8th International Conference on Energy Science and Applied Technology (ESAT 2023)

E3S Web of Conferences Volume 375 (2023)

Yichang, China
10-12 March 2023

Editors:

Zhigang Fang
Dongjuan Tong

ISBN: 978-1-7138-7238-2

Printed from e-media with permission by:

Curran Associates, Inc.
57 Morehouse Lane
Red Hook, NY 12571

Some format issues inherent in the e-media version may also appear in this print version.

For additional information, please contact EDP Sciences – Web of Conferences
at the address below.

EDP Sciences – Web of Conferences
17, Avenue du Hoggar
Parc d'Activité de Courtabœuf
BP 112
F-91944 Les Ulis Cedex A
France

Phone: +33 (0) 1 69 18 75 75
Fax: +33 (0) 1 69 28 84 91

contact-edps@webofconferences.org

Additional copies of this publication are available from:

Curran Associates, Inc.
57 Morehouse Lane
Red Hook, NY 12571 USA
Phone: 845-758-0400
Fax: 845-758-2633
Email: curran@proceedings.com
Web: www.proceedings.com

TABLE OF CONTENTS

Research on AGC Control Performance Optimization and Application of 135MW Coal-Fired Unit 1
Ding Yuming, Zhang Li, Cheng Weikun, Zhang Fang

Preparation of Polymeric Vesicles with Tertiary Amine and Epoxy Functional Groups Via Polymerization-Induced Self-Assembly 5
Zhang Fen, Yao Qian, Li Yantao, Niu Yanling, Chen Xiaoqi, Zhou Haijun

Study on Thermodynamic System Characteristics of Thermoelectric Units Under Variable Operating Conditions 10
Zhong Ziwei, Zhu Lingkai, Zheng Wei, Gong Zhiqiang, Shang Panfeng, Guo Junshan, Liu Jun, Zhang Yanpeng

Power Consumption Characteristics of Cement Industry and Parameter Analysis of Self Provided Power Plant 14
Zhu Lingkai, Wang Weishuai, Zhang Haijing, Zheng Wei

Application Research on Liner Tube Without Variable Diameter of Shale Gas Plunger Gas Lift Wellhead 18
Duan Chengzhi, Xu Haonan, Fan Jing, Wen Lujun, Ni Jianhui

Cause Analysis and Solutions About Abnormal Heating of Equipment Clamp 23
Zheng Jianjun, Xu Zhehao, Gong Weiwei

The Quantitative Research on Atmospheric Environmental Corrosion of Aluminum Alloy Products 27
Zhu Yuqin, Zhao Fangchao, Li Jiameng, Tan Tiantian, Dai Lu

Micro-Environment Corrosion Strength Based on Quartz Crystal Microbalance 33
Dai Lu, Zhu Yuqin, Qin Chaoxuan, Li Jiameng

Application Research of Computer Big Data Informatization Technology in Oilfield Mining Automation System 38
Shen Zhongmei

Response of Rhizosphere Microenvironment to Coupling of Climate Warming and Soil Pollution................. 43
Lei Na, Zhang Weihua

Numerical Investigation of Poroelastic Effects During Hydraulic Fracturing Using XFEM Combined with Cohesive Zone Model 47
Ren Xiaoyu, Guan Yueyang, Zhang Xiaofei, Liu Xiaobao, Liu Jie, Liu Kang, Luo Jiangwei, Shi Guiyuan

Research on the Deposition Mechanisms and Production Characteristics of Untabulated Reservoirs in the X Oilfield of the Songliao Basin 54
Wu Pan

Management Methods of Drilling Shut in 61
Wu Haiwei

Fixed Point Continuous Monitoring System of Oilfield Skeleton and Its Application.. 65
Du Kai

Application Strategy of Drainage Gas Production Technology in Shallow Gas Fields.. 69
Lv Na

Characteristics of Water Flooded Layer in Minghuazhen Formation 73
Li Zhonglin

Technology and Effect of Small Well Spacing Control Decline in Close Well Network 77
Gan Senlin

A New Method for Dynamic Prediction of Reservoir Pore Pressure in Development Wells 81
Gao Ying

Information Security Protection Technology of Station and Depot Industrial Control System of Oil Production Plant 85
Liu Tianze

Adjustment and Tapping Technology of Water Drive Reservoir in Late High Water Cut Period 89
Jia Peiyi

Influencing Factors and Measures of Water Drive Oilfield Development Effect 93
Gao Hang

Discussion on Technical Countermeasure of Improving Development Effect of Low Permeability Reservoir 97
Li Xiang

Analysis of Reservoir Geological Characteristics and Development Countermeasures 101
Zheng Hong

Adjustment of Infill in Transition Zone to Improve Oilfield Productivity 105
Long Xu

Analysis of Development Status and Adjustment Method of a Block 109
Han Hongzhi

Study on the Cause Classification and Prevention Methods of Casing Damage 113
Jia Fangfang

Study on the Cause of Casing Damage and the Feasibility of Comprehensive Prevention 117
Liu Shiping

Application of Close Cut Volume Fracturing in Tight Oil Horizontal Wells 121
Gao Yude

Discussion on Reservoir Description Method of Well Vibration Combined with Tight Well Pattern Area 125
Liu Pan

Analysis of the Exploitation Potential of Well DA1 129
Wang Lei

Study on Water Flooding Law in High Water Cut Stage of Complex Fault Block Reservoir 133
Chen Jing

Study on the Method of Remaining Oil Exploration in Thin Differential Reservoir 137
Gao Li

Discussion on the Function of Logging Technology in Geological Steering of Horizontal Well Drilling 141
Li Lin

Study on the Influence of Oilfield Water Injection on Casing Damage 145
Ding Xiangnan

The Optimization Method of Control Precision of Injection System in Well Metering 149
Zou Haochen

Washed Core and Tracer Tests Data Guide the Characterization of Single Sand Body in Low Permeable Reservoir 153
Qin Guoying

Discussion on the Countermeasure of Improving the Development Effect of Oil Well Dynamic Analysis 158
Jiang Meiyi

Causes and Treatment of Adhesive Damage in Polymer Flooding Injection Pipeline 162
Tian Long

A Certain Block of Oil and Gas Accumulation Understanding 166
Bao Yanfeng

Discussion on Reasonable Flow Pressure in Class I Block of a Oilfield 170
Lu Mingzheng

Residual Oil Start-Up Mechanism of Polymer/Surface Dual Flooding After Polymer Flooding in Medium-High Permeability Reservoir 178
Zhang Nan, Zhao Yuwen, Wang Wei, Wang Xiaoyan, Yan Xi, Yuan Guangyu, Lin Zeyu, Liu Yuanxian, Zhang Jing

Effect of Perforation Shear on Viscosity of Polymer Solution 185
Yan Xi, Wang Xiaoyan, Wang Wei, Lin Zeyu, Yuan Guangyu, Zhan Nan, Han Song, Cai Xidong, Wu Xing, Liang Yanbin

Analysis on Influencing Factors of Gully Control in Loess Plateau 191
Lei Na, Zhang Panpan

Capacity Analysis and Understanding of Pilot Test Well in Gas Storage A 195
Li Weiyan

Gas Distributor for Ultra-Large Air Blow-Out Bromine Extraction Plant 200
Chai Shujing, Zhang Wenyan, Yu Xiaoyu, Hao Xiaocui, Li Tao, Wang Zejiang, Liu Wei

The Vapor-Liquid Equilibrium of Bromine-Brine System at Atmospheric Pressure 204
Qian Jianting, Yu Zhaoling, Yang Xincheng, Zhang Xiaolong, Xu Paisheng, Liu Jundong, Liu Wei, Chai Shujing, Zhang Wenyan, Hao Xiaocui, Wang Yuqi, Zhang Qi, Huang Xiping

Preparation of High-Temperature Gels for Enhanced Oil Recovery Using Methyl Etherized Melamine-Formaldehyde Resin 208
Guo Hongbin, Ge Jijiang, Li Jiayi

Research on Foaming Agents for Gas Flooding in Medium-Temperature and High-Salinity Clastic Reservoirs 213
Li Kexin, Ge Jijiang, Li Longjie, Li Jiasu, Chu Pengju

Development of High Strength Polyvinyl Alcohol Gel for Water Shutoff in Carbonate Reservoirs 218
Chen Jin, Ge Jijiang, Wang Wenhui, Chen Pengfei, Chen Weixiong

Study on the Influencing Factors of Productivity Under Pressure Drive Development in Low Permeability Reservoir 223
Cui Chuanzhi, Wang Guobiao, Lu Shuiqingshan

Hot-In-Place Recycling Engineering Application of Low Penetration Aged Asphalt 228
Shi Ning, Xiao Hongyu, Xiao Jiang, Liu Qi

Study on the Method of Identifying Fractures by Ant Tracking Technique 233
Zhang Juan, Liu Ke, Han Xianbo

Study on the Influence of Different Key Parameters on EOR in LB Block 237
Wang Qunyi, Bi Yongbin, Zhang Yang, Cao Tongfeng, Liu Dong

Fine Dissection of Meandering River Sand Bodies Under Dense Well Nets 243
Liu Ye, Lu Hong, Hu Yun

Application of Multiphase Flow Fiber Optic Fluid Production Profile Testing Technology in W1 Well 249
Yu Fuji

Research Progress on Stability Analysis Methods of Granite Residual Soil Slope 254
Wang Zian, Zheng Chao, Zhang Le, Xiong Kangwei, Li Zhanlin

Preparation of Quaternary Ammonium Gemini Surfactant and Its Application in Insulator Cleaning 261
Wan Xinyuan, Xia Xiaojian, Chen Yiyang, Lin Deyuan, Hong Yicheng, Zhang Wenzhe, Hou Linxi

Development Prospect of Heat Supply and Peak Shaving of Nuclear Power Units in Shandong Power Grid 265
Zhu Lingkai, Zheng Wei, Zhong Ziwei, Gong Zhiqiang, Guo Junshan

Research on Technical Supervision of Steam Turbine in Nuclear Power Plant 269
Che Yongqiang, Chen Yisen

Application of Lidar in Comparison of Wind Speed and Wind Direction Meters in Wind Power Field 274
Wang Hua, Wang Jinshan, Tian Wenqiang, Bao Zhiqiang, Wang Bao, Gou Yanxu

Application of Mobile Vibration Online Comparison Sensor in Wind Power Field 278
Wang Hua, Wang Jinshan, Tian Wenqiang, Bao Zhiqiang, Wang Bao, Gou Yanxu

Study on Amplification Effect of Wind Turbine Blade Stopping Position on Vortex Induced Vibration of Tower 282
Wang Ziyue, Jiao Shoulei, Zhang Guowei, Chen Xiaoxu

Effect of Dam Removal of Small Hydropower Station on Migration and Distribution of Biogenic Elements: Progress and Prospects 286
Li Cai, Guan Linjie, Yang Kun, Zhang Li, Fan Qingsong

A Novel Doping Layer Strategy to Realize High Efficiency Laye-By-Layer Organic Solar Cells 290
Li Xinrui, Du Xiaoyang, Tao Silu

Unit Commitment Considering the Different Frequency Supporting Capability of Synchronous Generators 295
Zhang Jiaqi, Xu Guoyi, Li Beibei, Han Lei

Integrated Photo - Rechargeable Batteries: Photoactive Nanomaterials and Opportunities 300
You Liwen

Utilizing CO_2 from Refinery for Methanol and Electricity Coproduction: System Assessment 310
Ma Qian, Yu Xiaodan, Xia Song, Zhang Xu, Liu Anqi, Yuan Bo, Chen Xiaojun

Utilizing Carbon Dioxide from Refinery for Methanol and Electricity Co-Production: System Design and Economic Assessment 314
Ma Qian, Wang Yuxi, Zhang Xu, Yu Xiaodan, Liu Anqi, Yuan Bo, Chen Xiaojun

Calculation Model of Non-Nuclear Steam Rush in Steam Turbine of Nuclear Power Unit Construction and Application 318
Liu Junfeng, Kang Zhen, Wang Miaomiao, Chao Shima, Xi Jingbin, Wu Zhijun, Wang Wei, Zhang Zhenlu

An Improved Multi-Objective Gray Target Model for Project Resource Allocation Selection Research 323
Xu Yaqiang

Research Progress of Energy Equalization Topology of Power Lithium Battery Pack 329
Miao Yinbao, Zhang Wenhua, Liu Weihao, Kang Dongqi, Wang Shuai, Chen Zhe, Liu Jia, Chen Biaoxian, Zhu Leijing

Performance Test of a 660MW Thermal Power Unit Steam Turbine Based on the Background of "Double Carbon" 335
Zhong Ziwei, Zhu Lingkai, Zheng Wei, Gong Zhiqiang, Shang Panfeng, Guo Junshan, Liu Jun, Zhang Yanpeng

Application Research on Zero Output of Double Low Pressure Cylinder of 600MW Unit 339
Liu Jun, Zhu Lingkai

Research on Intelligent Regulation System of Solar Panels Driven by Low-Voltage Electric Energy............ 344
Zhang Wei, Lin Xuefeng

Economic Day-Ahead Scheduling of SOFC and Biomass-Based Integrated Tri-Generation Energy System Using Artificial Bee Colony Algorithm 348
Zhao Pengxiang, Zhu Jianjun, Zhang Yanru, Wang Qiang, Yang Xian, Cong Lin

Study on the Thermal Decomposition Reaction Process and Kinetics of SF6 and Tungsten 358
Xu Mengyuan, Lv Houhua, Liu Yunfang, Yang Ren, Han Yanhua, Xu Liuyu, Li Xu, Li Jiuhong, Zhang Xiao, Jian Yajun, Zhang Weiqiang

Study on Calculation of Inlet Steam and Inlet Water Flow of Drain Pump of Low Pressure Heater 364
Zhu Lingkai, Zheng Wei, Wang Qian, Zhong Ziwei, Gong Zhiqiang

A Novel Transmission Network Planning Framework Under the New Power System 368
Yan Chaochen, Tao Ye, Zhao Yu, Wang Jiaxing, Wang Yuqing, Zhu Jiang

Day-Ahead Load Optimal Distribution of Thermal Power Coupled Electric Energy Storage Frequency Regulation System Considering the Effect of Overtemperature 374
Zhou Xichao, Wang Nan, Li Zhen, Zhao Pengxiang, Cong Lin

Overview of Landfill Leachate Treatment Technology 381
Qin Ruqiong

Highly Active Cu-MFI Catalyst for Conversion of Furfuryl Alcohol to Pentanediols 385
Dai Dengfeng, Liu Dandan, Liu Yunqi

A Real-Time Corrosion Monitoring Method Based on Quartz Crystal Microbalance 389
Qin Chaoxuan, Dai Lu, Zhu Yuqin

The Detection Methods of Hydroxyl Radical: A Review 393
Huang Chenkun, Chen Qiuyue, Li Shengqiang, Wang Jian

Research on Time-Varying Economic Dispatch of Smart Grid Based on Lagrangian Pairing........................ 397
Zuo Ying

Application Prospect of High Efficiency and Energy Saving Bromine Extraction Technology in the Field of Bromine-Containing Waste Salt 401
Zhang Wenyan, Chai Shujing, Hao Xiaocui, Liu Wei, Wang Yuqi, Zhang Qi

Research on Energy-Fed Suspension Control System of New Energy Vehicles Based on Energy Flow Analysis 405
Ding Jiageng

Evaluation Model and Promotion Based on the Ecological Characteristics of Saihanba 409
Wang Bingqi

The Relation Between the CO_2 Concentration Levels and the Temperature 416
Lin Yuyuan, Zeng Zidong, Chen Lang, Deng Zifeng

A Light Pollution Risk Model Based on Improved Assessment and Prediction Methods............................... 421
He Ziyan, Guo Ruikun, Yang Haiwei

Energy Efficiency Optimization of Ships Based on Particle Swarm Optimization 425
Zhang Ailin, Liang Zhengxu

Diagnosis and Analysis of Thermal Power Plant Operation in Heating Season... 430
Zhang Yanpeng, Zhu Lingkai, Li Si, Shang Panfeng

Study on the Relationship Between Tea Concentration and Tyndall Effect 434
Yang Yifan, Su Xiaofeng, Xu Yongke, Jin Weigang, Shen Bangxing, Zhang Xinfeng, Lv Ruibo, He Yi, Guo Chunfeng, Li Yuefeng, Zou Jun

Basic Characteristic Analysis of Self-Provided Power Plant in Iron and Steel Industry 441
Zhu Lingkai, Zhang Haijing, Wang Weishuai, Wang Qian, Zheng Wei, Zhong Ziwei

On the Importance of Self Owned Power Plants Participating in the Adjustment of Supply and Demand of Power Network Under the Reform of Power System 445
Zhu Lingkai, Zhang Haijing, Wang Weishuai, Wang Qian, Zheng Wei, Zhong Ziwei

Test and Research on Load Capacity of a 210MW Unit in Heating State... 449
Zhong Ziwei, Zhu Lingkai, Zheng Wei, Gong Zhiqiang, Shang Panfeng, Guo Junshan, Liu Jun, Zhang Yanpeng

A 1000MW Ultra-Supercritical Secondary Reheat Unit Energy Consumption Analysis of Thermodynamic System... 453
Zhang Hu, Lao Jinxu, Cheng Huitao, Liang Kai

Analysis of Heating Transformation of Different Types of Thermal Power Units 458
Lao Jinxu, Zhang Hu, Zheng Wei, Shang Panfeng, Han Yue

Prospect of Peak Regulation Capacity Improvement Through Flexibility Transformation of Combined Heat and Power Units in Shandong Province 463
Gong Zhiqiang, Han Yue, Guo Junshan, Shang Panfeng, Zhu Lingkai

Research on Operation of Low-Pressure Cylinder Zero Output Heat Supply Retrofit Unit 468
Lao Jinxu, Zhang Hu, Han Yue, Shang Panfeng

Medium- And Long-Term Power Load Forecasting Model 472
Zhu Yuning, Xu Xiaoliang, Lai Ruoqi, Guo Hengqing, Jia Fan

Study on the Electrothermal Coupling Characteristics of Different Types of Thermoelectric Units 477
Zhong Ziwei, Zhu Lingkai, Zheng Wei, Gong Zhiqiang, Shang Panfeng, Guo Junshan, Liu Jun, Zhang Yanpeng

Introduction on Current Application of Voltage Transformer 481
Yang Jun, Fan Ziming, Yang Yue, Zhao Xiayao

Author Index

Preface

The 8th International Conference on Energy Science and Applied Technology (ESAT 2023) was held in Yichang City of China during March 10-12, 2023. ESAT 2023 organized by Wuhan University of Technology, co-organized by China Three Gorges University, Northeast Petroleum University, Huazhong University of Science and Technology, and sponsored by Asian Union of Information Technology.

ESAT conference is started in 2014, the aim of ESAT series conference is to provide a platform for researchers, engineers, and academicians, as well as industrial professionals, to present their research results and development activities in energy science, and its applied technology. It provides opportunities for the delegates to exchange new ideas and application experiences, to establish business or research relations and to find global partners for future collaboration.

This three-day conference was organized in technical association with the Asian Union of Information Technology and included invited keynote talks and oral paper presentations from both academia and industry. Each keynote speakers lasted for 30 minutes, including 5 minutes for Q&A. and for other sessions, there was 20 minutes for each presentation, including 5 minutes for Q&A. The themes covered in this proceedings are: Fossil Energy & Geological Engineering; Clean Energy and Electric Energy; Energy Sustainability & Energy-Related Environmental Science. The book is useful to professionals and scientists involved or interested in energy science and applied technology. Hopefully, all participants and other interested readers would benefit scientifically from the proceedings and also find it stimulating in the process.

We would like to thank all the author submitted papers to the conferences and thank all the reviewers for their time and effort in reviewing articles. Especially we would like to thank the organizing committee for their valuable advices in the organization and helpful peer review of the papers.

This 8th ESAT received 179 manuscripts, and 102 submissions had been accepted by our reviewers. By submitting a paper to ESAT conference, the authors agreed to the review process and understood that papers would undergo a peer-review process. Manuscripts were reviewed by appropriately qualified experts in the field selected by the conference committee, who took detailed comments and-if the submission was accepted-the authors would submit a revised version that took into account this feedback. All papers were reviewed using a double-blind review process: authors declared their names and affiliations in the manuscript for the reviewers to see, but reviewers did not know each other©s identities, nor did the authors receive information about who had reviewed their manuscript. The committees of ESAT 2023 invested great efforts in reviewing the papers submitted to the conference and organizing the sessions to enable the participants to gain maximum benefit.

Conference Organizing Committee

Statement of Peer review

In submitting conference proceedings to *Web of Conferences*, the editors of the proceedings certify to the Publisher that

1. They adhere to its **Policy on Publishing Integrity** in order to safeguard good scientific practice in publishing.
2. All articles have been subjected to peer review administered by the proceedings editors.
3. Reviews have been conducted by expert referees, who have been requested to provide unbiased and constructive comments aimed, whenever possible, at improving the work.
4. Proceedings editors have taken all reasonable steps to ensure the quality of the materials they publish and their decision to accept or reject a paper for publication has been based only on the merits of the work and the relevance to the journal.

Title, date and place of the conference

8th International Conference on Energy Science and Applied Technology (ESAT 2023)
March 10-12, 2023
Yichang City, China

Proceedings editor(s):

Zhigang Fang, Dongjuan Tong

Date and editor's signature March 17, 2023

17, avenue du Hoggar - PA de Courtabœuf – BP 112 - 91944 Les Ulis Cedex A (France)
Tél. : 33 (0)1 69 18 75 75 - Fax : 33(0)1 69 07 45 17 – www.edpsciences.org

Research on AGC control performance optimization and application of 135MW coal-fired unit

Yuming Ding, Li Zhang, Weikun Cheng and Fang Zhang

Huadian Electric Power Research Institute Co., LTD.Hangzhou, China

Abstract. This paper studies the corresponding performance optimization of unit coordinated control system. Combined with the actual operation of the unit, targeted optimization strategies are formulated and implemented for the main parameters such as active power, main steam pressure, feed water flow, main steam temperature and furnace pressure. After long-term tracking of operation and repeated adjustment of relevant control parameters, a more reasonable optimization effect is finally obtained, and the regulation performance of the unit coordinated control system meets the AGC input conditions.

1. Introduction

A 135MW unit of Xinjiang Huadian was originally a pure condensing unit, and the heating transformation was carried out in 2009. At present, the AGC regulation performance of the unit can not meet the requirements of Xinjiang dispatching and is often assessed by the power grid. Therefore, through loop optimization, improve the regulation quality of the coordinated control system and MCS subsystems, so that the AGC regulation performance of the unit can meet the assessment standards of AGC by Xinjiang dispatching control center (response time < 60 seconds, lifting rate > 1.5% PE / min), and can operate in AGC mode for a long time, And verified by test.

2. Anzlysis

2.1 The coal type

The section headings are in boldface capital and lowercase letters. Second level headings are typed as part of the succeeding paragraph (like the subsection heading of this paragraph). All manuscripts must be in English, also the table and figure texts, otherwise we cannot publish your paper. Please keep a second copy of your manuscript in your office. When receiving the paper, we assume that the corresponding authors grant us the copyright to use the paper for the book or journal in question. When receiving the paper, we assume that the corresponding authors grant us the copyright to use the paper for the book or journal in question. When receiving the paper, we assume that the corresponding authors grant us the copyright to use.

2.2 CCS parameters

The coordination control parameters do not match the current unit operation. Under the coordination mode, the boiler master controls the main steam pressure and the steam turbine master controls the unit load. The main control parameters of the steam turbine are poor and the load regulation is unstable. The main control parameters of the boiler are poor, the main steam pressure regulation is slow, the lifting load can not meet the dispatching demand, and the pressure deviation is large, which is easy to trigger the pressure pull back circuit, which is not conducive to the rapid load regulation.

2.3 Other automatic control parameters

Other automatic control system parameters cannot fully match the coordination effect. Such as water supply, primary air pressure control, air volume and oxygen content, coal, water and air cannot be added (reduced) quickly.

2.4 Equipment impact

If the load is increased rapidly, it is easy to cause overtemperature of boiler wall temperature.

2.5 Heating impact

With large heating units, the influence of heating and air extraction is not calculated in the coordinated control.

3. Optimization

3.1 parameter optimization

Optimize the static and dynamic parameters of boiler main control PID. See Table I~III for the comparison before and after parameter optimization.

.

* Corresponding author: yming-ding@chder.com

Table 1. Boiler main control PID parameters

Parameter optimization	static P1	static I1	static D1	dynamic P1	dynamic I1	dynamic D1
Before	0.6	250	0	0.3	350	0
After	0.8	210	0	0.3	320	0

Table 2. Boiler main control function

proportional variable parameter function (increase)								
Unit load X	0	80	90	100	110	120	130	135
coefficientY	0.9	0.9	0.9	0.9	0.9	0.9	1	1

Table 3. Boiler main control function

Integralvariable parameter function (increase)							
Unit load X	0	70	80	100	110	150	0
Coefficient Y	1	1	1	1	1.2	1.3	1

Optimize the main control PID parameters and dynamic parameters of the steam turbine. See Table IV and V for the comparison before and after parameter optimization.

Table 4. turbine main control PID parameters

Parameter optimization	static P1	static I1	static D1	dynamic P1	dynamic I1	dynamic D1
Before	0.4	15	0	0.4	15	0
After	0.45	30	0	0.45	18	0

Table 5. pressure pull back function

Integralvariable parameter function (increase)								
Unit load X	-10	-1.5	-1	-0.5	0.5	1	1.5	10
Coefficient Y	-0.9	-0.8	-0.7	0.0	0	0.7	0.8	0.9

The unit was originally a condensing gas generator unit. After heating transformation, the control performance of the unit changes. It needs to be optimized after heat supply accounting, which is divided into non heating sliding pressure curve optimization and heating sliding pressure curve optimization. Optimize the sliding pressure curve, change the input of the sliding pressure function from the main steam flow to the corresponding main steam pressure setting value under the comprehensive load command, set the third-order inertia link and output it as the final sliding pressure setting value after speed limit. See Table VI and VII below for the comparison before and after parameter optimization.

Table 6. Sliding pressure curve function (heating)

Before optimization						
Unit load X	0	70	90	100	115	150
Sliding pressure setting Y	8	9	10.8	12	13.2	13.5
Before optimization						
Unit load X	0	70	90	100	112	150
Sliding pressure setting Y	8	9	10.8	12	13.2	13.5

Table 7 Sliding pressure curve function (non-heating)

Before optimization						
Unit load X	0	70	90	100	115	150
Sliding pressure setting Y	8	9	10.8	12	13.2	13.5
After optimization						
Unit load X	0	70	90	100	115	150
Sliding pressure setting Y	8	9	10.8	12	13	13.2

3.2 Logical modification

(1) Modify the primary air pressure setting logic, optimize the primary air pressure setting curve corresponding to the coal volume command, and increase the manual correction and feedforward action correction.

(2) Modify the boiler main control feedforward, optimize the static feedforward function in the boiler main control feedforward, and add pressure deviation differential feedforward, pressure set value differential feedforward, dynamic load feedforward and desuperheating water action feedforward in the boiler main control feedforward loop; When the coal quantity of the coal feeder reaches the lower limit, the main control of the boiler will be locked and reduced. When the coal quantity of the coal feeder reaches the upper limit, the command of the boiler will be increased.The modification of boiler master feedforward logic mainly includes: Boiler master control: feedforward consists of four parts: load static feedforward (ff0), load dynamic feedforward , pressure deviation differential feedforward , pressure set value differential feedforward and desuperheating water feedforward .

(3) Increase the closed-loop control link of fuel master control, set the number of automatic coal feeders and coal feeding deviation, and correct the PID parameters.

(4) Increase the command of locking unit and steam turbine load with large main steam pressure deviation.

(5) Modify the unit instruction and add AGC allowable conditions and AGC switching logic.

Increase the function and screen for setting the upper limit of each coal feeder, and set the trigger logic for each coal feeder command to reach the upper and lower limits.

4. AGC load following test data

4.1 Analysis of AGC load following test data

At 17:26:38 on April 24, the load began to rise. The load change range was 76 MW ~ 101.5 MW, and the set rate was 2.1mw/min. At 17:39:18, the load change at this stage ended. The statistical data are shown in the table below.

Table 8. Dynamic deviation of various parameters of load rise and load response of unit

Adjust quality items	Adjust quality actual value	Adjust quality reference value (DL/657)	conclusion
Load variation	76→87.4→93.6→101.5MW	20MW	qualified
Load change rate	1.6%Pe/min	1.5%Pe/min	qualified
Load response delay	35s	< 60s	qualified
Main steam pressure	±0.6MPa	< ±0.4MPa	unqualified
Main steam temperature	AUTO: ±12℃ MAN: ±24℃	< ±10℃	unqualified
Drum water level	±27mm	< ±60mm	qualified
Furnace negative pressure	±152Pa	< ±200Pa	qualified

4.2 Response of AGC load reduction parameters

At 19:14:03 on April 24, the load was reduced. The load change interval was 132 MW ~ 99.3 MW, and the set rate was 2.1mw/min. At 19:30:51, the load change at this stage ended. The statistical data are shown in the table below.

Table 9. Dynamic deviation of various parameters of load reduction and load response of unit

Adjust quality items	Adjust quality actual value	Adjust quality reference value (DL/657)	conclusion
Load variation	132→122→113→99.3MW	20	qualified
Load change rate	1.6%Pe/min	1.5%Pe/min	qualified
Load response delay	38s	< 60s	qualified
Main steam pressure	±0.54MPa	< ±0.4MPa	unqualified
Main steam temperature	±28℃	< ±10℃	unqualified
Drum water level	±27mm	< ±40mm	qualified
Furnace negative pressure	±121Pa	< ±200Pa	qualified

4.3 Steady state deviation of unit

The statistical data of load deviation of the unit under stable working conditions under different loads are shown in the table below.

Table 10. Steady state deviation of unit

Load command	Load deviation	Adjust quality reference value	conclusion
134MW	±0.43MW	< 0.5%Pe	qualified
70MW	±0.35MW	< 0.5%Pe	qualified

4.4 AGC load following performance evaluation

After coordinated optimization of the unit, the unit was successfully put into the coordinated control system, and the application was transferred to AGC to carry out load following test. The load and main steam pressure of the unit were tracked well. The unit is capable of long-term stable operation in coordinated control mode and AGC mode.

5. Existing problems and suggestions

Due to the idle stroke (about 10%) of the air supply and liquid couple actuator, the air volume and oxygen control of the unit are still manually controlled. After the maintenance of the liquid couple actuator, put the air supply into automatic mode and optimize the oxygen tracking closed-loop control.

After optimizing the unit coordinated control system and main analog loop logic, the relevant parameters are optimized through variable load test, and the load following test is carried out under AGC working condition. The following conclusions are obtained: under variable load working condition, the unit variable load rate, load response time, drum water level The quality of furnace negative pressure regulation meets the requirements of DL / T 657-2015 code for acceptance test of analog control system in thermal power plant. During load change, the variation range of main steam pressure and main steam temperature is larger than the standard value, but it still meets the requirements of safe and stable operation and load change of the unit.

References

1. LI Bo-qun,DONG Hui,XIA Yi-ming,ZHAO Yong,MA Lei.Application on Neural PID Control of MN-AGC in Continuous Hot Strip Rolling[A]. Proceedings of the 33rd China Conference on Control and Decision Making (1) [C]. 2021
2. Li Sun,Donghai Li,Kwang Y. Lee,Yali Xue. Control-oriented modeling and analysis of direct energy balance in coal-fired boiler-turbine unit [J] . Control Engineering Practice . 2016.

3. Dan Ma,Jie Chen 0005. Delay Margin of Low-Order Systems Achievable by PID Controllers. [J] . IEEE Trans. Automat. Contr. . 2019 (5).
4. Wang Dangzhen, Zhao Chuanhao, Zhu Yingchun, Wei Ruixia. Analysis of factors affecting the comprehensive index of AGC [J]. Shandong Power Technology. 2016(12).
5. Liu Hao. Optimization of AGC control strategy for subcritical coal-fired units [D]. Shanghai Jiaotong University 2018.

Preparation of polymeric vesicles with tertiary amine and epoxy functional groups via polymerization-induced self-assembly

Fen Zhang*, Qian Yao, Yantao Li, Yanling Niu, Xiaoqi Chen and Haijun Zhou

Institute of Energy Resources, Hebei Academy of Sciences, 050081, Shijiazhuang, Hebei Province, China

Abstract. Polymeric vesicles with tertiary amine and epoxy functional groups were fabricated via the reversible addition-fragmentation chain transfer (RAFT) dispersion polymerization of 2-(diisopropylamino) ethyl methacrylate (DIPEMA) and glycidyl methacrylate (GlyMA) in an ethanol-water mixture by using poly (ethylene oxide)-4-(4-Cyanopentanoic acid) dithiobenzoate (mPEG45-CPADB) as the macro-chain-transfer agent (macro-CTA). Biocompatible polymer mPEG45 with molecular weight of 2000 and 4-(4-Cyanopentanoic acid) dithiobenzoate (CPADB) were used for the macro-CTA synthesizing via the esterification reaction. Investigation of the degree of polymerization (DP) of the hydrophobic block P(DIPEMA-co-GlyMA) and the solid content indicated that both have significant effect on the particle morphology and particle diameter. 1H NMR and FTIR analysis indicated the existing of tertiary amine and epoxy functional groups on the produced vesicles.

Keywords: Polymeric vesicles; polymerization-induced self-assembly; reversible addition-fragmentation chain transfer; dispersion polymerization.

1. Introduction

Nanoparticles have attracted significant attention due to their broad applications in areas of catalysis, drug delivery, imaging and nanoreactors.1-3 Polymerization-induced self-assembly (PISA) is a recently developed one-pot polymerization approach that can generate multiple morphologies such as micelles, as well as worm-like/rod-like and spherical vesicles at high yields and with high solid content (typically 10-50 wt%).4,5

The morphology control of block copolymers has always been a hot topic for research on PISA, but it is not simple to generalize the parameters that determine the particle morphologies produced via PISA. Generally the morphologies of the nanoparticles are thought to depend on the block copolymer composition and the solid content, and both of them influence the relative volume fractions of the constituent blocks (packing parameter P, P=v/(al), where v is the volume of the hydrophobic polymer chain, a is the optimal interfacial area per molecule and l is the hydrophobic length normal to the interface). Spherical particles were thought to be formed when P was ≤1/3, while worms and vesicles were expected for 1/3 ≤ P ≤ 1/2 and 1/2 ≤ P ≤ 1, respectively.7,8

In this study, we reported the copolymerization of 2-(diisopropylamino)ethyl methacrylate (DIPEMA) and glycidyl methacrylate (GlyMA) with poly(ethylene oxide)-4-(4-Cyanopentanoic acid) dithiobenzoate ($mPEG_{45}$-CPADB) as the macro-chain-transfer agent (macro-CTA) in an ethanol/water solvent for the preparation of polymeric particles with tertiary amine and epoxy functional groups. The synthesizing of the macro-CTA ($mPEG_{45}$-CPADB), and the effect of the degree of polymerization (DP) of the hydrophobic block P(DIPEMA-co-GlyMA) and the solid content on the particle morphologies were studied in detail.

2. Experiment

2.1 Materials

The α-methoxy-ω-hydroxypoly(ethylene oxide) ($mPEG_{45}$) with a number average molecular weight (M_n) of 2000 was purchased from Tokyo Chemical Industry Co. Ltd. and was used as received. The 4-(4-Cyanopentanoic acid) dithiobenzoate (CPADB), 4-(dimethylamino)pyridine (DMAP), and dicyclohexylcarbodiimide (DCC) compounds were purchased from Aladdin and also used as received. The monomers 2-(diisopropylamino)-ethyl methacrylate (DIPEMA, Aladdin, 97%) and glycidyl methacrylate (GlyMA, Aladdin) were purified by passing the compounds through a column of Al_2O_3 to remove the inhibitor prior to use. Initiator N,N'-Azobis(isobutyronitrile) (AIBN, Tianjin Chemical Reagent Co. Ltd.) was purified via recrystallization from ethanol.

* Corresponding author: xiaojj@hebust.edu.cn

2.2 Preparation of polymeric vesicles

mPEG$_{45}$-CPADB was used as the macro-CTA agent, and synthesized via an esterification reaction of mPEG$_{45}$ and CPADB at room temperature with DMAP as the catalyst and DCC as the dehydrant according to the reported method.[9,10] A typical protocol was performed as below: DIPEMA (0.3686 g, 1.5930 mmol), GlyMA (0.0614 g, 0.4319 mmol), mPEG$_{45}$-CPADB (0.0616 g, 0.0270 mmol), AIBN (0.0014 g, 0.0085 mmol), and a solvent (2.7937 g, with a mass ratio of ethanol/water=6:4) were added into a glass tube with a magnetic bar. This reaction mixture was degassed via N_2 purge in an ice bath for 30 min. After degassing by three pump-N_2 purge cycles, the glass tube was sealed under a vacuum, and then the sealed tube was placed in an oven at 70°C under magnetic stirring to polymerize for 7 h. The reaction mixture was quickly cooled to room temperature and then opened to air to quench the polymerization.

2.3 Characterization

The molecular weight (M_n) and molecular weight distribution (M_w/M_n) were measured by the gel permeation chromatograph (GPC, Waters 2695), and THF was used as the eluent and monodispersed polystyrene as the standards. The ^1H NMR spectra was done on a Bruker DMX500 spectrometer, where $CDCl_3$ was used as the solvent and tetra-methylsilane was used as an internal reference. The morphologies of the nanoparticles were characterized by transmission electron microscopy (TEM, JEM-2100Plus electron microscope), and stained with phosphotungstic acid before observations. All of the dynamic light scattering measurements were carried out on a commercial dynamic light scattering (DLS) spectrometer (PPS Z3000, Particle Sizing Systems, UK). The Fourier transform infrared (FTIR) spectroscopy was obtained using a PerkinElmer Frontier.

3. Results and discussion

3.1 Synthesis of macro-CTA mPEG45-CPADB

The synthesizing of macro-CTA mPEG$_{45}$-CPADB was conducted according to Scheme 1, and GPC and ^1H NMR analysis were done to characterize the produced macro-CTA, as shown in Figure 1. M_n of mPEG$_{45}$-CPADB was M_n=2631 with M_w/M_n =1.19 (Figure 1A, both M_n and M_w/M_n increased a little compared with mPEG$_{45}$), which showed relatively narrow M_w/M_n after esterification reaction. ^1H NMR spectroscopy result indicated that the esterification reaction was successfully done based on the integral values of signals of ester methylene protons at δ=4.3 (f). The average end-functionality of mPEG$_{45}$-CPADB was 98.3% by calculating the integral ratio of methoxy protons of mPEG at δ= 3.4 (h) and aromatic protons at δ=7.3-8.0 (a, b, c), as shown in Figure 1B.

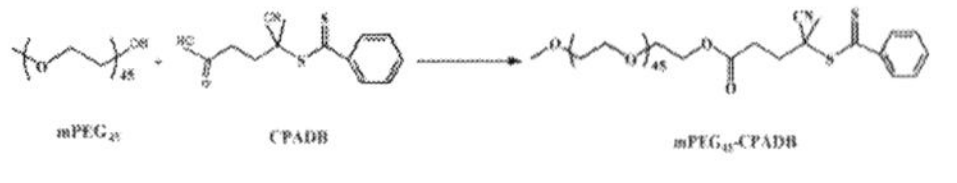

Scheme 1. Synthesis of mPEG45-CPADB macro-CTA.

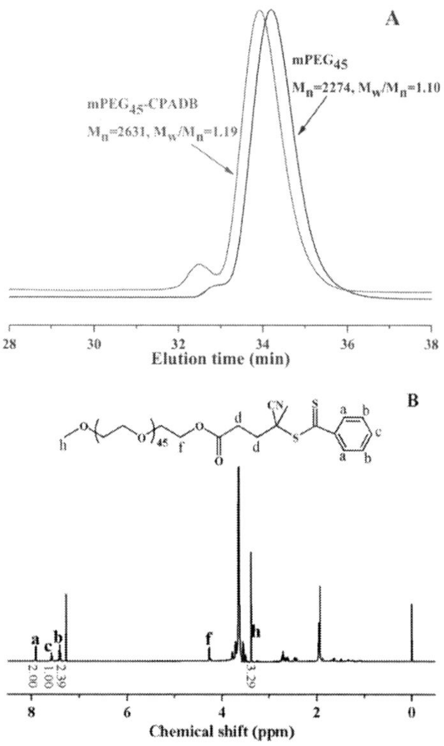

Figure 1. (A) GPC traces of mPEG$_{45}$ and mPEG$_{45}$-CPADB; (B) ^1H NMR spectrum of mPEG$_{45}$-CPADB in $CDCl_3$.

3.2 Effect of DP of hydrophobic block on polymeric vesicles

The mPEG$_{45}$-CPADB produced above was used as the macro-CTA to produce the functional polymeric vesicles, which also acted as the stabilizer to stabilize the produced nonoparticles. During the polymerization, the hydrophilic macro-CTA was chain extended with DIPEMA and GlyMA to form an insoluble block segment. Once the amphiphilic block copolymer chains grew to a certain length, they precipitated out and self-assembled to form nanoparticles, and the nanoparticles aggregated and finally were stabilized by the hydrophilic segment of the macro-CTA. The effect of DP of the hydrophobic block P(DIPEMA-co-GlyMA) on the particle morphology was evaluated.

A series of nanoparticles were prepared via changing DP of the P(DIPEMA-co-GlyMA) block from 40 to 100. Each sample was characterized by its morphology, hydrodynamic diameter and its distribution index (PDI). Figure 2 showed the TEM images of the produced nanoparticles, and Figure 3 showed the particle hydrodynamic diameter and its distribution index (PDI). When DP of P(DIPEMA-co-GlyMA) block was 40, spherical particles of 117 nm were formed, while vesicles could be obtained with DP of P(DIPEMA-co-GlyMA) block equal to or higher than 60. The DLS results of the particle hydrodynamic diameter in Figure 3 showed that the diameter of the particles increased from 117 nm to 593

https://doi.org/10.1051/e3sconf/202337501002

nm when DP increased from 40 to 100, while the PDI in range of 0.1-0.25. As the increasing of the DP of P(DIPEMA-co-GlyMA) block, more hydrophobic monomer was added to the polymerization system, the degree of swollen of the polymer chains would be increased, which helped the chain mobility and chain reorganization, and also promoted particle fusion, and thus facilitated the vesicle formation.

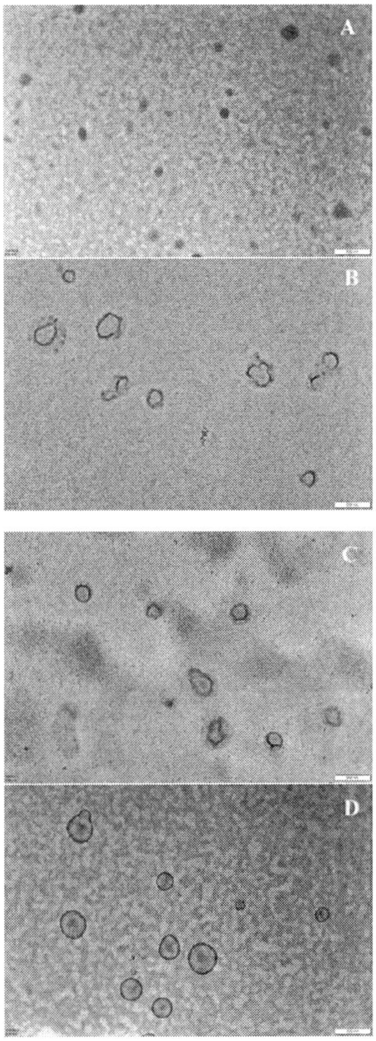

Figure 2. TEM images of mPEG$_{45}$-b-P(DIPEMA-co-GlyMA) nanoparticles prepared under different DP of P(DIPEMA-co-GlyMA) block, (A) DP=40, (B) DP=60, (C) DP=80 and (D) DP=100.

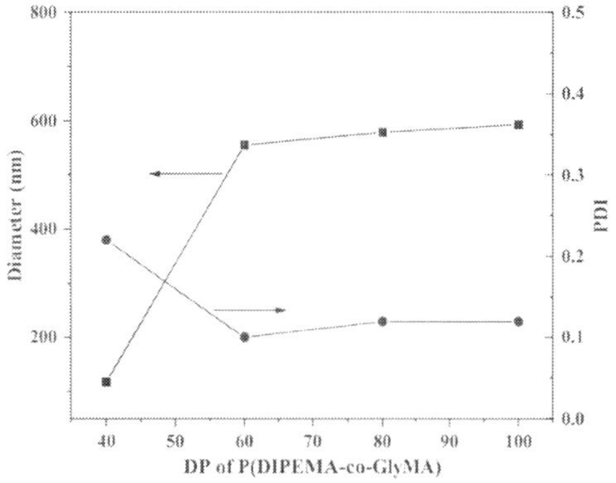

Figure 3. Hydrodynamic diameter and PDI of mPEG$_{45}$-b-P(DIPEMA-co-GlyMA) nanoparticles prepared under different DP of P(DIPEMA-co-GlyMA) block.

3.3 Effect of solid content on polymeric vesicles

The effect of solid content, which was one of the key factors that affected the morphology of the nanoparticles, was evaluated in this polymerization system. Figure 4 shows the TEM images of the nanoparticles produced with solid content changing from 5% to 20%, while Figure 5 illustrates the corresponding hydrodynamic diameter and PDI. It can be seen that pure vesicles could be fabricated when solid content increased to 10%, and the hydrodynamic diameter of the produced nanoparticles increased from 438 nm to 890 nm with solid content increased from 5% to 20%, while the PDI kept in range of 0.5-1.5. With a higher solid content, the number of nanoparticles generated in the nucleation stage would be larger, and the inelastic collision of the nanoparticles became more easily, and thus facilitated the formation of vesicles.

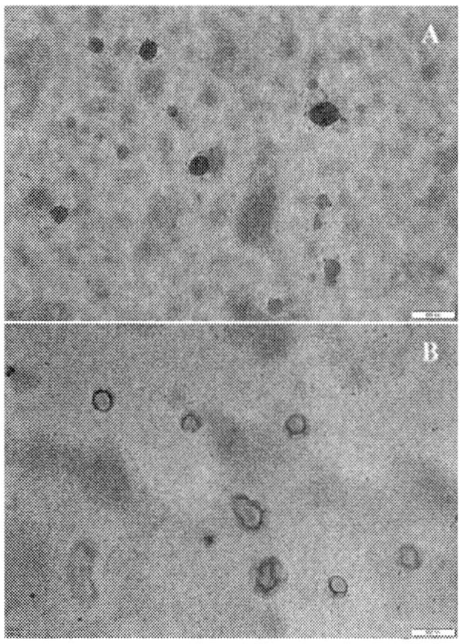

https://doi.org/10.1051/e3sconf/202337501002

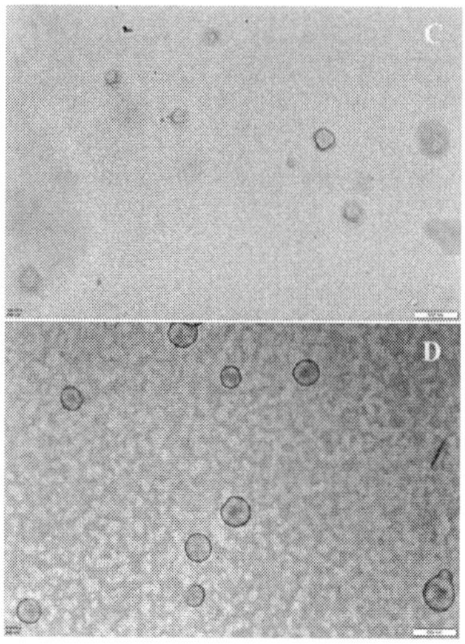

Figure 4. TEM images of mPEG$_{45}$-P(DIPEMA-co-GlyMA) nanoparticles prepared under different solid content, (A) 5%, (B) 10%, (C) 15% and (D) 20%.

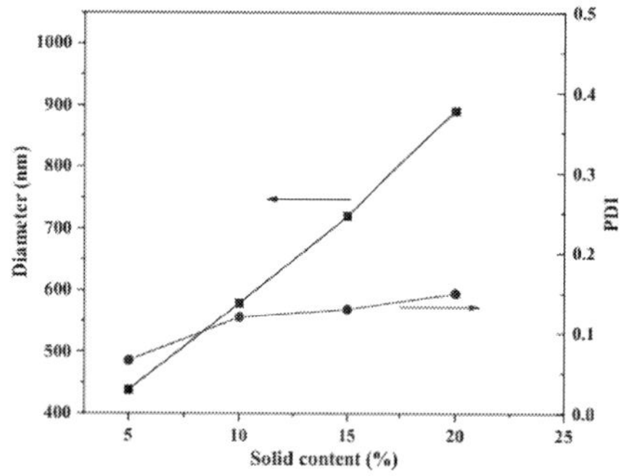

Figure 5. Hydrodynamic diameter and PDI of mPEG$_{45}$-b-P(DIPEMA-co-GlyMA) nanoparticles prepared under different solid content.

3.4 ^{1}H NMR and FTIR analysis of the polymeric vesicles

In current study, DIPEMA and GlyMA were used to copolymerize for the preparation of polymeric vesicles with the tertiary amine and epoxy groups via RAFT dispersion polymerization. The produced mPEG$_{45}$-P(DIPEMA-co-GlyMA) vesicles were isolated by freeze-drying. ^{1}H NMR spectroscopy and FTIR spectroscopy were employed to characterize the chemical structure of the produced block copolymers, as shown in Figure 6. From Figure 6A, the ^{1}H NMR spectra curve, the peak at δ 3.84 (d) attributed to -OCH$_2$- protons of PDIPEMA units, while the peaks at δ 3.2 (i), δ 2.63 and 2.82 (j) can be assigned to the protons of the epoxy group. And the absorption peak at 842 cm^{-1} in Figure 6B also indicated the existing of epoxy groups on the produced vesicles. The tertiary amine group on the PDIPEMA block could enable the vesicles with pH responsibility, and by using this epoxy functional groups on the vesicle membrane, the vesicles could be further functionalized or cross-linking of the vesicle membrane by reacting with other monomers, which would expand its potential application areas.

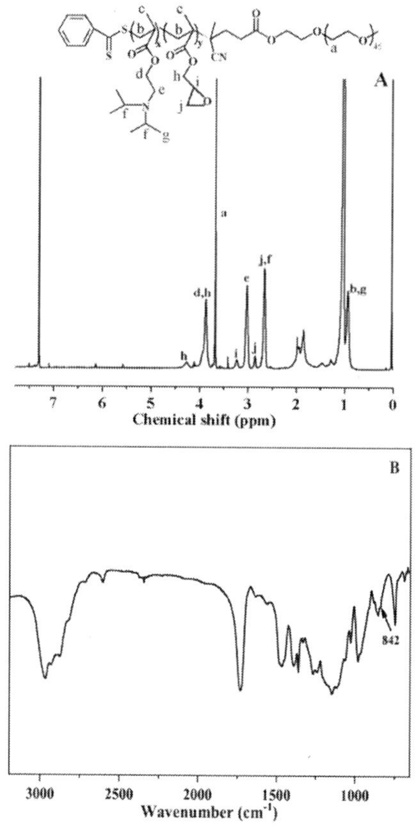

Figure 6. ^{1}H NMR (A) and FTIR (B) spectra of the mPEG$_{45}$-P(DIPEMA-co-GlyMA) vesicles

4. Conclusion

In summary, polymeric vesicles with tertiary amine and epoxy functional groups in range of 500-900 nm were fabricated by using mPEG$_{45}$-CPADB as the macro-CTA agent via the RAFT dispersion polymerization approach. DP of hydrophobic block P(DIPEMA-co-GlyMA) and the solid content both had significant effect on the morphology of the produced particles. Increasing DP of the hydrophobic block P(DIPEMA-co-GlyMA) to 60 or solid content to 10%, good shaped vesicles could be obtained, and the diameter of the vesicles increased accordingly based on the DLS results. ^{1}H NMR and FTIR analysis indicated the existing of tertiary amine and epoxy groups on the produced vesicles.

Acknowledgements

We gratefully acknowledge the financial support from the Hebei Academy of Sciences (22A04 and 22703) and Department of Human Resources and Social Security of Hebei Province (C20221059).

References

1. A. Lu, T. P. Smart, T. H. Epps, D. A. Longbottom, R. K. O'Reilly, l-Proline functionalized polymers prepared by RAFT polymerization and their assemblies as supported organocatalysts, Macromolecules, 2011, 44, 7233-7241.
2. X. L. Hu, Y. Q. Zhang, Z. G. Xie, X. B. Jing, A. Bellotti, Z. Gu, Stimuli-responsive polymersomes for biomedical applications, Biomacromolecules, 2017, 18, 649-673.
3. L. Qiu, C. R. Xu, F. Zhong, C. Y. Hong, C. Y. Pan, Fabrication of functional nano-objects through RAFT dispersion polymerization and influences of morphology on drug delivery, ACS Appl. Mater. Interfaces, 2016, 8, 18347-18359.
4. J. T. Sun, C. Y. Hong, C. Y. Pan, Recent advances in RAFT dispersion polymerization for preparation of block copolymer aggregates, Polym. Chem., 2013, 4, 873-881.
5. N. J. W. Penfold, J. R. Whatley, S. P. Armes, Thermoreversible block copolymer worm gels using binary mixtures of PEG stabilizer blocks, Macromolecules, 2019, 52, 1653-1662.
6. J. Rieger, Guidelines for the synthesis of block copolymer particles of various morphologies by RAFT dispersion polymerization, Macromol. Rapid Commun., 2015, 36, 1458-1471.
7. J. Z. Du, R. K. O'Reilly, Advances and challenges in smart and functional polymer vesicles, Soft Matter, 2009, 5, 3544-3561.
8. A. B. Lowe, RAFT alcoholic dispersion polymerization with polymerization-induced self-assembly, Polymer, 2016, 106, 161-181.
9. F. Zhang, Y. L. Niu, Y. T. Li, Q. Yao, X. Q. Chen, H. J. Zhou, J. J. Xiao, Fabrication and characterization of structurally stable pH-responsive polymeric vesicles by polymerization-induced self-assembly, RSC Adv., 2021, 11, 29042-29051.
10. X. F. Xu, C. Y. Pan, W. J. Zhang, C. Y. Hong, Polymerization-induced self-assembly generating vesicles with adjustable pH-responsive release performance, Macromolecules, 2019, 52, 1965-1975.

Study on thermodynamic system characteristics of thermoelectric units under variable operating conditions

Ziwei Zhong [1,*], Lingkai Zhu [1], Wei Zheng[1], Zhiqiang Gong [1], Panfeng Shang[1], Junshan Guo [1], Jun Liu[1], Yanpeng Zhang[1]

State Grid Shandong Electric Power Research Institute, Jinan, China

Abstract. The variable working condition of thermal system refers to that the working condition in the system changes, deviates from the design working condition or deviates from a certain reference working condition. There are various reasons for the change of thermal system's off design condition, including the change of unit heat load, the change of thermal system and equipment, and the change of steam parameters and reheat parameters. The purpose of variable condition calculation of thermodynamic system is to determine the steam parameters of each extraction port and exhaust end of the turbine and the corresponding parameters of the regenerative system. Its essence is to determine the new expansion process line and system parameters of the turbine. This is the basis for the safety and reliability analysis of the system under off design conditions and the calculation and analysis of economic indicators. Through the research on the electrothermal coupling characteristics of thermoelectric units under variable operating conditions, it can provide data support for the power grid dispatching department to master the electrical output characteristics of thermoelectric units in heating seasons.

key word: Thermoelectric unit; Off design condition calculation; thermodynamic system.

1. Introduction

The variable working condition of thermal system refers to that the working condition in the system changes, deviates from the design working condition or deviates from a certain reference working condition[1]. There are various reasons for the change of thermal system's off design condition, including the change of unit heat load, the change of thermal system and equipment, and the change of steam parameters and reheat parameters.

The purpose of variable condition calculation of thermodynamic system is to determine the steam parameters of each extraction port and exhaust end of the turbine and the corresponding parameters of the regenerative system[2]. Its essence is to determine the new expansion process line and system parameters of the turbine. This is the basis for the safety and reliability analysis of the system under off design conditions and the calculation and analysis of economic indicators.

Affected by the information segmentation between power grid and heat supply network, the traditional electric heat dispatching is to conduct real-time balancing in their respective regions[3]. At present, it is more and more difficult to balance power supply and demand, so it is necessary to carry out research and application promotion of electric heating coordination optimization technology from the perspective of energy supply system[4].

2. Study on off design calculation of thermodynamic system

2.1 Basic Steps of Off design Calculation

(1) Original calculation. According to the data obtained, the original working condition (i.e. the selected benchmark working condition) shall be calculated in detail and correctly by different methods (heat balance method, cycle function method or other methods), with the main purpose of obtaining detailed data required for subsequent calculation. These data mainly include steam extraction volume at all levels, flow rate at all levels, and main thermal economic indicators.

(2) Preliminary calculation. The variable (generally the flow of live steam) specified in the off design condition is directly substituted into the thermal system, while all other parameters are reserved as the values under the original working condition for the principled thermal system calculation of the whole plant. The primary function of the preliminary calculation is to provide calculation conditions and basis for the subsequent iterative calculation.

(3) Iterative calculation. Each iteration calculation repeats a process: all steam and water parameters of the new round and the flow rate of each stage of the turbine are

* Corresponding author: dkyzhongziwei@163.com

obtained through iteration from all steam and water parameters obtained from the previous calculation and the flow rate of each stage of the turbine.

Each iteration calculation will be carried out in two steps. The first step is preliminary calculation. The steam parameters (mainly extraction pressure and specific enthalpy) shall be corrected before this calculation by using the conclusions from the last calculation (the first iteration is the result of preliminary calculation). The specific correction method is shown below. The second step is basic calculation. Make use of the conclusions obtained from the preliminary calculation to carry out the calculation of the plant's principled thermodynamic system to prepare for the next iteration.

(4) End of iteration. If the calculation is carried out according to the above three steps, the results obtained will continue to approach the real situation, but an end condition needs to be given to complete the calculation[5]. Theoretically, the end condition of iteration calculation is that the steam parameters and flow rate obtained from the previous and subsequent two iterations are equal. However, considering the practicability of engineering problems, a minimum error is given in this project. When the calculation results of two iterations are less than the specified minimum error, it is considered that the calculation results meet the requirements and the iteration is completed.

2.2 Theoretical Basis of Off design Condition Calculation

(1) Determination of extraction pressure change

Flügel formula is the relationship between the turbine flow and the parameters before and after the stage. The expression is:

$$\frac{D_1}{D_{10}} = \frac{\sqrt{p_1^2 - p_2^2}}{\sqrt{p_{10}^2 - p_{20}^2}} \frac{\sqrt{T_{10}}}{\sqrt{T_1}} \quad (1)$$

In the formula:

D10, D1 -- stage flow before and after working condition change, t/h;

T10, T1 -- temperature before and after working condition change, ℃;

P10, p1 -- pre stage pressure before and after working condition change, MPa;

P20, p2 -- pressure after stage before and after working condition change, MPa

The formula is derived under the condition that the steam flow and flow area of the flow passage part of the steam turbine remain unchanged. For condensing units, it is considered that the regenerative steam extraction volume is approximately proportional to the heating steam flow, so Freugel formula can also be used. For heat supply extraction steam turbine, Freugel formula can also be used in sections to complete the calculation of the unit thermal system under variable conditions.

In case of variable working conditions, the influence of temperature correction term can be basically ignored. In general, p2/p1 between two steam extraction ports is very small, so equation (1) can be simplified as follows:

$$\frac{D_1}{D_{10}} = \frac{p_1}{p_{10}} \quad (2)$$

Through the calculation of the off design condition of the 135MW unit of Harbin Turbine Group Co., Ltd. and the 330MW unit of Dongfang Turbine Group Co., Ltd., and the comparison of the off design condition parameters in the thermal characteristics book, it is found that the accuracy requirements can be met by using Formula (2) to correct the extraction pressure.

(2) Determination of relative internal efficiency of steam turbine

In the approximate calculation of the variable working conditions of the steam turbine, the efficiency of the governing stage and the last stage varies greatly, while the internal efficiency of other pressure stage groups basically remains unchanged when the working conditions change, and the steam state line shows a process line of parallel movement. In the calculation of specific units, the internal efficiency of each unit can be verified through the thermal characteristics book, which can well prove the above theory.

The internal efficiency of the regulating stage can be calculated according to the flow efficiency curve of the manufacturer. In the absence of the manufacturer's curve, the internal efficiency curve of the regulating stage can be fitted according to the thermal test results.

The last stage efficiency curve can also be checked according to the characteristics of the manufacturer's variable working conditions[6]. For the final stage, the main loss is wet steam loss. The internal efficiency of the final stage can also be calculated from the change of wet steam loss. It is generally believed that the unit efficiency will decrease by 1% when the average steam humidity of the final stage group increases by 1%.

The detailed calculation of steam turbine condenser under off design conditions is very complicated, and the calculation formula using the simplified algorithm is as follows:

$$t_k = t_{w1} + \frac{D_k q_k}{C_W G_W [1 - \exp(-k_0 (-\frac{G_w}{G_{w0}})^{0.5} \frac{F_k}{C_w G_w}]} \quad (3)$$

In the formula:

Dk -- cooling water entering condenser, t/h;

Qk -- Condensation heat release of 1kg exhaust steam in condenser, t/h;

CW -- Specific heat of cooling water, J/(kg. C);

M -- cooling ratio

When the cooling water operates normally, the air partial pressure is very small and can be ignored. The calculated saturated water pressure corresponding to the saturated water temperature is the condenser pressure under variable working conditions.

For the calculation in this paper, since the influence of the condenser mainly lies in the final stage exhaust pressure, it has little influence on the overall off design condition calculation. At the same time, the condenser pressure is selected as the rated pressure of the condenser according to the thermal characteristics book under off design condition.

(3) Determination of Extraction Point Parameters and Hydrograph
In the calculation of the off design condition of the thermal system, the steam parameters at each extraction point are determined first and the process line of the operation condition change is drawn on the enthalpy entropy diagram of the water steam. The extraction pressure of each stage and the new exhaust pressure under the new working condition are determined by Freugel formula. Then determine the internal efficiency of each unit according to the determination method of internal efficiency of each unit.
The extraction enthalpy and exhaust enthalpy of each group can be further obtained by the obtained pressure of each extraction point and the efficiency value in each group. According to the definition of relative internal efficiency of steam turbine, the internal efficiency calculation formula of any stage group r is:

$$\eta_{oi(r)} = \frac{h_{i(r)} - h_{2(r)}}{H_{a(r)}} \qquad (4)$$

From this, the formula of extraction enthalpy at all levels can be obtained as follows:

$$h_{2(r)} = h_{1(r)} - H_{a(r)}\eta_{oi(r)} \qquad (5)$$

In the formula:
$h_{1(r)}$, $h_{2(r)}$ -- enthalpy value of steam at the inlet and outlet of stage r, kJ/kg;
$H_{a(r)}$—— isentropic expansion enthalpy drop of steam in steam turbine of stage r group, kJ/kg;

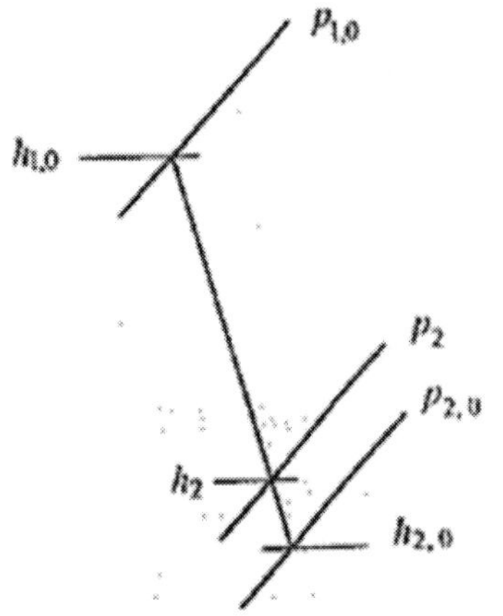

Figure 1 etermination of Specific Enthalpy at Inlet and Outlet of Unit

The peak shaving capacity of thermal power units is the highest and lowest electrical load that the unit can reach under a certain external steam extraction capacity. The limiting conditions for different types of thermal power units to carry the highest and lowest electrical loads are also different.

(4) Determination of water enthalpy at heater outlet
To determine the outlet water enthalpy of the heater, the variable working conditions of the heater shall be considered to determine the corresponding end difference and heat transfer coefficient. Considering the small change of end difference and the requirement of calculation speed, this paper does not consider the variable working condition of the heater when determining the steam and water parameters at each point of the thermal system. The upper and lower end difference of the heater is obtained from the thermal characteristics book and considered to be unchanged. According to the definition of heater end difference and the known outlet pressure of feed water pump and condensate pump, the outlet water temperature, outlet water enthalpy and outlet drain water temperature and enthalpy of heaters at all levels can be determined.

(5) Analysis of auxiliary steam and water components
Various auxiliary steam and water components in the thermal system mainly include the utilization of shaft seal steam leakage, steam used by steam extractors, valve stem steam leakage and blowdown expansion steam, etc. They should be correctly treated in the calculation of variable working conditions to ensure the reliability of the calculation. According to the theory in reference [2], the steam leakage parameters of the unit change with the change of the source parameters, and the steam leakage flow changes proportionally with the change of the live steam flow

3. Conclusion

In the analysis of Harbin Turbine 135MW unit, the heat supply and steam extraction together with the subsequent flow passage are treated as the fourth stage flow passage. It can be seen that when the extraction pressure rises, the internal efficiency of the unit decreases under the same flow rate; On the contrary, the efficiency in the unit increases.
There are few known data under heating and steam extraction conditions, but it can still be seen that under the same steam extraction pressure conditions, the internal efficiency of the fourth stage is approximately linear with the flow rate of the fourth stage. Therefore, the piecewise linear interpolation method is also used to solve the fourth stage internal efficiency under variable conditions.
The final stage wet steam loss is the main reason for the change of the final stage efficiency under off design conditions. Generally speaking, the unit efficiency will decrease by 1% when the average steam humidity increases by 1%. Because the change of the last stage steam humidity is caused by the change of the last stage flow rate in the variable working conditions, the relationship curve between the last stage flow rate and the internal efficiency is needed to find its corresponding relationship.

Acknowledgements

This work is supported by the science and technology project of State Grid Shandong Eectric Power Research Institute(ZY-2022-15), which studies Research on peak capacity improvement of thermal power unit based on electric heating coordination optimization under guaranteed supply demand.

References

1. Bu Yongdong, Yang Lijun, Du Xiaoze, etc. Numerical study on off design performance of power plant air cooling system [J]. Chinese Journal of Electrical Engineering, 2012, 32 (35): 8-13.
2. Lv Kai, Wang Hongyu, Zhou Jia, etc. Study on characteristics of electrothermal coal for cogeneration units [J]. Thermal Power Generation, 2018, 47 (5): 7-12.
3. Yang Bo, Li Zheng.Research on modeling method of thermal system leading factors of thermal power unit under off design conditions[J]. Chinese Journal of Electrical Engineering, 2005.
4. Song Binbin.State equation of thermal economy of thermal system of thermal power plant [J]. Science and Technology Innovation Guide, 2013, (4): 217-217.
5. Wang Wei.Analyze the application of thermal test in thermal power plants [J]. China Science and Technology Wealth, 2012, 000 (014): 214-214.
6. Yan Shui Bao.Energy saving principle and method of thermal system of power plant [M]. China Electric Power Press, 2007.

https://doi.org/10.1051/e3sconf/202337501004

Power consumption characteristics of cement industry and parameter analysis of self provided power plant

Lingkai Zhu [1], Weishuai Wang [2], Haijing Zhang [2], Wei Zheng [1]

[1] State Grid Shandong Electric Power Research Institute, Jinan, China
[2] State Grid Shandong Electric Power Company, Jinan, China

Abstract. Cement production is an industry that consumes a lot of electric energy. During the period of power shortage in China, the government encouraged multi-channel financing to run electricity to alleviate the power shortage problem of large industrial users. Therefore, many cement enterprises have established their own power plants. This paper analyses the basic electrical characteristics of the cement industry, classifies the loads of enterprises, and studies the regulation characteristics of various types of equipment.

1. Introduction

Self owned power plants are built by our country to alleviate the difficulty of power consumption in the era of power shortage. The first batch of self owned power plants was built by the government for specific industrial enterprises during the expansion of production scale, and the second batch was built at the weak stage of power infrastructure construction in China. In order to alleviate the problem of power shortage for large industrial users, the government encouraged multi-channel financing to run power plants and develop local self owned power plants of small thermal power types[1,2]. High power consumption industries include: electrolytic aluminum, calcium carbide, steel, cement and other enterprises. Among the high energy consumption users in some cities, the cement industry accounts for a large proportion. The survey shows that in 2011, among the 46 high power consumption industries of Jinzhong Power Supply Branch, 17 were cement, accounting for 36.96%[3].

The self owned power plant can provide a self balanced power supply system within the enterprise, and also provide a relatively stable and low-cost power supply for end users[4]. Some large cement plants are far away from the city, and there is no supporting power transmission and distribution system. For example, the 5000 t/d cement production line in Western Asia undertaken by Sinoma Construction Company is 80km away from the city, and there is no large power grid nearby. If the power supply is connected from the power grid, a 57 km power transmission and distribution system is required, and the power grid can only meet the discontinuous 10MW load capacity, and the unit price of industrial power is close to $1.2, so a self-contained power supply is built, The island operation mode is adopted[5].

2. Electrical characteristics of power plant basic production

The production process of cement industry mainly refers to ore mining, ore crushing, rotary kiln paper burning, clinker grinding, product packaging, and is often referred to as "two grinding and one burning". The cement industry mainly produces in the normal and low periods. The wind bone inversion is obvious, the nighttime load is high and relatively stable, the equipment operation cycle is long, the power load is high, and the power supply reliability requirements are high.

Typical daily load characteristic curve of cement industry is shown in Figure 1.

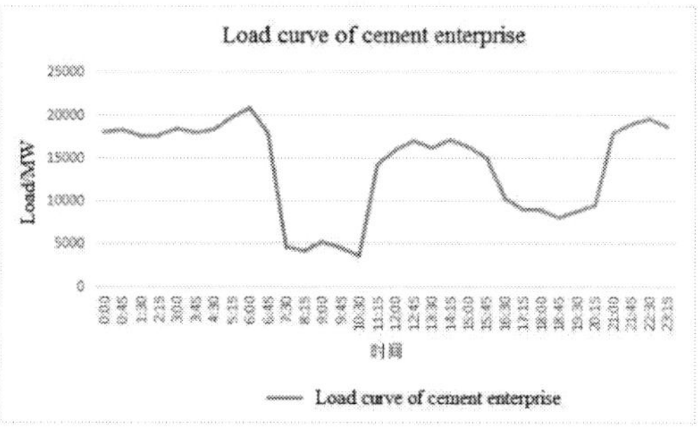

Figure 1. Typical Load Curve of Cement Enterprise.

---
* Corresponding author: zhulingkai@woyoxin.com

3. Equipment load characteristics of cement enterprises

The main production load of the cement industry accounts for 55%-60% of the total load, which is used for raw meal grinding, raw material burning and cement grinding. The main equipment includes raw meal mill, cement mill, ball mill, etc.

The auxiliary production load accounts for 15%-20% of the total load, which is used for the transmission of raw materials, the delivery of finished cement products, the preparation of intermediate products, etc. The main equipment are transmission fans, transmission belt motors, etc.

The safety guarantee load accounts for 8%-15% of the total load, which is used for the cooling of cement production process, lubrication of rotating equipment, etc. The main equipment includes cooling water pump, lubricating oil pump, etc.

Non productive load accounts for 2%-5% of the total load, including office electrical equipment, split and central air-conditioning equipment.

Equipment classification of cement industry is shown in Table 1 below.

Table 1. Classification of Equipment in Cement Industry.

Load Category		Load Ratio	Major Installation
Production Load	Safety Load	8%-15%	cooling water pump, lubricating oil pump, transmission equipment, etc
	Main Production Load	55%-60%	raw meal mill, cement mill, ball mill, rotary kiln, vertical kiln, etc
	Auxiliary Production Load	15%-20%	transmission fan, transmission belt motor, elevator, pulverized coal mill, etc
Non-production load		2%-5%	office lighting, office appliances, split and central air conditioners, blowers, etc

4. Analysis and Summary of Equipment Adjustment Ability

4.1 Potential analysis on power generation side of cement enterprises

In terms of heat and power cogeneration units, the heat and power cogeneration generator units used by cement enterprises generally adopt and are equipped with two types of steam extraction condensing steam turbine generator units and back pressure steam turbine generator units, which can provide heat and power for the process flow of cement enterprises. It can be seen from the production link that the heat load in the production process of the enterprise will not change, so the change of the power supply to the grid is only determined by the change of the load of electrical equipment.

In terms of waste heat and residual pressure utilization unit, there are two sources of high-temperature and high-pressure superheated steam for the generator unit. One is to concentrate the black liquor in the concentrator, and then enter the alkali recovery furnace for combustion, burning the organic components in the black liquor to generate heat energy, which is absorbed by the alkali recovery furnace to generate medium and high-pressure steam for power generation; The other is that the coal is transported to the circulating fluidized bed boiler for combustion, heating the deoxidized soft water in the circulating fluidized bed boiler and generating high-temperature and high-pressure superheated steam. The power generation capacity of this waste heat and residual pressure unit is determined by the waste liquid generated during drying in the production link, which is not flexible.

4.2 Potential analysis of power plant load side in cement enterprises

The continuous production of cement industry is to improve production efficiency, not limited by production process. It has interrupting potential in production process and can participate in power demand response. The main production loads of rotary kiln and vertical kiln account for 45% of the total load and 16% of the total adjustable load.

Load regulation of cement industry is shown in Table 2 below.

 https://doi.org/10.1051/e3sconf/202337501004

Table 2. Load Control in Cement Industry.

Load Category	Major Installation	Load Ratio	Regulatory Mode	Regulatory Time			Adjustable Load Ratio		
				Setup Time	Response Time	Resumption Operation Time	Individual Effect	Overall Proportion	Total
Main production load	rotary kiln	25%	automatic	2h	0.5h-2h	2h	40%		
	vertical kiln	20%	automatic	2h	0.5h-2h	2h	30%		
	raw meal mill	5%	automatic	1h	0.5h-2h	1h	20%	19%	
	cement mill	5%	automatic	1h	0.5h-2h	1h	20%		24%
	ball mill	5%	automatic	1h	0.5h-2h	1h	20%		
Auxiliary production load	conveyor belt motor	6%	direct (flexible)	minute grade	0.5h-4h	minute grade	50%	5%	
	air compressor	5%	direct (flexible)	minute grade	0.5h-4h	minute grade	50%		
	office lighting	<1%	direct (rigid)	second grade	0.5h-4h	second grade	<1%		
Non-production load	split and central air conditioning system	<1%	direct (flexible)	second grade	0.5h-4h	second grade	<1%	<1%	
	domestic electricity	<1%	direct (rigid)	second grade	0.5h-4h	second grade	<1%		

5. Development trend outlook

At present, under the national "double carbon" economic development goal, the government's management policy for self owned power plants is being tightened step by step, gradually stopping the disorderly development of self owned power plants. Due to the characteristics of multi energy cogeneration, self owned power plants generally have the effect of comprehensive utilization of resources. Encouraging self owned power plants to participate in the regulation of power grid supply and demand is of great significance in improving the efficiency of energy comprehensive utilization, improving the balance between supply and demand of regional power grids, and improving the energy consumption and environmental protection level of self owned power plants, which is also the direction of future development of self owned power plants.

Acknowledgments

This work was financially supported by The Science and Technology Project of State Grid Corporation of China - Research on Flexible Interactive Technology of Autonomous Power Plants for Market-oriented Supply and Demand Adjustment (SGSDDK00YJJS1900143).

https://doi.org/10.1051/e3sconf/202337501004

References

1. Yue Zhang, Zhongxian Dai, Jinli Lv, etc. Analysis on Related Policies and Development Problems of Self-provided Power Plant under Energy Transformation[J]. Electrical Technology and Economics, 2020, (4):65-66.
2. Yifan Huang, Haijing Zhang, Lei Wang, etc. Evaluation method for flexibility regulation of self owned power plants based on enterprise production characteristics [J]. Demand Side Management, 2021, 23 (1): 61-66.
3. Min Han, Zhengquan Tian. Research on Power Consumption in High Power Consumption Industries [J]. Caizhi, 2011, (1): 87-88.
4. Jiajia Yang. Main Problems and Policy Suggestions of Enterprise Owned Power Plants [J]. China Price Supervision and Inspection, 2012, (8): 13-14
5. Xianwei Wang, Wenjie Yin. Self owned coal-fired power plant in cement production line [J]. Cement Technology, 2013, (1): 87-88.

Application Research on Liner Tube without Variable Diameter of Shale Gas Plunger Gas Lift Wellhead

Chengzhi Duan [1], Haonan Xu [1], Jing Fan [1], Lujun Wen [2], Jianhui Ni [2]

[1]Sichuan Hwayoo Petroleum Drilling & Production Co., Ltd., Sichuan, China
[2]Chongqing Gas Field of PetroChina Southwest Oil and Gasfield Company, Chongqing, China

Abstract. Given that there is variable diameter in the shale gas plunger gas lift wellhead, which causes that the plunger cannot run up and down the gas well wellhead, affects the effect of plunger drainage gas production, the paper carries out research on liner tube without variable diameter of the shale gas plunger gas lift wellhead and installation methods, makes the diameter of the gas well wellhead consistent with that of the oil pipe, ensure that the plunger can be open up and down the wellhead of the Christmas tree, facilitate the periodic inspection and replacement operation of the plunger, and reduces the cost of plunger salvage operations. The successful application of the process plan in the shale gas well ** has achieved good drainage gas production effect, truly solved the variable diameter problem of shale gas plunger gas lift wellhead, and achieved the goal of once and for all.

Keywords: Plunger drainage gas production; shale gas; liner tube.

1. State of Shale Gas Plunger Gas Lift Wellhead Process

For shale gas development, the key technology is factory fracturing, and shale gas wells need to be fractured to transform the reservoir before they are put into production. In order to simplify the construction process and save investment costs OF gas well, well completion usually adopts fractured wellheads, namely fractured wellheads are used below the 1# main valve and will not be replaced, while conventional wellheads are replaced above the 1# valve, as a result, there is clash between the production tubular column and the gas wellhead diameter at the Christmas tress wellhead.

Shale gas** well adopts KQ180/78-105EE class gas production wellhead (big cross, 1#, 4# main valve: 105MPa, main diameter 180mm; small cross 105MPa, 78mm diameter; big and small cross connect in the middle by 78-105 to 180-105 conversion flange). The structure of production pipe column from bottom to top: one φ88.9mm sieve pipe (has bell mouth) x2.95m + one φ93mm rupture disk x one 0.24m + φ88.9mm 2 7/8 "EU-P x 2 7/8 "NU-B variable buckle 1 x 0.20m + one φ88.9mm plunger working cylinder x 0.31m + four hundred and thirty-three 2 7/8 "NU chamfered oil pipe x 4136.41m.

The well adopts 2 7/8" oil pipe, and the gas production wellhead adopts Christmas tree with 180mm main diameter, there is variable diameter at the wellhead. For this type of shale gas well, the existing plunger gas lift process usually only considers whether the plunger can be successfully dropped into the wellbore without considering whether the plunger can reach the blowout prevention tube, thus affecting the drainage gas production effect of plunger, make the plunger cannot run up and down at the wellhead, which is not conducive to periodic inspection and replacement of the plunger, and once the plunger is stuck or is replaced or inspected, wireline work must be carried out to salvage the plunger, which increase the cost of construction operations.

2. Design of Liner Tube without Variable Diameter

In order to ensure that the plunger can be open up and down the wellhead of the Christmas tree, facilitate the periodic inspection and replacement operation of the plunger, ensure the drainage gas production effect of plunger, and truly solve the technical difficulties of variable diameter of shale gas plunger gas lift wellhead. The gas production wellhead experts of our company carried out technology study and designed a liner tube without variable diameter of shale gas plunger gas lift wellhead and installation process, completely solve the variable diameter problem of shale gas plunger gas lift wellhead and achieve the goal of once and for all.

2.1 Design of Liner Tube

In order to adapt to the diverse working structure of the Christmas tree, the liner pipe is designed as straight liner pipe I, perforated liner pipe II, straight liner pipe III and bell mouth liner pipe IV, under the condition of ensuring the normal work of the Christmas tree, four sections of liner pipes are installed in the corresponding positions of the Christmas tree, as shown in Fig.1. The liner pipe is installed inside the channel of Christmas tree, achieve the same inner diameter of the production pipe column and the channel of Christmas tree, prevent the plunger from jamming when running up and down the wellhead, because the diameter of Christmas tress is much larger than the maximum outside diameter of the plunger, affecting gas well production.

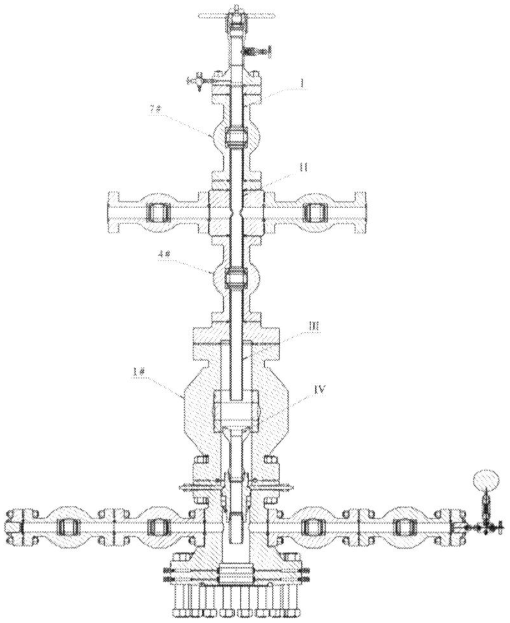

Fig.1 design diagram of the liner tube with variable diameter

The inner diameter of the production tubular column of the gas well is 62mm, and the maximum outer diameter of the plunger is 59.5mm, in order to achieve the same inner diameter of the production tubular column and the Christmas tree channel, and ensure the smooth work of the plunger up and down the wellhead, the inner diameter of liner is designed to be 62mm. In order to install the liner tube into the Christmas tree and ensure the strength of the liner tube, the outer diameter of the liner tube is 72 mm. For ease of installation and disassembly, each liner tube is installed on the threaded flange among valve bodies by threaded connection mode. When installing the liner, a certain distance is kept between the liner tube and the brake board to prevent the liner tube from affecting the normal use of the valve. Without affecting the normal work of the valve, the design length of the liner tube should be taken as round as possible in order to improve production efficiency.

The ends of liner pipe I, liner pipe II and liner pipe III are designed with 60° chamfer. The slight tilt will occur when the plunger moves up and down the wellhead, the ends of the liner tube are designed with 60° chamfer, when the plunger passes through the valve body, the plunger can smoothly enter the next liner by using the chamfer.

2.1.1. *Design of Straight Liner Tube I*

Straight liner pipe I is designed with a total length of 290mm and is installed to the upper flange of the 7# valve to the upper part of the valve plate. The inner diameter of the blowout prevention device is smaller than the inner diameter of the valve, in order not to affect the normal work of the blowout prevention device, the threaded flange is installed between the blowout prevention device and the 7# valve, and the installation thread of the liner pipe I is designed at the end. In order not to affect the normal on-off work of the 7# valve board, and ensure that the plunger can smoothly pass the valve, it is appropriate to minimize the distance between the liner I and liner II, the design spacing is 147.3mm. As shown in Fig.2, the liner tube I is installed in the threaded flange through the end thread, and the other end does not affect the normal work of the brake board.

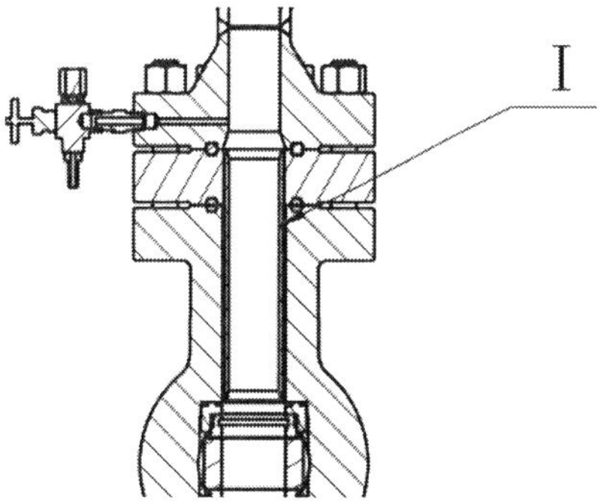

Fig.2 installation diagram of liner tube I

2.1.2. *Design of Perforated Liner Tube II*

The perforated liner II is installed from the lower part of valve seat 7 # to the upper part of valve seat 4 #. The liner tube II should not affect the normal on-off of the 7# valve board and the 4 # valve board, and the liner II should ensure that the plunger can pass through the valve smoothly, the spacing between liner tube II and liner tube I and liner tube III should be reduced as much as possible, the total design length of liner II is 835mm. For ease of the installation of liner II, the threaded flange is installed between the small cross and 7 # valve, and the installation thread of liner tube II is designed to be 225mm away from the upper end of liner II. Since the small cross channel is 449mm away from the upper end of liner tube II, a hole with diameter of 50mm is designed at 225mm away from the upper end of liner tube II to ensure that the produced fluid from the gas well can pass through the small cross channel smoothly. As shown in Fig.3, the installation position of liner tube II does not affect the normal work of valve 7 # and valve 4 #, and the small cross channel is opened to ensure that the liquid carried by the plunger can be sent out normally.

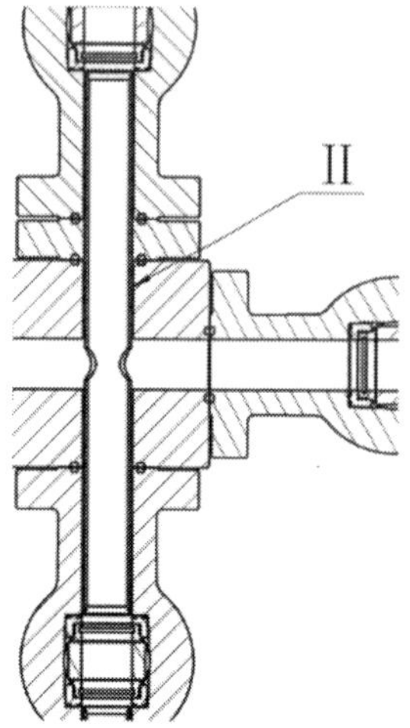

Fig.3 installation diagram of liner tube II

2.1.3. Design of Straight Liner Tube III

Straight liner III is installed from the lower part of valve seat 4 # to the upper part of valve seat 1 #. After liner tube III was installed, the normal on-off of 4 # board valve and 1 # board valve will not be affected, and liner tube III needs to ensure that the plunger can pass through the valve smoothly, the distance between liner III and liner II and liner IV should be reduced as much as possible, the total design length of liner III is 775mm. For ease of the installation of liner tube III, liner tube and thread are designed in the transformer flange, which is installed between 4 # valve and 1 # valve, the installation thread of liner tube III is designed to be 230mm away from the upper end of liner tube II. As shown in Fig. 4, the installation position of liner III does not affect the normal operation of valve 4 # and valve 1 #.

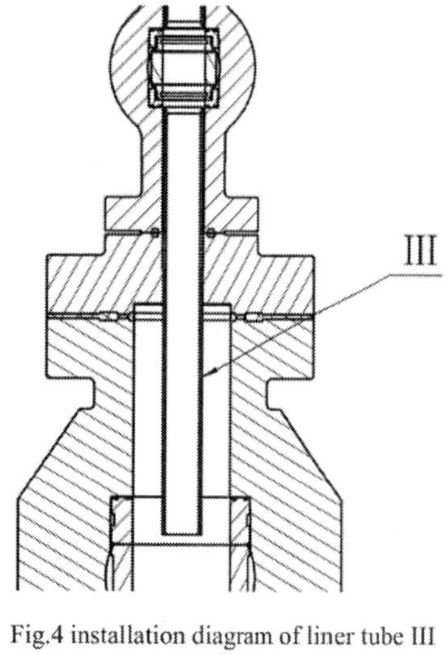

Fig.4 installation diagram of liner tube III

2.1.4. Design of bell mouth Liner Tube IV

The bell mouth liner tube IV is installed under the valve seat of 1 # valve to the oil pipe hanger. The lower end of liner IV is designed with tubing joint NU/EU thread to facilitate the connection of tubing hanger, and the position of liner IV should not affect the normal on-off of 1# valve ram, the total design length of liner tube IV is 379mm. Since the valve seat of 1# valve seat is larger than that of other valves, the installation space between liner tube IV and liner tube III is larger than that of other liner tubes, therefore, liner tube IV is designed to be bell mouth, so that the plunger can be reset in the inclined state by using the liner tube IV bell mouth slope, prevent the plunger from flying out of the liner tube due to inclination and getting stuck in 1 # valve, and ensure that the plunger can enter liner tube IV smoothly. Since the inner diameter of 1# valve channel is 89.5mm, the maximum outer diameter of the bell mouth is designed to be 80mm, while facilitating the installation of liner tube IV in 1 # valve, it ensures that the plunger can reach the bell mouth of liner tube IV even if it is tilted to the maximum angle, and it can be reset smoothly to enter to liner tube IV. The external part of bell mouth and the middle part of the liner tube are designed to be 60° chamfer, which reduces the weight of liner tube IV and weakens the stress concentration of liner tube IV, and the design is beautiful, the strength of liner tube IV is improved. The inner diameter of the bell mouth end of liner tube IV is designed with threads to facilitate the direct installation of liner tube IV at the tubing hanger by using the wellhead pickup device. As shown in Fig.5, the installation position of liner IV does not affect the normal work of valve 1 #, and can be installed on the tubing hanger.

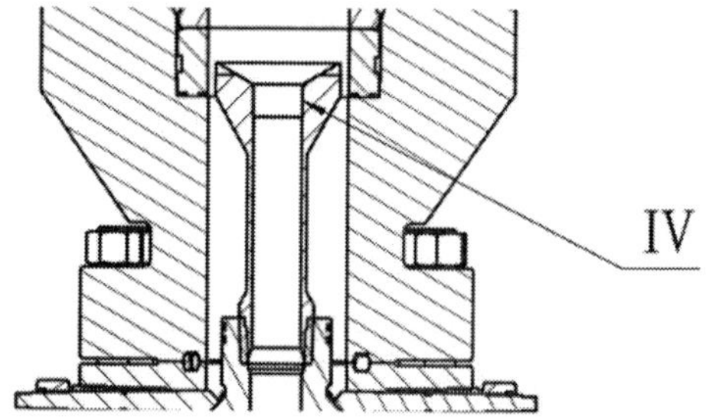

Fig.5 installation diagram of liner tube IV

2.2 Installation Method

According to the design of liner tube without variable diameter, combining the tool device of Christmas tree, the installation method of liner tube IV and other three sections of liner tubes with the pick-up device is designed. This installation process method needs to be implemented by team with corresponding qualification for pressure operation, and nonprofessional teams are strictly prohibited from construction.

(1) Shut in the well, close 1 # main valve and relieve pressure;

(2) Dismantle the wellhead process, namely the part above 1 # main valve;

(3) Install the liner tube IV to the lower end of the pressurized pick-up device, connect the flange, test the pressure of clean water at 30 MPa, and maintain the pressure for 15 minutes, there is no leakage outside the equipment, and the wellhead pressure drop should not be higher than 0.5 MPa;
(4) Open 1 # main valve, lower liner tube IV to hanger position, connect with hanger
(5) Remove the pick-up head and exit the pick-up head to the upper position of 1 # valve plate;
(6) Close 1 # main valve and open the pressure relief device and relieve the residual pressure at the wellhead;
(7) Remove the pressure pick-up device;
(8) Install the connecting flanges of 1 # and 4 # valves at the wellhead, replace the connecting lifting short joint (built-in liner III), and install 4 # valve and small cross;
(9) Install the small cross and 7 # valve to connect lifting flange (built-in liner tube II), and install 7 # valve;
(10) Install the upper part of 7# valve and connect lifting flange(built-in liner tube I);
(11) Close pressure test of 7 # valve, test pressure at 30MPa, no pressure drop and external leakage are qualified.

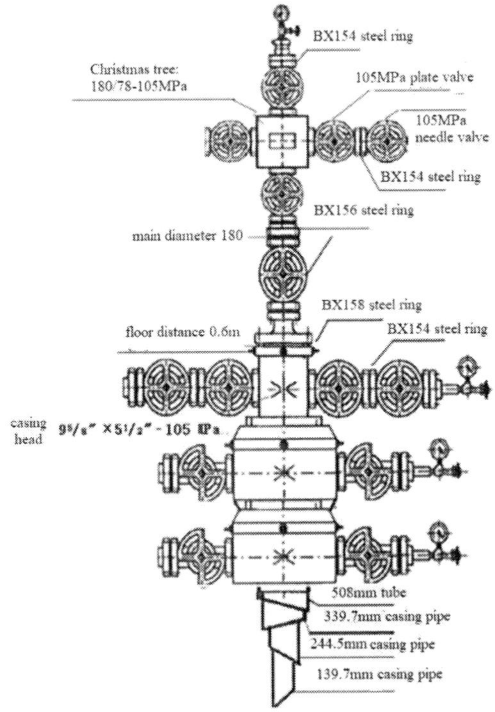

Fig. 6 schematic diagram of gas production wellhead equipment

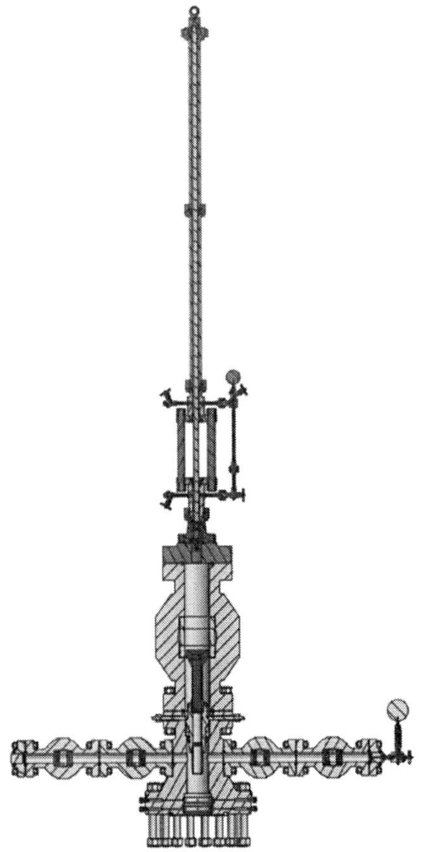

Fig.7 installation diagram of pick-up device under pressure and liner tube IV

3. Application Effect

Shale gas well ** is a shale gas development evaluation well/horizontal well deployed in the syncline structure of Fangjiagou of the southern Sichuan low fold belt in the Sichuan Basin, the well depth of drilled well is 5627m, the layer of drilled well is $S1_1$, and the well completion method is casing completion. It is put into production in April 2018, the early daily gas production is $6.17\times10^4 m^3/d$, daily water production is $8.1m^3$, casing pressure is 48MPa. By the beginning of 2022, the cumulative gas production is $2907\times10^4 m^3$, the accumulated water production is $10013m^3$.
In January 2022, the adjacent well was fractured before the plunger is put into operation, in order to avoid fracturing channel, the well was shut. As a result, the gas well was drowned, and it was difficult to resume production, intermittent well opening and liquid drainage into the air were adopted to maintain the gas well production, The daily output was $2.7707\times10^4 m^3 \downarrow 0.5478\times104m^3$.
In May 2022, the well was transformed by the plunger drainage gas production process, the plunger wellhead process of liner tube without variable diameter was adopted, which completely solved the variable diameter

https://doi.org/10.1051/e3sconf/202337501005

problem of the shale gas plunger gas lift wellhead, ensured that the plunger stably reached the wellhead blowout prevention tube, facilitated the periodic inspection and replacement of the plunger, and saved the salvage cost of plunger.

After the plunger is put into operation, the gas well has been running stably for more than 100 days, the plunger reaches the wellhead stably, reaching the detection rate is 100%, the gas well realizes automatic liquid drainage, the pressure difference of the oil jacket is 6.1↓1.3MPa, and the annular liquid level depth is 3363↓3643m, which effectively reduce the liquid accumulation in the gas well, the average daily gas production is $10000m^3$, production increases by 80%, the efficiency of drainage gas production is greatly improved, effectively reduce the cost of human management in the well station, and realize automatic drainage gas production in the gas well, green and environment-friendly production.

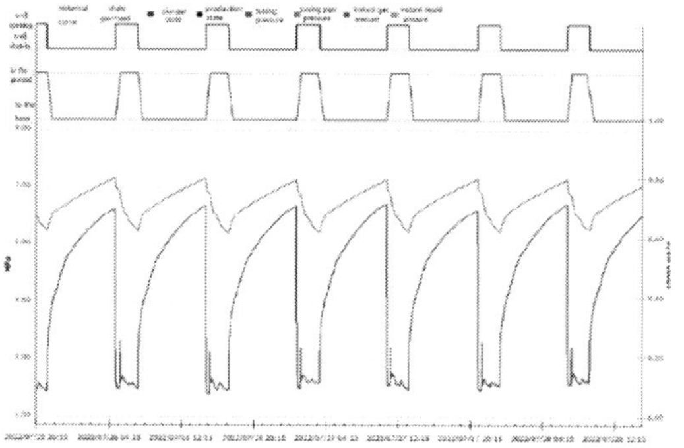

Fig.8 operation curve of plunger in shale gas well**

4. Conclusion and Recommendation

4.1 The liner tube device without variable diameter of shale gas plunger gas lift wellhead and its installation process completely solves the technical challenge: variable diameter of shale gas well plunger gas lift wellhead affects the drainage gas production effect of plunger;

4.2 The liner tube device without variable diameter of shale gas plunger gas lift wellhead and its installation process can ensure that the plunger reaches the ground blowout prevention pipe stably, facilitate the periodic inspection and replacement of the plunger, and save the salvage cost of plunger;

4.3 The plunger can run up and down periodically in the wellbore of shale gas well, which is the premise for plunger gas lift well to achieve better drainage gas production effect.

References

1. Guo Ruxiang, Xiang Xin, Li Yonglong, et.al. Discussion On the Key Technology of Plunger Gas Lift Drainage Gas Production Process [J]. China Petroleum and Chemical Standard and Quality, 2019(17):2.
2. Han Qianghui. Design And Application of a New Self-Buffered Plunger Gas-Lift Drainage Device for Horizontal Wells: A Case Study of the Changqing Gas Province, Ordos Basin [J]. Natural Gas Industry, 2016, 36(12): 67-71.
3. Jia Min, Li Jun, Li Nan. Progress and Application of Plunger Gas Lift Drainage and Production Technology [J]. Western Exploration Engineering, 2015, 27(7):4.
4. Tong Zheng, Pei Xiaohan, Shen Zejun, et.al. Plunger Lift and Monitoring System for Gas Wells Based on Deployment-Retrievement Integration [J] Natural Gas Industry, 2015, 35 (08): 44-50.
5. Yan Yingfei, Tang Yonghuai, Yao Jun. Application of the Plunger Airlift Draining Gas Recovery Technology in Yanchang Gas Field [J]. Petrochemical Industry Technology, 2020,27 (07): 80+86.
6. Feng Xiaoya. Process Design and Lifting Capacity Analysis of Plunger Gas Lift [D]. Changjiang University, 2020. DOI: 10.26,981/d.cnki.gjhsc.2020.000,336.
7. Liu Mengpeng, Gao Xiuli, Li Donglin, Hong Liang, Huang Shuang. Research on the Influence Factors of Return Pressure on Plunger Gas Lift Drainage Gas Production Technology [J]. Inner Mongolia Petrochemical Industry, 2021,47 (09): 75-77.
8. Yan Hong, Zheng Qingping, Shang Shaofen Development and Application of Shale Gas Well Drainage Production Process Chart in Southern Sichuan [C]//. Proceedings of the 32nd National Natural Gas Academic Annual Conference (2020). [Publisher unknown], 2020: 1579-1585. DOI: 10.26914/c.cnkihy.2020.065037.
9. Cheng Xingrong, Zhang Junxu, Xu Chao, Li Yaohui. Study on Operation Optimization of Plunger Gas Lift Drainage Gas Recovery Process [J]. Shanxi Chemical Industry, 2020, 40 (04): 120-121+138. DOI: 10.16525/j.cnki.cn14-1109/tq.2020. 04.41.
10. Ye Changqing, Xiong Jie, Kang Linjie, et.al. New Progress in R & D of Water Drainage Gas Recovery Tools in Sichuan Chongqing Gas Zones [J] Natural Gas Industry, 2015, 35 (02): 54-58.

https://doi.org/10.1051/e3sconf/202337501006

Cause Analysis and Solutions About Abnormal Heating of Equipment Clamp

Jianjun Zheng[1], Zhehao Xu[2], Weiwei Gong[1]

Inner Mongolia Power (Group) Co., Ltd., Inner Mongolia Power Research Institute Branch, Hohhot 010020, China
Inner Mongolia Power (Group) Co., Ltd., Inner Mongolia Power Economic and Technological Research Institute, Hohhot 010020, China
*Corresponding author e-mail: dkyzjj@163.com

Abstract: As a kind of link terminal in power grid system, clamp has become one of the key components and widely used for connecting the transmission lines and electrical equipment. However, the joint between the tubular structure and the terminal plate or the drainage plate is always a weak part, which is easy to produce defects and then accelerate the failure of the clamp during service. This is also limited the use space of clamp to a certain extent. In this paper, research status on heating causes of equipment clamps and the corresponding solutions are mainly summarized. And the application of disc shape memory alloy gasket as well as the design of mortise tenon structure clamp will fundamentally solve the problem of abnormal heating of wire clamp.

Keywords: Clamp; heating; locked; bolt; memory alloy; mortise tenon.

1. Introduction

Electric power is a basic system of the national economy and the most important part of the energy industry[1-3]. The production of the electric power is a continuous process including power generation, transmission, transformation and distribution, and problems occurred in any link will affect the whole system[4-5]. If these problems cannot be handled in time or reasonably solved, a series of chain reactions will be induced, causing a large-scale power failure of the system and leading to the expansion of the accident. Therefore, it is particularly important to ensure the safe and stable operation of the power system.

2. Importance of equipment clamps

Equipment clamp is a kind of link terminal in power grid system, which is one of the key components and widely used for connecting the transmission lines and electrical equipment [6-7]. The commonly used equipment clamps are divided into aluminum equipment clamps and copper aluminum transition equipment clamps, which are usually composed of the fastening strand part and the electrical equipment connection part. The former is mostly tubular structure, while the latter is terminal plate. In addition, SY type equipment clamp (Fig.1 (a)) is usually produced by welding aluminum pipe and terminal plate or integral casting, and NY type equipment clamp (Fig.1 (b)) is mainly used in power transmission lines and made of aluminum alloy, producing by welding the strain tube and the drain plate [8].

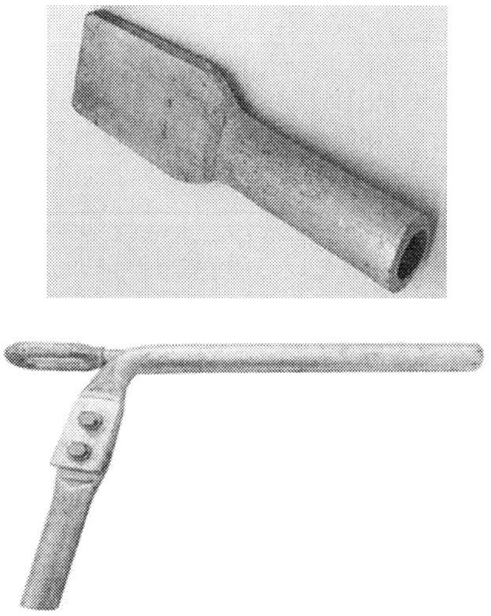

Fig.1 Common types of the equipment clamp

SY type (b)NY type

No matter whether welding or casting is adopted, the joint between the tubular structure and the terminal plate or the drainage plate is always a weak part, which is easy to produce defects and then accelerate the failure of the clamp during service, eventually leading to the interruption of electrical transmission and resulting in a relatively serious power grid accident[9-10]. In serious cases, it may also cause a large-scale power outage, and the resulting national economic losses are incalculable.

* Corresponding author: dkyzjj@163.com

3. Researches on heating causes of equipment clamps

The existing research shows that the temperature of the new clamp will be lower than the conductor temperature during operation. However, the clamp state is deteriorating constantly, when the deterioration reaches a certain degree, the clamp temperature will be higher than the conductor temperature, which will partly restrict the current carrying capacity of the line[11-12]. Therefore, solving this troublesome issue of equipment clamps has become one of the important ways to ensure the normal operation of lines and improve the current carrying capacity. However, at present, the research on different types of joints at home and abroad is mainly focused on the mechanical characteristics. Although some research has been carried out on the overheating problem, while most of them are conducted analysis on individual cases and few researches are provided the operable solutions. The research on overheating of wire clamps is generally attributed to improper installation process, corrosion of wire fittings and the effect of mechanical force [13]. And, there is a lack of systematic research on the mechanism and solution measures of abnormal temperature of wire joints, temperature rise characteristics as well as temperature distribution characteristics. Therefore, there is always a certain blindness in the treatment of wire clamps.

4. Solutions on heating of clamp

The heating of the equipment clamp is mostly caused by the loose connection bolts, the loose contact of the electrical contact surface, and the greatly reduced actual contact area, which leads to the rapid rise of the contact resistance. Therefore, in order to fundamentally reduce the heating of the equipment clamp, it is necessary to reduce the contact resistance of the connector connection during the use of the strain clamp [14]. In the past, when the power equipment has the defects of heating and power failure, method of comprehensive maintenance of power failure is always adopt. The heat of the power equipment is solved by polishing the drain plate, coating conductive paste, fastening bolts, and conducting auxiliary wire shunt [15, 16]. However, these maintenance methods can only temporarily solve the heating problem of the equipment, which belongs to the remedial measures after the event. Moreover, these treatment technologies are outdated, time-consuming and laborious, which cannot effectively reduce the heating fault defects. Therefore, a new type of material or equipment clamp must be developed and selected to increase the contact area at the connection of the drain plate to reduce the contact resistance and effectively prevent the heating fault.

4.1 Memory alloy gasket and its action mechanism

The joint position of the clamp is usually fastened with bolts, which belongs to fixed electrical contact. During the long-term operation of the clamp, the bolt loosens due to the vibration, leading to the change from the initial surface contact to the point contact between the drain plates and the increase of contact resistance. In addition, the small voids formed on the contact surface due to looseness and stagger will lead to metal oxidation, further increasing the contact resistance and causing abnormal heating. Too high temperature rise will lead to premature oxidation and corrosion at the bus connection, leading to the increase of contact resistance again. The increase of contact resistance will further lead to the increase of temperature rise of the drain plate, forming a vicious circle.

The connection will generate abnormal heat when the bolt loosens due to long-term operation. Because the memory alloy gasket has temperature sensitivity and automatic action execution, when the temperature reaches the deformation temperature point of the memory alloy gasket, the memory alloy gasket will change from martensitic state to austenitic state and return to the original state. The memory alloy gasket will produce a large restoring force during deformation, which will increase the pressure between the drainage plates, reducing the contact resistance and the temperature. Thus effectively avoiding the heating failure caused by abnormal temperature rise of the drainage plate contact surface.

Shape memory alloy gasket is made based on the shape memory function of the special alloy. At present, the shape memory alloy gasket is usually saucer shaped and have different diameters and thicknesses. When the memory alloy gasket is installed on the connecting conductor through a nut and a bolt, the gasket is in a flattened state. While, when the contact resistance increases due to the looseness of bolts due to the wind force, the drainage plate starts to heat up. When the temperature of the drain plate rises to the reverse martensitic transformation temperature of the memory alloy gasket, the shape of the gasket begins to recover. The shape recovery makes the bolt and nut compress the conductor with more and more pressure. When the bolt and nut are pressed again, the contact surface of the two conductors will be tightly pressed . Therefore, the shape memory alloy gasket increases the contact area of the joint when heating, effectively reducing the contact resistance and heat resistance of the joint.

Figure 2 shows the state of the disc shape memory alloy gasket after installation, the state in use and the shape recovery after the gasket temperature rises. It can be seen that once the temperature of the drain plate exceeds the martensitic reverse transformation temperature of the memory alloy gasket, the gasket will restore its shape and reimplements the fastening effect of the tension clamp bolt, increasing the contact area and reducing the contact resistance. Thus effectively controlling the heating of the drain plate.

Relevant standards requires that it is defined as a serious defect when the hot spot temperature of the connection between metal parts of clamp is between 90 °C and 130 °C. Therefore, the shape recovery temperature test and identification of memory alloy gasket shall be carried out under laboratory conditions, and appropriate memory alloy gasket could be selected according to the specific heating temperature of the equipment clamp.

https://doi.org/10.1051/e3sconf/202337501006

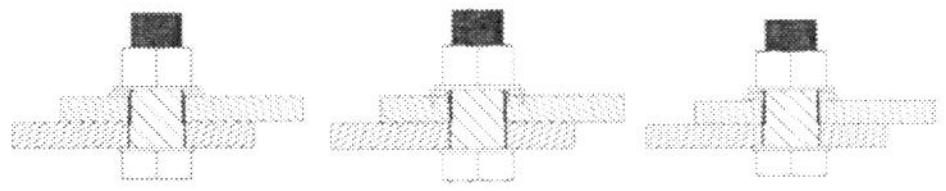

Fig.2 Working principle of disc shape memory alloy gasket

4.2 Structural improvement scheme

To meet the conductivity, the fundamental way to solve the heat problem is to improve the structure of the clamp. The author believes that there are mainly two aspects to improve the structure of the clamp. At first, the bolt connection can be reduced, and the crimping method shall be used as far as possible to avoid the increase of contact surface resistance caused by the loosening of bolts, so as to ensure the temperature rise of the strain clamp within the allowable range. Besides, double contact surface drainage clamp can be designed to increase the conductive area and effectively reduce the heating problem caused by rapid increase of contact resistance due to installation factors or loose bolts. According to the above ideas, clamps such as bolt free hydraulic, spiral, spiral connection and double contact surface drainage are produced as the times require. For new lines, it is appropriate to use bolted hydraulic strain clamp and spiral strain clamp, which can completely solve the hidden danger caused by the deterioration of electrical contact surface. Although the double contact surface drainage strain clamp is superior to the single contact surface drainage clamp, there is still a hidden danger of increased contact resistance. However, The mortise tenon drain plate uses the self-locking function of the dovetail wedge (Figure 7) to closely combine the upper and lower drain plates through the bite force, ensuring that the wire clamp can maintain a good contact state during use, and this connection method can also effectively increase the contact area between the upper and lower drain plates, avoiding the heating defect caused by the use of fastening bolts and their looseness. The author propose another design philosophy which is to combined the upper and lower drainage plates by the bite force from mortise tenon structure, as shown in Fig.3. The wire clamp with this structure can maintain a good contact state during use, and this connection mode can also effectively increase the contact area between the upper and lower drain plates, avoiding the heating defects caused by the use of fastening bolts and their looseness. This special mechanical connection structure can maintain a good contact state during use of the wire clamp, and this connection mode can also effectively increase the contact area between the upper and lower drain plates, avoiding the heating defects caused by the use of fastening bolts and their looseness.

Fig.3 Structure diagram of dovetail type drain plate

5. Inclusion

In general, the fastening effect of the drain plate bolt can be re realized by restoring the shape of memory alloy gasket due to the increase of contact area, the reduction of the contact resistance and the inhibition of the further temperature increase of the clamp, which can fundamentally solves the abnormal heating problem of the equipment clamp. Besides, multi surface contact structure is adopted between the upper and lower drainage plates of bolt free tenon and mortise equipment clamp, which greatly increases the electrical contact area.

The self-locking function of dovetail tenon structure can reduce the use of hot-dip galvanized bolts, improving the anti loose as well as anti galloping performance of equipment clamps and solving the overheating problem of previous clamps in the transmission operation process, which is of great significance to ensure the reliability and safety of substation equipment operation. The development of tenon and mortise type equipment clamp can realized the structural optimization of the drain plate and further improved the conductivity of the drain plate. After further structural optimization, this type of equipment clamp can also be considered to be widely used in power transmission line projects and produce significant economic benefits.

Acknowledgments

This research was financially supported by 2022 Youth Science and Technology Project of Inner Mongolia Power Research Institute Branch (Grant No. 2022-QK-09).

References

1. O.A. Makarov, E.A. Barbashina. Analysis of the problems of modern electric power industry and strategic ways to resolve them in accordance with the concept of the energy strategy to 2035. Proceedings of the Voronezh State University of Engineering Technologies, 2016, 2: 366-373.
2. N.H. Liao, Z.H. Hu, Y.Y Ma, et al. Review of the short-term load forecasting methods of electric power system. Power System Protection and Control, 2011, 39(1): 147-152.
3. G. Cardoso, J.G. Rolim, H.H. Zurn. Application of neural-network modules to electric power system fault section estimation. IEEE Transactions on Power Delivery, 2004, 19(3): 1034-1041.
4. X. Zhang, J. Song, Z. Li. Design and implementation of virtual simulation experiment for generation, transmission and application of electric power//[C]. 11th International Conference on Education Technology and Computers, 2019.
5. B. Chen, Q. Chen. The whole-process reliability evaluation of power system including generation, transmission, transformation and distribution//[C]. 5th International Conference on Electric Utility Deregulation and Restructuring and Power Technologies (DRPT), 2016.

6. X.B. Wu. One development of rotary clamp equipment of ceramic pipe workpiece[J]. Manufacturing Technology&Machine Tool, 2010, 32(4): 145-148.
7. Z.P. Li Numerical analysis and computation of a multi-order laminated microstructure[J]. Applied Mathematical Modelling, 2009, 33: 81-91.
8. I. Suda, M. Hajika, Y. Nishiba, et al. Simple and rapid method for the selective detection of individual lipoxygenase isoenzymes in soybean seeds. Journal of Agricultural and Food Chemistry, 1995, 43(3): 742-747.
9. Z.L. Li, G.X. Liu, M.J. Jian, et al. Detection and cause analysis of structural defects in lead clamp weld of transformer//[C]. IOP Conference Series: Earth and Environmental Science, 2019, 267(042062): 1-6.
10. Y. Shi. Analysis of suspension clamp crack accident of longyanxia-dashimen overhead grounding line II[J]. Qinghai Electric Power, 2008, 27: 31-34.
11. B. Zhang, J. Liang, C. Fang, et al. Study on temperature rise characteristics of insulation jumper clamp for 10kV distribution line//[C]. 2022 IEEE International Conference on High Voltage Engineering and Applications (ICHVE), 2022.
12. W.U. Gang, R. Wang, G.E. Haitao , et al. Effect of corrosion layer coverage rate on clamp on temperature rise of strain resistant clamp joint[J]. Heat Treatment, 2017, 32(05): 20-25.
13. A.S. Timothy, M.P. John. Hardness and stress-corrosion of rubber dam clamps[J]. Journal of Endodontics, 1997, 23(6): 397-398.
14. Z. Liu, J. Wu, X. Liu, et al. Influence groove clamp the contact resistance factors[J]. Electrical Engineering, 2016, 10: 132-134.
15. S. Zheng, Z. Yang, Z. Gao, et al. The development of tool for adding and fastening U-bolts of transmission line suspension clamp[J]. North China Electric Power, 2017, 2: 8-12.
16. S. Li, C.S. Jiang, Y.D. Zhu. A novel shunt clamp design and application[J]. Northeast Electric Power Technology, 2016, 37(7): 51-53.

The Quantitative Research on Atmospheric Environmental Corrosion of Aluminum Alloy Products

Yuqin Zhu[1,2], Fangchao Zhao[1], Jiameng Li [1], Tiantian Tan [1], Lu Dai [1]

[1] Southwest Technology and Engineering Research Institute, Chongqing 401329, China
[2] School of Oceanography, Zhejiang University, Hangzhou 310058, China.

Abstract: In this paper, we conduct quantitative research on the atmospheric environmental corrosion of aluminum alloy based on the atmospheric environmental factors. we apply the elastic network regression method to construct a regression model for corrosion rates based on the identified damage factors, which allows us to calculate the dependent variable as long as we know the main environmental factors in a specific year. Finally, we introduce a measure of atmospheric environment corrosion based on the established regression which can characterize the severity of corrosion with different transformations.

Keywords: Aluminum alloy; Environmental adaptability; Random forest; Elastic network; Atmospheric environmental corrosion measure.

1. Introduction

According to GJB 4239 "General Requirements for Equipment Environmental Engineering"[1] and GJB 6117-2007 "Equipment Environmental Engineering Terms"[2], environmental adaptability of a product refers to the ability of a product to maintain normal function and performance in all predetermined environments, and the requirements for environmental adaptability[3] are the requirements of a product's tolerance to environmental stress.An extensive analysis of typical cases of products' environmental adaptability shows that the environmental adaptability of different products varies with the environment, thus the environmental adaptability level of a product should be considered systematically when choosing the materials[4]. In this paper, A regression model for the corrosion of aluminum alloy in the atmospheric environment is constructed. we define a new measure of atmospheric corrosion to reflect the severity of environmental conditions.

2. Variable Selection

This paper selects seven aluminum alloy materials, 2B06-T4, 2D12-T4, 3A21-H24, 5A05-O, 7A09-T6, 7B04-T6 and 5A90-T3S. The environmental corrosion of aluminum alloy materials during storage is quantified based on the factors influencing the average corrosion rate. We define the independent variables X_1 (temperature), X_2 (relative humidity), X_3 (Cl^- deposition rate), X_4 (pH of rainwater), etc., to represent the environmental factors that have effect on the corrosion of aluminum alloy material.

TABLE 1 gives the atmospheric average corrosion rates of the seven aluminum alloys in four typical environments (Jiangjin, Wanning, Mohe and Beijing). The dependent variable Y considered in this paper is the ratio of the average corrosion rate to the mean of each group. If the ratio is less than 1, it can be interpreted as "being averaged", while if the ratio is greater than 1, it indicates that the original average corrosion rate has made contributions to the final average. We construct a regression model of Y using the selected environmental factors as the independent variables. We finally construct an environmental corrosion measure H to characterize the severity of the environmental damage, which is a function of Y. The larger H, the more severe the environment will be, and thus the more severe corrosion will be if the aluminum alloy material is exposed to this environment.

Table 1. Average corrosion rates of seven aluminum alloys in different localities

Locality	Year	2B06-T4	2D12-T4	3A21-H24	5A05-O	7A09-T6	7B04-T6	5A90-T3S
Jiangjin	2011	0.508	0.967	0.522	0.479	2.115	0.295	0.099
	2012	1.284	0.197	0.135	0.359	0.643	0.179	0.163
	2013	0.389	0.416	0.646	0.414	0.389	0.176	0.112
	2014	0.22	0.208	0.453	0.223	0.508	0.073	0.142
Wanning	2011	2.075	2.737	2.251	0.861	15.417	3.208	0.865
	2012	1.31	1.999	0.969	0.509	6.734	2.324	1.25
	2013	1.125	1.62	1.491	0.696	5.034	—	0.846
	2014	1.449	1.137	—	—	6.327	1.97	1.03
	2015	—	1.347	1.561	0.75	5.667	1.509	0.67
Mohe	2011	0.001	0.141	—	0.441	0.126	0.09	0.001
	2012	0.018	0.021	0.026	0.244	0.062	0.014	0.031
	2013	0.061	0.029	0.041	0.181	0.045	0.025	0.064
	2014	0.084	0.105	0.304	0.088	—	0.05	0.044
Beijing	2011	0.34	0.573	0.17	0.271	0.612	0.333	0.181
	2012	0.165	—	—	—	—	0.126	—
	2013	0.109	—	—	—	—	0.114	—
	2014	—	0.326	0.051	0.169	0.332	—	0.086

3. Regression Model for Atmospheric Environmental Corrosion

For aluminum alloy products, this paper aims at establishing a regression model for the atmospheric environmental corrosion with independent variables X_i and response variable Y, i.e.,

$$Y = Xw + \varepsilon \tag{1}$$

where the first component of the vector $X = (1, X_1, \cdots, X_n)$ represents the constant term, $w = (w_0, w_1, \cdots, w_n)$ are the regression coefficients, and $\varepsilon \sim N(0,1)$. The basic linear regression method uses the least square method to estimate the coefficients $w = (w_0, w_1, \cdots, w_n)$ as follows:

$$\min_w f(w) = \sum_{i=1}^{m}(y_i - x_i^T w)^2 = (y - Xw)^T(y - Xw) \tag{2}$$

where x_i, y_i are the i-th observation values of the independent variables and the response variable.

In this paper, we use 14 environmental variables excluding the chemical components as the independent variables for modeling, i.e., SO_2 deposition rate, temperature、relative humidity、pH of rainwater、solar radiation, Cl^- of rainwater, rain days, atmospheric pressure, NH_3, Cl^- deposition rate, snow days, average wind speed, fog days, and H_2S。

The variance inflation factors given in Table 2 suggest that there is serious multicollinearity between independent variables. Therefore, if simple linear regression is performed directly, the estimation results are not reliable and the variance of the parameter estimators will be too large.

Table 2. VIFs of independent variables

Independent variable	VIF	Independent variable	VIF
H_2S	36.91	Solar radiation	190.89
NH_3	234.72	Atmospheric pressure	1006.33
SO_2	217.31	Average wind speed	336.90
Temperature	2480.46	Rain days	211.35
Relative humidity	710.45	Fog days	161.28
pH of rainwater	521.87	Snow days	240.52
Cl^- of rainwater	84.85	Cl^- deposition rate	208.28

Elastic Network Regression is another regularized regression method[4], which combines the L_1 and L_2 penalties of the lasso and ridge methods. Its loss function is as follows,

$$f(w) = \sum_{i=1}^{m}(y_i - x_i^T w)^2 + \rho\alpha\sum_{i=1}^{n}|w_i| + \frac{1-\rho}{2}\alpha\sum_{i=1}^{n}w_i^2 \tag{3}$$

where $\rho\alpha$ is an important parameter to determine which regularization method dominates in the end. Elastic network regression is quite useful when there is a group of highly correlated variables. The value of λ obtained by cross-validation[9] is 0.140766, which gives us the following model,

$$Y = 2.813209 + 0.028445x_{Temperature} + 0.000239x_{Relative\ humidity} - 0.466436x_{pH\ of\ rainwater} + 0.101853x_{Average\ wind\ speed} + 1.111378x_{Cl^-\ deposition\ rate} \tag{4}$$

For which, the mean square error is 0.16, and the goodness of fit is 97.80%.

Based on this model, the value of Y of a region can be obtained only based on the annual average values of temperature, relative humidity, pH of rainwater, average wind speed, and Cl^-deposition rate. The larger the value of Y, the more severe the corrosion of aluminum alloy products.

Based on the established model, we provide the values of Y in four different regions in Table 3 and plot the trend of each region in Figure 1.

https://doi.org/10.1051/e3sconf/202337501007

Table 3. The value y of Y in four regions

Region	Year	y	Region	Year	y
Jiangjin	2011	1.010229	Mohe	2011	0.079678
	2012	0.952298		2012	0.118373
	2013	0.952537		2013	0.127745
	2014	0.914403		2014	0.206743
Wanning	2011	1.765198	Beijing	2011	0.384689
	2012	1.527571		2012	0.351415
	2013	1.494026		2013	0.295100
	2014	1.424120		2014	0.361261
	2015	1.386805			

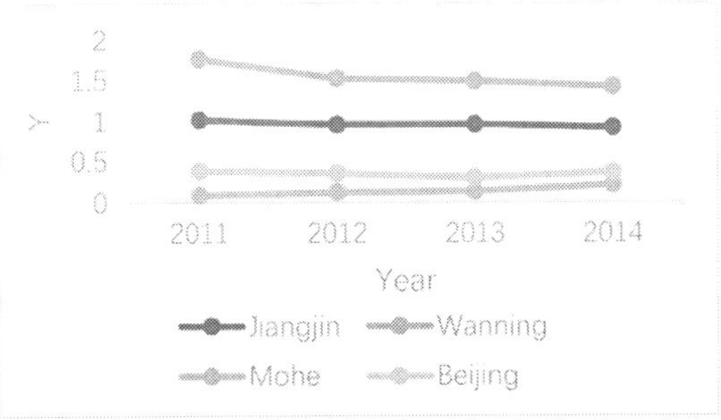

Fig. 1 Value trend chart of each region in the first four years

For each year, it can be seen that the relationship among the values of Y in the four regions is $Y_{B2} > Y_{B1} > Y_{B4} > Y_{B3}$. In other words, for aluminum alloy materials, the environment in Wanning is the worst, followed by that in Jiang, Beijing. In addition, except for Mohe, the values in other regions show a decreasing trend year by year. This may be because the strengthening of environmental protection measures in China in recent years has improved the atmospheric environment.

4. Environmental Corrosion Measurement

According to the corrosion regression model of aluminum alloy material in the atmospheric environment provided in Eq (4), we can calculate Y of a region in a certain year when the independent variables are known, which allows us to give a general judgment of the severity of the environment. The larger the value of Y, the harsher the environmental conditions for the aluminum alloy products are. However, to have a more uniform and intuitive understanding, this paper constructs the environmental corrosion measurement based on the dependent variable. We consider using the quadratic fraction transformation[5,6]:

$$h(y) = \frac{y^2}{1+y^2}, \quad (5)$$

or the logit transformation[10]:

$$h(y) = \frac{e^y}{1+e^y}, \quad (6)$$

to obtain a measurement of atmospheric environmental corrosion within [0,1], which allows us to better analyze the severity of the environment. A value of 0 indicates that the environment is the best for aluminum alloy products, and a value of 1 indicates that the environment is the harshest.

Moreover, we introduce two undetermined parameters α_0 and α_1 ($\alpha_1 > 0$) such that the transformed environmental corrosion measurement $H(Y,\alpha_0,\alpha_1)$ is a strictly increasing function of the dependent variable Y. Thus, a large $H(Y,\alpha_0,\alpha_1)$ also indicates a harsh environment. Specifically, the measure is given by,

$$H(Y,\alpha_0,\alpha_1) = \frac{(\alpha_0+\alpha_1 Y)^2}{1+(\alpha_0+\alpha_1 Y)^2} \quad (7)$$

or

$$H(Y,\alpha_0,\alpha_1) = \frac{e^{\alpha_0+\alpha_1 Y}}{1+e^{\alpha_0+\alpha_1 Y}}, \quad (8)$$

4.1 Quadratic fraction transformation

When applying the atmospheric environmental corrosion measurement, we hope that it can distinguish different typical environments as well as possible. Therefore, when determining the parameters α_0 and α_1, this paper tries to maximize the difference between the two regions with the smallest difference in the measurement of atmospheric environmental corrosion among the four regions. Let the environmental corrosion measure in Jiangjin, Wanning, Mohe, and Beijing be denoted by $H_1(Y_1,\alpha_0,\alpha_1)$, $H_2(Y_2,\alpha_0,\alpha_1)$, $H_3(Y_3,\alpha_0,\alpha_1)$, $H_4(Y_4,\alpha_0,\alpha_1)$, or abbreviated as H_1, H_2, H_3, H_4, there comes a constrained non-linear programming problem:

$$\begin{aligned} \max \quad & \min\{|H_1-H_2|^2,|H_2-H_3|^2,|H_3-H_4|^2,|H_4-H_1|^2\} \\ & +\lambda_1\sum_{i=1}^{4}|H_i|^2+\lambda_2\sum_{i=1}^{4}|1-H_i|^2 \\ s.t. \quad & \begin{cases} \alpha_1 > 0 \\ \alpha_0 \in \mathbb{R} \quad \text{(Field of real numbers)} \end{cases} \end{aligned} \quad (9)$$

Where $\lambda_1\sum_{i=1}^{4}|H_i|^2+\lambda_2\sum_{i=1}^{4}|1-H_i|^2$ denotes the penalty term, and λ_1, λ_2 are penalty coefficient satisfying $\lambda_1 > 0$ and $\lambda_2 > 0$.

Then the sequential quadratic programming method can be used to convert the problem into the following standard form:

$$\begin{aligned} \min \quad & -\min\{|H_1-H_2|^2,|H_2-H_3|^2,|H_3-H_4|^2,|H_4-H_1|^2\} \\ & -\lambda_1\sum_{i=1}^{4}|H_i|^2-\lambda_2\sum_{i=1}^{4}|1-H_i|^2 \\ s.t. \quad & \begin{cases} -\alpha_1 < 0 \\ \alpha_0 \in \mathbb{R} \end{cases} \end{aligned} \quad (10)$$

The values of H_1, H_2, H_3, H_4 are relevant to the specific year corresponding to environmental factors in the area, whereas Jiangjin and Wanning are different in the time dimension. Thus, we average Y_i of each region to get $\overline{y}_i$, and let $H_i(Y_i,\alpha_0,\alpha_1)=H_i(\overline{Y}_i,\alpha_0,\alpha_1)$. The results are given in Table 4.

Table 4. Corrosion measurement expressions and $\overline{y}$ for four regions

Locality	$\overline{y}$	H
Jiangjin	0.957367	$H_1=\dfrac{(\alpha_0+0.957367\alpha_1)^2}{1+(\alpha_0+0.957367\alpha_1)^2}$
Wanning	1.519544	$H_2=\dfrac{(\alpha_0+1.519544\alpha_1)^2}{1+(\alpha_0+1.519544\alpha_1)^2}$
Mohe	0.133135	$H_3=\dfrac{(\alpha_0+0.133135\alpha_1)^2}{1+(\alpha_0+0.133135\alpha_1)^2}$
Beijing	0.348116	$H_4=\dfrac{(\alpha_0+0.348116\alpha_1)^2}{1+(\alpha_0+0.348116\alpha_1)^2}$

Let the penalty factors $\lambda_1=0.00001$, $\lambda_2=0.0001$, taking the values in the table into the nonlinear programming problem of Eq. (10) gives

$$\begin{cases}\alpha_0=-0.020926\\ \alpha_1=1.497601\end{cases} \tag{11}$$

Therefore, the measurement formula of atmospheric environmental corrosion under quadratic transformation is:

$$H(Y)=\frac{(1.497601Y-0.020926)^2}{1+(1.497601Y-0.020926)^2} \tag{12}$$

These four regions represent four typical atmospheric environments in China, namely, the humid and hot marine atmospheric environment, the sub-humid and hot industrial atmospheric environment, the cold rural atmospheric environment, and the warm and semi-rural atmospheric environment. The atmospheric environmental corrosion measurements for the four regions are given in Table 5.

Table 5. Environmental corrosion measurement in four regions after quadratic fraction transformation

Locality	Atmospheric environment	H
Jiangjin	Sub-humid and hot industrial atmospheric environment	0.6662
Wanning	Humid and hot marine atmospheric environment	0.8356
Mohe	Cold rural atmospheric environment	0.0309
Beijing	Warm and semi-rural atmospheric environment	0.2003

4.2 Logit transformation

Similarly, we establish the environmental corrosion measurement based on the Logit transformation[11,12]

$$H(Y,\alpha_0,\alpha_1)=\frac{e^{\alpha_0+\alpha_1}}{1+e^{\alpha_0+\alpha_1}},$$

To determine the parameters α_0 and α_1, we need to solve the following constrained non-linear programming problem.

$$\begin{aligned}\max\quad & \min\left\{|H_1-H_2|^2,|H_2-H_3|^2,|H_3-H_4|^2,|H_4-H_1|^2\right\}\\ & +\lambda_1\sum_{i=1}^{4}|H_i|^2+\lambda_2\sum_{i=1}^{4}|1-H_i|^2\\ s.t.\quad & \begin{cases}\alpha_1>0\\ \alpha_0\in\mathbb{R}\end{cases}\end{aligned} \tag{13}$$

Where $\lambda_1\sum_{i=1}^{4}|H_i|^2+\lambda_2\sum_{i=1}^{4}|1-H_i|^2$ denotes the penalty term, and λ_1, λ_2 are penalty coefficient satisfying $\lambda_1>0$ and $\lambda_2>0$. We can transform the problem into the following standard form:

$$\begin{aligned}\min\quad & -\min\left\{|H_1-H_2|^2,|H_2-H_3|^2,|H_3-H_4|^2,|H_4-H_1|^2\right\}\\ & -\lambda_1\sum_{i=1}^{4}|H_i|^2-\lambda_2\sum_{i=1}^{4}|1-H_i|^2\\ s.t.\quad & \begin{cases}-\alpha_1<0\\ \alpha_0\in\mathbb{R}\end{cases}\end{aligned} \tag{14}$$

In TABLE 6, we provide the average values $\overline{y}_i$ and the corresponding measurements $H_i(Y_i,\alpha_0,\alpha_1)=H_i(\overline{Y}_i,\alpha_0,\alpha_1)$ for each region.

Table 6. Environmental corrosion measurement in four regions after logit transformation

Locality	$\overline{y}$	H
Jiangjin	0.957367	$H_1=\dfrac{e^{\alpha_0+0.957367\alpha_1}}{1+e^{\alpha_0+0.957367\alpha_1}}$,
Wanning	1.519544	$H_2=\dfrac{e^{\alpha_0+1.519544\alpha_1}}{1+e^{\alpha_0+1.519544\alpha_1}}$,
Mohe	0.133135	$H_3=\dfrac{e^{\alpha_0+0.133135\alpha_1}}{1+e^{\alpha_0+0.133135\alpha_1}}$,
Beijing	0.348116	$H_4=\dfrac{e^{\alpha_0+0.348116\alpha_1}}{1+e^{\alpha_0+0.348116\alpha_1}}$,

The sequential quadratic programming (SQP) algorithm[13,14] is used to solve the optimization problem (14), for which, we consider the following Lagrange function,

$$\begin{aligned}L(\alpha_0,\alpha_1,\lambda)=&-\min\left\{|H_1-H_2|^2,|H_2-H_3|^2,|H_3-H_4|^2,|H_4-H_1|^2\right\}\\ &-\lambda_1\sum_{i=1}^{4}|H_i|^2-\lambda_2\sum_{i=1}^{4}|1-H_i|^2+\lambda\cdot(-\alpha_1)\end{aligned} \tag{15}$$

Where λ is Lagrange multiplier. Based on this operation, this paper transforms an optimization problem with two variables and one constraint condition into an extreme value problem of a system of equations with three variables without any constraints. Next, we consider the quadratic approximation of the Lagrange function to improve the similarity of the quadratic programming subproblem, which makes the optimization problem with strong nonlinearity can also be calculated.

Let $\lambda_1=0.001$, $\lambda_2=0.0001$, we can get:

$$\begin{cases}\alpha_0=-1.944134\\ \alpha_1=3.647075\end{cases}$$

The environmental corrosion measurement with the transformation $h(y) = e^y / (1 + e^y)$ is given by

$$H(Y) = \frac{e^{-1.944134+3.647075 \cdot Y}}{1 + e^{-1.944134+3.647075 \cdot Y}}, \quad (16)$$

Table 7. Environmental corrosion measurement in four regions after logit transformation

Locality	Atmospheric environment	H
Jiangjin	Sub-humid and hot industrial atmospheric environment	0.8245
Wanning	Humid and hot marine atmospheric environment	0. 9733
Mohe	Cold rural atmospheric environment	0.1887
Beijing	Warm and semi-rural atmospheric environment	0.3375

For this transformation, the environmental corrosion measures of all four regions also have the $H_2 > H_1 > H_4 > H_3$. The results show that the environment of Wanning is the worst among the four regions, followed by Jiangjin, which is consistent with the results in Section 4.1.

The related reference[15] demonstrates that Jiangjin is perennial and cloudy. Long-term exposure of metal materials to a humid atmosphere increases the thickness of the surface water film, impeding the passage of oxygen, thus reducing the corrosion rate of the metal materials. While in Wanning where the measured value of the corrosion rate is the largest, the metal materials are exposed to the humid and hot marine atmospheric environment for a long time. The high temperature and humidity as well as the erosion of seawater will accelerate the corrosion of metal materials. Among the four regions, Mohe's atmospheric environmental corrosion measure is the smallest, where the metal materials are in the cold rural atmospheric environment. The low temperature and humidity will reduce the corrosion of metal materials. All these findings based on our model are in accordance with the objective law.

As shown in Section 4, the atmospheric environmental corrosion measures constructed based on the quadratic power transformation or logit transformation can all be used to evaluate the effects of typical atmospheric environments on the corrosion of aluminum alloy material in China. In this section, we explored the corrosion measures of seven typical atmospheric environments, i.e., humid and hot marine atmospheric environment, humid and hot rainforest atmospheric environment, sub-humid and hot industrial atmospheric environment, cool plateau atmospheric environment, warm and semi-rural atmospheric environment, dry and hot desert atmospheric environment, and cold rural atmospheric environment.

Referring to the atmospheric environment data of the seven atmospheric environments in 2016, this paper predicts the dependent variable values of Y in the corresponding environment based on the established model, as shown in TABLE 8.

Table 8. Typical environmental data and dependent variable values for 2016

environment	Te mpe ratu re	Relati ve humi dity	pH of rainw ater	Avera ge wind speed	Cl^- depos ition rate	Y
Humid and hot marine atmospheric environment	25.6 25	84.50	5.57	2.642	0.841 158	2. 1 6 7 4
Humid and hot rainforest atmospheric environment	21.4 67	77.58	6.15	0.850	0.004 75	0. 6 6 5 6
Sub-humid and hot industrial atmospheric environment	19.6 07	77.50	5.40	1.050	0.001 642	0. 9 8 0 2
Cold and warm plateau environment	8.61 7	40.18	6.32	2.408	0.003 117	0. 3 7 0 4
Warm and semi-rural atmospheric environment	12.7 83	46.51	6.09	1.912	0.293 433	0. 8 7 0 1
Dry and hot desert atmospheric environment	12.3 33	37.10	7.18	2.858	0.543 875	0. 7 2 0 6
Cold rural atmospheric environment	2.44 2	66.17	5.83	2.533	0.003 325	0. 3 0 4 3

The atmospheric corrosion measures for the seven typical atmospheric environments based on both the quadratic power and the logit transformations are given in TABLE 9 and TABLE 10, respectively.

Table 9. Measurements of Seven Typical Atmospheric Environments after Transformation

Typ e	Humid and hot marine atmos pheric enviro nment	Humid and hot rainfor est atmos pheric enviro nment	Sub-humid and hot industr ial atmos pheric enviro nment	Cold and warm platea u enviro nment	Warm and semi-rural atmos pheric enviro nment	Dry and hot desert atmos pheric enviro nment	Cold rural atmos pheric enviro nment
$H(Y)$	0. 9123	0. 4878	0. 6768	0. 2217	0. 6218	0.5283	0. 1590

Table 10. Measurements of seven typical atmospheric environments after logit transformation

Ty pe	Humid and hot marine atmosp heric enviro nment	Humid and hot rainfor est atmosp heric enviro nment	Sub-humid and hot industr ial atmosp heric enviro nment	Cold and warm plateau enviro nment	Warm and semi-rural atmosp heric enviro nment	Dry and hot desert atmosp heric enviro nment	Cold rural atmosp heric enviro nment
$H(Y)$	0.9974	0.6186	0.8363	0.3558	0.7737	0.6646	0.3027

5. Summary

From the results in the above two tables, we can see that in 2016, the corrosion severity of the seven typical atmospheric environments has the following relationship: Humid and hot marine atmospheric environment > Sub-humid and hot industrial atmospheric environment > Warm and semi-rural atmospheric environment > Dry and hot desert atmospheric environment > Humid and hot rainforest atmospheric environment > Cold and warm plateau environment > Cold rural atmospheric environment. According to the long-term monitoring data of environmental factors and the field exposure test results of the seven environments[20], the order of environmental corrosion severity is: Marine atmospheric environment > Sub-humid and hot industrial environment > Dry and hot desert environment > Cold rural environment and Cold and warm plateau environment. So the results based on the atmospheric environment corrosion measurement in this paper are in line with the objective law.

References

1. Yaochang Zhu, Jianyong Sun. GJB 4239-"General Requirements for Material Environmental Engineering" and Its Application Analysis (I) (in Chinese) [J]. Aviation Standardization and Quality, 2006(05): 32-37.
2. GJB 6117-2007, Materiel environmental engineering term[S].
3. Zeqi Xu, Shiyan Zhang, Weifang Xuan. Environmental Worthiness Evaluation of Equipment (in Chinese) [J]. Equipment Environmental Engineering, 2012, 9(1):54-58.
4. Bowei Wu. Research on Sufficient Dimension Reduction Method Based on Regression Tree (in Chinese) [D]. East China Normal University, 2020.
5. Khaleghi S , Givehchi S , Karimi S . Multivariate correlation analysis and its application in environmental analysis[J]. World Applied Programming, 2013, 242(1):5-9.
6. Xiaoguang Wang and Junhui Fan. Variable selection for multivariate generalized linear models[J]. Journal of Applied Statistics, 2014, 41(2) : 393-406.
7. ISO 9223-2012, Corrosion of Metals and Alloys-Corrosivity of Atmospheres-Classification, Determination and Estimation[S].
8. Liyuan Wang, Xiutong Wang, Haofen Sun, Baorong Hou. Study of SO2 Influence on Metal Corrosion in Atmospheric Environment (in Chinese) [J]. Equipment Environmental Engineering, 2011, 8(02): 62-66.
9. Junxin Hu, Gongjie Zhang. K-Fold Cross-Validation Based Selected Ensemble Classification Algorithm [J]. Bulletin of Science and Technology (in Chinese), 2013, 29(12): 115-117.
10. Wenke Huang. The Diagnosis and Process Solutions in Multicollinearity of Multiple Regression Model (in Chinese) [D]. Harbin Institute of Technology, 2012.
11. Liu, Dacheng Tao. Review on Recent Method of Solving Lasso Problem (in Chinese) [J]. Journal of Data Acquisition & Processing, 2015, 30(01): 35-46.
12. Boya Wu. Research on New Elastic Network Algorithm for Cluster Analysis (in Chinese) [D]. Beijing University of Civil Engineering and Architecture, 2020.
13. Ying Li, Xuan Zhou, Huifang Meng, Haohua Wang. A Robust Logistic Regression with Exponential Squared Loss (in Chinese) [J]. Natural Science Journal of Hainan University, 2019, 37(04): 287-291.
14. Cong Ma, Zhe Liu, Xiaoxian Zhen. Combining penalty function with sequential quadratic programming method for norm minimization(in Chinese)[J]. Computer Engineering and Applications, 2013, 49(18): 212-216.
15. Laizheng Luo, Yong Xiao, Yan Su, Xiaofeng Li, Yong Zhong. Corrosion Behavior of 7050 High-strength Aluminum Alloy in Four Typical Atmospheric Environments in China (in Chinese) [J]. Equipment Environmental Engineering, 2015, 12(04): 49-53.

 https://doi.org/10.1051/e3sconf/202337501008

Micro-environment Corrosion Strength Based on Quartz Crystal Microbalance

Lu Dai *, Yuqin Zhu, Chaoxuan Qin, Jiameng Li

Southwest Technology and Engineering Research Institute,401329 Chongqing, China

Abstract. Equipment service environment often has a certain corrosive, accelerate equipment corrosion, affect the integrity of equipment. In order to accurately monitor the micro-environment corrosion strength of the equipment, the multi-channel quartz crystal microbalance calibration test and the crystal vibration sensor temperature supplement test were carried out to realize the frequency-quality signal conversion of the quartz crystal vibration sensor and improve the accuracy of data acquisition of the device. Using a multi-channel quartz crystal microbalance, For an equipment micro-environment under the hot and humid ocean environment, Packaging box under warm and wet acid rain environment, For 30h, respectively, 65h Continuous in situ monitoring, The test data showed that the mass gain of a 30h copper crystal vibration sensor is 0.5 μg, The rate of mass change was 0.017 μg/h, High corrosion strength, Metal materials in this area are prone to corrosion, Key protection is required; The 65h mass temperature increase: 0.022 μg, The rate of mass change was 0.000034 μg/h, Low corrosion strength, The internal metal material is corroded slowly.

Keywords. Multi-channel quartz crystal microbalance, micro-environment corrosion strength, in-situ monitoring.

1. Introduction

Equipment service environment often has a certain corrosive, accelerate equipment corrosion, affect the integrity of equipment. The researchers often use GB/T 19292.1-2018[1] for typical atmospheric environment corrosion grade evaluation, specific method for steel, aluminum, zinc, copper four metal standard metal pieces to different natural environment, calculate the different metal corrosion rate within a year, according to the value of the natural environment rating, this method is widely used, can implement, but not in accurate quantitative evaluation of micro-environment corrosion strength in a short time. However, compared with the natural environment, the equipment micro-environment is more suitable for the equipment, and plays a vital role in the integrity of the equipment[2], and should be regarded as the monitoring object.

Quartz crystal microbalance is a high-precision test device, with simple structure, online, real-time dynamic monitoring characteristics, is widely used in biology, chemistry, physics and many other fields[3].In the biological field for nucleic acid aptamer screening and detection[4], and protein adsorption detection research[5][6]. In the field of environmental monitoring, organic toxic gases and liquid pollutants such as organic aldehydes, organic acids and dioxins can be detected[7][8], as well as corrosive gases such as SO2 and NO2[9][10]. In the field of electrochemistry, it is used to study the interaction between common-rail/non-co-rail polyelectrolyte pairs, final membrane components and viscoelastic behavior of self-assembled films[11], and the dissolution behavior of pure magnesium sensors in dilute sodium chloride electrolytes[12].

This project takes advantage of the rapid response of quartz crystal vibration sensor to the environment to carry out real-time monitoring of the comprehensive corrosion concentration for 65h and 30h for the packaging cylinder and a wheeled equipment micro-environment respectively. By analyzing the data obtained, the corrosion rate of the crystal vibration sensor in the sampling period is obtained, and the quantitative evaluation of the corrosion ability of the micro-environment is obtained.

2. Experiment

2.1 Materials and instruments

Experimental materials: $AgNO_3$, KCl, Na_2SO_4, Chengdu Cologne Chemical Co., Ltd.; 49U, copper coated crystal, Nanjing Meixuan Radio and Television Materials Co., Ltd.

Experimental instrument: 5 μL microprototype, Shanghai High Pigeon Industry and Trade Co., Ltd.; Quttro-S field, Thermo, TT711-5 High and low

* Corresponding author:

https://doi.org/10.1051/e3sconf/202337501008

temperature test box, Chongqing Harding Environmental Test Technology Co., Ltd.; 53220A frequency meter, USA; calibration module, multi-channel quartz crystal microbalance, laboratory self-research;ME204E electronic balance,METTLER.

2.2 Crystal frequency-corrosion quality calibration test

The crystal mass-frequency relation is shown in Sauerbrey equation[13] as shown in equation (1).

$$\Delta m_f = -\frac{\rho_q \mathrm{h}_q}{f_0}\frac{\Delta f}{n} = -C\Delta f \qquad (1)$$

In this equation $C = \rho_q h_q / (n f_0)$, ρ_q is quartz crystal density, h_q *is* quartz crystal vibration thickness, f_0 is the quartz crystal vibration base frequency, n is the general frequency number, for the AT cut quartz crystal, n=1,3,5,.......

The purpose of the calibration test is to find the C value of the Sauerbrey equation. The specific method is: the first step is to prepare 5~7 concentration gradient $AgNO_3$ with distilled water in a sufficient amount of 50mL beaker3, KCl, Na_2SO_4 Salt solution was 5mL each and diluted 1000-fold, See Table 1 for details; the second step, At room temperature, Open the self-developed calibration module (equivalent to the device single-circuit test assembly), Preheat for 10 min after power, Connect the vibration signal end to the frequency meter, Record the crystal vibration base frequency; the third step, The diluted salt A solution prepared in step 1 to the crystal vibration surface with a micro-sampler, Drying at room temperature for 2 to 3 h, Record the frequency after the constant vibration frequency within 15 min, After that, when cleaned with distilled water, drying to constant weight, The same step was used to measure the frequency of the remaining B, C, D, E, F and G, on the basis of $\Delta f = f_1 - f_0$. The fourth step, according to the Sauerbrey equation, when the crystal vibration plating film is uniform, the Δf is only m related, fitting the Δf to the actual salt mass Δm, to obtain the relationship between the crystal vibration frequency change Δf and the crystal vibration mass change m. The materials and equipment required for calibration are shown in Figure 1.

Table 1. Type and concentration of calibration solutions

	Concentration of calibration solutions (mg/mL)						
	A	B	C	D	E	F	G
KCl	20.84	31.36	57.4	79.98	104.44	139.86	207.82
Na_2SO_4	22.92	67.32	102.32	141.88	181.38	230.1	276.98
$AgNO_3$	24.06	79.26	126.88	203.32	283.98	317.56	——

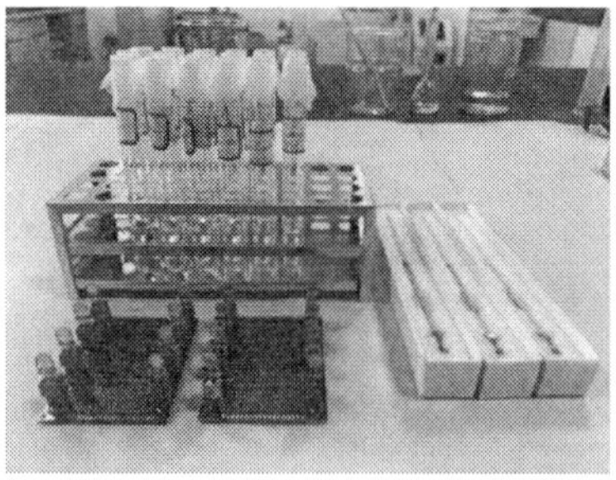

(a)Calibration Materials

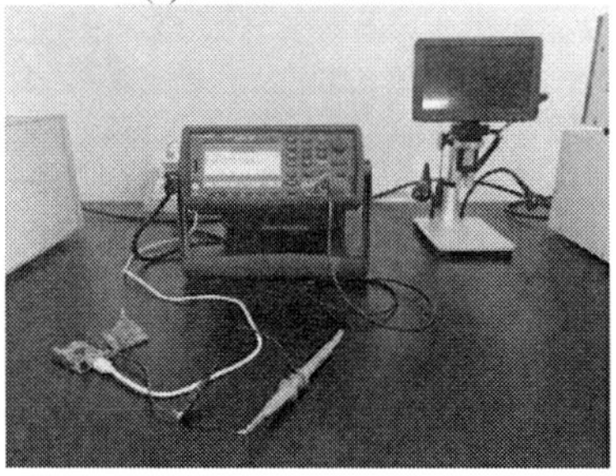

(b) calibration data acquisition

1Figure 1. Calibration experiment

2.3 Temperature fill test

Temperature and humidity of environmental factors will have a certain impact on the oscillation frequency of QCM sensor, especially the frequency change of 1℃ and about 1Hz[14]. In order to obtain the accurate frequency change value of each coating crystal vibration, the relevant study of the device temperature compensation is done, as shown in Figure 2. Specific method: place the device in the temperature control box to slowly cool down to-15℃ for 20 min; then to 1℃/ min to-10℃/ min for 20 min; to-5℃ at 1℃/ min for 20 min; to 0℃ for 1℃/ min for 20 min; and then to 45℃, and finally slow down to room temperature.

Figure 2. Temperature supplement test

2.4 Micro-environment corrosion and corrosion data collection

The acquisition micro environment mainly includes two micro environments: packaging box and wheel equipment base, as shown in Figure 3. The packaging box used for micro-environment data collection is mainly the packaging box of an equipment under warm and wet acid rain environment and the upper end micro-environment of a wheeled car base under hot and humid ocean environment. When collecting data, the device is first connected with the upper computer program, and the sampling frequency is set to 10 times/s, the sampling time is set to 65h, and the wheeled equipment is set to 30h. After the device is set in the corresponding micro-environment, the packaging box should close the upper cover for collection.

(a)Box micro-environment

(b) Wheel equipment micro-environment

Figure 3. micro-environment data acquisition

3. Results and Discussion

3.1 Crystal frequency-corrosion quality calibration result

A total of 40 copper coated crystal vibration calibration data were measured. According to the data integration and analysis, the copper crystal vibration test results are shown in Figure 3. Δf and Δm showed an obvious linear relationship: $\Delta f=644.2\cdot\Delta m$, and the goodness of fit is 0.982, which conforms to the Sauerbrey equation. The crystal vibration sensor frequency signal can be converted into the corrosion quality signal to directly reflect the crystal vibration corrosion situation, and the corrosion strength of the micro-environment can be monitored through real-time signal processing and transformation.

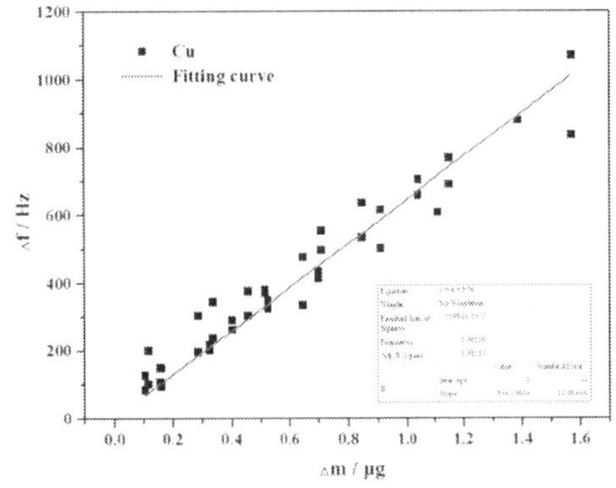

Figure 4. Mass-Frequency data calibration

3.2 Temperature compensation results

The quartz crystal vibration temperature frequency curve of copper coated crystal vibration as measured by the high and low temperature test box is shown in Figure 4. Copper coating crystal vibration frequency is negatively correlated with temperature, and temperature decrease frequency increases and temperature increase frequency decreases. Temperature change-15 ℃ ~45 ℃ , the frequency change of 78.4Hz, the temperature of every change of 1 degrees Celsius, the frequency change of about 1.3Hz, the temperature measurement and supplement results will be written to the self-developed quartz crystal micro-balance program, the subsequent data acquisition will be automatically compensated, to improve the accuracy of the data acquisition of the device.

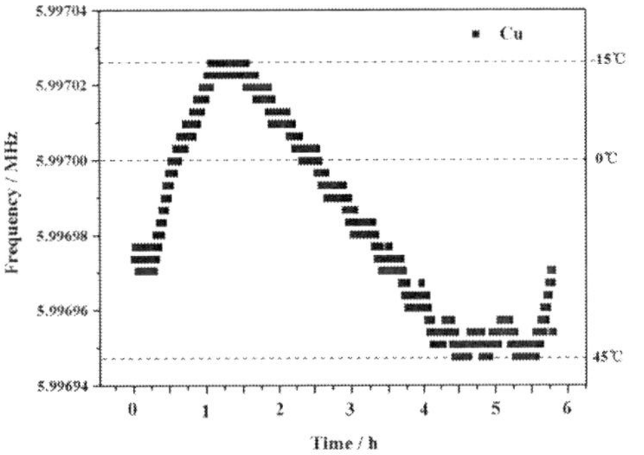

Figure 5. Installation temperature supplement

3.3 Analysis of micro-environmental corrosion data

The two micro-environment corrosion data are shown in Figure 5. According to Figure (a), the change of copper crystal vibration quality in the wheeled equipment micro-environment for 30h is about 0.5μg, and the mass change rate is 0.017g/h. From Figure (b), the packaging box micro-environment is 65h, the change of copper crystal vibration quality is about 0.022μg, and the mass change rate is 0.00034μg/h. Through the data comparison, it can be seen that the crystal vibration quality of the wheeled vehicle change is 50 times that of the packing box, and the concentration of corrosion medium is much higher than that of the packing box. Except for the corrosion concentration difference of two micro-environmental corrosion, It can also be seen that the packaging box sensor is in a relatively stable environmental quality in a constant speed increase state, The micro-environment of HAINAN station wheeled armored vehicle is semi-open, The environment changes greatly, Causes the sensor mass change to fluctuate, See Figure 10 (a) for the data, When the trial lasted for 22h, The mass change reached a peak within 24h, About 0.3μg, The following 2h showed a significantly decreasing trend, This is due to the high micro-ambient air humidity, When the air temperature is below the crystalline vibration surface temperature, Poor condensation on crystal vibration sensor[15], As the air temperature rises, Evaporation of the surface condensate droplets, The crystal recovered to the initial weight.

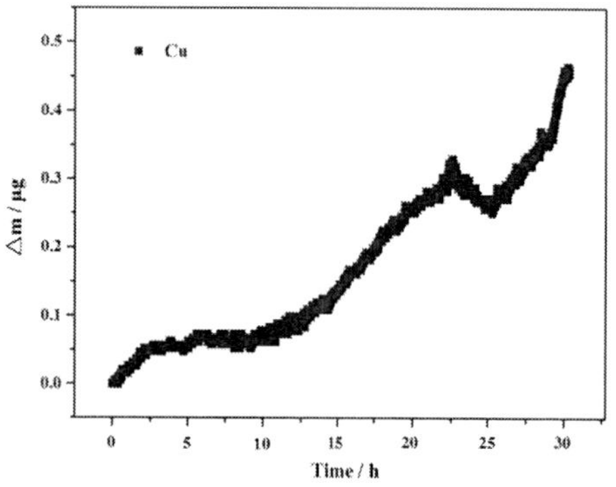

(a) Wheel, equipped micro-environment

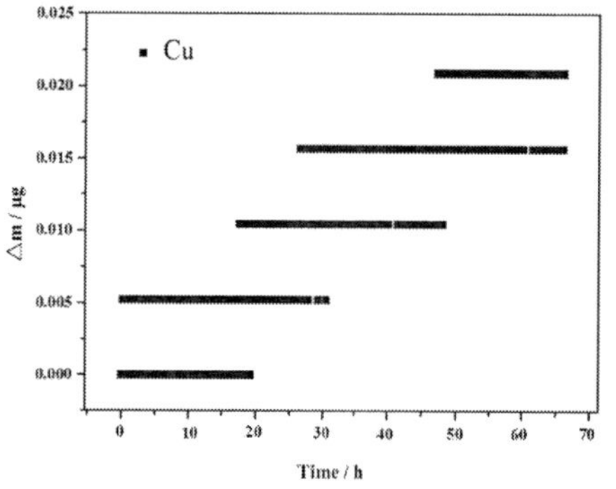

(b) packaging micro-environment

Figure 6. Results of two micro-environment data acquisition

3.4 Analysis of crystal vibration and corrosion mechanism

The microscopic analysis of quartz crystal vibration sensor as shown in figure 6, after the test copper crystal vibration coating metal layer cracks and corrosion serious parts peeling, through the EDS spectrum analysis, weak parts, the main elements of oxygen, copper elements, the total content of 25.54%, 64.77%, can be speculated as copper metal corrosion oxidation product CuO, consistent with copper element corrosion oxidation products.

(a) SEM photograph

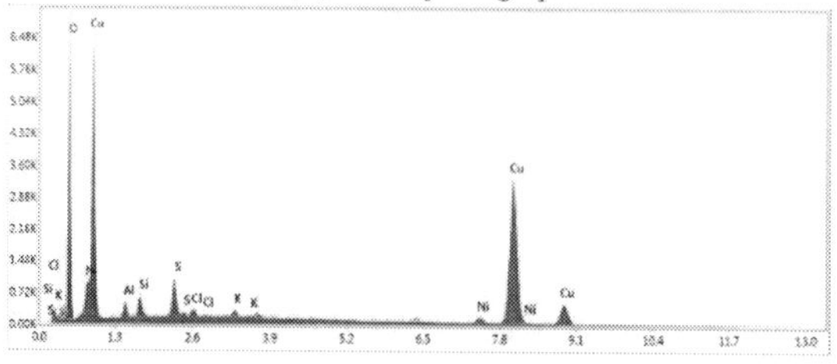

(b)EDS analysis results

Figure 7. Microscopic analysis of the quartz crystal vibration sensor after corrosion

4. Summary

This paper uses the laboratory self-developed multi-channel quartz crystal micro-balance to monitor the equipment micro-environment corrosion monitoring. Test data show that a wheeled micro-environment 30h copper vibration sensor mass weight of 0.5μg, the mass change rate of 0.017μg/h, high corrosion strength, metal materials prone to corrosion; 65h copper vibration sensor mass weight of 0.022μg, mass change rate of 0.000034μg/h, low corrosion strength, slow internal metal corrosion. micro-environment corrosion monitoring data can provide a reliable basis for the service, storage and daily maintenance of equipment.

References

1. Corrosion of GB / T 19292.1-2018 metals and alloys — Part 1: Classification, determination and assessment.
2. Silver N A, Gaebel W.Facilities Environmental Severity Classification Study[J]. Leidos, February, 2017.
3. Zou Chunyang, Wang Bowei, Yan Xueqing, He Xian, Wu Wenjuan. Research Progress in the application of Quartz Crystal Microbalance (QCM) [J]. Journal of Zhejiang A&F University, 2020,37(05):1006-1013.
4. Wu Liting, Su Xue, Lin Junsheng. Research Progress of Quartz Crystal Microbalance Sensor Based on Aptamer [J]. Biotech Bulletin, 2018,34(09):97-103.
5. Chen Zhuoyue, CAI Yulong, Zhang Cheng, CAI Hu. Application of quartz crystal microbalance for protein adsorption detection [J]. China Medical Device Information, 2014,20 (08): 55-57+62.

6. Fu Yakang, Zhao Yuqiang, Weng Jie, Liu Yaowen. Progress in the characterization of protein adsorption on biomaterial surfaces [J]. International Journal of Biomedical Engineering, 2019(03):250-257.
7. Zhu Zhonghao, Hao Yanfei, Sun Mingying, Pang Xiangyang, Liu Zhigang, Zhu Jianqiang. Online monitoring of organic pollutants based on quartz crystal balance and micro-nano optical fiber [J]. Photonic Journal, 2016,45 (09): 135-140.
8. KUROSAWA S, PARK J W, AIZAWA H, et al.Quartz crystal microbalance immunosensors for environmental monitoring [J]. Biosens Bioelectron, 2006,22(4):473-481.
9. Wang Fengping, Yan Chuanwei, Zhang Xueyuan, Du Yuanlong. Quartz crystal microbalance (QCM) and its application in the research of atmospheric corrosion [J]. Chemical Bulletin, 2001(06):382-387.
10. Forslund M, Leygraf C.A quartz crystal microbalance probe developed for outdoor in situ atmospheric corrosivity monitoring[J].Journal of the Electrochemical Society, 1996, 143(3): 839.
11. DENIZ M, DELIGOZ H.Flexible self-assembled polyelectrolyte thin films based on conjugated polymer: Quartz Cristal Microbalance Dissipation (QCM-D) and cyclic voltammetry analysis [J]. Colloids Surfaces A Physicoche} c Eng Aspects, 2019,563:2061:216.
12. RALSTON K D, THOMAS S, WILLIAMS G, et al.An electrochemical quartz crystal microbalance stu dy of magnesium dissolution [J].Apple Surface Sci, 2016, 360:
13. Zhang illumination. Quartz crystal microbalance [M]. Beijing: Science Press, 2015.1.
14. Mo Chentian. Research on QCM detection technology of corrosive gases in cultural collection [D]. Zhejiang University, 2018.
15. Luo Xuanguo, Xia Lijian. Research on anti-condensation technology of Electrical Equipment [J]. Renewable Energy, 2014,32(04):489-492.

Application research of computer big data informatization technology in oilfield mining automation system

Zhongmei Shen

Exploration and Development Research Institute of Daqing Oilfield Co Ltd., Daqing 163712, China

Abstract. Science and technology are developing continuously, and oilfield exploitation has gradually realized informatization and automation. In order to meet the needs of the oilfield mining automation system and to better monitor the data in the oilfield mining process, the computer big data information technology application system has become an indispensable part of the oilfield mining operation enterprise. In this paper At the beginning, the status quo of the oilfield mining automation system is analyzed. In this paper, the role of the computer big data information technology automation system in oilfield mining is analyzed. This paper discusses the key points of the development and design of the automation system system. The ultimate purpose of this paper is to provide certain ideas for oilfield mining automation.

Keywords: Computer big data;Information technology;Oilfield exploitation automation system.

1. Introduction

The oilfield mining automation system application system is a system that collects and summarizes oilfield mining data, analyzes, organizes, warns[1] and applies. This system plays an important role in troubleshooting oilfield mining equipment failures and improving oilfield mining safety levels. The application system of the oilfield exploitation automation system has epoch-making significance at present, which marks that the oilfield exploitation, construction and maintenance have entered a new period. Information technicians of oil companies should devote more energy to the research[2] and development of system functions and contribute to the realization of modern oilfield construction.

2. Status Quo of Oilfield Exploitation Automation System

Mastering the situation and data in the process of oilfield exploitation has become an important basis for the decision-making of the oil company management. Various sensors and analyzers installed on the oilfield exploitation site, downhole, and on the equipment all provide support for the collection and summary[3] of oilfield exploitation data. The basis for the management to make decisions is more scientific and reasonable, so the application system of the oilfield exploitation automation system has become an important part of the modern oilfield construction. The oilfield exploitation automation system has received extensive attention from the management[4] of oilfield exploitation enterprises. At present, oilfield exploitation data cannot support managers and technicians to use mobile communication terminals to view data at home, which makes the drawbacks of the production automation system application system prominent. Information technicians need to further optimize the production automation system, and the system needs to obtain better The stability and quickness of the system, the scope of application of the system is further expanded, and the system can provide[5] greater technical support for oilfield development.

3. Functional Analysis of Petroleum Exploitation Information Management System

Based on the author's many years of practical operation experience, this paper derives the main functions of the oil exploration information management system through a detailed analysis of oil exploration, oil exploration, equipment management and other personnel. These functions include oil exploration information management, oil exploration progress management, Oil extraction equipment management, etc. These functions provide users with a powerful operational support to ensure real-time grasp of oil extraction information.

3.1 Petroleum Exploration Information Management

The main purpose of oil exploration is to find and identify oil and gas resources. Oil companies use various advanced technologies and methods to understand the geological situation. Technicians are fully aware of the oil reserves. Technicians can also evaluate and analyze the prospect of oil and gas exploration. The technicians determine the favorable areas for oil and gas accumulation, the technicians find an area for storing oil and gas, the technicians find out the situation and production capacity of the oil and gas layers, and the technicians identify the relevant products for the country to increase the crude oil reserves. Petroleum exploration information management can provide users with strong operational support. Petroleum exploration information management records the exploration status of oilfield resources, which is convenient to determine the optimal path for obtaining resources and storage.See Figure 1.

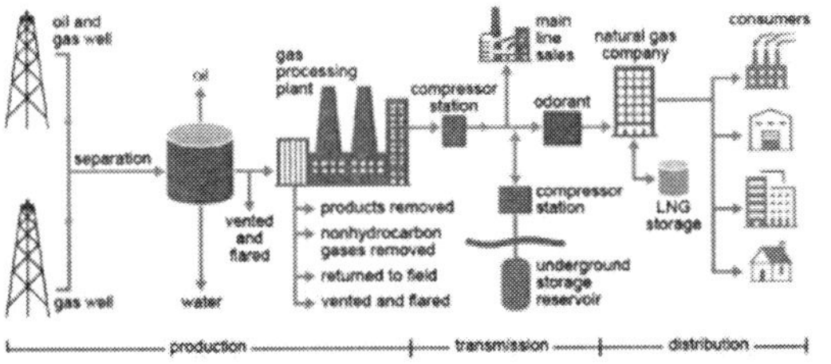

Fig. 1. Research on the Function of Petroleum Exploitation Information Management System

3.2 Oil exploitation progress management

In the process of oil extraction, each oil field involves a very wide range. In order for oil companies to better grasp the development of oilfields, the management of oil development progress can record and share the development progress according to the data collected by people, which has an important role and significance.

3.3 Management of oil extraction equipment

There are a lot of oil extraction equipment, and technicians need to purchase equipment, repair faults, scrap equipment, and manage equipment in the process of oil extraction. In oil extraction, if the oil extraction personnel need new equipment, they can apply for purchase at this time, and the oil company will provide the extraction tools in time. The fault maintenance can record the use time, the number of faults, and the fault maintenance of each equipment; the equipment scrap can provide equipment exit management for oil exploration, ensure the timely scrapping of equipment and improve the details of oil exploration; equipment classification management can provide users with powerful This can further improve the convenience and automation of equipment operation.

4. Key technologies of oilfield informatization construction

4.1 IoT Technology

The Internet of Things technology belongs to a modern management technology. When using the Internet of Things technology, oil company technicians first extensively collect the latest technical information to ensure that information entry and information perception can be performed in oilfield mining operations, construction and other links, and the Internet of Things technology will gradually transform it into digital information. Afterwards, technicians should also reasonably apply the IoT management center to ensure the rationality of data control and data management. Only by combining the processed data and information can the managers of the oil company make correct decisions and instructions, and the managers can direct the information to be transmitted to the relevant institutions of production and operation in a timely manner.

4.2 Computer big data information technology

In order to realize oilfield informatization, oil companies must reasonably apply computer big data informatization technology. Oil companies ensure the security of computer big data informatization. The computer big data informatization technology reasonably applies virtual resources, and its goal is to quickly improve the efficiency of information processing to achieve information integration. The characteristics of computer big data informatization technology are fast information transmission, which can quickly complete the information mapping of the actual situation and upload it to the corresponding field area. Technicians build models with the help of Java to ensure that the final system conforms to actual needs. In the application of computer big data information technology, the technology is highly dependent on the three basic technologies. One is information collection technology and information storage technology. In the new era, the amount of information has increased significantly, and in the information collection process, the requirements of information collection technology on hardware equipment and facilities have gradually increased. On the one hand, technicians should actively improve the accuracy of information, to ensure the feasibility of the information entry scheme and the integrity of the information collection device. A large-scale storage base station can also be established to ensure that the storage space meets the requirements of the amount of information in the new era. In addition, technicians should also consider the supporting role of CPU technology in computing. The second is information transformation technology. The process of information scheduling and information processing will inevitably involve the conversion of virtual information and physical resources.

4.3 Setting up electronic duty

Electronic road cards are installed in relatively important production intervals, and electronic road cards can intelligently monitor the time when vehicles arrive at the oilfield well site. After the electronic road card takes pictures of the passing vehicles and then identifies the information, the electronic road card accurately determines the vehicle information and time for tracking. Such information technology not only realizes the intelligent function of electronic duty, but also effectively improves the management of roads near the oilfield well site, so that the oilfield has a certain safety.See Figure 2.

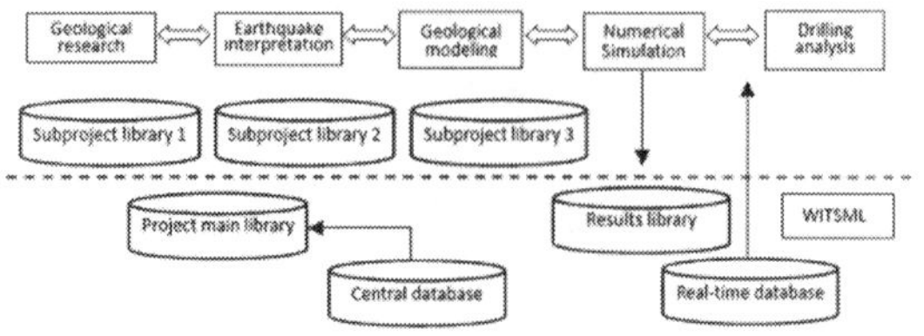

Fig. 2. The key technology of oilfield informatization computer construction

4.4 Using information technology to set up an alarm system

In the information technology of the oil field, the alarm system is also an intelligent technology. Oil companies collect information data from oilfield well sites through self-inspection technology. Information technology sets up an alarm system to diagnose the situation in production, and then analyzes and optimizes parameters. The information technology sets the alarm system to judge the fault of abnormal data, conduct regular inspections, make alarms and provide corresponding measures and suggestions. The information technology sets the alarm system to provide a guarantee for safe operation.

4.5 Application of digital skid-mounted booster integrated device

In order to realize the mixed output of oil and gas from pump to pump, oil companies need to heat, pressurize, control and buffer the oil and gas mixture. The oil company integrates these steps into one place through an intelligent control system, and Oil companies can operate these processes remotely. The oil company optimizes the entire process to digitize and integrate this set of procedures.

4.6 Using information technology to automatically pitch and remotely allocate water

The oil company uses the key technology of information technology to install the automatic ball pitching device in the oilfield well site. The number of balls loaded at a time is about 13. The technicians complete the pitching according to the set pitching time and collect the pressure and flow of the water source well, the pressure and the pressure of the water injection well. The technical personnel can remotely control the electric submersible pump of the water source well, and the water injection volume of the water injection well can be adjusted remotely, thereby improving the production efficiency.

5. The role of oilfield exploitation automation system in oilfield exploitation

The oilfield production automation system is an oil production engineering optimization analysis software for oil companies to monitor oil well conditions, analyze faults, evaluate oil displacement effects, and analyze oil production energy consumption. Energy consumption analysis of oil well power grid: The quality and efficiency of power supply are directly related to the production cost and assessment indicators of the enterprise; the system provides analysis data of reactive power, harmonics and faults of oil and water well power supply lines and power supply transformers to avoid high energy consumption operation of power supply lines; oil and water wells Monitoring and analysis of the reactive power compensation equipment of the power supply transformer, the software actively reports the name and location of the reactive power compensation equipment running inefficiently or in failure. Oil well electricity safety early warning: It is difficult to monitor the working conditions of oil wells with electricity, leakage, and low insulation resistance, and the resulting safety accidents occur from time to time; the system provides fault early warning and alarm instructions such as line leakage, short circuit, open circuit, and equipment live The company can better prevent the occurrence of vicious accidents caused by such hidden dangers. Remote working condition monitoring of oil wells: Oil wells are widely distributed, complex terrain, inconvenient transportation in rainy and snowy weather, changes in oil well parameters or fault shutdown, and it is difficult for well patrol personnel to detect in time; the system provides remote working condition monitoring (data acquisition, remote start and stop of wells)) and fault alarms, thereby speeding up troubleshooting time, reducing the number of well inspections; improving production efficiency and reducing production costs.

5.1 Update data in real time

The data generated in the process of oilfield exploitation is in a state of constant change. If the data update period is long, the data cannot provide early warning of oilfield exploitation failures in time, nor can it meet the needs of management and technical personnel for real-time data, and the oilfield exploitation automation system The powerful and rigorous data processing system of the application system makes it possible to update the data of the oilfield portal website in real time. The oilfield exploitation automation system application system combines database technology with implementation processing technology. The system transmits data through SCDMA and establishes a mapping relationship with PEOFFICE data items. The system collects, analyzes, processes and feeds back data in a relatively short period of time. Form single-well ground power map, pump working condition map, and pump efficiency analysis map. These maps meet the needs of management and technical personnel for real-time data. This processing method requires strict process control to be realized in the case of massive data. During the operation of the real-time database, information technicians need to pay attention to its operation status when the data flow is too large, so as to avoid the problem of difficult data management. In comparison, the real-time database is more in line with the needs of modern oilfield exploitation than the traditional database.

5.2 Timely fault warning

Fault early warning is one of the very important functions of the application system of the oilfield mining automation system. The fault early warning is mainly realized through the system's real-time data collection and monitoring and analysis. The fault early warning system is widely used in oilfield portals, urban pipe network end management systems, etc. In the field, the fault early warning system can effectively monitor the flow parameters, pressure parameters, temperature parameters and other information in the process of oilfield exploitation and oil transportation. The system obtains possible fault information through analysis and processing, and then synchronously responds to the monitoring system and gives early warnings.See Figure 3,

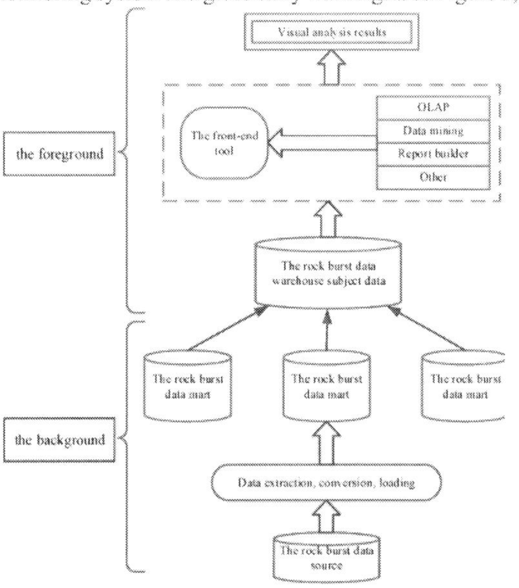

Fig. 3. Fault early warning is the oilfield mining automation system

5.3 Automatically provide processing comparison of historical data

The oilfield production data in the traditional data application system mostly exists in the form of Excel. The data stored in this way is not conducive to the application of the management and technical personnel, and cannot intuitively reflect the changes in the data trend. The traditional data application system limits the Management efficiency of oilfield mining enterprises. The development and implementation of the application system of the oilfield exploitation automation system allows the technical personnel and the management to process and compare the historical data. The oilfield exploitation automation system reflects the data trend in the form of charts. The oilfield exploitation automation system reduces the time consumed by manual data processing. The mining automation system improves the efficiency of oilfield mining data processing, thereby improving the efficiency of oilfield mining management. Application strategies of key technologies of oilfield informatization In order to ensure the steady progress of construction of major and detailed oilfield projects, oil companies give full play to the advantages of key informatization technologies.

5.4 Strengthen the key technology research of informatization

At present, information technology emerges in an endless stream. Oil companies should strengthen the research on Internet of Things, computer big data information technology, big data and other technologies based on their own information construction. Oil companies formulate practical implementation plans based on the actual work of oilfield exploitation. On the one hand, oil companies should always pay attention to the development of key information technology, establish good cooperative relations with well-known domestic universities and research institutes, clarify the construction goals and key technologies of "digital oil fields", and strive for breakthroughs in key technologies. At the same time, do a good job in the integration research between key information technology and existing hardware equipment, do a good job in upgrading the functions of existing software and hardware equipment, and reduce the application cost of key information technology. See Figure 4

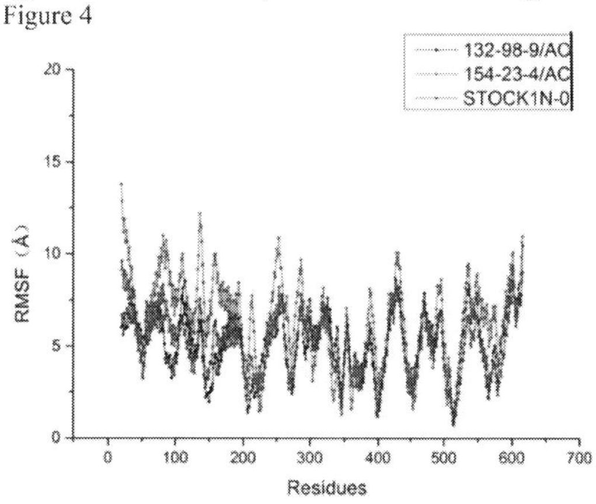

Fig. 4. Fault warning is a comparison of advantages of oilfield mining automation system

On the other hand, some large oilfield companies have achieved remarkable results in informatization construction. Therefore, oil companies should visit and study with similar enterprises to understand their achievements in informatization construction and the goals of future informatization construction. In particular, companies need to understand the breakthroughs made by similar companies in the Internet of Things and computer big data informatization technology. Oil companies can learn from their own actual situations. At the same time, in terms of key technologies, oil companies can also establish cooperative relationships with well-known domestic oilfield companies to jointly develop and research information technology to ensure that the Internet of Things and computer big data information technology can be better applied to oilfield exploitation.

5.5 A good job in the deployment of informatization construction

At present, with the rapid development of information technology, oilfield enterprises should do a good job in the rational deployment of informationization construction. Oilfield enterprises should ensure that information technology software and hardware not only meet the current work needs, but also have redundancy to facilitate the expansion of software and hardware performance in the future. When deploying informatization construction work, oilfield enterprises should pay attention to the implementation of the following contents: On the one hand, oilfield enterprises should do a good job in designing all parts of the informatization system. As we all know, computer big data informatization technology not only has strong professional knowledge, but also requires a reasonable network architecture. Computer big data informatization technology requires a reasonable arrangement of hardware to ensure information transmission and processing speed. Therefore, the oil company can establish a good cooperative relationship with the domestic well-known network service provider, and the oil company entrusts it to be responsible for the information construction service work. At the same time, the oil company clearly explained the informatization construction needs to the network service provider to ensure that the informatization construction plan is scientific and reasonable.

6. Conclusion

This paper comprehensively studies the application and development of oilfield exploitation technology in oilfield automation system. This paper carefully analyzes the practicability, real-time, easy maintenance and other functional characteristics of oil well monitoring system. Petroleum enterprises should continuously optimize the application of computer big data information technology in the oilfield exploitation automation system. Only when oil companies continuously improve their mining automation technology, can oil companies improve the efficiency, stability and safety of oilfield mining and promote the stable and healthy development of the energy industry.

Acknowledgements

Natural Science Foundation.

References

1. Ge Zhucong. The International Petroleum and Petrochemical New Technology, New Process and New Equipment Promotion Conference was held. Dongying Daily.Vol.13(2019) No.9, p.101-105.
2. Wang Hui. Design of Oilfield Exploitation Automation System, Xidian University, Vol. 20(2020) No.31, p.142-146.
3. Lei Xue. Key technologies of oilfield informatization and its application effects. China Management Informatization, Vol. 27(2019) No.46, p.204-208.
4. Wang Lijun. Research and application of key technologies in intelligent oilfield construction. China Management Informatization, Vol. 43(2018) No.14, p.168-171.
5. Duan Hongjie, Ma Chengjie, Liu Huanzong. Research and application of key technologies of oilfield informatization. Petroleum Science and Technology Forum, Vol. 25(2018) No.29, p.67-71.

https://doi.org/10.1051/e3sconf/202337501010

Response of rhizosphere microenvironment to coupling of climate warming and soil pollution

Na Lei [1,3,4] and Weihua Zhang[2,3,4*]

[1]Institute of Land Engineering and Technology, Shaanxi Provincial Land Engineering Construction Group Co., Ltd, Xi'an, China
[2]Shaanxi Key Laboratory of Land Consolidation, Xi'an 710075, China
[3]Key Laboratory of Degraded and Unused Land Consolidation Engineering, the Ministry of Natural Resources, Xi'an, China
[4]Shaanxi Provincial Land Consolidation Engineering Technology Research Center, Xi'an, China

Abstract. With the development of industrialization, the pollution of heavy metals in soil, such as lead (Pb) and cadmium (Cd), has become increasingly serious. On the other hand, the century old climate warming is considered to be the result of the significant increase in the concentration of greenhouse gases, such as CO2, in the atmosphere. These changes will affect the function and stability of the terrestrial ecosystem. Therefore, in recent years, there are more and more studies on the effects of heavy metal pollution and elevated atmospheric CO2 concentration on plants. But most of them focus on the study of the aboveground part of herbaceous plants under the action of single factor, and the study of the soil microenvironment of woody plants under the combined action of these environmental factors is relatively less. This study summarized the response of rhizosphere microenvironment to the coupling of climate warming and soil pollution, in order to provide theoretical reference for improving soil quality and ecological environment.

Key word: Climate warming; Soil pollution; Rhizosphere microenvironment; Coupling effect

1. Introduction

The global warming is largely due to the increase of atmospheric CO_2, N_2O, CH_4 and other greenhouse gas emissions generated by the burning of a large number of fossil fuels. Since mankind entered the era of industrialization in the 1970s, the concentration of CO_2 in the atmosphere has been increasing year by year [1]. It is estimated that the concentration of CO_2 in the atmosphere will double by the end of the 21st century [2].The IPCC (Intergovernmental Panel on Climate Change) climate assessment report points out that the global average temperature has increased by 0.3-0.6°C in the past 100 years [3]. By the end of this century, it is predicted that the earth's surface temperature will continue to rise by 1.4-4.8 °C [4], which will directly lead to climate anomalies and exacerbate natural disasters. Heavy metal pollution in soil is becoming increasingly serious in today's society [5, 6]. At present, the problem of heavy metal pollution in soil generally focuses on the pollution of significant elements such as Cd, Hg, Cr, As, Pb, among which Pb and Cd are the most prominent ones. Pure soil Pb or Cd pollution is rare, and they often exist together [7], which will pose a great threat to the safety and stability of plant rhizosphere soil ecosystem [18]. When the accumulation of heavy metals in the soil exceeds the self-purification carrying capacity of the soil, it will have a great toxic effect on aboveground plants and other soil microorganisms, which will not only reduce the quality and yield of crops, but also affect the food safety of agricultural and sideline crops through the food chain, affecting people's health [8, 9]. Both of them are major ecological issues that have become the concern of human society today, but the research on the impact of their coupling effect on soil is rare. This study will highlight the impact of their coupling on soil ecological environment, providing theoretical reference for improving soil quality and ecological environment.

2. Impact of climate warming on soil rhizosphere microenvironment

2.1 Research progress on the influence of atmospheric CO2 on root exudates

The composition and content change of root exudates is the most direct and obvious response of plants to environmental stress. The increase of atmospheric CO_2 concentration has become the most important environmental problem facing the society today. Research shows that the increase of atmospheric CO_2 concentration affects both the aboveground parts of plants and the root environment of plants [10]. With the increase of atmospheric CO_2 concentration, the photosynthesis of plants is enhanced, and the input of organic compounds to underground roots is increased, thus stimulating the

* Corresponding author: 619648133@qq.com

growth and secretion of plant roots. Because root exudates can reflect the response of plants to environmental stress and are the link between soil and plants, they have become the focus of scientific research.

According to previous studies, the increase of atmospheric CO_2 concentration can promote the secretion of amino acids and organic carbon by soil roots, further affect the quantity and activity of rhizosphere soil microorganisms, and promote the symbiosis between plant roots and mycorrhizal fungi, resulting in significant changes in plant rhizosphere soil microenvironment. The changes in the number and morphology of roots with the increase of atmospheric CO_2 concentration may help plants to absorb more nutrients and water under environmental stress, so as to better adapt to the environment after the increase of atmospheric CO_2 concentration.

2.2 Research progress on the impact of atmospheric CO_2 on soil microorganisms

In recent years, the increase of atmospheric CO_2 concentration has become a global environmental problem, which will inevitably affect soil microorganisms. The increase of atmospheric CO_2 concentration mainly affects plant growth and indirectly affects soil microbial activity [11]. The increase of atmospheric CO_2 has different effects on soil microbial community structure, metabolic activity, microbial flora and activity [12]. The increase of atmospheric CO_2 concentration can stimulate microbial activity to a certain extent, and microbial population, quantity, composition and activity change with environmental change. Among them, the number of living cells of microorganisms is the most sensitive biological indicator to environmental change.

Although there are more and more studies on the response of soil microbial biomass and microbial population to atmospheric CO_2, most of the studies are still focused on the impact of a single environmental factor on microorganisms, and there are still few reports on the interaction conditions between the increase of atmospheric CO_2 and the combined pollution of heavy metals in soil. The response of soil microorganisms to environmental changes is a very important link in the process of ecosystem feedback, and is the key to comprehensively evaluate the impact of environmental changes on terrestrial ecosystems, which needs further research.

2.3 Research progress on the effect of atmospheric CO_2 on soil enzyme activity

Soil enzymes are biologically active proteins in soil. Through the biochemical transformation of a variety of organic substances and organic residues entering the soil, the components of the ecosystem are functionally linked, thus maintaining the relative stability of soil biochemistry, which can be used as one of the indicators of soil fertility. In recent years, the impact of elevated atmospheric CO_2 concentration on soil enzyme activity has attracted extensive attention of researchers. The final research results usually show significant promotion, but there are also different conclusions. In general, the increase of atmospheric CO_2 concentration increases the number and activity of most enzymes, but the experimental results are often different due to different research methods, research time, research environment and enzyme types.

3. Impact of soil pollution on soil rhizosphere microenvironment

3.1 Research progress on the effect of heavy metals on root exudates

Plant roots play a very important role in the biosphere biosphere interface. When the soil environment changes or is stressed, roots will actively release specific secretions. Root secretions change the bioavailability of heavy metals by changing the pH value of rhizosphere soil and chelation, thereby affecting the absorption of heavy metals at the root interface and reducing the toxicity of heavy metals to plant roots.

Under the combined action of Pb, Cd and Cd/Pb, the types of root exudates change to varying degrees, which is mainly related to the complexation reaction between root exudates and heavy metals. The complexation ability is related to the nature of heavy metals themselves. The complexation ability of wheat root exudates to Pb is greater than that of Cd. When wheat plants are polluted by Pb, Cd and Cd/Pb, the amount of sugar secreted by wheat roots increases, and tartaric acid is newly added in root exudates. Citric acid is only detected in root exudates polluted by Cd and Cd/Pb, indicating that heavy metals will have an impact on wheat root exudates in terms of both species and quality.

3.2 Research progress on the effects of heavy metals on rhizosphere soil microorganisms

Soil microorganisms are the most active members in the terrestrial ecosystem, and also the important driving force and main participants in the decomposition of soil organic matter and the transformation and circulation of soil nutrients. They promote the material cycle and energy flow of the ecosystem and maintain the normal operation of the ecosystem. The species, quantity, distribution and community composition of rhizosphere soil microorganisms are affected by soil nutrient status, pH, texture, temperature, humidity and aeration. Soil microorganisms need energy and nutrients in their life activities, so any factors affecting the soil environment will have an impact on soil microorganisms to a certain extent.

In recent years, the pollution of heavy metals in soil has become more and more serious, which inevitably changes the living environment of soil microorganisms, thus affecting soil microorganisms in varying degrees. The influence of heavy metal pollution on soil microorganisms is related to its species and concentration, soil physical and chemical properties, and single or combined heavy metal pollution. Heavy metal ions in the environment can form coordination complexes or chelates with various organic polymers and anions, which are easily adsorbed on the surface of minerals and organic substances and accumulated in sediments, thus affecting the bioavailability of heavy metals in soil, and also having a significant negative impact on soil microbial biomass and activity, thereby affecting the stability of soil ecological structure and function.

3.3 Research progress of heavy metal pollution on soil enzyme activity

Heavy metals can interact strongly with proteins and various enzymes in the soil, so that their biological effects are inhibited and inactivated, and they can also be enriched in the soil. If the ion level of heavy metals reaches or exceeds the physiological limit that plants can tolerate, it will cause toxic effects. Soil enzymes are sensitive to the toxicity of heavy metal ions, and the determination of soil enzyme activity is helpful to determine the degree of heavy metal pollution in soil and the ability of plants to repair contaminated soil [13].

In many cases, soil enzyme activity will decrease with the aggravation of soil heavy metal pollution. Under the stress of heavy metals, the molecular structure of soil enzymes is destroyed, leading to the inactivation of enzymes. However, when the content is low and does not reach the level of pollution, it will activate soil enzymes. This is mainly because they promote the coordination and binding of enzyme molecules with the substrate, thus enhancing the enzyme activity.

4. Response of rhizosphere microenvironment to coupling of climate warming and soil pollution

Soil microorganism is a biological indicator of soil health, which has an important impact on soil fertility. The growth and activity of soil microorganisms are affected by environmental conditions such as soil pH, humidity, temperature, external interference, etc., which can sensitively indicate changes in climate and soil environmental conditions. Therefore, human industrial and agricultural development activities inevitably change the living environment conditions of soil microorganisms. The results of this study showed that the increase of atmospheric CO_2 concentration and the coupling of soil Cd/Pb pollution had a significant impact on the microbial activity of plant rhizosphere soil. Soil microbial activity is very sensitive to atmospheric CO_2 concentration, and the increase of atmospheric CO_2 concentration has a positive effect on the activity of bacteria, fungi, actinomycetes, microbial biomass C, microbial quotient and total microorganisms.

This may be due to the increase of atmospheric CO_2 concentration will bring more organic substrates to rhizosphere microbial metabolism, thus stimulating rhizosphere microbial activity and improving microbial activity. Under Cd or Pb stress, bacteria, actinomycetes, microbial biomass C, microbial quotient and total microbial activity were inhibited. The inhibition increased with the increase of Cd concentration, but the number of fungi did not change significantly, which may be due to the lower sensitivity of fungi to heavy metals than bacteria and actinomycetes. Root exudates interact with rhizosphere soil microorganisms. Root exudates are an important medium to connect plants and root soil microorganisms. Changes in plant root exudates can affect the activities of microorganisms. Root exudates a large amount of sugars, amino acids and other organic substances provide important nutrients and energy materials for the growth and reproduction of plant rhizosphere soil microorganisms, enabling microorganisms to multiply in large quantities.

5. Conclusion

(1) The increase of atmospheric CO_2 concentration promoted the total soluble sugar, amino acids, organic acids and total phenolic acids secreted by roots of Robinia pseudoacacia seedlings. Under Cd/Pb compound condition, the stimulation of Robinia pseudoacacia root system was enhanced. Compared with the effect of Cd alone, the content of total soluble sugar, amino acid and organic acid increased significantly, while the content of total phenolic acid decreased. Under the coupling of high concentration of atmospheric CO_2 and soil Pb and Cd pollution, the content of total soluble sugar, amino acid, organic acid and total phenolic acid secreted by plant roots is higher than that of heavy metals or CO_2 alone.

(2) Compared with the effect of Cd or Pb alone, the soil enzyme activity under Cd/Pb combined pollution decreased, indicating that the toxic effect of combined pollution on soil biological activity was enhanced. Soil urease activity, dehydrogenase activity β The activities of glucosidase and invertase were significantly higher than those of Cd, Pb and Cd/Pb alone, indicating that high concentration of atmospheric CO_2 could reduce the inhibition of heavy metals on soil enzymes.

(3) At the same Cd, Pb and Cd/Pb pollution levels, compared with the atmospheric CO_2 concentration, the increase of CO_2 concentration significantly increased the number of bacteria, fungi and actinomycetes in the rhizosphere soil of Robinia pseudoacacia seedlings, and the microbial biomass C, microbial quotient and total microbial activity also gradually increased. CO_2 has a certain role in promoting the survival and development of soil microorganisms. On the whole, the increase of atmospheric CO_2 concentration can mitigate the adverse effect of soil Cd and Pb pollution on plant rhizosphere soil microecology.

Acknowledgments

This work was financially supported by by the Fundamental Research Funds for the Central Universities, CHD (300102352503), Xi'an Science and Technology Plan Project, China (22NYGG0001).

References

1. E.G. O'Nell, Responses of soil biota to elevated atmospheric carbon dioxide, Plant Soil. Commun. 165 (1994) 55-65.

2. S. Solomon, D. Qin, N.M. Man, The scientific basis: Contribution of working group of the fourth assessment report of the intergovernmental panel on climate change, Cambridge University Press, 2007.
3. S.M Qiu, P.L. Gu, H. Hao, Analysis on the change and control of CO_2 emissions from energy consumption, Journal of Coal Industry. Commun. 27 (2002) 412-416.
4. IPCC. Third Assessment Report Climate Change, Cambridge: Cambridge University Press, 2001.
5. Q.X. Tang, X. Miao, Current status of cadmium pollution in soil and research progress in remediation, Environmental Engineering. Commun. 31 (2013) 747-750.
6. B. He, Z.J. Yun, J.B. Shi, Research progress of heavy metal pollution in China:Sources,analytical methods,status,and toxicity, Chinese Science Bulletin, Commun. 58 (2013) 134-140.
7. Z.Q. Meng, Environmental Toxicology, Beijing: China Environmental Science Press, 2000.
8. G. Gadd, M. Metals, minerals and microbes: geomicrobiology and bioremediation, Microbiology. Commun. 156 (2010) 609-643.
9. A.Z. Ren, Y.B. Gao, S. Liu, Effects of Cr, Cd, Pb Stress on Several Physiological and Biochemical Indexes of Vegetable Leaves, Journal of Applied and Environmental Biology. Commun. 6 (2000) 112-116.
10. J.A. Ryan, H.R. Pahren, J.B. Lucas, Controlling cadmium in the human food chain: a review and rationale based on health effects, Environmental Research. Commun. 28 (1982) 251-302.
11. Y.L. Ma, K.Y. Wang, Q. Sun, Effect of elevated atmospheric CO_2 concentration on plant roots, Journal of Ecology. Commun. 26 (2007) 1640-1645.
12. Y. Li, G.H. Huang, Y. Shi, Effects of elevated atmospheric CO_2 concentration on farmland soil microorganisms and related factors, Journal of Applied Ecology. Commun. 14 (2003) 2321-2324.
13. J.W. Liu, A.Q Shan, X.M. Chuai, Study on the effect of heavy metal ion pollution on soil enzyme activity, Pollution Prevention Technology. Commun. 21 (2008) 19-21.

 https://doi.org/10.1051/e3sconf/202337501011

Numerical investigation of poroelastic effects during hydraulic fracturing using XFEM combined with cohesive zone model

Xiaoyu Ren [1], Yueyang Guan [2], Xiaofei Zhang [3], Xiaobao Liu [3], Jie Liu [3], Kang Liu [4], Jiangwei Luo [5], Guiyuan Shi [5]

[1] The Second Gas Production Plant of Petro China Changqing Oilfield Company, Yulin, China
[2] CNPC Engineering Technology R&D Company Limited, Beijing, China
[3] The Second Gas Production Plant of Petro China Changqing Oilfield Company, Yulin, China
[4] CNPC Bohai Drilling Engineering Company Limited No.4 Drilling Engineering Company, China
[5] China University of Petroleum (Beijing), China

Abstract. Hydraulic fracturing is an effective technology used to stimulate hydrocarbon production from unconventional reservoirs. Numerical simulation of the hydraulic fracturing process plays a key role in the design of hydraulic fracturing. However, most of existing models are overly simplified by neglecting the poroelasticity of rock matrix, which may have a substantial effect on the growth of hydraulic fractures, especially in oil and/or water saturated formations. In this study, we developed a fully coupled finite element model for hydraulic fracture propagation in permeable formations by combining XFEM technique with cohesive zone model, and used it to investigate the effects of poroelasticity on the geometry evolution of single fracture which initiated from an injection point. Fluid flow within the fractures, Darcy flow within the rock matrix, hydraulic fractures propagation, and fluid leak-off into the formation are simultaneously taking into account in the model. Then, the model is validated by comparing the results with available analytical solutions. To understand and quantify the poroelastic effects on the propagation of hydraulic fracture, several cases with different matrix permeability, leak-off coefficients, and bulk modulus of pore fluid are performed. The simulation results show that the total volume of leakage is controlled by the combined action of matrix permeability and leak-off coefficient. The fracture aperture and length decrease with the increase of matrix permeability or leak-off coefficient, while as the bulk moduli of pore fluid increases, the fracture aperture and length tend to increase.

Keywords: Poroelastic effect; Hydraulic fracturing; XFEM; Cohesive zone model.

1. Introduction

Hydraulic fracturing is an effective technology used to stimulate hydrocarbon production from unconventional reservoirs, in which one or more fractures are driven by the internal pressurized fluid (Bunger and Cardella, 2015). The growth of hydraulic fracture in porous media is a complex, multi-physics problem, which can be generally considered as the following coupled processes (Adachi et al., 2007; Detournay, 2016): (i) rock deformation caused by the fluid pressure on the fracture surfaces, (ii) fluid flow within the fracture, (iii) fracture propagation, and (iv) fluid leak out of the fracture into the formation matrix. Accurately predicting the growth behavior of hydraulic fracture is necessary to design a suitable fracturing scheme, which requires a fully coupled numerical model that can simultaneous account for the aforementioned processes.

In the early times, analytical and semi-analytical methods (Detournay, 2004; Nordgren, 1972; Perkins and Kern, 1961) have been developed to understand the evolution of variables of interests, such as fracture aperture, fracture length, and injection pressure. Due to the inherent limitation of analytical methods, these solutions, however, are usually limited to some simple situations such as constant injection rate and linearly elastic medium. To simulate the growth behavior of hydraulic fracture in more sophisticated situations, a large number of numerical approaches have been proposed in literatures, including the boundary element method (Gordeliy and Detournay, 2011; Long and Xu, 2017; Zhang and Jeffrey, 2008), the distinct element method (Damjanac and Cundall, 2016; Han et al., 2015), the finite element method (Carrier and Granet, 2012; Guo et al., 2015; Li et al., 2017a; Li et al., 2017b), the finite discrete element method (FDEM) (Lisjak et al., 2017; Yan et al., 2015), the discontinuous deformation method (DDA) (Choo et al., 2016), and the extended finite element method (XFEM) (Mohammadnejad and Khoei, 2013; Salimzadeh and Khalili, 2015). Although significant progress has been made, there are still some critical points that need to be elucidated. For example, most of existing models use

Carter leak-off model to characterize the velocity of fluid leak out of fracture, but neglect the poroelasticity effect of rock matrix, which may have substantial impacts on the growth of hydraulic fracture, especially in oil and/or water saturated formations.

To understand and quantify the poroelasticity effect on the evolution of hydraulic fracture, this paper develops a fully coupled XFEM-based cohesive model which is able to take into account several crucial physical aspects during hydraulic fracture propagation in porous media, including fracture propagation, fluid flow within fracture, fluid leak-off, pore fluid flow, and rock deformation. In the following, the governing equations used in the model are provided first. Next, this model is validated against analytical solutions. Finally, parametric sensitivity analyses are performed to investigate the impact of poroelastic effect on the growth of hydraulic fracture.

2. Governing equations

2.1 Deformation of porous medium

The rock formation is assumed to be a homogeneous, isotropic, and linear poroelastic medium. For quasi-static conditions, the stress equilibrium equation can be written as

$$\nabla \cdot \boldsymbol{\sigma} + \mathbf{F} = 0 \tag{1}$$

where $\mathbf{F}$ is the body force vector, and $\boldsymbol{\sigma}$ is the total stress tensor. Assuming that the rock matrix is fully saturated by single-phase fluid and obeys poroelasticity theory, the effective stress of rock matrix reads (Detournay and Cheng, 1993)

$$\boldsymbol{\sigma}' = \boldsymbol{\sigma} + \alpha p \mathbf{I} \tag{2}$$

where $\boldsymbol{\sigma}'$ is the effective stress, p is the pore pressure, $\mathbf{I}$ is the second-order unit matrix, and α is the Biot coefficient, which is defined as

$$\alpha = 1 - K / K_s \tag{3}$$

where K and K_s are the bulk modulus of porous rock and of the rock matrix grain, respectively. The effective stress is related to the matrix strain as

$$\boldsymbol{\sigma}' = \mathbf{D}\boldsymbol{\varepsilon} \tag{4}$$

where $\mathbf{D}$ is the elastic stiffness matrix, and $\boldsymbol{\varepsilon}$ is the strain tensor.

The pore fluid flow obeys mass balance equation as

$$\frac{\partial(\rho_f \phi)}{\partial t} + \nabla \cdot (\rho_f \phi \mathbf{v}_f) = 0 \tag{5}$$

where ρ_f is pore fluid density, ϕ is the rock porosity, $\mathbf{v}_f$ is the fluid flow velocity of the pore fluid. The Darcy's fluid velocity is given by

$$\mathbf{v}_d = \phi(\mathbf{v}_f - \mathbf{v}_s) \tag{6}$$

in which $\mathbf{v}_s$ is the solid phase (rock matrix) velocity, and $\mathbf{v}_d$ is the Darcy's fluid velocity, can be defined as

$$\mathbf{v}_d = \frac{\mathbf{k}_m}{\mu_f}(\nabla p + \rho_f \mathbf{g}) \tag{7}$$

in which $\mathbf{k}_m$ is the permeability of the rock matrix, μ_f is the fluid viscosity, and $\mathbf{g}$ is the vector of gravitational acceleration.

2.2 Fracture fluid flow

Assuming that the hydraulic fracture is filled with a Newtonian fluid, the mass balance equation for the fluid flow within the fracture reads (Hibbitt et al., 2016)

$$\dot{w} + \nabla \cdot q + (q_t + q_b) = 0 \tag{8}$$

where w is the fracture width, q is the tangential fluid flow velocity within the fracture, q_t and q_b are the fluid leak-off into the formation across top and bottom fracture surfaces, respectively.

The tangential fluid flow velocity obeys cubic law as

$$q = -\frac{w^3}{12\mu_f}\nabla p_f \tag{9}$$

in which p_f is the fluid pressure within the fracture.

The normal fluid velocity per unit area (q_t and q_b) can be described as

$$\begin{cases} q_t = c_t(p_f - p_t) \\ q_b = c_b(p_f - p_b) \end{cases} \tag{10}$$

where p_t and p_b are the pore fluid pressures in the rock matrix abut to the top and bottom surfaces, respectively. c_t and c_b are the "leak-off coefficients" of top and bottom surfaces, respectively. Different with the classical Carter's model, c_t and c_b here characterize the resistance of fluid across the fracture surfaces. The leak-off velocity is controlled by the rock matrix permeability if c_t and c_b are large enough.

2.3 Fracture propagation

XFEM technique, implemented in Abaqus, is used in this paper to simulate the hydraulic fracture propagation in porous media. The fracture propagation behavior is described by cohesive zone model with linear traction-separation response, as shown in Fig. 1. Damage is assumed to initiate when a quadratic interaction function involving nominal stress ratios reaches 1.0, which can be expressed as (Hibbitt et al., 2016)

$$\left(\frac{\langle t_n \rangle}{t_n^0}\right)^2 + \left(\frac{t_s}{t_s^0}\right)^2 + \left(\frac{t_t}{t_t^0}\right)^2 = 1 \tag{11}$$

where t_n, t_s, and t_t are the nominal traction in the normal and two local shear directions, respectively; t_n^0, t_s^0, and t_t^0 represent the cohesive strength in the normal and two local shear directions, respectively.

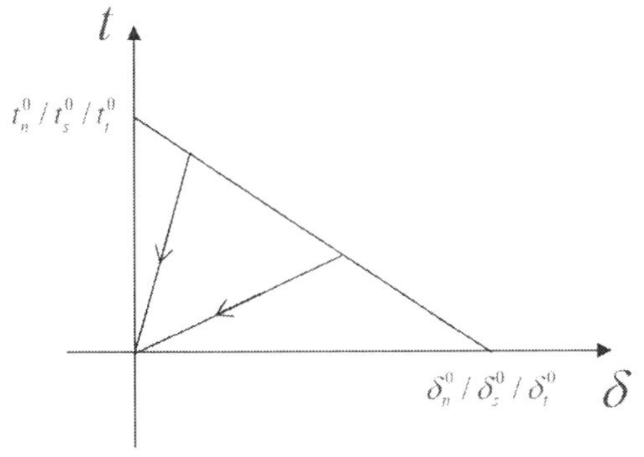

Fig. 1 Linear elastic traction-separation law

Damage evolution of cohesive zone model is described by progressive degradation of cohesive strength as

$$t_0 = \begin{cases} (1-D)T_n & T_n \geq 0 \\ T_n & T_n < 0 \end{cases} \quad (12)$$

$$t_{s/t} = (1-D)T_{s/t} \quad (13)$$

where $T_{n/s/t}$ are the stresses in the corresponding directions predicted by linear traction-separation law without damage. D is the scalar damage parameter.

3. Model validation

In this section, Terzaghi's problem and KGD problem are performed to verify the capability of the XFEM-based cohesive zone model on modeling the process of fluid diffusion and the growth of hydraulic fracture in rock matrix, respectively.

3.1 Terzaghi's problem

Consider a water-saturated, plane strain porous column, with the height of 100 m, as shown in .Fig. 2(a). The top surface is drained and other surfaces are undrained boundary conditions. The normal displacement of left, right, and bottom boundaries are fixed. A load P_L acts instantaneously on the top boundary at initial time. The relevant parameters of this case are shown in Table. 1 and the analytical solution are provided in Appendix A.

Table. 1 Input parameters for the Terzaghi's problem

Parameters	Value
Elastic moduli, GPa	1.0
Poisson's ratio	0.0
Bulk moduli of grain, GPa	50.0
Bulk moduli of pore fluid, GPa	2.2
Permeability, D	1.0
Fluid viscosity, mPa·s	1.0
Load P_L, MPa	1.0

Fig. 2(b) shows the evolution of pore pressure along the length of the column at different time, and Fig. 2(c) shows the vertical displacement of top boundary. The numerical results agree well with analytical solutions, which indicates that the XFEM-based cohesive model has the capability to model the fluid diffusion of rock matrix.

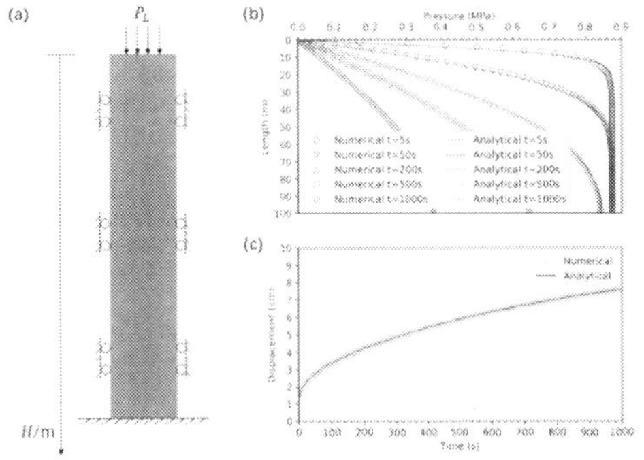

Fig. 2 (a) Sketch of the Terzaghi's problem; (b) Evolution of pore pressure along the length of the column; (c) Vertical displacement of top boundary

3.2 KGD problem

As shown in Fig. 3(a), a bi-wing fracture is created by injecting a Newtonian fluid into an impermeable medium. Because the typical growth of hydraulic fracture in an unconventional reservoir is under viscosity dominated, the parameters used in this case were tuned to this regime, as provided in Table. 2. The details of semi-analytical solutions are shown in Appendix B.

Fig. 3 (b) and (c) show the temporal evolution of injection pressure and mouth aperture. The agreement between numerical results and analytical solutions is very good, which proves the adequacy of the XFEM-based cohesive zone model on modeling the growth of hydraulic fracture.

Table. 2 Input parameters for the KGD problem

Parameters	Value
Elastic moduli, GPa	10.0
Poisson's ratio	0.2
Mode-I fracture energy, J/m^2	10
Mode-II fracture energy, J/m^2	300
Tensile cohesive strength, MPa	0.2
Shear cohesive strength, MPa	0.2
Injection rate, m^3/s	5×10^{-4}
Fluid viscosity, Pa·s	2.0
Injection time, s	150.0

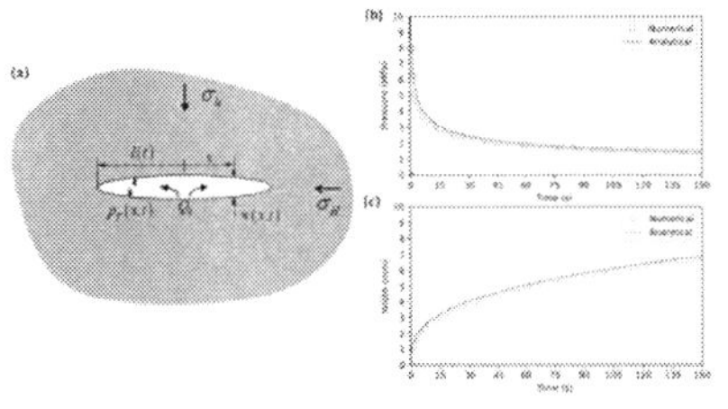

Fig. 3 (a) Sketch of the KGD problem; (b) Temporal evolution of injection pressure; (c) Temporal evolution of mouth aperture

4. Poroelastic effect on hydraulic fracture propagation

This section aims to investigate the impact of poroelastic effect on the propagation behavior of hydraulic fracture. To this end, a series of parametric analyses for KGD problem are performed in this section. As shown in Fig. 4, a 2D plane-strain fracture is propagated by injecting a viscos fluid from the initial crack. The dimension of the domain is 200×50 m, and the length of the initial crack is 0.4 m. The left boundary of the model is symmetric boundary condition, the normal displacements of other external boundaries are fixed, and a constant pore pressure boundary condition is also imposed on these boundaries. The input parameters for the base case of parametric analysis are listed in Table. 3. Additionally, in each parametric analysis, only the investigated parameter is changed while all other parameters are kept same with the values from the base case.

Table. 3 Input parameters for the base case

Parameters	Value
Elastic moduli, GPa	20.0
Poisson's ratio	0.2
Mode-I fracture energy, J/m^2	10
Mode-II fracture energy, J/m^2	300
Tensile cohesive strength, MPa	0.2
Shear cohesive strength, MPa	0.2
Permeability, D	1.0
Leak-off coefficient, m/(Pa·s)	1.0×10^{-5}
Bulk moduli of pore fluid, GPa	2.2
Initial pore pressure, MPa	20.0
$\sigma_H/\sigma_h/\sigma_v$, MPa	30/27/35
Fluid viscosity, Pa·s	0.1
Injection rate, m^3/s	5×10^{-4}
Injection time, s	80.0

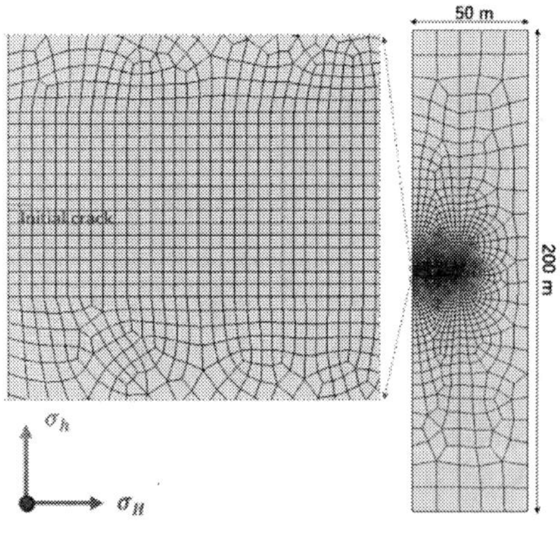

Fig. 4 The geometry and mesh of the KGD problem

4.1 Results of the base case

Fig. 5 shows the final fracture geometry and pore pressure distribution of the base case. The fracture grows to a length of 6.2 m at the end of the simulation (t=80 s). The leak-off coefficient in this case is large enough and the rock permeability dominates the fluid leak velocity. The simulation results for temporal evolution of net injection pressure and aperture at the injection point are shown in Fig. 6. The maximum aperture at the end of simulation is about 1.35 mm.

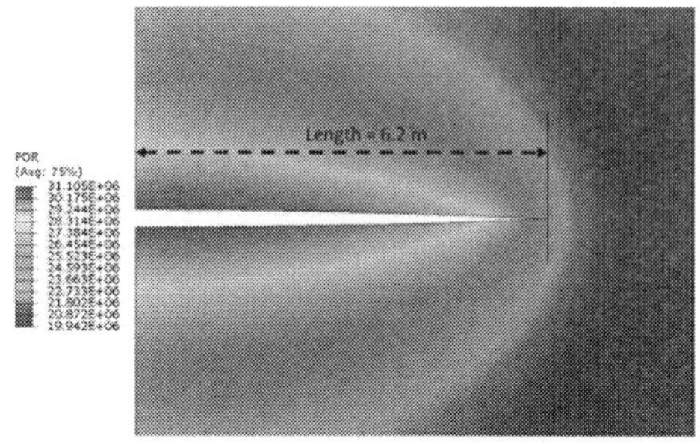

Fig. 5 Fracture geometry and pore pressure distribution of the base case at the end of the simulation

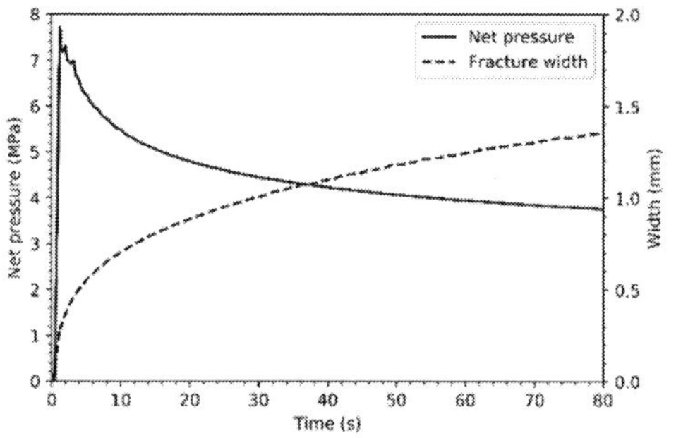

Fig. 6 Temporal evolution of net injection pressure and aperture at the injection point

4.2 Effect of permeability

As an important parameter for porous media properties, permeability is expected to significantly influence the pore pressure distribution around the hydraulic fracture, and consequently affects the fracture evolution. Thus, we perform a sensitivity analysis by varying permeability in this section. Assume there is no filter cake developed on the fracture surfaces. Fig. 7 depicts the fracture profile and pore pressure distribution at the end of the injection for different rock permeability. It readily observed that fracture length increases monotonically as permeability decreases. Large permeability leads to high fluid loss volume, which indicates that the leak-off in a system with large leak-off coefficient is controlled by rock permeability.

https://doi.org/10.1051/e3sconf/202337501011

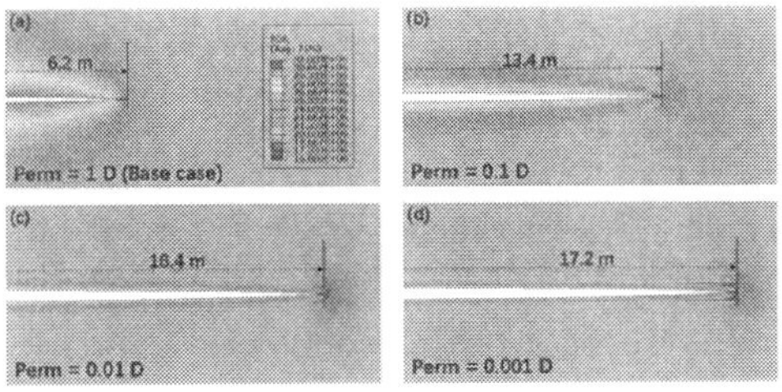

Fig. 7 Effect of permeability on the fracture profile and pore pressure distribution

Fig. 8 shows the effect of permeability on the net pressure and aperture at the injection point as time elapse. It can be observed that the net pressure of high permeability case is much larger than the low permeability case, while the fracture aperture of high permeability case is much lower than the low permeability case. This is due to the large back-stress induced by poroelastic effect in high permeability case, which in turn reduces the aperture.

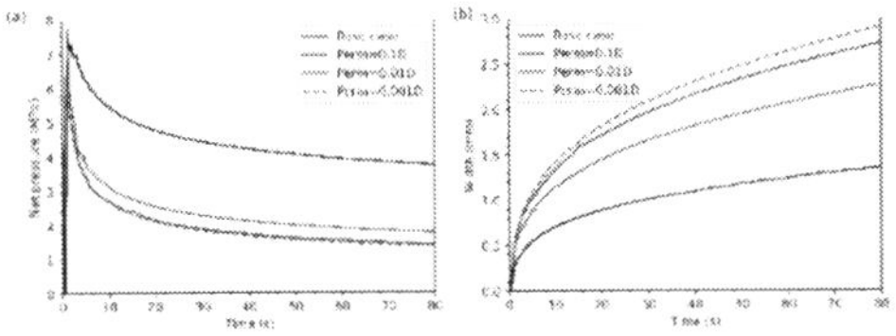

Fig. 8 Effect of permeability on the net pressure and aperture at the injection point

4.3 Effect of leak-off coefficient

Leak-off coefficient in this work characterizes the resistance of filter-cake during the fracture propagation. Fig. 9 shows that the impact of leak-off coefficient on the growth of hydraulic fracture is substantial. Fluid loss is controlled by permeability in base case while it is controlled by the leak-off coefficient in Fig. 9(d) and consequently more fluid is utilized to drive the fracture propagation.

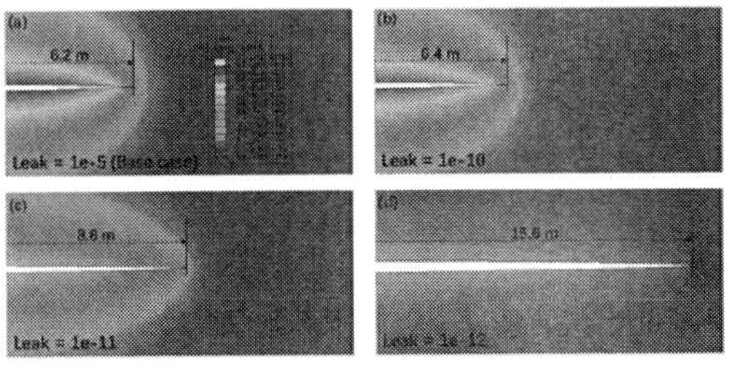

Fig. 9 Effect of leak-off coefficient on the fracture profile and pore pressure

Fig. 10 shows the temporal evolution of net pressure and aperture at the injection point for different leak-off coefficient. The results indicate that lower leak-off coefficient results in lower propagation pressure and higher fracture aperture. This is because the back-stress induced by poroelastic effect is lower.

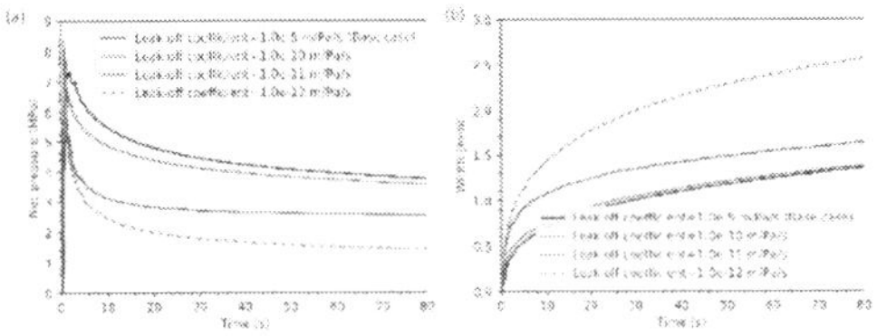

Fig. 10 Effect of Leak-off coefficient on the net pressure and aperture at the injection point

4.4 Effect of bulk modulus of pore fluid

Bulk moduli of pore fluid is one of factors determining the dissipation rate of pore pressure, which may influence the hydraulic fracture propagation. However, to the authors' knowledge, there is no previous studies have investigated its influence on the fracture propagation. Fig. 11 shows the fracture profiles for different bulk modulus of pore fluid. The results indicate that low bulk moduli of pore fluid tends to suppress the increase of the pore pressure around hydraulic fracture, which leads to high leak-off velocity and low fracturing fluid efficiency.

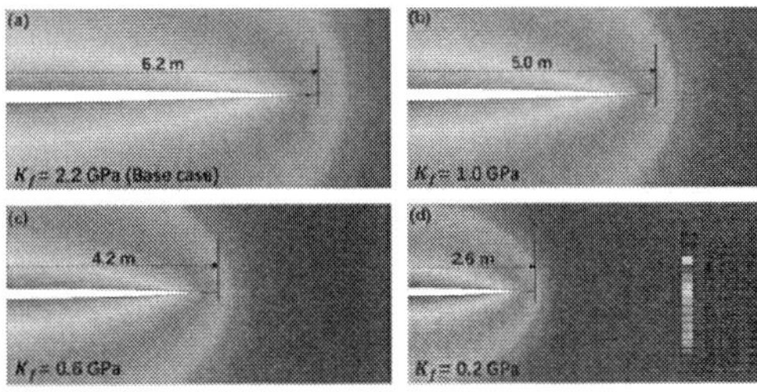

Fig. 11 Effect of fluid bulk moduli on the fracture profile and pore pressure

5. Conclusions

In this paper, a fully coupled finite element model is proposed to investigate the influences of poroelastic effect on the growth of hydraulic fracture, and the model was validated against available solutions. The simulation results show that the total volume of leakage is controlled by the combined action of matrix permeability and leak-off coefficient. The fracture aperture and length decrease with the increase of matrix permeability or leak-off coefficient, while as the bulk moduli of pore fluid increases, the fracture aperture and length tend to increase.

Acknowledgements

This work is financially supported by National Natural Science Foundation (No.11502304), and National Science and Technology Major Project (2016ZX05058002-006).

Appendix A. Terzaghi's 1D consolidation problem(Wang, 2004)

Pore pressure evolution:

$$p(z,t)=\frac{4}{\pi}p_0\sum_{m=0}^{\infty}\frac{1}{2m+1}\exp\left[\frac{-(2m+1)^2\pi^2ct}{4L^2}\right]\sin\left[\frac{(2m+1)\pi z}{2L}\right] \quad (A.1)$$

Vertical displacement:

$$u(z,t)=c_M p_0\left\{(L-z)-\frac{8L}{\pi^2}\sum_{m=0}^{\infty}\frac{1}{(2m+1)^2}\exp\left[\frac{-(2m+1)^2\pi^2ct}{4L^2}\right]\cos\left[\frac{(2m+1)\pi z}{2L}\right]\right\}+u_0 \quad (A.2)$$

where $p_0(z)=\frac{\alpha M}{K_u+4\mu/3}P_L$; $u_0(z)=\frac{1}{K_u+4\mu/3}P_L(L-z)$; $M=\left[\phi\beta+(\alpha-\phi)c_{br}\right]^{-1}$ is the Biot modulus; c_{br} is the solid grain compressibility; K_u is the undrained bulk modulus; c_M is the vertical uniaxial compressibility; c is the consolidation coefficient.

Appendix B. Viscosity dominated KGD problem

A bi-wing fracture is created by injecting an incompressible Newtonian fluid into rock. The rock is assumed to be impermeable and under a plane strain condition. The fluid injection rate is Q_0. Fracture half-length $l(t)$, fracture aperture $w(x,t)$, and net pressure of the fracture $p(x,t)$ will be determined, where x is the distance along the fracture and t is injection time. The analytical solutions of this problem are provided by Detourney (Detournay, 2004) as functions of the injection rate Q_0 and three material parameters E', K', and μ',

$$E'=\frac{E}{1-\nu^2};\quad K'=4\left(\frac{2}{\pi}\right)^{1/2}K_{IC};\quad \mu'=12\mu; \quad (B.1)$$

where K_{IC} is the fracture toughness, E is the Young modulus, ν is the Poisson ratio, and μ is the fluid viscosity.

The solutions provided by Detourney (Detournay, 2004) can be expressed as:

$$l(t)=L(t)\gamma(t);\quad w(x,t)=\varepsilon(t)l(t)\Omega(\xi,t);$$
$$p(x,t)=\varepsilon(t)E'\Pi(\xi,t) \quad (B.2)$$

where $\xi=x/l(t)$ is the scaled coordinate $(0\le\xi\le1)$; $\varepsilon(t)$ is a small dimensionless parameter; $L(t)$ is length scale; γ, Ω, and Π are dimensionless fracture length, aperture, and net pressure, respectively.

In viscosity-dominated propagation regime, the aforementioned dimensionless parameters can be expressed as follows:

$$\varepsilon(t)=\left(\frac{\mu'}{E't}\right)^{1/3};\quad L(t)=\left(\frac{E'Q_0^3}{\mu'}\right)^{1/6}t^{2/3} \quad (B.3)$$

$$\gamma_{m0}=0.616 \quad (B.4)$$

$$\bar{\Omega}_{m0}(\xi)=A_0\left(1-\xi^2\right)^{2/3}+A_1^{(1)}\left(1-\xi^2\right)^{5/3}+B^{(1)}\left[4\sqrt{1-\xi^2}+2\xi^2\ln\frac{1-\sqrt{1-\xi^2}}{1+\sqrt{1-\xi^2}}\right] \quad (B.5)$$

$$\Pi_{m0}^{(1)}=\frac{1}{3\pi}B\left(\frac{1}{2},\frac{2}{3}\right)\left[A_0\,{}_2F_1\left(-\frac{1}{6},1;\frac{1}{2};\xi^2\right)+\frac{10}{7}A_1^{(1)}\,{}_2F_1\left(-\frac{7}{6},1;\frac{1}{2};\xi^2\right)\right]+B^{(1)}\left(2-\pi|\xi|\right) \quad (B.6)$$

where $A_0=3^{1/2}$; $A_1^{(1)}=-0.156$; $B^{(1)}=0.0663$; B is the Euler beta function; and ${}_2F_1$ is a hyper-geometric function.

References

1. Adachi, J., Siebrits, E., Peirce, A. and Desroches, J., 2007. Computer simulation of hydraulic fractures. International Journal of Rock Mechanics and Mining Sciences, 44(5): 739-757.
2. Bunger, A.P. and Cardella, D.J., 2015. Spatial distribution of production in a Marcellus Shale well: Evidence for hydraulic fracture stress interaction. Journal of Petroleum Science and Engineering, 133: 162-166.
3. Carrier, B. and Granet, S., 2012. Numerical modeling of hydraulic fracture problem in permeable medium using cohesive zone model. Engineering Fracture Mechanics, 79: 312-328.
4. Choo, L.Q., Zhao, Z., Chen, H. and Tian, Q., 2016. Hydraulic fracturing modeling using the discontinuous deformation analysis (DDA) method. Computers and Geotechnics, 76: 12-22.
5. Damjanac, B. and Cundall, P., 2016. Application of distinct element methods to simulation of hydraulic fracturing in naturally fractured reservoirs. Computers and Geotechnics, 71: 283-294.
6. Detournay, E., 2004. Propagation regimes of fluid-driven fractures in impermeable rocks. International Journal of Geomechanics, 4(1): 35-45.
7. Detournay, E., 2016. Mechanics of Hydraulic Fractures. Annual Review of Fluid Mechanics, 48: 311-339.
8. Detournay, E. and Cheng, A.H.-D., 1993. Fundamentals of poroelasticity, Chapter 5 in Comprehensive Rock Engineering: Principles, Practice and Projects, II, pp. 113-171.
9. Gordeliy, E. and Detournay, E., 2011. Displacement discontinuity method for modeling axisymmetric cracks in an elastic half-space. International Journal of Solids and Structures, 48(19): 2614-2629.
10. Guo, J., Lu, Q., Zhu, H., Wang, Y. and Ma, L., 2015. Perforating cluster space optimization method of horizontal well multi-stage fracturing in extremely thick unconventional gas reservoir. Journal of Natural Gas Science and Engineering, 26: 1648-1662.
11. Han, Y., Hampton, J., Li, G., Warpinski, N.R. and Mayerhofer, M.J., 2015. Investigation of Hydromechanical Mechanisms in Microseismicity Generation in Natural Fractures Induced by Hydraulic Fracturing. SPE Journal.
12. Hibbitt, H., Karlsson, B. and Sorensen, P., 2016. Abaqus Analysis User's Manual Version 2016, Dassault Systèmes Simulia Corp, Providence.

13. Li, Y. et al., 2017a. Numerical simulation of limited-entry multi-cluster fracturing in horizontal well. Journal of Petroleum Science and Engineering, 152: 443-455.
14. Li, Y., Deng, J.G., Liu, W. and Feng, Y., 2017b. Modeling hydraulic fracture propagation using cohesive zone model equipped with frictional contact capability. Computers and Geotechnics, 91: 58-70.
15. Lisjak, A. et al., 2017. A 2D, fully-coupled, hydro-mechanical, FDEM formulation for modelling fracturing processes in discontinuous, porous rock masses. Computers and Geotechnics, 81: 1-18.
16. Long, G. and Xu, G., 2017. The Effects of Perforation Erosion on Practical Hydraulic-Fracturing Applications. SPE Journal, 22(02): 645 - 659.
17. Mohammadnejad, T. and Khoei, A.R., 2013. An extended finite element method for hydraulic fracture propagation in deformable porous media with the cohesive crack model. Finite Elements in Analysis and Design, 73: 77-95.
18. Nordgren, R., 1972. Propagation of a vertical hydraulic fracture. Society of Petroleum Engineers Journal, 12(04): 306-314.
19. Perkins, T. and Kern, L., 1961. Widths of hydraulic fractures. Journal of Petroleum Technology, 13(09): 937-949.
20. Salimzadeh, S. and Khalili, N., 2015. Fully coupled XFEM model for flow and deformation in fractured porous media with explicit fracture flow. International Journal of Geomechanics, 16(4): 04015091.
21. Wang, H.F., 2004. Theory of linear poroelasticity with applications to geomechanics and hydrogeology.
22. Yan, C., Zheng, H., Sun, G. and Ge, X., 2015. Combined Finite-Discrete Element Method for Simulation of Hydraulic Fracturing. Rock Mechanics and Rock Engineering, 49(4): 1-22.
23. Zhang, X. and Jeffrey, R.G., 2008. Reinitiation or termination of fluid-driven fractures at frictional bedding interfaces. Journal of Geophysical Research: Solid Earth, 113(B8).

 https://doi.org/10.1051/e3sconf/202337501012

Research on the Deposition Mechanisms and Production Characteristics of Untabulated Reservoirs in the X Oilfield of the Songliao Basin

Pan Wu

No.4 Oil Production Plant of Daqing Oilfield Company Ltd., Daqing 163511, China

Abstract. In the late stage of the ultra-high water cut development, the untabulated reservoir (not included in the submitted reserve sheet) is considered an important part of production sustainability in the X oilfield. According to the coring well data of 5 wells in the eastern district, the untabulated reservoir is associated with higher heterogeneity. Specifically, it develops not only the conventional low and ultra-low permeability sands but also medium-high permeability sands with thickness below 0.2 m. Given this, the research investigates the sedimentary models, spatial distribution, and heterogeneity of the different types of untabulated sand bodies, based on the core analysis. Moreover, the physical model water flooding experiments are carried out and the production characteristics of the different sand bodies in the untabulated reservoir with the different recovery approaches are clarified. The findings of this research effectively solve the development contradictions of the untabulated reservoir and lay down the foundation for the further potential tapping of the untabulated reservoir.

Keywords: Untabulated reservoir, sedimentary model, planar distribution, heterogeneity, recovery situation.

1. Introduction

Reservoir rocks of the Songliao Basin are derived from sediments of the fluvial-delta system of the intra-continental shallow-water lacustrine basin. The prolonged complex deposition is associated with not only sediments of high-energy environments but also those of low-energy environments with an insufficient supply of clasts. Such sediments of low-energy environments are mainly muddy siltstone characterized by the presented hydrocarbon occurrence of the oil patch and oil immersion, and located in the top, bottom, surrounding, and inside of the high-energy sediments. The corresponding reservoir rocks present degraded physical properties and are typically excluded from the effective reservoir rock. Therefore, they are not included in the reserve sheet and are referred to as the untabulated reservoir [1].

The X oilfield lies in the southern Songliao Basin and is associated extensive development of untabulated reservoirs. One drilled well encounters 48.6 independent untabulated sands on an average basis, with the sand thickness of 28.3 m, and the reserves of the untabulated reservoirs account for 17.9% of the total oil initially in place of the region. In 2015, five sealed-coring wells are placed in the eastern X oilfield, within 1000 m perpendicular to the provenance. The core samples reveal that the average permeability of the untabulated sands is 12.5×10^{-3} μm^2 and the average porosity is 20%. Further classification by permeability shows that the untabulated reservoirs are associated with an expanded range of air permeability. The maximum air permeability of the samples reaches 1300×10^{-3} μm^2, representing the high-permeability sandstone, while the lowest permeability is less than 10×10^{-3} μm^2, representing the ultra-low permeability sandstone.

Upon the arrival of the ultra-high water cut stage, the thin poor reservoirs, dominated by the untabulated reservoirs, have become the main target for water flooding-based reserve potential tapping in the X oilfield. As the low-efficient and insufficient circulation problems of injected water intensify, the development contradiction of the untabulated reservoirs becomes increasingly prominent. Given these, this research investigates the heterogeneity of the untabulated reservoirs, based on the samples of the five coring wells, in an attempt to clarify the sedimentary model and planar distribution of sand bodies with varying levels of permeability in the untabulated reservoirs. Moreover, the physical model flooding is carried out to reveal the production characteristics of untabulated reservoirs. Efforts of this research are made to further provide theoretical and practical references for the potential tapping of untabulated reservoirs.

2. Heterogeneity of untabulated reservoirs

2.1 Core characteristics of untabulated reservoirs

Two types of sandstone are identified in the untabulated reservoirs, according to the core characteristics of the coring wells. One is the low and ultra-low permeability sandstone, with permeability below 50×10^{-3} μm^2, which are classified as the conventional untabulated reservoir. It is seen with a predominance of muddy siltstone or silty mudstone, and diverse oil-containing attributes, mostly of oil patch and oil stain. Moreover, it features numerous inclusions of muddy strips or aggregates, and can be subdivided into three groups, namely strip-like, thin-bedded, and patchy. The other is the untabulated sand with permeability above 50×10^{-3} μm^2 and called the favorable band. Its top and bottom contact with the low and ultra-low permeability stands, and the lithology is siltstone or fine sandstone. The oil-containing attribute is typically identified as oil immersion, with that of some samples reaching the oil-rich level. The oil occurrence of this type is more similar to that of the effective sandstone reservoir, and yet the corresponding thickness is less than 0.2 m, which results in failure of identification in well logs and subsequent inclusion into untabulated reservoirs (Fig. 1).
Among the 522 untabulated samples in total collected from the Sa-Pu oil layers of the five coring wells, the conventional untabulated samples account for 44.8% of the total, constituting the main permeability range. However, the sand samples of the favorable bands also hold the shares of 29.4% (Table 1), which indicates that the interior of the untabulated reservoirs is associated with high intra-layer heterogeneity, due to the extensive development of favorable bands.
The water washing observation of the coring wells demonstrates that the development of favorable bands does not only affects the intra-layer heterogeneity of the untabulated reservoirs but also the production of these reservoirs. Most of the favorable bands have been through wash washing at medium-intensive levels, while the conventional untabulated samples with permeability below 50×10^{-3} μm^2 are mostly free of water washing. Therefore, it is suggested that the presence of favorable bands in the untabulated reservoirs is a major factor impacting the hydrocarbon recovery of untabulated sands.

Figure 1. Core characteristics of untabulated reservoirs

Table 1. Permeability statistics of the five coring wells with high coring density

Type	Permeability Range ($\times10^{-3}$μm^2)			
	1-10	10-50	50-500	>500
Sample proportion (%)	44.8	25.7	25.7	3.7
Layer proportion (%)	36.2	24.8	32.6	6.4

2.2 Micro pore structure characteristics of untabulated reservoirs

Comparison of the mercury intrusion curves of the favorable bands and conventional untabulated reservoirs (Fig. 2) shows that the conventional untabulated reservoir is found with a lower maximum mercury injection saturation, lower residual mercury saturation, higher mercury withdrawal efficiency, and preferred median-saturation pressure. In other words, the conventional untabulated reservoir, compared with the favorable band, has smaller pore-throat radii and higher displacement pressure.

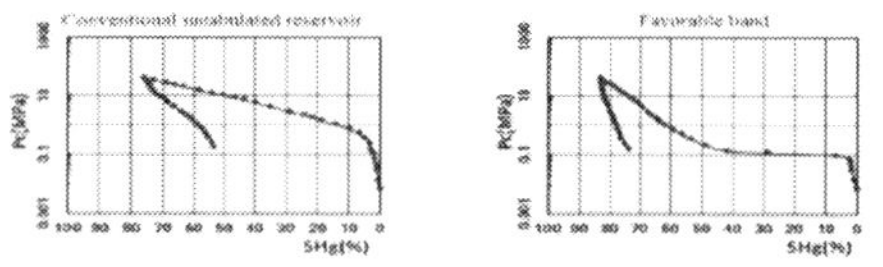

Figure 2. Comparison of the mercury intrusion curves of the conventional untabulated reservoir and favorable band

11 samples of the on-sheet and untabulated reservoirs in the X oilfield are put through the constant-rate mercury intrusion tests (specifically, 3 for the tabulated reservoir (in other words, included in the reserve sheet), 4 for the favorable band, and 4 for the conventional untabulated reservoir), which implies considerable differences of throat radii of the different samples (Fig. 3). In terms of the samples of the tabulated reservoir, the throat radius range is expanded and the proportion of larger throats is higher, with the radius peak generally occurring above 7 μm. For the favorable band, the throat radius distribution is closer to that of the on-sheet reservoir, with the radius distribution peak occurring in 5–7μm. At last, the conventional untabulated reservoir is found with a narrow range of the throat radius, with the peak generally below 5 μm.

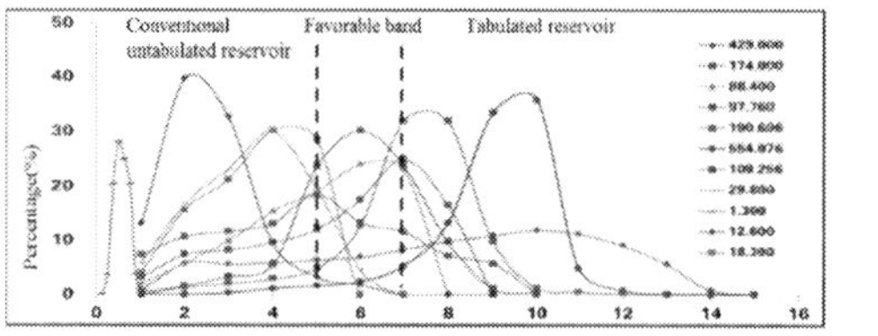

Figure 3. Throat radius distribution of core samples with varied permeability

https://doi.org/10.1051/e3sconf/202337501012

3. Deposition mechanisms of untabulated reservoirs

3.1 Sedimentary models of varied types of sands

The formation and distribution of untabulated reservoirs are controlled by the original sedimentary environment and associated conditions. The deltaic distributary plain facies mainly develops the (fluvial) channel sand, with sufficient clast supply. The occurring of the untabulated reservoir is mostly attributed to breach and overbank floods. Due to the high-energy sedimentary environment, the untabulated reservoir typically contains multiple medium-high permeability banded sands. Moreover, the untabulated sand body is seen with rapid lateral thinning and, because of the quick horizontal decline of the sedimentary environment energy and a lack of clastic supplement. The outward extension of the inter-channel sand body from the channel is short. Hence, in a planar view, the medium-high permeability untabulated sand presents itself as a sand body attached to the effective (on-sheet) sand body in the forms of edging and bridging. The coring well data shows that the farthest extension of medium-high permeability sand from the effective sand is only 25 m, which is rather difficult to identify in the case of the current well pattern.

The inner front of the delta is subjected to the joint effects of the river and lake and typically develops large-area sheet sands. The lake is found with high lateral continuity of energy. Therefore, the massive coarse-grained clasts, entrained and supplied by the river, deposit with decreasing thickness outward from the two river banks, with the degrading lake energy, which results in the successive formation of the fluvial channel sand, sheet sand, and medium-high permeability untabulated sand. As for the fine-grained clasts, the corresponding sediments reach farther, with the help of waves, and the connected low-ultra low permeability untabulated sands are formed. In the deltaic inner front sedimentary environment, the extension of the medium-high permeability sand is larger—the coring well data shows that the farthest extension of the medium-high permeability sand from the effective sand is 200 m.

The outer front facies is mainly affected by the lake. In such a low-energy environment lacking clastic supply, the untabulated reservoir mainly presents itself as the low and ultra-low permeability sand. The development area and probability of medium-high permeability sands are higher in the outer front I zone than in the outer front II zone. The untabulated reservoir in the latter is found with a predominance of connected low and ultra-low permeability sands and sporadic isolated medium-high permeability sands (Fig. 4).

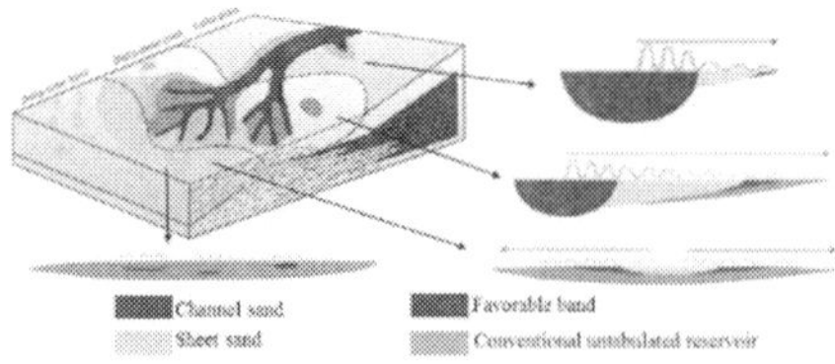

Figure 4. Sedimentary models for different sedimentary environments

3.2 Planar distribution characteristics of untabulated reservoir sands

According to the planar development location of untabulated sands and the planar sedimentary facies distribution, 5 planar development types of untabulated reservoirs are identified, namely the edging type (bridging with the channel sand), the embedding type (embedded into the main and non-main sands), the extending type (that is no less than 200 m away from the effective sand), the connected type (connected untabulated sands), and the isolated type (untabulated sands scattered in mudstone).

Analysis of the corresponding untabulated sand types of the collected samples shows that the development proportion of the favorable band is the highest in the edging-bridging type of the underwater distributary fluvial channel sand and reaches 88%, associated with only the conventional untabulated reservoir accounting for only 12%. The development proportion of the favorable band in the untabulated sand embedded into the main or non-main thin sands reaches 87%, with the corresponding proportion of the conventional untabulated reservoir of 13%, similar to the case of the edging-bridging type. The favorable band proportion is 75% in the extending type, with the conventional untabulated sand proportion of 25%. The least proportion of the favorable band is found in the connected type (only 15%), with the corresponding conventional untabulated sand reservoir proportion reaching 85%. In summary, the development probability of favorable bands ranks from high to low as the edging type, embedded type, extending type, and the connected type.

The water absorption statistics show that for the above four planar types of untabulated sands, the water absorption ratios of favorable bands rank as the edging type, embedded type, extending type, and connected type, which is consistent with the development probability regularity of favorable bands. The concentrated zone of favorable bands in the untabulated reservoir has higher produced degrees. Therefore, the favorable band is vital for further potential tapping of the remaining oil.

The sedimentary environments and planar distribution locations of conventional untabulated reservoirs and favorable bands (both of the untabulated reservoir) are clarified via investigation of the sedimentary patterns and planar distribution characteristics of untabulated reservoirs. The favorable band mainly occurs in the distal sedimentary environment of the delta inner front and the proximal sedimentary environment of the delta outer front. The sands of the favorable band mainly contact with the

https://doi.org/10.1051/e3sconf/202337501012

fluvial channel sand, and main and non-main thin sands. As for the conventional untabulated reservoir, it mainly occurs in the distal zone of the outer front, and some in the medium zone of the outer front. The conventional untabulated reservoir is dominated by the lake and typically occurs as a connected areal development or contacting with the mudstone.

4. Production characteristics of untabulated reservoirs

To clarify the production behaviors of the untabulated reservoir sands of various types with varied injection-production modes, five physical model simulation experiments and numerical simulations are performed for the various types of untabulated sands. Sample #1 is designed to simulate the separate recovery of the conventional untabulated reservoir. Sample #2, for the parallel recovery of the favorable band and conventional untabulated reservoir; Sample #3, for the parallel recovery of the favorable band and conventional untabulated reservoir, with only the latter perforated; Sample #4, for the cascade recovery of the favorable band and conventional untabulated reservoir; Sample #5, for the recovery of the pinch-out favorable band. The experimental preparation and sample treatment (mainly including the drying, vacuuming, oil and water saturation, parameter calculation, and finally water flooding test) are done in accordance with the Chinese national standard SY/T 5358-2002.

4.1 Results of the physical model and numerical simulations

The total recovery factors of the five physical models (five recovery approaches) are all above 40%, and yet the recovery of Samples #2 and #3 are associated with considerable production interference harmful to the effective exploitation of the conventional untabulated reservoir. The recovery approaches of Samples #4 and #5 are found in favor of producing from the conventional untabulated reservoir.

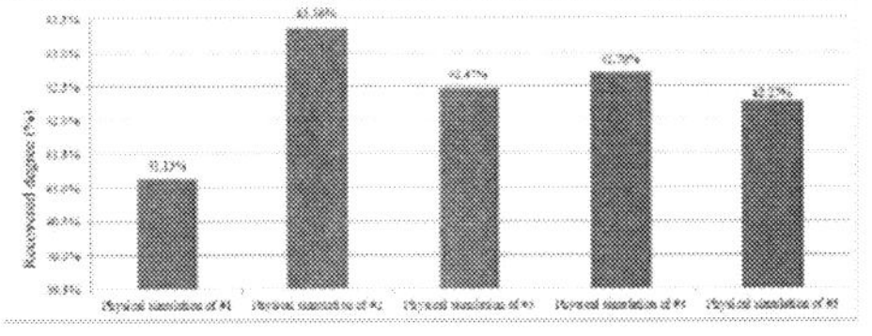

Figure 5. Histograms of recovery factors of the physical models

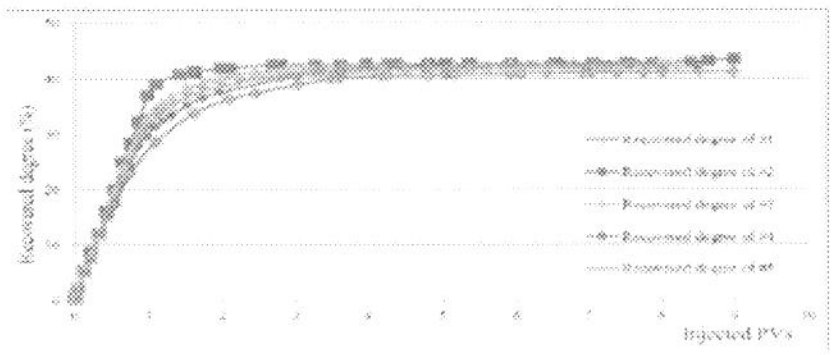

Figure 6. Comparison of the recovered degrees of the physical models

The numerical simulation (Figs. 7 and 8) demonstrates that the highest recovery factor is attributed to the parallel recovery of the favorable band and conventional untabulated reservoir (Sample #2) and reaches 43.25%; however, this recovery approach suffers from intensive interlayer interference, which leads to the limited recovery of the conventional untabulated reservoir (with the contributions of only 23.8% to the total). Moreover, the least recovery of the conventional untabulated reservoir with the contributions of only 17.8% is associated with the recovery approach of Sample #3, in which the flow rate and streamline indicate occurring of cross flow and the production is mainly supplied by the non-perforated favorable band (the recovered degrees of the conventional untabulated reservoir are low). The recovery approaches of Samples #4 and #5 are in favor of production stimulation of the conventional untabulated reservoir. The recovery factors of the cascade recovery for the physical and numerical simulation reach 42.7% and 42.10%, respectively. The corresponding flow rate and streamline demonstrate that these two approaches help to promote the recovery of the conventional untabulated reservoir.

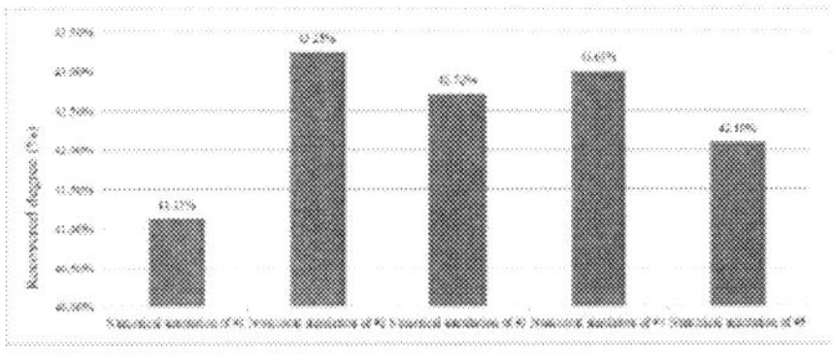

Figure 7. Histograms of recovery factors of the numerical simulation

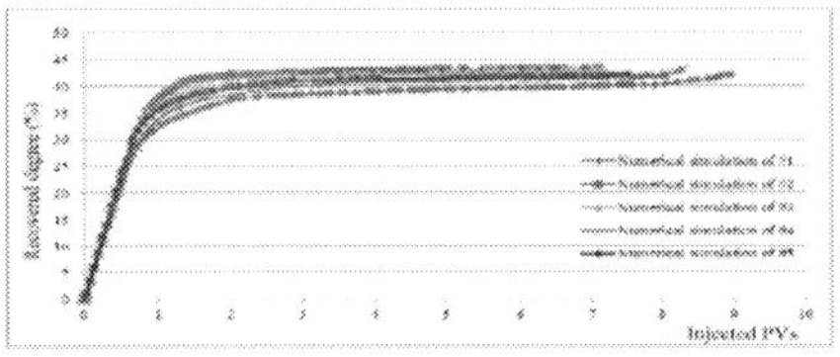

Figure 8. Comparison of recovered degrees of numerical simulation

The physical and numerical simulations demonstrate that the separate recovery of the conventional untabulated reservoir can present the total recovery factor above 40%, with the slow growth of water cut, and the production of the conventional untabulated reservoir is highly disturbed by the favorable band. In the untabulated reservoir containing the favorable band, the injected water always finds the dominant channel via the favorable band, regardless of whether the favorable band is perforated or not. Under such circumstances, it is difficult to exploit the conventional untabulated reservoir. In the cascade arrangement of the favorable band and conventional untabulated reservoir, the recovery performance of the evenly-displaced favorable band side is better, and yet the contributions of the conventional untabulated reservoir are lower than the injection of the conventional untabulated reservoir. It should be noted that in the case of either the cascade arrangement or parallel arrangement of sands, the simultaneous recovery of the favorable band and conventional untabulated reservoir, the contributions of the favorable band always exceed 60%, which make it the main contributor in the untabulated reservoir.

4.2 Numerical simulation of the production interference in parallel recovery

To determine whether or not the joint recovery of the favorable band and the conventional untabulated reservoir is associated with interference, and if any, how strong it may be, a parallel recovery experiment is designed to calculate the separate contributions of the two types of untabulated reservoirs. The samples of the favorable band and conventional untabulated reservoir are separately put into different core holders and then flooded in a parallel arrangement. The liquid, oil, and water production are measured in the outlets also in a separate fashion to compute the respective contributions to the total liquid and oil production.

The model parameters for the production interference simulation of the parallel recovery are listed below (Fig. 9). The model parameters are 4.5×4.5×30 cm (for each core sample), with the porosity of 18%–23%, the permeability of 17.8–80 mD, the pore volume of 182.25 cm^3, oil initially in place 100.23 cm^3, and the initial oil saturation of 55%. The injection rate during flooding is 2 mL/min.

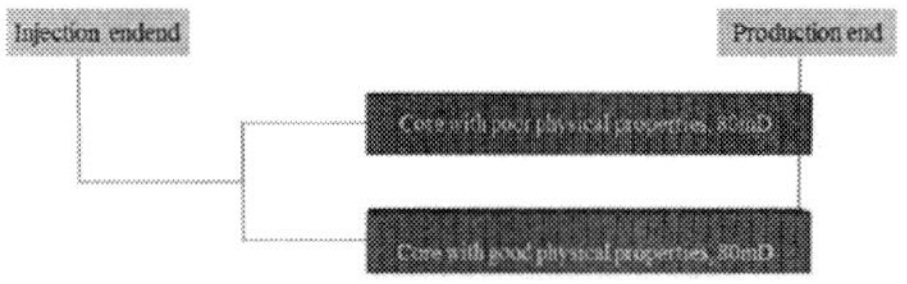

Figure 9. Schematic diagram of the parallel recovery

The simulation ends, as the water cut reaches 98%. By the end of the simulation, the ultimate recovery factor is 41.12%. As shown by the recovered degree versus injected water (expressed as multiples of pore volumes, injected PV), When the injection volume is from 0 to 1 PV, it is the rapid growth of the recovered degrees. From 1 PV to 3 PV, the recovered degree climbs up steadily. After 3 PV, the injection is low-efficient or ineffective. The favorable band presents considerable interference. After the water-free production, the flow rate of the favorable band is far higher than that of the conventional untabulated reservoir. The produced oil is mainly from the favorable band (contributing 87.5%; Figs. 10 and 11). This result is similar to those of numerical simulations of Samples #3 and #2, which demonstrates that the joint recovery in such reservoirs is unfavorable for the exploitation of the conventional untabulated reservoir.

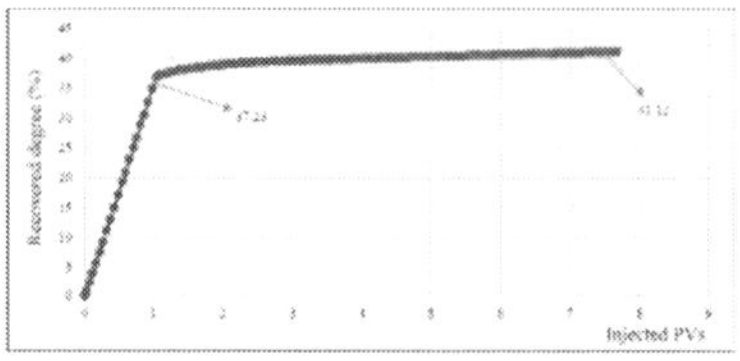

Figure 10. Recovered degrees of numerical simulation

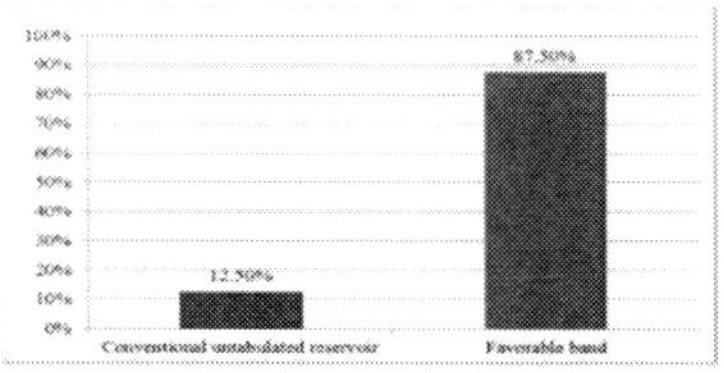

Figure 11. Contributions of the conventional untabulated reservoir and favorable band

The conventional untabulated reservoir is severely impacted by the favorable band. The actual production is much more complex than the simulated one and therefore it is safe to say that it is highly difficult to effectively recover the conventional untabulated reservoir in the case of joint recovery. In theory, the cascade recovery of the conventional untabulated reservoir and favorable band in which the injection well is placed in the untabulated reservoir is expected with better recovery performance. Nonetheless, the actual production suffers from severe interlay interference and it is impossible to build the separate well pattern for recovery. Hence, it is recommended that the development of the untabulated reservoir should preferably recover the un-produced favorable band.

5. Conclusions

Under the guidance of the above findings, the type-specific adjustment of the untabulated sands (that are not included in the reserve sheet) of the D well district is implemented. For the favorable band with permeability above 50×10^{-3} μm^2, the adjustment combines the injection rate enhancing and control. For the highly-permeable layer of the favorable band that is already producing, the ineffective injection is controlled after testing and adjustment. For the non-producing favorable band with remaining oil potential, the refined layer division is performed for recovery. Moreover, for the low-permeability untabulated sand with the permeability of (10–50) $\times10^{-3}$ μm^2, the injection-production adjustment is optimized to improve the pressure gradient. At last, for the ultra-low permeability untabulated sand with permeability below 10×10^{-3} μm^2, an attempt is made for the precise-control fracturing that highlights targeted reservoir stimulation. The adjustment of the D well district results in effective control of water injection and liquid production and the adjustment performance is considerable, considering the cumulative oil gain of 1286000 tons. The development adjustment performance of the D well district validates the identified deposition mechanisms and production characteristics of the various untabulated reservoir sands.

(1) The untabulated reservoir presents high heterogeneity. A considerable discrepancy is found between the micro-pore structures of the conventional untabulated reservoir and favorable band. The physical properties and pore-throat characteristics of the favorable band, compared with those of the conventional untabulated reservoir, are closer to those of effective sands of the tabulated reservoir (included in the submitted reserve sheet).

(2) The deltaic distributary plain facies is highly subjected to the high-energy river and the resultant untabulated reservoir largely contains favorable bands. Due to the joint effects of the river and lake, in the delta inner front facies, the clastic sediments gradually thin outward from both sides, and the favorable band mainly develops within 200 m from the effective sand. As for the conventional untabulated reservoir, it can reach farther away, with the help of waves, and deposit in a connected pattern. Affected by the lake, the medium zone of the outer delta front representing a low energy environment is seen with the predominance of the conventional untabulated reservoir, and the favorable band is found with limited development in the proximal zone of the outer delta front.

(3) In a planar view five types of untabulated reservoir sands are identified, namely, the edging, embedded, extending, connected, and isolated types. The favorable band mostly occurs as the edging, embedded, and extending types, while the conventional untabulated reservoir generally presents stable connected areal development and is mainly of the connected and isolated types.

(4) The favorable band severely interferes with the production of the conventional untabulated reservoir and highly impacts the recovered degree of the latter. The contributions of the favorable band to production are much higher than those of the conventional untabulated reservoir. In other words, the favorable band is the main producer of the untabulated reservoir.

Acknowledgements

Funded by the 13th Five-Year-Plan National Science and Technology Major Project "Development of Large Oil and GasFieldsandCoalbed Methane—The Demonstration Area of the Water Flooding Water Controlling and Efficiency Enhancement of the Eastern Xing-6 District" (Grant No. 2016ZX05054-002)

References

1. Xu Chaopeng. Method Of Enhancing The Developed Degree Of The Untabulated Reservoirs. Petroleum Geology and Oilfield Development in Daqing 33(4):57-61(2014)
2. Liu Ding-ceng, Wang Qi-min, Li Bo-hu. Development of multi-sandstone reservoirs in Daqing oil field. Petroleum Industry Press(1996).
3. Gao Da-peng, Ye Ji-gen, Hu Yong-le, et al. An evaluation model for stratified water injection in multilayer reservoirs. Petroleum Exploration and Development 42(6), 711-715(2015).
4. Fu Guang-qun, Wei Ze-gang, Guo Hong-wei, et al. Characteristics and recovery method of untabulated reservoirs in Xingnan oilfield. Journal of Daqing Petroleum Institute27(4), 15-17(2003).
5. Liu Yi-kun, Wang Feng-jiao, Hu Chao-yang, et al. Study on oil / water percolation in thin and poor pay zone. Special Oil and Gas Reservoirs20(5), 98-101(2013).
6. Gao Da-peng, Ye Ji-gen, Li Qi, et al. Anindependent development method of low permeability oil thin layers with extreme high water cut in Changyuan Structure, Daqing oilfield. Oil and Gas Geology38(1), 181-188(2017).
7. Zhang Dong, Di Jjing-dong, Li Jie, et al. Effective development technology of unbooked reservoirs in Daqing La-Sa-Xing oilfield. Petroleum Geology and Oilfield Development in Daqing28(5), 112-118(2009).
8. Qi Chun-yan. Adjustments on thin and poor reservoir and unworkable reserves in La-Sa-Xing oilfield. Petroleum Geology and Oilfield Development in Daqing29(2), 64-69(2010).
9. Xu Xiao-ping. Micro-pore structures of the untabulated reservoir and the characteristics of polymer injection. Petroleum Geology and Oilfield Development in Daqing31(4), 135-138(2012).
10. Xu Pengchao. Method of enhancing the developed degree of the untabulated reservoirs. Petroleum Geology and Oilfield Development in Daqing33(4), 57-61(2014).
11. Cui Yue, Shi Jingping. Experiment on the seepage features for the untabulated reservoirs in Daqing oilfield. Petroleum Geology and Oilfield Development in Daqing36(2), 69-72(2017).
12. Mechanism of CO2 enhanced oil recovery in shale reservoirs[J]. Hai-Bo Li, Zheng-Ming Yang, Rui-Shan Li, Ti-Yao Zhou, He-Kun Guo, Xue-Wei Liu, Yi-Xin Dai, Zhen-Guo Hu, Huan Meng. Petroleum Science.2021(06)
13. Jin-you DAI, Jing YAO, Meng KONG, Jing WANG. New Method to Quantitatively Describe Remaining Oil by Dynamic Data[A]. Proceedings of 2016 International Conference on Electrical Engineering and Automation (ICEEA2016) [C]. 2016
14. Yu Qi-tai. A Study On Remaining oil[J]. Petroleum Exploration and Development,1997,24(2):46-50
15. Quick assessment to ascertain technical rational well spacing density in artificial water flooding oilfield[J] . Hu Gang, Tian Xuanhua, Liu Quanwen, Yi Linzi, Liu

https://doi.org/10.1051/e3sconf/202337501012

Dawei, Li Pengchun. Journal of Petroleum Science and Engineering . 2021

16. The study on exploitation potential of original low-oil-saturation reservoirs[J] .Sun Yingying, Zhang Shanyan, Wei Xiaofang, Sang Guoqiang, Minghui Zhou, Huang Jia, Yang Jiru. Petroleum Research . 2020 (1)
17. Effect of displacement rates on fluid distributions and dynamics during water flooding in tight oil sandstone cores from nuclear magnetic resonance (NMR) [J] . Meng Chen, Jiacai Dai, Xiangjun Liu, Yan Kuang, Zhongtao Wang, Shunchao Gou, Minjun Qin, Min Li. Journal of Petroleum Science and Engineering . 2020 (C)
18. Research the Electromagnetic Detecting Electrical Parameters Through the Casing[A]. Proceedings of IEEE 2011 10th International Conference on Electronic Measurement & Instruments (ICEMI'2011) VOL.04[C]. 2011
19. Simulation of Residual Oil Distribution Based on 2D-network Model[A]. 2010 Information and Management Sciences--Processing of the Ninth International Conference on Information and Management Sciences[C]. 2010
20. Sedimentary characteristics and pattern of distributary channels in shallow water deltaic red bed succession: A case from the Late Cretaceous Yaojia formation, southern Songliao Basin, NE China[J] . Li Zhang, Zhidong Bao, Luxing Dou, Dongsheng Zang, Shuwei Mao, Jian Song, Jiahong Zhao, Zecheng Wang. Journal of Petroleum Science and Engineering . 2018
21. Fluviatile fining-upwards cycles and the magnitude of palaeochannels[J] . M. R. Leeder. Geological Magazine . 1973 (3)
22. Modeling of CO2 solubility in crude oil during carbon dioxide enhanced oil recovery using gene expression programming[J] . Alireza Rostami, Milad Arabloo, Arash Kamari, Amir H. Mohammadi. Fuel . 2017
23. Capillary driven flow in nanochannels – Application to heavy oil rheology studies[J] . Saeed Mozaffari, Plamen Tchoukov,Ali Mozaffari, Jesus Atias, Jan Czarnecki, Neda Nazemifard. Colloids and Surfaces A: Physicochemical and Engi . 2017
24. Relationship between bulk foam stability, surfactant formulation and oil displacement efficiency in porous media[J] . Kofi Osei-Bonsu, Paul Grassia, Nima Shokri. Fuel . 2017
25. Numerical simulations of seepage flow in rough single rock fractures[J].Qingang Zhang, Yang Ju, Wenbo Gong, Liang Zhang, Huafei Sun. Petroleum . 2015
26. Impact of pore-scale three-phase flow for arbitrary wettability on reservoir-scale oil recovery[J] . Adnan Al-Dhahli, Sebastian Geiger, Marinus I.J. van Dijke. Journal of Petroleum Science and Engineering . 2014
27. A New Method base on Water Drive Curve for Determining the Reasonable Injection-Production Ratio for High Water-Cut Oilfield[J]. ZHANG Zhang, QUAN Honghui, XU Zhongbo, HE Xinrong, LI Jun. Acta Geologica Sinica (English Edition) . 2015(S1)
28. Calculation on a reasonable production/injection well ratio in waterflooding oilfields[J] . Cunyou Zou, Yuwen Chang, Guohui Wang, Lifeng Lan. Petroleum Exploration and Development Online . 2011(2)
29. Cui Yue. Experiment on the seepage features for the Untabulated Reservoirs in Daqing Oilfield. Petroleum Geology and Oilfield Development in Daqing36(02),69-72(2017)
30. Xiao Lu-chuan, Cao Wei-fu , Xu Yun-ling. Characteristics of non-Darcy flow velocity of oil-water phases in extra-low permeability reservoirs. Petroleum Geology and Oilfield Development in Daqing26(5),61-63(2007)

Management Methods of Drilling Shut in

Haiwei Wu

The Third Oil Production Plant of Daqing Oilfield Limited Company, Daqing, Heilongjiang, China

Abstract. The Daqing oilfield was put into development in 1960. At present, it is in the late stage of ultra-high water cut period. Various displacement modes coexist, and the reservoir development and management is complex and difficult. In order to control the decline of production, the number of new drilling wells increases year by year, the surface and underground conditions are complex, and the management is difficult. The drilling shut in affects the production year by year. The formation pressure decreases greatly during the drilling shut in process, which affects the development situation. In order to reduce the impact of drilling shut in on production, control the decline of formation pressure, and explore the management methods and methods of drilling shut in. In order to reduce the impact of drilling shut in on production and control the decline of formation pressure, we explore the management methods of drilling shut in.

Keywords: Development situation; Drilling shut in; Management methods.

1. Introduction

A target management project team with the chief geologist as the project manager and the participation of relevant personnel from the geological research institute, the process research institute, relevant departments of plant authorities and operation area has been established, and an "integrated" management guarantee system of reservoir engineering, oil production engineering and surface engineering has been established, forming a long-term management mechanism. At the same time, optimize the division of labor, decompose the goals vertically and horizontally, achieve the vertical decomposition in place, expand horizontally to the end, and decompose and implement various goals to all units and individuals, forming a situation in which everyone has a heavy burden and indicators on everyone. At the same time, we optimize the division of labor and decompose the objectives vertically and horizontally, so as to achieve the vertical decomposition in place and horizontal expansion to the end. We have broken down and implemented various goals to all units and individuals, forming a situation in which everyone has a heavy burden and everyone has indicators.

2. Preparation Before Drilling in Advance

2.1 Before drilling, the engineering organization is meticulous, and the well location survey is carried out in advance

The geological research institute actively organized, conducted well site survey half a year in advance, implemented the cushion work of 315 low-lying wells, and prepared the drilling operation plan in advance. According to the field survey results, we completed the overall planning of drilling rig operation in advance, optimized the preparation of drilling operation scheme, and defined the drilling time of each well. The drilling disclosure meeting shall be organized half a year in advance, with the participation of the drilling company, the production and operation department of the plant, the planning institute and other departments, to preliminarily discuss and plan the drilling rig operation line and power erection line and formulate the next work plan. [1]

2.2 The ground is paved in place

The planning department shall actively coordinate and issue the temporary plan for paving earthwork in advance. The production and operation department shall organize the earthwork paving team to complete the earthwork paving and pipe laying in advance according to the preliminary well location survey results.

* Corresponding author: whw227@petrochina.com.cn

2.3 The lines are erected in place

The production and operation department shall comprehensively organize the disclosure and implement the material plan. The supervision and management center systematically organizes the construction team, completes the stringing work in advance, and constructs the temporary transformer outlet and high-voltage line to meet the drilling needs. [2]

3. Innovation of Drilling Operation

There are many influencing factors involved in drilling off optimization. Combined with the actual drilling operation of our plant, drilling off optimization is mainly carried out from four aspects: drilling off distance, drilling off time, drilling operation and drilling off recovery.

3.1 Drilling shut in distance is standardized

In order to ensure the normal operation of drilling, shut in and depressurization shall be implemented for water injection wells within a certain range around the well shut in. Combined with the previous shut in practice experience, the wellhead residual pressure of water injection well before drilling is adjusted from 2MPa to 3Mpa. The standard of well control and injection shut in is adjusted from 450m to 300m. For the layer system well pattern with injection production well spacing less than 300m, the drilling shut in well spacing can be adjusted to not less than the injection production well spacing. If there is pressure relief at the production well point between the water injection well and the new drilling point, the water injection well may not be closed. For water injection wells where the residual pressure at the wellhead still fails to meet the standard requirements for a long time, the existing injection control methods shall be used. At the same time, the reasons for the failure shall be analyzed to determine the corresponding injection control standard. [3]

3.2 Drilling shut in time is personalized

The number of shut in days is adjusted from 15 days to optimize the number of shut in days according to reservoir conditions. Based on the actual monitoring pressure data of more than 300 wells, the pressure drop trend is analyzed, and the shut in days before drilling shut in of each well pattern are counted and analyzed. From the analysis results, if the pressure drops below 3MPa after shut in, the drilling requirements can be met. The average shut in days are 6.9 days for the basic strata with well-developed reservoir conditions and high permeability and tertiary oil recovery class I and II reservoirs with well spacing of 125m. Primary infill wells with medium and low permeability, relatively good reservoir conditions and low formation pressure and class I oil reservoir of tertiary oil recovery have a well spacing of 250m, and the average shut in days are 9.4 days. For secondary and tertiary infill wells with poor reservoir conditions, high formation pressure and slow pressure drop after shut in, the average shut in days is 14.6 days.

According to the numerical simulation results, if the pressure drops below 3Mpa after shut in, it can meet the drilling requirements. The average shut in days are 6.0 days for the basic strata with well-developed reservoir conditions and high permeability and class I and II reservoirs of tertiary oil recovery with well spacing of 125m. Primary infill wells with medium and low permeability, relatively good reservoir conditions and low formation pressure and class I oil reservoir of tertiary oil recovery have a well spacing of 250m, and the average shut in days are 9.0 days. For secondary and tertiary infill wells with poor reservoir conditions, high formation pressure and slow pressure drop after shut in, the average shut in days is 16.0 days.

According to the above two methods, a "personalized" drilling shut in scheme is designed to shorten the drilling shut in time. The specific principles are as follows:

(1)The injection wells are shut in 7 days in advance for the basic strata of water flooding.

(2)The injection wells are shut in 10 days in advance for the primary infill strata of water flooding.

(3)The injection wells are shut in 15 days in advance for the secondary and tertiary infill strata of water flooding.

(4)The injection wells are shut in 7 days in advance for class I and II reservoirs of tertiary oil recovery with well spacing of 125m.

(5)The injection wells are shut in 10 days in advance for class I reservoirs of tertiary oil recovery with well spacing of 250m.

3.3 Drilling rig operation is integrated

According to the well location distribution, centralized drilling, overall promotion, local optimization and piece improvement are carried out to reduce the impact of drilling and shorten the drilling construction cycle on the premise of ensuring the drilling quality. According to the load of the drilling line and the control analysis of the impact of drilling, we arrange a reasonable number of drilling rigs for drilling construction. The construction period of a single well is calculated as 7 days, which can calculate the pure drilling period required to complete all drilling tasks in the block. In addition, according to the impact of shut in time on production and pressure (Fig.1), the impact of formation pressure and production after 38 days of shut in of water injection well is increased. Therefore, if the drilling shut in time is controlled within 38 days through reasonable rig operation, It can greatly reduce the impact of drilling shut in on formation pressure and production. [4]

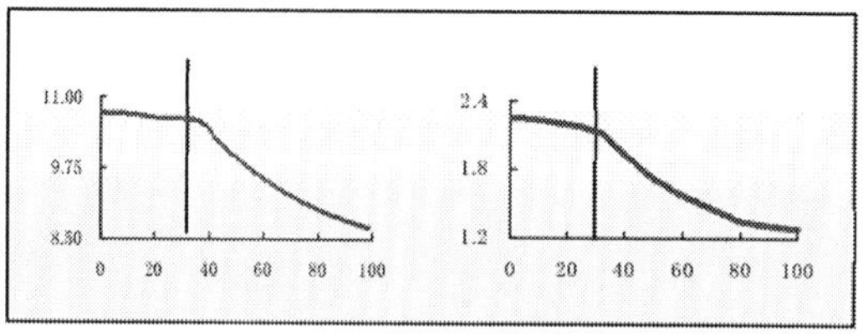

Fig.1 Effect of shut in time on pressure and production

3.4 Drilling shut in recovery scheme is refined

Drilling shut in is not only a factor affecting production, but also a potential that can be optimized. The step-by-step recovery scheme shall be determined according to the water flooding interpretation data of new wells and the comprehensive water cut changes of surrounding oil production wells during drilling shut in. For the oil wells with high liquid production, the water cut drop value is larger after drilling shut in, while the water cut drop value of the oil wells with low liquid production is smaller, which is mainly caused by the large drop range of high-yield liquid and high aquifer in the wells with high liquid production, and the small drop range of low-yield liquid and low aquifer. This reflects the great influence of interlayer differences. At the initial stage of drilling, the water injection pressure decreases and the water injection volume increases for the well-developed reservoir with good connection with the oil well, while after drilling, the water injection pressure decreases but the water injection volume does not increase for the reservoir with poor water absorption and poor connection with the oil well, which indicates that the pressure of the well-developed reservoir with good connection decreases more than that of the poor reservoir. This proves that the water cut of the oil well with high-yield liquid decreases greatly after the drilling shut in of the water injection well. Therefore, we follow the following principles when recovering from drilling. For the water cut descending well area, and the water flooding interpretation data of the new wells is interpreted as a high water flooded layer, we mainly control the injection of high aquifer, which is restored to 80% in the second step of water injection wells and 100% in the second step of polymer injection wells. For the water cut rising well area, and the water flooding interpretation data of the new wells is interpreted as medium and low water flooded layer, we mainly improve the injection of low aquifer, which is restored to 120% in the second step of water injection wells and 120% in the second step of polymer injection wells. For the water cut stable well area, the water injection volume should be kept stable, which is restored to 80% in the second step of water injection wells and 100% in the first step of polymer injection wells. For the low-pressure and under load well areas, the main task is to speed up the recovery of production, and the water injection volume is restored by 100% in the first step. [5]

4. Institutionalization of Process Management

Process management highlights institutionalization. We have established the management regulations on drilling shut in of injection well, check the process nodes every week, include them in the assessment, report them at the production and operation meeting, and require timely rectification. We have formulated management regulations for well opening and well shut in and established procedures. The drilling team of the Geological Research Institute shall provide the drilling well number, drilling time and completion time to the relevant business departments of the Geological Research Institute. The relevant business departments and offices shall prepare the drilling shut in plan and drilling open plan, which shall be dispatched to the relevant operation areas, so that the special work can be managed by a specially assigned person, so as to realize no more shut in of one well and no more shut in for one day. We have formulated the management regulations of electric pump well, established the operation table of electric pump well in drilling shut in area. First of all, we should verify the well conditions and data to ensure that they are accurate, and the parameter adjustment equipments such as electric pump nozzle and frequency conversion are fully prepared. The second is process management. When the water well is shut in, we shall reduce the oil nozzle in time, strengthen the liquid level monitoring, increase the oil quantity times, track and adjust the oil nozzle in time. When the oil nozzle is adjusted to the minimum, the electric pump well is still under load. We shut in the well after wax scraping and well washing according to the electric pump management measures. The third is to resume production. Before well opening, we shall clean and scrape the well according to the electric pump management method, use the daily current card within one week after well opening. We formulate the oil measurement scheme and liquid level test scheme according to the production situation, and switch to the conventional management after the operation is stable. [6]

5. Targeted Site Supervision

The supervision and inspection on drilling site highlight pertinence. As the drilling operation is 24-hour construction, it is difficult to supervise. In order to strengthen supervision and management, we adopt targeted on-site supervision methods. In the drilling stage, we implement three monitoring measures. The first is to monitor the performance of drilling fluid in strict accordance with the design. The second is to monitor the wellbore quality and observe whether there are special conditions. The third is to monitor the wiper trip and reaming to ensure the borehole quality. The first is to check the pressure test report of incoming casing to ensure the compressive strength of casing. The second is to check the position and quantity of centralizers to ensure the centering of casing. The third is to check the casing appearance and casing running torque to ensure the casing quality. In the cementing stage, we should control the speed, density and cleanliness. The first is to control the displacement speed to 1.5m/s to ensure the displacement efficiency. The second is to ensure that the density of cement slurry is strictly in accordance with the design value. The third is to keep the drilling fluid circulating before cementing to improve the cleanliness of the wellbore. In addition, we should control the amount of pad fluid, injected mud and displacement fluid. 4m3 clear water is generally used as the pad fluid to fully isolate the drilling fluid and cement slurry. According to the well depth and casing structure, the injection mud volume shall be implemented in strict accordance with the design volume. The displacement fluid volume is generally 3m3-5m3 clear water. Through the above methods, clear objectives and key control are achieved to ensure that the

 https://doi.org/10.1051/e3sconf/202337501013

drilling team is implemented in place according to the design, so as to improve the drilling quality.

6. Conclusion

Based on the results and discussions presented above, the conclusions are obtained as below:
(1) The preparation before drilling is the basis for realizing the overall promotion of the drilling rig and reducing the impact of production.
(2) The drilling shut in limit is applicable to large-scale regional drilling. Since there are relatively few wells drilled and shut in for scattered wells, and the pressure drop is relatively slow, the shut in distance and shut in days should be appropriately increased.
(3) Drilling shut in stage is an effective opportunity to understand the difference of formation physical properties and residual potential.

References

1. Liwei Zhou, Kai Cui, Xuejing Ma, Ruoxia Du. An Effective Way to Optimize Drilling Method. Oil and Gas Field Surface Engineering, 2004.
2. Yajuan Zhang. Study on Adjustment before Drilling and Recovery after Drilling. Oil and Gas Field Surface Engineering, 2011.
3. Youzhi He. Study on Pressure Relief Method of Drilling Shut in[J]. China Petroleum and chemical industry standards and quality, 2011. China Petroleum and Chemical Industry Association.
4. Jinming Ji. Study on Drilling Drop Law of Dense Well Pattern. Petroleum and Petrochemical Energy Conservation, 2014.
5. Jurong Wang. Discussion on the Ways to Reduce the Influence of Drilling off in the First Fault Block Area of Pubei. Science and Technology Innovation Guide, 2013.
6. Jicheng Zhang, Jun Zhang, Bingyu Lv. Production Change Law in Drilling Shut in Area [J]. Oil and Gas Industry, 2020.

https://doi.org/10.1051/e3sconf/202337501014

Fixed point continuous monitoring system of oilfield skeleton and its application

Kai Du

No.4 Oil Production Plant Daqing Oilfield Co., LTD, Daqing 163511, Heilongjiang, China

Abstract: Aiming at fixed monitoring well point in oil field monitoring deployment chaotic, block monitoring percentage change is big, suitable for lack of contrast data, and poor adaptability to changes in the cost problem, plane layout, well pattern deployment, give attention to two or morethings injection-production differences, etc., to determine the skeleton designated strip continuous monitoring, use the statistical methods such as cluster analysis, deep learning, combining with the field dynamic data changes, To optimize the deployment of well points, determine the principle of "skeleton well fixed continuous monitoring, non-skeleton well according to demand", establish skeleton fixed continuous monitoring system, achieve the purpose of saving monitoring costs and improving the accuracy of reservoir evaluation. Research results show that the skeleton fixed-point continuous monitoring system to meet the needs of overall reservoir evaluation, more through rules deployment of a certain percentage of the skeleton well point and the fixed location monitoring of the stability can be compared to the continuity of the data, ensure the evaluation precision reservoir framework, at the same time using the non-framework change monitoring well balance monitoring proportion, key and the key blocks of costs change with a strong adaptability. Through research and field practice, it has been proved that the skeleton fixed-point continuous monitoring system combines oilfield monitoring deployment with traditional geological profile analysis mode, and realizes the fusion application of three system frameworks: dynamic monitoring, geological static and numerical simulation, which is helpful to promote the process of precise grid management of reservoir.

Key words: Skeleton monitoring; Fixed point continuity; Cluster analysis; Monitoring ratio; Well point selection.

1. Introduction

According to the Outline of Oilfield Development Management of CNPC, the dynamic monitoring of oilfield development should follow the design scheme of development unit, and the number of monitored Wells should meet the requirements of reservoir development dynamic analysis as the principle. In accordance with the principle of point-surface combination and highlighting key points, the general and typical blocks are combined, and the fixed and non-fixed well points are combined. The selection of monitoring well sites should be representative and continuous. The traditional monitoring system gives priority to meet the needs of adjustment, the fixed monitoring well points are disordered, the monitoring proportion of each block changes greatly, the applicable continuous contrast data is insufficient, and the application level of results is poor, which cannot meet the needs of the overall reservoir evaluation. With the deepening of oilfield development, the distance between injection and production Wells decreases and the density of well pattern increases gradually. The monitoring system is difficult to adapt to the dual restrictions of monitoring ratio and cost, which increases the difficulty of monitoring deployment. Since the "13th Five-Year Plan", the concept of oilfield development has changed from fine to precise. It is necessary to integrate monitoring results with other types of data more closely, and carry out accurate evaluation of reservoir production from whole to local, macro to micro [1-4]. Based on the above requirements, the framework fixed-point continuous monitoring system was built, which fully considered the representativeness and continuity of monitoring well points, combined point and surface, highlighted key points, and accurately reflected the reservoir development situation.

2. Determine the framework of continuous monitoring system

In hydrogeology, Marine geology and other observation work [5], some key locations are often selected for fixed-point continuous observation to meet the overall evaluation and continuous comparative analysis of data.

With the deepening of the grid management reform of the social governance system, the advantages of management methods are gradually highlighted. Therefore, the two modes should be integrated into oilfield monitoring work to promote grid management of oilfield monitoring, adapt to cost changes and meet development needs.

The basic well pattern deployment in the old area of Daqing Placanticline adopts the row and row cutting mode, that is, three rows of oil Wells are deployed in the middle of two rows of Wells, so that the horizontal and longitudinal well arrangement mode is mostly used in the later oil-wells deployment in the pure oil area, which also promotes the fine division of blocks and the construction of multi-turn infill well pattern to take well arrangement as the boundary. Based on the principle of considering plane layout, well pattern deployment and regional injection and production differences, the relationship between overall deployment and unit management is balanced, and the skeleton strip is established: horizontal well row is the base oil well row, vertical uniform coverage is given priority, and all well patterns are deployed, and the width refers to the common multiple of well spacing.

By comparing the relationship between the number of skeleton Wells and the total number of Wells with historical data, the proportion of skeleton Wells is always stable. Therefore, the continuous monitoring and interpretation results of skeleton well points can characterize the overall development situation of the reservoir to a certain extent.

3. Determine the reasonable monitoring proportion within the framework

The framework fixed-point continuous monitoring system should not only ensure the orderly and continuous monitoring of skeleton Wells, but also keep the reasonable monitoring proportion of skeleton Wells and ensure the accuracy of reservoir evaluation framework. Therefore, it is very important to clarify the reasonable monitoring proportion in the framework for system construction.

3.1 Cluster analysis method [6]

Center clustering method is a common algorithm in cluster analysis. It takes finding the center of each cluster as the basic task, and classifies the points in a certain range into the center cluster. K-means algorithm is one of the representative algorithms of center clustering. It takes the iteration process as the main line and the positive integer K as the hyperparameter, and updates the centers of K cluster classes in each turn. For the sample set in a given space, the initial center points of K family clusters are selected randomly or with input sample features. Then, in each iteration process, the distance between samples and related centers is calculated and classified into the nearest class cluster. After the new class cluster is obtained, the sample average value is calculated, which is used as the next iteration center to continue the calculation, and then the sample category division is realized.

3.2 Determine the reasonable monitoring proportion

Daqing changyuan old USES encryption rounds on gradually in-depth development of the reservoir development technology, each set of well pattern development objects intertwined overlap, plus local horizon seal oil and water Wells, take the set of methods of uniform distribution pattern already can't satisfy the needs of representative, therefore decided to adopt the method of clustering center, to belong to cluster the oil and water Wells.

(1) Single well feature scoring method [7]

Oil Wells can be divided into underground reservoir, wellbore structure, three components of the ground parameters, the use of these factors affecting on liquid water injection and production capacity for each hit a comprehensive score well, that is, from all kinds of attribute data for public factor, when the common factor contribution rate reaches a certain level, the composite scores represent the characteristics of oil and water Wells. The spatial correlation structure of all injection Wells can be obtained by using the single well score and well location coordinates, and the Wells with similar structures can be grouped into the same cluster.

(2) Determine the reasonable monitoring proportion

The interpolation error of each class cluster is obtained according to the variational function of single well score, well location coordinates and Kriging interpolation. The number of sample class clusters is drained according to a certain rule, and the mean square deviation of class cluster value is calculated. The curve of attribute change rate with proportion of well number can be drawn.

Table 1 Some data after cluster analysis and calculation

Class number of clusters	proportion (%)	clustering	
		Mean square error	Degrees of freedom
38	5%	7.807	37
76	10%	3.855	75
114	15%	2.559	113
153	20%	1.903	152
191	25%	1.522	190
229	30%	1.268	228
267	35%	1.087	266
305	40%	0.951	304
343	45%	0.846	342
382	50%	0.759	381
420	55%	0.690	419
458	60%	0.633	457
496	65%	0.584	495
534	70%	0.543	533
572	75%	0.506	571
610	80%	0.475	609
649	85%	0.446	648
687	90%	0.422	686
725	95%	0.399	724
763	100%	0.38	762

As can be seen from the curve, with the continuous improvement of the monitoring proportion, the slope of the curve gradually decreases and tends to be stable. The mean square deviation 1.5 is taken as the limit, and the reasonable monitoring proportion of the clear block is 25%-35%. The reasonable monitoring density of different blocks is different, and the monitoring density of skeleton bands in different blocks should also be differentiated. Therefore, the proportion of fixed-point continuous monitoring of skeleton strip should be consistent with the reasonable monitoring proportion of the block, and the monitoring proportion of non-skeleton well and skeleton non-fixed-point well should be adjusted to adapt to the change of monitoring cost.

4. Determine the method of monitoring well site selection within the framework

According to the principle of cluster analysis, the number of monitored Wells can be clarified, but the well points with the most monitoring value in the cluster cannot be selected. Therefore, the local injection-production relationship should be given priority in the selection process of monitoring well points to meet the needs of dynamic adjustment.

4.1 Selection method of injection profile logging well point

The MONITORING WELL point OF injection profile SHOULD be THE WELL with strong water injection sensitivity and great influence on the surrounding oil Wells. Therefore, the deployment of monitoring well point should be optimized from this perspective. The factors that determine the production are related to the distribution of reservoir physical parameters, reservoir fluid characteristics, well pattern forms and injection-production parameters. All parameters have complex historical dependence, so convolutional neural network (CNN) is suitable to solve this problem [8].

Based ON THE HISTORICAL PRODUCTION DATA, A single WELL history FITTING model IS established with an oil well as the center. By changing the water injection volume of an injection well, the oil well production changes are induced and the oil-well connectivity relationship is determined.

When the water injection volume of a well changes and the oil well production changes greatly, the water injection sensitivity of the well is strong. Based on the comprehensive judgment of oil and Wells in each cluster area and the influence of injection Wells on oil Wells in this cluster area, the water injection sensitivity of injection Wells in the region is ranked, the priority of injection profile monitoring is clarified, and the skeleton fixed point of continuous monitoring well point is determined. Due to the variation of dynamic parameters, the calculation result is short-term, so it needs to be recalculated every year to obtain a new sensitive sequence to guide the change of non-skeleton well monitoring deployment.

4.2 Well point selection method for well test monitoring

Injection-production ratio is one of the main influencing factors of formation pressure variation, which reflects the change of injection-production ends. Therefore, the concept of apparent injection-production ratio is used to evaluate the regional extent of the same cluster to measure the deployment and turnover of monitoring well sites.

Based on the classification results of different clusters obtained from cluster analysis, the concept of quasi-deviation of apparent injection-production ratio was introduced to calculate the apparent injection-production ratio of a single cluster, and the standard deviation of apparent injection-production ratio within the block was calculated. As a function of the number of class clusters, it is found that when the classification proportion reaches a certain degree, the variation range of standard deviation gradually decreases, which means that it has little influence on the representation of the overall pressure law of the block, so the monitoring proportion can be further optimized.

Theoretically, the central well within each type of cluster is regarded as a typical oil well, and the monitoring result represents the bottom pressure value of this type of cluster. However, due to the extensibility of oil reservoir properties and the timeliness of dynamic parameters, it is necessary to formulate the change mechanism of monitoring well:

I. The pressure measuring point should be set at the critical point because the injection-production ratio in the adjacent area is significantly different; If similar, it can reduce the number of manometric well points.

II. Pressure measuring well points should be added in areas with abnormal injection-production ratio; The monitoring well point can be reduced appropriately according to the injection production ratio area.

5. Examples of application

Based on the above methods, A skeleton fixed-point continuous monitoring system was established, which was exploratory applied in A development zone of Daqing Oilfield. The injection profile monitoring ratio and well test monitoring ratio were kept at about 27% and 12%, both lower than the planned ratio of 5%.

The PRACTICAL APPLICATION SHOWS THAT THE RESULTS OBTAINED FROM THE monitoring deployment meet the demand of block development and adjustment, and the utilization rate of data directly applied to injection-production adjustment keeps above 70%. At the same time, this method can further integrate monitoring deployment with geological framework model and digital modeling framework model, break through the traditional monitoring mode, combine with geological profile analysis, accurate digital modeling results, and further improve the accuracy of reservoir evaluation.

6. Conclusion

(1) The fixed-point continuous monitoring system of oilfield skeleton adopts big data statistical algorithms such as cluster analysis and convolutional nerve to start from the macroscopic deployment of oilfield monitoring, meet the principles of "Outline of Oilfield Development Management", and optimize the monitoring well points.
(2) Reservoir evaluation is an important basis for oilfield development adjustment, and monitoring data is one of the main factors affecting the evaluation results. Therefore, monitoring deployment should not only meet the daily production requirements, but also lay the foundation for accurate reservoir evaluation.
(3) Combined with the fixed-point continuous monitoring system of oilfield skeleton and the development adjustment needs, the evaluation data body should be further enriched, the evaluation method should be improved, and a set of integrated development monitoring and evaluation technology system should be established to adapt to the evaluation of reservoirs in the late stage of ultra-high water cut and accurately tap potential.

References

1. Wenzhi Mu. Evaluation method of production condition of heterogeneous oil reservoirs in ultra-high water cut period [J]. Journal of Southwest Petroleum University (Science & Technology Edition), 2014, 36 (2) : 128132.
2. Qu Yongchun. Reservoir condition evaluation technology based on multiple injection profile logging and its application [J]. Petroleum Geology & Oilfield Development in Daqing, 2016,35 (5) : 152155.
3. Gong Bao, Li Songyuan, Tian Xiaodong, et al. System engineering method for evaluating development effect of water drive oilfield [J]. Petroleum Geology & Oilfield Development in Daqing, 2015,34 (5) : 5863.
4. Yin Xiaozhe, Wang Jiaxu, Wu Qiong, et al. Comprehensive evaluation method of water flooding control degree in multi-layer sandstone oilfield [J]. Petroleum Geology & Oilfield Development in Daqing, 2015,34 (5) : 64-67.
5. Zhou Yangxiao, Li Wenpeng. Optimal design method of regional groundwater level monitoring network [J] Hydrogeology and Engineering Geology, 2007,34 (1) : 1-9
6. Yin Xiaotian, Zhan Gaofeng. Overview of clustering analysis algorithms and comparison of their applicability [J]. Science and Technology Information, 2018,33:230-232
7. PEI Jianya, Yang Jinghai, Yang Yunjie, et al. Petroleum Geology & Oilfield Development in Daqing, 2018,37 (5) : 130-134
8. WANG Huijun, Zhao Guiping, Li Liang, et al. TOC prediction model of argillaceous source rocks based on Convolutional neural network (CNN) : A case study of Hangjin Qi area, Ordos Basin [J]. Journal of University of Chinese Academy of Sciences, 2020,1:103-112

Application Strategy of Drainage Gas Production Technology in Shallow Gas Fields

Na Lv

Geological Research Institute of The Ninth Oil Production Plant of Daqing Oilfield Co., Ltd., Daqing, Heilongjiang, China

Abstract. During the use of gas wells, the production time must be increasing. If reasonable measures are not taken for a long time, the bottom hole pressure will be reduced. If the gas flow in the gas well can no longer drive the water at the bottom to the ground, it will lead to the accumulation of water at the bottom of the well and produce back pressure, which will have a very obvious impact on the production of the gas well until the gas well finally can not meet the normal production requirements. This paper first analyzes the water production law and productivity details of shallow gas fields, then analyzes the diagnosis mode of wellbore effusion, and finally puts forward the practical application mode and strategy of drainage gas production technology, hoping to provide a reasonable reference for the optimal development of related work.

Keywords: Drainage gas production; Wellbore unblocking; Shallow gas field

1. Introduction

In the production process of gas wells in shallow gas fields, liquid will be produced when gas is produced. If the bottom hole pressure is high enough, the liquid will be driven to the ground, but if the pressure is insufficient, the liquid will remain at the bottom of the well, which is a common phenomenon of gas well effusion. If this phenomenon is not treated for a long time, it will continue to accumulate and eventually form a high liquid column, accompanied by the phenomenon of well water back pressure. The impact of these pressures on the gas well will reduce its self injection capacity and eventually lead to shutdown. In addition, if the gas-liquid mixing occurs in the gas well, it will also lead to the reduction of the purity of the produced natural gas.

2. Study on water production law and productivity of shallow gas fields

2.1 Water production mode and type

The first is the water source of gas wells in shallow gas fields [1]. The periodic characteristics of condensate water yield are not obvious. The influencing factor closely related to the water yield is the condensate water gas ratio. What will affect this parameter is the temperature and pressure of the gas reservoir. Obviously, the condensate water yield will show a relatively different water yield state according to different gas reservoir conditions. The primary layer is also a common water source. It is easy to find the law when this water source gets water. In the relevant process from the start of operation of shallow gas fields to stable production and production reduction, the water volume is always decreasing. The water channeling in the interlayer will not occur in the non exploitation stage of the gas reservoir, but the bottom hole pressure will decrease after a period of gas reservoir exploitation. In this case, the water will cross the interlayer and finally enter the wellbore with the help of pressure. In addition, the edge water source with random water output exists in the whole process of gas well production and operation. Compared with the water output time of interlayer water, the water output time of edge water is usually relatively late, and the specific factors affecting the time are uneven factors such as pressure and energy conversion[2].

The second is the type analysis of produced water. The main content of this work is to analyze the properties of produced water, in which the formation saturated steam water content is the key parameter to be calculated. The water content in the formation refers to the saturated equilibrium water vapor in the formation under the influence of formation temperature and pressure. Conventional gas reservoir formation is often accompanied by a large amount of formation water. Therefore, it is normal for the gaseous fluid of gas reservoir to contain water vapor, and the state of water vapor is also saturated. In fact, the water vapor content is closely related to reservoir temperature, pressure and gas. The relationship between temperature and water vapor content is positive, while the relationship between pressure, heavy hydrocarbons in gas and water vapor content is negative. There are various ways to confirm its content, among which the simulation of Mcketta-wehe belongs to the common algorithm, and the algorithm diagram coefficients are shown in Tab.1.

Tab.1 Coefficients of simulated Mcketta-wehe algorithm graph

Pressure (MPa)	a_0	a_1	a_2
4	-24.7175	0.1120	-1.0085×10^{-4}
6	-25.1163	0.1128	-1.0264×10^{-4}
10	-25.4407	0.1133	-1.0425×10^{-4}
15	-22.6263	0.0973	-8.4136×10^{-4}
20	-22.1364	0.0946	-8.1751×10^{-4}
30	-20.4434	0.0851	-7.0353×10^{-4}
40	-21.1259	0.0881	-7.4510×10^{-4}

Finally, it is the reference basis for judging the produced water type of gas reservoir and gas well, and the specific contents are shown in Tab.2.

Tab.2 Judgment standard of produced water types of gas reservoir and gas well

Source of water	Mineralization degree (mg/L)	Water gas ratio (t/m³)
Condensate water	< 10000	≤0. 13
Formation water	≥10000	> 0. 13

2.2 Water production law and characteristics

The analysis of water production law and characteristics of shallow gas fields should be combined with actual cases. Through the analysis of water gas ratio and mineralization degree of a shallow gas field, the gas wells can be divided into gas wells of producing condensate water, gas wells with liquid production signs and gas wells of producing formation water. The first is the gas well of producing condensate water. The water gas ratio and mineralization degree of this gas well are relatively stable, and even if they change, the synchronization degree is relatively high, which means that the formation belongs to homogeneous formation. The second is the gas well with liquid production signs. The mineralization degree and water gas ratio of this gas well do not meet the judgment standard of the gas well of producing formation water, and a certain index of the gas well in this case always shows an upward state. Finally, the water gas ratio and mineralization degree of gas wells of producing formation water will change synchronously. It may also be that the change of water gas ratio is slower than mineralization degree , or the water gas ratio is always rising, which is normal.

2.3 Influence of water production on gas well productivity

In the late stage of development, the gas wells in shallow gas fields will be affected by formation pressure and the threat of formation water entering wellbore, resulting in water production from gas wells. In view of this situation, technicians need to study the actual inflow performance characteristics of gas water co-production wells [3], and work on the management and evaluation of these wells. There are three conventional working methods, namely, converting water production into gas production, single-phase treatment, and dividing the gas reservoir into two overall regional ranges [4]. It should be noted that the two-phase flow area in this way is to artificially reduce the formation permeability. In fact, it is still a single-phase treatment method.

3. Wellbore effusion diagnosis

3.1 Theoretical analysis of gas well effusion

Wellbore effusion actually refers to the problem that the formation produced liquid cannot be carried out of the well due to insufficient gas energy. This situation will inevitably lead to the slow accumulation of liquid in the well, and finally form an obvious additional liquid column pressure, resulting in the reduction of gas well production and work efficiency. If this situation continues to develop, the fluid accumulation in the well will continue to increase, and the gas well will eventually stop production. At the initial stage of production, the gas well is in the stage of the strongest energy and can carry the liquid out of the gas well through its own strength. The gas rises rapidly at this stage, and there is basically no effusion in the wellbore. After the gas well is put into production for a period of time, the energy of the gas reservoir will also have an obvious decreasing trend, and the effusion gradually begins to accumulate in the wellbore.

3.2 Evaluation method

In shallow gas wells, there are mainly the following evaluation methods for gas well effusion. One is the IP and IC index method of gas wells, which needs to calculate the IP and IC values of gas wells under different water production conditions. In this way, technicians can systematically master and judge the index changes in different effusion wells, and master the effusion in gas wells at the same time. However, it should be noted that this method can not achieve the effect of prediction in advance, and can only be used as a judgment method.
The other is the field test method, which contains more technical details. The flow pressure detection method needs to draw up the test curve through the field test of the flow pressure value. If the curve itself has an inflection point, it indicates that there is effusion in the gas well, and the technicians can realize the purpose of depth confirmation of the enterprise interface with the help of the inflection point. This technology is a common and mature technology, which is applicable to a variety of gas wells. The detected data are very accurate and can accurately judge the actual situation of effusion in gas wells.

4. Field application of drainage gas production process

4.1 Optimized drainage gas production tubes

This technology needs to select the tubes with reasonable size for use, which is also the key core of the application of this technology. It should be noted that the smaller the tubes diameter, the greater the resistance of the fluid in the tube. Under the same wellhead pressure and flow pressure, serious energy loss will occur. Taking a gas well as an example, after comprehensively considering the friction resistance of the tubes and the liquid carrying effect, the technicians believe that the use of 2-3/8' tubes can meet the actual requirements of liquid carrying of the gas well. For the gas well with obvious decline in production capacity, the use of 4-1/2' tubes can not only optimize the liquid carrying effect, but also improve the stable production capacity.

4.2 Using foam method to assist drainage gas recovery

This kind of gas recovery technology is in fact a kind of auxiliary exploitation technology of gas field development. With the aid of equipment, surfactant is added into the casings, and the surfactant reacts with water to produce foams. A large number of foams significantly reduce the overall density. This technology is very convenient to use and can produce high economic benefits. It is a common technology in drainage gas production.
The key point of the use of this technology is the selection of foaming agent. Technicians can evaluate the performance of foaming agent by stirring method. When adding 0.5% foaming agent to each 100ml, the foam volume will be detected immediately after mixing 1min, and the time of taking out the liquid in 50ml foams is observed and recorded. This is the half-life period of the foam. The larger the foam volume and the longer the half-life period[5], the better the foam effect.

4.3 Drainage gas production by gas lift

The application of this technology needs to rely on external gas sources. Generally, nitrogen or carbon dioxide will be used. After pressurization with compressors and other equipments, the gas can be compressed and polymerized into high-energy gas and added to the wellbore, so as to reduce the overall density and improve the efficiency of carrying liquid out of the well. The use of this technology requires the addition of drainage assistance process on the basis of the original gas well process, and finally forms the annular space of the gas well. After the gas-liquid mixture is produced through the oil tubes, it can be directly used as the gas source.

4.4 Drainage gas production by mechanical pumping

The application of this technology requires the use of deep well pumps to repeatedly pump the effusion in the wellbore to the surface, and then use mechanical equipments to assist in drainage. The advantage of this technology is that it can start and stop work at any time, change the liquid level under the influence of different working systems, and finally act on the wellbore liquid column to produce back pressure on the bottom of the well. It is a mining method with high rationality.

5. Conclusion

Based on the results and discussions presented above, the conclusions are obtained as below:
(1) Drainage gas production itself is one of the secondary gas production technologies. The working principle is to use a variety of technologies to work together to reduce the bottom hole flow pressure, and can also form an obvious production pressure difference to lift the bottom hole liquid.
(2) In order to alleviate the problem of water production of gas wells and improve their liquid carrying capacity, technicians need to select a drainage gas production technology more suitable according to the actual situation, and increase the system operation stability and work efficiency. In addition, it also has a very obvious effect on the control of operation cost and the improvement of economic benefits.

References

1. [Qiaorong Feng, Gaijiang Liu. Water Production Characteristics of Low Permeability Tight Gas Reservoir and Application Analysis of Drainage Gas Production Technology. J. Petrochemical Technology, 2021, 28(07): 136-137.

2. Haidong Xue. Study on Combined Technology of Bottom Hole Purification, Drainage Gas Production in Middle and Shallow Layers in Western Sichuan. D. Southwest Petroleum University, 2019.
3. Lei Tang, Guangliang Lu, Xue Zhang, Haoru Wang, Jing Liu. Application of Vortex Tube Drainage Gas Production Technology in West Sichuan Gas Field. J. China Petroleum and Chemical Industry Standards and Quality, 2018, 38(21): 149-150.
4. Jianfeng Zhao, Dong Liu, Dan Fang, Chenglin Zhang. Discussion on Drainage Gas Production Technology of Shallow Gas Wells. J. Science and Technology Wind, 2017(11): 109.
5. Ping Li. Study on Drainage Gas Production Technology of Shallow Horizontal Wells in Western Sichuan. D. Southwest Petroleum University, 2016.

Characteristics of water flooded layer in Minghuazhen Formation

Zhonglin Li

Institute of Geology, The Fourth Oil Production Plant, Daqing Oilfield Co., LTD Heilongjiang Daqing 163000 China

Abstract: In the process of oilfield exploitation, the water-flooded layer in the Minghuazhen Formation cannot be accurately identified, which has a great impact on the oilfield exploitation efficiency. If the water-flooded layer marker in the Minghuazhen Formation cannot be fundamentally defined, the oilfield may not be able to carry out normal exploitation operation. Based ON this, THIS PAPER expounds the basic characteristics of oil accumulation in the water-flooded layer of Minghuazhen Formation, puts forward the characteristics of the water-flooded layer, and analyzes the factors of low resistance for reference.

Key words: Oilfield Minghuazhen Formation; Water flooded layer; Characteristics analysis

1. Introduction

Once the reservoir is in service, it may be exposed to production water or groundwater from the reservoir, which can lead to waterflooding and formation of flooded zones. In order to maintain a stable reservoir pressure in the subsurface and generate sufficient fluid from the well, oil fields usually inject water into the Wells surrounding the production well after production, forming a pressurized network of production Wells. This measure of "filling oil" is the supplement of formation energy and plays an important role in improving production well productivity. However, water injection cementing measures have some modification effect on the original rock. Lead to changes in the content of material, electricity and oil and gas in primary strata. Therefore, the characteristics of water-flooded layer in Minghuazhen Formation should be further analyzed.

2. Overview of water flooded zone in oil field

In order to increase the recovery rate of oil fields, water injection is now a common way to produce oil fields, as shown in the following figure:

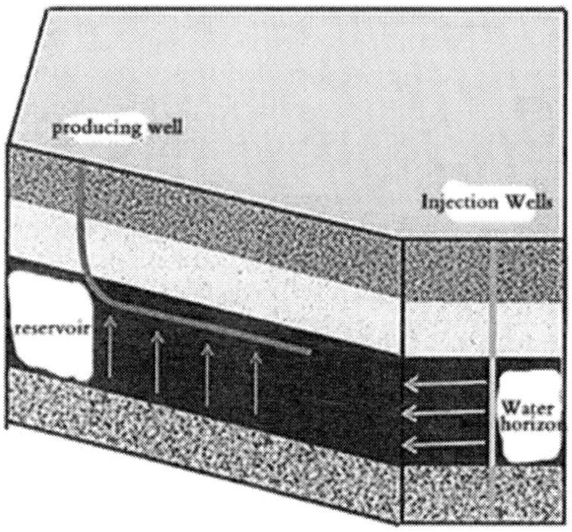

Figure 1: Schematic diagram of oilfield water injection

When the oil field is flooded, the physical properties of the deposit will change. Permeability, oil removal efficiency, resistivity, oil-water distribution, porosity, moisture, and temperature will vary. The quality of flooded layer is also affected by floods in different periods. Upon flushing, clays and clays degrade the formation. The saturation of bound water decreases with the increase of mean particle size. In some highly heterogeneous reservoirs, pumping methods, rates, and techniques vary due to uneven permeability and porosity. This process is mainly done by injecting water into highly permeable areas, which is very rapid. When the oil in the pore space is almost completely purified, most of the oil is mainly stored in the weak permeability zone. In the oil injection industry, oil removal is most common in large fractures, while large amounts of oil are retained in smaller Wells.

Therefore, areas with low porosity and low permeability become "potential reservoirs" for high oil production [1].

3. Basic characteristics of oil accumulation in water-flooded layer of Minghuazhen Formation

With the long-term development of the deposit and the continuous movement of water injection into the deposit, some changes have occurred which are obviously different from the original state of the stratum. Its main features are as follows:
(1) Increase the saturation of water. With the increase of flood intensity, the initial critical water saturation (equivalent to correlation saturation) gradually increases to any maximum value. (2) Change the physical characteristics of the layer. Flushing water washes or fills the clay on the surface of sandstone or sandstone particles, thereby increasing the pore radius (mainly the pore size). As a result, permeability increases significantly after extensive washing. At the same time, the water content of the rock changes, the hydrophilicity becomes hydrophilic, and the saturation index n also changes. However, flooding can also cause some clay minerals to swell, clogging pores and reducing the permeability of the producing formation. In general, water injection increases porosity; (3) Formation resistivity variation. At the beginning of filling, the collector resistivity decreases with the increase of water saturation. (4) The distribution of inundation area is closely related to the sedimentary rhythm. The upper part of the museum is the Yun dam and the point bar. When water is squeezed, a large channel is formed in the lithologic part of the lower reservoir, leading to local early water injection. (5) Reservoir inhomogeneity causes flooding and the difference between them. Due to the difference of reservoir material properties, reservoir heterogeneity is caused, and channels are formed in the middle and high strata of the reservoir, while residual oil is relatively abundant in the low-permeability parts and thin ground layers [2].

4. Characteristics of water flooded layer in oil field

4.1 Petrological characteristics

The core of Minghuazhen Formation is siltstone and fine sandstone, and the main mud is cement. Due to the relatively new age, shallow burial depth, less compaction and consolidation, the rock is relatively free. The latent layer is mainly composed of fine grained and medium grained sandstone with good separation and high structural maturity. This group of sands has many lenses with flat distribution, limited dispersion, large thickness variation and high heterogeneity. Analysis of rock logging data shows that the oil and gas in the reservoir are mainly fluorescent. Petrology has a certain correlation with gas. The higher the sandstone thickness, the better the oil and gas. And vice versa; The lower the shale content, the better the performance, and vice versa. The analysis of rock mineral composition shows that the average content of cement in underwater sandstone is 28.3%, of which clay cement accounts for 86.4% and calcium 8.1%. The content of clay cement depends on burial depth and decreases from top to bottom. The analysis of minerals in clay cement shows that these minerals are divided into illite, kaolinite, chlorite and illite/lignite mixture. The statistical analysis shows that clay minerals are mainly mixed line and kaolinite, accounting for 73.8% and 14.3%, respectively. In the mixed layer of montmorillonite, the average proportion of overall mixing is 81.2%.

4.2 Physical characteristics

The data of rock pore permeability test show that the reservoir porosity of Minghuazhen Formation ranges from 9.2% to 42.5%, with an average of 33.2%. The average pore size of 7.56μm permeability coefficient varies greatly, generally between 10x10-3μm~2~7500×10μm², the average permeability is about 500x10μm², heat collection performance has high, medium pore and high permeability. Generally good, the porosity is generally more than 30%, the permeability is generally more than 300×10.3μm-2, the porosity and permeability are relatively high. The analysis shows that the porosity types of sedimentary sandstone are mainly primary fractures, and the secondary porosity is small. The larynx is mainly composed of plates, layers and tubes. The physical and physical properties of the Minghuazhen Formation have two characteristics: with the increase of sedimentary depth, the physical properties of the rocks change; With the decrease of shale content, both shale quality and quality have a positive physical tendency to decrease. The analysis of the relationship between sandstone permeability and shale content shows that with the increase of shale content, the permeability value gradually decreases from high to low, and the decreasing trend is obvious. It can be seen that shale content has a great influence on the physical properties of water-flooded layers in Minghuazhen Formation [3].

4.3 Log characteristics

During the development and exploitation of oil fields, the curve logging data are mainly affected by the following factors: the oil content and water content of the reservoir change with the change of the exploitation degree; Formation salinity changes with the change of water; The physical and chemical properties of rocks vary with the amount of water under hydraulic impact. The vertical pressure of a well depends on the irregular pressure in the reservoir. These factors mainly reflect the changes of spontaneous potential, resistivity, acoustic wave propagation time, density, PFT and C/O ratio. The deeper the development of the deposit, the higher the degree of mineralization of the deposit, the higher the total salt content in the pore, and the higher the irrigation amount. However, this rule does not exist when the irrigation water is fresh. When water enters a certain level through chemical reactions such as ion exchange, mineralization is reduced. Due to the increase of water supply, the reservoir rocks are affected by different degrees of water, and the porosity structure of the reservoir changes to a

 https://doi.org/10.1051/e3sconf/202337501016

certain extent, so as to improve the permeability and porosity of the reservoir. When water does not flow in, the pressure in the flooded layer is relatively lower. However, vertical formation pressures must be different due to the limitations of the use of collection nets, sandstone petrology, and existing enclosed emissions. The pressure coefficient of the original formation is higher than that of the sedimentary layer. Neutron gamma logging uses neutron gamma radiation to measure the amount of hydrogen and helium in the formation. As the amount of water injected increases, so does the mineral and chlorine content of the aquifer. The gamma-ray measurements of the seeds will also increase proportionally to the size of the unsubmerged collector.

4.4 Distribution law of water-flooded layer

Extensive geological research and production practice at home and abroad show that submerged strata do not mean that the entire strata have no potential. On the contrary, determining the distribution pattern of the submerged layer is very important for predicting the remaining oil distribution. The comprehensive analysis of more than 20 oil Wells in an oil field shows that the latent layer in the plane is mainly distributed in the well network development area. Only a few are new or incomplete; The remaining LPG in submerged formation is mainly distributed at the "waterline", which is the main direction of water injection. Therefore, accurate analysis of the waterline injection direction of the production well pattern is very important to determine the distribution pattern of the latent layer. From vertical aspect, there are obvious urban sinter inundation strata in this oilfield, which are mainly concentrated in the oil well section with settlement depth less than 1400 meters, showing the characteristics of surface inundation. In sand, water deposits are widely distributed at the bottom of rhythmic sandstone strata. The analysis shows that scintillation strata are mainly formed in fluvial sediments. The rock at the base of these rocks is usually well separated and has a high degree of structural maturity. The physical properties are generally high porosity and high permeability. The results show that high porosity, high reservoir permeability and good permeability are prone to the formation of flooded and submerged layers. Therefore, the distribution of submerged strata is closely related to petrology and sedimentary factors.

4.5 Logging mode characteristics of flooded layer

In order to interpret the flooded zone log, multiple factors become more and more difficult as the drilling depth deepens. Pool logging interpretation: During the pouring process, the moving oil is washed away by water, but in some places, even after pouring, the moving oil is not washed away, but only "washed away". Therefore, mineral saturation is an important criterion to evaluate the degree of submergence and mineral saturation. The residual saturation and correlation saturation can be measured by analyzing the liquidity problem. (1) The theoretical approach to fluidity analysis is based on the dynamics of mixed liquids, allowing the sum of fluid flow rates in the reservoir from different phase drift equations. Under constant differential pressure, the oil, gas and water contents and their industrial properties are determined by phase permeability, permeability profile and liquid properties. It's all about effective penetration. Each phase liquid is measured in terms of the volume of the rock, which depends on the viscosity, humidity, and porosity of the rock; (2) With the help of natural potential, the unevenness and filling degree in the formation are also very uneven, and it is easy to deviate from the base naturally. The main cause of the potential baseline shift is local salinity variation due to drainage and desalination. The base shift is determined by the formation salinity ratio before and after pouring. This process is mainly an ion exchange between water injection and melting, so the formation resistance increases accordingly. Due to the low permeability of the formation, the ion exchange time is prolonged. The closer the ion exchange rate is to the reservoir, the faster the transmission speed of the potential bottom to the reservoir is. Advanced carbon and oxygen recording methods are based on the principle of fast inelastic neutron scattering. In this method, the neutron source is a controlled neutron accelerator that forms high-energy neutrons through hydrogen-sensitive reactions.

5. Low resistance factor analysis

Based on the study of low resistance deposits in other areas, the causes of low resistance deposits are analyzed. Geologists usually attribute the low resistance of a stratum to rocks. There are two "internal causes", such as porosity and hydraulic properties, and three "external causes", such as formation distribution in adjacent layers, formation salinity and drilling fluid properties. In the context of the formation of large areas of sedimentary layers and oil reservoirs, low resistance oil reservoirs may be formed in areas with low fluid dynamic capacity, low amplitude and low oil saturation [4].

5.1 Internal factors

The low resistivity of Minghuazhen Formation has two internal factors: (1) High lithology fineness, high clay content, more clay minerals, and high formation rate of bound water, which lead to the increase of rock conductivity and the decrease of formation resistivity; In particular, the logs show a decrease in the natural gamma potential relative to the aquifer, while the resistivity between the normal formation and the aquifer increases by less than a factor of two. This is caused by the formation of low resistance reservoirs in the area; 2) The relatively weak material and complex pore structure means that the total oil content is reduced, resulting in lower reservoir resistance.

5.2 External factors

The three external factors leading to low resistivity of Minghuazhen Formation are as follows: (1) weak reservoir. In formation logging, due to the limited vertical resolution of the logging instrument, the logging value is affected by the upper and lower layers (clay or aquifer),

 https://doi.org/10.1051/e3sconf/202337501016

resulting in low formation resistance. Considering that sediments in curved rivers are easy to form thin reservoirs, this is also one of the main reasons for the formation of low specific oil-resistant reservoirs in this area. (2) In areas with small tectonic amplitude, reservoirs are low in height and easy to form in the same reservoir. The resistance of the formation depends on the aquifer, resulting in low resistance; There are many oil, gas and water systems in the west flank of Port of Spain oilfield, and the separation of oil and water is not uniform. The local development of the same reservoir is one of the reasons for the low resistance. (3) For drilling reasons, some Wells use high-conductivity saline solutions. As it enters the formation, the resistance of the formation decreases.

5.3 Conclusion analysis

From the above analysis, it can be seen that the distribution of the Minghuazhen Formation with low resistivity is largely related to sedimentary factors such as petrology and sedimentary environment. The analysis shows that the sedimentation of Shimminghua Formation is fluvial curve deposition, the reservoir is positive circulation, and the recombination of sedimentary cycle is mainly heterogeneous. The distribution of KMG low resistivity strata is related to the sedimentary period. In the upper part of the positive circulation and the bottom of the reflux, due to the thin sandstone layer and thin rock, the shale content is relatively increased, and the bound water is more saturated, forming the stratum with lower specific resistance [5].

6. Conclusion

In a word, the following conclusions can be drawn from the analysis of water-flooded layer characteristics of Minghuazhen Formation in the above oilfield:
(1) There is a nonlinear relationship between water saturation and resistance, and an inverse relationship between resistance and humidity; (2) The curves of spontaneous potential and induced resistance of submerged layer changed obviously;
(3) Use low salinity dielectric logging to measure oil content.
(4) For Wells with uncertain formation salinity, formation saturation can be obtained by carbon and oxygen logging to determine water flooding and water flooding levels;
(5) RMT measures the reservoir mainly through two models to assess the degree and nature of the flooded layer.

References

1. Li Min, Tang Xinping. Science and Technology Innovation and Application, 2021,11 (18) : 27-29.
2. Chen Xin, Fan Xiangyu, Wang Zhaofeng, Dai Xiongjun, Chen Kegui, Liu Hui. Logging response characteristics and waterlogging analysis of water-flooded zone in Aryskum oilfield [J]. Progress in Geophysics, 202, 37(2) : 561-569.
3. Shen Xiaoxiu, Zhang Jieshu, Miao Feifei, Han Jianbin, Liu Wenchao. Petroleum Geology and Engineering, 2020,34 (1) : 55-60. Reservoir characteristics of the lower member of Minghuazhen Formation in Huanghekou Sag and their influence on productivity [J].
4. Hou Dongmei, Zhou Junliang, Zhao Junshou, Zhang Xiaolong, Quan Bo. Sedimentary characteristics of shallow water delta in the west end of Bonan low Uplift: A case study of the lower member of Neogene Minghuazhen Formation in area B [J]. Fault Block Oil and Gas Field, 2021, 1028 (002) : 205-211.
5. Li Jian, Qu Sa, Zhang Haina, Shen Huilin, Hu Chenbin, Sun Qipeng. Well logging comprehensive evaluation method of Chengdao Oilfield saltwater flooded layer [J]. Well Logging Technology, 222, 46(3) : 304-310.

 https://doi.org/10.1051/e3sconf/202337501017

Technology and effect of small well spacing control decline in close well network

Senlin Gan

The Third operation area of No.1 Oil Production Plant of Daqing Oilfield Co., LTD; heilongjiang daqing 163000, China

Abstract: In order to further improve oil recovery, water flooding was adjusted for three times from 2009 to 2012. The spacing of injection-production Wells was shortened from 250 meters to 106 meters, and the density of well pattern increased from 90.7 to 196 per square kilometer. This leads to the problems of "rapid rise of water cut, large difference in oil formation production, and large production decline in block u". In recent years, we by fine geologic research, to improve the poor thin oil layer use condition, controlling water cut rising velocity, as the key point, the energy recovery network continuously grope for dense well spacing control regulation technology, gradually formed dense well network from decreasing "three-excellent" control technology, through the effective implementation, improved the block development effect, achieve the goal of stabilizing oil water control.

Key words: small well spacing; Natural decline; Stable oil and water control; Control technology

1. Introduction

After three encryption adjustments, the number of small Wells accounted for 85%. On the longitudinal, average interval is 198 meters long, single well interval number is 5.8, and a big difference between layers, check with the percent of pass is 10% lower than conventional well, lead to a use of sandstone percentage is only 25%, after rising water cut, inefficient invalid circulation to further expand, injection-production in the development of dynamic high speed, low degree of use, water cut rises quickly, the characteristics of big production decline, Water injection adjustment and on-site management is more difficult.

Stabilize in order to further extend the production period, to achieve the goal of stabilizing oil water control, through the comprehensive analysis of dynamic and static data, first looking for remaining oil, and at the same time for different sedimentary microfacies in the reservoir and the status of the use of the single sand body, water flooded, optimization measures, aimed at the prominent contradiction development problems in production process, summarizes the mining practice of the process of recognition, To solve the influencing factors, to find out the potential of a single well, to promote the fine analysis, fine potential exploration, fine measurement and adjustment, according to this development idea, gradually developed and formed the "three best" regulation technology of tight well network and small well spacing control decline.

2. "Three optimal" control technology for tight well network and small well spacing

In the double extra-high development stage, in view of the contradictions such as low use degree of small well spacing, difficult well selection measures, and rapid water cut rise, the precision water injection and water control are deeply promoted, and the "three-best" water control and efficiency improvement technology is developed

2.1 Water injection program optimization and adjustment technology

One is the combination of fine geological research results, do fine water allocation scheme. Adhering to the concept of "precision is potential, precision is potential", we have carried out solid precision mapping work, deepened the digital and modeling understanding of the research and development block modeling, and the research object has changed from oil formation to sedimentary unit and then to the internal configuration of single sand body, so as to realize the progress of remaining oil characterization from precision to precision. The distribution characteristics of remaining oil in the late stage of ultra-high water cut are revealed, and the main potential objects are identified precisely.

Second, it is combined with the use of refined layer segment, do fine water injection stratification scheme. Considering the long distance from the well to the well and the number of layers, the analysis method combining dynamic and static data was used to introduce the coefficient of permeability variation based on the "7788" subdivision limit standard adjustment technology, and the layers with similar coefficient of variation were combined. Through the reorganization, the water absorption difference between the layers was significantly reduced.
Thirdly, the optimal stop-control and recovery plan is made based on the numerical well test research results. With the decrease of pressure and the redistribution of oil and water, it has become a potential reservoir for the recovery of water injection. The numerical well test simulation method is used to study the variation law of bottom hole pressure and guide the design of injection distribution scheme for long stop zone. On the basis of theoretical understanding, comprehensive analysis of well pressure level, injection-production connectivity, remaining oil distribution characteristics, effectively adjust the small layer injection.
Fourth, the combination of round - round inspection and intelligent test and adjustment, do accurate water injection program implementation. The polling inspection and adjustment technology was adopted: a two-month cycle was carried out to optimize the test cycle, so as to ensure the encryption and adjustment of key Wells and stable Wells as required, and effectively improve the qualified rate of inspection and distribution on the basis of reducing the workload. Adopting intelligent measurement and adjustment technology: In 2016, the field test was first carried out in the narrow well spacing area. Based on the understanding of the results obtained, the intelligent measurement and adjustment technology was comprehensively promoted in the area, which achieved a good effect of remote real-time monitoring and adjustment of stratified water quantity, a leap in the quality of water injection, greatly reducing the test workload and reducing labor cost.

2.2 Measure scheme optimization design technology

For fracturing, the difficulty of selecting well, clear "potential match technology, combination of conventional and unconventional" working train of thought, on the basis of analysis of remaining oil potential, the potential for different types of targeted measures tapping technology, by doing conventional technology, do best unconventional technology, realize all kinds of effective measures to potential release. Especially for small well spacing block action potential, potential in alignment with low water cut Wells, on the basis of the conventional fracturing, by summing up, widen the fracturing optimization design, the wall pressure combined with exploration technology, smaller size fracturing technology, coiled tubing fracturing process four potential direction, in terms of unconventional fracturing, reverse pressure, displacement process, Through the detailed design and individual design of fracturing schemes according to different reservoir development conditions and development conditions, the potential space is effectively broadened by applying various process combinations.

2.3 2Ineffective injection and production optimization control technology

For a dramatic increase in late high water cut oil liquid ratio, invalid injection-production contradictions prominent problems, such as using tracer flow simulation software and the mathematical model research, preliminary screening invalid injector-producer layer, combination and utilization of water saturation, injection-production ratio parameter identification standard preliminary judgment may be ineffective circulation well layer, analysing the injection-production relationship, sandbody connecting relations, injection-production profile, Judge invalid loop identification result.

3. Practical application effect

Through the implementation of "three excellent" regulation technology, the oilfield development effect has been significantly improved
It is output, decrease get effective control
From the five - year production changes, the annual oil decline significantly reduced.

Table 1 Production of water flooding in recent five years

year	2017	2018	2019	2020	2021
Annual oil production (ten thousand tons)	44.02	37.73	33.73	30.21	28.12
Production decline (%)	18.8	14.3	10.6	10.4	6.9

Table 2. Natural decline of water drive in recent five years

year	2017	2018	2019	2020	2021
Natural decline (%)	15.1	12.8	12.7	10.9	10.8

Table 3. Natural decline of stratified systems

year	2017	2018	2019	2020	2021
MrPortuguese formation (%)	15.0	11.8	10.2	8.5	8.4
High platform sub-layer system (%)	20.0	16.7	15.6	12.0	11.7

Second, the rise rate of water cut is effectively slowed down
The regional water cut is stable as a whole, and the rising rate of water cut is controlled.

Table 4. Variation of water cut in water drive year

year	2017	2018	2019	2020	2021
In aqueous (%)	96.31	96.59	96.86	97.11	97.35
Rise in water content (%)	0.37	0.3	0.3	0.25	0.24

Table 6. Variation of appreciation over water in stratified systems

Year	2017	2018	2019	2020	2021
Mr Portuguese formation (%)	0.32	0.25	0.23	0.22	0.21
High platform sub-layer system (%)	0.41	0.36	0.31	0.28	0.27

Third, the injection-production capacity is balanced and the formation pressure is stable

The formation pressure is stable, and the overall submergence decreases by 51m, but the overall structure tends to be reasonable.

Table 7. Formation pressure variation

Year	2017	2018	2019	2020	2021
Injection-production ratio	1.17	1.2	1.2	1.14	1.16
Formation pressure (MPa)	9.45	9.57	9.62	9.68	9.66

Table 8. Variation of Wells with different levels of water flooding

classification	Number of Wells in 2020 (Wells)	Number of Wells in 2021 (Wells)	difference
submergence <100m	0	1	1
100m≤submergence < 300m	421	541	120
300m≤submergence < 500m	189	78	-111
submergence≥500 m	71	48	-23

Fourth, the use of oil reservoirs has been effectively improved

The water absorption ratio of off-sheet sandstone increased by 11.59 percentage points.

Table 9. Triennial water absorption ratio change

year	2017	2018	2019	2020	2021
Water absorption ratio once every three years (%)	62.18	69.85	70.14	72.86	75.08

4. Conclusion

(1) The late stage of extra high water cut is still an important stage to tap potential. Accurate characterization of structure, reservoir and remaining oil is an important basis to ensure the release of small-scale potential.

(2) Through the integration and application of conventional technologies and the demonstration and application of new technologies, the potential space of narrow well spacing is broadened, and the supporting technology and model of tight well network and small well spacing in the late stage of extra high water cut is formed.

(3) Based on the fine anatomy of tight well pattern reservoir, through the optimization and adjustment of water injection, the qualified rate of water injection is improved and the quality of injection is improved by making full use of intelligent separate injection and round check and adjustment, and the comprehensive potential mining adjustment of small well spacing is formed to improve the water injection efficiency.

(4) The combination of dynamic and static is used to establish the automatic identification of invalid circulation and the potential exploitation measures of the combination of dredging and blocking in the aspect of injection and production, forming the integrated adjustment technology of identification of inefficient invalid circulation and water control and potential exploitation.

References

1. Wang Lixia. Improvement method of water flooding technology in high water cut stage of complex fault block reservoir in Huabei Oilfield [J]. Petroleum Knowledge, 2019(4):2-3.
2. Zhao Beichen. Coupling mechanism and parameter optimization of water flooding in fault block reservoir [D]. China University of Geosciences (Beijing), 2020.
3. Wang Nian-ming, GUO Hua-gu, JIANG Ming-sen. Research on oil stability and water control technology in middle and high water cut stage of complex fault block reservoirs: A case study of Yuejin No.2 Donggao Oilfield [J]. China Petroleum and Chemical Standards and Quality, 2013(11):1-3.
4. Wu Gang. Water flooding enhanced Oil recovery technology of fault-block reservoir in Shengli Oilfield [J]. China Chemical Trade, 2015(9)23-24.

5. Liu Xishun. Research on remaining oil distribution and potential exploitation strategies of extremely complex fault block reservoirs in high water cut development period [J]. Xinjiang Oil and Gas, 2008, 4(4):4-5.

A new method for dynamic prediction of reservoir pore pressure in development Wells

Ying Gao

Institute of Geology, No.4 Oil Production Plant, Daqing Oilfield Company, Daqing, Heilongjiang 163511, China

Abstract: The rapid development of social economy in our country, the industry in innovation and development of the growing demand for oil, promote people to constantly increase the intensity of oilfield development, and oilfield development time prolonged, and the application of water injection production in the process of development, cause of formation pore pressure is more and more complex, as the new adjustment well drilling work more difficult. Therefore, the research on dynamic prediction methods of reservoir pore pressure in development Wells is intensified, which can predict reservoir pore pressure in development Wells in a short time, and the pore pressure can be predicted more accurately through the comparative analysis of predicted and measured pressure, providing strong support and guarantee for the design and construction of drilling engineering.

Key words: Development well reservoir; Pore pressure; Dynamic prediction; The new method.

1. Introduction

Formation pore pressure prediction is generally in combination with acoustic and resistivity logging data and corresponding calculation, and only in the existing drilling, or goal well drilling can only be carried out under the condition of existing around, and in the process of actual prediction to comprehensively consider the p-wave velocity equation of stress, but there exist certain disadvantages to the these equations, The predicted structure is affected by the rate of compaction trend. Therefore, it is difficult to ensure the accuracy of the normal compaction trend velocity calculation when the shallow velocity logging data are incorrect or incomplete. To solve this problem, a new dynamic pore pressure prediction method is applied to further improve the accuracy of pore pressure prediction.

2. A new method for dynamic prediction of pore pressure

2.1 Calculation principle

Eaton Method is commonly used in the calculation of formation pore pressure curve, and the calculation formula is:

$$\sigma = (OBG-PPG_N)\left(\frac{dco}{dcn}\right)^E$$

Among them , σRepresents the effective stress equivalent density (g/cm^3) ;

OBG Represents the pressure equivalent density of the overlying rock (g/cm^3) ;

PPG_N Represents the equivalent density of formation pore pressure (g/cm^3) ;

Dco Represents the acoustic time difference of undercompaction;

dcn Represents the acoustic time difference of the normal compaction curve;

E It represents the power index, which is a specific parameter that is mainly taken in combination with experience, laboratory specimen tests and engineering measurement information data。

The acoustic time difference of normal formation compaction curve and related parameters of geological deposition history play an important role in the calculation of Eaton method [1]. Due to the relatively complex sedimentary history, the information of sedimentary history is lacking in many cases. Therefore, the normal formation compaction curve needs to be calibrated according to the actual situation observed in existing drilling and drilling engineering. When has the anticline structure, and at the same time orogeny caused by horizontal extrusion effect is significant, under the condition of local stress is affected by the structure is very big, causing the overlying rock pressure coefficient value and density of integral calculation data have large deviation, the information obtained from this kind of cases, the accuracy of dynamic pore pressure prediction results under the influence of a certain extent, It shows a decreasing trend. Therefore, the results of numerical calculation of overlying rock pressure based on three-

https://doi.org/10.1051/e3sconf/202337501018

dimensional geostress field are used as OBG, and the pore pressure is calculated and analyzed in detail using Eaton formula to effectively solve this problem. The pore pressure results obtained will be included in the influence of anticlinal structure on overlying rock pressure. Therefore, this dynamic pore pressure prediction method is more accurate than the conventional overlying rock pressure prediction method based on density integration with depth. When the other parameters in Eaton formula are set to, E is 3, the normal hydrostatic pressure density is PPGN=1.03 g/cm3, and the information data applied to calibrate the curve for the well already drilled or the pore pressure already drilled around the target well are measured formation pressure coefficient (PP), overflow drilling fluid density value and other measured data.

2.2 Calculation process

The calculation flow of the new pore pressure dynamic prediction method applied in this paper mainly includes the following three steps:

(1) Three-dimensional seismic wave interpretation structure to establish a geological model, this step mainly uses three-dimensional geological analysis software, combined with block seismic information data, to create the geological structure of each layer of the block [2].

(2) In the process of single well geomechanical analysis of the upper strata of adjacent Wells, the deep strata of some development Wells lack relevant logging information data. Therefore, the single well logging information data of the drilled upper strata can be analyzed in detail to obtain the magnitude and direction of the principal component of ground stress in the upper strata. In the process of this analysis, the relevant personnel can calculate pore pressure, overlying rock pressure coefficient and other parameters based on the actual drilling information, such as transverse wave, longitudinal wave, density log and other relevant logging information data of acoustic logging. The minimum horizontal principal stress among the three components of ground stress can be calculated according to Poisson north, elastic modulus and other relevant parameters, and the relevant parameter values in the work can be effectively calibrated and calculated with the help of floor drain test. This analysis process is relatively complicated. In the process of calculating pore pressure using acoustic logging information data, due to the lack of analytical function relationship between acoustic wave and pore pressure, semi-empirical and semi-analytical empirical formula method is usually used to calculate pore pressure in detail.

In addition, the value process of the maximum horizontal principal stress in the detailed analysis of geomechanics of a single well by relevant staff is complicated, and its value cannot be directly calibrated by ground drain test [3]. In the process of drilling engineering construction, the indirect calibration of maximum horizontal principal stress is usually carried out by matching the curve of collapse pressure and equivalent drilling fluid density of real drill. In this process, the matching mainly means that the collapse pressure should not be less than the real drilling fluid at the choke point, and the collapse pressure should not be greater than the real drilling fluid at the safe drilling depth point under complex conditions such as no jam.

There is usually a certain relationship between the characteristics of the upper and lower crustal stress distribution. When the upper reverse fault stress occurs, the transverse extrusion coefficient of the lower crustal stress will be affected to a certain extent and gradually increase. When the upper ground stress is under normal fault stress, the transverse extrusion coefficient of the lower ground stress will gradually decrease. At the same time, when the overlying rock pressure is subjected to strong transverse extrusion effect, the effect of transverse deformation is relatively large. Therefore, in the process of predicting and analyzing the geo-stress of the lower strata, the geo-stress value of the upper strata can be regarded as an effective reference data [4].

(3) The horizontal tectonic stress coefficient is defined based on the fault morphology and structural characteristics, and then the magnitude of the horizontal tectonic stress is determined. The main purpose is that when the transverse extrusion effect is relatively strong, the overlying rock pressure value is larger due to the influence of the transverse deformation effect, which leads to the corresponding increase of pore pressure. In general, the minimum horizontal principal stress coefficient corresponding to the reverse fault is not less than 1, and the overlying rock pressure in the principal stress component corresponding to the normal fault is not less than the maximum horizontal principal stress. In the process of evaluating the horizontal tectonic stress coefficient, the empirical method can be applied to determine the tectonic stress coefficient of the specific location of the target well in combination with the value of the tectonic stress coefficient obtained from the analysis of adjacent Wells with similar structures and the relevant experience. In addition. The trial-and-error method has good applicability in the value of horizontal tectonic stress coefficient, mainly by setting different values of tectonic stress system, and observing in detail whether the pore pressure obtained by it is compatible with the density curve of drilling fluid.

(4) Solve the three-dimensional fine stress field. The fineness of the calculation process is supported by the following factors: stratigraphic structure; The initial input parameters, such as pore pressure and three principal stress components, are combined with the geomechanical analysis of a single well. Due to the lack of logging information data of deep drilling strata of development Wells, the geomechanical analysis of single well is limited to the upper strata. The measured information data mainly include the minimum horizontal stress point obtained by the floor drain test and the maximum horizontal principal stress direction determined by the image logging [5]. In the calculation process, the gravity load, initial ground stress and displacement constraint boundary conditions are appropriately added to the 3D block finite element model.

(5) Predict the pore pressure. In the process of pore pressure prediction, the acoustic time difference of the formation should be calculated in detail by combining the

velocity of the formation, and the rock density of the formation should be calculated in detail by combining the acoustic time difference with the Gardner empirical formula. In the process of pore pressure prediction, the most conventional method is to calculate the overlying rock pressure accurately in combination with the compaction density. The new dynamic prediction method of pore pressure proposed in this paper is mainly to establish a three-dimensional fine ground stress field of the block, calculate the finite element values in detail, and obtain the overlying rock pressure. At the same time, the pore pressure curve is calculated in detail by Eaton formula combined with the overlying rock pressure and acoustic time difference.

3. Engineering case analysis

For example, a block contains multiple Wells such as well X7 that have been drilled with the upper reservoir as the target reservoir. This paper focuses on the formation pressure analysis of the lower source-reservoir-cap combination under 6000m in well X1. The upper X7 well has been drilled to a depth of 3,000-4,000m. The fault forms in this block mainly include reverse faults and thrust faults, indicating that the horizontal crustal stress component has a high value. Meanwhile, in the process of pore pressure prediction and calculation, the pore pressure data already drilled in the upper part should be combined, and the specific analysis and calculation are as follows.

3.1 Construct the model and boundary conditions

3D fine stress field modeling is a new technology of finite element numerical calculation, which requires the application of seismic wave information data to create an accurate geological model. On this basis, the existing rock and stratum logging information data is used to provide a perfect elastic model and other material parameters for the model construction. The finite element model of Abaqus in this block is shown in FIG. 1. The finite element model includes 370 000 nodes and about 90 000 C3D20R-20 node high-precision elements.

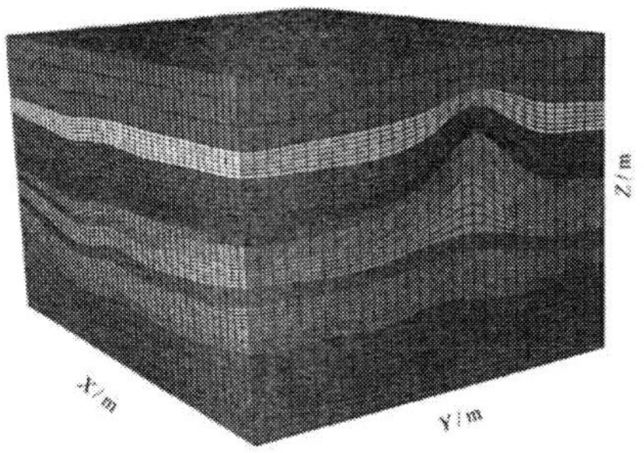

Figure 1: finite element model diagram of Abaqus

The Abaqus finite element model of this block was simplified. The changes of material properties on both sides of the fault were comprehensively considered, and the simulation of the fault plane itself was ignored. The fault plane was not considered as the fracture plane. There is gravity load in the model, and the calculation is mainly aimed at the static ground stress field. In the actual calculation and analysis process, there is no relative movement between the two ends of the fault, which proves that the simplified treatment of the model is scientific and effective. The single well ground stress field analysis result of well X7 in this block is used as the initial value of the model ground stress field. In the input process, the lateral pressure coefficient is directly converted through the single well ground stress field analysis result.

3.2 Pore pressure of X1 well based on 3D in-situ stress and zone velocity

Combined with the numerical solution of three-dimensional fine geostress field, the stress component value of the geostress field is extracted from the target well trajectory points and input into the single well geostress analysis tool. Then, based on the given layer velocity, the pore pressure of the three-dimensional geostress field is predicted for well X1, and relevant data are obtained as shown in Table 1. Through practical analysis, it is found that the rock density calculated by acoustic empirical formula is better than the density logging data, which fully shows that a layer velocity curve is reasonable and effective. Therefore, the pore pressure prediction results of the X1 well profile based on the numerical solution of the 3D in-situ stress field and the layer velocity are obtained. There is no significant difference between the pressure curves of the single well and the three-dimensional overlying rock, and thus there is little difference between the two pore pressure curves. The upper pore pressure is between 1.2g/cm3 and 2.2g/cm3, and the surrounding fault fracture zone has the phenomenon of drilling fluid leakage, and the pressure plugging operation is carried out. The predicted pore pressure value of the lower undrilled formation is between 1.9g/cm3 and 2.02g/cm3. The bottom strength obtained by acoustic wave calculation is relatively high, and the collapse pressure is not greater than the pore pressure value. The lower limit of the drilling fluid density window in this depth section is the pore pressure value, and the upper limit is the minimum stress curve. The normal fault stress and strike-slip fault stress state of the minimum principal stress curve represent the minimum horizontal principal stress. The stress state of reverse fault mainly represents the pressure coefficient of overlying rock. Combined with the prediction, the pore pressure distribution curve was obtained. During the actual drilling operation, the corresponding gas display was obtained by the gas logging in this depth section, which fully proved that the pore pressure at this position was close to the drilling fluid density, indicating that the predicted pore pressure distribution curve was very accurate.

Table 1: Formation pore pressure prediction table for well X1

Well depth formation	Interval velocity (m.s^{-1})	Acoustic time (μs.ft^{-1})	Rock density (g.cm^{-3})	Equivalent density (g.cm^{-3})	
1200 Tower group kosi	2000	170	210	0.8	
1800 Sha wan group	3400	150	220	1.0	
2400 Anjihaihe Formation	3500	200	200	1.9	

4. Conclusion

This paper mainly based on three dimensional stress field, extracted from seismic data volume of interval velocity method, in view of the development well reservoir pore pressure prediction calculation, based on the analysis of the process of 3 d geostress in single well ground stress results as an input parameter, the real drilling and drilling fluid density curve as a calculation, calibration of geostress component values of key reference data. The drilling fluid density curve is used to calibrate the value range of pore pressure indirectly to further improve the accuracy of pore pressure prediction results. In addition, this paper a block X1 well drilling practice application of the new method of pore pressure prediction, the prediction of pore pressure and drilling fluid density window, and the existing drilling section of drilling fluid density data of drilling engineering information related to material were analyzed, obtained the same between measured and predicted pore pressure and drilling fluid density, It is proved that the pore pressure prediction method proposed in this paper is reasonable and feasible. Due to the relationship between the tectonic stress system and the local anticlines, the value of the tectonic stress system changes with the change of the location in the space of the three-dimensional block, which increases the accuracy of pore pressure prediction. It needs to be studied and improved continuously in the calculation and analysis.

References

1. Dong Pinghua, Zhang Lei, Lin Hai, et al. Unconventional Oil and Gas, 2020, 7(2):7-7.
2. Li Wentuo, Li Zhong, Wu Tao, et al. Journal of Chongqing University of Science and Technology: Natural Science Edition, 2020, 22(2):5-5.
3. Shi Mengxuan, Liu Zhizhi, Yang Xuefeng, et al. Review and prospect of geophysical logging prediction techniques for formation pore pressure [J]. Progress in Geophysics, 2020, 35(5):9-9.
4. WANG Yi. Accurate calculation method of formation pore pressure in fracture zone of shale gas reservoir [J]. Petroleum Drilling Techniques, 2020, 48(3):6-6.
5. Chen Ming, Yu Liwen, Qin Rui, et al. A new method for predicting pore pressure in shale formation [J]. Complex Oil and Gas Reservoirs, 2021, 14(3):5-5.

https://doi.org/10.1051/e3sconf/202337501019

Information security protection technology of station and depot industrial control system of oil production plant

Tianze Liu

The Sixth Operation Area of the NO.2 Oil Production Plant of Daqing Oilfield Co., LTD. Daqing Heilongjiang 163000, China

Abstract: With the rapid development of Internet technology, as well as the development of informatization and industrialization in oil production enterprises, industrial control system of information security protection technology can make the efficiency improved, but once the invasion by the virus, can cause the control system failure, pressure, temperature, flow, liquid level, such as measuring index will appear problem, chain alarm detection system failure, Or the solenoid valve and power supply of various instruments are not supplied. Once a major hazard source is out of control, it will cause serious accidents and endanger the safety of operators and the ecological environment.

Key words: information security; Industrial control system; Protection technology.

1. Introduction

With the rapid development of industrial information, industrial control system and public network, office network, Internet and other forms of connection, making the industrial control system from a relatively closed, isolated state to the transition of interconnection. As the traditional DCS, PLC, RTU and other industrial control systems are increasingly open, general, standardized, so that the Trojan horse, virus and other security problems in the enterprise's control systems continue to spread. Stuxnet in 2010, Night Dragon in 2011, Duqu, Nitro, and Flame in 2012 all illustrate information security issues in industrial control systems.

2. Current situation and characteristics of information security of industrial control system in oil production plant

2.1 Status quo of information security of station and depot industrial control system of oil production plant

In the early stage, the control system of oil production station is designed for closed special environment, which is only suitable for the production of single machine or process section, and there is no network with the outside world. In the design, implementation and deployment of the system, usability, functionality and real-time as the main indicators, network attack, information security and other aspects are less considered.

With the in-depth combination of computer, network technology, especially information technology and industrialization, industrial control systems increasingly use general protocols, hardware, software, etc., so that their openness is constantly improved, but also under the threat of the following aspects: (1) malware attack; (2) major industrial control processes are destroyed; (3) Improper operation of industrial control data: (4) Illegal access to the function of industrial control system.

At present, most of the industrial control systems used in oil production plants were developed many years ago, and most of them use special software, hardware and communication protocols. They only pay attention to their performance and reliability when designing, but ignore the demand for safety measures. With the application of Internet technology in industrial control system more and more widely. Industrial control systems also have many weaknesses that make them vulnerable to attack. The safety defects of industrial control system and their influencing factors are as follows: the system safety strategy and management system are not perfect; The proliferation of common protocols and applications; No antivirus software is installed on the console of the control system, or the virus library of the antivirus software is not updated in a timely manner. The introduction of general Ethernet technology makes the industrial control network face the threat of Ethernet communication. Many hardware platforms that use Windows PCS as workstations for engineers are as compromised as general PCS.

https://doi.org/10.1051/e3sconf/202337501019

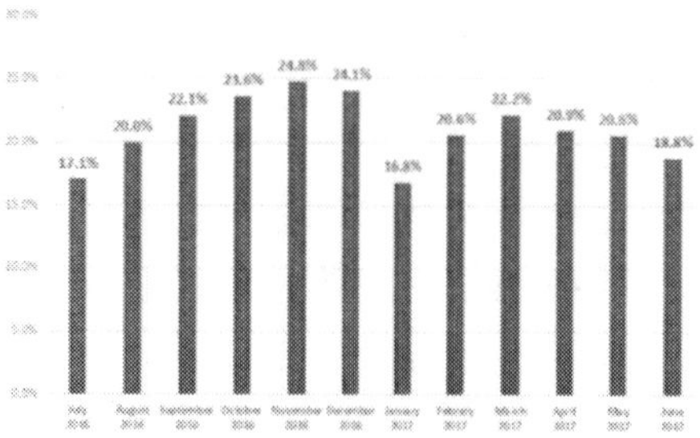

Fig. 1 Safety threat of industrial control system in oil production plant

2.2 Characteristics of station and depot industrial control system of oil production plant

(1) The system has good closure. Industrial control system is an important business subsystem in the management network of oil production enterprises. It involves many important technologies and sensitive information in the production process, so it can only be properly disclosed.
(2) The system needs high ease of use and real-time performance. Industrial control systems must ensure sustainable use. Stability of system access. System characteristics. Safety protection techniques for special industrial control systems. At the same time. Most real-time reaction times are within 1 millisecond [1].
(3) Industrial communication protocol and TCP/IP communication protocol. There are two kinds of protocols in industrial control system, one is industrial control protocol and the other is TCP/IP, when industrial communication protocol and TCP/IP coexist. Some protocols are not designed with security requirements in mind, so security solutions must be targeted.
(4) The safety relationship between the work network and the industrial control system is closely related. With the deep integration of the two industries, more and more information is exchanged between the industrial control production network and the office network, and the interrelation between the two needs higher security.
(5) The system has been maintained for a long time. Industrial control system is closely linked with equipment, production, generally using DCS. DCS. DLL/0 embedded equipment, specifically for site control. In a long-term and stable industrial control system, if there is no certain impact on its normal production, it will not be easily adjusted and changed [2].

3. Demand analysis of industrial control system information security

Information security of industrial control system is a complex system engineering, involving technology, products and systems, but also related to the safety management of enterprises. Compared with traditional IT information security, the information security of industrial control system has its own characteristics. It is mainly reflected in the degree of attention to security. IT security is generally office-specific, starting with confidentiality, integrity, and ease of use.
The goal of safety protection is field DCS, PLC, real-time network communication, industrial control applications, the main features are as follows:
(1) To ensure 24/7/365 availability, continuous operation and access to industrial control systems must be ensured;
(2) Guarantee the operation of the system;
(3) Ensure real-time data transmission;
(4) whether the system and data are complete;
(5) Adopt open standards to achieve tankless communication and functional interaction;
(6) Use common components as the basis for automated solutions;
(7) In order to ensure the real-time monitoring of production, communication between the office network and the production network is not interrupted;
(8) Prevent wrong operation and intentional damage;
(9) Protection of patented technology;
(10) Safety record and change management, etc.
The information security of industrial control system must first ensure the high availability of the system, then the integrity, and then the confidentiality. The main requirements for information security of industrial control systems are as follows:
(1) Improve the design of the project. Construction of the industrial control system is currently underway. In the design of the system, there are also many uncertain security risks, such as: which process equipment needs to be installed separately, which equipment needs to be installed to strengthen the information security of the third party, these are empirical judgments, there is no standard answer.
(2) Improve the design of the whole system. The quality of industrial control system also has advantages and disadvantages. In the early stage of the project, it is necessary to go through a series of design, installation and debugging, which are completed by the supplier. Supplier's technical level and experience. The degree of perfection of the system obtained also varies. The difference in the participation of user engineering designers and related professional and technical personnel will also have a certain impact on the quality of the system and produce certain security risks.
(3) Improve the daily operation and maintenance of equipment. After the completion of the industrial control system, it is necessary to carry out daily maintenance, regular inspection and a series of work to make it continue to play a role. The quality problems of these works will cause the risk of Angin.
(4) Information assurance measures need to be taken. The traditional industrial control system is generally closed, and the function is relatively single. With the advancement of technology. The control system is gradually opening up, making the connection between industrial equipment more convenient. Due to the lack of corresponding security countermeasures against viruses and hackers, it is particularly important to establish systematic information security measures.
(5) New challenges need to be faced. With the acceleration of the two integration process, the emergence of new computer technology and network technology,

industrial control system through a variety of interface technology, the realization of industrial control system network structure. With the new situation of industrial control system information security, it is urgent to formulate industrial control system network planning and specification, in order to ensure the sustainable development of industrial control system.

4. Industrial control system protection strategy construction

4.1 Overall architecture design of defense policies

The protection strategy of the industrial control system of oil production enterprises can be divided into: safety protection of production management, production monitoring and on-site equipment protection; Based on the design ideas of boundary control, host monitoring and internal audit, the multi-level protection of the whole industrial control system is realized. It also unified logs at all levels, analyzed and evaluated the information security situation of the automatic network, realized real-time monitoring of the Internet of Things, and effectively dealt with it [3].

4.2 Security Protection at the Production Management Layer

Three industrial control firewalls are set between the DCS control cabinet and the PC host in three processing stations, and three industrial control firewalls are set between the DCS of three processing stations and the switch of the automation center in the work area. In the SCADA system of the production monitoring center, a set of industrial control firewall is configured, and it is composed of the central control switch. DCS and SCADA systems can be protected in the area to avoid data sensitivity and leakage. By using the whitelist function on the industrial control firewall, you can actively protect the security software or programs in the industrial control network, thus ensuring safety and reliability. The firewall enables abnormal message detection and abnormal traffic detection to implement passive defense and prevent land attacks such as eardrop. ping deah attacks.

4.3 Security Protection on the production monitoring layer

Unified management of all industrial control host protection software clients in the oil production plant, status monitoring of each system, policy allocation and release, collection, summary, update and synchronize the white list data of each independent customer, unified audit of each customer, and data analysis [4].

In industrial control client hosts configured on host protection, and carries on the unification of protection, established a complete set of security maintenance plan, strategy and planning, to strengthen the daily safety training, strict control of process, prevent the internal attack, achieve the purpose of protect the terminal, is not going to solve the technical problem in view of the present, take effective management measures. It effectively improves the safety of the industrial control system. The information security of industrial control system is a dynamic process, it must run through the whole life cycle of industrial control system, and constantly improve and improve.

4.4 Onsite Device layer Security protection

The normal section can be read on a computer without an industrial control system; It can be viewed and used normally only after the host is installed with the industrial control host reinforcement and the corresponding policies are enabled at the same time, so as to achieve "regional exclusivity" [5]. Because of the use of high-density commercial encryption algorithm, it can effectively prevent data leakage. Security USB flash drive and industrial control host enhance the management of mobile storage media and their own security, can effectively prevent Trojan horses, viruses and other intrusions into the artificial control host. To ensure the safety of the industrial control host. Strengthen the management of industrial control medium, it is forbidden to use U disk for data copy and transmission, only use the security U disk for data copy, to prevent the misuse of U disk caused by the virus into the industrial control network and spread, improve the security level.

5. Industrial control system information security protection solutions

Security work of modern industrial control system. Strengthen the construction of the company's security strategy system, strengthen the security awareness of employees, strengthen security protection from different levels, different areas, different functions, different priorities, through the deployment of multifaceted security policies and processes including physical security. Firewalls are configured to protect different security devices. VPN technology is used to enhance communication between devices, patch management, configuration account management, and malware detection and protection to ensure the security of key industrial control processes and applications.

(1) Establish a boundary of physical protection: Physical security is a prerequisite for the information security of the control system. It first ensures that the system is in a controllable physical environment, and then establishes a security isolation system between the enterprise office network and the industrial production network [6].

(2) Safety strategy and procedure: Formulate a complete safety strategy and procedure. Set up the audit system between the industrial control network and the industrial control network, strictly implement the access control of the industrial control network, the operation of the audit engineer workstation,

(3) Set up safety monitoring device. Secure the boundaries of the system; It is divided into different Anjin areas according to its use and communication services, and isolation facilities such as firewalls and security gateways are set up at the junction of Anjin areas.

(4) Use VPN technology for communication protection of each control device: configure VPN in each security device to protect the communication of each unit, ensure the authentication, integrity and confidentiality of the communication, prevent illegal communication.
(5) Strengthening and updating: real-time updating and system strengthening of engineer station and application server. Only vendor-specific protocol data is protected, preventing all communications that do not comply with standards and legal requirements.
(6) Account management: establish industrial control applications, such as DCS, PLC, etc., establish account management, and strengthen the access control of passwords.
(7) Malware protection technology: not only in the industrial control network on the PC host virus monitoring, but also in its entrance to set the antivirus gateway.

6. Conclusion

Industrial control system information security is not only a technical problem, but involves the management process, architecture technology, products and other aspects, the need for management, operators, integrators, component suppliers and other multi-party efforts to improve the information security awareness of all personnel, fully aware of the importance of information security. To ensure the safe operation of industrial control system, it is necessary to continue to improve it.

References

1. Wu Jun, Zhang Xinzheng, Li Jun, et al. Research and application of information security protection technology for station and depot Industrial control System of Oil production Plant [J]. China Management Information Technology,2015,18(19):91-93.
2. YAO Xiangzhen, Zhao Zitong, ZHOU Ruikang, et al. Interpretation of National Standard of Information Security Technology Industrial Control System Information Security Protection Capability Maturity Model [J]. Automation Expo, 222,39(7):48-52.
3. NI Min, Fan Jing, Li Chenguang, et al. Review on Information Security Protection Technology of Industrial Control System [J]. Journal of Yunnan Minzu University (Natural Science Edition),2020,29(6):619-627.
4. TAN Xinzhang. Research on Information Security Protection Technology of Industrial Control System [J]. Computer Programming Skills and Maintenance, 2019(7):145-146,165.
5. LIU Yong. Research on network protection technology of industrial control system information security [J]. Digital User,2019,25(34):53.
6. Zhu Shishun, Huang Yibin, Zhu Yingfei, et al. Research on Key Technologies of Industrial Control System Information Security Protection [J]. Electric Power Information and Communication Technology, 2013,11(11):106-109.

 https://doi.org/10.1051/e3sconf/202337501020

Adjustment and tapping technology of water drive reservoir in late high water cut period

Peiyi Jia

Institute of Geology, No. 2 Oil Production Plant, Daqing Oilfield Co., LTD., Daqing, Heilongjiang 163000, China

Abstract: The implementation of oilfield development technology is closely related to geological conditions. Scientific and reasonable exploitation strategy is of great significance to improve oilfield economic benefits and technical utilization rate. At present, the development of oilfield is in the middle and late stage, and there are still many problems in technology, which need to be adjusted and innovated urgently. According to the geological conditions of oil fields and the current intensive exploitation technology, it is pointed out that in the late development of high water cut, it is an important method to improve the economic benefits of oil fields by using the potential tapping technology. In ORDER TO REALIZE THE TRANSFORMATION OF oil and gas exploration technology, THE key LIES in increasing research and development, accelerating technology innovation and application, and pushing oil field development into a new development track.

Key words: High water cut; Late water flooding oil; Adjust the latent technology

1. Introduction

With the increase of the number of Wells and the improvement of the extraction depth, the distribution of remaining oil is highly dispersed, and the potential benefit of each measure decreases year by year. The imbalance between storage and production is quite prominent, and the substitution ability of resources is poor. With the development of high water cut oilfield in late stage, the production decreases and the pressure of water cut increases gradually. The contradiction of large formation pressure difference is becoming more and more prominent, the structural adjustment is difficult, and the geological conditions involved in the development and analysis are becoming more and more complex, so the fine geological survey results and manual analysis can no longer meet the requirements. In this context, the oilfield put forward multidisciplinary integrated research, leading the field at home and abroad. Through research to guide practice, practice to transform knowledge, multidisciplinary research results reproduce history, quantify the distribution of remaining oil, and optimize the water drive adjustment program.

2. Study on geological characteristics of continental sandstone oil fields

First, continental sandstone oil fields are distributed in a wide range, mainly in fluvial sedimentary areas, delta areas and coastal areas. Based on known oil reserves, it is large, estimated to be several billion tons or more.

Secondly, the formation fluidity of continental sandstone reservoir is poor, the thickness of oil layer is larger, the number of sandstone layers in the reservoir is more, and its density characteristics are different. The oil in the reservoir is of great quality difference, but the overall viscosity is high, and its freezing point is low, the proportion is high, so it has great development value.

Thirdly, due to the low activity of the boundary water in the continental sandstone oil area, coupled with the lack of natural energy, the saturation pressure of the formation is low, which leads to the formation of the stingy cap.

Due to the improvement of oilfield development degree, the influence of regional environment, climate and other factors, its geological conditions have changed, and the sediment in the stratum has also changed, therefore, the analysis of the stratum of the oilfield, the adjustment of its technical program has a very important role.

3. Problems faced in the late stage of high water cut in waterflooding oilfield

Based on the analysis of the geological characteristics of the oil field and the actual production status of the oil field, it is proposed that the development of the oil field should be carried out in stages and layers, and the technical adjustment and improvement should be carried out in each stage. In the early stage of development, most of the basic well pattern is used, with large spacing between Wells and thick layers. With the development of oil field in the late stage, the water cut continues to increase, the water injection technology has been strengthened, and the simple enhanced water injection technology in the development of oil field also encountered many problems. First, as the water injection oilfields at later stage of high water cut, water cut sharply increases, the demand for water also will increase, therefore, the water injection and waterflooding are increasing year by year trend, this phenomenon in many Wells in daqing oil field are reflected, while take the corresponding measures to curb, but with the deepening of the development, the problem also intensified [1].

Second, the continuous injection caused great damage to the equipment. In the stage of high water cut, high pressure water injection and complicated non-underground operation are required, resulting in a large number of casing damage. In the long-term high pressure and frequent use of oil well casing, casing damage is increasing, and many casing are nearly to the limit.

Third, oil fields are becoming more expensive to develop. The first is the loss of equipment, which must be continuously maintained and replaced, increasing the demand for funds; Secondly, in the late development of high water cut, not only the water injection pipeline but also the oil transmission pipeline should be adjusted. Therefore, it is necessary to carry out a lot of surface engineering transformation and introduce the corresponding equipment. Thirdly, since the operation cost of oil field is closely related to water content, the effective control of water content must be established when the cost control is carried out. With the increasing energy consumption of oil extraction, through the investigation of oil fields across the country, it is found that the energy consumption per ton of oil is increasing, and the energy consumption is also increasing.

4. Development technology adjustment strategy in high water cut late stage of waterflooding oilfield

This article aims at our country's oil field entering the late exploitation demand and the existing problem, proposes two countermeasures to the perfection of the development technology system of current oil field high water cut.

4.1 Technical application of periodic water injection

In this paper, in the study of the late development technology of high water cut in foreign countries, a kind of technology, namely circulating water injection technology, is proposed, which can effectively improve the displacement effect and enhance the recovery efficiency. To adjust the reservoir fluid condition by periodic water injection is to inject water into the oil field according to the change of water content of oil Wells, so as to achieve the purpose of improving the recovery efficiency. Circulating water injection technology is the most effective method in the development of high water cut oil Wells, but it still needs further exploration and improvement. In addition, in the late development of high water cut, modern technological means should be reasonably used to improve the development process, such as gel and fracturing, so as to establish a supporting technical system suitable for the development of this region.

4.2 Adjustment of management mode focusing on single well management

In the early stage of oilfield development, we pay attention to the overall effect, so the technical scheme is based on blocks. This management way has a very good effect in the early stage. However, due to the scattered distribution of remaining oil in high water-cut oil Wells, and the remaining oil is mostly concentrated between Wells, it is necessary to pay more attention to the remaining oil reservoirs in the later development of high water-cut oil fields, and the well spacing will also change greatly, so it is necessary to focus on a single well. On the premise of improving the development effect of a single well, the technical scheme of a single well and well group should be adjusted to optimize the streamline of a single well and improve the recovery efficiency of a single well so as to improve the economic benefits of the whole block [2].

4.3 Research on technical limits of subdivided water injection

In order to effectively control high water-content invalid water injection, increasing the effective degree of water flooding, increasing the oilfield development, this paper segment segmentation boundary are studied, water injection and discusses the layer between the small number, thickness and interlayer inhomogeneity and development status, the connection between found in high water cut development stage, the permeability grade difference, the effect of Duan Youxiao thickness is reduced. Four indexes, including the number of oil layers, the coefficient of variation between layers and the thickness of sandstone, have significant influence on the utilization of oil and gas. Through analysis, the reasonable stratification parameters and various parameters within stratification are determined, which can be applied to different strata and blocks.

(1) In the longitudinal direction, reservoir injection conditions have the greatest influence on reservoir production. In THE WATER-INJECTION LAYER, THE LOWER THE NUMBER OF SHEETS IS, THE HIGHER THE USE DEGREE IS, AND THE FINER THE THEORY IS, BUT DUE TO THE LIMITATION OF THE WATER-INJECTION TECHNOLOGY, THE MORE REASONABLE NUMBER OF SHEETS CAN BE OBTAINED BY STATISTICS AND ANALYSIS. The linear regression method is used to study the relationship between water absorption of single sand body and single sandstone.

(2) On the plane, injection-production relationship is an important factor affecting reservoir production, and the role of reservoir can be fully played by increasing the number of interconnected Wells between Wells. In the design of the water-injection well area division scheme, the number of single reservoir, thickness of single reservoir and characteristics of reservoir should be optimized longitudinally, and the development of upper

https://doi.org/10.1051/e3sconf/202337501020

reservoir, sand body contact relationship, water flooding control degree, and injection-production relationship of well group should be fully considered to determine the reservoir characteristics and water injection intensity reasonably.

4.4 Water drive adjustment in the late stage of high water cut

4.4.1 Develop block adjustment potential

According to the quantitative analysis and numerical simulation of remaining oil, the vertical stratification standard of each layer is determined. According to the relationship between the overall production of single layer and the comprehensive water content, each layer is divided into strengthening layer, control layer and equilibrium layer, and the existing water content and the degree of exploitation are taken as the reference. In the plane, according to the use of the situation is divided into strengthening parts, control parts and balance parts, and determine to meet the conditions. According to the clear injection and production conditions of the oilfield, combined with the relationship between water cut and formation pressure, numerical simulation method is adopted to formulate the regulation principle of water flooding in the late stage of high water cut, that is, according to the requirements of "three extraction, three control and one stability", the corresponding adjustment of oil and water Wells is carried out [3].

4.4.2 Developing individual well adjustment potential

(1) Reasonable determination of the range of water injection regulation. In order to protect the casing, the limits of water injection regulation can be reasonably determined by statistical analysis based on the measured data. According to the limit that the annual variation range of rock pressure is 0.5 MPa, water injection is adjusted for different types of sand bodies. The water extraction capacity of the outer edge facies sand body is not more than 20%, while that of the inner edge facies and distributive plain facies is not more than 30%, and the control water volume is 15% and 25%, respectively. (2) Reasonably determine the lower limit of flow pressure. In order to avoid the degassing of reservoir, the minimum allowable flow pressure under different water cut and different pressure is given based on the flow man dynamic equation of oil well. (3) Reasonable determination of the scope of adjustment measures. The numerical simulation method is used to find out the indicators of formation selection, such as fracturing, acidification and water plugging, when the output rate of each measure well exceeds 2. According to the above criteria, combined with the single well dynamic, static data and measured data, the single well was adjusted.

4.4.3 Optimize the adjustment scheme and improve the development benefit

After the implementation of the potential, the production decline and water cut increase were strictly controlled from the adjustment of injection-production system and injection-production structure. In order to complement the difference between the measured conventional capacity and the preset index, the scheme is optimized.
(1) Optimization of injection and production system. Its main objective is to optimize the injection and production relationship of a single sand body, reorganize the well pattern under the existing well pattern conditions, and use the numerical simulation method to select the scheme with the best, maximum exploitation degree, the lowest water cut and the highest input-output ratio [4]. (2) Optimization of injection and production structure. The adjustment and economic effect of different schemes are analyzed, and the best economic effect is obtained, which is optimized and optimized.

5. Evaluation method

In the late stage of high water cut, due to the influence of many factors and the complex relationship between them, the effect of each factor can not be accurately judged, and the overall effect can not be expressed by a single factor. On the basis of "fuzziness", the fuzzy comprehensive evaluation method is used to evaluate it.
(1) Mine statistical analysis. According to the recent statistics of oilfield, at present, many reservoirs in the eastern area of our country have entered the high water cut later period, and at present, the natural decline rate of composite fault block reservoir is about 11.7%. Based on the above data and the current industry norms, the criteria for the natural decline rate of the integrated reservoir and fault block reservoir in the late stage of high water cut are obtained (see Table 1).

Table 1 Evaluation criteria of natural decline rate in the late stage of high water cut

Reservoir types	Natural decline rate %		
	Excellent	Good	bad
ready	≤10	10~20	>20
block	≤15	15~25	>25

(2) Establish the index. If m indices are selected for evaluation, then these indices can be constructed as a set of indices. F={F1, F2, Fm}.
(3) Establish evaluation set. The evaluation set represents the final evaluation of several levels, such as A={A1,A,... Ak}, indicating that there are k evaluation levels [5]. If the rating is good, medium, or poor, k=3. The three levels were standardized, among which the positive score was 1, the bad score was 0, and the medium score was 0-1.
(4) Establish the weight set. There are many ways to determine the weight set, such as statistical method, analytic hierarchy process and so on. The AHP method is used to stratify the problem, establish a stratified model, and gradually distribute the weight from top to bottom. The weighted assignment vector for this set of factors is W = {W1, W2... Wm).
(5) Establish fuzzy matrix. Due to the numerous influencing factors, a single evaluation index can only reflect the local effect but not completely represent the

https://doi.org/10.1051/e3sconf/202337501020

overall situation. If there are n evaluation indexes, whose index set is F and evaluation set is A, then the evaluation matrix P=(rij) mxn, then (F, A, P) is the judgment space. If there are 12 evaluation indicators, then n=12.

6. Effect evaluation

(1) The water injection structure is improved. The WELL pattern DEVELOPED IN THE early STAGE HAS good utilization conditions, high water content and good water injection effect, which can be reduced by 7.83 percentage points. In ADDITION TO THE WELL PATTERN AFFECTED BY DRILLING PIPELINE, THE WELL PATTERN IMPLEMENTED IN RECENT YEARS CAN ALSO INCREASE THE WATER INJECTION BY 7.83 PERCENTAGE POINTS, SO THAT THE UTILIZATION RATE OF RESERVOIRS IS IMPROVED AND THE development BENEFIT IS OBVIOUSLY improved. In the pumping profile of the injection well, it is found that the stratum utilization thickness increases by 0.88 percentage points, the proportion of water-absorbing sand body increases by 0.88 percentage points, and the outer layer and inner layer increase the most [6]. (2) The liquid production structure was adjusted. According TO THE METHODS OF profile CONTROL, subDIVISION, periodic water INJECTION, WATER SHUtoff and fluid drawDOWN, THE CONTROL of profile control, SUBdivision, periodic water injection, WATER SHUtoff and fluid DRAWdown were carried out for more than 92% WATER-bearing reservoirs. For Wells with low water cut, fluid production can be increased by increasing water injection and taking measures to increase production, so as to effectively adjust fluid production.

7. Conclusion:

In the stage of high water cut, strengthening multidisciplinary research and promotion can provide reliable technical support for water injection regulation. The results can quantitatively evaluate the occurrence state of remaining oil in the late stage of high water cut, so as to determine the idea of water flooding control. It is a prerequisite for the optimal measure scheme to study the adjustment boundary suitable for water flooding measures in the late stage of high water cut. Multidisciplinary numerical simulation technology is an effective method to optimize water flooding adjustment scheme and improve mining efficiency.

References

1. LIU Lisha. Research on potential exploitation technology of water drive reservoir adjustment in late high water cut stage [J]. China Science and Technology Expo, 2015,0(14):183.
2. Research on water flooding technology of Daqing Oilfield in late high water cut stage [Z]. Petrochina Daqing Oilfield Co., LTD., 2001.
3. Feng Yonghua. A review of water-drive seepage characteristics and potential exploitation techniques in late high water-cut sandstone reservoir [J]. China Chemical Trade,2021(34):64-66.
4. Ding Shuping, Dong Qing, Ren Haizhen, et al. Fault block Oil & Gas Field,2009,16(4):100-101.
5. Zhang Yunyan. Study on multi-disciplinary comprehensive exploitation potential adjustment of high water cut reservoir in late development period [D]. Beijing: China University of Petroleum (Beijing), 2005.
6. YU Shanshan, Jiang Tao, Li Qirui, et al. Oil and Gas Well Testing, 2016,25(5):9-11. (in Chinese)

Influencing factors and measures of water drive oilfield development effect

Hang Gao

Institute of Geology, No. 2 Oil Production Plant, Daqing Oilfield Co., LTD., Daqing, Heilongjiang 163000, China

Abstract: Oil field is a very important non-renewable resource in China, which plays a very important role in economic strategy. In the process of oil field development, water flooding technology can effectively improve the production quality, provide a strong technical support for our oil field exploitation, and help to promote the sustainable development of oil field enterprises. However, there are many influencing factors in the exploitation process of water flooding oilfield, so the workers need to deeply explore the root cause of the problem from different angles and all aspects, and take scientific and reasonable countermeasures to improve the exploitation effect of water flooding oilfield. This paper mainly introduces the factors affecting the development effect of water flooding oilfield and puts forward the corresponding solutions. At the same time, combined with practical cases, the measures to improve the development effect of water flooding oilfield are analyzed in detail.

Key words: Water drive oil field; Development effect.

1. Introduction

With the rapid development of petrochemical industry, the demand for oilfield development is also gradually increasing, which leads to the increasing pressure of oilfield development. For oilfield enterprises, how to improve the exploitation efficiency of oilfield has become one of the urgent problems to be solved. In order TO improve the quality and efficiency of oil field exploitation, oil field enterprises should set up a professional exploitation team, scientifically evaluate the results of water flooding oil field exploitation, try to find out every influencing factor and take solutions, so as to create more economic benefits for oil field enterprises.

2. Factors affecting the development effect of water drive oilfield

2.1 Liquid withdrawal rate

The quality of oilfield development is closely related to the rate of fluid production. In general, the faster the rate of fluid production, the higher the quality of the oilfield. But in the process of collecting liquid, sometimes it will be affected by the water cut stage. For example, in the stage of low water cut, the influence on the liquid collection speed is relatively small; However, if the water cut period enters the middle and high stage, the collection speed of liquid will be seriously affected [1]. Moreover, when the water-cut period enters the ultra-high stage, the liquid collection speed will gradually decrease on the basis of the liquid collection speed in the middle and high stage.

2.2 Permeability

When the rate of water cut rise is slowing down or remains constant, if the permeability increases, the development effect will be enhanced. It can be seen that if the development effect of water drive oilfield is not very ideal, it may also be caused by permeability. The staff should make an accurate judgment based on the actual situation.

2.3 Dip

The formation dip Angle has a great influence on the rise of water content. Generally, the smaller the dip Angle is, the lower the obstruction of water content will be, leading to an increase in the rise rate, and it is difficult to improve the rate of oilfield development. According to the research analysis, when the dip Angle of the local layer is around 1.8° to 4.8°, it is better to ensure that the oil can be more effective in the ultra-high water stage.

2.4 Heterogeneity

China's land area is broad and terrain conditions are very different, so the geological conditions and terrain conditions in the oil area are very different, even the geological conditions are the same, sometimes the oil quantity inconsistency will occur. At this time, for the area with serious heterogeneity problem, if the permeability is not very strong, it will easily cause a lot of bad influence on the development effect of water drive oilfield.

3. Specific measures to solve the factors affecting the development effect of water drive oilfield

3.1 Improve the well pattern

Well pattern is one of the most important links in the development process of water flooding oil fields, which can better ensure the effectiveness of well pattern distribution and strive for more economic benefits for oil field development while ensuring the production efficiency [2]. Therefore, when using water flooding technology for oilfield development, workers should attach great importance to the perfection of well pattern. First of all, it is necessary to ensure that oil reservoirs of similar nature are combined in the same production layer as far as possible in the process of oilfield development, so as to ensure that the same results can be obtained after using water injection method. Secondly, it is necessary to ensure the stability of oil field production speed. At this time, it is necessary to ensure that each independent reservoir has the minimum reserves. By constantly improving the rationality of well pattern distribution, the exploitation area of the oil field is improved, so as to better guarantee the production quality of water flooding oil field. Finally, the staff should strengthen the statistical analysis of the original well pattern, encrypt the development of oil and water Wells whose economic benefits are not very obvious, improve the speed, and extend the service life of old oil Wells.

3.2 The old well is sidetracked

For the area with relatively high structure, it is very likely that the retention problem will occur. If effective methods are not taken in time to dredge, the later oilfield development work will not be smooth. At this time, the staff can sidetrack the old well, which can not only reduce the cost input, but also make reasonable use of resources. Sidetracking of old Wells is often used to reduce the impact of obsolete Wells on normal Wells. In particular, casing damage Wells and fallen Wells, which are difficult to be reused after repair, can be used as the main body of sidetracking, which can improve the production efficiency, enable them to be recycled, scientifically control oil reserves, and provide better service for people.

3.3 Using water flooding method to drive water washing mining

The recovery factor formula of the reservoir produced by water flooding method is ER=ESEd, where ER is the recovery factor, ES is the sweep coefficient, and Ed is the oil displacement efficiency. It can be seen from the formula that the sweep coefficient and oil displacement efficiency are one of the two important factors affecting the water flooding recovery efficiency. The workers can improve the recovery efficiency through various measures. However, the recovery factor is not an immediate parameter. The recovery factor can only be determined after the reservoir has been developed, so the displacement process must be considered. At this time, R0=ESEfEd, where R0 refers to the degree of recovery and Ef is the degree of washing. According to the formula, the degree of water washing is 0 when the reservoir is put into production, and the degree of water washing also increases with the deepening of the waterflood development process. When all the movable oil in the formation is driven out, the degree of water washing reaches 100%. $E_f = \frac{S_w - S_{wc}}{1 - S_{wc} - S_{or}}$, According to the formula, the degree of water washing is 0 when the reservoir is put into production, and the degree of water washing also increases with the deepening of the waterflood development process. When all the movable oil in the formation is driven out, the degree of water washing reaches 100%. According to R0 formula, it can be judged that there are three main methods to improve the degree of oilfield recovery, which are expanding sweep, deepening water washing and improving oil displacement efficiency. In this case, the water drive reservoir can be divided into three areas, namely, unwashed, weak washed and strong washed, as shown in FIG. 1, where the outermost red part is unwashed, the middle pink part is weak washed, and the innermost side is strong washed.

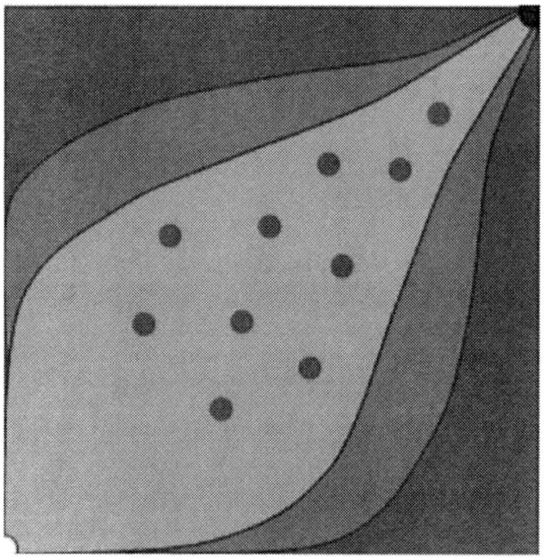

Fig. 1 Washing degree distribution diagram

The unwashed area refers to the area that has not been affected. The way to develop the remaining oil in this area is to expand the sweep and improve the sweep coefficient Es. In weak washing area, the water flooding time is relatively short, the degree of washing is also very low, and the remaining movable oil is more. The method of deepening washing can be used to develop the remaining oil and improve the degree of washing Ef. The water flooding time in the strong water washing area is long, the degree of water washing is high, and the amount of remaining oil is small, which is basically discrete oil. In this case, the recovery effect can be improved by improving the water flooding efficiency, that is, improving the Ed. Waterflood development of oil fields is mainly to increase the sweep, so as to improve the recovery factor.

3.4 Strengthen the operation effect of liquid extraction

To strengthen liquid extraction operation, the electric pump is mainly used to regulate the reasonable control of the production pressure, so as to reduce the decline rate of the oilfield and improve the oilfield production efficiency. When using this method, attention should be paid to selecting a flat place as the liquid extraction site and selecting an electric pump with appropriate power [3]. In addition, water with different temperatures can be used to complete water flooding. As shown in Table 1, water with different temperatures can affect the original oil saturation and oil displacement efficiency, which can be selected according to the actual situation.

Table 1 Effects of different displacement modes

Displacement way	Original oil saturation (%)	Residual oil saturation (%)	Oil displacement efficiency (%)	The final displacement (PV)
55°CWater drive	63.9	28.3	55.7	6.427
120°CWater drive	61.8	24.1	61.1	5.475
200°CWater drive	55.8	19.2	65.6	4.441
270°CWater drive	54.1	16.4	69.7	3.481

3.5 Reinforced cement to block

Due to the large difference of geological environment and not very uniform, it is easy to encounter collapse and damage in the process of oilfield exploitation, which will inevitably increase the difficulty of exploitation and even lead to safety accidents. At this time, the staff can pass the refined level of cement blocking method, so as to avoid the waste of oil, damage to resources, reduce the quality and efficiency of oil production. There are three main ways of cement sealing, the first is the use of expansion agent, the ring welding in the horn position, and then filled with expansion agent, and quickly add water, so as to play a blocking role. However, this method has high cost and will increase the economic burden of oilfield enterprises. The second is to drill holes in the cement cake surface, with drilling tools to squeeze it, until the surface of the mud, the quality of the way is very good, the disadvantage is in the process of sealing need to use a large amount of cement as a basis, increase the strength of work. The third is to bridge in the table set, the staff in the table set under the ring, and then perfusion small stones, so that it forms a new cement cake surface, the cost is low, and the operation steps are simple, but also more commonly used.

4. Actual case analysis

4.1 Geological situation

The fault block of an oilfield tends to the northwest. The western part is relatively open and the structure is very gentle with a dip Angle of 1.8°, while the eastern part is steep with a dip Angle of 4.8°. The main oil bearing zone of the oilfield is Sha3 to Sha2 member, with an area of 3.71 square kilometers and oil reserves of 1,808 *104 tons. The Wells range from -1765 m to -2120 m and can be divided into eight groups with an average thickness of 40 m.

4.2 Analysis of influencing factors of water flooding oil field

In this oilfield, there are four factors that affect the production effect. The first one is the influence of formation dip Angle. As shown in Figure 2, the smaller the formation dip is, the faster the water cut rises, the lower the recovery degree and the worse the recovery effect.

The second factor is the rate at which water is extracted. The faster the rate at which water is extracted, the faster the rise in water will be in a reservoir with a certain inclination Angle. However, for different water cut stages, the impact intensity also has certain differences. The impact is small in the period of low water cut, increases in the period of medium and high water cut, and decreases in the period of extra high water cut.

The third is the influence of permeability. Based on the actual situation of this oilfield, it can be concluded that the higher the permeability is, the higher the degree of recovery will be, the slower the rising rate of water cut will be, and the better the development effect will be [4].

The fourth is the influence of heterogeneity. This oilfield is seriously affected by heterogeneity, and the permeability of the same sandstone group can be tens of times different. Different sedimentary rhythm has a great influence on the effect of water flooding development. However, with the growth of time, the water flooding effect of medium permeability is the worst, while the water flooding effect of small permeability is very ordinary.

4.3 The measures

According to the analysis of reservoir distribution law in this oilfield, it is found that the remaining oil is mainly concentrated in the place with relatively stable structure, and there is a small amount of water-bearing remaining oil in the area without well control. In addition, there are no production Wells in the low level area of the western structure because of the shielding effect of cutting water injection.

The first point is to improve the well pattern. In the non-well control area in the west, a total of 5 Wells have been developed. The oil reservoirs are 37 meters deep in 12 layers, and the oil and water reservoirs are 76.5 meters deep in 23 layers. A total of 24,168 tons of oil was added. The second point is that lateralization of old Wells can be used to restore reserve control by lateralization of abandoned Wells in the stagnant flow area with high structure. For example, No. 61-58 well was scrapped previously due to casing changes and downhole falls. At this time, by using sidetracking, the workers encountered four oil layers with a total of 17.8 meters and seven low-flooded layers with a total of 32.1 meters. At the initial stage of production, 67 tons of oil could be pumped every day with a water content of 34%, resulting in a total oil increase of 16,384 tons.

The third point is to strengthen the liquid extraction. In the area with relatively gentle oilfield structure, the workers use the electric pump to extract liquid and reduce the pressure of the oil field to expand the sweep volume of water flooding. The initial oil production increased from 967 tons to 3,364 tons, with a cumulative increase of 17,462 tons, which effectively controlled the decline of the oilfield.

The fourth point is cement sealing. For the problem of water pipe channeling out of the fault block caused by the large number of oil bearing layers in the longitudinal direction and the low degree of low permeability thin layer, the workers can use ultra-fine cement to seal the borehole and then adopt heavy shooting measures [5]. A total of eight Wells were plugged in the field, with an initial oil gain of 56 tons per day for a cumulative oil gain of 7,452 tons.

4.4 Effect Analysis

In this oilfield, through the judgment and treatment of the factors affecting oilfield production, the daily oil production increased from 256 tons to 323 tons, the cumulative oil increase was 65,466 tons, the new geological reserves were 30*104 tons, the recoverable reserves were 47.3*104 tons, and the recovery rate increased from 49% to 52%.

5. Conclusion

The oil field is a very important resource for our country, in the oil field development process, water flooding is the very common and very effective way. The rational use of water flooding method can improve the oil exploitation efficiency and create more economic profits for the enterprise. However, in the use of water flooding technology, there are some factors that will affect the quality and efficiency of water flooding production fields. It is necessary for staff to conduct reasonable research and analysis on each factor and formulate corresponding measures to improve the production quality. According to the actual situation of the oilfield, the staff should make a comprehensive analysis and evaluation, put forward scientific solutions, and make more contributions to the steady development of the oilfield enterprises.

References

1. SUI Xin. Factors influencing development effect of water flooding oilfield and countermeasures [J]. Chemical Engineering and Equipment, 2021(04):80+70.
2. ZHAO Chongzhi, Tang Hai, Li Minglin, Yang Zhibin, Yi Fan. Analysis of influencing factors and treatment countermeasures of water flooding development effect in Block X [J]. Shandong Industrial Technology, 2019 (02) : 98-99:.
3. HUANG Rui. Principle and application condition analysis of periodic water injection to improve water flooding effect [J]. Yunnan Chemical Industry, 2019, 46(01):48-49.
4. Zhao Jun, Weng Dali, Chen Ping, et al. Laboratory experimental study on periodic water injection of heterogeneous reservoir in high water cut stage [J]. Applied Science and Technology, 2019, 46(06):92-95.
5. WANG Zenglin. Research and practice of stratified water injection technology in Shengli Oilfield [J]. Petroleum Geology and Recovery Efficiency, 2018, 25(06): 5-10.

 https://doi.org/10.1051/e3sconf/202337501022

Discussion on technical countermeasure of improving development effect of low permeability reservoir

Xiang Li

Institute of Geology, No. 2 Oil Production Plant, Daqing Oilfield Co., LTD., Daqing, Heilongjiang 163000, China

Abstract: Low PERMEABILITY RESERVOIR IS AN IMPORTANT MATERIAL FOUNDATION FOR STABLE PRODUCTION IN OLD AREAS, BUT DUE TO THE limitations OF reservoir physical properties, exploitation methods and other factors, the late production effect is poor, therefore, from the reservoir and development approach, the technical measures to improve oil recovery in low permeability reservoir are discussed. Based on the analysis of the factors influencing the development effect of double low reservoir and the geological characteristics of reservoir, from the viewpoint of seepage mechanics, this paper systematically classifies five indexes, such as main pipe radius, flow percentage, starting pressure gradient, clay mineral content and crude oil viscosity. The five-element combination evaluation method is used to optimize the low permeability point reservoir, and on this basis, favorable oil and gas accumulation areas are found. According TO THE DEVELOPMENT status of low permeability reservoir, reasonable injection-production network and well type are demonstrated to break the bottleneck of conventional low permeability water injection and improve the development benefit of low permeability reservoir.

Key words: low permeability reservoir; Reservoir grading evaluation; Reasonable well pattern; Natural gas flooding.

1. Introduction

In recent years, the development of low permeability reservoirs has become the main source of domestic oil resources, and the proportion of oil production is increasing year by year, its importance is increasingly prominent. With the development of low permeability oil field, China has already formed the situation of low recovery rate and low recovery rate at present. Combined WITH THE TYPICAL BLOCK, THE EXPLOITATION STATUS AND THE PROBLEM OF "DOUBLE LOW" OIL FIELD ARE IMPLEMENTED, THE TECHNICAL BOTTLENECK OF "DOUBLE LOW" OIL FIELD IS carried out, THE IN-DEPTH research IS carried out, THE practical AND feasible development methods are formulated, and THE injection and production well pattern is constantly improved. A set of technical means for economical and efficient utilization of low permeability and double low permeability reservoirs to improve ultimate recovery is established, and the development plan of low permeability and low efficiency blocks is guided and completed, which provides a reference for the efficient development of similar low permeability reservoirs.

2. An overview of low permeability reservoirs

2.1 Low permeability meaning

It turns out that "low permeability" refers to a reservoir with poor permeability and mainly refers to the concept of oil reservoir. At present, with the progress of industrial production technology, low permeability reservoirs have been widely used in low permeability reservoirs, low permeability reservoirs and so on. Because low permeability oil and gas fields have smaller Wells, weak fluid and permeability, reservoir void, low productivity and other problems, only through the reservoir transformation, to ensure the normal operation of oil and gas fields.

2.2 Division of low permeability reservoirs

At present, there is no clear definition and standard for low permeability oil field in the world, which is determined by each country according to its existing oil reserves and technical status, according to its existing crude oil reserves and technical status. In China, low permeability oil fields are divided into three types according to their average permeability according to the actual situation and production characteristics in China.

https://doi.org/10.1051/e3sconf/202337501022

Generally, the permeability of low permeability oil field is the largest, and oil Wells can basically meet the production requirements, but the production is not high, so fracturing measures must be taken to achieve greater stimulation effect. The permeability of ultra-low permeability oil field is relatively normal, and the water content is also relatively high, but in order to achieve normal recovery effect, it must carry out large-scale fracturing transformation. Such fields generally have no natural yield because their reservoirs are so dense, their fluid volumes are small, and their bound water content is so high that they are generally not exploited.

2.3 Characteristics of low permeability reservoirs

Low permeability oil fields generally have the characteristics of low abundance, low permeability and low productivity of single well. Compared with medium and high permeability oil fields, low permeability oil fields have low permeability, poor fluid characteristics, small pore throat diameter, and "differential pressure for starting production", which requires a lot of flow resistance and pressure. The second characteristic is poor continuity, because the spacing between Wells is generally too large, the development scale of sand body is not large, and the water flooding control is difficult. The third problem is that after the discovery of water in the low-permeability reservoir, the indicators of crude oil extraction and oil production have fallen sharply, which seriously affects the stable production of the oilfield. Thirdly, for the enrichment degree of low permeability reservoirs, fracturing and other stimulation measures must be taken due to low oil saturation, low production capacity and insufficient production capacity.

3. The main problem of low permeability reservoir development

The effect of water injection in low permeability reservoir is not obvious, and the influence is great. It usually does not appear obvious change until half a year after water injection. It is an objective law that oil production decreases with water content increasing after water injection. Medium - and high-permeability reservoirs have a higher yield index after water injection, which can be stabilized or reduced by increasing production. However, the fluid production index of low permeability reservoir will decrease sharply after water injection, and the fluid production index is about 40% when the water content is above 60%. At the same time, low formation pressure makes production difficult, resulting in increasing production and decreasing production in low-permeability Wells after water discovery.

The PROBLEMS existing in the development of water injection in oil field are that the formation pressure of injection well rises rapidly, the oil pressure and oil production decrease sharply, and finally the water injection, oil production, exploitation rate and recovery rate are all poor, resulting in no injection or production. In fact, these problems occur in all low permeability reservoirs, some of which are particularly severe, causing production to grind to a halt.

4. Study on countermeasures to improve the development effect

4.1 Study the distribution of fractures and geostress field, and reasonably arrange well patterns

4.1.1 Study the direction of fracture formation extension

One is to track natural fractures, and the other is to create new fractures in the rock, which are parallel to the direction of the maximum principal stress in the current stress field. If a transverse fracture can be formed, it will help expand the oil spill area and further increase the production of the well.

4.1.2 Make full use of reservoir numerical simulation technology

Through the collection and analysis of a large number of data, the mathematical model of reservoir is established, and the dynamic of reservoir is corrected, and the historical fitting is carried out, so as to provide a new method for the future dynamic of reservoir, and improve the economic benefit of development.

4.2 Reservoir protection technology should be applied to reduce reservoir damage

In order to reduce the damage of drilling operations to the formation, polymer amine salt mud system, underbalanced pressure drilling technology and shielding temporary plugging technology are used in the drilling process. The construction technology mainly adopts various construction methods, such as negative pressure perforation, reasonable use of well washing fluid in different construction stages, lower oil and gas reservoir protector, etc., in order to reduce the damage to the reservoir during construction. In oil well exploitation, suitable auxiliaries should be selected according to reservoir characteristics. In ORDER TO REDUCE RESERVOIR DAMAGE, VARIOUS MEASURES SHOULD BE TAKEN, SUCH as ANTI-SWELLING technology IN advance, fine treatment of water injection and modification of water injection.

4.3 Early water injection

Since the natural energy of low permeability oil fields is generally low, their elastic recovery and dissolution rates are also low, so early water injection and maintenance of formation pressure must be adopted to enhance the recovery. For fields with high elastic properties and abnormal pressure, recovery can be enhanced by delayed water injection and enhanced oil recovery. Through the investigation of low permeability oil fields in China, it is found that permeability coefficient and porosity decrease with the increase of overburden pressure, and this change is an irreversible process. In order to maintain the low pressure of the low permeability reservoir, the porosity and permeability of the reservoir are avoided to decrease sharply, and the permeability of the reservoir is good.

Table 1 Basic parameters of water flooding core

NO	The length of the core (cm)	Core diameter (cm)	porosity (%)	Gas logging permeability (x 10-3m^3) μ
1	3.20	2.49	11.94	0.263
2	4.10	2.50	12.50	0.500
3	4.10	2.49	13.20	1.560
4	4.15	2.49	12.70	7.500
5	3.16	2.48	13.70	10.80

4.4 Use fracturing technology to improve the productivity of low permeability reservoirs

Low permeability reservoirs have low natural productivity and usually cannot meet the requirements of industrial production. Therefore, fracturing is the core technology to develop low permeability reservoirs. The optimal design technology of "integral fracturing" has surpassed the conventional single-well oil augmentation and injection augmentation technology, which is a great development of hydraulic fracturing technology at home and abroad in recent years. It plays an important role in the whole oilfield development plan [1].

At present, the fracturing processes for low permeability reservoirs mainly include current limiting fracturing, ball-throwing multilayer fracturing, packer multilayer fracturing, CO2 fracturing, high-energy gas fracturing, composite fracturing technology, etc. After fracturing technology modification, reservoir permeability is mainly caused by fracture forming mechanism, and a number of long and narrow sand filling fractures are formed in the bottom hole area. Due to the opening of reservoir, the formation permeability is increased, the pressure and direction of reservoir fluid are changed, the injection-production relationship of well pattern is optimized, and the production of oil Wells is greatly increased. Practice shows that there are three factors that can affect the fracture of sand-filled body [2]. Permeability mechanism of reservoir fluid: The formation of sandstone fractures changes the reservoir near the bottom of the well from radial flow to unidirectional flow, and flows downward to the borehole through the fractures. In particular sedimentary facies bands, the fractures extend from the middle to the high seepage phase, or to the deep pay zone and natural fractures. Due to the plugging effect, a variety of factors such as entering fluid will be encountered in all aspects of drilling, oil testing, production, etc., resulting in formation plugging, especially in low-permeability reservoirs, which will lead to the decrease of formation permeability and thus production. The occurrence of fracturing fractures can not only alleviate the plugging in the near-well area, but also improve the permeability of the reservoir, so that it has a better stimulation effect.

4.5 Use acidification technology to improve the productivity of low permeability reservoirs

As an adjunct technology, acidification can increase the production of a well at a specific well location. The improvement of reservoir permeability by acidification technology can be divided into three types: conventional acidification, pickling and fracturing acidification. Acidification is through acid chemical dissolution and extrusion formation of certain minerals in the dung and formation, formation seam hole to expand, extension, communication or in formation with diversion ability of crack formation, so as to restore and improve the formation permeability, reduce the resistance of formation, increase the power of the injection Wells, achieve the goal of increasing oil and water Wells [3].

4.6 Pressurized water injection profile control technology to improve reservoir permeability

The causes of water injection difficulty in medium and low permeability reservoirs are very complex. In addition to the pollution and water quality in deep drilling, there are mainly two reasons: first, the formation condition is poor, the formation permeability is low, the porosity is small, and the pipe diameter is small; Second, for reservoirs with special high-pressure characteristics, the energy and pressure released in the process of opening the reservoir will lead to formation "plastic change", which is very unfavorable to water injection. In VIEW OF THE UNDERINJECTION WELL CAUSED BY THE ABOVE MENTIONED CAUSES, IT is difficult to obtain good results by adopting chemical transformation measures. After investigation, a new technology suitable for high-pressure water injection and profile control technology is put forward. The pressure can be directly raised to 40 MPa according to the requirement of 30 MPa water supply.

4.7 Using aerodynamic deep penetration augmentation technology to improve reservoir permeability

4.7.1 Mechanism of action

Put various reagents in the well bottom, will mix together all sorts of different drugs, the formation of strong chemical reaction, high temperature and high pressure gas formation, formed a huge crack in the strata, under the effect of repeatedly, the fracture will extend to the deeper, so as to improve the permeability of formation, to eliminate formation pollution [4].

4.7.2 Improve formation characteristics

This method is suitable for medium and low permeability reservoirs, and is also suitable for oil and gas production decrease due to the blockage of impurities such as asphalt gel and insoluble organic salt, and water injection pressure increase resulting in the reduction of water injection. The technology has a variety of functions such as fracturing, pyrolysis plugging, acid plugging, etc., which can eliminate the pollution of the reservoir near the bottom of the well, increase water injection and volume, increase production, improve formation permeability, and thus enhance recovery efficiency.

4.8 Well spacing test is carried out to explore reasonable development mode of low permeability reservoir

At present, the phenomenon that the water injection well does not inject water and the high-pressure zone is formed is common in low permeability oil fields. Because the oil well is in the low pressure area, it cannot be exploited, resulting in the production of the oil field into a passive, even into paralysis. To solve this contradiction, the key lies in the appropriate reduction of well spacing and the reasonable increase of well pattern density. Only in this way, can a set of effective driving force system be formed to ensure the effect of water injection in the oilfield, ensure its stable production, and increase the oil recovery. When laying the development well pattern, it is necessary to take the whole reservoir as the research object, fully consider the match between in-situ stress direction and hydraulic fracture, and make full use of the potential of hydraulic fracture to achieve the purpose of efficient well pattern mining [5].

4.9 Main Application Technologies

In the early stages of development, the formation pressure is high and the production can be elastic, so the traditional method can achieve better production. During the middle and late stages of development, as the formation pressure decreases and the average fluid level of the well decreases, the application of deep pumping technology can increase the recovery efficiency, thereby increasing the economic value of the reservoir.

4.9.1 Supporting technology of deep well pump

Rod PUMP DEEP PUMPING technology, equipped with speed regulating motor, can be IN the production process, according to the change of moving liquid level to adjust the flushing, determine a reasonable working mode, coordinate the relationship between oil supply. The use of long stroke pump deep pumping process, can increase stroke, reduce stroke, so that the production pressure difference is stable.

5. Conclusion

Reservoir protection is very important to improve the early productivity of low permeability oil Wells. Reasonable well pattern design is the premise of achieving good results, and appropriate supporting technology is also the key to ensure the stability of low permeability oil fields. Integrated fracturing, acidification and other techniques are the main means of developing low and medium permeability reservoirs. In oilfield DEVELOPMENT, IT IS NECESSARY TO comprehensively CONSIDER FRACTURING, acidification, injection addition, aerodynamic deep penetration injection, AND well pattern and water injection, so as to realize the comprehensive utilization of the oilfield. In medium and low permeability reservoirs, high pressure water injection and fine water quality treatment should be strengthened to ensure the normal operation of injection Wells, timely replenish formation energy, and increase the formation pressure of the block. No matter in low permeability oil reservoir or using aerodynamic deep pressure technology, the plugging of oil well can be realized.

References

1. Yao Jun, Ge Haijiang, Yang Dazhong, et al. Discussion on development technology of low permeability reservoir [J]. Inner Mongolia Petrochemical Industry,2008,34(11):97-99.
2. Zhu Weijun. Study on remaining oil distribution and potential exploitation in low permeability reservoir with high water cut stage [D]. Shaanxi: Xi 'an Shiyou University,2013.
3. Ma Chao. Study on reservoir sedimentary facies and reservoir development characteristics of Upper fourth Member of Shahejie Formation in Wangjiagang Oilfield [D]. Shandong: China University of Petroleum (East China),2011.
4. LEI Chuanling. Research on technical countermeasures to improve development effect of low permeability reservoir [J]. Science and Technology Innovation, 2020(31):18-19.
5. Shi Xiuxiu, Yang Rongrong. Technology and application to improve waterflood development effect of low permeability carbonate reservoir [J]. Chemical Engineering Design Communication, 2017, 43(11):255.

Analysis of reservoir geological characteristics and development countermeasures

Hong Zheng

NO4. Oil Production Plant of Daqing Oilfield Company Limited, Daqing Heilongjiang 163511, China

Abstract: Under the background of the continuous improvement of the scale of modern petroleum engineering construction in our country, the level of oilfield development technology has been improved in an all-round way, which provides sufficient resources for the development of society. In the process of oilfield development, it is necessary to do a good job in the analysis of the geological characteristics of the reservoir to clarify the basic geological characteristics of the reservoir, based on which the development of the oilfield can improve the scientific work of reservoir development. Therefore, it is necessary to strengthen the analysis of geological characteristics and strengthen the innovation of exploration and development technology. Therefore, this paper will conduct in-depth research and analysis on the geological characteristics of the reservoir and development countermeasures, and summarize some measures based on practical experience, in order to be helpful to the related engineering.

Keywords: Reservoir field; Geological features; Development strategy; Developing technology; Optimization measures.

1. Introduction

Oil and gas resources are the basic energy in social production, but the total reserves of oil and gas resources are insufficient, and the development of many oil and gas reservoirs is difficult. In order to ensure the sufficient energy of the society and the national energy security, we must do a good job in oil and gas reservoir development. Oil and gas storage of geological environment is complex and variety, and in the process of reservoir development, will continue to consume the energy, the development of the current way to complement formation energy, so in order to improve the efficiency of reservoir development, need to combine the basic conditions of reservoir geological characteristics, optimize the development mode, strengthen geological exploration of stored oil resources, Under the condition of accurately mastering the geological characteristics of the reservoir, it can provide sufficient guarantee for the subsequent reservoir development operation.

2. Analysis of reservoir geological characteristics

2.1 Geological characteristics of the

The reservoir geological characteristics of in certain oilfield, mainly for the delta facies, delta plain subfacies include distributary channel and distributary interchannel, including underwater distributary channel subfacies and delta front position, front mud, mouth bar, between tributary bay, etc., the small layer structure characteristics of different areas to study the reservoir with high similarity, in a good inheritance.

2.2 Reservoir characteristics

Through the exploration of the characteristics of the reservoir, it is clear that the reservoir is mainly composed of mineral cementation and particle cementation, and the arrangement of mineral cementation particles can directly affect the oil quality. In this reservoir, the overall sorting of the reservoir throat is poor, the structural quality of the throat is poor, has the characteristics of low permeability, the degree of weathering of the oilfield is medium, the overall particle sorting way is good, it is a high-quality reservoir area. In the exploration of the reservoir, the reservoir of the reservoir thickness of about 13.8 km, 0.68 MD, permeability reservoir in the region is suitable for the low permeability limited mining technology, through scientific optimization of recovery techniques, effectively improving the quality of oil field development, in the discovery in the process of oilfield development, however, there are water crack problems during water injection, Therefore, in order to give full play to the advantages of low permeability development technology, it is necessary to optimize the development technology based on the geological characteristics of the reservoir to ensure the improvement of reservoir development efficiency and quality [1].

3. Analysis of difficulties and influencing factors in reservoir development

3.1 Difficulties in reservoir development

Because the reservoir lithology density, seepage resistance is larger, and the characteristics of the pressure conduction ability is poor, has its own rimland flexibility is poor, lack of natural resources, the characteristics of leading to a low permeable reservoir has a larger penetration resistance, so in the process of the special low permeability oil reservoir energy consumption too fast, need a lot of energy in the mining, In addition, with the development of ultra-low permeability reservoir, the overall productivity of the oilfield will gradually decrease. Compared with other oilfield types, the attenuation rate of ultra-low permeability reservoir is faster, so the exploitation effect of ultra-low permeability reservoir is poor, and it is difficult to maintain stable exploitation. Which is a super-low permeable lithologic reservoir, the reservoir belongs to the typical low pressure and physical property is poor, because the oil region development scope is larger, ready for a large-scale oilfield, on the longitudinal development of many sets of reservoir, the plane has good sand body connectivity, reservoir scale to large, early try oil yield is higher, producing fluid volume dropped from 5.90 m3 / d in the early to the current 2.07 m3 / d, The water cut increased from 5% at the beginning to 30% at present, and the daily production of a single well decreased from 4.10t /d at the beginning to 1.20t/d at present. The reservoir was developed for a long time, most of the reservoir was injected later, and the well pattern was locally imperfect, resulting in a rapid drop in formation pressure.

3.2 Analysis of influencing factors of development

In the development process of this reservoir, the design adopts the water injection development mode, but in the actual development process will be affected by a number of factors, including the following: (1) water quality factors. Water quality is a key factor in the application of water injection development technology. If the water quality is poor, it will have a direct impact on the exploitation. However, in most low permeability reservoirs, due to the corresponding water quality problems, the production work is more difficult. During the water injection recovery period, the water quality is not up to the standard. There are many impurities in the water, which will have a great negative impact on the downhole pipe string and reduce the water absorption effect of the reservoir, which is not conducive to the smooth implementation of the exploitation work. (2) Water injection pattern factors. In the process of low permeability oil reservoir, in order to guarantee the production efficiency, improve to the requirement of mining technology, the traditional pattern of water injection is difficult to achieve good results, there exists a problem of energy shortage and low efficiency is, so you need to do optimization of water injection mode, but the current unreasonable parts used in oil field water injection mode, make the efficiency of low permeability oil reservoir is low [2]. (3) Test factors. In low permeability reservoir development process, the water injection development technology can reach the effect, directly affecting reservoir development effect, so usually need to do a good job of testing water injection development technology, provide data support for the subsequent application, but some early test of field work and test the lack of management, the cause is not in conformity with the test results and the actual situation, Thus, the subsequent water injection recovery effect is insufficient, so it is necessary to strengthen the preliminary test optimization of water injection development technology.

4. Analysis of reservoir development strategy

Can clear the combination of the above analysis, the characteristics of the reservoir with low permeability, so need to be combined with geological characteristics, with the development of science and technology, to ensure the quality of the overall development work, to strengthen the analysis of geological features, in accurately on the basis of geological characteristics, through the optimization of the development process, can effectively improve reservoir development efficiency and quality. Therefore, this paper combined with relevant practical experience, according to the basic characteristics of the reservoir, summed up a number of development countermeasures.

4.1 Technical feasibility analysis

In the process of the reservoir development, combined with the feature of the basic characteristics of reservoir, water injection development technology, it is proposed the main on the basis of developing unit division, for the separate development of the reservoir, reservoir boundary tracking, design reasonable injection-production well pattern, improve the reservoir sand body, injection-production relation between optimization of injection-production parameters, overall waterflooding reservoir. With the help of natural energy, the ultimate recovery factor is relatively low, while water injection artificially replenished energy to maintain reservoir pressure has the advantages of sustainable high and stable production, high displacement efficiency, relatively high ultimate recovery factor and good economic benefits. In this block, water injection Wells decline more slowly than Wells without water injection, and currently produce significantly more than Wells with natural capacity. Therefore, it can be seen that the waterflooding development method of low permeability reservoir is an effective method. In this reservoir, the content of reservoir intercaulk is 10-15%, and the intercaulk is dominated by clay minerals such as chlorite and kaolin, while other minerals are relatively few. Therefore, there will be no minerals that expand and clog the formation when exposed to water, and the reservoir sensitivity is weak. Thus, it provides a good

 https://doi.org/10.1051/e3sconf/202337501023

application condition for water injection development technology [3].

4.2 Application design optimization of water injection development technology

The reservoir is the main problem is the local injection-production well spacing is not perfect, because of the large longitudinal reservoir development result in injection-production corresponding, to finish Good injection-production well spacing, increase the well by the effect direction, continue to increase the intensity of oil well transfer, increase the well the effect direction, the formation of "weak point and strong plane" injection-production pattern, maintain formation pressure, increase the degree of water drive. Measures such as "shutdown, adjustment, union, diversion and blocking" are taken to further improve the injection and production well pattern and improve the corresponding degree of single sand body. Only by realizing reasonable injection and production well pattern in the plane and achieving water injection balance in each layer in the longitudinal direction can we ensure good and sufficient water injection and ensure scientific, effective and sustainable production operation of the oilfield.

The principle of water injection design is to give full consideration to the distribution of sedimentary facies, sand body and oil reservoir, to expand the affected area of water injection and improve the degree of water driving based on reasonable water injection pressure and the principle of balance of injection and production. In order to balance formation energy and prevent over-rapid flooding of production Wells, it is necessary to determine a reasonable injection-production ratio, gradually restore and balance formation pressure in the oil area, and determine the injection-production ratio in stages according to the recovery of formation pressure in the oil reservoir. The general principle is as follows: In the early stage of water injection, due to the serious formation energy deficit and low fluid production of the oil well, it is meaningless to design the water injection volume completely in accordance with the principle of balance of injection and production. On the premise that the water absorption capacity of the formation is sufficient and the oil well is not flooded, water injection volume should be as large as possible to restore the formation pressure at a fast speed. Therefore, the initial injection-production ratio is set as 1.7 ~ 2.0 compared with the experience of similar fields. To water injection later period, formation pressure level is higher, production pressure differential can enlarge appropriately to improve liquid producing capacity, the water injection main purposes for strengthening regional overall water cut rising velocity control, and prevent the well water flooded, so on the basis of lower injection-production ratio with water, the design of 0.8-1.1, in the whole process of water injection development, in order to make regional plane by equilibrium, The overall distribution of formation pressure is uniform, and the water injection in a single well must be adjusted according to the production dynamics. The general principle is that the intensity of water injection can be appropriately increased for the well groups with no or no obvious effect of long-term water injection, while the intensity of water injection can be strictly controlled for the well groups with Wells that are flooded or have high overall water cut [4].

In terms of water injection pressure adjustment, for the most part need to make sure that all reservoir or reservoir water at the same time, and ensure that ensure the injection-production balance or accomplish a certain production index as the prerequisite, ensure that water injection technology can achieve pressure range at the same time, to ensure that the injection pressure limit shall not exceed the fracture pressure of reservoir rocks, thus to realize optimization water injection pressure, It is the key to improve the technical level of water injection development and has a good effect in this reservoir.

4.3 Reservoir development quality control optimization measures

In the process of reservoir development, in order to ensure the application effect of water injection development technology, it is necessary to do a good job in technical application quality control, which mainly adopts the following control methods: (1) Optimization of water quality management. In the application process of water injection development technology, in order to improve the efficiency and quality of oil recovery, it is necessary to do a good job in water injection water quality management, otherwise it will have a direct impact on water injection oil recovery efficiency. With the development depth of low permeability reservoir gradually increasing, the difficulty of exploitation is gradually increasing. In order to ensure the development effect of low permeability reservoir, it is necessary to strengthen water quality management and reasonably deal with diversified and complicated water quality. For example, the backwashing frequency control technology of sewage treatment tank can be adopted. Through the advantages of frequency control and automation technology, the efficient treatment of sewage can be realized, so as to improve the water quality and avoid the negative impact on the exploitation of low permeability reservoir. In the application process of this technology, it is necessary to ensure the standard operation of the flushing pump and improve the accuracy of the pump running speed control during the operation of the system, so as to improve the effect of water injection oil recovery and reduce the influence of the complex environment of low permeability reservoir on oil recovery, which is an effective optimization measure for the application of water injection development technology. (2) Optimization of water injection development mode. Water injection technology in low permeability reservoir is affected by the model, so it is necessary to optimize the water injection recovery model according to the actual situation of low permeability reservoir. In the water injection oil recovery mode, the centrifugal pump is the core equipment. It is necessary to clarify the curve characteristics of the centrifugal pump, comprehensively optimize the equipment power, displacement and other technical parameters, and adjust the technical parameters according to the characteristics of low permeability reservoir to ensure that the operation efficiency of the centrifugal pump can be improved. Reciprocating pump as an important infrastructure during water flooding oil recovery, the efficiency will be affected by the influence of the pump valve, so you need to do the optimization of the overall system, to optimize the water injection pipeline, the reciprocating pump and integration, to ensure that it can have enough power, combined with the operation parameters of accurate calculation, can make the system running effect, Thus, the optimization of water injection oil recovery mode is realized, and the goal of improving the efficiency of water injection oil recovery is realized [5]. (3) Test management optimization. Effect to enhance early water injection oil production testing, it is necessary to do a good job of testing management, innovation for the test system and related technology, intelligent technology can be used, for example, through the intelligent technology to build perfect testing system, test system needs to be included in the underground test, the ground water injection test, etc., so that they can get more comprehensive test information, Realize the high efficiency transfer of test information. In ground water injection test system, intelligence technology is used to analyse the information and data collection, to ensure that the test system running continuity, so that they can get more accurate test results, improve test efficiency, make it meet the demand of the implementation of the low permeability reservoir water flooding oil recovery technology, can effectively improve mining efficiency and security.

5. Conclusion

To sum up, this paper briefly expounds the address of a reservoir characteristics and the basic situation of the reservoir characteristics, and the difficulties of reservoir development work and detailed analysis of the factors affecting technology application, and finally puts forward several development countermeasures and measures, hoping to have some reference for oilfield development and help enhance the technical level of reservoir development, Promote the development technology level of petroleum engineering.

References

1. Jia Zhiwei, Ban Lijia, Zou Yihu. Reservoir geological characteristics analysis and development strategy study [J]. Petroleum and Petrochemical Materials Purchase, 2022,000(016):2-2.
2. Liu Yao, Liu Xiao, Zheng Huimin. Analysis of geological characteristics of low permeability reservoir and discussion on development strategies [J]. Market Weekly • Theoretical Edition, 2020, 000(002):3-3.
3. Yu Xiyan, Xiao Lihua, He Nengxin. Discussion on reservoir geological characteristics and development strategies [J]. Science and Technology Wind, 2020(14):1-1.
4. Tan Tao, Guo Chen, Chen Yong, et al. Study and practice on Enhanced Oil Recovery Mechanism of high temperature and high pressure fracture-vuggy reservoir by N_2 flooding [J]. Reservoir evaluation and Development, 2020, 10(2):5-5.
5. Zhang Haifeng, Xu Qianwen, Zhu Hongmei, et al. Discussion on reservoir geological characteristics and development strategies [J]. Petroleum and Petrochemical Materials Purchase, 2021,12(002):1-1.

Adjustment of infill in transition zone to improve oilfield productivity

Xu Long

The Third Oil Production Plant of Daqing Oilfield Limited Company, Northeast Petroleum University, Daqing, Heilongjiang, China

Abstract. Affected by the high viscosity of crude oil, the large injection-production well spacing, the poor injection condition, and the low oil recovery rate, the overall development effect of transition zone is poor, besides, the low produced degree and the poor utilization of thin oil layer makes transition zone have some potential excavation. Under the condition that the current conventional measures to improve block development are not clear, it is necessary to further optimize the injection-production system. Based on the principle of "Well pattern rules, Well spacing optimization, Injection-production perfecting, Facilitation of adjustment", layer combination is thus optimized. This paper analyzes the effect of encryption adjustment and tracking adjustment.

Keywords: Transition zone; produced degree; encryption adjustment.

1. Introduction

The transition zone was put into development in 1971 and is currently divided into four sets of strata well pattern mining. The basic well pattern develops the main reservoir the first to the third strips which adopts 300m four-point well pattern; 250m inclined linear well pattern was used in exploitation of C Oil Layer in the fourth stripes which adopts in 1989. The area well pattern of 173m four-point method was used to encrypt and adjust the first to the second strip C and P oil layers in 1996. The well pattern of 250×250m five-point method was used to develop polymer in P formation in 2000, a series of strata complementary tests were carried out in the first to the second strip in 2010. With the method of filling holes corresponding to the encryption well and the foundation well, reducing well spacing to improve injection-production relationship of narrow sand body, improving the degree of water flooding control and multi-directional connectivity ratio, development effect of water flooding in transition zone is improved.With the continuous development of oil fields, the effect gradually decreases. Because the third and fourth zones are affected by factors such as high crude oil viscosity, large injection-production well spacing, poor injection condition and low oil recovery rate, the overall development effect is poor. The third and fourth zones have low recovery degree, poor production degree of thin and poor reservoirs, and there is a certain potential tapping space. Facing the current situation that conventional measures are not obvious to improve block development, it is necessary to further optimize the injection-production system.Based on the general principle of "well pattern rules, well spacing optimization, perfect injection-production and favorable adjustment", optimize the combination of strata series, reduce the well spacing of high viscosity reservoirs and maximize the recovery factor. In the third and fourth zones, new wells were drilled between rows of old wells and between wells, and the old wells were transferred to injection, forming a regular 150m five-point area well pattern. A total of 106 new wells (69 oil production wells and 37 water injection wells) were drilled, and 23 wells were transferred to injection. According to the numerical simulation results, combined with the current exploitation situation of old wells, the waterflooding development index after encryption adjustment is predicted. It is predicted that the recovery ratio of water flooding will increase by 3.08 percentage points, the geological reserves of the adjustment area will be 389.41×10^4t, and the recoverable reserves will increase by 11.99×10^4t.

2. Main problems in the development of transition zone

The third and forth stripes in the eastern transition zone of S Development Zone were put into development in 1971. The third stripes were mined by four-point method with 300m well spacing, and the fourth stripes were mined by oblique linear well pattern with 250m well spacing. At present, it has entered the stage of extra high containing water and the comprehensive containing water is as high as 96.0%[1]. Due to the high viscosity of crude oil and low well pattern density, the problems of low water flooding control, poor injection-production condition and

* Corresponding author: xulong1357@petrochina.com.cn

https://doi.org/10.1051/e3sconf/202337501024

low recovery degree are mainly manifested in the following three aspects:

Table 1. Single well production classification table of the third and fourth stripes.

Stripes	classification	Wells proportion	daily liquid production	daily oil production	water content ratio	flow pressure	submergence depth
the third stripes	<1	70.6	13.6	0.5	96.33	5.76	450
	2~3	29.4	75.5	2.6	96.60	5.41	444
	total	100.0	31.8	1.1	96.52	5.64	448
the fourth stripes	<1	80.0	7.1	0.3	95.62	3.87	193
	1~2	10.0	50.5	1.7	96.70	5.82	474
	2~3	10.0	63.2	2.2	96.50	5.08	345
	total	100.0	17.0	0.6	96.27	4.40	265
the third and the fourth stripes	<1	74.1	11.0	0.4	96.15	5.18	364
	1~2	3.7	50.5	1.7	96.70	5.82	474
	2~3	22.2	73.4	2.5	96.58	5.36	428
	total	100.0	26.3	0.9	96.46	5.27	390

Firstly is high viscosity of crude oil, large injection-production well spacing, poor water injection. In the eastern transition zone, the viscosity of underground crude oil in the third stripes is 20.9mPa.s, and the viscosity of ground crude oil in the fourth stripes is 40.7mPa.s, which is 32.6mPa.s higher than that in the pure oil zone. The third and fourth stripes are mined by well pattern with 250~300m well spacing. Affected by the high viscosity of crude oil and the large injection-production well spacing, the starting pressure of the reservoir is relatively high. At present, the starting pressure of the third and fourth stripes is 12.5MPa and 13.4MPa respectively, which results in the non-water absorption ratio of the third and the fourth stripes reaching 43.5%.

Secondly is the low well pattern density, the low degree of water flooding control. A set of strata development is deployed in the third and fourth stripes of the eastern transition zone. The third stripes are mined by four-point method with 300m well spacing, and the fourth stripes are mined by oblique linear well pattern with 250m well spacing. The well pattern density is only 14.3 per km^2, which is 26.3 per km^2 and 31.2 per km^2 lower than the pure oil area and the first and second strips respectively.

The degree of water flooding control and the proportion of multi-directional connectivity are low. The degree of water flooding control is 75.8%, 10.6 percentage points lower than that of the first and second strips, 18.2 percentage points lower than the pure oil area, while the proportion of multi-directional connectivity is 18.3%, 18.4 percentage points lower than the first and second strips, and 42.3 percentage points lower than that of the pure oil area. It is necessary to further adjust the well pattern to improve the degree of water flooding control and the development effect of the block.

Thirdly is low single well production, high containing water, high and uneven flow pressure level. The average daily liquid production of single well in the third and fourth stripes of the eastern transition zone is 26.3t, the daily oil production is 0.9t, and the comprehensive containing water is 96.46%, of which the proportion of oil wells with daily oil production less than 1.0t accounts for 74.1%. Among the third stripes, the average daily liquid production of single well is 31.8t, daily oil production is 1.1t, and the comprehensive containing water is 96.52%. The fourth stripes' average daily liquid production is 17.0t, daily oil production is 0.6t, comprehensive containing water is 96.27%. The flow pressure level of the third and the fourth stripes reaches 5.27MPa, of which the proportion of wells above 5MPa reaches 50%. The flow pressure level of the third stripes is 5.64MPa, while more than 5MPa wells accounted for 57.1%, the flow pressure level of the third stripes is 14.4MPa, while more than 5MPa wells accounted for 33.3%. The pressure imbalance between strips and wells is serious[2].

In conclusion, the overall development effect of the third and the fourth stripes of the transition zone is poor due to the high crude oil viscosity, large injection-production well spacing and poor injection-production conditions. Besides the adjustment of conventional water flooding is difficult. It is necessary to further optimize the injection-production system[3].

3. Potential for well pattern adjustment of the third and the fourth stripes

First and foremost, flooding data show that the thin and poor layers are less flooded. From the interpretation results of the new well flooded layer drilled in 2015 in the weak alkali ternary test area of the second type oil layer in the eastern transition zone, all the water has been seen, and the proportion of water flooded thickness reaches 99.28%, among which the proportion of low unflooded thickness is 7.63%, and the proportion of medium water flooded thickness is 40.94%. which means there is a certain adjustment potential. According to the flooding condition of different thickness oil layers, the effective thickness more than 1.0m oil layer has higher flooding degree, the ratio of low unflooded thickness is 2.73%, and the ratio of medium flooding thickness is 35.59%. The water-flooded degree of layer with effective thickness less than 1.0m is low, and the proportion of low unflooded thickness is 28.30%, and the proportion of medium water flooded thickness is 63.55%. From the different reservoir groups, the degree of flooding in the first group is relatively low, the ratio of low and no water flooded is 29.36%, the ratio of medium flooding is 60.61%, the degree of flooding in the CII and CIII groups is relatively high, the ratio of low unflooded is 4.83%, the ratio of medium flooding is 38.75%.

Secondly, numerical simulation results show that, recovery of the third and the fourth stripes is low, the total recovery is 28.0%, while recovery of CII is 31.1%, correspondingly higher. The remaining geological reserves still account for 69.7% of the total remaining geological reserves, which is the main potential tapping object[4].

4. Changes of indicators after layer adjustment

The third and the fourth stripes is poor due to the high viscosity of crude oil, the large injection-production well spacing, the poor injection condition, and the low oil production rate. At the same time, the recovery degree of the third and the fourth stripes is low, and the utilization

https://doi.org/10.1051/e3sconf/202337501024

degree of thin and poor reservoirs is poor, so there is a certain potential space. Faced with the current situation that the conventional measures to improve block development are not obvious, it is necessary to further optimize the injection-production system[5].

Based on the overall principle of "Well pattern rules, Well spacing optimization, Injection-production perfecting, Facilitation of adjustment", layer system is well optimized. The final adjustment scheme is that a regular well pattern of 150m five-point method area is formed by the combination of new drilling among the old rows, new wells, old wells and old diversion wells in the third and the fourth stripes. A total of 106 new drilling wells(69 production wells and 37 injection wells) and 23 diversion wells are implemented.

After encryption adjustment, the adjustment well area achieves the following effects:

First of all, the degree of water flooding control has been significantly improved. Water drive control degree of the third and the fourth stripes increased from 75.8% to 96.2%, and the multi-directional connectivity ratio increased from 18.3% to 66.8%, which increased respectively by 20.4 and 48.5 percentage points. Numerical simulation predicted that water flooding recovery increased from 32.74% to 35.82%, which increased by 3.08 percentage, while predicted recoverable reserves increased 11.99×10^4t[6].

Secondly, the water injection condition is improved significantly. At present, the daily water injection rate in the north of Eastern transition zone reaches 5184m3, of which the daily water injection rate in the third and the fourth stripes reaches 1468m3, which is 1050m3 higher than that before adjustment. The proportion of non-suction wells decreases from 43.5% to 19.6%, which is 23.9 percentage lower. The injection pressure of single well also decreases from 12.6MPa to 12.0MPa, which is 0.6MPa lower than that before adjustment.

Thirdly, both single well production and oil recovery rate increases. At present, the daily liquid production in the north of Eastern transition zone reaches 31.5t and daily oil production is 1.2t, which is 5.2t and 0.3t higher than those before adjustment respectively, and the water content ratio decreases by 0.23 percentage. Among them, the third and the fourth stripes' old wells increase oil by 0.1t and water content ratio decreases by 0.3%. From the point of view of production classification, the proportion of less than 1t decreased from 74.1% to 66.7%, which decrease by 7.4 percentage, and the proportion of 1 to 2t increased by 9.6 percentage.

Fourthly, the flow pressure level is well controlled and the distribution is more balanced. After the strata adjustment, the flow pressure levels of the third and the fourth stripes decreased from 5.27MPa to 4.33MPa. Among them, the flow pressure of the old wells in the third and the fourth stripes decreased from 5.27MPa to 4.32MPa which decreased by 1.04MPa. From the point of view of the strip, the flow pressure of the third stripes decreased from 5.64MPa to 4.49MPa, and the flow pressure of the fourth stripes decreased from 4.40MPa to 4.12MPa, indicating that the pressure difference between the strips is reduced. From the classification of single well flow pressure, the proportion of wells in the reasonable well area between 3~5MPa increased from 50% to 76.9%, which increased by 26.9%. The pressure difference between wells is gradually reduced, and the pressure distribution is more balanced.

5. Complementary tracking adjustment after hierarchy adjustment

In the first place, the adjustment of water injection wells in the adjustment area. A total of 32 wells were adjusted for the water injection well, the water injection volume increased by 88m3, the single well of the surrounding 53 wells increased by 0.11t, and the water content ratio decreased by 0.12 percent. Among them, 20 wells were subdivided, 37 layers were increased after adjustment, the daily water injection rate increased by 121m^3, and the single well oil production of 42 wells increased by 0.13t, and the water content ratio decreased by 0.14 percentage. In addition, four old wells were matched by the new-old water injection relationship, and the water injection in the old direction was controlled to 76m^3. The 11 production wells around increased oil by 0.05t individually, and the water content ratio decreased by 0.07 percentage.

The second is to take out methods to improve oil and water wells. In order to ensure inject smoothly, Acidizing was carried out in 15 wells when putting into production. In addition, in order to ensure enough water injection, 16 wells around were taken measures, including 2 old well acidified, 1 old well and 8 new wells fractured, which correspondingly adding daily injection to 28m3, 183m3 and 230m3.The average daily oil increment per well is 0.15t, and the water content ratio decreases by 0.18 percentage. When the new wells were put into production, 9 wells were fractured by limiting flow method. In addition, 24 wells' production parameters are reasonably adjusted, and 4 wells' pump setting depth are deepened for low submergence wells.

Third, through the above work, the oil production of new well increased from the initial 27t to the current 34.6t. water content decreased from 97.18% to 96.05%, with a decrease of 1.13 percentage points. The oil well opening rate has always remained above 90%.

6. Conclusion

First, physical properties of crude oil in transition zone is poor and the potential tapping effect of conventional measures is not ideal, reasonable well pattern adjustment can better improve the development effect of transition zone.

Second, timely subdivision adjustment and matching of old and new wells are needed to effectively slow down the rising rate of water content ratio in the area of strata adjustment.

References

1. Yu Yue, S.: Research on encryption adjustment scheme of three to four belts in eastern transition zone.

https://doi.org/10.1051/e3sconf/202337501024

Inner Mongolia petrochemical industry. Z1 , 139–141 (2015).

2. Wang Zedong. Study on four out-of-band expansion and encryption methods in the western transition zone of Daqing Sanan Oilfield . Journal of Yangtze University (self-edited edition) 14(23), 61–63 (2017).
3. Liu Wenhui, Xu Yongchang. On bio-thermal catalytic transition zone gas. Petroleum exploration and development 32(4), 123–125 (2005).
4. Hu Fangting. Su Jianwen . Lu Qiang. Zhongsheng Wang. Liu Peng. Tong Gang. Optimum selection of geometric dimensions of drill pipe thickening transition zone. Oil field machinery 36(5),. 77–79 (2007).
5. Liu Wenhui. Geochemical characteristics of natural gas in transition zone. Geochemistry 24(4) ,11–13 (1995).

Analysis of development status and adjustment method of a block

Hongzhi Han

Geological room of Geological Research Institute of the fifth oil production plant of Daqing Oilfield, Daqing, China

Abstract. Since a certain block was put into development, it has always put improving the injection production relationship and improving the ultimate recovery factor in the first place. Through a series of infill adjustment, supplement and extended drilling, the injection production relationship of single sand body has been gradually improved. In high water cut period, the remaining oil distribution has been quantified through joint multidisciplinary research, and the fluid production structure has been adjusted by combining the oil well injection, hole filling, drilling and other measures, Hard and stable production in high water cut stage is realized. However, due to the small scale of sand body development and fault development, the contradiction between single sand body injection and production is still prominent. From the perspective of development effect of Gaotaizi old area, compared with Xingnan development area, the degree of recovery at the same water bearing stage is obviously low, and the development effect is poor. At present, the comprehensive water cut has reached 92.77%, and the recovery percentage is only 28.05%. The research shall be based on the results of fine reservoir description, guided by the results of multidisciplinary research, and focus on the comprehensive analysis and sorting of the injection production relationship of single sand body through the re understanding of geology, so as to find the adjustment potential of supplement and injection conversion. Through the adjustment of the injection production system, increase the water injection well points, expand the swept volume, improve the degree of water drive control, so as to increase the formation pressure and improve the development effect of the block.

Key words: water drive control degree; Current situation of injection production; Single sand body.

1. Background and purpose of the thesis

An oilfield is a heterogeneous oilfield of continental fluvial delta sedimentation, mainly composed of straight distributary channel sand bodies. This type of sand body has small development scale, low drilling rate, strong heterogeneity on the plane, and low well pattern control. Due to the small scale of sand body development, fault development and imperfect injection production relationship of single sand body, the contradiction is prominent: Problem 1 is affected by the development characteristics of sand body, the contradiction between imperfect injection production relationship is prominent, with a large decline, and the adjustment is difficult; Problem 2 The formation pressure level of a certain block is low and the development effect is slightly low. Therefore, it is necessary to analyze the development status of a block and study the adjustment potential, optimize and implement the potential of the adjustment plan, and predict the single well productivity and effect. In order to increase the water injection well points, expand the swept volume, improve the water drive control degree, increase the formation pressure, improve the injection production relationship, and achieve the purpose of improving the block development effect.

2. Establish a reasonable number water ratio of oil wells

According to the injection production balance principle, under the condition that the injection production pressure difference is basically stable and the total number of wells in the oilfield is certain, the reasonable oil-water well number ratio to obtain the maximum liquid production of the oilfield is equal to the square root of the ratio of the water absorption index of the water injection well to the liquid production index of the oil well, namely:

$$Row=\sqrt{\frac{Jw}{Jl}}$$

Where Row is the reasonable ratio of oil wells to water wells, dimensionless Jw is the water absorption index of water wells, $m^3/(d.Mpa)$; Jl - Fluid production index of oil well, $m^3/(d.Mpa)$

* Corresponding author: hhongzh1i@petrochina.com.cn

https://doi.org/10.1051/e3sconf/202337501025

3. Investigation and study on injection production status in a block

The oil-bearing area of a block is 47.8km2. As of December 2018, there are 559 oil wells, including 383 oil production wells and 176 water injection wells. The well pattern density is 11.69/km^2, and the oil-water well ratio is 2.2. Up to now, the cumulative crude oil production is 690.86 × 10^{4}t, accumulated water injection 4205.80 × 10^{4}t, cumulative injection production ratio 1.20, recovery percent 28.05%, oil recovery rate 0.28, and annual oil production 6.45 × 10^{4}t, the daily oil production of each single well is 0.84t/d, and the comprehensive water content is 92.77%. Affected by the development scale of faults and sand bodies in Gaotaizi old area, the contradiction between injection and production is outstanding. Seen from the water drive control degree table of different thickness levels, one-way and two-way connectivity is dominant, and the proportion of connected layers below two directions (including two directions) is 54.9%. The connectivity ratio of three-way sandstone is only 12.3%. From the water drive control degree table of the subarea, the water drive control degree of Block B is slightly higher than that of Block M, and the proportion of connectivity above three directions in a block is slightly lower. On the plane, an oilfield subdivides sedimentary microfacies, including distributary plain facies, which can be subdivided into channel sand, interriver thin layer sand, off surface reservoir and pinch out area. The inner front facies can be subdivided into underwater distributary channel sand, interdistributary thin layer sand, off surface reservoir and pinch out area. According to the size and shape of sand body distribution in each unit, the sand bodies of an oilfield can be divided into the following four types (see Table 1).

From the point of view of the degree of water drive control of oil layers in a block, the latter three sand bodies are relatively low in water drive control, which is the focus of treatment. From the perspective of water drive control degree in the subarea, the water drive control degree of sandstone in an old area and Block B is similar, and the unidirectional connection proportion is high; From the point of view of the degree of water drive control of oil layers, the degree of perfection of single sand bodies of different sedimentary types is also different. It is believed that there is still a certain potential to improve the injection production relationship of single sand bodies.

Table 1 Sedimentary Types of Oil Layers in an Old Area

Sand body classification	Number of units	Channel sand drilling rate (%)	Sedimentary unit
Straight distributary sand body with main belt	6	23.0	5, 6, 7, 8a, 9, ,10
Straight distributary sand body without main belt	4	21.1	2a, 3, 4, ,8b
Intermittent narrow strip straight distributary sand body	3	14.8	1b, 2b, 11
Stable distribution of sheet sand	2	7.3	1.1, 1a

4. Single sand body potential analysis

For oil wells that have been perforated, the potential of single sand body injection and production is not perfect or only one direction is effective due to poor water well sand body, excessive well spacing, non perforated water wells, poor water absorption and other reasons. The potential can be tapped through water well fracturing, hole patching and other measures; For oil well perforation, if the water well is not perforated or there is no water well, it will lead to recovery without injection. The direction of tapping the potential is to supplement the well or increase the water injection well points. For the imperfect single sand body caused by interlayer interference, plane interference and intra layer interference, the potential can be tapped by means of subdivision adjustment of water wells, plane adjustment of oil wells and profile control of water wells. For the non injection and non production caused by the non perforated oil well and the non perforated water well, the injection and production relationship can be improved through the simultaneous perforation of oil and water wells; For non perforated oil wells and non water well type non injection and non production, water injection well points can be increased and the oil wells can be re perforated at the same time; If the oil well is not perforated, and there is a water well, and there is no production due to perforation, the oil well can be re perforated. Seen from the types of imperfect injection and mining in an old area (see Table 2), there are mainly four types: no injection and no mining, in layer, fault edge and water absorption difference. Since most of the non injection and non production layers are oil-water same layer without perforation of oil and water wells, such as PI8, PI9, PI10, PI11 and other sedimentary units, they are not analyzed. Fault edge, injection without mining and mining without injection are the key points of this injection production system adjustment.

Table 2 Imperfection of Single Sand Body in an Old Area

Potential type	Number of layers		Sandstone		Valid	
	Number of layers(Nos.)	Proportion (%)	thickness (m)	proportion (%)	thickness (m)	proportion (%)
Injection without mining	136	4.51	250.5	4.21	181.2	4.25
Mining without injection	368	12.19	639	10.74	451.1	10.57
No injection and no mining	296	9.81	471.2	7.92	320.3	7.51
Interlayer interference	338	11.20	386.5	6.50	185.5	4.35
Intraformative	637	21.11	1864.8	31.35	1528.6	35.83
Planar interference	286	9.48	356.6	5.99	173	4.05
Fault margin	609	20.18	1305.3	21.94	923.6	21.65
Isolated sand body	2	0.07	2.7	0.05	1.8	0.04
Riverside	214	7.09	517	8.69	415.5	9.74
Water absorption difference	59	1.95	56.6	0.95	31.4	0.74
other	73	2.42	98.4	1.65	54.7	1.28

5. Research on the adjustment mode of imperfect injection production of single sand body

5.1 Evaluation on the perfection degree of single sand body in a block and analysis on the cause of single sand body imperfection

In the late period of high water cut, the remaining oil is mainly distributed between wells, and the direction of improving oil recovery should be to find sand bodies or local enrichment zones in sand bodies, loss zones of sand bodies and well patterns, local remaining oil enrichment zones caused by fault shielding, and the remaining oil between wells that can not be used well despite well pattern control. For this reason, the first main means is to drill various types of scattered adjustment wells, which are used to improve the sand body injection and production system, and to excavate the remaining oil in the fault and sand body corners; Second, sidetracking horizontal wells, directional deviated wells, branch wells, etc. are used to exploit oil enrichment zones between wells, especially near oil wells due to changes in reservoir physical properties, and can also be used for oil retention areas caused by fault shielding; The third is to apply conventional sidetracking to renew wells to solve the problem of incomplete injection and production caused by various reasons.

Focusing on sorting out the single sand body injection production relationship in the well area near the fault, such as production without injection, production with less injection and more production, the well area with adjustment potential is classified, and effective adjustment measures are formulated to improve the injection production relationship and tap the residual potential.

5.2 Residual oil retention area near the fault

In the vicinity of faults, it is easy to form remaining oil enrichment areas. For this situation, in combination with the results of fine reservoir description and multidisciplinary research, additional wells are arranged at appropriate locations to tap the remaining potential of the oil layer.

For example, the well block A34-29 is located near the fault. The fault is relatively complex. The production of oil well A35-28 in the well block is low and no fault is encountered. The distance between the well A34-29 and the oil well A35-28 is 360 meters, and the remaining oil around the fault is large. Based on the results of fine reservoir description, the sand body development is predicted. In order to reduce the injection production well spacing and excavate the remaining oil in the river near the fault, it is proposed to arrange the supplemental oil well A34-28 on the connecting line between Well A35-S27 and Well A34-29, 200 meters away from Well A34-29. Based on the fine reservoir description results, the drilling thickness of the supplementary well is predicted. The drillable sandstone thickness is 11.2m, the effective thickness is 6.4m, and the recoverable reserves can be increased by 0.5158×10^4t.

6. Formulate adjustment ideas, implement the potential of adjustment plans, and demonstrate research methods

6.1 Adjustment of injection production relationship

In order to further improve the development effect of an old area, it is mainly to apply the results of fine reservoir description to find the supplementary potential in the well areas near the fault and with great changes in the development of sand bodies. At the same time, for the well areas with large ratio of oil wells to water wells and imperfect injection and production of single sand bodies in the plane, the injection and production relationship is improved by means of oil well conversion.

6.2 Formulate adjustment principles

1. For areas with no water injection direction or multiple well areas with less water injection and production, improve the injection production relationship of the well area by drilling supplementary wells or transferring oil wells to water injection;

2. For the well area with more injection and less production and lack of recovery well points, the injection production relationship in the well area shall be improved by drilling additional oil wells;

3. After re understanding the structure through combination of well and earthquake, for the well area where the injection production relationship changes greatly due to fault changes, the injection production relationship in the well area can be improved by drilling additional oil and water wells or transferring oil wells to injection;
4. The thickness of sandstone encountered during single well drilling of supplementary well shall not be less than 5.0m, and the effective thickness shall not be less than 1.0m;
5. To prevent casing damage, no water injection wells shall be arranged within 100m from the fault;
6. The transfer well shall be no less than 100m away from the fault, and the increased recoverable reserves shall be more than 1500t.

6.3 Implement adjustment potential, prepare adjustment plan and demonstrate it

After the well location is preliminarily determined, the drillable thickness of a single well shall be predicted in combination with the development of sand bodies in the supplementary well area, the production status of single sand bodies in the well group shall be analyzed in combination with the fine mapping results, and the initial water cut and productivity shall be predicted in combination with the multidisciplinary research results. Finally, the economic benefits of the single well shall be evaluated. Plan the well locations that meet the well layout requirements.
According to the above ideas, Gaotaizi Old Area has preliminarily determined 22 injection wells adjusted by the injection production system (see Table 3); 8 internal supplementary wells (see Table 4), including 5 oil wells and 3 water wells; Renew 3 wells; There are 33 wells in total. After adjustment, the thickness ratio of connected sandstone in an old area has increased by 5.3 percentage points, the effective thickness ratio has increased by 3.8 percentage points, and the recoverable reserves are predicted to increase by 10.15×10^4t, the ratio of oil and water wells has reached 1.90 from 2.18, which is closer to the ideal value of 1.60, fully demonstrating the rationality of this research method.

Table 3 Well Layout Results of Injection Production System Adjustment Scheme

Number of design wells			Renewal well (water)	Average adjustable thickness of single well		Single well increase Recoverable reserves (10^4t)	Forecast Well productivity (t/d)	estimate Initial moisture content (%)
oil well	well	Subtotal		Sandstone (m)	Valid (m)			
5	3	8	3	8.3	6.0	0.46	2.2	70

Table 4 Statistics of Oil Well Reinjection in Injection Production System Adjustment Scheme

Number of wells converted from oil wells (mouth)	Average perforation thickness of single well		Increased recoverable reserves of single well (10^4t)	Estimated daily water injection of single well (m^3)
	Sandstone (m)	Effective (m)		
22	10.7	7.3	0.2294	30

7. Conclusion and understanding

1. The exploration of injection production balance principle is a necessary means to establish a reasonable well number ratio for reservoir development;
2. The application of reservoir related software, such as PREP2.0 and numerical simulation, assists in the implementation of adjustment potential and prediction of development indicators;
It is an effective way to improve the development effect of the block to excavate the remaining oil in the partial well block with imperfect injection production relationship through adjustment schemes such as supplement and injection conversion.

References

1. Lei Chen,Jianjun Xu,Shuang Chen,Hui Yang,Linhu Liu. Power Quality Disturbance Signal Denoising Based on Overcomplete Representation. IEEJ TRANSACTIONS ON ELECTRICAL AND ELECTRONIC ENGINEERING.2022; 17: 544–555
2. Xu Jianjun, Huang Lida, Yan Limei, Yi Na. Insulator Self-Explosion Defect Detection Based on Hierarchical Multi-Task Deep Learning. Transactions of China Electrotechnical Society, 2021,36(07):1407-1415.
3. Yan Limei,LIU Yongqiang,XU Jianjun,et al.Broken string diagnosis of composite insulator based on Grabcut segmentation and filler area discrimination.Power System Protection and Control,2021,49(22):114-119
4. Yi, Q. Wang, L. Yan, et al., A multi-stage game model for the false data injection attack from attacker's perspective. Sustainable Energy Grids & Networks 28 (2021).
5. Na Yi,Jianjun Xu,Limei Yan,Lin Huang. Task Optimization and Scheduling of Distributed Cyber-physical System Based on Improved Ant Colony Algorithm. Future Generation Computer Systems, 109(Aug. 2020),134-148.
6. Yang Zhao, Jianjun Xu, Jingchun Wu. A New Method for Bad Data Identification of Oilfield Power System Based on Enhanced Gravitational Search-Fuzzy C-Means Algorithm. IEEE Transactions on Industrial Informatics. VOL. 15, NO. 11, NOVEMBER 2019 5963-5970

 https://doi.org/10.1051/e3sconf/202337501026

Study on the cause classification and prevention methods of casing damage

Fangfang Jia

No. 5 Production Plant of Daqing Oil Field Daqing, Heilongjiang 163513, China

Abstract: The casing damage wells in non oil layers are mainly distributed in the N2 and SI-SII oil shale interlayer, and the casing damage types are mainly deformation and dislocation. The main reason is that the injected water enters into the oil shale, the cementing quality is poor, and it is easy to form an invasion water area after invading along the channeling channel, which reduces the stability of the formation. Under the effect of geostress, the force generated by the formation sliding along the weak surface is directly applied to the casing, causing shear casing damage; The casing damage in the oil reservoir is mainly concentrated in SII1-5 layers, which is analyzed as a result of high pressure water injection and large difference in plane and interlayer pressure; The casing damage on the fault plane is mainly caused by slip caused by water softening on the fault plane and fault activity caused by excessive difference in pore pressure of oil layers on both sides of the fault. This paper expounds the measures taken in daily management from the aspect of casing damage prevention, which provides a basis for the prevention and control of casing damage wells in the future.

Key words: cause of casing damage well; Prevention and control of casing damage well.

1. General situation of casing damage

Since its production and development in 1970, Block A can be roughly divided into three stages, namely, water injection and flowing production stage, flowing to pumping stage, and formation subdivision and infilling adjustment stage. In the layer system subdivision and adjustment stage, it is divided into primary encryption adjustment, secondary encryption adjustment and tertiary encryption adjustment. From the change curve of the number of wells with damaged casing over the years, it can be seen that from 1978 to 1981, the number of wells with damaged casing increased year by year. From 1982 to 1986, the number of wells with damaged casing was stable at about 10. From 1986 to 1990, the number of wells with damaged casing decreased from 8 in 1986 to 3 in 1990. Therefore, before 1990, the basic well pattern went through the production stage of flowing, flowing, pumping and overall pumping, while the casing damaged well has gone through the process of rising, stabilizing and falling since it was discovered in 1978. Since 1990, infilling adjustment has been started. After infilling adjustment, with the increase of the number of wells with damaged casing in primary wells, secondary wells and tertiary wells, the number of wells with damaged casing is generally on the rise.

2. Causes of casing damage

2.1 Analysis of casing damage well in non oil layer

The casing damage in non oil layer is mainly concentrated in N2 and SI-SII interbeds. It is believed that water inflow from oil shale is the main factor causing casing damage.

1) Water ingress into the oil shale of N2 causes casing damage around the well

The lithology of the marker bed at the bottom of N2 is mainly dark grey, grey black and black grey massive mudstone and mud shale with weak sound. The rock contains rich ostracods and phyllopoda fossils, which are mostly distributed along the layers, but most of them are unevenly distributed. Some layers are densely distributed, while some almost do not contain fossils. The phyllopoda fossils are also enriched into layers locally. There are few fossils in the shale, the foliation is developed, the inner layer is flat, and sporadic fish fragment fossils can be found locally. Silt is contained locally in mudstone, horizontal bedding is mostly developed, rock section is also relatively neat, and lithology of the whole section is characterized by weak hard lithology.

The study shows that there is a weak lithologic interface in the fossil belt corresponding to the third peak of the electrical logging curve at the bottom of N2 Member. After the injected water is immersed along the channeling

* Corresponding author: 458116900@qq.com

channel, it is easy to form an immersion area, which reduces the stability of the formation. Under the action of in-situ stress, the force generated by the formation sliding along the weak surface is directly applied to the casing, causing shear casing damage.

A total of 67 casing damage wells were found in the standard layer of N2, accounting for 21.27% of the non oil layer. In the first half of 2013, two casing damage wells were found in the standard layer of N2: Well A is a well in Block A. During the overhaul in April 2013, 861.3m casing was found to be broken. The casing damage layer is N2, with a minimum diameter of 92mm. At present, water is injected after the overhaul. In order to verify whether there is water inflow into the standard layer of N2, the water injection wells around the well block have been investigated to determine the water immersion range of N2. A total of 5 wells have been inspected; Well B is an oil well in Block A. During profile control in January 2013, it was found that the casing at 717.53m was broken. The casing damage layer was at Section N2, with a minimum diameter of 104mm. The well was not overhauled successfully and is now to be scrapped.

2) The injected water channeled to the interlayer oil shale of SI-SII along the bedding, resulting in casing damage

Horizontally, there are two or three obvious peaks in the SI-SII interlayer, which is an oil shale with high abundance of ostracoda, phyllopoda and other fossils, brittle texture and developed bedding. The analysis shows that the main reason is the casing damage of oil and water wells in the oil shale after water inflow along the bedding. The causes of water ingress may include:

(1) The cementing quality is poor, and the casing is damaged due to the external immersion of injected water. According to statistics, there are 158 wells with casing damage in SI-SII interlayer, 61 wells with good cementing quality, accounting for 38.6%, and 97 wells with poor cementing quality, accounting for 61.4%.

(2) Fracturing of SII1 layer may cause injected water to move up along the longitudinal fracture, causing casing damage of SI-SII interlayer.

(3) The casing damage of adjacent wells is caused by water inflow at the interlayer of Well SI-SII.

Therefore, it is believed that the main reasons for casing damage of SI-SII interlayer are the poor quality of water well fracturing and cementing of SII1 interlayer.

2.2 Analysis of casing damage wells in oil layers

1) High pressure water injection easily leads to casing damage

Long term high-pressure water injection in water injection wells is the main cause of casing damage in oil layers. From the perspective of the two blocks that have experienced high pressure water injection, the casing damage ratio is relatively high, while the casing damage ratio of the blocks that have been put into development at the same time and have low water injection pressure is very low. It can be seen that high pressure water injection is an important cause of casing damage, and the well pressure drop of top fracture pressure water injection is an important means to reduce casing damage.

2) Uneven plane pressure distribution and large interlayer pressure difference easily lead to casing damage

Due to the influence of development well pattern, production method and development strata, the development block forms a high and low pressure area on the plane, which makes the distribution of formation pressure uneven on the plane and between layers. Due to the change of the maximum and minimum horizontal principal stress distribution, the circumferential stress distribution of the casing is uneven. When the change of this force exceeds the anti extrusion or anti shear strength of the casing, the casing is damaged.

With the continuous deepening of oilfield development, the conditions of the adjusted target oil layer become worse, and the shale content in the oil layer increases. In the oil field developed by water injection, it is easy to cause the pore plugging of the oil layer, which reduces the water injection volume. High pressure injection is required to meet the needs of oilfield development. With the continuous increase of water injection pressure, the pressure difference between different sub layers is increasing, and the difference of formation pressure is an important reason for casing damage.

3. Study on Prevention and Treatment of Casing Damage Wells

3.1 Prevention and treatment methods for N2 and SI-SII interbeds in non oil bearing section

(1) For the water injection wells around the newly added N2 bottom casing damaged wells, water injection shall be controlled or stopped immediately, and then the well condition shall be investigated. For the casing damaged wells, water injection must be resumed after major repair. The casing damage at this position brings the greatest harm to oilfield development, which is the most important part of casing damage prevention and control work, and also the focus of casing damage analysis.

(2) For the casing damage at the oil shale of N2 and the interlayer between SI-SII, if the casing protection packer is not sealed, the first stage packer shall be installed at the casing damage position to prevent casing damage at the same position of the surrounding oil and water wells. Investigation was made on the casing damage of the packer in N2 Member and SI-SII Sandwich. The investigation results show that 21 of the 54 wells have been scrapped due to multi-point casing damage and serious dislocation. The remaining 33 wells were fully investigated and found to be overhauled. According to horizon analysis, there are 5 wells in N2 and 28 wells in SI-SII interlayer. There are 26 wells with eccentric passing and 7 wells with smooth tube passing. In view of the actual situation, for the wells that currently use packers, put dead nozzles, and the last operation time is before June 2012, all the wells have been re sealed, with a total of 20 wells.

https://doi.org/10.1051/e3sconf/202337501026

(3) Investigate the scrapping method of scrapped wells for casing damage at non oil layer N2 oil shale and SI-SII interlayer, and scrap the well used micro expansive cement scrapped with heavy mud.
(4) When new wells are arranged in the casing damage area of the standard layer of N2 and the SI-SII interlayer, p110 casing is run in the standard layer of N2 and the SI-SII interlayer to improve the casing shear strength and prevent casing damage.
(5) To prevent casing damage caused by water entering the standard layer in the free section of N2 section. During the completion of new wells, a part of the cement return height is selected to be higher than the standard layer of N2 for future comparison and analysis.
(6) In order to prevent casing damage caused by mud water inflow expansion, anti expansion agent is added to newly injected water injection wells to reduce mud water immersion expansion.

3.2 Prevention and treatment methods for oil layers

1) Reduce water injection pressure
High pressure water injection is an important cause of casing damage, and the well pressure drop of top fracture pressure water injection is an important means to reduce casing damage.
2) Adjust the high and low pressure blocks to reduce the plane and interlayer pressure difference
In order to avoid casing damage of oil layer caused by abnormal pressure, the monitoring data, oil and water well dynamic and static data, etc. are used to comprehensively analyze the injection and production status of single sand layer and determine the abnormal pressure well layer. In 2013, we will focus on the adjustment of water injection scheme and periodic water injection to control the water injection of SII1- SII5 and SII9-SII11 layers with relatively high formation pressure, appropriately increase the water injection volume of the main oil layers and some main thin sand layers of PI Formation, and strive to balance the plane pressure between layers. At the same time, the newly discovered casing damage wells in oil layers are analyzed and treated. In view of the actual situation that the pressure difference between layers and planes is large, it is planned to reduce water in the high-pressure layer.

4. Conclusion

(1) At the bottom of N2 in Block A, casing damage is easy to occur at the position with high fossil content and the highest apparent resistivity. The positions where casing damage is easy to occur in oil layers are mainly concentrated in the positions where lithology changes. The casing damage is the result of mechanical shearing, and the casing damage type is mainly shear casing damage and there is casing damage shear band.
(2) Water injection with excessive reasonable pressure can lead to casing damage of water injection wells in the overpressure water injection well section, and the greater the overpressure amplitude, the faster the casing damage of oil and water wells. Considering the influence of various factors, the reasonable water injection pressure limit of the oil layer should be controlled within 0.5MPa lower than the overlying rock pressure of the oil layer.
(3) The genetic type of abnormal high pressure layer is mainly caused by the imperfect injection production relationship of single sand body. With the continuous infilling and adjustment of the well pattern, the mining object gradually becomes worse, and the distribution of high-pressure layer also changes with the change of the adjusted target layer.
(4) The prevention and control measures for casing damage at the bottom of N2 in Block A should mainly focus on improving the cementing quality of new wells and the key prevention measures in key areas.
(5) The formation pressure on both sides of the fault shall be controlled reasonably and monitored in time to prevent casing damage caused by excessive differential pressure on both sides exceeding the critical value.
(6) For abnormal water injection wells near the fault, timely test and adjust. If the adjustment is not successful, shut in the well in time for operation to prevent the injected water from entering the fault to further expand the immersion area and aggravate casing damage.
(7) For casing damage wells that have been at the breakpoint, pay close attention to other wells with the same breakpoint on this fault, and reduce the formation pressure difference in time to prevent casing damage in pieces.

References

1. Lei Chen,Jianjun Xu,Shuang Chen,Hui Yang,Linhu Liu. Power Quality Disturbance Signal Denoising Based on Overcomplete Representation. IEEJ TRANSACTIONS ON ELECTRICAL AND ELECTRONIC ENGINEERING.2022; 17: 544–555
2. Xu Jianjun, Huang Lida, Yan Limei, Yi Na. Insulator Self-Explosion Defect Detection Based on Hierarchical Multi-Task Deep Learning. Transactions of China Electrotechnical Society, 2021, 36(07):1407-1415.
3. Yan Limei,LIU Yongqiang,XU Jianjun,et al.Broken string diagnosis of composite insulator based on Grabcut segmentation and filler area discrimination.Power System Protection and Control,2021,49(22):114-119
4. Yi, Q. Wang, L. Yan, et al., A multi-stage game model for the false data injection attack from attacker's perspective. Sustainable Energy Grids & Networks 28 (2021).
5. Na Yi,Jianjun Xu,Limei Yan,Lin Huang. Task Optimization and Scheduling of Distributed Cyber-physical System Based on Improved Ant Colony Algorithm. Future Generation Computer Systems, 109(Aug. 2020),134-148.

https://doi.org/10.1051/e3sconf/202337501026

6. Yang Zhao, Jianjun Xu, Jingchun Wu. A New Method for Bad Data Identification of Oilfield Power System Based on Enhanced Gravitational Search-Fuzzy C-Means Algorithm. IEEE Transactions on Industrial Informatics. VOL. 15, NO. 11, NOVEMBER 2019 5963-5970

Study on the cause of casing damage and the feasibility of comprehensive prevention

Shiping Liu

No.4 Oil Production Plant, Daqing oilfield company limited, Daqing, China

Abstract: Through the study of casing damage in oil and water wells, the causes of casing damage are found out. At the same time, a set of systematic prevention and control methods are summarized based on the fine geological research results and the combination of dynamic and static conditions. The source of water inflow is cut off in time, effectively preventing the further spread of casing damage area, which plays an important role in controlling casing damage speed and improving the development effect of casing damage area.

keywords: Causes of casing damage Study on prevention and control methods cognition.

1. Cause analysis of casing damage

In only two years, 15 oil and water wells have casing damage, with casing damage rate of 19.7%, including 11 of 24 water injection wells, with casing damage rate of 45.8%, which shows that the casing damage situation in this area is very serious.

The reservoir exploitation in the casing damage area is mainly based on the off surface reservoir, and a small part is the main thin layer sand on the surface and the non main layer sand on the surface. The oil layer is poorly developed, and the pressure loss of injected water during seepage is large.

From the perspective of casing damage, casing damage wells are concentrated, and the casing damage layer is mainly I-II interlayer, and the casing damage depth is near the perforation top boundary. Through analysis, it is believed that the main reasons for casing damage are as follows:

1.1 Casing damage is induced by injected water flowing into non oil layer along well section with poor cementing quality

The reservoir development is dominated by Group II. According to the statistics of isotope data of water injection wells in recent two years, Layer II1-4 has a high water absorption ratio and is the main water absorption layer of the whole well. If the cement bonding quality outside the pipe at the top boundary of the perforation is poor and there is a channeling channel, the injected water in layer II1-4 will directly enter the mudstone to form a flooding area to induce casing damage. According to the statistics of the secondary acoustic changes of the three casing damaged water injection wells, 28.7% of the wells are from the cement return height to the bottom boundary of the measurement, and 10.2% of the wells are from the top boundary of the perforation with worse cementing quality (see table). It can be seen that the hidden danger of cementing quality is the main factor leading to casing damage in this area.

Table 1. Statistics of Cementing Quality of Three Casing Damaged Water Injection Wells by Secondary Acoustic Variation Measurement

Number of wells	Segmentation	Total thickness (m)	Cementation			
			Medium cementation (m)	Poor cementation (m)	Medium to poor cementation	
					Subtotal (m)	proportion (%)
3 ports	I Top to perforation top boundary	41.8	21.1	6	27.1	64.8
	Perforation top boundary to II4 bottom	48.8	29.8	14	43.8	89.8
	Cement returns to the bottom of measurement	1064.8	386	372.7	758.7	71.3

1.2 Casing damage is related to lithology

From the lithological characteristics of 0-I and I-II interbeds, 0-I interbeds are gray black mudstones, with three layers of low-quality oil shale or ostracode mudstones in the middle, I-II interbeds are gray black ostracode mudstones at the top, gray black mudstones at the middle, and a layer of dark gray ostracode layer and a layer of marlstone at the bottom. The volume expansion of the interbed mudstones can reach about 7% after the montmorillonite content is relatively enriched and water is absorbed, Therefore, after absorbing water, the mudstone expanded with water will inevitably increase its plasticity, reduce its ability to withstand vertical rock pressure, and generate creep pressure on the casing, which will damage the casing.

1.3 Delayed or delayed scrapping of staggered water injection wells is also the cause of casing damage in this area

The water injection wells M and N are staggered at the I-II interlayer successively. It can be seen from the micro structural map that M well is at the structural high point and N well is at the structural low point. Through analysis, it is believed that the water injection well M at the structural high point was not found in time after the staggered fracture, and the later scrapping was not timely. The injected water continued to enter the formation along the staggered fracture, resulting in the staggered fracture of the adjacent water injection well N at the same layer.

2. Study on the feasible method of casing damage prevention

In view of the casing damage situation in this area, we have deeply studied the causes of casing damage in each well, applied fine geological research results, summarized and implemented comprehensive prevention and control measures in combination of static and dynamic, and achieved some results.

2.1 Use various monitoring means to find and cut off the source of water inflow in the shortest time to prevent further spread of casing damage.

A comprehensive prevention and control plan was developed for timely finding the causes of casing damage wells, including 22 wells for isotope production, 2 wells for noise leak detection, and 13 wells for casing damage re zoning in the casing damage area. After the implementation of the plan, 11 wells were found to be unable to be pulled out, and the wells were shut in immediately to reduce harmful water injection. Through careful analysis, the water inlet wells * and * * were found and cut off in time to prevent further spread of casing damage in this area.

2.2 Maintain block pressure balance and prevent casing damage at oil reservoir

In order to maintain the pressure balance in this area, 9 wells were overhauled and opened in a timely manner, and the repaired water injection wells have polished the II1-4 layers to prevent the injection water from channeling upward, coordinate the supply and drainage of oil and water wells, improve the injection production system, and reduce the hidden danger of casing damage. At the same time, the three water injection wells that failed to be overhauled were scrapped in time to prevent further expansion of casing damage.

2.3 Strengthen on-site inspection of water injection wells from the ground to eliminate hidden dangers of casing damage.

Regularly check the admission of on-site data of water injection wells, the implementation of the system by post workers, improve their awareness of casing damage protection, earnestly implement the casing protection laws and regulations, and timely solve the problems found in the inspection. Strictly adhere to the reporting system of abnormal wells, immediately test the D well with abnormal pressure and water volume, and find that the pipe string fails, and timely operate to find that the well cannot be pulled out.

2.4 Improve cementing quality and prevent new water source

The use of casing external packer above the perforation top boundary can more effectively block the water channeling due to poor cement bonding. There are 7 downhole cement surface control tools in the secondary wells in the casing damage area this year. In addition, in order to improve the anti extrusion ability of the casing itself, N80 high anti extrusion casing is run into all the oil and water wells that are injected later. In addition, it is planned to conduct secondary acoustic variable cementing quality logging for the water wells in this area, shut down the water injection wells with poor cementing at the top of perforation according to the logging results, and repair some water injection wells with poor cementing by using cement sheath repair technology, so as to reduce the reduction of the degree of reservoir production caused by the shutdown or shut in.

2.5 Investigate the immersion area, and stop the injection of the water injection well II1-4 layers in the casing damage area in time to prevent the injection water from channeling upwards

According to the statistics of the water injection intensity of 24 whole wells, the effective water injection intensity of 18 water injection wells in Layer II1-4 is relatively high. In order to prevent the injection water from channeling upward, the water immersion area is judged according to the transverse map of the new wells and the electrical logging curve. At the same time, the layer II1-4 of the water injection wells is exposed in time in combination with the isotope data, and the wells with abnormal noise curves are subdivided. In addition, 2 staggered oil wells around the staggered water injection well continue to produce and relieve pressure, so as to narrow the scope of the immersion zone. The pressure relief measures were taken for 4 oil wells by injecting the I-II interlayer.

3. Conclusion and understanding

Casing damage will damage the well pattern and affect the oilfield development effect. In order to fully understand the cause of casing damage and clarify the ideas and direction of casing damage control, this paper draws the following conclusions through the analysis and research on casing damage situation in this area:
3.1 Poor cementing quality and mudstone swelling are the main reasons for casing damage in this area;
3.2 The casing damage at non oil layer can be avoided to some extent by using the cement surface control tools and running the high anti extrusion casing. For the casing damage wells at the newly emerging I-II interlayer, the investigation of the well conditions of water injection wells should be strengthened to reduce the hidden dangers of casing damage;
3.3 For the damaged casing area, it is necessary to timely scrap the water injection well project that has not been wrongly broken during the overhaul, continue the production and pressure relief of the wrongly broken oil well, or select the adjacent oil well to re shoot the I-II interlayer for pressure relief, which can effectively control the further spread of the immersion area;
3.4 It is necessary to strengthen the dynamic monitoring work in the casing damage area, grasp the trend of casing damage at any time, and take preventive measures.

References

1. Lei Chen,Jianjun Xu,Shuang Chen,Hui Yang,Linhu Liu. Power Quality Disturbance Signal Denoising Based on Overcomplete Representation. IEEJ TRANSACTIONS ON ELECTRICAL AND ELECTRONIC ENGINEERING.2022; 17: 544–555
2. Xu Jianjun, Huang Lida, Yan Limei, Yi Na. Insulator Self-Explosion Defect Detection Based on Hierarchical Multi-Task Deep Learning. Transactions of China Electrotechnical Society, 2021,36(07):1407-1415.
3. Yan Limei,LIU Yongqiang,XU Jianjun,et al.Broken string diagnosis of composite insulator based on Grabcut segmentation and filler area discrimination.Power System Protection and Control,2021,49(22):114-119
4. Yi, Q. Wang, L. Yan, et al., A multi-stage game model for the false data injection attack from attacker's perspective. Sustainable Energy Grids & Networks 28 (2021).
5. Na Yi,Jianjun Xu,Limei Yan,Lin Huang. Task Optimization and Scheduling of Distributed Cyber-physical System Based on Improved Ant Colony Algorithm. Future Generation Computer Systems, 109(Aug. 2020),134-148.

6. Yang Zhao, Jianjun Xu, Jingchun Wu. A New Method for Bad Data Identification of Oilfield Power System Based on Enhanced Gravitational Search-Fuzzy C-Means Algorithm. IEEE Transactions on Industrial Informatics. VOL. 15, NO. 11, NOVEMBER 2019 5963-5970

 https://doi.org/10.1051/e3sconf/202337501028

Application of close cut volume fracturing in tight oil horizontal Wells

Yude Gao

No.10 Oil Production Plant, Daqing Oilfield Co., LTD., Daqing 166405, China

Abstract. S tight reservoir is rich in oil and gas and has good development potential. However, the reservoir lacks good physical property, strong heterogeneity, tight lithology and complex reservoir formation conditions, and the fracturing operation is characterized by lack of good fracture complexity and difficulty in initiation. For the geological characteristics of the reservoir and the difficulties of reconstruction, we actively follow the principle of fracture-controlled reserves, broaden the understanding of unconventional volume fracturing, and integrate cutting fracturing technology in the long-term practice process. The combination of cluster perforation, plug segmentation, reverse mixed injection and other methods can fully improve the fracture initiation and extension effect of multiple clusters. The combination of sand addition technology and smooth water to replace guar gum, promote the laying quality of fracture proppant to be fully improved, promote the formation to further achieve the energy storage and increase goal, tight oil layer to achieve the volume transformation goal, and promote the long-term, stable effect of fracturing reconstruction.

Key words: Close cutting; Volume fracturing; Horizontal well.

1. Introduction

In order to make full contribution to the stability and efficiency of tight oil formation development, different technologies should be used together. For tight oil horizontal Wells, preparatory treatment should be carried out before formal production, and volume fracturing is a key step. This work covers cutting and fracture-net fracturing modes. The fracture-net fracturing technology has a good application effect in the large brittle reservoir, and the cutting volume pressure has a good application effect in the small brittle reservoir, while the tight oil reservoir is less brittle, so the cutting pressure technology can be adopted [1].

2. Specific application of medium volume pressure technology in horizontal well

2.1 Specific application of subdivision cutting technology in horizontal well medium volume pressure technology

The S reservoir is an extremely low and low-porosity permeability reservoir with starting pressure in the range of 0.017MPa/m -- 0.14MPa/m, which is higher than other reservoirs of the same type. As starting pressure increases, oil flow decreases and the reservoir is highly dependent on fractures. The S reservoir mainly adopts the form of two wing fractures. Based on increasing the number of artificial fractures per unit volume, the flow distance between bottom fluid and fractures can be shortened by subdividing and cutting the reservoir, which can fully increase the production volume of the reservoir and increase the production of a single well.

In addition, the starting pressure of BKQ formation on the north slope of S reservoir is 0.14MPa/m, and that of BKQ Formation on the west slope is 0.017MPa/m, and there is an order of magnitude difference between them. Therefore, the Arctic slope BKQ reservoir should have a denser cut density, and the distance between artificial fractures in the horizontal section should be smaller than that in the western slope BKQ reservoir, so the fracturing process should be designed in a differentiated way [2].

The oil saturation of the north slope BKQ reservoir is simulated under different conditions of 1000m horizontal interval, 300m distance of well development and fracture spacing. See the picture below.

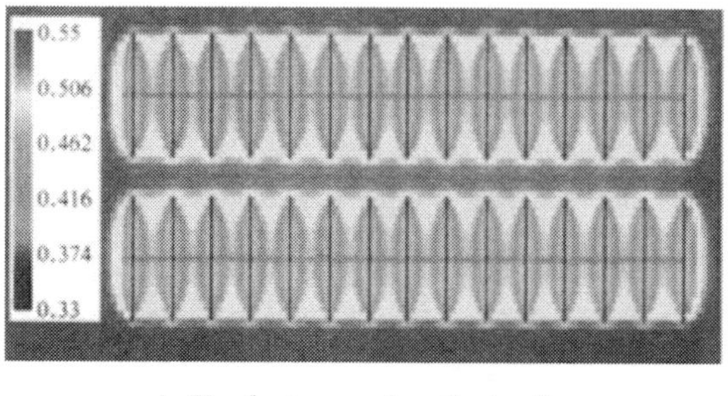

A: 70m fracture spacing oil saturation

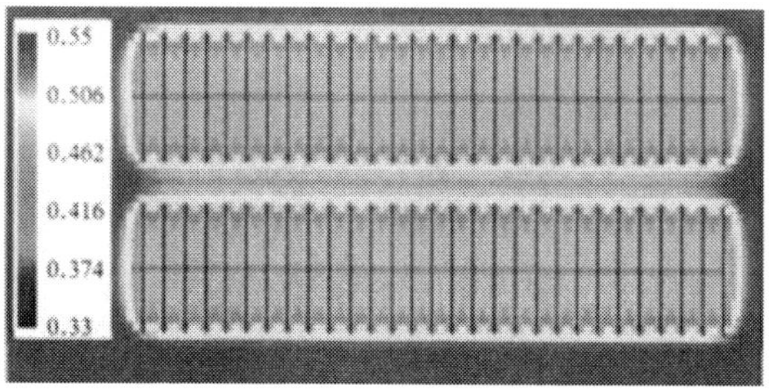

B: 70m fracture spacing oil saturation

Figure 1 Oil saturation at various fracture distances after 3 years

It can be found from the figure above that the effective control area is located around the main peak when the spacing is 70m, and there is a lack of good effective mobilization between different fractures. When the fracture distance is 35m, the fracture control area is expanded, the overall oil saturation in the reservoir is reduced, and the operation of jointing is controlled. Through production simulation, it is found that when the starting pressure is not taken into account, the production of a single well can be effectively increased when different fractures change from 70m to 35m.

Therefore, it was determined to use the cementing bridge cluster perforation and pressure division technique. The number of perforations in the interval was 2 to 3 clusters, and the fracture distance in the north slope BKQ reservoir was shortened from 70m to 30m, and the fracture distance in the west slope BKQ reservoir was shortened to 40m. The microseismic survey showed that the fracture morphology is mainly banded, and the sweep bandwidth can fully cover the fractured well interval, so as to fully improve the fracture complexity target and volume reconstruction target.

2.2 Large displacement reverse mixed joint making process

(1) Fracturing pumping technology. The fractures in S reservoir have large roughness, and the fracture extension path is tortuous, and the proppant movement in the fracture lacks good regularity, so it is difficult to simply introduce the "slippery water + sand" technology. According to the physical simulation data of S reservoir, in the fracture initiation stage and the initial propagation stage, gravel will have an influence, and the viscosity of slippery water is low. Therefore, the pre-fracture formation with large displacement can further improve the fracture complexity in the well zone. Therefore, the complex fracturing technology of "guar gum + slippery water" was selected in the initial stage of horizontal well S development to construct fractures with good complexity and further enhance the reconstruction effect. According to the field practice, frequent overpressure and high pressure can not effectively improve the construction displacement in the process of joint construction with slippery water, which will not only affect the construction process, but also affect the actual cracking effect of artificial cracks. The main reason is that the leakage of slippery water is large and the viscosity is low, and the fractures in the well zone of conglomerate formation are narrow. In addition, the physical simulation found that the cluster perforation and shallow deep hole may lead to the increase of fluid friction, while the liquid in conglomerate has a circumferential flow phenomenon, which promotes the complexity of fractures near the wellbore. Through the combined action of the two factors, the construction process always maintains a high pressure, unable to raise the displacement smoothly, and thus difficult to fully extend the artificial cracks.

Therefore, combined with the advantages of various fracturing fluids, 200mPa·s cross-linked Guar gum was selected to effectively improve the cracking effect and efficiency of the main joint by taking advantage of the low filtration loss and high viscosity of Guar gum during the fracture initiation period. At the same time, it was combined with slug treatment to build a good seam width and reduce the construction pressure. After the construction displacement meets the requirements, 5.5mPa·s slippery water is selected for joint construction. The reverse mixing technology of "Guar gum sand carrying + smooth water seam making + Guar gum seam making" can not only effectively extend the main crack, but also effectively connect the pore space of the main crack wave. According to the field practice results, the fracturing displacement of guar gum and smooth water can exceed $8m^3/min$ and $9.8m^3/min$ by using the above technology, which fully reduces the construction difficulty, improves the fracture initiation and extension effect, and further improves the contact area between reservoir and fracture. In 2018, the construction displacement was continued to increase, and the displacement reached $12m^3/min$ under the condition of 5000m horizontal Wells and 3 clusters of perforation, realizing the limit of wellbore construction [3].

(2) continuous + slug combination sand adding technology. With the increase of grain size fluctuation and content of gravel in conglomerate formation, the roughness of fracture surface also increases, which makes the placement of proppant more complicated than that of shale formation. In addition, the Young's modulus of S reservoir is in the range of 20GPa -- 25GPa, Poisson's ratio is in the range of 0.22 -- 0.25, and the formation is soft, resulting in proppant embedment, which has a certain impact on the conductivity level of artificial fractures. Therefore, this kind of reservoir needs to lay proppant strictly.

In S reservoir, there are depressions, and some fan bodies are developed. Due to certain differences in reservoir formation mechanism, reservoir characteristics and reservoir characteristics of different slopes are also quite different. The BKQ reservoir in West Slope 18 is dominated by mudstone with small sand span, which enables effective placement of proppant in the reservoir. The sand body of the second member of BKQ in the back slope 131 well area has a large span. The first member of Baier Member of the upper part of the sand body is an oil reservoir, while the second member of the middle and lower part of the sand body is a tight non-oil reservoir. There is no significant difference in lithology between the two layers, and there is no stress or lithology between each layer. With conventional fracturing techniques, the longitudinal extension of the artificial fracture is not controlled, which can affect effective propping during proppant settlement.

To solve the above problems, with the help of sand carrying performance test analysis, the proppant settlement law of various sand carrying liquids and the settlement law of various injection methods were revealed, and then with the help of sand adding process optimization, proppant placement was effectively optimized to provide a good guarantee for placement concentration.

First, the law of proppant settlement under various sand carrying liquid conditions. For low-viscosity liquids, dynamic sand processing is generally carried out with the help of the turbulent action of fracturing fluid to simulate the displacement condition of $4m^3/min$. When injected into sub-fractures, the proppant will settle in fractures. Only after the displacement is adjusted to $8m^3/min$ can the settlement be effectively alleviated. Compared with low viscosity liquid, high viscosity liquid has more prominent sand carrying capacity, and will not easily appear settlement based on low displacement conditions.

Second, settlement law based on different injection mode conditions. Analysis of proppant deposition of low viscosity fluid based on various injection conditions. In the case of continuous injection, the proppant rapidly settles and becomes a berm that cannot be fully injected into the distal fracture. The slug injection method is selected. After the front slug completes the injection operation and becomes a sand bank, the rear slug will gradually advance the injection [4].

Under the condition that there is no stress shielding at the bottom of the reservoir and the lithology is not obvious, the method of "slug carrying sand + slippery water" is selected in the pre-forming link of the reservoir to construct the shielding layer at the bottom to avoid the fracture derivatives.

In the phenomenon of proppant embedment, the high sand ratio of ggel was selected for continuous sand addition to fully strengthen the proppant distal injection level, increase the proppant placement thickness per unit area, and fully improve the fracture conductivity level.

Therefore, the organic combination of sand addition technology and large-displacement reverse-mixed fracturing technology can fully improve the fracture-forming effect of conglomerate formation. When guar gum is selected, the crack initiation effect can be fully improved. After meeting the design displacement requirements, smooth water is selected to increase sand concentration according to the step mode for sand addition treatment. When the slug sand content is 25%, the high sand ratio of ggel is selected to carry out continuous sand addition operation. The application of this technology in field practice can not only have outstanding adaptability, but also can carry out construction activities smoothly.

3. Field application status

After 2016, 86 secondary volume fracturing processes were applied in the S reservoir. Selecting the cementing plug cluster perforation process not only ensured that the requirements of horizontal well reconstruction in the 6000m and 2000m horizontal sections were fully met, but also ensured that the maximum length of the horizontal section was 2022m, the maximum depth was 6008m, and the distance between fractures was shortened from 70m to 30m, further improving the effect of reservoir reconstruction. The proportion of slackwater was increased to 70%, and a large amount of liquid was injected into the formation to ensure that the formation energy was maintained. The maximum fracturing volume and support dose reached more than $4000m^3$ and $2400m^3$ [5].

On this basis, the production of horizontal Wells in tight reservoirs has been effectively broken through after reconstruction treatment, and the stable production capacity has been continuously improved. At present, the cumulative production of horizontal Wells in S formation has reached 86, all of which are capable of self-injection production. The daily oil production of more than 86% of the peak well is more than 40t and the maximum oil production is 117.7t, and the cumulative oil production of a single well exceeds 8589t in 300d and 28500t in 800d, reflecting stable and sustainable production capacity. See table below.

Table 1 Peak daily oil production of horizontal Wells in S reservoir

Oil production	proportion
≤10t	14%
40—60t	60%
60—80t	14%
80—100t	2%
100—120t	10%

Table 2 300d oil production of horizontal Wells in S reservoir

300d Oil production	Distribution ratio
$0.2—0.4*10^4t$	3%
$0.4—0.6*10^4t$	27%
$0.6—0.8*10^4t$	24%
$0.8—1.4*10^4t$	35%
$1.4—2.0*10^4t$	11%

4. Conclusion

In summary, for the reconstruction work of S conglomerate reservoir, combined with the concept of "fracture-controlled reserves", cutting volume pressure and reverse mixed fracture-building technology are proposed to fully improve the operation effect of tight reservoir horizontal Wells. This technology is highly targeted and operable. Volume fracturing + level and technology can fully improve the development quality of tight reservoir. It is also necessary to combine the theoretical research of reducing cost, increasing benefit subjects, improving nonlinear seepage flow and conglomerate fracture expansion mechanism, continuously optimize the fracturing scale and fracturing materials, and fully strengthen the production efficiency of horizontal Wells in tight oil reservoirs.

References

1. Zhang Yanming, Close cut volume fracturing technology and scale application of tight gas horizontal well in Sulige. Changqing Oilfield Company, petrochina Co., LTD., Shaanxi Province,2019-04-01.
2. Zhang Hongtao, Wang Xianjun, Wang Wei, Li Yonghuan, Gu Mingyong. Close cut volume fracturing technology for Songbei tight oil [C]//.2021IPPTC International Petroleum and Petrochemical Technology Conference Proceedings, 202:508-516
3. Chen Chaofeng, Wang Bo, Wang Jia, Xu Shifu, Qin Yingmin, Li Xuebin. Horizontal well fracturing technology in the second class lower sweet spot of Jimsar shale oil [J]. Petroleum Drilling Techniques, 2021, 49(04):112-117.
4. Huang Yue, Jin Zhirong. Study on close volume fracturing strategy of shale oil in Huazhuang Block [J]. Petroleum Geology and Engineering, 2022, 36(05):96-100.
5. Liu Honglei, Xu Shengqiang, Zhu Biwei, Zhou Linbo, Huang Yajie, Li Baolin. Research and practice of volume fracturing technology for intersalt shale oil [J]. Special Oil and Gas Reservoirs,2022,29(02):149-156.

https://doi.org/10.1051/e3sconf/202337501029

Discussion on reservoir description method of well vibration combined with tight well pattern area

Pan Liu

Institute of Geology, No.1 Oil Production Plant, Daqing Oilfield Co., LTD., Daqing, Heilongjiang 163000, China

Abstract. In the process of rapid social and economic development, the demand for the oil of related industries is more and more large. The continuous increase of oil exploitation is causing some oil fields to gradually enter the period of special high water cut development and increasing the difficulty of oil exploitation. Therefore, how to improve oil recovery is the key question that relevant personnel think and solve. Based on this, this paper mainly describes the channel sand body description technology combined with well shock, and analyzes the reservoir description method combined with well shock combined with the general situation of an oilfield working area, aiming at improving the prediction accuracy of sand body in dense well pattern area and effectively guiding the development and adjustment of potential excavation work.

Key words: Tight well pattern area; Well shock combination; Reservoir description method.

1. Introduction

The application of the well vibration combination technology in the dense well pattern area can more accurately describe the inter-well sand body, accurately combine the channel sand body, effectively improve the reliability and effectiveness of various work, and provide strong support and guarantee for the later oilfield development and production. In addition, the method of combining well seismic with reservoir description in dense well pattern can recombine abandoned channel, correct various types such as variable sand body in channel, main channel shape and swing, and describe channel sand body more accurately, effectively avoiding the uncertainty of reservoir description by only using well data.

2. Analysis of channel sand body description technique combined with well shock in dense well pattern area

2.1 Section extraction and selection

The equal-scale subdivision method based on the top and bottom of the reflection marker layer applied in the time domain can effectively characterize the relative equal-size of the interface of the small layer. Meanwhile, in the actual operation, the method of correcting well curve, extracting frequency conversion and variable wavelet can be used in the dense well pattern area to form a synthetic record, and the more precise calibration and final zone calibration can be applied in the combination of well vibration and marker layer. In addition, the staff realized equal proportion subdivision according to the marker layer to analyze the sedimentary characteristics of the target layer in detail. Local adjustment of stratigraphic level is targeted, and fine isochronous stratigraphic framework close to sedimentary unit level is created at the same time to ensure the extraction accuracy of the target layer as slice [1].

With the help of a variety of dense well pattern area information, the specific thickness of sandstones drilled at well points and the amplitude energy of stratum sections are analyzed in detail. Generally, about 74% of the range of medium-strong amplitude area is distributed in the well points of a class of sandstones, the thickness of which is no less than 2 meters, so as to determine that the well shock matching relationship is good. Ensure that the optimized strata sections have good applicability in channel sand body characterization.

2.2 Channel sand body characterization

The seismic data show a higher plane resolution, and the boundary and strike of the channel sand body can be identified more accurately. Based on the level of seismic identification and current technology, Continuous research and practice have formed a variety of fine description methods of well seismic combination reservoir, such as well point microfacies control, collaborative analysis of seismic sedimentology plane prediction, fine tracking and characterization of key target sand bodies, and separate treatment of different kinds of sand bodies, which promote the identification of river channel sand bodies and realize the semi-quantitative description of "seismic trend guidance".

(1) Composite river channel. Combined with the sedimentary study of the new period, it is clearly found that for the composite channel, there is not only one abandoned pattern of meandering ring, but also the abandoned pattern formed by multiple abandoned meandering ring in the area of high curved composite channel. It is generally difficult to identify the point dam body of the channel in the same period on the plane. In addition, based on the analysis of diagenetic conditions, it is difficult to accurately determine whether the multi-stage channel plane is cross-cutting and overlapping, or a single channel has undergone multiple channel changes [2]. However, based on the analysis of slice attributes, it is difficult to define the key regularity of the composite channel sand body morphology under the influence of high value surrounding rock and multi-point dam plane convergence, etc., and it is relatively easy to define the boundary of point dam and the contour of variable sand body. (1) A comprehensive analysis of the seismic trend information, to determine the size of the point dam, the specific spread of abandoned river in the meansional zone. (2) Reasonable application of optimal seismic attribute slices can more accurately define the abandoned channel strike and spreading point in relevant areas, and accurately depict the channel trend within the range of large high-value sandstone. (3) Guide the plane combination of internal point dam and abandoned channel in the composite channel, and then show a good response to the abandoned channel in the sand in the seismic inversion profile, further define the boundary points between Wells, constantly optimize the understanding of plane combination, and then effectively complete the fine characterization of well seismic combination reservoir.

(2) Narrow river channels. The application of seismic attribute slicing and inter-well inversion profile can effectively improve the certainty of the boundary of the channel sand body. At the same time, the application of frequency division attribute slicing can effectively guide the characterization of the channel sand body, fully implement the narrow and small channel, realize the effective combination of points and lines, draw the boundary line more precisely, and define the spreading point of the narrow and small channel. For example, analyze the actual situation of one of the small layers (as shown in FIG. 1), clarify the specific strike and spread point of the narrow and small channel in "looking for sand in the mud", clarify the boundary line of the channel between Wells with sudden change of seismic attribute value (as shown in FIG. 1A), and effectively correct the amplitude slice by combining with the actual well point curve. The boundary and strike of narrow and small river channels are more accurately reflected (as shown in Figure 1B). In addition, in profile, combined with the actual situation of inter-well sand body connectivity in seismic inversion profile, the boundary points between narrow and small river channels were further defined (as shown in FIG. 1C).

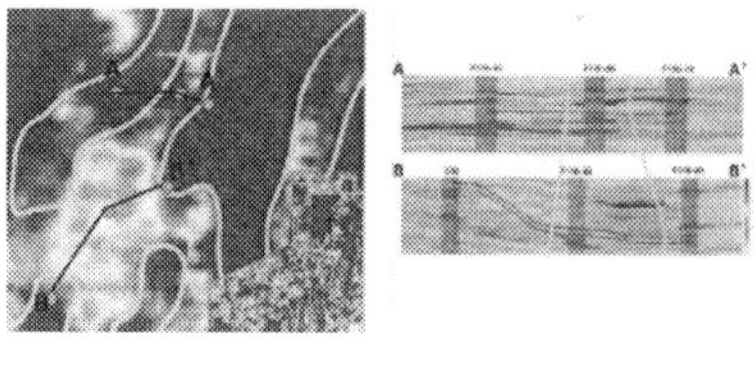

(A) Looking for sand in the mud (B) The inversion profile confirms the connectivity of sand bodies between Wells

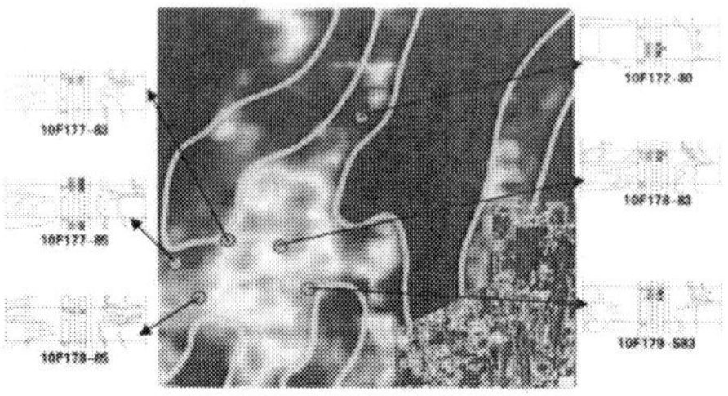

(C) Well point verification

Figure 1: Combination of well and seismic characterization of narrow and small channel sand bodies

3. The significance of fine characterization of reservoir combined with well vibration

On the one hand, well seismic combined with reservoir fine characterization to achieve effective integration of seismic data and geological information, and constantly improve and perfect the previous model mapping method. The traditional mapping method is mainly based on the microfacies of modern deposition and outcrop, and the planar combination of well points. However, the combination of well seismic and reservoir description method can organically combine seismic and well information, give full play to the application value of seismic information, provide more complete and reliable model information for fine characterization of channel sand bodies, and obtain the result of fine characterization of channel that is most similar to the underground reality [3].

On the other hand, the method of reservoir characterization based on well vibration has changed greatly, which further realizes the fine characterization of channel sand body. In practical application, this method is mainly combined with 3D seismic data to correctly interpret the specific strike and distribution characteristics of the channel sand body in space. At the same time, combined with the specific analysis of well information, it can be effectively transformed into the channel distribution characteristics of different sedimentary units, so as to more accurately identify the channel boundary on the plane and more accurately define the channel boundary points on the profile. Therefore, on this basis, effective combination of well point microfacies plane and section is carried out to create a 3D geological model meeting the actual conditions of the well, which effectively solves the problems of unclear boundary and multiple solution of plane combination in the characterization of well channel sand body by traditional model mapping method, and comprehensively improves the prediction accuracy of inter-well channel sand body.

4. Overview of tight well pattern area

The east-west fast fault of the first block North of an oilfield is mainly located in the middle of the Short-sleeve anticline structure in the north of the oilfield, and the oil-bearing area is about 17.2km2. This study area is mainly located in the frequent Baidong area along the shoreline of the lake, where there are various types of oil deposits. Meanwhile, channel sandbodies, estuary bar sandbodies and thin bedded sandbodies interlace with each other, showing serious heterogeneity, and channel sandbodies are key reservoirs. Reservoir porosity is between 23% and 25%, permeability is about (300-1500) $\times 10^{-3}\mu m^2$, present in the hole, the permeability characteristics of composite moisture content is above 90%, belong to determining phase.

5. Application of reservoir description method in combination with well shock

5.1 Conditions of Use

Compared with the logging method, the seismic method has a significant advantage in the coverage density on the plane. However, the current seismic identification ability and technical conditions in the practical application process tend to cause the seismic longitudinal resolution is relatively low, about 10 meters. The well shock coincidence rate of different types of channel sandbodies in the study area was analyzed, mainly the well points with sandstone thickness of no less than 2 meters, which were located within the range of medium-strong amplitude attribute. It was obviously found that the well shock coincidence rate was not less than 50% and the ratio of layers was not less than 65% in meander river, delta distributary plain and channel sandbodies. However, the well seismic coincidence rate of the delta far bank and outer front facies sandbodies is relatively low, which fully indicates that the seismic data have good applicability in the prediction of meandering river, delta distributary plain and inner front facies channel sandbodies.

Channel microfacies distribution direction, width, fork Angle, width-thickness ratio and other parameters in different development zones are different to some extent, so it is necessary for staff to conduct statistical analysis and create reservoir parameter database, as shown in Table 1.

Table 1: River channel parameters of a and b dense well pattern development zones

block	Mean channel width (m)	Mean channel thickness (m)	Channel width to thickness ratio	Angle of bifurcation (°)	Extension azimuth (°)
a	56.1	1.4	37.34	39.1	31-38
b	90.3	2.1	40.90	38.2	32-44
On average	82.4	1.7	39.20	36.6	32-39

5.2 Improve the prediction accuracy of inter-well sand body

The application of well vibration combined with reservoir description method can more accurately depict different types of channel sand bodies, fully reflect the sedimentary features of sand bodies in various sedimentary environments, so that people can more correctly understand the distribution characteristics of sand bodies and effectively approach the actual underground conditions. In addition, the staff confirmed through well inspection that the combination of well shock and well sand body prediction, clear boundary and well data study results are more reliable, mainly in the following aspects:

(1) Channel sand body boundary position transformation. Combined with the depositional facies zone diagram of the well, it is clear that the channel boundary is mainly in the central part between the two Wells, and the channel width is predicted to be about 110 meters. However, combined with the analysis of seismic attribute slices, the location about 30 meters south of the well is the attribute mutation position, and then it is determined that the width of the river channel presents a rapid change rate, while the width of the vertical channel strike is about 50 meters. The detailed analysis of the logging characteristics of the reserved posterior well shows that the spontaneous potential presents a straight curve. However, the microelectrode curve mainly presents slight fluctuation without significant difference, thus determining that the well is a pinch-out microphase, which fully indicates that the application of the well shock combined with the sedimentary facies zone diagram can effectively enhance the identification accuracy of the boundary position of the channel sand body between Wells [4].

(2) Newly added channel sand body. Combined with the depositional facies zone diagram of the well, it is determined that two Wells in the study area are not channel microfacies, so it is determined that there are corresponding channel microfacies between the Wells. However, the analysis with the help of seismic attribute section shows that there is a certain low value range of seismic attributes between the two Wells, and then the well seismic combined depositional facies zone diagram is used to effectively identify the channel microfacies about 40 meters wide between the two Wells. A detailed analysis of the detailed logging characteristics of the reserved post-test well shows that the spontaneous potential presents a bell-shaped curve, and the curve amplitude difference at the bottom of the microelectrode is relatively high. The comprehensive determination that the well belongs to the channel microfacies fully indicates

that the application of the combination of well shock and sedimentary facies zone diagram can effectively improve the effective identification accuracy of narrow and small channel sand bodies between Wells [5].

(2) Added microphase inside and outside the surface and pinch-out. According to the sedimentary zone facies diagram of the Wells, it is mainly determined that two Wells in the study area are the same channel, and there is no surface or inside-out microfacies between the Wells, showing good connectivity. With the help of seismic attribute slices, it can be clearly found that there are attribute mutation areas behind the two Wells, and then according to the combination of well seismic sedimentary facies map, the existence of about 50 meters wide pinch-off microfacies between the two Wells can be effectively identified. In addition, by analyzing the logging characteristics of the reserved post-test well, it is obviously found that the spontaneous potential mainly presents a flat curve, while the microelectrode does not have a curve amplitude difference. Therefore, it is determined that the well belongs to the pinch-out microfacies, which fully indicates that the application of the well shock combined with the sedimentary facies map can comprehensively improve the identification accuracy of the newly added inner and outer channel sand bodies and pinch-out microfacies between Wells.

6. Conclusion

In the practical application of the method of combining well seismic with reservoir description, it can effectively give full play to the maximum application value of the dense well pattern data and seismic data in the development block, reduce the subjectivity in the study of sedimentary facies with the help of well data, reduce the multi-solution of the prediction of inter-well channel sand body with seismic data, and effectively improve the accuracy of the sedimentary microfacies map.

References

1. ZHAO Mingzhu. Quantitative prediction method of reservoir thickness based on well seismic combination [J]. Journal of Xi 'an Shiyou University (Natural Science Edition), 2021, 036(006):16-22,55-55.
2. Wei Feng, Xue Guoqing, Xiong Cong, et al. Classification and comprehensive evaluation of facies controlled reservoir in sparse well area combined with well earthquake: A case study of Zhujiang Formation in WC-A Oilfield, Zhujiang Estuary Basin [J]. Journal of Oil and Gas Technology, 2020, 42(2):10-10.
3. NA Xiaoxun. Application of well seismic combined statistical inversion method in thin layer sand body prediction [J]. Journal of Heilongjiang University of Science and Technology, 2020, 030(005):572-576.
4. TAN Zhaobin. Reservoir description method based on well vibration in dense well pattern area [J]. China Metal Bulletin, 2020(6):2-2.
5. LI Wenlong. Reservoir description method based on well vibration in dense well pattern area [J]. Petroleum Geology & Oilfield Development in Daqing, 2021(4):6-6.

https://doi.org/10.1051/e3sconf/202337501030

Analysis of the exploitation potential of well DA1

Lei Wang

Geological Management Office, Geological Technology Research Institute, Gas Production Company of Daqing Oilfield Co., LTD. (Gas Storage Company), Daqing Heilongjiang 163000, China

Abstract: DA1 well exploited two sets of reservoirs, volcanic rock of Ying1 member and conglomerate of Ying4 member. In the initial stage of exploitation, the well had a high productivity. After an operation in 2010, the production situation of the well changed dramatically. In this paper, through the comparison of the production situation before and after the operation, the geological situation and the adjacent well, the operation pollution situation of the well is analyzed, and the exploitation potential of the well is analyzed, the measure scheme of plugging removal is put forward, and the stimulation prediction and economic benefit evaluation of the well are carried out, proving that the measure is effective and feasible.

Key words: Operation pollution; mining potential; plugging removal.

1. Gas well overview

The DA1 well is a development well with a west-dipping monoclinic structure and two sets of volcanic rocks in the first member and conglomerate in the fourth member. From March 25, 2007 to May 28, 2007, the well was tested after pressure. The 158 and 161 I zones achieved open flow of 63,000 m3 / day and 49,000 m3 / day respectively, indicating high gas well productivity. The well was put into operation on November 13, 2007, with a cumulative production of 68 million cubic meters of gas and 8,937 cubic meters of water up to now. In 2010, the well experienced an operation. Due to the pollution, the production situation of the well changed dramatically after the operation, and the wellhead pressure and daily gas production decreased significantly.

2. Operation pollution analysis

From September 2010 to October 2010, DA1 well was overhauled and plugging due to the high safety risk due to the pressure of the technology sheath. During the operation, the well was killed for several times and there were large losses, so it was decided to restore the process and maintain the production.

2.1 Comparison of production conditions

Before the operation, the wellhead pressure of this well is 18.1MPa and the daily gas volume is 70,000 m3. After the operation, the wellhead pressure drops sharply to 6MPa and the daily gas volume drops to 38,000 m3, while the water production only drops 0.8 m3 and the water-gas ratio increases. As shown in FIG.1, the current wellhead pressure of this well is only 5.4MPa and the daily gas volume is 11,000 m^3.

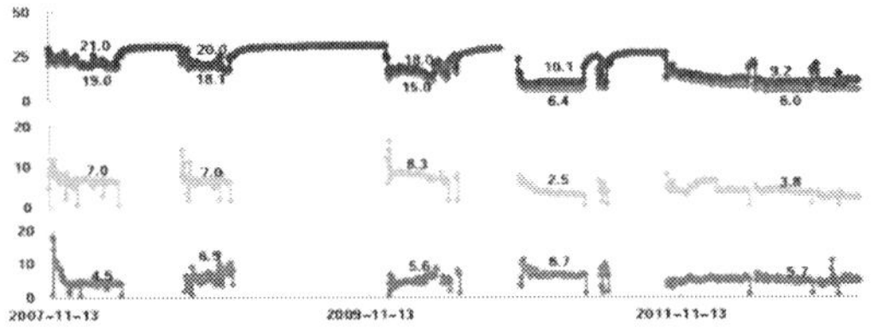

Fig. 1 Production curve of well DA1

2.2 Contrast of geological condition

The current average open flow rate of well DA1 is 14,500 m3 /d, which is 76% lower than the original average open flow rate. Before the operation, the formation pressure dropped slowly, producing 43.99 million cubic meters of gas per unit of pressure drop. After the operation, the pressure dropped 21.7%, and the gas production per unit of pressure drop dropped to 12.9 million cubic meters. The well control dynamic reserves have decreased significantly from 520 million cubic meters before the operation to 360 million cubic meters today.

2.3 Near well contrast

A longitudinal comparison between this well and two non-kill Wells in the same block during the same period shows that the production situation of the non-kill Wells DA2 and DA3 did not change significantly before and after the operation, as shown in Figure 2 and Figure 3, while the productivity of the DA1 well declined sharply after the killing operation.

 https://doi.org/10.1051/e3sconf/202337501030

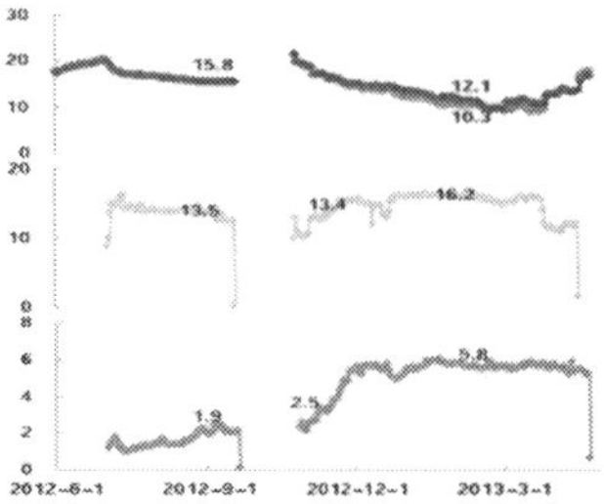

Fig. 2 Production curve of well DA2

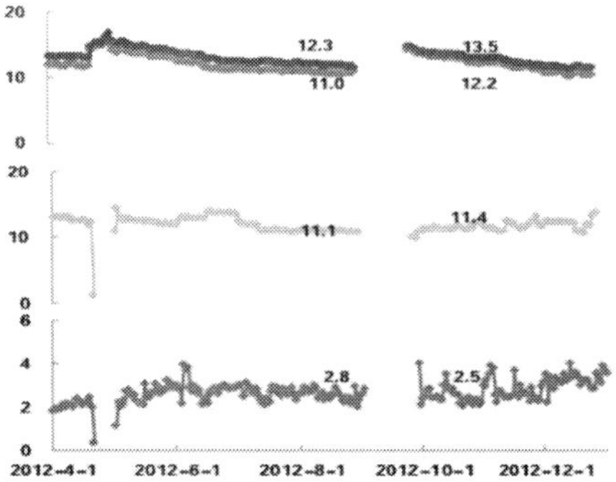

Fig. 3 Production curve of well DA3

2.4 Analysis of operation pollution

A total of 521 square meters of killing fluid were injected in the operation, and the loss reached 188 square meters. The types of killing fluid include Nacl killing fluid, solidified water, gel and solidless killing fluid. The damage factors of this well can be summarized into five categories according to the damage types and causes of oil and gas formations, as shown in Table 1 [1]. It causes blockage of part of the orifice throat and pollutes the formation. At the same time, it causes blockage of the seepage channel, which makes it difficult for gas to enter the wellbore and reduces the gas production of a single well. Second, the well mining volcanic rock, conglomerate, is a low porosity and low permeability reservoir, in the long-term kill fluid immersion, strong clay expansion, reduce the permeability of the reservoir, resulting in formation pollution; Thirdly, in the production process of this well, there is a two-phase flow of gas and water. Due to the different working conditions between the stratum and the bottom hole, part of the phenomenon of "calcium scale" may appear, leading to the increase of the skin coefficient near the well. Fourth, during the operation, a certain amount of plugging agent was added for plugging, and the construction amount reached about 80m3. The formation pollution caused by plugging agent was one of the important factors leading to the decrease of gas production in the well. Fifth, in the production after the operation, the water gas ratio of the well increased, and the test data showed that there was fluid accumulation at the bottom of the well, which resulted in the water lock effect in the production process.

Table 1 Shows the types and causes of oil and gas formation damage

type		Element of injury			Mechanism of damage
		Reservoir (internal cause)	Formation fluid (internal cause)	External fluid (external cause)	
Water sensitivity		Montmorillonite, illite/montmorillonite mixed layer, hydrated mica, illite, chlorite, kaolinite	—	Fresh water working fluid	Lattice expansion, dispersion, migration
Salt sensitivity				High salinity working fluid	Dispersion and migration
Acid sensitivity	HCL	Chlorite, montmorillonite, mixed layer, kaolinite, iron calcite, iron dolomite, biotite, heavy minerals containing iron	—	Hydrochloric acid	Fe (OH)$_3$ precipitates and the acid-etched particles are released and transported
	HF	Calcite, dolomite, clay, mica, feldspar, quartz	K^+、Na^+、Ca^{2+}、Mg^{2+} in formation water	Hydrofluoric acid	Fluosilicate, fluoaluminate precipitate, CaF_2 precipitate
Alkali sensitivity (PH>8)		All kinds of clay, feldspar and quartz	Ca^{2+}、Mg^{2+}、CO_3^{2-}、HCO_3^-、SO_4^{2-} in formation water	High OH^- Concentration working fluid	Silicate precipitation, inorganic mineral precipitation, dispersion, migration, lattice expansion
Velocity sensitive		Clay minerals, microcrystalline quartz, microcrystalline feldspar,	—	Strong mining makes formation fluid flow at	Dispersion and migration

		microcrystalline carbonate minerals, fibrous anhydrite, amorphous materials		high speed	
Sand production		Formation particles, clastic particles, loose sandstone formation	—	Strong mining makes formation fluid flow at high speed, and the amount and concentration of acid are too large	Weak consolidation, destruction of rock cementation, particle migration
scaling	inorganic	Carbonate, part of clay, salt crystals	Salt in formation water	High PH, different ion types and content, production caused by temperature, pressure changes	$CaCO_3$、$CaSO_4 \cdot 2H_20$ Inorganic salt precipitation
	organic	—	Paraffin and bitumen in crude oil	Gas injection, acid solution, steam inhalation	The solution - precipitation equilibrium is disrupted
External blockage		—	—	A solid phase carried by a foreign liquid	Solid ions were deposited and bridged in the pore throat

3. Analysis of exploitation potential

3.1 Gas well potential analysis

The current well controlled dynamic reserves of this well are 360 million cubic meters, the cumulative gas production is 68 million, the recovery degree is only 18.33%, the remaining reserves are large, the potential of the gas well has not been fully played, and the subjective ability to explore potential. This well is located in the high part of local structure, with favorable structural position, good reservoir property, porosity development and strong storage and permeability capacity. After measures, it is expected to restore high productivity. The physical conditions of this well are similar to those of well DA4 and well DA5, as shown in Figure 4. However, these two Wells have higher productivity and good production situation under the condition of no operation pollution, and their recovery degree is higher than that of this well.

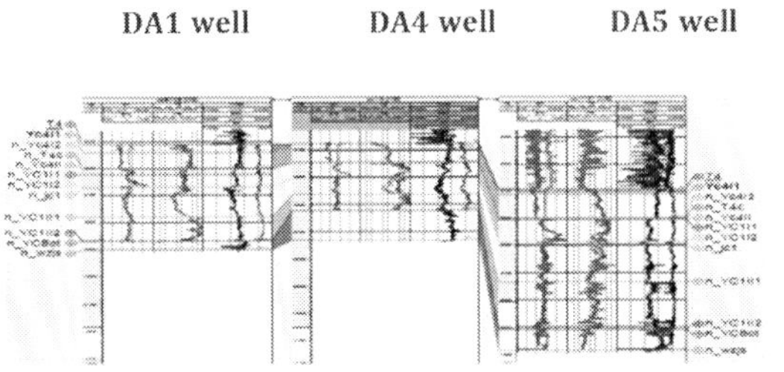

FIG. 4 Stratigraphic correlation between well DA1 and adjacent well Yingcheng Formation

3.2 Analysis of measures

In 2013, the 13Cr anti-corrosion tubing was replaced without killing the well in DA1 well, and no resistance was encountered in all tests, indicating that the wellbore conditions were perfect and the technological conditions were met with measures. Refracturing and plugging are two common ways to stimulate such contaminated gas Wells.

3.2.1 Feasibility analysis of refracturing

Through the statistical analysis of refracturing Wells in DA block, it can be seen that refracturing can increase the production of gas Wells to some extent, but the fracturing cost is large, the payback period is long, the construction process of fracturing is multiple, the construction is difficult, and the risk of construction failure is relatively high. In addition, DA1 well is originally self-injection after pressure, and the fracture characteristics after fracturing are obvious, so it is not necessary to conduct secondary fracturing.

3.2.2 Feasibility analysis of unplugging

Commonly used unplugging methods are: Organic acid deplugging, inorganic acid deplugging, surfactant deplugging and alcohol deplugging. Take DS1 well as an example, before deplugging, the skin coefficient is large, the reservoir pollution is serious, and the daily gas production is low. However, after deplugging, the skin coefficient decreases, the pollution is relieved, the daily gas production increases significantly, and the deplugging effect is good. It is suggested that well DA1 be de-plugged to restore gas well productivity.

3.3 Gas well stimulation prediction and economic benefit evaluation

3.3.1 Production increase forecast

Because of the particularity of gas Wells, the wellhead pressure before and after the measurement is difficult to be consistent. Therefore, it is more representative to count the production as open flow when calculating the stimulation prediction. According to formula (1), the stimulation factor was 1.92 times and the stimulation period was 500 days.

$$J = qsc/ qsco \quad (1)$$

Where: J -- yield increase multiple, dimensionless;

qsc——Test output after construction, 104m3/d;

qsco——Test output before construction, 104m3/d。

3.3.2 Evaluation of economic benefits

According to formula (2), (3) and (6), it can be calculated that the payback period of plugging removal in well [2]DA1 is 3.3 months, the net present value is 2,300.15 million yuan, the internal rate of return is 30%, the net present value is greater than 0, and the internal rate of return is greater than 10%. The scheme is feasible.

$$T=G/g \quad (2)$$

Where: T -- payback period, month;

G——Project investment: ten thousand yuan;

g——Net increase profit, ten thousand yuan/month。

$$NPV=NPV1-NPV0 \quad (3)$$

Where: NPV——Actual cash inflow, ten thousand yuan;

NPV1——Cash inflow from increased production, ten thousand yuan;

NPV0——Investment in stimulation measures is converted into cash flow, ten thousand yuan。

$$NPV0=V0(1+i/12)n \quad (4)$$

$$NPV = \sum_{j=1}^{n} \left[V_j (1 + \frac{i}{12})^{(n-1)} \right] \quad (5)$$

Where: V j——Net production increase revenue for month j, ten thousand yuan;

i——Bank annual interest rate, dimensionless;

n——Effective period for increasing production, month;

V0——Direct investment in stimulation measures, ten thousand yuan.

$$IRR = a + [NPV_a/(NPV_a - NPV_b)] \times (b - a) \quad (6)$$

Where: IRR——Internal rate of return, %;

a、b——Rate of discount, a>b;

NPVa——When the discount rate is a, the net present value that you calculate is going to be positive;

NPVb——When the discount rate is b, the net present value that you calculate is going to be negative。

4. Conclusion and Understanding

4.1 Well DA1 is a combined production well of volcanic rock and conglomerate of Yingcheng Formation. It has good reservoir properties, high original open flow and high daily gas production at the initial stage of exploitation.

4.2 In 2010, the treatment technology of DA1 well was set with pressure, and the plugging failed, resulting in serious loss of kill fluid, resulting in reservoir pollution and sharp decline in productivity. However, the degree of recovery was low, the gas well itself had good physical properties, favorable structural position, and had potential enhancement ability.

4.3 It is suggested to restore the production capacity of gas well DA1 through plugging relief. This paper forecasts the production increase of this well after plugging measures. According to the forecast, the production increase multiple of this well after plugging measures is 1.92 times, the net present value is 2.30.15 million yuan, and the internal rate of return is 30%.

References

1. Wen Hao. Oil test operation technology. Petroleum Industry Press, 2002.4
2. Li Runzi. Dynamic analysis of oil and gas field gas reservoir and new technology of gas well stimulation and high production. Petrochemical Press, 2008.8 (5)
3. Wang Lixia. Improvement method of water flooding technology in high water cut stage of complex fault block reservoir in Huabei Oilfield [J]. Petroleum Knowledge, 2019(4):2-3.
4. Zhao Beichen. Coupling mechanism and parameter optimization of water flooding in fault block reservoir [D]. China University of Geosciences (Beijing), 2020.
5. Wang Nian-ming, GUO Hua-gu, JIANG Ming-sen. Research on oil stability and water control technology in middle and high water cut stage of complex fault block reservoirs: A case study of Yuejin No.2 Donggao Oilfield [J]. China Petroleum and Chemical Standards and Quality, 2013(11):1-3.
6. Wu Gang. Water flooding enhanced Oil recovery technology of fault-block reservoir in Shengli Oilfield [J]. China Chemical Trade, 2015(9)23-24.
7. Liu Xishun. Research on remaining oil distribution and potential exploitation strategies of extremely complex fault block reservoirs in high water cut development period [J]. Xinjiang Oil and Gas, 2008, 4(4):4-5.

https://doi.org/10.1051/e3sconf/202337501031

Study on water flooding law in high water cut stage of complex fault block reservoir

Jing Chen

Institute of Geology, No.5 Oil Production Plant, Daqing Oilfield Co. LTD, Daqing 163513, Heilongjiang, China

Abstract: In recent years, with the progress of science and technology and the development of times, it has brought great opportunities for the development and protection of Tibetan resources in our country. Many reservoirs are located in very complex fault blocks with small fault blocks and many faults, coupled with strong heterogeneous media. After years of waterflooding development, they have entered the high water cut oil recovery stage. In view of this, this paper will focus on the analysis of the water drive law in the high water cut stage of complex fault block reservoirs, and carry out specific strategy optimization for increasing production, in order to promote the exploration of the law of high water cut area in complex fault block reservoirs.

Key words: Complex fault block reservoir; High water cut stage; Water drive law.

1. Introduction

At present, many reservoirs are complex fault block reservoirs, which have entered the late stage of development. Because the structure is very complex, and the waterflood development history is long, and the relationship between oil and water is very complex, also unconsciously aggravate the difficulty of development, but also cause the decline of production. By carefully analyzing the factors of oilfield development effect, understanding the existing problems, and timely taking targeted measures to stabilize production, we can better explore the water drive law of complex fault block reservoirs in the high water cut period. The combination of dynamic and static, and the use of relevant advanced information technology to conduct reservoir numerical simulation, understand the direction of the source water, summarize the law of its water area, and understand the distribution of remaining oil, can provide clear guidance for the formulation of subsequent ideas [1].

2. Geological survey of complex fault block reservoir

The overall development of a complex fault block reservoir is located in the ascending arm structure of fault No. 1, which grows in the northwest direction and is partially affected by the lateral blocking of volcanic rocks, forming a tectonic closed circle. The east-west faults gradually show a complicated trend. The average size is 0.75 square kilometers. The viscosity is 1.8mPas, belonging to a conventional thin oil reservoir [2].

Table 1 Statistical table of water flooding velocity of shallow strata in complex fault block reservoir 1-29 over the years

time	The first year	The second year	Third year	The fourth year	The fifth year	The sixth year
Water drive velocity (m/d)	1.76	1.91	1.93	1.95	2.01	2.10

3. Rules and specific research methods of water flooding in middle and high water cut stage of complex fault block reservoirs

3.1 Rule analysis

For reservoirs in high water cut development period, when faced with such special circumstances, it will intensify the difficulty and intensity of excavation, so it is necessary to seek effective operation rules. The first point is the law of plane waterflooding. The contradictions in reservoir plane of this complex fault block are more due to the control of the original advantageous permeability channels and other aspects, coupled with the influence of injected water inrush, so that the overall permeability shows a gradually increasing trend, and the channel velocity increases year by year and the average channel velocity even reaches 2.1M/D. In addition to the plane Internet alarm is not perfect area so that the water area is relatively weak, Erjing network perfect area when the water area is relatively strong. The second point is about the rule from the shallow water area. Because the permeability changes among the different layers in the reservoir of this complex fault block have great differences, it shows that the degree of water area also has obvious differences. Because water flow in a water drive reservoir will bypass the permeability zone, the high permeability zone with low channel resistance or some larger pores, the canal with high permeability will have a faster speed, and the water balloon with the highest permeability will have a slower speed [3]. In addition, different water injection methods also have a great influence on the effect of profile control. The third point is that the law of the water zone in the reservoir is complicated, and the fault block reservoir is mainly based on complex rhythm, and the logging curve will show different shapes, and be affected by the reservoir, physical property and other aspects. From the large physical property difference in the section and the high permeability at the bottom, the non-uniform medium strong water area may cause the internal oat surface plane to be in a contradictory state, which is difficult to transform. However, with the continuous maturation and development of the reservoir in the later stage, the pressure and permeability between each small layer increased significantly, resulting in a great change in the number of water-absorbing layers. And the long-term impact of rain erosion and other aspects will cause the middle and lower part of the reservoir is easy to deformation and form high mountain channels, such as water, the main deep channel in the lower part of the sweep or enrichment from the top of the remaining oil of the middle and strong water rock at the bottom. In order to further achieve the purpose of controlling water and stabilizing oil as well as the development effect of lifting cake, in normal production, the staff need to combine the actual distribution of water area on the plane, readjust the flow between production Wells and injection Wells, realize the allocation of resources, and form a state of uniform area on the plane. After a period of comprehensive treatment, it was found that the annual oil production increased by 0.65 million tons, the water driven reserves were 74,000 tons, and the degree of water driven increased by 0.2%, which decreased by 5.63% compared with the same period last year. The increase of the entire water cut was controlled below 2%, indicating that ideal effects were achieved [4].

3.2 Specific research methods

The first point is for the water zone speed discrimination. Due to the heterogeneity of reservoir and other factors, the velocity of the channel on the plane is not mature and uniform. The dynamic connectivity between water injection and surrounding production Wells, as well as the advancing direction and actual velocity of injection can be more objectively and clearly reflected in the relevant construction data. However, with the continuous injection of Wells, the difference of permeability and pressure often leads to the formation and development of large historical permeability channels. When the water content of oil Wells continues to rise, the dominant seepage channels in general high-yield areas will become more mature [5]. Subsequently, after analyzing the dynamic production data of a single well in this complex block, the staff found that the production characteristics of oil and water Wells on the plane were not consistent, mainly because the water content distribution of oil Wells on the plane was not uniform, and the proportion of oil Wells exceeding 60% was increasing year by year, so as to ensure the stability of oil production every year. The industrial chain of oil well will be continuously improved with the increase of water cut, which causes the aggravation of plane contradiction virtually. According to relevant data statistics, when the water content of oil well is greater than 40%, it is easier to create a superior seepage channel. At the same time, because the internal neck water flooding will be affected by rhythm and other aspects, the impact of load rhythm bottom and middle is more serious. The variation of interlayer water zone is more affected by pressure and permeability. Specifically, with the continuous exploitation of post, the contradictions between canals began to increase. Through the general analysis of the water injection well profile over the years, it is found that the water absorption layer presents a trend of decreasing year by year, and the phenomenon of single-layer main water absorption is more and more serious. Moreover, according to the analysis of relevant data, once the interval is greater than 5MPA, it is a multi-layer production well, and the pressure difference caused by the contradiction between the interval will be more significant.

The second point is about the research method of reservoir numerical simulation. This method is a kind of high quality reservoir numerical simulation technology based on theory and practical means of introducing advanced information technology. It can well solve a series of complex problems in the process of oilfield development. By simulating the variation of the saturation of melaleuca reservoirs in areas 1 to 29, the specific water flooding rule can be obtained. After specific research, it is found that the internal contradiction of LiuXiaozeng is the most serious in the section, mainly because the reservoir is dominated by the composite rate reservoir, the difference of physical properties in the reservoir is large, the non-uniform medium is strong, and the long-term impact of rain erosion. It is easy to form a deep channel in the middle and lower part of the reservoir. In addition, the effect of in-layer utilization in some reservoirs is not ideal, and the oil collected in the top and middle of the rhythm layer, plus the existence of advantageous penetration channels, leads to an intensified ineffective circulation of water between the injection well and the production well, which reduces the efficiency of water area spread and leads to the intensified contradiction between layers.

4. Study on stable production technology of complex fault block reservoir in middle and high water cut stage

Firstly, the waterflood project should be implemented to improve the effectiveness of waterflood development and improve the recovery efficiency by improving the well pattern. Dynamic detection, seismic and other related technical means are used to deepen the research on the connectivity of complex fault block reservoirs, optimize well patterns, and promote the effective improvement of double-direction benefit efficiency and corresponding rate of oil and water injection and production. Through the bridge matching commissioning linkage and injection eccentric technology, it can carry out matching commissioning linkage and electrodynamic inspection and sealing, and has the linkage inspection ability of commissioning injection polymerization well. At the beginning of the test, the number of intelligent metering oil and water Wells with throttling core is 8. Currently, they are used on 5 caliber, all of which have normal metering and fishing capabilities. In order to cater to the multi-layer detailed injection on the basis of thin reservoir and thin interlayer, the integrated sealing technology was used in the test to promote the realization of the goal of fine water injection. At the same time, the water drive wave and volume should be improved through the optimization of deep profile control. Focusing on the significant effect of water flooding, the large pore generated by long-term water injection is selected with higher strength and less damage to the bottom layer for the well area with obvious interzonal conflict. Based on the test results of the dominant front, for the direction of water injection near the fracture, the oil well is prefered with significant effect and the water cut is increased quickly. For Wells with low water injection pressure, profile control should be implemented to improve the direction of water flow, increase the volume of water injection and water injection wave, reduce the ineffective circulation of water injection, and improve the development efficiency of water flooding. Through the implementation of dynamic water transfer method, the effect of unstable water flooding can be brought into play. Through the simulation study of flow line simulation parameters, pulse irrigation, plane water transfer, intermittent irrigation and corresponding control can be carried out according to the basic principles of seepage dynamics, thus promoting the improvement of well group water flooding efficiency and reducing the decline of production. Based on the changes in the beneficial Wells and the timely description of the flow field changes, periodic irrigation and dispensing perfusion can be used to reduce the injection volume in the tight flow line oil-well zone and the supplementary injection in the sparse flow line oil-well zone. The well group decline has been effectively solved.

The second point is the scientific development of oil and gas fields by using multiple dynamic monitoring methods. Orderly monitoring is an important method to have a reasonable understanding of oil and gas reservoirs, so as to correctly evaluate the quality of oil and gas reservoirs and promote the effective increase of recovery. In the process of studying the whole oil field natural gas resources, the injection profile and production of oil field gas popularization are studied. The most important method of exploration well production is to detect the saturation of reservoir, test casing damage, monitor the ability of cementing, etc., which also involves testing the well pressure, recovering the pressure drop, detecting the edge, measuring the pulse in the interference well, can avoid the situation of ineffective water injection. As the development cycle of water injection continues to increase, the problem of well underinjection is becoming more and more serious, because of the low water quality, repeated well opening, high sensitivity to rock formation and other factors caused by the problem of rock formation pollution is becoming more and more serious. By using the coefficient of skin in well test studies, the fault blowdown situation is evaluated. The problem of under-injection in well can be solved reasonably by using acidizing process interpretation and formation energy can be reasonably supplemented. It is necessary to reasonably increase the injection water distribution in the weak water reservoir and non-mainstream line, so as to increase the injection and production pressure difference and increase the sweep range. For Wells in normal state, it is necessary to ensure the stability of water allocation, so as to maintain the stability of injection and production pressure difference and promote the steady growth of water content.

5. Conclusion

In a word, after an objective analysis of the water flooding laws of some alpine channels in many complex fault block reservoirs, the differences between different layers can be understood, so that the permeability, grade difference detection and pressure level difference in the reservoir are also decisive factors to determine whether profile control can be real and effective. The implementation of interval recombination according to its profile situation can achieve the effect of improving production and technology. With the continuous development of reservoir exploitation technology, we can further analyze the characteristics and distribution law of oil reservoir distribution, further optimize the reservoir exploitation process, effectively deal with the contradictions between various oil layers, effectively do all the work in the high water cut development stage, properly solve the relationship between different layers, and then achieve good exploitation effect.

References

1. Wang Lixia. Improvement method of water flooding technology in high water cut stage of complex fault block reservoir in Huabei Oilfield [J]. Petroleum Knowledge, 2019(4):2-3.
2. Zhao Beichen. Coupling mechanism and parameter optimization of water flooding in fault block reservoir [D]. China University of Geosciences (Beijing), 2020.

https://doi.org/10.1051/e3sconf/202337501031

3. Wang Nian-ming, GUO Hua-gu, JIANG Ming-sen. Research on oil stability and water control technology in middle and high water cut stage of complex fault block reservoirs: A case study of Yuejin No.2 Donggao Oilfield [J]. China Petroleum and Chemical Standards and Quality, 2013(11):1-3.
4. Wu Gang. Water flooding enhanced Oil recovery technology of fault-block reservoir in Shengli Oilfield [J]. China Chemical Trade, 2015(9)23-24.
5. Liu Xishun. Research on remaining oil distribution and potential exploitation strategies of extremely complex fault block reservoirs in high water cut development period [J]. Xinjiang Oil and Gas, 2008, 4(4):4-5.

https://doi.org/10.1051/e3sconf/202337501032

Study on the method of remaining oil exploration in thin differential reservoir

Li Gao

Technical Management Office of the Fifth Operation Area, No. 2 Oil Production Plant, Daqing Oilfield Co. LTD, Daqing, Heilongjiang: 163000, China

Abstract. It is difficult to excavate the remaining oil in the thin differential reservoir and the oil-water reservoir. The remaining oil can be excavated by means of static and dynamic analysis method and micro-structure method to continuously improve the recoverable benefit of the oilfield. Based on this, this paper studies the potential excavation methods of the remaining oil in the thin differential reservoir.

Key words: Oil-water layer; Thin differential reservoir; Remaining oil.

1. Introduction

By studying the sedimentary characteristics of thin oil layer, we can fully understand the characteristics of thin oil layer sand body and the distribution law of remaining oil, and do well in the preparation of drilling plan. According to the sedimentary characteristics of different thin reservoirs, selecting the right type of completion and perforation fluid can effectively develop thin reservoirs and fully excavate the remaining oil.

2. Fracturing technology of remaining oil in oil-water reservoir and thin differential reservoir

2.1 Flow-limiting fracturing completion technology

For new Wells that have not been perforated, the flow-limiting fracturing and completion technology can be used. Perforation is also an important part of the flow-limiting pressure completion technology. The flow-limiting fracturing and completion technology mainly controls the diameter of the fracturing purpose or the number of perforated perforations in order to maximize the displacement construction. Relying on the number of layers that have been previously opened, the bubbles generated by absorbing fracturing fluid continuously increase the pressure at the bottom of the hole along the module, so that the fracturing fluid can be divided and then a larger area of fracturing can be opened. The objective of treating all reservoirs during construction is to maintain fractures in each zone through Gaza.

2.2 Thin interval balanced flow limiting fracturing technology

In order to excavate the thin reservoir without perforation and the thin differential reservoir with high aquifer, the thin interlayer balanced linear flow method is mainly used. The purpose of protecting the thin interval during the fracturing process is to keep the fracture horizontal and to keep the staggered fracture harvest zone within the actual reservoir. The high aquifer and the high aquifer are continuously stabilized under the same pressure system by using the thin interval equilibrium flow limiting method, where the high aquifer and the fracture target are shot apart and then placed within the same fracture zone and re-fractured. At this time, the pressure above and below the thin layer is in a relatively balanced state, and will not be affected by the temperature difference between the cement ring and the thin layer, so as to protect the thin layer.

2.3 Ball-drop multi-fracture fracturing technology

In order to ensure the orderly production of the well after fracturing, it is necessary to plug the high aquifer continuously after the pressure. In the process of fracturing and reconstruction of the multi-thin and differential oil layer, which cannot be divided by the splitter, the ball-drop multi-fracture fracturing technology can be reasonably used to optimize. When the first layer of the oil layer is pressed open, it is reasonable to apply the gaza to maintain the fracture, and then press the high-strength plugging agent and fracturing slurry into the pressure fracture, so that the opening position of the fractured oil layer can be integrated, the suction channel can be blocked, and the direction of the fracturing fluid

https://doi.org/10.1051/e3sconf/202337501032

can be constantly pushed to continue the pressure and pressure opening of the target layer. Through this cycle, all the target layers achieve pressure balance.

2.4 Positioning and balancing fracturing techniques

In conventional perforated Wells, thin compartments and balanced linear flow are properly fractured to transform poorly permeable zones. In this process, the main application is positioning balanced fracturing technology, positioning balanced fracturing technology is mainly used in the downhole, that is, the use of positioning balanced pressure splitter, the use of positioning balanced pressure splitter in the telephoto cylinder and pressure balancer and sand blasting body, etc., to control the specific position of the target layer of pressure and control the number of vacuole holes. In this way, the target layer can be moderately aspirated to realize the location of cracks.

2.5 High sand ratio wide short fracture fracturing technology

The pressure balancer is similar to the actual quasi-high water content sandblasting body, which can not only contain the liquid, but also generate unsupported fractures, ensuring that the pressure target and high aquifer are always maintained within the same pressure range, thus ensuring that the thin barrier is always stable. The high sand ratio wide and short fracture pressure technology is mainly used in the construction of high sand ratio fracturing. The corresponding fracturing of pipe string can improve the concentration of mortar to a large extent, especially the concentration of tailings. In the process of using the flow limiting method for fracturing, the radius of the propped fracture should be controlled at about 10 meters in a reasonable way, mainly to manage and meet various requirements in the process of high sand ratio construction to the maximum extent.

2.6 Plug fracturing technology

In the process of converting stratified fracturing into multiple single pressures, the reasonable use of plug fracturing technology can continuously improve the multi-zone pressure difference ratio and meet the requirements of large span, which has a higher comprehensive advantage. The reasonable design of the bridge plug through finite element optimization can optimize the field application effect, the reasonable application of the bridge plug fracturing technology, there is no problem of clamping, can meet the specific pressure of height difference and other requirements of multi-level. At the same time, because the sand is not carried to the casing, the use of plug fracturing technology can greatly prevent casing impact and damage.

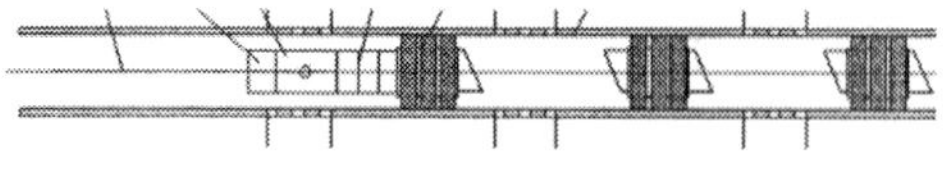

Figure 1. Plug fracturing technology

2.7 Hydraulic fracturing Technology

In the process of hydraulic fracturing technology, the reasonable use of pressure technology can better optimize the fracturing effect. The practical requirements of the application are to make reasonable use of the reservoir for dynamic simulation to predict the oil and gas production that can be met and achieved under different conductivity and fracture length, and then combine the data already used and measured to construct the corresponding fracture length to maximize the economic benefit and calculate the cost required for the actual operation. To improve the overall level of work. In the process of analyzing the corresponding influencing factors of fracturing construction, the most important thing is the fracturing fluid used and the actual performance. Nowadays, China has developed fracturing fluids with strong sand carrying capacity and good shear sensitivity, so the actual application cost and operation safety are guaranteed to a certain extent.

3. Ways and methods of exploring remaining oil in oil - water and thin - difference reservoirs

3.1 Dynamic and static analysis

In order to prevent the expansion of clay minerals in thin oil layer and the dispersion and migration of formation particles, filter water suitable for this area was selected as the gas liquid of the perforating fluid, and clay stabilizer and stabilizer were added to inhibit the migration of formation particles, and the density of perforating fluid was adjusted according to the actual downhole conditions. In addition, studying the mineral sensitivity of the reservoir, the compatibility between the thin reservoir and various perforating fluids and selecting the appropriate perforating fluids can reduce the damage to the thin reservoir. Nowadays, the field usually chooses to use water-based fracturing fluid for fracturing, and usually uses the way of differential treatment and cycle contrast to compare the reservoir of lithologic structural oil. By measuring the scenic line, the contrast of the reservoir is further refined, and it is refined into a single reservoir. After the oil layer comparison, the thick coating is made according to the target layer to find out the inflammatory annihilation line, so as to obtain the optimal oil bearing area distribution map of the target layer. Then the actual collected development data are analyzed and sorted out to draw the specific distribution of the remaining oil and find the oil well with the most development value and economic benefits. Through scientific and reasonable way to excavate the actual residual potential and exploitable situation of thin inserted oil layer and oil-water layer, through this way can better provide a certain guarantee for the actual development of oil layer and the reasonable development of subsequent oil layer.

3.2 Electrical plate method

Logging interpretation can be used to analyze the content of sandstone oil and gas layers with poor permeability. Generally speaking, the interpretation of the content is relatively low. The main reason for this is that when argillaceous sandstone and silty sandstone are used as oil storage layers, the actual sandstone particles forming the formation are smaller in size and larger in area. And the actual inclusion of clay minerals in the rock that can be filled will cause the actual crevices to narrow. Sometimes the radius value will be very small, more than 10, which will also lead to the increase of the curvature of the micro-gap, the continuous development of the gap, with high hydrophilicity, but with very low permeability. Therefore, the high bound water content is actually proposed based on the main production layer of argillous sandstone and siltstone, which has the characteristics of high availability. The actual form of pore water is unable to flow, bound water. Therefore, even if the saturation of water has reached about 65%, there may still be a situation that only oil production cannot produce water. The rational use of microstructure method can better optimize the overall structure of oil field.

3.3 Microstructural method

The oil field itself is constantly undergoing subtle changes. The height difference between oil layers should always be controlled within 10 meters, and the scope of the structure should also be controlled within 0.3 square kilometers. The use of forward microstructures is actually to make the top of the reservoir undulation actually higher than the surrounding sedimentary bodies, and the forward microstructures can better collect oil and gas.

3.4 Sedimentary microfacies method

After water injection development, when the water saturation is relatively high, the forward structure will also become the main accumulation area and component of the remaining oil. The reasonable use of reservoir sedimentary crisis method can better optimize the corresponding configuration. The use of such a way is also based on the thin layer dust machine between diverges. The distribution form and actual thickness of the sedimentary branches of the delta front and the actual oil storage physical oil storage zone will also be different due to different image zones. Usually, the thin and sharp layers of distributive yarns may be tightly attached to the perimeter of the river sand, or may be distributed in the crossing zone of the river, or even distributed separately. The sand bodies are connected with each other, and different modern oil storage properties will also be different. Water injection is carried out in the middle of the plane, along the sand mouth of the river dam and the underwater distributary channel, so that the phenomenon of flooding continues to worsen, and the hookah of the thin layer yarn in the front will also appear lighter.

4. Countermeasures for remaining oil exploitation in oil-water reservoir and thin differential reservoir

First of all, for areas with many faults, there is no way to arrange injection Wells around the faults, and faults have the function of blocking, so there may be unsound housing and remaining oil. Therefore, oil, water change, injection or drilling supplementary Wells can be used reasonably to balance the relationship between injection and production, and the remaining oil can be excavated reasonably. Secondly, for the environment with poor relative conditions of thin differential oil reservoir, scenic spots with good residual potential can reasonably use pressure or optimize fracturing technology according to the actual situation to continuously tap the residual pressure potential and improve the thickness of the oil reservoir. However, the actual effect of using the corresponding way to apply the pressure of thin differential oil reservoir may be unsatisfactory. So the top layer needs to be chosen carefully.

Finally, the special technology of thin and differential oil reservoir should be used positively and reasonably. First of all, the test is carried out in the outer layer. When the well pattern density is relatively small and the scenic spot drilling is encrypted, only the surface layer is excavated. The test block is set, and the drilling with good residual potential is encrypted with water injection. In order to optimize the availability of the outer surface, define the injection pressure and the reasonable well tool, the use of advanced oil displacement agent, to find a certain way for the subsequent exploitation of the reservoir.

5. Conclusion

As can be seen from the above, in the late stage of development, the corresponding old oil fields mainly have fewer well tools and low oil storage and production. Therefore, it is necessary to optimize the development mode constantly to better replace the production capacity of thin and poor oil reservoirs and water-oil reservoirs. In this case, the actual remaining oil distribution of thin differential reservoir and the distribution of oil-water reservoir are usually affected by microstructure and sedimentary microdirection. Therefore, we can reasonably find and excavate the remaining oil and oil-water reservoir of thin differential reservoir by using microstructure method, sedimentary microdirection method, dynamic and static analysis method and telecom drawing board method. The corresponding practice shows that these methods can continuously improve the reservoir benefit and stimulation efficiency.

References

1. WANG Changxu. Development and application of continuous closed coring tool for horizontal Wells in thin reservoir [J]. China Petroleum Machinery, 201,49(11):45-49. (in Chinese)
2. Pan Linhua, Wang Haibo, He Jiayuan, Li Fengxia, Zhou Tong, Bruce Lee. Numerical simulation of simultaneous fracturing interference in multiple horizontal fractures in shallow thin differential reservoir [J]. Journal of Northeast Petroleum University,20,44(06):114-124+12.
3. Bai Zhenqiang, Wang Qinghua, Li Ya. Distribution of micro remaining oil in reservoir after polymer

flooding in sandstone oilfield [J]. Petroleum Geology & Oilfield Development in Daqing,2021,40(04):101-106.

4. Cao Dandan, Wu Dan, Sun Ji. Research on development and adjustment of Chang 2 thin differential reservoir in Yucha area [J]. Inner Mongolia Petrochemical Industry,20,46(09):120-122.
5. WANG Hongli. Application of ultra-short radius horizontal well technology in thin differential reservoir [J]. West China Exploration Engineering, 20,32(03):105-106.

 https://doi.org/10.1051/e3sconf/202337501033

Discussion on the function of logging technology in geological steering of horizontal well drilling

Lin Li

Geological Research Institute of the Third Oil Production Plant of Daqing Oilfield Co. LTD, Daqing, Heilongjiang 163000, China

Abstract: Logging technology is the key technology used in modern petroleum resources exploitation. Reasonable application of logging technology is conducive to efficient exploitation of petroleum resources. Therefore, the application of logging technology in oil exploitation has always been the key research target of oil well exploitation technology. This paper discusses the function of logging technology in geological steering of horizontal well. This paper discusses the logging technology and the geological steering work of horizontal well, discusses the specific application and due function of the logging technology. In the process of discussion, in order to ensure the practical research of this paper, a good application process of the logging technology in the drilling geological steering of a horizontal well is put forward.

Key words: logging technology; Horizontal well; geosteering.

1. Introduction

Petroleum resource is an important resource used in current social development and widely used in social production. Therefore, under the current social development background, petroleum enterprises are stepping up the research of drilling technology, constantly improving the application efficiency of drilling technology, and improving the efficiency of oil resource exploitation. For example, horizontal well is a kind of special drilling technology and is very important in the production of oil drilling. Under the background of continuous technical research, relevant experts put forward the geological guidance in horizontal drilling technology, and also combined with the application of logging technology to ensure the optimal efficiency of horizontal drilling.

2. Discussion on logging technology

Logging technology is the key technology used in oil well exploitation. In the process of application, the technology is mainly used to observe, collect and store the data of solid, liquid and gas in the process of oil drilling by rock and ore analysis, physics, chemistry and other related methods. According to the relevant data of logging, the geological profile can be made and the resource evaluation and analysis of oil and gas reservoir can be completed, which can provide guidance for oil exploitation technology. In addition, the completion of logging protection can also provide technical guidance for drilling engineering and other oilfield construction projects.

In the process of specific application of logging technology, core logging, cuttings logging and other related technologies can be completed. The logging technology of physics principle includes fluorescence logging and nuclear magnetic resonance logging. Chemical principles of logging technology I oh rock pyrolysis logging, light hydrocarbon logging and other related technologies, different logging process implementation according to the actual situation of different oil fields. However, under the current social development background, in order to ensure the high accuracy of logging and meet more needs, logging technology began to develop in the direction of comprehensive. For example, the application of modern logging technology in the process has begun with integrated logging instrument. The logging instrument integrates drilling time logging, gas logging, drilling fluid logging and other logging methods into one, enabling measurement while drilling, remote data acquisition control and other functions, thus improving the efficiency of comprehensive logging.

Comprehensive logging instrument has comprehensive function in the process of application. Generally, comprehensive logging instruments are used in the application of modern logging technology. The instruments mainly include winch sensor, acoustic sensor, pressure sensor, casing pressure sensor, electric torque sensor, H_2S sensor, etc., which can collect a variety of information and improve the working efficiency of logging.

3. Discussion on geosteering of horizontal Wells

Geological steering of horizontal well refers to the exploitation of horizontal well by geological steering method.

First, a horizontal well is a special well that has a maximum inclination of 90° or close to it (generally not less than 86°) and maintains a certain length of horizontal section in the target zone. Sometimes a well can be "turned up" at an Angle of more than 90° for a particular need. In general, horizontal Wells are used in thin or fractured reservoirs to increase the exposed area of the reservoir.

Secondly, geosteering is specifically in the process of drilling implementation, formation parameters and well trajectory can be measured and data acquisition, after the completion of acquisition, log curve is drawn, using the log curve to complete drilling control, to realize drilling construction.

Geological steering of horizontal Wells is to carry out drilling records and complete real-time drilling guidance in the process of horizontal well drilling to ensure the best results of drilling construction. At present, our country has been discussing the technology of horizontal well geo-oriented drilling. For example, rotary geosteering drilling system is the research and development direction of modern petroleum enterprises. On November 18 this year, the horizontal geosteering drilling system achieved new success in the direction of development. The cost is 50% lower than that of the same type of international products. The simultaneous hanging technology can withstand downhole high temperature and pressure of 165 degrees Celsius and 140 mpa, with a wide range of adaptability, multi-parameter measurement, measurement, recording and guiding complementary, precise trajectory control, obvious cost advantages, etc.

4. Discussion on the application and function of logging technology in geosteering

In the process of horizontal well drilling, the application of geosteering technology is very important. It can be found through the study of geosteering technology that the specific application of logging technology in geosteering is actually to use advanced logging technology to measure relevant parameters, obtain data, and conduct downhole mining guidance, so as to ensure that various technologies reach the highest standard and improve the drilling effect of horizontal Wells.

4.1 Specific application of logging technology

First of all, logging technology is used to select drilling methods during the implementation of geosteering in horizontal Wells. The application of drilling method is very important in drilling operation. However, in the implementation process of traditional horizontal well operation, the application effect of logging technology is relatively poor, and the selection of drilling methods is relatively extensive, which is difficult to form a good effect, affecting the entire production. In the case of gesteerable drilling with logging technology, in the actual construction operation, the personnel need to investigate the geological environment, logging and reservoir volume of the area where the horizontal well is located, and then obtain the investigation data and data, and then use these data as the basis to reasonably design the drilling path and provide guidance for the subsequent drilling. However, in the traditional geosteering process, it is impossible to accurately analyze well trajectory and reservoir location.

Modern logging technology has been developed into a comprehensive logging technology that can complete multiple information acquisition of horizontal Wells, so as to implement comprehensive logging control and ensure the best drilling process implementation. In the application process, logging technology can be used to complete a number of horizontal well information acquisition, so as to ensure the best drilling process implementation.

For example, in the process of logging technology application, cuttings logging, core logging and other methods can be used to complete the collection of horizontal well related data, and then select the drilling method. The application of cuttings logging technology is mainly to directly use the drilling process of energy-saving logging, after drilling according to the color of cuttings and related parameters to choose the best drilling method. In addition, the cuttings in the reservoir can be stained with oil, which can be used to complete the comprehensive analysis of oil and gas resources. For example, in the course of the study, the relevant experts found that the oil-bearing cuttings are generally brown gray and brown yellow, but also give off the initial aroma of crude oil, and depending on the weight of the cuttings, it can be specific to determine the level of oil in the formation. Cuttings logging method is a relatively stable and reliable logging method, which can be used in the background of geosteering implementation of horizontal Wells to select the drilling method, which is more in line with the drilling requirements of horizontal Wells, as well as the drilling process quality, to ensure the best drilling results.

Secondly, the application of logging technology in geo-guided drilling of horizontal Wells can make drilling adjustment. Guided by the relevant data obtained by logging technology, drilling control of horizontal Wells can be completed, thus ensuring that logging achieves the best objectives. In the process of drilling implementation, the comprehensive logging technology can be used to collect related parameters.

For example, the study found that the integrated logging technique was significantly more dominant than the actual drilling with LWD parameters. In the application study of horizontal well logging technology, it is found that the well interval is 189-191m water and the average drilling time is 16min. In general, gamma logging while drilling is used. Under these parameters, the drilling time is 16X13min, and the natural gamma value begins to decline after drilling is completed. The integrated logging technology can be shorter than the integrated logging technology 188min, proving the efficiency of the

https://doi.org/10.1051/e3sconf/202337501033

integrated logging technology announcement. The resistivity logging while drilling technology also has a lag of 135min compared with logging to ensure the best drilling results. The following is a comprehensive analysis and research on the application of comprehensive drilling technology to ensure the best drilling results [1].

The application of comprehensive logging technology can also optimize and supervise geological logging parameters in horizontal Wells. In the process of gesteerable drilling in traditional horizontal Wells, it is difficult to accurately describe the data and grasp the data imaccurately, so it is difficult to judge the relevant technical parameters of drilling, which affects the specific implementation of drilling technology. Therefore, in the process of drilling technology application, the underground structure of some horizontal Wells is also relatively complex, which is affected by geology to some extent. By applying comprehensive logging technology, comprehensive parameters of logging can be collected, data analysis and protection can be done well, and real data and effective mining technical guidance can be provided for horizontal Wells [2].

For example, modern horizontal drilling is beginning to use new integrated logging instruments, which are able to complete multiple data acquisition while completing multiple data acquisition with high-precision horizontal logging tools. Taking SK-DMS logging while drilling instrument for horizontal well logging in some oilfields in China as an example, the instrument is equipped with 3Q04/3Q05/3Q06 field fast chromatograph to realize high reliability storage of data and provide guarantee for post-analysis. It can withstand 1h in 950°C high temperature environment and 10h in 260 °C medium temperature environment. In addition, the maximum operating temperature of the instrument is 150°C and the maximum pressure is 135MPa, which is suitable for exploration and development of deep and ultra-deep Wells. In the process of specific application, the geological parameters, data parameters and other parameters of the oil well can be collected to understand the geological information, to ensure that the collection work to achieve the ideal effect, but also to improve the application effect of logging. Table 1 below shows the specific working parameters of the equipment.

Table 1 Comprehensive application parameters of logging technology

Technical specification	Specific parameter
Indoor positive pressure	50Pa～120 Pa
Indoor operating temperature	0°C～30°C
Power input	220V/380V/440V/480V ±5% 3P
frequency	50Hz/60Hz±5%
power	20KVA

4.2 Application effect of logging technology in geosteering logging

The proper function of logging technology in geosteering logging has been proved by time and has a good effect. The following is an analysis of the application of logging technology in horizontal well geosteering logging.

The application of well logging technique in geologic steering well is able to improve the accuracy of geological steering data. Modern logging technology includes different technologies such as physics, chemistry and lithology. Different technologies obtain related parameters according to different scientific principles. The parameters obtained by scientific principles are more accurate. In particular, under the current technical background, the application of comprehensive logging technology can realize the accurate collection of a number of technical parameters, which can be stored and analyzed after completion of collection, so as to provide accurate data for geosteering and ensure the best results of geosteering logging [3].

In addition, the late time can be encrypted during the implementation of horizontal well logging. On the one hand, the late time of the comprehensive logging instrument belongs to the category of calculation, which needs to provide a theoretical value, but because of the complexity of the external environment, it is difficult to determine the theoretical value. Therefore, there is often a large difference between the late time of the logging instrument and the calculation time of the theoretical value, which requires the prediction of the late time to be encrypted to ensure the accuracy of the cuttings retrieval time, which has a positive promoting effect on the project [4].

5. Empirical analysis

In the process of research, this paper analyzes the logging data of a horizontal well, so as to complete the mining guidance. For example, in the process of the horizontal well logging application, the formation treatment method of logging has been applied to carry out the logging correlation study. In the drilling design of 63# of this well, comparing the logging data of adjacent well with that of this well with reference data of adjacent well, the thickness and pure thickness were converted, and real-time comparison of drilling process was realized. Finally, it was confirmed that the top of the target layer of this well was -1722m, 5m lower than the original design, and logging analysis was carried out according to this parameter. Ensure the best effect of technology application [5].

6. Conclusion

This paper summarizes the specific application of logging technology in horizontal well geo-guided drilling, hoping that the research in this paper can be helpful to the application of horizontal well geo-guided drilling.

https://doi.org/10.1051/e3sconf/202337501033

References

1. Jiao Xuesong. Discussion on how to apply reservoir prediction and monitoring technology in geosteering drilling of horizontal Wells [J]. Petrochemical Technology, 2020, 27(2):2-2.
2. Peng Ye Realgar for Liang Li Shifang Jia Hao Fang Tieyuan Liu Fangfang. Comprehensive application of various logging techniques in the optimization of horizontal well perforating interval [J]. Logging Engineering, 2022, 33(1):60-64.
3. Yan Ronghui, Huang Zijian, Fang Tieyuan, et al. Application of logging technology in geological steering of shale oil Wells [J]. Logging Engineering, 2020, 31(3):5-5.
4. FANG Yunfeng. Discussion on key technologies of geosteering logging for horizontal Wells [J]. West China Exploration Engineering, 2020, 32(2):2-2.
5. GUO Zhengfeng. Application of logging technology in geosteering of offshore oil drilling [J]. Chemical Engineering Management, 2021(8):2-2.

Study on the influence of oilfield water injection on casing damage

Xiangnan Ding

No.3 Oil Production Plant of Daqing Oilfield Company Ltd, Daqing 163113, China

Abstract: The casing damage of oil and water wells has seriously affected the normal production of the oilfield. This paper analyzes the high occurrence of casing damage in Saertu oil layer, and finds that there is a contradiction between the water injection production of the wells in Saertu formation and the prevention and control of casing damage. Although the wellhead oil pressure can be guaranteed to be lower than its fracture pressure during production, the injected energy at the wellhead includes pressure potential energy and kinetic energy. Because the fracture pressure of Saertu formation is relatively low, the total injected energy is often higher than its fracture pressure, resulting in water injection at an excess fracture pressure. Unreasonable water injection will lead to the destruction of the underground rock stratum, which acts on the casing and forms casing damage.

1. General situation and statistical law of casing damage in Sabei Development Zone

As of February 2022, there are 2023 casing damaged wells in Sabei Development Zone, including 878 oil wells and 1145 water wells.

1.1 Main types of casing damage

From the perspective of casing damage types, the vast majority of casing damage wells are stagger and deformed, including 796 stagger wells and 880 deformed wells, with a total of 1676 wells accounting for 82.85% of the total casing damage wells. Both dislocation and deformation are expressed in the form of displacement, that is to say, most casing damage is caused by mechanical casing damage caused by the release of formation stress.

1.2 Main layers of casing damage

From the perspective of casing damage horizons, casing damage horizons are mainly concentrated in Nen 2 member and Saer layer system, in which 524 wells in Nen 2 member account for 31.26%, and 935 wells in Saer layer system account for 55.79%.

Table 1 Statistics of Casing Damaged Wells in Sabei Development Zone

Stratigraphy	Number of wells	Proportion (%)
N2	524	31.26
S	935	55.79
P	118	7.04
G	62	3.70
Others	37	2.21

Casing damage of the second segment of the Nenjiang River usually occurs in a certain area and in a relatively short period of time, with concentration in both time and space. The horizontal bedding of oil shale in the standard layer of Nenjiang Formation 2 is developed. When the injected water enters, it will diffuse along the densely superposed paleontological fossil bedding plane, forming a soaking water area, which will crack the mudstone in pieces. When the rock is subjected to shear stress, the stress will be released along the fossil weak surface of the standard layer, leading to casing damage in pieces.

Sal layer is a production layer, and its casing damage is mainly related to water injection. This paper mainly analyzes the relationship between water injection production and casing damage.

2. Water injection and casing damage

The steel grade strength of the casing itself, the consolidation strength of the casing and the rock stratum, and the rock mechanical properties of the rock stratum are the basic factors for casing damage. Generally, these factors are relatively stable, and the casing strength can offset the formation stress, which is the guarantee for normal production. However, in the process of oilfield development, water injection production is required, which will inevitably affect the underground rock stratum. Under low water injection pressure, the rock stratum is generally stable and will not affect the casing. However, once the water injection pressure reaches a certain level, the underground rock stratum will be damaged, causing displacement, and a huge radial force on the casing will cause deformation or dislocation of the casing. In order to prevent casing damage, the injection pressure at the bottom of the well must be lower than the overlying rock

* Corresponding author: 236324253@qq.com

pressure of the formation. Since only wellhead equipment can be controlled during production, the production site will adjust the oil pressure through the oil pressure gate at the wellhead to ensure that it is lower than the fracture pressure for water injection.

3. Wellhead injection energy analysis

3.1 Wellhead process and principle of water injection well

During production, the water injection pipe network system will maintain a high pressure. At the wellhead end, the high pipe network pressure will be converted into a low injection pressure through an oil pressure gate to achieve constant pressure and quantitative water injection. The pipe network pressure at the wellhead end is called the pump pressure. The injection pressure is called the oil pressure through the pump pressure gauge. The oil pressure gauge shows that between the pump pressure gauge and the oil pressure gauge is the oil pressure gate that regulates the oil pressure. Generally, the pump pressure is more than 2MPa higher than the oil pressure.

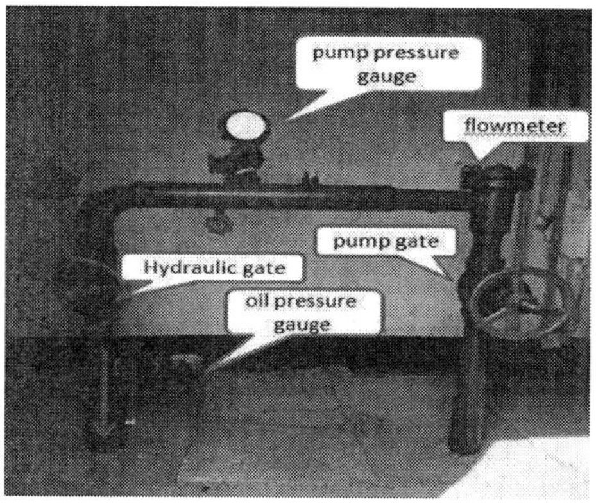

Fig. 1 Flow Chart of Well Head Control System

3.2 Principle of energy conservation in pipe flow system

According to the principle of energy conservation, the pipe flow energy at the oil pressure gauge should be equal to the pipe flow energy at the pump pressure gauge minus the pipe loss between two points. The distance between the pump pressure gauge and the oil pressure gauge is very close, and the pipe loss along the way can be ignored, only the gate pipe loss. The daily injection volume of the water intake well is 100 m3, the pipe diameter R is 0.05 m, the pipe flow velocity can be calculated as 0.6 m/s, the wall roughness of the old steel pipe n is 0.012, and the pipe loss when the gate is only 1/4 open can be calculated. Calculation of pressure loss per unit length:

$$I = v^2 / \left(\frac{R^{\frac{4}{3}}}{n^2} \right)$$

I —— Pressure loss per unit length

v —— Pipe flow velocity

n —— Pipe wall roughness

R —— caliber

Equivalent length of pipe diameter under different opening degrees of gate:

Table 2 Relation between gate opening degree and equivalent length of pipe loss

Valve opening size	Gate equivalent length (Multiple of pipe diameter)
Fully open	7
3/4	40
1/2	200
1/4	400

It can be seen from the calculation here that when the gate is 1/4 open, the loss is only 0.28P, which can be completely ignored. Generally, when the gate is more than 1/2 open, the pipe loss will be smaller. It can be considered that the pipe flow energy at the pump pressure gauge does not decrease after reaching the oil pressure gauge. The pipe flow energy in front of and behind the oil pressure gate is unchanged, which conforms to the Bernoulli equation of capacity conservation law of the pipe flow system:

$$C = P + \frac{1}{2}\rho v^2 + \rho g h$$

P —— Pressure potential energy

$\frac{1}{2}\rho v^2$ —— kinetic energy

$\rho g h$ —— Gravitational potential energy

C —— Pipe flow energy constant

The sum of pressure potential energy, kinetic energy and gravity potential energy at any point in the pipe flow is constant, and the pipe flow energy at the pump pressure gauge and the oil pressure gauge is equal.

$$P_1 + \frac{1}{2}\rho {v_1}^2 + \rho g h = P_2 + \frac{1}{2}\rho {v_2}^2 + \rho g h$$

P_1 —— Pump pressure

P_2 —— oil pressure

v_1 —— Speed in front of oil pressure gate

v_2 —— Speed behind oil pressure gate

Taking a well with a daily injection of 100 m3 as an example, the flow rate is 0.6m/s, assuming that the pump pressure drops by 2 MPa, the equation is:

$v_2 = 63.2\text{m/s}$

After passing through the oil pressure gate, although there is a pressure loss of 2MPa, the pipe flow speed has changed from 0.6m/s to 63.2m/s, greatly increasing. The oil pressure gate only regulates the pressure. When the injected energy is not reduced, the total energy before and after the oil pressure gate is the same, but the situation is different. The injected energy at the oil pressure gauge includes pressure potential energy and kinetic energy, that is, the sum of oil pressure and kinetic energy at the oil pressure gauge. Therefore, the total injection energy at the wellhead is close to the pump pressure, which is much larger than what we think.

4. Contradiction of water injection well production

4.1 Analysis of water injection before and after pump pressure drop

In oilfield water injection production, the problem of low pump pressure will be encountered. After the pump pressure drops, the injection volume of the well will often drop, and the injection allocation cannot be completed. In this case, the field workers will open the oil pressure gate to increase the oil pressure. At the same time, in order to prevent casing damage, the oil pressure should be lower than the fracture pressure. This method can increase injection volume to a certain extent, but there are still many wells unable to complete injection allocation. Take North 3-But-6 - Lateral 453 of water injection well in North 3rd East Block of Sabei Development Zone as an example:

Table 3 Injection Status before and after Pump Pressure Reduction of B2-D3-CS453

Well name	Rupture pressure (MPa)	Production status	Pump pressure (MPa)	oil pressure (MPa)	Water distribution (m^3)	Injection volume (m^3)
B3-D6-CS453	11.2	normal	13.4	10.5	65	67
		Low pump pressure	12.4	11.1	65	30
		After recovery	14.0	10.4	65	66

During normal production, the pump pressure is 13.4 MPa, and the oil pressure is 10.5 MPa, which can complete the injection allocation. When the pump pressure drops to 12.4 MPa, even if the oil pressure is increased to 11.1 MPa, the injection allocation cannot be completed. Under the condition that all parameters of the same well remain unchanged, the water injection rate is proportional to the injection energy, and the decrease of water injection rate indicates that the injection energy is reduced at this time. Before the oil pressure gate, the pipe flow speed is very slow, and the injected energy here only includes the pressure potential energy, that is, the pump pressure. After passing the oil pressure gate, the pipe flow speed is greatly increased, and the front and rear pressure differences are converted into the kinetic energy of the pipe flow. The pipe flow energy here includes the pressure potential energy and the kinetic energy. After the pump pressure decreases, the total injected energy decreases. Although the pressure potential energy at the oil pressure gauge remains unchanged or even increases, the kinetic energy decreases significantly. The total injected energy can no longer meet the water injection requirements, resulting in a decrease in water injection volume.

4.2 Comparison of injected energy and fracture pressure

The pressure difference between pump pressure and oil pressure will be converted into kinetic energy and retained in the pipe flow system. That is to say, in addition to the 11.1 MPa pressure potential energy, 1.3 MPa kinetic energy will exist in the pipe flow system after the pump pressure of Well Bei 3-Ding 6-Skew 453 decreases. When this energy enters the formation through the downhole string, it will be converted into pressure potential energy again (conversion efficiency is unknown). The pressure potential energy of 11.1MPa is already the top fracture pressure. With the kinetic energy of 1.3MPa, the total action energy on the formation must exceed the fracture pressure by 11.2MPa. After the well pump pressure drops, water injection cannot be completed at the top fracture pressure. It can be inferred that the injection energy at this time is lower than that in normal production. Although the injection pressure is only 10.5MPa, much lower than the fracture pressure of 11.2MPa, in fact, in addition to the 10.5MPa pressure potential energy, there is also 2.9MPa kinetic energy. The total injection energy is much higher than the fracture pressure.

4.3 Overbreak pressure water injection of the system wells in Saer Formation

This phenomenon of water injection under super fracture pressure is common in the wells of the Sal layer system. The Sal layer system is shallowly developed with low fracture pressure, which is obviously different from the other two layers. At present, the water injection pipe network system is the same for the water wells of the three development strata, and the injection energy is similar. The average pressure at the wellhead of the pipe network system in Sabei Development Zone is about 14.5MPa, while the average fracture pressure of the Sal Formation is only 12.4MPa, and the average injection energy is 2.3MPa higher than the fracture pressure. This high-energy water injection method of super rock formation fracture pressure will inevitably cause certain damage to the underground rock strata.

Table 4 Comparison of Injection Pressure Status of Stratified Systems and Number of Casing Damaged Wells in Sabei Development Area

Stratigraphy	Pump pressure (MPa)	Rupture pressure (MPa)	Pressure Difference (MPa)	Number of wells with damaged casing	Proportion (%)
S	14.7	12.4	2.3	935	83.86
P	14.6	14.0	0.6	118	10.58
G	14.5	14.9	-0.4	62	5.56

Statistically, it is found that the injection energy of the Sal layer system exceeds the average fracture pressure of the layer system by 2.3MPa, and the casing damage rate is particularly high, accounting for 83.86% of the casing damage wells in the oil layer; The injection energy of

Putaohua Formation is 0.6 MPa higher than its average fracture pressure, and its casing damage rate is greatly reduced compared with that of Sal Formation, accounting for only 10.58% of the casing damage wells in the oil layer; The injection energy of Gaotai sublayer is 0.4MPa lower than its average fracture pressure, and its casing damage rate is less. It can be seen that the casing damage occurrence rate of a set of layers is closely related to the difference between its injection energy and fracture pressure. The more the injection energy exceeds the fracture pressure, the easier it is to damage the formation and cause casing damage. It also confirms the correlation between the degree of the super fracture pressure of the water injection well and the number of wells where casing damage occurs. Therefore, the main reason for the high casing damage occurrence rate of the Sal layer system is its unreasonable injection status.

5. Conclusion

(1) Casing damage in production horizon is mainly concentrated in Sal layer system, and most of the casing damage types are stagger and deformation. The main reason for casing damage is that there is extrusion and shearing of rock stratum to casing at the bottom of Sal layer system.

(2) The oil pressure at the wellhead of a water well does not represent the injected energy. The total injected energy includes pressure potential energy and kinetic energy, which are much larger than what is generally believed.

(3) Overbreak pressure water injection is a common phenomenon in the system wells of Sa'er Formation, and the main reason for the high casing damage is the unreasonable injection condition.

References

1. Jin Haiying. Production Performance Analysis of Oil and Gas Wells [M]. Petroleum Industry Press, 2010. 4
2. Zhang Zhaoshun, Cui Guixiang. Fluid Mechanics [M]. Tsinghua University Press. July 2015
3. Li Wenkui. Theoretical analysis and damage repair technology of downhole casing string [M]. Petroleum Industry Press, 2010
4. Lin Yuanhua. Casing working mechanics [M]. Petroleum Industry Press. 2016. 4
5. Lian Zhanghua. In situ stress and casing damage mechanism [M]. Petroleum Industry Press. 2009
6. Zhang Xiaoyu, Zhao Guozhen, Zhang Xianpu, Research Survey and Development Trend of Casing Damage at Home and Abroad [J]. Petroleum Machinery. 1996.9, 52-54.
7. Wang Yuezhi. Calculation of anisotropic formation stress and prediction of deep hole formation fracture pressure [J]. Journal of Rock Mechanics and Engineering. 1998.1, 36-42
8. Zheng Junde, Wang Wenjun. Influence of high-pressure water injection on sandstone interval interpolation [J]. Petroleum drilling and production technology. 1997.1.64-68
9. Shen Chuanhai. Cause Analysis of Wang Zhen Ring Casing Damage and Comprehensive Treatment Technology [J]. Petrochemical Application. 2007.26 (5): 47-50
10. Xie Hui, Wang Wei, Wen Zhigang. Practice and understanding of casing damage well treatment in Longdong Oilfield [J]. Journal of Jianghan Petroleum Institute, 2005.27 (3): 548-549
11. Xu Dashu, Xiao Baojun, Chen Yunliang. Analysis on casing damage mechanism and countermeasures of oil and water wells in Jiangsu Oilfield. Drilling and Production Technology. 2006.29 (3): 66-68

 https://doi.org/10.1051/e3sconf/202337501035

The optimization method of control precision of injection system in well metering

Haochen Zou

The seventh operation area, the No.6 oil production plant, Daqing Oilfield Co., Ltd., Daqing, China

Abstract: Based on the analysis of the current situation of single well metering, data acquisition and control mode of oilfield injection system, the data acquisition and control mode of polymer injection volume and dilution water injection volume of single well were optimized and improved respectively. The automatic control of injection volume of single well was realized, the injection accuracy and time effect of metering and control were enhanced, and the work intensity of employees was reduced. Meanwhile, the polymer and dilution water injection volume and the abnormal alarm function of water metering regulation were increased. These make the operation of flow regulator, pump and other equipment in injection station more reasonable, and improve the operation life and efficiency of system equipment.

Keywords: Injection accuracy; data acquisition; well metering.

1. Field automatic control injection accuracy

The situation of 34 injection stations in Oilfield at the injection site was investigated. 10 wells were selected for each station, and 340 single wells were selected for automatic control. Before optimization and improvement, the injection volume of single wells was completely controlled by automatic control. Compared with the injection allocation value of the scheme, the average compliance rate of the injection scheme of the injection station after 24 hours of continuous production was 81.79%, the coincidence rate with the injection scheme of polymer flooding injection station required by the management is ± 5%, with a difference of 13.21%.Therefore, it is still necessary to manually regulate the injection volume of a single well. The survey data of some polymer injection wells are shown in Table 1.

Table 1 questionnaire of automatic control production before optimization and improvement of some injection stations

Well number	injection medium	Instantaneous injection (m3/h)	9:00 Table base (m3)	11:00 Table base (m3)	14:00 Table base (m3)	Difference between two table bases (m3)	5-hour automatic control error rate%
1# injection well	water	1	12551.8	12553.8	12556.8	5	0
	mother liquor	0.46	824.8	825.7	827	2.2	-4.35
2# injection well	water	0.96	9356.6	9358.6	9361.4	4.8	0
	mother liquor	0.29	625.6	626.1	627	1.4	-3.45
3# injection well	water	0.83	6462.1	6463.8	6466.3	4.2	1.2
	mother liquor	0.21	491.1	491.5	492.1	1	-4.76
4# injection well	water	0.79	6157.7	6159.2	6161.5	3.8	-3.8
	mother liquor	0.25	711.9	712.4	713.2	1.3	4

2. Optimization and improvement of injection metering control technology

2.1 Workflow and algorithm of automatic control of polymer injection volume

Firstly, manually input the injection volume of single well scheme into the main control computer, and PLC automatically collects the actual injection volume fed back by polymer flow meter through 485 communication,

* Corresponding author: zouhaochen@petrochina.com.cn

calculates the actual injection volume of polymer in the configuration software, and runs the program regularly.(the actual injection volume in this period is obtained by subtracting the accumulated volume base of the polymer instrument being recorded from the accumulated volume base of the polymer instrument recorded last time) the main control computer regularly compares the actual injection volume with the set injection volume. When there is a deviation between the actual injection volume and the set injection volume, the control program of the main control computer calculates the injection volume value to be corrected through the formula. This calculation process works circularly on the main control computer, constantly refreshing and generating new set injection volume.

Calculation formula for automatic control of polymer injection volume:

$$S_n = S_{n-1} - (\frac{V_n - V_{n-1}}{T}) + F \qquad (1)$$

S_n - the nth set injection volume (m^3/h);
S_{n-1} – the n-1th set injection volume (m^3/h);
V_n - base number of the nth polymer flow meter (m^3);
V_{n-1} - base number of n-1th polymer flow meter (m^3);
F - injection volume of scheme (m^3/h);
T - calculation cycle interval (H).

2.2 Method for judging rationality of setting value of polymer injection volume

For the rationality of the set value of polymer injection volume, the following design is carried out.

Determination of the upper limit value of polymer injection volume: when the new set volume is greater than the maximum displacement (P_{max}) of the pump, the equipment cannot adjust and compensate the difference of the previous period back to the injection volume of the scheme in the next period, indicating that the equipment is not operating normally, so it is necessary to alarm and remind the on-site operators for maintenance and inspection. That is, when $S_n > P_{max}$, the alarm will be prompted.

Determination of the lower limit of polymer injection volume setting: due to the maximum working displacement (P_{max}) of the injection pump, when the set displacement is less than 40% of the actual maximum displacement, the pump operation frequency will be less than 20Hz, resulting in the motor being lower than the reasonable operation frequency. It is necessary to alarm and remind the on-site operators for maintenance and inspection. That is, when $S_n < 40\% P_{max}$, the alarm will be prompted.

Upper limit alarm formula of polymer injection volume setting:

$$P_{max} - [(S_{n-1} - \frac{V_n - V_{n-1}}{T}) + F] < 0 \qquad (2)$$

Polymer injection lower limit alarm calculation formula:

$$(S_{n-1} - \frac{V_n - V_{n-1}}{T}) + F < 40\% P_{max} \qquad (3)$$

P_{max} - actual maximum displacement of pump (m^3/h).

2.3 Optimization of PLC polymer instantaneous flow data acquisition mode

At present, the PLC data acquisition polling method is used for periodic acquisition, and the data processing method is to directly apply the collected data and participate in the control. In this way, it is easy to collect the instantaneous peak, trough value and system noise of the on-site flowmeter. If the collected abnormal data is directly involved in the internal control of PLC, it will cause the instantaneous fluctuation of the frequency converter and affect the stable injection and the service life of the equipment. In view of this problem, the PLC data acquisition mode is optimized, and the mode of directly participating in the control of the original data is changed to participating in the calculation after data statistics.

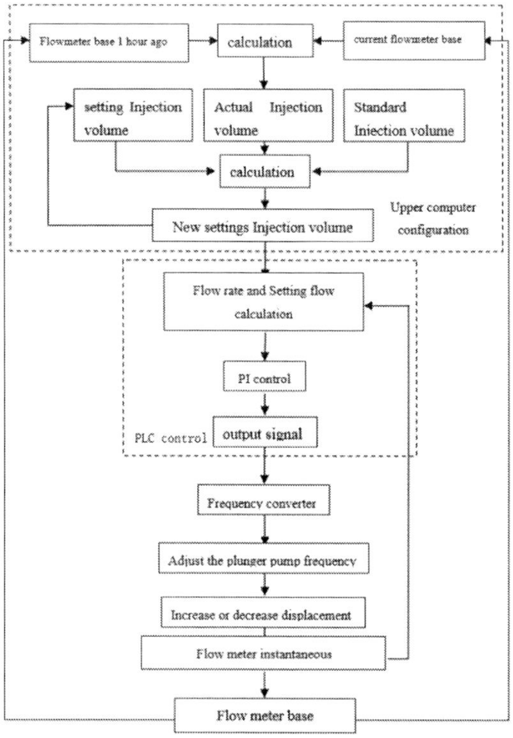

Fig. 1 work flow chart after optimization and improvement of polymer regulation and metering control system

3. Study on the reasonable range of allowable error of flow controller

According to the on-site staff who do not perform automatic adjustment when they reflect the slight difference of the field flow controller, carry out field test, and determine that the cause of the problem is the "dead zone" setting range of PID control module (the dead zone is the common name of the maximum allowable error). The normal "dead zone" control is to reduce the number of actuator actions, (generally, a "dead zone" is set for PID regulator. In ± "dead zone", the input deviation is considered to be 0. When the dead zone is exceeded, the input deviation will be calculated from 0. The dead zone can effectively reduce the number of actuator actions and ensure the equipment used.) However, because the dead zone is set too large, the adjustment accuracy is reduced, so the "dead zone" range of flow controller is tested on site, and the reasonable "dead zone" range is determined to be 0.05 m^3 / h.

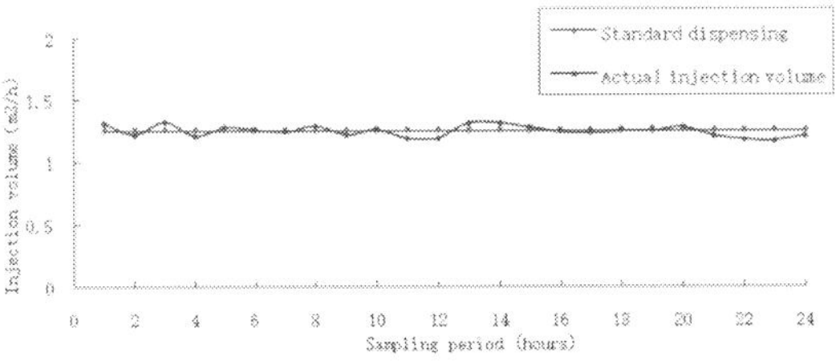

Figure 2 regulation of setting dead zone range 0.05 m^3/h

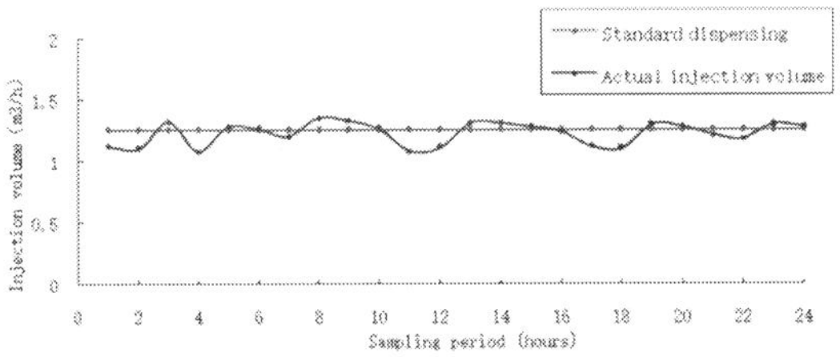

Figure 3 regulation of setting dead zone range 0.1 m^3/h

According to the research results of the optimization of the dilution water injection metering control system, the flow chart of the dilution water metering control system is given, and the program is written and implemented on site.

4. Application effect

The improved metering control system will be optimized and field test will be carried out in three injection stations in the oilfield. Similarly, the injection plan of three injection stations will be injected for 24 hours continuously. The average compliance rate of injection schemes of three injection stations is 98.36%. The injection scheme of polymer flooding injection station which fully meets the management requirements is within ± 5%, and it can achieve automatic control without manual intervention.

Table 2 questionnaire of automatic control production after optimizing and improving injection wells in three injection stations

Well number	Injection	Instantaneous injection (m3/h)	9:00 Table base (m3)	11:00 Table base (m3)	14:00 Table base (m3)	Difference between two table bases (m3)	5-hour automatic control error rate %
1# injection well	water	0.83	8737	8738.64	8741.17	4.17	0.48
	mother liquor	0.21	1055	1055.45	1056.07	1.07	1.90
2# injection well	water	1.17	15759	15761.34	15764.86	5.86	0.17
	mother liquor	0.5	2216	2217.00	2218.50	2.50	0.00
3# injection well	water	1.29	16916	16918.60	16922.50	6.50	0.65
	mother liquor	0.58	2614	2615.19	2616.93	2.93	0.46
4# injection well	water	1.75	17146	17149.52	17154.79	8.79	0.46
	mother liquor	0.54	1428	1429.10	1430.68	2.68	-1.05

5. Conclusion

(1) Recognizing the existing problems of metering control at this stage, new improvement schemes are proposed for the injection volume control of polymer and dilution water respectively, and the precise regulation of injection volume of polymer injection well based on base matching is proposed to effectively improve the metering control capacity of injection station.

(2) Aiming at the problems existing in the metering control, the optimization and improvement are carried out, and the interconnection among the flow controller, dilution water metering flowmeter, frequency converter, mother liquor metering flowmeter, PLC and main control computer is realized, which provides a feasible path for the signal transmission and linkage control between the control units in the automatic control system of the injection station.

(3) By setting the maximum control range of the hourly injection volume of each single well, judge the injection volume automatically written in every hour, give an alarm when the hourly injection volume is too large, and remind the staff on duty to deal with it, so as to realize the

optimization of the alarm prompt of insufficient metering control of single well pump efficiency in the injection station.

(4) Through the optimization and improvement of the injection automatic control system of polymer injection station, the accuracy of unmanned automatic injection has been improved from 81.79% of the original scheme to 98.36% of the current scheme, and the quality of polymer injection has been well guaranteed.

References

1. Wang Gang Research on remote centralized control management in industrial automation system [J]China informatization, 2013,5:42-48
2. Zhao Yuanjun Development and application of general server OPC [D]Shanghai: Shanghai Jiaotong University, 2004
3. Ding Yuhui Lou Wenfa Li Junhong Integrated automation system of coal mine based on iFix. China's high-tech enterprises, 2014,3:20-26.
4. Li Yun; Wang Wei Application of flow meter in automatic control system [J]Chemical design communication. 2016042,91
5. in TianjinDesign and Research on automatic control system of polymer injection device [D]China University of petroleum. October 2008

Washed core and tracer tests data guide the characterization of single sand body in Low Permeable Reservoir

Guoying Qin

Tenth Production Division of Daqing Oilfield Company, PetroChina, Daqing 163000, Heilongjiang, China

Abstract. A Oilfield is Fluvial Facies Low Permeable Reservoir, main layer is Low bend distributary channel sand body deposition, In order to better understand the distribution of single sand body and remaining oil in the reservoir, inspection well coring and tracer tests have been carried out successively in the same well area of the oilfield, based on these data, it is found that, the actual distribution of single sand body is very different from the original description in both vertical and plane. Discovered by the inspection well core data, the scale of the vertical single-phase body was too large. It was originally considered to be the mudstone interlayer inside the 0.2m single sand body, which is actually the interlayer between different single sand bodies, and this situation was confirmed by the trace data five years later. The distribution position and scale of single sand body on the plane are also different from the original description. It was thought that the oil and water well sand body with good internal connection in the same single sand body is actually not connected. Therefore, the experience of these data on vertical and horizontal single sand body identification is summarized to guide the subsequent single sand body characterization of the oilfield, so as to better guide the oilfield to describe the remaining oil and adjust and tap the potential.

Key words: Low Permeable Reservoir; Single sand body; Fluvial Facies.

1. Preface

In order to find the remaining potential in the next step of oilfield A, according to the sedimentary characteristics of the main oil layer of oilfield A is the low-bend distributive channel facies, the flood surface is mainly used as the reference standard layer, the elevation difference is compared, and the correlation methods such as sedimentary microfacies orientation are used [1-6]. For the deposited complex meander sand body, the single sand body identification methods [6-12], such as abandoned channel, sand body thickness change and log curve shape change, are mainly used to characterize the single sand body. On this basis, the core water washing data and tracer data of A closed core inspection well located in the same well area of A main block in oilfield A were analyzed. It was found that some single sand bodies were either not divided in the previous vertical comparison or not characterized in the previous identification of flat single sand bodies. Therefore, this paper further discussed the vertical and plane characterization of single sand bodies.

2. Core water washing and trace data guide vertical characterization of single sand body in low permeability reservoir

Well J1 is A closed coring inspection well drilled in one of the main blocks of Oilfield A. In the C1 layer of the well, there is a weakly washed sand body (FIG. 1) separated by about 0. 2 m of argillaceous rock above the unwashed sand body. The mudstone thickness of about 0. 2 m is consistent with the mudstone intercalation between the lateral deposits of the point dam, which was originally classified as such. Reference 6 points out that "mudstone interlayers between the lateral deposits usually only develop in the middle and upper part of the point dam, namely the upper two-thirds of the point dam thickness, which has a strong blocking effect on the fluid flow. Due to the non-presence of late flood peak erosion in the lower part, the connected body is formed in the lower part, which is an important channel for fluid migration "[6]. This indicates that the lateral interlayer only blocks the displacement in the middle and upper part of the sand body, but does not block the displacement in the lower part of the sand body, and the lower part of the sand body should be washed first. The above condition is inconsistent with the condition of washing the upper part of C1 sand body and not washing the lower part of C1 sand body in well J1, indicating that the argillaceous rock is not the lateral mudstone inside the single sand body, but the interlayer between different single sand bodies.

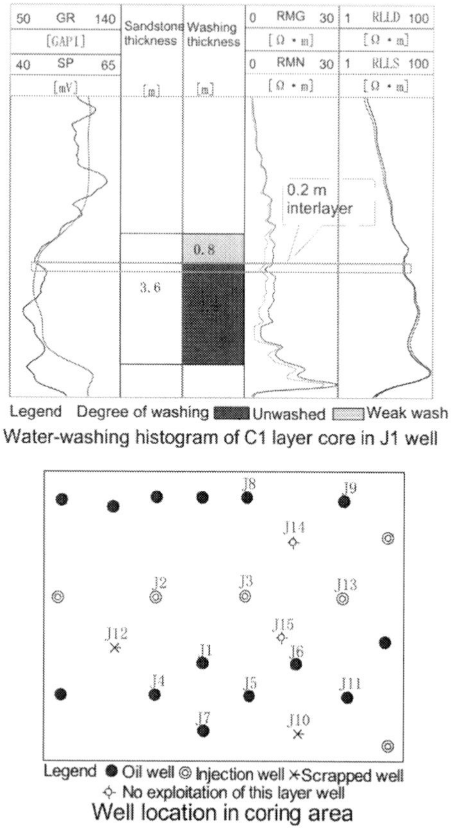

Fig. 1 Water washing histogram of C1 layer core in J1 well and well location in coring area

After J1 was drilled, J5 was injected and J1 was used as an oil well for production. Five years later, tracer was injected into two Wells J3 and J5 in J1 well area of C1 formation, and there were 6 directions for detection (see Figure 2). The detection time in 5 directions ranged from 8. 3 hours to 92 hours, and the detection time from J3 well to J1 well was 878 hours. The time of seeing agent in Yuanda and other directions is consistent with the weak water washing condition of the upper sand body of C1 during coring in well J1. Meanwhile, it is confirmed that the argillaceous rock between the upper sand body and the lower sand body of C1 in well J1 is not the lateral mudstone inside the single sand body, but the interlayer separating the upper and single sand body. Otherwise, if it is the lateral mudstone inside the single sand body, the tracer percolates through the bottom of the sand body of C1, and the maximum time is 92 hours according to other direction of agent detection in the same well area. In particular, the detection time from J5 to J1 is 14. 3h, which is similar to the injection-production distance from J3 to J1. The target direction of the tracer is the C1 sand body of J1, and the detection time from J3 to J1 is not much longer than 14. 3 hours but 878 hours.

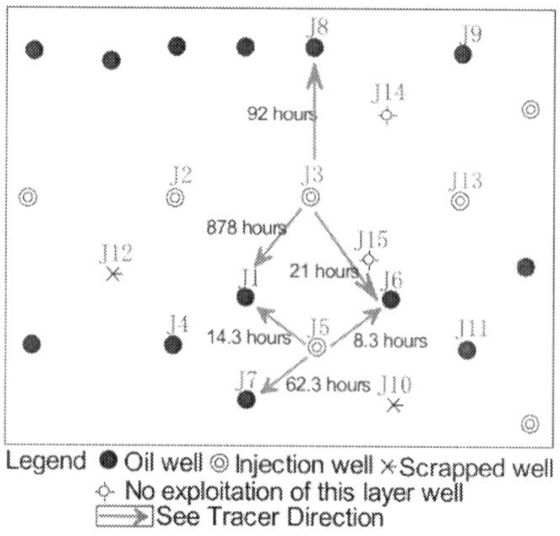

Fig. 2 Well location and tracing time diagram of J 1 well after coring five years

For the sand bodies of Wells J1, J2, J3, J4 and J5 in the C1 layer, according to the existing interpretation criteria of sand bodies in oilfield A, the degree of silt return is not enough to divide them into two independent single sand bodies in the C1 layer. Therefore, they are initially classified as one single sand body in the vertical comparison (see Figure 3). A number of literatures have emphasized the problem that reservoir correlation may be too finely divided [2, 5]. The overview of literature 7 is as follows: "In the process of small-bed correlation, if the series gully associated locally with the point dam, or even every bank overflow in the natural dike or fracture fan is taken as a single sand body and the main sand body is forcibly split, then the seepage barrier inside the main sand body will be forcibly split. It will seriously impede the later research on the law of oil and water movement "[7]. The core washing data of C1 layer and the tracer data of this well area show that the upper sand body and the lower sand body in C1 have completely different seepage barriers. J1 is based on this barrier, and the rest of the Wells are based on the flood surface closest to the barrier on the logging curve (see Figure 3). It is necessary to subdivide the argillous return near the top boundary of layer C11 and layer C12 (see Figure 3) into layers C11 and C12, so as to more accurately describe the true displacement seepage relationship, and thus more accurately guide the subsequent description of remaining oil and adjustment of excavation potential.

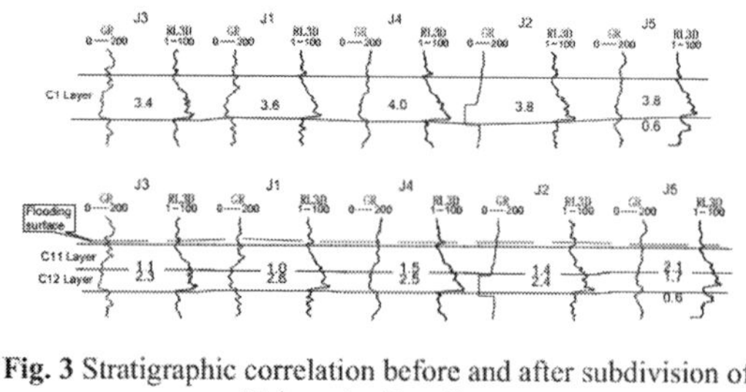

Fig. 3 Stratigraphic correlation before and after subdivision of C1 layer in J1 well area

3. Core water washing and trace data guide the plane characterization of single sand body in low permeability reservoir

Before subdivision of C1 layer in J1 well area, sand bodies were distributed in several Wells in the lower right area (see Figure 4-C1 layer). After subdivision into layers C11 and C12, multiple Wells in the lower right part of layer C12 have sand body distribution (see Figure 4-C12), while multiple Wells in the lower right part of layer C11 have sand body distribution (see Figure 4-C11). It shows that sand bodies only develop in the C12 layer, but not in the C11 layer, which further proves that the C1 layer in well J1 is subdivided into different single sand bodies in the C11 layer and the C12 layer.

In the initial vertical comparison of C1 layer, the division of single sand body was too large, and the different single sand bodies in the upper and lower stages were divided into the same single sand body. Compared with other well sand bodies, the sand body of J12 was the same as the bottom of other well sand bodies, and the sand body only developed at the bottom, with mud at the top and small thickness (see Figure 4-C1 layer). The well sand body is considered to be the abandoned channel sand body of the bar sand body at the point of Wells J1, J2, J3, J4 and J5 on the right (see layer 4-C1). After C1 layer is subdivided into lower C12 layer and upper C11 layer (see Figure 4-C12 layer and Figure 4-C11 layer), the average thickness of single well sandstone in C11 layer is 1. 0m, and that in C12 layer is 2. 0m. The average thickness of single well sandstone in C11 layer is 1. 0m smaller than that in C12 layer, and the hydrodynamic force in C11 layer is weaker than that in C12 layer during deposition. Accordingly, the porosity and permeability of layer C11 is weaker than that of layer C12, and the fluid preferentially passes through the well-permeated channel. Except for the longer 878 hours of observation time, which is the percolation time of layer C11 (the C11 layer of J1 core was washed and the C12 layer was not washed), the observation time of other tracers should be its percolation time in layer C12, otherwise it should also be the same as that of layer C11. The direction of the absence of agent indicates that C11 and C12 layers are absent of agent.

Layer C12 is completely based on core washing data and tracer data, and the direction without data is regarded as sand body connectivity. Single sand body of this layer is described (see Figure 4-C12 layer). During coring, the core of J1 was not washed, and the tracer injection occurred in 14. 3 hours from J5 to J1, so J3, J2 and J1 were not connected. The abandoned channel of C1 was added to C12 in this direction. J5 and J1 were connected, and the two Wells were located in the same point dam. After 8. 3 hours from J5 to J6, J5 and J6 are connected. The abandoned channel predicted by C1 in this direction is cancelled in C12. After 21 hours from J3 to J6, J3 and J6 are connected. The abandoned channel predicted by C1 in this direction is cancelled in C12. From well J5 to well J4, there is no fluid, and the unpredicted abandoned channel of C1 layer is added to layer C12 in this direction. The predicted abandoned channel in C1 between J3 and J8 was cancelled in C12. No fluid was found from J3 to J9, and the direction of abandoned channel in layer C12 was different from that predicted in layer C1.

The single sand body of layer C11 is characterized mainly based on core washing data and tracer data, and the direction shown in no data is based on the thickness change of sand body (see Figure 4-C11 layer). During coring, the core of well J1 was washed, and the agent was found from well J3 to well J1, so there was no abandoned river channel between well J3 and well J1. Three abandoned channels were identified based on the absence of agent in Wells J5 to J4, J5 to J11 and J3 to J9. Five abandoned channels were predicted based on thickness changes in the other directions.

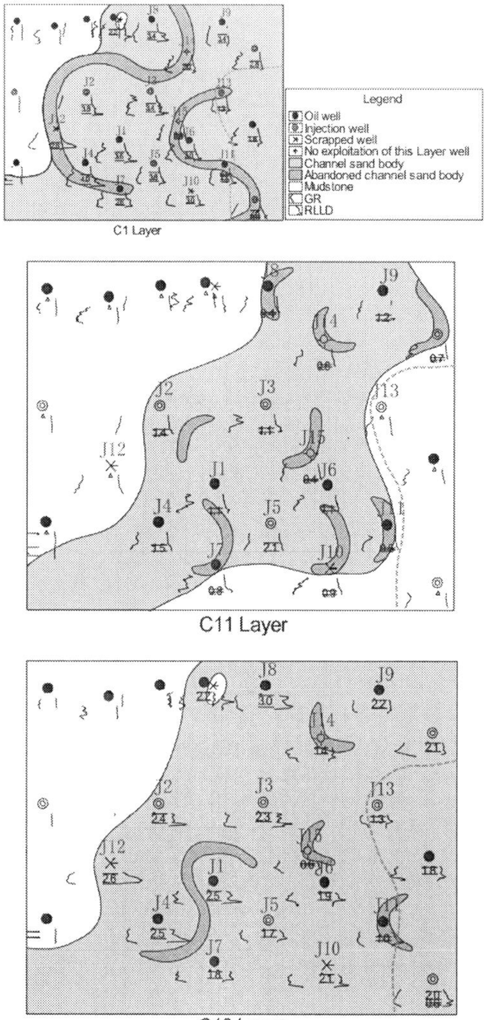

Fig. 4 Facies maps of C1 layer and C11 layer and C12 layer in J1 well area

In general, after vertical redivision and plane recharacterization, the vertical thickness and plane width of single sand body became smaller. In fact, many of the previously predicted abandoned river channels did not exist, many of the previously predicted abandoned river channels did exist, and the distribution direction of many of the previously predicted abandoned river channels changed.

The distribution of single sand bodies in fluvial facies is complex and variable. Based on the characterization results of single sand bodies guided by the above core and tracer data, it is found that the following points should be paid attention to when logging data of well area is used alone to characterize single sand bodies:

First, when the spacing between different vertical single sand bodies is too small, it is easy to be mistaken as the interlayer inside the single sand body, which will lead to the wrong judgment of the vertical seepage relationship, and also lead to the subsequent plane single sand body characterization errors based on the division of vertical single sand bodies.

Second, when the well is drilled into the abandoned channel sand body (the abandoned channel sand body is obviously thinner than the normal sand body, and the sand body is developed at the middle bottom or bottom, and the top is mud), it is easy to determine the existence of the abandoned channel in the well, but it is difficult to determine the distribution direction of the abandoned channel. If the wrong judgment is made, the understanding of the connectivity relationship of the sand body will be wrong, and the adjustment of potential excavation will be misleading. For example, J6 is closest to J15, and the curve is bell shaped. It is easy to predict J6 as the downstream end sand body of the point dam near the abandoned river channel [7], that is, it is predicted that the abandoned river channel bends towards J6, and it is considered that J6 and J15 belong to the same point dam, and J15 and J3 belong to different DAMS. In this way, J3 well and J6 well will be separated by abandoned river channel and not connected, which is inconsistent with J3 well to J6 well. Therefore, J15 well should be bent to J13 well so as not to block the communication between J3 and J6 well. Another example is the abandoned river channel at J11 well in the C12 layer. If J5 well bends to the left instead of to the right, then J5 well and J11 well become connected, which is inconsistent with the lack of agent from J5 to J11 well. Another example is the abandoned river channel of well J14 in the C12 layer. When it bends to well J9, well J14 is connected to well J9, but not to well J3. In this way, well J9 is not connected to well J3, which is consistent with well J3 to J9. If the abandoned channel bends to well J3, well J14 is connected to well J3 and is not connected to well J9, then well J9 is also not connected to well J3. Although it is consistent with well J3 to J9, a little extension of the abandoned channel spreading in this direction will block the communication between well J3 and well J8, which is inconsistent with well J3 to J8, so it is not depicted in this direction. If the abandoned channel bends towards well J13, just as the abandoned channel of J15 does not block the connection between well J3 and well J6, the abandoned channel has no blocking effect on well J3 to J9, which is not consistent with the absence of agent from well J3 to well J9. If the abandoned river bends to well J8, it is more likely to block the communication between well J3 and well J8, which is inconsistent with the detection of agent between well J3 and well J8. The abandoned channel may also have other bending directions. Only with tracer data can we determine that the abandoned channel of well J14 bends to well J9, which is the most consistent with the actual direction. If we only rely on well logging data of the well area, without tracer data, it is difficult to determine that the abandoned channel bends to well J9, which is the most consistent with the actual direction.

Third, when the sand body thickness of three adjacent Wells in the same straight line in the region presents thick-thin-thick distribution, and the sand body of the middle well becomes thinner. The thin sand body does not correspond to the middle bottom or bottom of the thick sand body on both sides, but to the middle top or top of the thick sand body on both sides, and the bottom is mud, then the middle well is not abandoned river channel sand body. There may be intact mudstone near the intermediate well sand body, which has not been washed out by the river and is another sand body boundary to prevent the effect of water flooding. It is necessary to carefully predict the location, distribution direction, extension length, and even the presence or absence of the boundary of the single sand body. If the prediction is wrong, the actually connected direction will be judged as disconnected, or the disconnected direction will be judged as connected. For example, in the connection direction of Wells J1, J5 and J10 in the C12 layer, the thickness of the three Wells is 2. 5 meters, 1. 7 meters and 2. 1 meters respectively. If it is predicted that the boundary of single sand body is between J5 and J1, it is inconsistent with the detection of agent from J5 to J1. If the boundary of single sand body extends between J5 and J6, it is inconsistent with the detection between J5 and J6. If the boundary of single sand body extends between J5 and J7, it is inconsistent with the detection from J5 to J7. When J10 is shut in, the connectivity between J5 and J10 cannot be determined. It can only be predicted that there may be a single sand body boundary between J5 and J10, and the direction is more likely to bend to J10, so as to avoid the contradiction between J5 and J6, J5 and J7, as well as between J5 and J6 and J7. Only with tracer data can it be determined that the single sand body boundary does not exist between J5 and J1. Without the tracer data, it is difficult to determine whether the single sand body boundary does not exist between J5 and J1, and whether the single sand body boundary exists between J5 and J10, depending on the well logging data of the well area. It is difficult to determine whether the single sand body boundary extends between J5 and J6 and between J5 and J7.

Fourthly, the sand bodies of oil and water Wells in different DAMS are not absolutely disconnected, but also well connected. For example, in layer C12, the connection line of well J7, well J5, well J6, well J3 and well J8 is followed. Assuming that the reverse flow time of the tracer is equal to the original flow time, the total flow time of the tracer in this route is 183. 3 hours. The connection between any adjacent Wells on this route is good. The length of the line is far beyond the width of a single point dam in the same area, so there must be some adjacent Wells located in different dam sand bodies.

Fifth, when the thickness of each local well sand body is the same, it is difficult to determine whether and where the boundary of single sand body exists, and its distribution direction and extension length between these Wells based on logging data alone. For example, the abandoned river channel between well J1, J2 and J3 in layer C12 was determined only under the instruction of checking well core water washing data and tracer data. It is difficult to predict it, as well as their distribution direction and extension length, based on logging data alone without checking well core water washing data and tracer data.

4. Conclusion

The correct characterization of vertical single sand body in low permeability reservoir is the basis for the correct characterization of single sand body on the plane. It is necessary to avoid the vertical division of single sand body. When the vertical division of single sand body is conducted according to the logging curve, it is necessary to realize that the small argillaceous return of about 0. 2m can also be used as the separation layer of upper and single sand body and become the seepage barrier of upper and single sand body.

(1) In the well area of low permeability reservoir, according to the well logging data, argillus return occurs simultaneously in the position with the same elevation difference between a part of Wells in a small layer and the nearest reference standard layer. In the other part of Wells that can be connected together, only the sand body below the argillus return develops, but not above the argillus return. It is necessary to recognize the possibility that the upper and lower argillaceous sand bodies in the previous part of the well are different single sand bodies.

(2) The comparison between the single sand body characterized by combination of logging, core and tracer data and the single sand body characterized by single logging data shows that the distribution of single sand body in low permeability reservoir is complex and changeable, and there are many uncertainties in the characterization of single sand body in low permeability reservoir by single logging data.

References

1. Yu Xinghe. Existing problems and sedimentogenesis-based methods of reservoir characterization during the middle and later periods of oil-field development[J]. Earth Science Frontiers, 2012, 19(2): 1-14.
2. Qiu Yinan, Zhang Zhisong, Tang Meifang, et al. The detailed correlation of fluvial sandbody reservoirs[J]. Petroleum Exploration and Development, 1987, 14(2):46-52, 9.
3. Zhao Hanqing. Formation correlation of fluvial-deltaic deposition in Daqing oil field[J]. Petroleum Geology & Oilfield Development in Daqing, 1988, 7(4):25-31.
4. Deng Hongwen, Wang Hongliang, Yan Weipeng, et al. Architecture model of sequence stratigraphy in fluvial facies[J]. Acta Sedimentologica Sinica, 2004, 22(3):373-379.
5. Qu Fang, Chen Qinghua, Lian Chengbo. Discussion on the method for the subdivision and comparison of fluvial reservoir[J]. Journal of Xi'an Shiyou University(Natural Science Edition), 2008, 23(1):17-21.
6. Zhang goodwill Research on point dam identification and building structure of Fuyu oil layer in Chang 43-39 block [D] Northeast Petroleum University, 2011.
7. Chen Xinkai, Liu Jingyan, Chen Cheng, et al Single sand body identification of meandering river under subdivision of main configuration elements [J] Acta sedimentologica Sinica, 2020, 38 (1): 205-215.
8. Ma Shizhong, Yang Qingyan Sedimentary model, three-dimensional configuration and heterogeneous model of meandering point dam [J] Acta sedimentologica Sinica, 2000, 18 (2): 241-247.
9. Yan Baiquan, Ma Shizhong, Wang long, et al Physical simulation of formation and distribution law of residual oil in meandering point dam [J] Geoscience frontier, 2008, 15 (1): 65-69.
10. Wang Mingchuan, Zhu Weiyao, Dong Weihong, et al Internal configuration of point bar type thick reservoir unit in meandering river and its influence on remaining oil distribution [J] Oil and gas geology and oil recovery, 2013, 20 (3): 14-17.
11. Ma Shizhong, Wang Yibo, Li Shikui, et al Internal building structure of single sand body and its control mechanism on residual oil [a] Abstracts of the 8th Academic Conference on paleogeography and sedimentology [C]. 2004.
12. He Miao, Jin zhenkui, Li Tingdong, et al Characterization of fluvial facies single channel sand body -- Taking the Triassic Kexia formation in the first middle area and the third 4 area in the northwest margin of Junggar Basin as an example [J] Science and technology Herald, 2014, 32 (6): 17-24.

 https://doi.org/10.1051/e3sconf/202337501037

Discussion on the countermeasure of improving the development effect of oil well dynamic analysis

Meiyi Jiang

Dynamic Room of Geological Research Institute, No.6 Oil Production Plant, Daqing Oilfield Company, Heilongjiang 163000, China

Abstract: With the rapid development of society and economy, higher requirements are put forward for oil and gas exploitation. Therefore, effective measures to improve the development effect should be analyzed with oil and gas well dynamic analysis as the entry point. This paper first introduces the definition of oil and gas well dynamic analysis, then introduces the well dynamic analysis and measures to improve the exploitation effect, then introduces the injection well dynamic analysis, and finally explores through the analysis and study of each production link of oil and water Wells, analyzes and determines the actual situation of underground advantage. The aim is to find out the law of oil and water movement between each reservoir, and take corresponding adjustment measures constantly to ensure a higher recovery rate of oil and gas Wells.

Key words: oil-well dynamic analysis; Development effect; discuss

1. Introduction

Well dynamic analysis, also known as production dynamic analysis, includes water injection dynamic analysis and well dynamic analysis, and is also an important part of oilfield production engineering. Oil well dynamic analysis mainly uses a variety of detection types and production data of oil and gas Wells to analyze changes in output, pressure and water content. Through detailed data research and analysis, we can understand the water level of oil and water underground, judge the downhole technology, and determine the rationality of its production process and production technology. The dynamic analysis of water injection well mainly refers to the analysis and research when changes are found in the aspects of water injection and water absorption capacity and wellhead pressure, and to judge the actual situation of downhole operation. The main function of dynamic analysis of oil and water Wells is to discover possible problems through dynamic analysis and comparison of Wells and oil Wells in the production stage, and constantly adjust injection and production to ensure that Wells and oil Wells are in a relatively stable condition in oil production, water injection, pressure and water cut, so as to develop the reservoir reasonably.

2. Dynamic analysis of oil and water Wells

In general, a single well is mainly composed of the wellbore surface and underground, and the combination of production management and underground analysis, and then through the analysis of the ground, and then the actual situation of the wellbore, and finally explore the recent underground conditions, through the progressive method of layer by layer, can effectively improve the dynamic analysis and development level of oil and water Wells. Combining economic benefits with Wells and oil Wells, through analysis and research, strong rationality and feasibility measures can be put forward without affecting the overall mining efficiency and quality, which can maximize the productivity of oil Wells and improve the economic benefits of oil field enterprises. In the process of crude oil exploitation, oil Wells are the main way to extract crude oil from the surface [1]. The process involves oil being pressured through fractures in the rock into the well and then down the wellbore to the wellhead. Ultimately, the oil is delivered centrally to the gathering station, so the well production process also consists of wellbore, reservoir and surface phases to enable dynamic analysis to ensure reservoir quality and safety.

2.1 Ground management

In the production process of oil field enterprises, the management of oil well surface is mainly divided into wax removal, hot washing and casing pressure selection. The main function of paraffin removal is to ensure that the oil flow is unimpeded, so as to minimize the frequency of paraffin removal hot wash, and control the sleeve pressure, so that the dynamic liquid level is at a relatively level, in order to ensure the pumping force. When the casing pressure is too high, the dynamic liquid level in the annular gap of the casing will be reduced, which will lead to the exhaust gas into the deep well pump, resulting in gas invasion, affecting the pump efficiency and reducing the production of the well. In severe cases, air locking may even occur. Therefore, based on the actual situation, part of the gas in the tube should be discharged accordingly, effectively reduce the casing pressure, raise the hydrodynamic level, and effectively avoid the waste gas into the pump.

2.2 Well shaft performance analysis

The change of wellbore dynamics can be caused by a variety of factors, the most common of which is (1) formation blockage, which is usually caused by damage to the formation around the bottom of the well, such as silt, or sand production, resulting in sand buried in the formation, which prevents the smooth flow of oil into the wellbore and ultimately leads to a decline in production. The wellhead pressure decreased significantly, and so did the casing pressure. By monitoring the bottom hole sand samples and sand surface, the main analysis objective is to determine whether there is a large well production and cementing of the reservoir, as well as changes in the characteristics of the drilling fluid, such as relative density coefficient, loss volume, permeability duration, leakage, etc., to ensure the safe operation of the newly commissioned well. In order to ensure the safe operation of the old working well, the level of operation, the performance of the kill fluid, and other possible causes of formation damage should be tested. For example, if the local formation of wax, glue, salt, scale, etc., may lead to perforation or reservoir penetration surface obstruction; (2) The size of the nozzle should not be too large, because too large nozzle will lead to the increase of gas-oil ratio and the increase of water content too fast, thus reducing the production, and may lead to sand and other problems when the oil reservoir pressure drops seriously [2]. As the wellhead oil and casing rise, sand and water in the reservoir rise, causing hydraulic pressure to drop and casing pressure to stabilize or decrease. If the nozzle becomes smaller, its ability to carry water becomes worse, which will cause water at the bottom of the well, so that the oil and casing pressure decreases, the production decreases, and the water content decreases. (3) Other factors. Because the oil supply capacity is not enough, oil well intermittent work, the condition is: when oil, oil pressure will drop sharply, and it makes oil well production capacity and pressure change dramatically, oil, casing pressure will rapidly reduce or improve, resulting in oil well production capacity greatly reduced.

2.3 Submergence (hydrodynamic level) analysis

In the production process of oil well, the moving fluid level mainly refers to the depth of the liquid level in the annular space of casing and tubing, while the subduction degree mainly refers to the difference between the moving fluid level and the depth of the pump. The analysis of the specific changes of the dynamic fluid level can further reflect the actual changes of the formation. Hydrodynamic level rise is mainly due to pump parameters smaller, reservoir pressure rise, pump frame change. Before the deep well pump can begin work, it is necessary to solve the screen, anchor, sand anchor and Vallhole under the pump equipment obstruction, and ensure adequate displacement. This requires the deep well pump to go deep enough below the liquid level to form a type of pressure head to reduce the obstruction to the pump and to avoid degassing the oil so that the well can be produced smoothly. Once the head is too small, it will affect the satisfaction coefficient of the pump; Once the head is too large, the wreck will be too large, affecting the smooth production of the well. The increase of the pumping unit's load will lead to energy consumption. Therefore, according to the successful experience of some oil fields at home and abroad, the sinking degree should be controlled within 200-300m for the oil well with thick oil, high water content and large production capacity, and the electric pump well should also be maintained within 200-300m, but the specific depends on the characteristics of the oil well.

2.4 Analysis of subsurface performance of oil well

Initial formation pressure is an important reference value before reservoir development. But when the well is shut down for a while, the pressure measured is no longer the same. In contrast, for Wells with heterogeneous multiple zones, the measured static pressure is uniform across the formation. The driving mode and mining rate are the key factors affecting the rock pressure in the waterflood oilfield. In a waterflood oilfield, water content changes obviously, and different water states will lead to different water increase rates. Generally speaking, in the middle and low water cut stage (before 60%), with the increase of water content, the water rise rate will gradually accelerate; However, in the high water cut stage (after 60%), the rate of water rise will gradually slow down with the increase of water content. By analyzing the water cut of the well, we can better understand the properties and distribution of the reservoir and understand the connectivity between the well and the water well. In addition, we can also determine the cause of water cut rise, and put forward timely adjustment suggestions and measures. The amount of fluid a well produces depends on its ability to extract fluid under certain water conditions and on the pressure difference it is subjected to. With the increase of water content of oil Wells, the liquid recovery index will also increase year by year [3]. However, when the reservoir pressure remains stable, the production pressure differential of the well will continue to increase, resulting in a gradual decrease of the production pressure differential. Therefore, in the middle and high water cut stage, although the fluid recovery index of the gusher well is increased, the fluid production is hardly increased or slightly decreased due to the decrease of working pressure.

3. Dynamic analysis of injection well

Water injection well is an important production technology, its main purpose is to inject water from the wellhead to the reservoir through the water distribution, in order to meet the needs of current oilfield development and achieve the balance of injection and production. Its main function is to ensure the development of oil field, but also can reasonably improve the benefit of oil field. In order to effectively improve the quality of water injection, layered water injection can be used to ensure that different layers can meet the requirements, so as to achieve the effect of water absorption balance. However, during the actual operation, because of the different water injection volume changes, may lead to the completion of the mismatch of the injection volume. Therefore, the relevant personnel should analyze the changes of water absorption of each layer in real time, and compare the oil reservoir pressure, water cut changes in the surrounding production Wells and the production rate. Through dynamic analysis and research on various factors, timely optimization and improvement of water injection volume can effectively improve water injection efficiency. Fundamentally improve water drive recovery [4].

3.1 Analysis of reservoir condition of injection well

The physical properties of oil reservoir and petroleum are the key factors affecting the dynamic analysis of injection well. Therefore, it is necessary to fully understand the geological, natural and physical characteristics of oil reservoir, including lithology, infiltration rate, oil characteristics, pressure, as well as the number of perforated layers, thickness and perforation condition, so as to better grasp the water absorption status of oil reservoir.

As a result of unqualified quality, reservoir perforation and clay imbalance, percolation surface obstruction, biochemical sediment blocking reservoir gap, and bacteria propagation and blockage, resulting in a sharp increase in the water pressure of the well, a sharp decline in water absorption index, significantly reduced water injection, and even completely unable to fill. Tests should be carried out in time to determine the root of the problem and take appropriate measures. Drainage and stabilization of mudstone should be performed prior to the injection of Wells to ensure that the injected water quality is up to standard, which is the key to successful irrigation.

3.2 Variation of water injection

Due to the combined needs of reservoir injection-production balance and reservoir energy conservation, the injection allocation of each well must be strictly regulated. Overinjection or underinjection can lead to disruption of the injection-production balance. Therefore, during the analysis of injection volume, it is necessary to carefully check whether there is water nozzle obstruction, poor water quality, unreasonable selection of killing fluid and other reasons according to the system test resources measured in the previous injection well, so as to ensure the accuracy of injection volume. If the flow rate exceeds the normal range, a test should be performed to check if the downhole packer is faulty. There is no closure of the water base, there is no spout, etc. By analyzing the causes of overinjection and underinjection, propose effective solutions, such as adjusting the water volume, replacing the nozzle, reconfiguring the test or increasing the water volume, to ensure that the water flow in the injection well is as consistent as possible.

3.3 The water absorption of each layer changes

By means of differential well temperature logging or isotope logging, the absorption profile data of the output well are analyzed, so as to understand the absorption between different layers and observe its change trend in history. In this way, the absorption differences between different layers can be better understood and more effective application techniques can be developed more effectively. To maximize the area covered by injected water, ensure that each layer absorbs water evenly as much as possible.

In order to ensure the balance of injection and production in an oilfield, the amount of water injected into the ground should be equal to the amount of fluid produced. This requirement must be met whether it is an entire well, a stratification or a single sand body. However, due to the heterogeneity of the reservoir, the uneven water absorption in plane, between and within the reservoir results in the unbalanced reservoir pressure. In the interlayer, the distribution of reservoir pressure is unbalanced, resulting in interlayer interference. Within the reservoir, the difference of water drive state and reservoir pressure makes the distribution of remaining oil more fragmented, especially in the late period of high water cut, and these situations become more complicated [5]. Therefore, according to the analysis of injection-production ratio and reservoir pressure, the injection allocation between zones and planes should be adjusted as much as possible. In addition to the usual adjustment of the nozzle in the wellbore to change the interzonal water absorption, a series of measures to transform the reservoir should be carried out if necessary.

3.4 Wellbore performance analysis

Based on the dynamic analysis, we should assess whether downhole tools such as water distributors, packers, and taps are performing to relevant standards and observe how they perform during the injection process. If the cause of failure or state change occurs in the above parts, we should respond by observing the wellhead oil pressure and system measurement indicators curve. If part of the fault occurs, the water flow will be unstable. Therefore, it is necessary to check and analyze the failure factors regularly and take preventive measures to solve them. Significant changes in oil pressure and water flow can occur when a tubing perforation is lost, a first-stage packer is damaged, or cement is routed outside the sleeve. If the reservoir above the stage 1 packer absorbs more, a strong casing pressure increase, oil pressure decrease, and water injection increase can occur. When stage 2 and 3 packers are damaged, oil pressure decreases and water injection increases. If the water nozzle is blocked or out, the oil pressure and water volume will change greatly, the oil pressure will increase, and the water injection will decrease, which indicates that the water nozzle has been blocked. When the oil pressure decreases and the amount of water injected increases, it indicates that the water nozzle has come out.

4. Conclusion

To sum up, in order to adjust the differences between different oil formations, low-pressure and low-aquifer injection can be carried out, water injection profile can be adjusted, and the vertical production profile can be more consistent. In addition, combining with the exploitation condition of oil sands and water flooding condition, some layered water injection Wells can be repored, injection and production system can be improved, or water injection well points can be added, so as to achieve better results and improve the fluid production of oil Wells. Improve the exploitation condition of oil sand body, improve the exploitation efficiency of oil sand body and realize the effective utilization of oil and gas resources. As a result of increased pressure, more productive potential can be developed.

References

1. LIU Lu. Application of on-line continuous fluid level monitoring in oil and water well dynamic analysis [J]. Chemical Engineering and Equipment,2021(11):240-241.
2. Yang Lei. Application of oil-well dynamic monitoring data in oilfield development dynamic analysis [J]. China Petroleum and Chemical Standards and Quality, 201,41(20):57-58.
3. Shu Nan. Application of oil-well dynamic monitoring data in oilfield development dynamic analysis [J]. China Petroleum and Chemical Standards and Quality, 201,41(16):130-131. (in Chinese)
4. Yang Haiwei. Application of oil-well dynamic monitoring data in oilfield development dynamic analysis [J]. Chemical Engineering and Equipment,2020(04):58-59.
5. Zhang Runjie. Application of oil-well dynamic monitoring data in oilfield development dynamic analysis [J]. Chemical Engineering Design Communication,2019,45(01):48.

Causes and treatment of adhesive damage in polymer flooding injection pipeline

Long Tian

No.6 Oil Production Plant No.7 Operation Area, Daqing Oilfield Co., LTD., Daqing Heilongjiang 163000, China

Abstract. The viscosity of polymer liquid is an important parameter used in oilfield. Understanding the loss rule and regularity of polymer liquid will help to improve the distribution and water injection quality of polymer liquid and ensure the effective distribution of polymer liquid. According to the oil-water viscosity data of the previous polymer flooding fluid, it is found that the oil-water viscosity is within the range of 5%-20%, with an average of 9.8%. The reasons of single well pipeline viscosity are complex and varied, which are closely related to reducing substances, types and quantities of bacteria, fouling oil suspensions on the inner wall of pipeline, pipeline length and running time. Therefore, from the aspects of Fe2+, pipeline running time and so on, the influence factors and prevention methods of single well pipeline viscosity loss were studied, and the corresponding treatment measures were worked out, and good results were obtained.

Key words: injection pipeline; Polymer; Adhesion loss

1. quotation

The principle of the polymer zone is to increase the hydraulic viscosity through the polymer and thus achieve better recovery. Due to various factors, the viscosity of polymer solution decreases and viscosity loses in the process of water injection, which seriously affects the production effect of oilfield. Therefore, it is very necessary to study the polymer flooding problem in the process of polymer flooding.

2. The key factor of single well viscosity loss

2.1 Distribution law of single well pipeline viscosity loss

According to the research on polymer flooding oil system over the years, it is found that the viscosity of crude oil in polymer flooding oil system is 5% to 15%, with an average of 9.8%. According to different materials, different diameters and different running times, the viscosity loss of single well pipelines at several injection stations in similar regions was calculated. Through the analysis of the test data, the following rules are obtained: under the same working condition, the adhesion loss rate of carbon steel pipe is 2%-3% higher than that of glass steel pipe; The viscous resistance of each pipeline increases with the increase of pipeline length. The results show that the viscosity loss of carbon steel pipe increases with the extension of service time, but the viscosity loss of FRP pipe does not change significantly. The viscosity loss of pipe has little relation with pipe diameter.

2.2 Causes of single well pipeline viscosity

In order to find out the cause of pipe adhesion, the pipes of various materials were selected, and the single well pipe was washed with water to wash away a lot of glue and strips in the pipe. Through comparative analysis of imported and exported water samples, it is found that the concentration of iron ions increases and decreases significantly, as well as the concentration of iron [1]. Among them: after the glass steel wire by washing, the concentration of hot iron almost did not change; Carbon steel pipe after the early cleaning, iron content increased greatly. Through the analysis of the experimental results, it is found that the residual iron ion and gum in the polymer solution are the important factors leading to the viscosity loss.

3. Mechanism of iron ion degradation

This project intends to analyze the decrease of high viscosity property caused by Fe ion in the pipeline through laboratory tests, clarify the influence of Fe ion on the viscosity property of polymer, clarify the mechanism of Fe ion reducing the viscosity property of polymer, and find the way to reduce the influence of Fe ion on the viscosity of polymer.

3.1 Influence of Fe2+ on viscosity of polymer solution without oxygen

In an open system, Fe2+ is easily oxidized to Fe3+ by air, which makes the interaction between Fe2+ and the polymer inadequate. To achieve this goal, a polymer solution with a concentration of 1000 mg/L was prepared in the laboratory with simulated water. The FeSO4 was added under anaerobic conditions, stabilized for a period of time, and the viscosity value was measured. As can be seen from Table 4, when the concentration of Fe2+ is less than 10 mg/L, the influence on the viscosity change of polymer solution is no more than 10% under closed anaerobic conditions.

3.2 Influence of Fe2+ on viscosity of polymer solution under aerobic condition

In the tertiary recovery process, most of which takes place in an open system, Fe2+ is easily oxidized to Fe3+ by atmospheric oxygen. Although sealing measures are taken in some areas, oxygen can still fuse with the solution to gradually oxidize Fe2+ to Fe3+, which is very different from the effect of Fe3+ alone and Fe2+ in the absence of oxygen [2]. In the validation laboratory open system, a polymer solution with a concentration of 1000 mg/L is prepared with simulated water, and FeSO4 is added for 3 hours, and the viscosity is measured. The oxidation treatment from Fe2+ to Fe3+ has a great damage to the viscosity of the polymer solution. Under the concentration of 1 mg/L Fe2+, Has caused viscosity loss in a quarter of the solution.

3.3 Influence of Fe3+ on polymer viscosity

When Fe2 (SO4)3 was added to the polymer solution, it was observed that the viscosity of the polymer solution did not change much with the addition of Fe3+.

3.4 Influence mechanism of iron ion on polymer viscosity loss

Fe2+ is the only known substance that can dissolve polymers close to water in a matter of seconds. Among them, the degradation of macromolecules caused by Fe2+ oxidation to Fe3+ is the most important reason leading to the decrease of solution viscosity. The polyacrylamide solution is very stable in the absence of oxygen, even in the presence of Fe2+ and FeS. However, in actual production and operation, due to the presence of dissolved oxygen and Fe, the viscosity of the polymer system will be reduced.

4. Methods to reduce pipeline viscosity

4.1 Oxygenation

To change Fe2+ from diluted water to Fe3+, oxygen exposure equipment was installed at the water injection station to measure Fe2+ before and after oxygen exposure prior to mixing with the polymer. The experimental results show that appropriate increase of oxygen content in wastewater can effectively inhibit the adverse effects of different kinds of fungi on the viscosity of polymer materials in wastewater. From the perspective of oxygen aeration effect, the concentration of Fe2+ can be reduced to 1 mg/L, achieving the goal of oxygen aeration.

4.2 Flushing Pipeline

The multi-channel cleaning of single well pipeline can effectively eliminate the adhesion of reducing substances such as colloidal substances in pipeline to polymer, and can effectively reduce the influence of crude oil, suspended matter and various ions on the viscosity of polymer solution. In order to facilitate operation and save investment, the injection station uses an established well cleaning process and uses diluted water to quickly clean the injection line. Although this method is very difficult to operate, its effect is not significant, and only reduces viscosity to a small extent. Therefore, for a single well pipeline with a viscosity loss rate of more than 10%, It is necessary to adopt more efficient cleaning methods.

4.3 Clean water Rinse

When cleaning the pipeline, in order to ensure the effect of cleaning, first of all, the station process is improved, so that it can be connected with the cement car, and cleaning under clear water at 60°C. As far as scour effect is concerned, the scour effect of clean water + washing aid is the best, followed by that of clean water under pressure. In actual production, using cement truck to pull clean water for cleaning is not only time-consuming and time-consuming, but also unable to meet the needs of large-scale pipeline cleaning [3]. Therefore, the current clean-out process should be used to increase the pressure and flow rate gradually, using diluted effluent initially, so that post-wash cementation is greatly reduced.

4.4 Hole jet flushing

Cavitation jet cleaning technology is a new technology suitable for oil pipeline. It has the characteristics of fast flushing speed and obvious anti-rust effect on oil and water pipeline. First, the mechanism of stomatal ejection. Cavitation jet is a physical fluid dynamics that takes advantage of "hole effect". Driven by high pressure water flow, a low pressure area is formed in the pipeline. The microbubbles in these low pressure areas are rapidly compressed and ruptured, and powerful microjets are ejected instantly to clean the dust in the pipeline. Second, the scour effect of hole jet. In this area, through hole jet cleaning of single well pipeline, the efficient cleaning of oil-water interface is realized, and the anti-fouling ability of oil-water interface is improved.

5. Research on cohesive loss control method of polymer flooding

Viscosity loss refers to the viscosity loss caused by mechanical, chemical and biological degradation of polymer materials in injection molding. The viscosity of the system can be improved and the development effect can be improved by the viscosity of the polymer. Through years of research and practice, the causes and changes of viscosity in polymer solution during injection distribution were clarified, and a series of viscosity control measures were put forward, so that the viscosity of polymer flooding can be effectively controlled.

5.1 Understanding of cohesive loss rules of polymer flooding in Lamadian Oilfield

In order to understand the viscosity loss law of polymer flooding, it will have certain effects on polymer molecular weight, polymer solution concentration, dilution water quality and maturity time, etc. In this study, we will study the influencing factors of the above aspects, so as to better understand the viscosity loss change law of different injection systems. In this way, the management of adhesion failure can be carried out more clearly. Laboratory test results show that in the same molecular weight, the same aqueous solution, under the same shear force, the viscosity retention rate of the solution is higher, and its shear resistance is stronger. This is because at a higher concentration, the repulsive force between molecules increases, causing the molecular chain to curl, thus reducing the mechanical properties [4]. The mixing tank is made of polymer matrix, and there is almost no adhesion phenomenon during mixing, but the mixing time is too long, so that the macromolecules are completely developed, and its shear resistance decreases.

5.2 Viscous loss control technology of polymer flooding in Lamadian Oilfield

By studying the causes and influencing factors of gelling phenomenon, the prevention and control measures of gelling phenomenon are explored, so as to realize the effective control of gelling phenomenon and ensure the injection of gelling phenomenon with high quality and high efficiency. When determining the concentration of the polymer parent solution, the concentration of the polymer parent solution should be increased as much as possible according to the operating parameters of the polymer injection pump and the design concentration of the individual well, so that the viscosity phenomenon can be fully controlled. Combined with the transport capacity of the preparation station and the transport time in the pipeline, the mother liquor in the mature stage can be reduced as far as possible, and the overall control of viscosity loss can be achieved. On the basis of clarifying the causes and rules of adhesion loss of all kinds of joints, formulate the adhesion loss control criteria of the joints, divide the joints into A. B. C three categories, adopt the "guarantee A. According to the overall idea of "controlling B and treating C", targeted grading and benchmarking should be carried out to improve the treatment efficiency of joint adhesion damage. The viscous-loss treatment method of "flush-washer-separation-repair" has been popularized in Daqing oilfield for many years, but its implementation specifications are still unclear. Therefore, this project proposes a set of viscous-loss treatment measures specifications based on "flush-washer-separation-repair" from three aspects of influencing factors, rule changes and control specifications, in order to enhance the applicability of measures and improve the treatment effect of measures.

6. Important types and influencing factors of oilfield polymer flooding and water scaling

6.1 Calcium Carbonate

Calcium carbonate ($CaCO_3$) is one of the most common scaling substances in offshore crude oil production, and its solubility in water is very low. The substance is made by combining an ion called carbonate or a hydroate with an ion called calcium.

6.1.1 Effects of carbon dioxide

In general, the number of carbonate ions is much smaller than that of carbonate ions, so it can be regarded as a deposit dominated by calcium carbonate deposits. Increasing the concentration of CO_2 can reduce the amount of calcium carbonate deposition. In oilfield water, the lower the concentration of CO_2, the more calcium carbonate precipitation. The increase of CO_2 concentration and water pressure can lead to the increase of CO_2 concentration in water.

6.1.2 Influence of pH value

There are usually different carbonic acids in oilfield water. Due to the difference of pH value, the concentration ratio of $CO_2+H_2CO_3$, HCO_3^- and CO_3^{2-} in water is also different at equilibrium. It can be seen from the pH that the higher the pH in water, the more calcium carbonate will precipitate; Otherwise, it will not be easy for sediments to form.

6.1.3 Influence of Temperature

Temperature is another major factor affecting the scale formation of calcium carbonate. The higher the temperature, the greater the normal state of the greater the solubility of most salts in water. Calcium carbonate scaling occurs easily when high temperature injection Wells are injected.

6.1.4 Influence of total pressure

The dissolution of calcium carbonate will be beneficial when pressure is increased and, conversely, will encourage precipitation. In a single phase, from a thermodynamic point of view only, the dissolution degree of calcium carbonate increases as the pressure increases.

6.1.5 Influence of salts dissolved in water

The solubility of calcium carbonate increases with the increase of salt content in aqueous solution with different concentration. The solubility of calcium carbonate in water is determined by its total solubility in water, the greater the total solubility. In general, an increase in pH, as well as a decrease in salt content, increases the scaling tendency of calcium carbonate with increasing temperature.

6.2 Magnesium Carbonate

When the scale samples brought back from the offshore platform are analyzed, it is often found that there is a small amount of magnesium carbonate scale in the calcium carbonate scale. The precipitation process is as follows: $Mg^{2+}+2HCO_3^- \rightarrow MgCO_3\downarrow+CO_2\uparrow+H_2O$ (6). With the increase of temperature, the decrease of partial pressure of CO2, the increase of pH value and the decrease of salt, and thus the increase of scaling tendency of magnesium carbonate. However, the solubility of CaCO3 is much lower than that of CaCO3, and CaCO3 will precipitate first when present conditions change. In the process of offshore oil platform exploitation, a large amount of seawater is often used, and the Mg ion content in the ocean is relatively high, which will be mixed with the well water to a certain extent. If the well water is relatively alkaline, it is very likely to generate magnesium carbonate scale. Magnesium hydroxide is easily hydrolyzed to magnesite in aqueous solution, and its hydrolysis process is as follows: $MgCO_3+H_2O\rightarrow Mg(OH)_2+CO_2\uparrow$(7). Its solubility in aqueous solution is low and decreases with the increase of temperature.

6.3 Calcium Sulfate

Calcium sulfate is another common form of scale found in offshore oil fields. The crystallization rate of calcium carbonate is higher than that of calcium sulfate. Calcium sulfate scale does not dissolve easily under acidic conditions, so it is difficult to remove.

6.3.1 Influence of Temperature

Below 38°C, the solubility of gypsum gradually increases with the increase of temperature, but above 38°C, it gradually decreases.

6.3.2 Influence of dissolved salts in water

When NaCl or other salts that do not contain calcium ions and sulfate ions are present in water, the solubility of calcium sulfate or anhydrous calcium sulfate increases when the concentration is less than 150 g/L, and the increase of salt content leads to the decrease of solubility of calcium sulfate.

6.3.3 Influence of pressure

The solubility of all scale in water increases with the increase of pressure.

6.3.4 Influence of pH value

pH has little effect on the solubility of calcium sulfate.

6.3.5 Iron compounds

In the analysis of scales, trace amounts of ferric compounds are often detected. The water containing CO2 will erode iron and generate ferrous, and the precipitation of iron is related to the pH of the system. When the pH is greater than 7, iron is easily precipitated. The water contains H2S, which forms iron-based sulfides due to iron erosion. When iron and oxygen come into contact, corrosion forms such as $Fe(OH)_2$ and $Fe(OH)_3$ are produced, and precipitates are produced.

7. Concluding remarks

In summary, the material, length and running time of the pipeline will have a direct effect on the viscosity loss of the polymer. In the process of surface distributed hydraulic transportation, the content of iron ion should be reduced by appropriate measures such as oxygen. The influence of impurities such as glue in the pipeline on polymer viscosity can be reduced by regular pipeline cleaning. For the pipelines with small viscosity loss, "high temperature water + lotion" can be used for cleaning. For the pipelines with large viscosity loss and more dirt, the method of hole jet can be used for cleaning.

References

1. Zheng Bin, Qiao Fang, Sun Fenglin, Zhang Jiguang, Liu Fushuang, Jiang Guanghong, Cao Xin. Analysis and treatment of pipeline scaling in polymer flooding injection process [J]. Chemical Industry Management,2022(17):131-134. (in Chinese)
2. LI Yuehua. Causes and treatment methods of viscous damage in polymer flooding injection pipeline [J]. Oil & Gas Field Surface Engineering,2022,41(02):17-21.
3. LI Dong. Discussion on countermeasures to reduce viscosity loss by surface injection process in oilfield [J]. Petrochemical Technology,2021,28(11):148-149.
4. ZANG Minghao. Study on viscosity change of displacement fluid in pipe fitting of injection distribution system and low shear sampling device [D]. China University of Petroleum (Beijing),2020.

A certain block of oil and gas accumulation understanding

Yanfeng Bao

No.5 Production Plant, Daqing Oil Field, China

Abstract: The block is located in the southern part of an oil field. A grooved area held by three structures. The blank area is 41.6 square kilometers. It accounts for about 20% of the blank area. The faults in this area are more developed, mainly in the NNW direction, Permeability 0.7~1542mD, The average is 275.1mD, the porosity is 11.1~27.8%, the average is 23.2%, the average effective thickness is 2.2m, and the river channel width is 100-300m. The three-dimensional seismic data were collected in 2004, the accuracy of the survey network was 20m×40m, the number of coverage times was 66 times, the main frequency was 39-42Hz, the surface element was 10m×10m, and the data was finely processed and explained in 2008.

Keywords: Oil and gas accumulation; Sedimentary facies; Trapping.

1. Analysis of accumulation conditions

Oil and gas accumulation is inseparable from two factors, one is static geological factors, including source rocks, reservoirs, caps and overlying strata; The second is dynamic geological action which includes trap formation and hydrocarbon formation, transport and aggregation. These geological elements and processes must have an appropriate spatial and temporal configuration to convert the organic matter in the source rock into oil and gas, and then form oil and gas reservoirs. These concepts are very consistent with the research idea of "raw storage cap transport circle protection" oil and gas bearing system proposed by petroleum geologists in China. In the following analysis, the accumulation control factors of this block are analyzed from two aspects: static geological elements and dynamic geological effects.

1.1 Raw oil conditions

Sufficient oil source is the basis of reservoir formation, Qingshankou Formation is mainly lacustrine sedimentary oil layer, Qingshankou Formation sedimentary large section of mudstone is not only regional cap, but also good source rock. The overlying grape flower oil layer is the result of upward accumulation of oil from the green section after a large amount of oil is discharged from the source rock, under the joint matching of fluid potential and fault, and continues to move laterally after entering the grape flower oil layer. The dark mudstone of the green section of this block develops, and the predicted thickness is between 50~80m, and the average thickness is 70m. The source rocks of this block are developed, providing sufficient oil and gas for the grape flower oil layer.

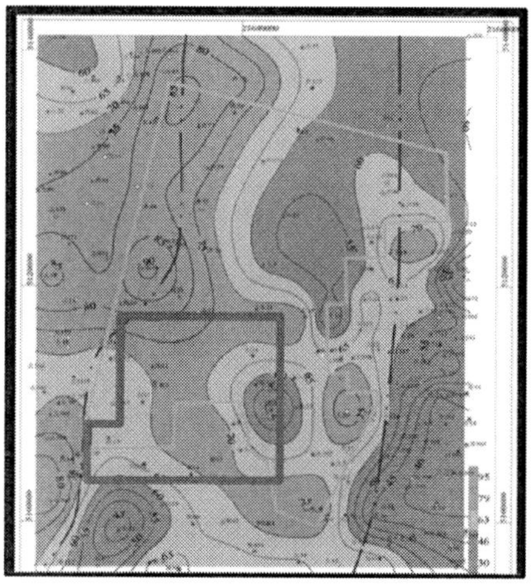

Fig. 1 Cyan dark mudstone thickness prediction map

1.2 Reservoir Conditions

The grape flower oil layer is located in the sedimentary environment of the delta front, and it can be seen from the prediction results of sedimentary facies that the entire grape flower oil layer is a set of reservoirs deposited in a large water retreat process, mainly developing underwater diversion channels, mat-like sand, side beaches and inter-stream bay sedimentary microfacies, the sand body is thicker and the sand quality is purer, the physical properties and connectivity are better, and the sand body is mainly flaky or nearly north-south strips.

In addition, the grape flower oil layer is the product of the short lake retreat in the depression stage of the lake basin, and the source of material mainly comes from the north and northwest of the block, and the main reservoir is the sediment of river channel sand and side beach. The thickness of the oil layer is about 50~70m, the scale of the sand body contained is large, the number of development layers is small, the thickness of single sandstone is generally less than 3m, the lateral connectivity of the sand body is good, and it is lens-like and short axial distribution. From the perspective of the physical characteristics of the reservoir, the porosity distribution range of the samples of the grape flower oil layer was 4~26%, and the average was 15.0%, which did not change much overall, and it belonged to the medium pore reservoir. The permeability distribution range of grape flower oil layer samples is generally about 0.1~300mD, which is a low permeability reservoir, so the grape flower oil layer is a mesoporous and low permeability reservoir.

1.3 Trap conditions

There are various types of traps in this area, including tectonic traps, lithological traps and composite traps, and the formation time, type of trap, and distance from the oil source all determine the effectiveness of the trap. In terms of formation time, the end of Nenjiang-Mingshui is the formation and shaping period of various types of traps in this block, and it is also the maturity and expulsion period of source rocks of the Qingshankou Formation, and the two match in time and are effective traps.

1.4 Migration Conditions

In terms of oil-generating environment, the Qingyi Formation is the source rock layer in this area, and according to the previous research, it is known that the block is accompanied by three fold movements at the end of the Nenjiang Formation, the end of the Mingshui Formation and the end of the Old Tertiary period, and at the same time, the arrival of the peak period of the Qingyi Oil-producing Rock, the peak period of production and oil discharge, under the action of the regional stress field, a large number of early and middle faults are revived and opened during the migration period, which has become the most favorable channel for migration. Through the plane and sectional comparison, a total of 83 oil source faults were developed in this block, and the whole area was developed, which could effectively communicate the lower source rocks and sufficient oil sources.

The grape oil layer needs a medium-term fault to play a communication role, supplemented by the disintegration surface between the Qingshankou Formation and the bottom layer of the Yaojia Formation, so that the oil generated by the Qingshan Pass Formation crosses the mudstone sedimentary strata of the Qingyi Section II and III, and under the guidance of local structure, a longitudinal and horizontal net blanket-like channeling mode is formed to reach the Grape Flower Oil Layer, and the relatively good physical conditions in the oil layer also create conditions for lateral transport. The lateral conduction channel is a fault-sand composite channel. In the eastern tectonic belt, north-north-west migration is the mainstay, and the north-north-west fault and sand body form a reverse trending fault-sand body transmission system. In the western part of the block, the lateral migration direction is mainly in the near east-west direction and the north-north-west direction, and the forward fault-sand body form is transmitted.

2. Oil-water distribution characteristics

On the plane, according to the analysis of the existing data, there are 24 exploration and evaluation wells in the blank area, a total of 12 oil tests, 5 industrial oil flow wells, 4 low-yield oil wells, 3 water wells, and the highest daily oil production well, with a daily oil output of 11.664t. Five of the obtained industrial flow wells are located in the northeast slope of the blank depression area, which is inside the tectonic trap. The oil test wells located in the structural depression are mainly based on water layers.

In the longitudinal direction, from the analysis of the drilling situation of 22 wells that have been drilled, each single well is drilled with sandstone, and the interpretation results are oil layer, same layer, water layer and a small amount of dry layer. The average sandstone thickness is 16.2m, but only 9 wells have been drilled effectively, averaging 1.58m. The relationship between oil and water is complex, and the oil-water layer is interlayered, and the inverted relationship between the oil and the water appears under the oil.

Tab 1 Statistics table of the number of oil test layers

Layer number	Total number of layers (pcs)	Number of oil outlet layers (pcs)	Proportion (%)
PI1.1	5	1	20.0
PI1	3	0	0.0
PI2	1	0	0.0
PI3	5	1	20.0
PI4	2	1	50.0
PI5	2	0	0.0
PI6	4	2	50.0
PI7	3	2	66.7
PI82	3	1	33.3
PI9	2	1	50.0
PI10	1	0	0.0
PI11	1	1	100.0
total	27	9	33.3

Tab 2 Test oil thickness statistics table

Layer number	Total thickness (m	Oil thickness (m)	Proportion (%)
PI1.1	12.2	3	24.6
PI1	4.4	0	0.0
PI2	1	0	0.0
PI3	6.3	1.5	23.8
PI4	4	3	75.0
PI5	2.6	0	0.0
PI6	12.4	4.2	33.9
PI7	10.6	4	37.7
PI82	9.2	3.2	34.8
PI9	5.2	1.6	30.8
PI10	2.4	0	0.0
PI11	1.2	1.2	100.0
total	71.5	21.7	30.3

3. Conclusion

Through the analysis of the main controlling factors of reservoir formation in this block, the oil and gas plane migration direction is moved from the high potential energy area to the low potential energy area, that is, the middle low-lying area is transported to the surrounding high parts, and the reservoir is formed by effective trapping and occlusion in the migration direction. The faults on the west and southeast sides of the lower part are consistent with the oil and gas migration direction, the fault occlusion effect is not good, and the angle between the northeast fault trend and the oil and gas migration direction is large, which is easy to form lithology-fault reservoirs.

(1) Oil and gas are transported from low-lying areas to high parts, and after encountering structural traps on the migration path, structural reservoirs are easy to form;

(2) There are faults blocking the oil and gas migration channel, and the sand body and fault are well matched, which is easy to form fault-lithological reservoirs;

(3) Oil and gas are transported from low-lying areas to high parts without obstruction, and oil and gas are transported until they are blocked.

Therefore, accurate understanding of the structure and sand body and their matching relationship is the key to drilling high-yield evaluation wells.

References

1. Cui Baowen, et al. Progress and recommendations of oil exploration in the northern part of the Songliao Basin [J]. Daqing petroleum geology and development, 2018.
2. Xu Jianjun, Huang Lida, Yan Limei, Yi Na. Insulator Self-Explosion Defect Detection Based on Hierarchical Multi-Task Deep Learning[J]. Transactions of China Electrotechnical Society, 2021, 36(07): 1407-1415.
3. Limei, LIU Yongqiang, XU Jianjun, et al.Broken string diagnosis of composite insulator based on Grabcut segmentation and filler area discrimination[J].Power System Protection and Control, 2021, 49(22):114-119
4. Yi, Q. Wang, L. Yan, et al., A multi-stage game model for the false data injection attack from attacker's perspective. Sustainable Energy Grids & Networks 28 (2021).
5. Na Yi, Jianjun Xu, Limei Yan, Lin Huang. Task Optimization and Scheduling of Distributed Cyber-physical System Based on Improved Ant Colony Algorithm. Future Generation Computer Systems, 109(Aug. 2020), 134-148.
6. Yang Zhao, Jianjun Xu, Jingchun Wu. A New Method for Bad Data Identification of Oilfield Power System Based on Enhanced Gravitational Search-Fuzzy C-Means Algorithm. IEEE Transactions on Industrial Informatics. VOL. 15, NO. 11, NOVEMBER 2019 5963-5970

7. Jing Han, Xi Wang, LiMei Yan, Aida Dahlak, et al. Modelling the performance of an SOEC by optimization of neural network with MPSO algorithm. International Journal of Hydrogen Energy, Volume 44, Issue 51, 22 October 2019, Pages 27947-27957. https://doi.org/10.1016/j.ijhydene.2019.09.055
8. HU Jianyi. Frontiers of petroleum geology. Beijing: Petroleum Industry Press, 2002.
9. FU Xiaofei et al. Tectonic characteristics of slope in western Songliao Basin and control of hydrocarbon accumulation[J].Chinese Journal of Geology, 2007, 42(2): 209-222.

 https://doi.org/10.1051/e3sconf/202337501040

Discussion on Reasonable Flow Pressure in Class I Block of A Oilfield

Mingzheng Lu

No. 10 Oil Production Plant of Daqing Oilfield Co., Ltd, Daqing, Heilongjiang, China

Abstract. Reasonable flow pressure is the premise to ensure the maximum production capacity of oil wells. After years of development and adjustment, the adaptability of pump depth and flow pressure needs to be further studied, so as to formulate a reasonable flow pressure under the current development situation. This paper studies the relationship between flow pressure and production in a class of blocks by using theoretical research and data probability distribution method, determines the current reasonable flow pressure system in line with the actual situation, and proposes the test potential wells for adjustment

Key words: Reasonable flow pressure; Probability distribution; Reasonable pump depth.

1. Questions raised

A oilfield belongs to low-ultra-low permeability oilfield. After years of development, the productivity level of the old wells in the oilfield has gradually declined. At present, the reasonable flow pressure at the development stage needs further study. It is necessary to further carry out the research on the reasonable flow pressure at the medium-high water cut stage, gradually carry out the matching and adjustment of the reasonable flow pressure of the oil well, and give full play to the maximum production capacity of the oil well

2. Theoretical method for determining reasonable flow pressure

Well flow pressure is the back pressure of the wellbore fluid column, and its value directly affects the percolation capacity and productivity of the reservoir, pump efficiency and system efficiency. In the early stage of oilfield development, the determination of reasonable flow pressure was deeply discussed, and the determination formula of reasonable flow pressure was given[1].

2.1 Determine the minimum allowable flow pressure of the well

2.1.1 Calculation of relative flow capacity of liquid and oil phases

Due to low reservoir permeability, poor water injection efficiency, low productivity, and well flow pressure lower than saturation pressure, if crude oil is degassed, the flow capacity of oil phase will change. The bottomhole fluid flows as oil, gas and water three-phase, and the relative flow capacity of the oil phase can be expressed as:

$$K_o = \frac{V_o}{V_m} \tag{1}$$

Where $V_m=V_o+V_g+V_w$, When one ton of surface oil is produced, the volume flow of oil, gas and water under bottom hole conditions can be calculated by the following formula:

$$V_0 = B(P_b) - \beta(P_b - P_{wf}) \tag{2}$$

$$V_g = 0.1033\ \frac{2T}{293}\frac{a}{D_o p_{wf}}(p_b - p_{wf}) \tag{3}$$

$$V_w = \frac{f_w}{1-f_w} \tag{4}$$

Where: $B(p_b)$—Volume conversion coefficient of crude oil under saturation pressure, m^3;

β--Change rate of crude oil volume conversion coefficient, m^3/Mpa;

P_b—Saturation pressure, MPa;

P_{wf}—Flow pressure, MPa;

a—Natural gas dissolution coefficient, $m^3/(m^3.Mpa)$;

T—Absolute temperature of oil reservoir, K;

Z—Natural gas compression coefficient;

D_o—Surface crude oil density, t/m^3;

f_w—Water cut of oil well, decimal.

Substitute (2), (3) and (4) into formula (1), Then the relative flow capacity of oil phase in the reservoir near the bottom of the well can be obtained:

* Corresponding author: lumingzheng@petrochina.com.cn

$$K_o = \frac{1 - f_w}{1 + (1 - f_w)R} \quad (5)$$

Where: R - gas-oil ratio at the outlet of oil layer near the bottom of the well $R = \frac{V_g}{V_o} = 0.1033 \frac{2T}{293} \frac{a}{D_o p_{wf}} (p_b - p_{wf}) / [B(pb) - 3(p_b - p_{wf})]$ m^3/t.

If the bottom hole pressure is greater than the saturation pressure, then R=0. At the same time, the relative flow capacity of the liquid phase can be obtained:

$$K_L = \frac{1}{1 + (1 - f_w)R} \quad (6)$$

It can be seen from the analysis of (5) and (6):

① When the crude oil near the bottom of the well is degassed, the relative flow ability of the oil phase will decrease, that is, $K_o<1$. And with the increase of gas-oil ratio, the decline rate of relative flow ability of oil phase increases. At the same time, the relative flow capacity of the liquid phase also decreased;

② With the increase of water cut in oil wells, the relative flow ability of oil phase decreases and the relative flow ability of liquid phase increases.

2.1.2 Establishment of inflow performance equation of oil well

According to the above analysis, when the bottom hole flow pressure is lower than the saturation pressure, the inflow performance equation of the oil well can be obtained by modifying the initial flow capacity of the oil well with (5) and (6):

For oil phase:

$$q_0 = \frac{J_0(1 - f_w)}{1 + (1 - f_w)R}(P_R - P_{wf}) \quad (7)$$

For liquid phase:

$$q_L = \frac{J_0}{1 + (1 + f_w)R}(P_R - P_{wf}) \quad (8)$$

Where: J_0—Production index, t/d·MPa;

R—Gas-oil ratio at the outlet of oil layer near the bottom of the well, m^3/t;

P_R—Formation pressure, MPa.

It can be seen from the two formulas (7) (8) that when the flow pressure is lower than the saturation pressure, when it is greater than a certain value, the liquid production will increase with the decrease of the flow pressure; When it is less than this value, the liquid production will decrease with the decrease of flow pressure, which is the minimum allowable flow pressure of the oil well.

2.1.3 Calculation of minimum allowable flow pressure of oil well

Figure 1 is the dimensionless inflow performance curve of oil wells with different water cut based on the calculation results of (7) (8).

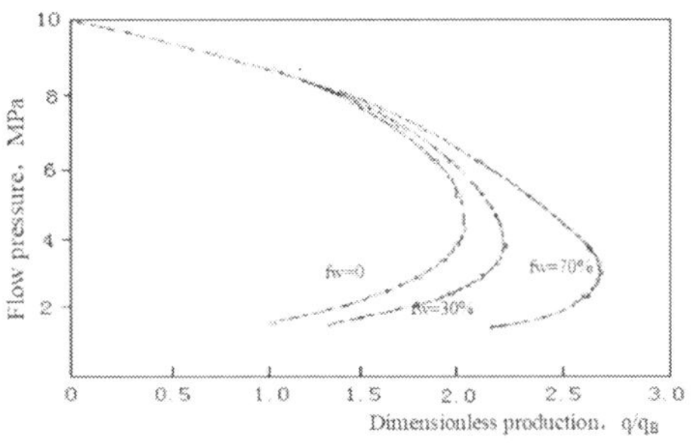

Fig. 1 Inflow performance curve of oil wells with different water cut

The curve can be divided into two parts: straight section and curve section. Within the range of straight section, the relative oil production index is stable and the flow conforms to Darcy's law.

There are two characteristic points in the curve part: the first is the starting point of the straight line starting to bend. The flow pressure at this point is equal to the saturation pressure. After the flow pressure is lower than this point, the oil production index decreases and the production growth rate slows down; The second is the maximum production point. The pressure corresponding to this point can be called the minimum allowable flow pressure of the oil well. After the flow pressure is lower than this point, the output begins to decrease.

The main reason for this situation is that after the flow pressure drop reaches a certain level, there is a three-phase flow of oil, gas and water near the bottom of the well, resulting in a sharp decline in the flow capacity of the oil phase. The contribution of the production pressure difference to the production has been less than the impact of the decline of the production index on the production.

Calculate the first derivative of equation (8) and make it zero to obtain the lowest allowable flow pressure equation:

$$Lp^4_{wfmin} + Mp^3_{wfmin} + Np^2_{wfmin} + Qp_{wfmin} + W = 0 \quad (9)$$

Where: $L=\beta^2 . j_L$

$M=2(a-cf_o)\beta . j_L$

$N=(3bf_o+ap_R)\beta . j_L + (a-cf_o)(a-\beta p_R) \ j_L$

$Q=2bf_o j_L(a-\beta p_R)$

$W=-bf_o j_L a p_R$

Among the above $c = \frac{a}{D_o}\bar{t}$ $b = \frac{1}{D_o} a p_b \bar{t}$; a= B(pb)-βpb

because $\beta(p_b-p_{wf}) << B(p_b)$, so the second item at the right end of formula (2) can be omitted, so that formula (9) can be simplified into the following form:

$$(1-n)p^2_{wfmin} + 2np_b p_{wfmin} - np_b . p_R = 0 \quad (10)$$

The formula for calculating the minimum allowable flow pressure of oil well is obtained by solving the above formula: $\bar{t} = 0.1033 \frac{ZT}{293}$

https://doi.org/10.1051/e3sconf/202337501040

$$P_{wf\ \min} = \frac{1}{1-n}\left[\sqrt{n^2 P_b^2 + (1-n).nP_b.P_R} - nP_b\right] \quad (11)$$

among: $n = \frac{\alpha \bar{t}}{B_0}(1-f_w)$

2.2 Determine the suction inlet pressure of oil extraction pump

Based on the analysis of the working conditions of the pumping unit itself, the minimum allowable flow pressure limit of the oil well is usually described by the bottomhole degassing amount. With the rise of water cut in oil wells, water cut has also become an important factor that cannot be ignored in calculating the minimum allowable flow pressure limit. It can be seen from this that the pressure formula at the suction inlet of the oil well pump is:

$$P_\lambda = \frac{M(1-f_w)}{f_w + B_o(1-f_w)} \quad (12)$$

$$M = \left(\frac{1}{G}-1\right) \times 3.53 \times 10^4 TZ\left(S_{gi} - S_g\right)$$

Where: P_λ—Pump suction pressure, MPa;
G—Volume percentage of gas and liquid, decimal;
Sgi—Original dissolved gas-oil ratio, m3/t;
Sg—Dissolved gas-oil ratio of suction pressure, m3/t;

2.3 Determine reasonable flow pressure

For the reasonable flow pressure of the pumping well, it is necessary to ensure the condition of the minimum allowable flow pressure of the reservoir, and to ensure that the pump can work under the optimal suction pressure. Since the pump suction pressure converted to the flow pressure in the middle of the reservoir is less than the minimum allowable flow pressure, the pump suction pressure can be used as the reasonable flow pressure. The calculation formula shall be expressed as follows:

$$P_R = \frac{M(1-f_w)}{f_w + B_o(1-f_w)}$$

The calculated average reasonable flow pressure of oilfield A is 2.51MPa (see Table 1 for details).

Table 1-A Calculation table of reasonable flow pressure and reasonable submergence of oil field

classify	Middle depth (m)	Average pump depth (m)	Average pump efficiency (%)	Dynamic fluid level (m)	Submergence degree (m)	Actual flow pressure (MPa)	Reasonable flow pressure (MPa)	Reasonable submergence (m)	Difference between theory and practice	
									Flow pressure (MPa)	Submergence degree (m)
Strata 1	494.7	491.1	19.98	458.0	33.1	1.00	1.52	119.1	0.50	86.0
main	1045.6	1039.7	21.3	966.1	73.6	1.31	2.34	205.1	1.00	131.5
S 1-2	10055 4.9	984.6	17.34	912.2	72.3	1.20	2.72	221.6	1.50	149.3
Wing	1082.2	1032.2	16.25	970.0	62.2	1.32	2.63	190.8	1.30	128.6
Strata 3	1248.9	1156.5	14.19	1068.8	87.6	1.03	2.55	152.9	1.50	65.3
average	1051.5	1003.1	18.31	942.2	60.9	1.30	2.51	169.4	1.20	108.5

3. Study on the relationship between oil production level and flow pressure

Due to the different geological characteristics and development characteristics of various blocks in Oilfield A, whether the actual production level of the oil well is compatible with the reasonable flow pressure system established at the initial stage needs to be further discussed. The following work will be carried out to determine the reasonable flow pressure system of the block by means of data statistics and probability distribution, and adjust the flow pressure by adjusting the pump depth and pumping parameters.

3.1 A-block

Considering the current poor oil production capacity of oil wells, the vast majority of oil wells are in the state of insufficient liquid supply, the distance between the dynamic liquid level and the pump suction is relatively close, or has reached the pump suction, and the criterion of the dynamic liquid level data, the distance between the pump depth and the middle depth of the oil layer is used to calculate the bottom hole flow pressure, and the scatter point of the relationship between the initial production and flow pressure, as well as the scatter point of the relationship between the current production and flow pressure, is established.

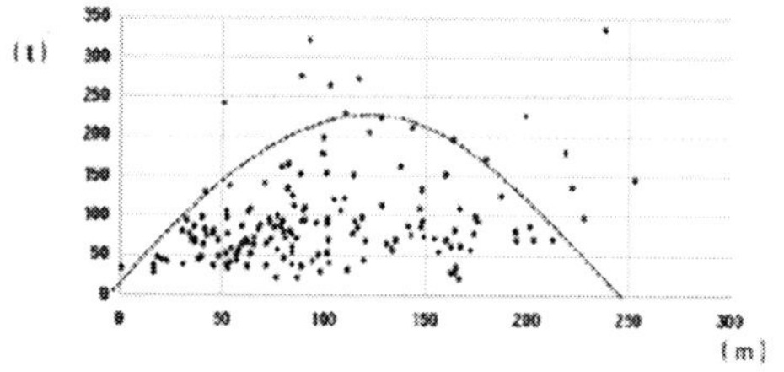

Figure 3-Scatter diagram of relationship between initial liquid production and pump depth - oil layer distance in Block A

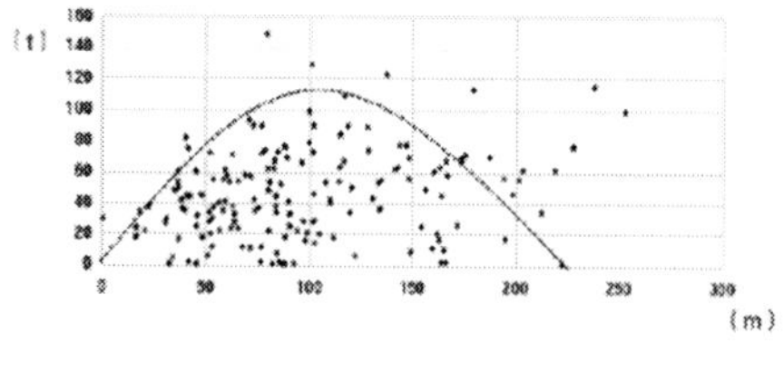

Figure 4-Scatter diagram of relationship between initial oil production and pump depth - oil layer distance in Block A

From the relationship between the initial liquid production of Block A and the distance between the pump depth and the oil layer, when the distance is less than 100m, the liquid production is significantly reduced, which means that too small flow pressure will reduce the liquid production capacity of the oil well, and when the distance is more than 150m, the liquid production has an obvious downward trend, which means that too large flow pressure will cause the oil production capacity of the oil layer to be suppressed. From the perspective of scatter probability, when the distance is 100-150m, the liquid production capacity of the oil well is the largest.

From the perspective of oil production capacity, the oil production capacity is the largest when the distance is 100m-120m. From a comprehensive point of view, when the distance is 100m-120m, the oil well is considered to have the largest production capacity.

At this time, the flow pressure is the pressure generated by the distance between the length of the liquid column and the submergence degree of the liquid column in the middle depth of the oil layer. Generally, the submergence degree is about 40m, the depth of the liquid column is 140m-160m, and the initial water content is calculated as 20%, and the flow pressure is about 1.4Mpa.

Table 2 - Relationship between liquid column depth and flow pressure

containing water(%)	20	20	20	20	20	20	20	20
Liquid column height(m)	100	110	120	130	140	150	160	170
pressure(Mpa)	0.88	0.96	1.05	1.14	1.23	1.31	1.40	1.49

At the current development stage, the comprehensive water content of Class I block is about 65%. Taking Block A as an example, by December 2022, the comprehensive water content of the block will be 67%. The reasonable pumping depth is obtained when the maximum production capacity is obtained by plotting the scattered points of liquid production, oil production and pump depth - medium depth.

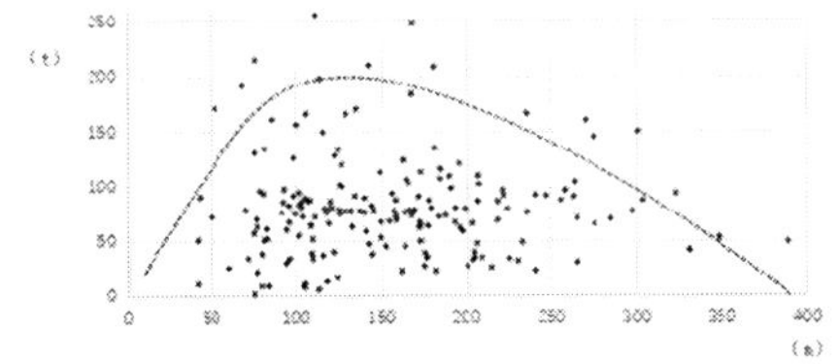

Figure 5-Scatter diagram of the relationship between current liquid production and pump depth - oil layer distance in Block A

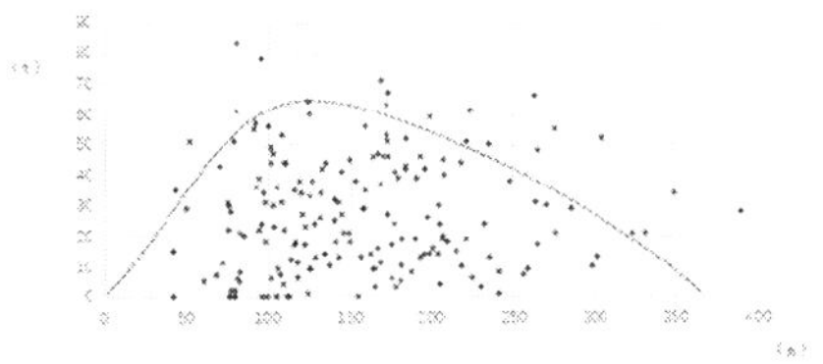

Figure 6-Scatter diagram of the relationship between current oil production and pump depth - oil layer distance in Block A

From the current production capacity and distance of Block A, the maximum production capacity is reasonable depth, which is about 10m smaller than the initial distance, and the reasonable pump depth is 90m-110m. Due to years of development, the formation pressure has decreased, the production pressure differential has decreased, and the oil production capacity of the reservoir has decreased. To maintain a high production capacity, the flow pressure also needs to be adjusted downward to maintain a high production pressure differential.

Table 3 - Relationship between liquid column depth and flow pressure

containing water(%)	67	67	67	67	67	67	67	67
Liquid column height(m)	100	110	120	130	140	150	160	170
pressure(Mpa)	0.94	1.03	1.12	1.22	1.31	1.41	1.50	1.59

According to the submergence of 40m, the liquid column height is 130m-150m, and the reasonable flow pressure is 1.22Mpa-1.41Mpa.

3.2 B-block

According to the current production level of Block B and the scattered points of the relationship between the pump depth and the middle depth of the reservoir, the current reasonable pump depth parameters are obtained.

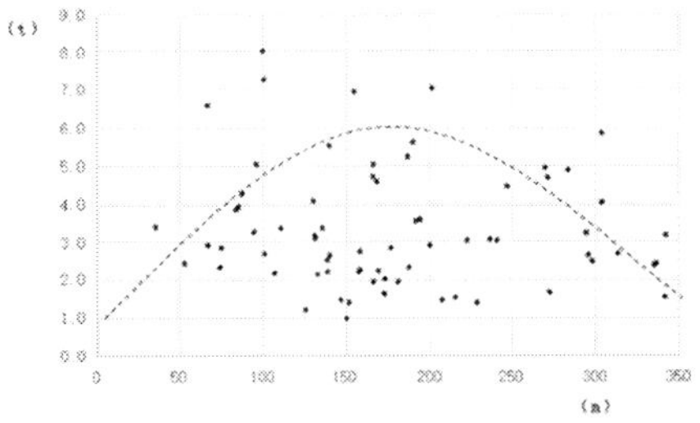

Figure 7-Scatter diagram of the relationship between current liquid production and pump depth - oil layer distance in Block B

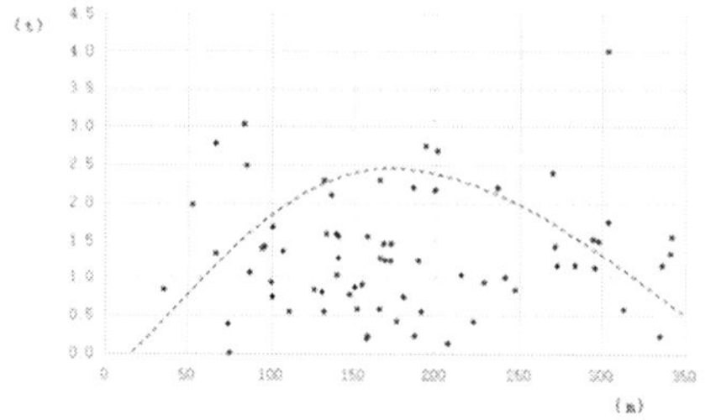

Figure 8-Scatter diagram of the relationship between current oil production and pump depth - oil layer distance in Block B

According to the scattered point relationship, when the distance between the middle depth of the reservoir and the pump depth is 150-200m, the production level of the oil well is the highest.

In Block B, the current comprehensive water content is 61%, and the reasonable flow pressure is estimated according to the submergence degree of 40m.

Table 4 - Relationship between liquid column depth and flow pressure

containing water(%)	61	61	61	61	61	61	61	61
Liquid column height(m)	170	180	190	200	210	220	230	240
pressure(Mpa)	1.58	1.67	1.77	1.86	1.95	2.04	2.14	2.23

The reasonable flow pressure level of Block B is 1.77Mpa - 2.23Mpa, with an average of 2.0Mpa.

3.3 C-block

According to the current production level of Block C and the scattered points of the relationship between the pump depth and the middle depth of the reservoir, the current reasonable pump depth parameters are obtained.

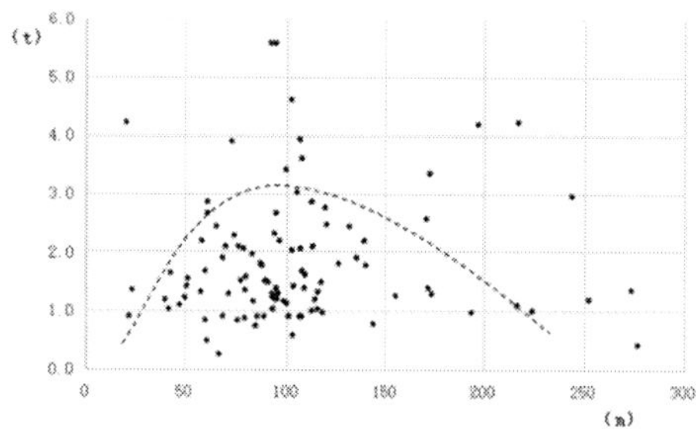

Figure 9-Scatter diagram of the relationship between current liquid production and pump depth - oil layer distance in Block C

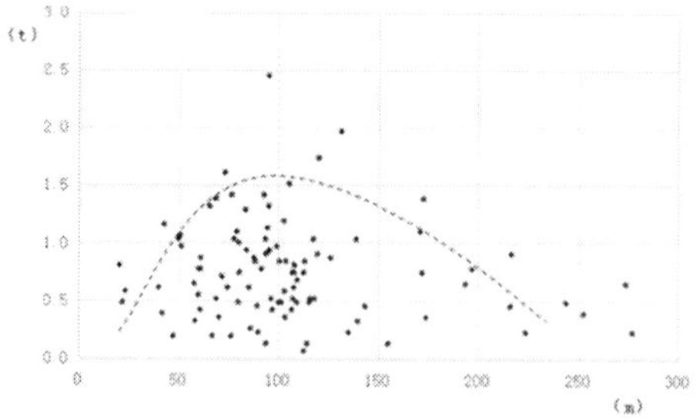

Figure 10-Scatter diagram of the relationship between current oil production and pump depth - oil layer distance in Block C

According to the scattered point relationship, when the distance between the middle depth of the reservoir and the pump depth is around 100m, the production level of the oil well is the highest.

In Block C, the current comprehensive water content is 59%, and the reasonable flow pressure is estimated according to the submergence degree of 40m.

Table 5- Relationship between liquid column depth and flow pressure

containing water(%)	59	59	59	59	59	59	59	59
Liquid column height(m)	100	110	120	130	140	150	160	170
pressure(Mpa)	0.93	1.02	1.11	1.20	1.30	1.39	1.48	1.57

The reasonable flow pressure level of Block C is 1.3Mpa.

3.4 D-block

According to the current production level of Block D and the scattered points of the relationship between the pump depth and the middle depth of the reservoir, the current reasonable pump depth parameters are obtained.

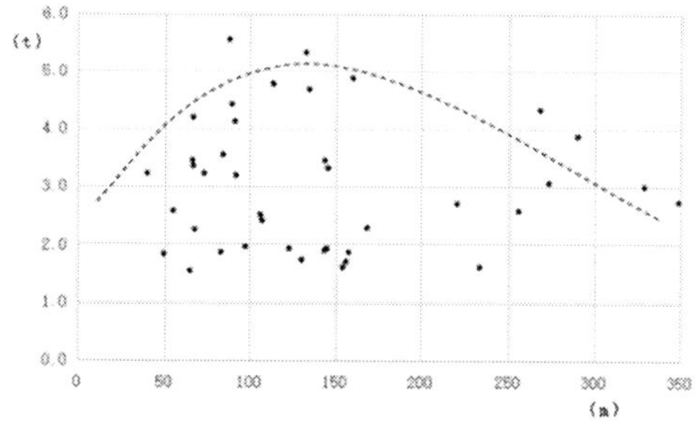

Figure 9-Scatter diagram of the relationship between current fluid production and pump depth - oil layer distance in Block D

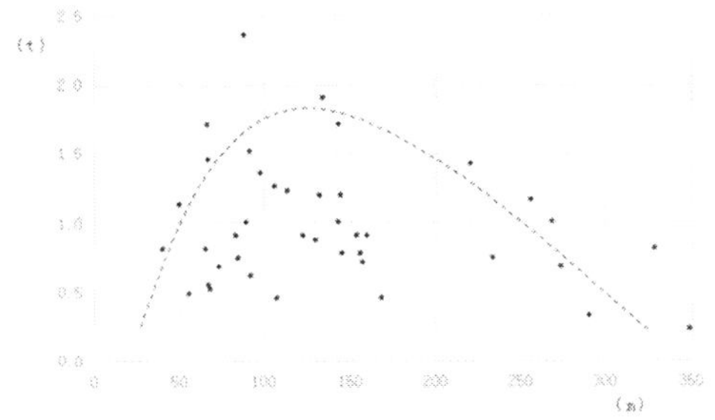

Fig. 10-Scatter diagram of the relationship between current oil production and pump depth - oil layer distance in Block D

According to the scattered point relationship, when the distance between the middle depth of the reservoir and the pump depth is around 120m, the production level of the oil well is the highest.

In Block D, the current comprehensive water content is 67%, and the reasonable flow pressure is estimated according to the submergence degree of 40m.

Table 6- Relationship between liquid column depth and flow pressure

containing water(%)	67	67	67	67	67	67	67	67
Liquid column height(m)	110	120	130	140	150	160	170	180
pressure(Mpa)	1.03	1.12	1.22	1.31	1.41	1.50	1.59	1.69

The reasonable flow pressure level of Block D is 1.5Mpa.

3.5 Total of Class I blocks

According to the current production level of Class I block and the scattered points of the relationship between the pump depth and the middle depth of the reservoir, the current reasonable pump depth parameters are obtained.

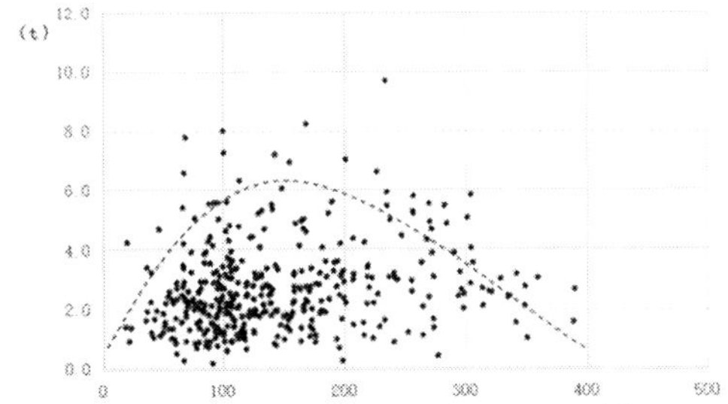

Figure 11 - Scatter diagram of the relationship between current liquid production and pump depth - oil layer distance in Class I block

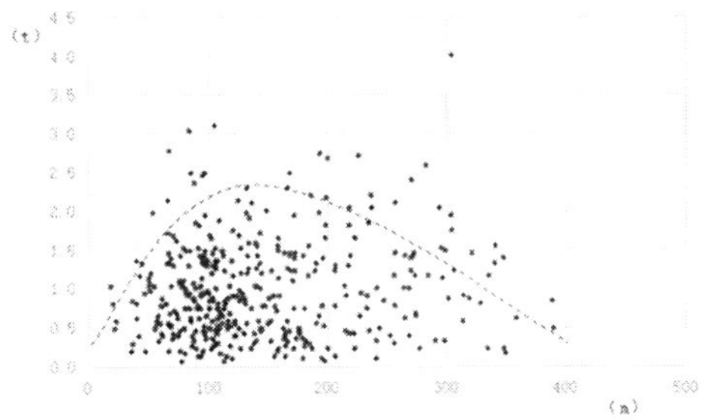

Figure 12 - Scatter diagram of the relationship between current oil production and pump depth - reservoir distance in Class I block

According to the scattered point relationship, when the distance between the middle depth of the reservoir and the pump depth is around 130m, the production level of the oil well is the highest.

In Class I block, the current comprehensive water content is 63%, and the reasonable flow pressure is estimated according to the submergence degree of 40m.

Table 7- Relationship between liquid column depth and flow pressure

containing water(%)	63	63	63	63	63	63	63	63	63
Liquid column height(m)	120	130	140	150	160	170	180	190	200
pressure(Mpa)	1.12	1.21	1.30	1.40	1.49	1.58	1.68	1.77	1.86

The reasonable flow pressure level of Class I block is 1.58Mpa.

4. Current situation of flow pressure in Class I block of oilfield

According to the above research results, there are 395 oil wells in Class I block, with an average flow pressure of 1.81Mpa, which is slightly higher than the reasonable range of flow pressure. From the statistical results of flow pressure classification, the overall adjustment potential is large.

Table 8-A Classification Statistics of Flow Pressure Status in Class I Block of Oilfield

Flow pressure classification	Number of wells	Flow pressure (Mpa)	Daily liquid (t)	Daily oil production (t)	liquid production (%)	oil production (%)	Proportion of wells (%)
(0.0,1.0)	35	0.82	58	27	5.5	7.0	8.9
[1.0, 1.4)	105	1.21	224	107	21.2	27.7	26.6
[1.4, 1.8)	93	1.57	269	85	25.5	22.1	23.5
[1.8, 2.5)	91	2.10	264	90	25.1	23.4	23.0
[2.5, 3.0)	37	2.75	134	41	12.7	10.6	9.4
[3.0,10.0)	34	3.60	104	35	9.9	9.2	8.6
	395	1.81	1053	386	-	-	-

There are 93 wells in the reasonable range, accounting for 23.5% of the wells and 25.5% of the liquid production.

5. Optimization and effect prediction of potential wells for next adjustment test

5.1 Pump hanger deepening potential

Taking Block A as the test area, through screening the medium and low water cut oil wells with too shallow pumping depth, 17 wells with deepening potential are initially selected, and the deepening pumping test will be carried out next, and the flow pressure will be adjusted to 1.3Mpa - 1.4Mpa.

Table 9-Potential wells for flow pressure adjustment in Block A

Potential well	liquid (t)	Oil (t)	fw (%)	Pump depth (m)	Medium depth of oil layer (m)	Dynamic fluid level (m)	Current flow pressure (Mpa)	Adjust pump depth (m)	Deepen the distance(m)	Adjusted flow pressure (Mpa)
A01	1.1	0.6	44.1	802.96	1067.8	747.7	2.90	960	157.04	1.34
A02	1.1	0.9	20.6	720.56	925	679.3	2.15	810	89.44	1.36
A03	1.1	0.4	62.9	801.64	1011.3	755.0	2.39	900	98.36	1.41
A04	1.5	0.7	54.3	601.19	1116.5	565.4	5.07	1000	398.81	1.44
A05	1.5	0.8	45.8	562.75	951.7	517.0	3.95	840	277.25	1.38
A06	1.6	0.6	60.8	833.16	1040	784.0	2.38	933	99.84	1.37
A07	1.7	0.8	52.8	740.22	972.8	700.7	2.50	870	129.78	1.31
A08	1.7	1.1	38.9	900.65	1105.9	862.6	2.19	1000	99.35	1.31
A09	2.0	1.2	38.7	638.53	1134.2	597.1	4.83	1020	381.47	1.39
A10	2.1	1.4	31.3	752.07	1016.9	706.3	2.76	910	157.93	1.31
A11	2.1	0.9	56.9	840.63	1154.7	792.5	3.35	1050	209.37	1.34
A12	2.1	1.4	33.8	600.21	948.9	558.8	3.49	840	239.79	1.33
A13	2.2	1.3	40.3	702.47	921.1	654.0	2.41	810	107.53	1.36
A14	2.4	1.4	41.9	859.88	1129.7	813.8	2.86	1020	160.12	1.35
A15	2.5	1.5	43.0	762.05	1083.4	720.0	3.29	970	207.95	1.39
A16	2.6	0.9	66.3	797.44	1116.4	749.4	3.43	1010	212.56	1.37
A17	2.9	1.3	54.4	700.87	985.8	658.2	3.02	880	179.13	1.34
Av	1.9	1.0	46.8	742.2	1040.1	697.8	3.12	930.8	188.6	1.36

Effect prediction: in Block A, the average formation pressure is 8.1Mpa, the average flow pressure of 17 potential wells is 3.12Mpa, the adjusted average flow pressure is 1.36Mpa, the production pressure difference rises from 5.0Mpa to 6.7Mpa, and the average daily oil production is 1.0t. Under the condition of unchanged oil production index, the average daily oil production of a single well can increase by 0.36t, and the average daily oil production of 17 test wells is expected to increase by 6.1t. According to statistics, there are 129 adjustment potential wells (flow pressure greater than 2Mpa) in Class I block. The current average flow pressure is 2.73Mpa, and the adjusted flow pressure is expected to be 1.6Mpa. At present, the formation pressure is 8.0Mpa, and the production pressure difference is increased from 5.27Mpa to 6.4Mpa. At present, the average daily oil production is 1.1t. With the oil production index unchanged, the average daily oil production of a single well is expected to increase by 0.24t, and the cumulative daily oil production level is expected to increase by 30.4t.

Table 10 - Potential well for flow pressure adjustment in Class I block

Number of potential wells	Liquid (t)	Oil (t)	Fw (%)	Pump depth (m)	Medium depth of oil layer (m)	Dynamic fluid level (m)	Current flow pressure (Mpa)	Adjusted flow pressure (Mpa)	Deepen the distance (m)
129	3.2	1.1	66.8	788	1027	734	2.73	1.60	121

5.2 Pump hanging lifting potential

Take Block C as the potential area for lifting and pumping, select 12 potential wells, and adjust the flow pressure to around 1.3Mpa.

Table 11-Potential wells for flow pressure adjustment in Block C

Potential well	liquid (t)	Oil (t)	fw (%)	Pump depth (m)	Medium depth of oil layer (m)	Dynamic fluid level (m)	Current flow pressure (Mpa)	Adjust pump depth (m)	Lifting distance(m)	Adjusted flow pressure (Mpa)
C01	1.4	1.0	28	957.9	1037.8	926.4	0.99	912.9	45.0	1.46
C02	1.5	1.0	31	945.9	1023.6	907.9	1.03	900.9	45.0	1.45
C03	1.2	1.0	17	1009.5	1059.4	972.5	0.76	949.5	60.0	1.31
C04	0.9	0.5	42	936.3	1005.3	901.3	0.94	891.3	45.0	1.39
C05	1.2	0.9	20	946.2	1030.2	909.5	1.06	901.2	45.0	1.48
C06	1.5	1.0	38	999.0	1050.4	958.8	0.82	939.0	60.0	1.36
C07	0.3	0.2	28	930.4	997.3	902.2	0.84	880.4	50.0	1.39
C08	0.7	0.6	16	927.4	1012.6	893.5	1.04	897.4	30.0	1.35
C09	0.5	0.4	11	936.5	997.4	902.7	0.82	886.5	50.0	1.30
C10	1.3	1.0	20	1008.7	1102.0	970.2	1.15	978.7	30.0	1.43
C11	0.9	0.5	44	909.9	989.7	876.2	1.03	869.9	40.0	1.45
C12	0.8	0.5	35	950.0	1010.1	914.2	0.86	900.0	50.0	1.34
Av	1.0	0.7	27.5	954.8	1026.3	919.6	0.94	909.0	45.8	1.39

Effect prediction: the average daily oil production of potential wells is 0.7t, the flow pressure is increased from 0.94Mpa to 1.39Mpa, and the distance from the middle depth of the reservoir to the pump is increased from 72m to 117m. From the perspective of scatter relationship, the daily oil production is expected to increase by 12%, and the total daily oil production is increased by 1.1t.

6. Sum up

With the adjustment of oilfield development, the flow pressure system established in the early stage of oilfield development gradually cannot adapt to the current production capacity of the reservoir.

First, with the extension of the development period, the formation pressure decreases year by year, and the production pressure difference decreases gradually, and the oil production capacity of the reservoir is not fully developed.

The second is that the depth of the oil well pump is not adjusted in time, and the pump depth is too shallow, resulting in high oil level and high flow pressure, which limits the production capacity of the reservoir; When the pump depth is too deep and the liquid supply capacity of the oil layer is insufficient, the flow pressure drops rapidly, resulting in serious degassing of the near-well oil layer, and the rapid decline of the crude oil flow capacity, resulting in the reduction of the production capacity. Therefore, further reasonable pump depth field tests need to be carried out.

The third is to use the probability distribution analysis of the field data to draw a conclusion that the reasonable level of flow pressure is more suitable for the actual production. The field test can be further carried out to observe the effect of the test well.

Fractures are developed in Class I block of A oilfield, and the adaptability of theoretical flow pressure level needs further study. Based on data statistics and probability distribution, this paper determines the overall reasonable flow pressure level of each small block and a class of blocks. According to the adjustment effect of potential well test, the next promotion work will be carried out step by step. It is the most economical, direct and effective way to improve the production capacity of oil wells by properly adjusting the pump depth and production parameters and matching the reasonable working system. The research and test results can be replicated and popularized in the whole oilfield.

References

1. Wu Shang.Determination of Reasonable Flow Pressure in Chaoyanggou Oilfield
2. He Qihou. A new method for calculating flow pressure of pumping wells [J] Petroleum drilling and production technology 1985
3. Gao Xing.The importance of maintaining reasonable flow pressure of oil wells and its influencing factors [J] China Petroleum and Chemical Standards and Quality 2012 (05)
4. Li Fayin.Research on productivity prediction method of low flow pressure oil wells [J] Petroleum drilling and production technology 1991 (02)
5. Wang Li.Study on the relationship between flow pressure and production of oil wells in Sashi transitional zone [J] Inner Mongolia Petrochemical 2013 (20)
6. Ju Yanqin, Lu Baohua. Study on reasonable flow pressure of Yingdong Gaotaizi [J] China Petroleum and Chemical Standards and Quality 2013 (12)

Residual oil start-up mechanism of polymer/surface dual flooding after polymer flooding in medium-high permeability reservoir

Nan Zhang[1,2,3,*], Yuwen Zhao[1], Wei Wang[1,2,3], Xiaoyan Wang[1,2,3], Xi Yan[1,2,3], Guangyu Yuan[1,2,3], Zeyu Lin[1,2,3], Yuanxian Liu[4], Jing Zhang[4]

[1] Oil Production Technology Institute, Dagang Oilfield, CNPC, Tianjin, China
[2] Key Laboratory of Nano-chemistry, CNPC, Tianjin, China
[3] Enterprise Key Laboratory of Tertiary Oil Recovery and Oilfield Chemical, CNPC, Tianjin, China
[4] Exploration and Development Research Institute, Tuha Oilfield, CNPC, Xinjiang, China

Abstract. The remaining oil distribution and displacement mechanism of poly/surface dual flooding stage in medium and high permeability reservoir are not clear. In this paper, the reservoir conditions in Gangxi Area 3 of Dagang Oilfield were taken as the research object. Based on nuclear magnetic resonance and large model physical simulation experiment, the characterization of micro and macro residual oil start-up mechanism and enhanced oil recovery experiment were carried out respectively after polymer flooding poly/surface binary flooding. The experimental results suggest that polymer-surfactant flooding could increase the oil displacement efficiency by 12-29% after polymer flooding, and the contribution of porous to recovery degree is in the order of medium pore, large pore, and small pore. When the surfactant concentration is higher than 0.25%, the increase of oil displacement efficiency tends to be stable, and further enhancing oil recovery requires the expansion of sweep volume. Physical model experiments results show that medium permeability layer has the highest contribution to the recovery degree in the chemical flooding stage, and the displacement mechanism is mainly based on the synergistic effect of expanding sweep volume and improving displacement efficiency. The contribution of low permeability layer to recovery degree is next, and the displacement mechanism is mainly to expand sweep volume. The high permeability layer has the lowest contribution to the recovery degree, and the oil displacement mechanism is mainly to improve the oil displacement efficiency.

Keywords: Polymer-surfactant flooding, Nuclear magnetic resonance, Recovery factor, Starting mechanism.

1. Introduction

In order to solve the problem of low water flooding recovery in medium and high permeability reservoirs, polymer flooding technology has been applied to the old oil fields represented by Daqing, which has achieved a better effect of enhanced oil recovery [1-3]. As the development continues and the injection and production Wells form advantageous channels, the polymer application becomes less effective and new measures to enhance oil recovery are needed. Polysurface binary flooding is an innovative technology after polymer flooding, which can expand the sweep volume of polymer and improve the washing efficiency of surfactant [4-6]. The high permeability reservoir in Dagang oilfield is rich in reserves, but the phenomenon of inter injection and production Wells is obvious. it has reached the development stage of "double high" (water cut is higher than 60%, and the degree of recoverable reserves is higher than 60%). At present, the application of polymer flooding or polymer surface binary flooding technology has achieved obvious effect of oil increase and precipitation [7-9]. With the development of polymer flooding, it is difficult to determine whether continuous injection of chemical systems is effective or not, and the contribution degree of polymers and surfactants in various core pores. Therefore, it is necessary to carry out relevant research to support the design of oilfield post-polymer flooding development scheme.

At present, nuclear magnetic resonance technology is widely used in the industry to study the distribution of remaining oil in core pores at different displacement stages, and certain research results have been achieved. Fan Meng, aiming at the development of polymer flooding in high permeability heavy oil reservoirs, based on the nuclear magnetic resonance technology, analyzed that the oil signal of medium and large pores decreased after polymer flooding, and the polymer significantly improved the sweep degree of small pores and the basic law of displacement efficiency of medium and large pores

* Corresponding author: cy_zhangnan@petrochina.com.cn

[10]. Liu Zheyu made clear the difference between single-mode and complex mode pore structure cores through the in-situ NMR displacement experiment of natural cores, and the main reason for the utilization of residual oil in large and medium pores to enhance oil recovery in binary composite flooding [11]. Using nuclear magnetic resonance technology, Zhang Zhentao et al. compared the emulsification effect of chemical flooding on crude oil in each pore and the effect of reducing interfacial tension, and analyzed that the injection of surfactant after gas flooding has a higher degree of use on small pores [12]. Sun Chen used ultraviolet fluorescence and heavy water nuclear magnetic resonance technology to analyze all the displacement test cores for conglomerate reservoirs, to clarify the utilization of microscopic remaining oil in pore throat of different sizes, and to give the application limit template of chemical flooding in different types of conglomerate reservoirs [13]. It is found in literature research that there has been a clear explanation for the characterization of microscopic residual oil utilization in homogeneous core, but homogeneous core is difficult to represent the permeability change of the actual reservoir. It is necessary to associate the NMR micro-characterization with the macroscopic physical experiment to clarify the contribution of each system to the oil recovery in binary flooding, so as to provide reference for the actual oilfield development.

2. Conditions of experiment

2.1 Material of experiment

The polymer was partially hydrolyzed polyacrylamide BHHP-112 provided on site, with a relative molecular weight of 25 million, hydrolysis degree of about 25% and solid content of 88%. The surfactant is DWS-3 nonionic surfactant with an effective content of 40%. The experimental oil was the crude oil from the third section of Gangxi in Dagang Oilfield. The simulated oil was prepared with dehydrated crude oil and kerosene in a certain proportion. The viscosity was 17mPa·s at 53°C. In the process of experiment, the mother liquor of polymer and surfactant was prepared by using the sewage produced from the oil well and diluted to the concentration required by the experimental scheme. The salinity of the produced sewage was shown in Table 1.

Table 1. The ion composition of the sewage in the Gang Xi

Type of ions	K^{+}+Na^{+}	Mg^{2+}	Ca^{2+}	Cl^{-}	SO_4^{2-}	HCO_3^{-}	CO_3^{2-}	Total salinity
Content(mg/L)	2043	36	39	1347	0	3126	135	6750

The NMR experimental core is an artificial homogenous columnar core with quartz sand epoxy resin cement, and the relevant parameters are shown in Table 2. The large-scale physical displacement experimental core is a longitudinal three-layer heterogeneous high-pore core with a size of 80 cm×80 cm×4.5cm. The gas permeability of the high, middle and low layers is $3000\times10^{-3}\mu m^2$, $900\times10^{-3}\mu m^2$ and $300\times10^{-3}\mu m^2$, respectively, and the porosity is 32%. Electrode layout: 36 pairs are distributed in each layer, 108 pairs in total, for oil saturation monitoring, one injection and one production well pattern. The model is shown in Figure 1.

Table 2. Physical property parameters of experimental cores

Plan	Length (cm)	Diameter (cm)	Porosity (%)	Permeability ($\times10^{-3}\mu m^2$)	Oil saturation (%)
1	8.45	2.52	30.00	636	72.83
2	8.29	2.52	29.59	770	72.85
3	8.45	2.52	28.77	675	72.52
4	8.12	2.52	29.98	748	73.69
5	8.03	2.52	30.01	705	73.92

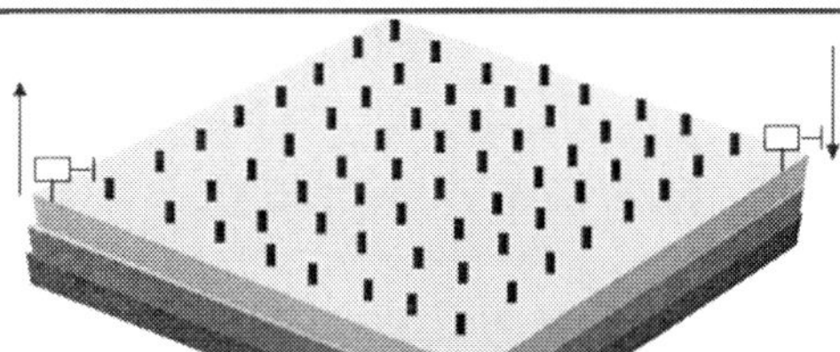

Figure 1. Physical experimental model of flat plate

2.2 Experimental instruments and procedures

2.2.1 Instruments and equipment

BROOKFIELD DV-II+Pro Brinell viscomter was used to test the viscosity of the system, and TX-500 rotary drop interfacial tensimeter was used to test the interfacial tension of the system. Related supporting instruments and equipment include HJ-6 multi-head magnetic agitator, electronic balance (JA2003A, Accuracy is 1mg), beaker, test tube and Brock nuclear magnetic resonance instrument (Fourier 80, Germany Brock Magnetic Resonance Division)

A displacement experimental device, including SG83-1 incubator, advection pump (LB-1), pressure sensor, core gripper, hand pump and intermediate vessel, was used to evaluate the oil displacement effect in core of polymer dual system.

2.2.2 Experimental procedures

(1) The permeability of the core is measured by gas;
(2) After the core is pumped out for 2 hours, the site sewage is saturated and the porosity is measured;
(3) The saturated sewage model was placed in a thermostatic box at 53°C for more than 12 hours;
(4) When the model is saturated with oil, the oil floods until there is no water at the outlet of the model, the original oil saturation is determined, and the T2 spectrum is measured by core scanning.

(5) After water flooding to 100% water content at the outlet of the model, T2 spectrum was measured by core scanning.
(6) The surfactant and polymer mother liquor were prepared according to the experimental scheme, and stood at room temperature for 12 hours. Then the required binary composite system was prepared, and the binary composite system was placed in a constant temperature box for 2 hours;
(7) Chemical flooding was carried out according to the experimental scheme;
(8) After all the chemical slug was injected, water flooding was followed, and the water cut at the outlet reached 100%. The core was scanned and the T2 spectrum was determined;
The injection rate was 0.6mL/min during the above experiment, and the changes of oil saturation were monitored after saturated oil, water flooding, poly flooding and binary composite flooding.
All the experiments were carried out at 53°C in Gangxi Reservoir.

3. Results and Discussion

3.1 Comparison of oil displacement effect

The oil displacement experiment schemes include: water flooding to 100% water cut at the outlet, polymer (1500mg/L) flooding 3PV, poly/surface binary flooding 3PV, and subsequent water flooding to 100% water cut. Among them, the binary system: the polymer concentration is 1500 mg/L, 2000mg/L, 2500mg/L, and the surfactant concentration is 0.15%, 0.25%, 0.30%. The results of oil displacement experiment are shown in Table 3.

Table 3. Experimental data of oil displacement in binary system with different concentrations

Plan	Concentration of polymer (mg/L)	Concentration of surfactant (%)	Viscosity (mPa·s)	Interfacial tension (mN/m)	Oil recovery factor (%)			
					Water	Polymer	Two element	EOR
1	1500	0.15	39.67	2.03×10^{-2}	30.05	10.30	12.06	52.41
2	2000	0.15	71.23	3.97×10^{-2}	29.98	9.90	16.95	56.83
3	2500	0.15	117.33	5.56×10^{-2}	30.05	10.13	19.35	59.53
4	2000	0.25	70.40	8.92×10^{-3}	30.17	9.72	27.14	67.03
5	2000	0.3	70.82	7.80×10^{-3}	30.20	9.89	28.07	68.16

The final oil displacement efficiency of columnar core flooding experiment is 52.41%~68.16%. After polymer flooding, the oil displacement efficiency was increased by 12.06~28.07%. It can be seen that only relying on polymer flooding to improve the oil displacement efficiency is only about 10%, and the improvement range is limited. With the implementation of binary flooding and the increase of concentration, the final oil displacement efficiency is gradually improved. When the surfactant concentration is constant, the final oil displacement efficiency increases with the increase of polymer concentration, which indicates that the high viscosity of polymer can effectively expand the sweep volume and improve the utilization degree of crude oil. When the polymer concentration is constant, the final oil displacement efficiency increases with the increase of surfactant concentration. The increase of surfactant concentration can effectively reduce the oil-water interfacial tension, increase the capillary number, reduce the adhesion of crude oil on the rock surface, and improve the oil displacement efficiency. Therefore, the improvement of oil displacement efficiency in the binary system is generally the result of the synergistic effect of the expanded sweep volume of polymer and the improvement of the washing efficiency of surfactant.
It is worth noting that when the surfactant concentration reaches a critical value, the increase of oil displacement efficiency becomes stable. For example, compared with scheme 4, the polymer concentration of scheme 2 is the same, the surfactant increases from 0.15% to 0.25%, the oil displacement efficiency increases by 10.19 percentage points, and the interfacial tension decreases from 10^{-2} to 10^{-3}, indicating that the DSS-3 surfactant can reduce the capillary force and improve the oil displacement efficiency. The comparison between scheme 4 and scheme 5 showed that although the surfactant increased by 0.05%, the oil displacement efficiency increased by less than 1 percentage point. It shows that 0.25% surfactant can basically exert the maximum effect of oil washing in the affected area.

3.2 Comparison of recovery degree

NMR instrument mainly reflects the fluid in large and small pores through the length of time component of T2 distribution spectrum [14-16]. The relaxation time of fluid in core pores mainly depends on the strength of fluid molecules under the applied force of pore solid surface [17-18]. When the force exerted by the pore solid surface is strong, the fluid is in a bound or immobile state, and the

T2 relaxation time is small on the NMR. On the contrary, the fluid in the large pore, when the force on the solid surface of the pore is weak, this part of the fluid T2 relaxation time is larger. Therefore, NMR T2 spectrum can be used to analyze the occurrence of crude oil in core pores, and give quantitative evaluation parameters such as oil saturation and utilization degree at each stage of displacement.

According to the NMR data, the oil phase recovery degree of each interval after water flooding and chemical flooding is shown in Table 4. According to the different signal amplitude of T2 spectrum NMR in different states, the size of pore throat was divided into three sections. Signal values with T2 relaxation time less than 10ms reflect small pores; signal values with T2 relaxation time between 10-50ms reflect medium pores; signal values with T2 relaxation time greater than 50ms reflect large pores.

Table 4. Oil recovery degree in different intervals of T2 spectrum after water flooding/binary flooding (%)

Plan	All intervals		<10ms			10-50ms			>50ms		
	Water	Binary	Water	Binary	Growth	Water	Binary	Growth	Water	Binary	Growth
1	30.05	52.41	1.42	10.53	9.11	15.16	22.64	7.48	13.47	19.24	5.77
2	29.98	56.83	1.40	10.12	8.72	15.13	26.12	10.99	13.45	20.58	7.13
3	30.05	59.53	1.42	9.16	7.74	15.16	29.44	14.28	13.47	20.93	7.46
4	30.17	67.03	1.46	11.23	9.77	15.22	31.34	16.12	13.49	24.45	10.96
5	30.16	68.16	1.46	11.64	10.18	15.22	31.94	16.72	13.49	24.58	11.09

Compared with schemes 1, 2 and 3, it can be seen that under the condition of constant surfactant concentration (0.15%), with the increase of polymer concentration, the recovery degree of the middle pore increases rapidly, the large pore only increases slightly, and the small pore decreases slightly. The analysis shows that in the polymer/surface binary flooding after polymer flooding, the polymer mainly uses the remaining oil in the middle pore which has not been affected. Due to the relatively high utilization degree of macropores in the early stage and insufficient development potential in the later stage, the increase of recovery degree of binary flooding is low. Due to the increase of polymer concentration, its hydrodynamic size increases, its injection in small pores decreases, and the recovery degree of binary flooding decreases slightly.

Compared with schemes 2, 4 and 5, it can be seen that under the condition of a constant polymer concentration (2000mg/L), with the increase of surfactant concentration, the recovery degree of each pore is improved, and the increase order is as follows: medium pore > large pore > small pore. The analysis shows that the surfactant can improve the oil washing efficiency in the affected area, and the middle and large pore sweep volume is relatively high, so the oil washing capacity of surfactant can be fully brought into play. Similarly, due to the relatively high utilization degree of large pores in the early stage, the increase of recovery degree of dual flooding is less than that of middle pores.

3.3 Contribution rate of large/medium/small pores to recovery degree

The contribution rates of different pore intervals to the improvement of recovery degree in the binary composite flooding stage are shown in Table 5.

Table 5. Contribution rate of binary flooding to the improvement of recovery degree in different pore intervals

Plan	Recovery degree contribution rate		
	Small pore (%)	Medium pore (%)	Large pore (%)
1	40.74	33.45	25.81
2	32.49	40.95	26.56
3	26.26	48.44	25.31
4	26.51	43.74	29.74
5	26.80	44.01	29.19

As can be seen from Table 5, except for Scheme 1, the contribution rate of middle pores is the largest in the binary compound flooding stage, which is all above 40% or even 48.44%, while the contribution rate of small pores and large pores is relatively low. It can be seen that the binary composite flooding mainly displaces the remaining oil in the middle pores, and small pores and large pores can also achieve a certain effect. Due to the high degree of recovery in the water flooding stage, the contribution rate of the binary compound flooding stage is reduced. In Scheme 1, the polymer concentration is low, the molecular dynamics radius of the system is relatively small, and it is easy to enter small pores, so the recovery degree of small pores contributes relatively high. However, due to the low concentration of poly/surface binary, the oil displacement effect on the large pores is relatively insufficient, so the final oil recovery rate is low.

3.4 Polymer/surfactant contribution to recovery

A large-scale physical model was used to conduct oil displacement simulation experiment. The steps included water flooding to 98% water cut, 1500mg/L polymer flooding to 0.25PV, binary composite flooding to 0.4PV, and subsequent water flooding to 98% water cut. The binary flooding scheme was 0.2PV (Cp=2000mg/L, Cs=0.25%) + 0.2PV (Cp=1500mg/L, Cs=0.2%)

https://doi.org/10.1051/e3sconf/202337501041

combined slug. The contribution rates of mobility control and increase of capillary number to recovery after polymer flooding were analyzed. Table 6 and Table 7 respectively compare the statistical table of stage recovery degree results and the contribution rate of each layer recovery degree in chemical flooding stage.

Table 6. Statistical table of recovery degree at each stage

Plan	Porosity (%)	Oil saturation (%)	Oil recovery factor (%)		
			Water	Chemical drive	Total
6	29.55	77.48	37.09	33.15	70.23

Table 7. Contribution rate of recovery degree of each layer in chemical flooding stage

Plan	Contribution rate of recovery degree of each layer in chemical flooding stage (%)		
	Low permeability layer	Medium permeability layer	High permeability layer
6	42.36	51.34	6.30

As can be seen from Table 6 and 7, chemical flooding can further increase the oil recovery rate by 30 percentage points on the basis of water flooding, indicating that the remaining oil after water flooding still has great development potential. During chemical flooding, the contribution rate of recovery degree of middle and low permeability layer is relatively high. The analysis shows that the high permeability layer has developed most of the crude oil in the water flooding and polymer flooding stage, which has formed the dominant channel, and the remaining oil potential is insufficient. On the contrary, the middle and low permeability layer is continuously deployed in the binary flooding stage, and the recovery degree changes greatly in this stage.

The remaining oil saturation electrode test results of each layer at the end of each displacement stage are shown in Figure 4. In the figure, ▲ is the injection well and ● is the production well.

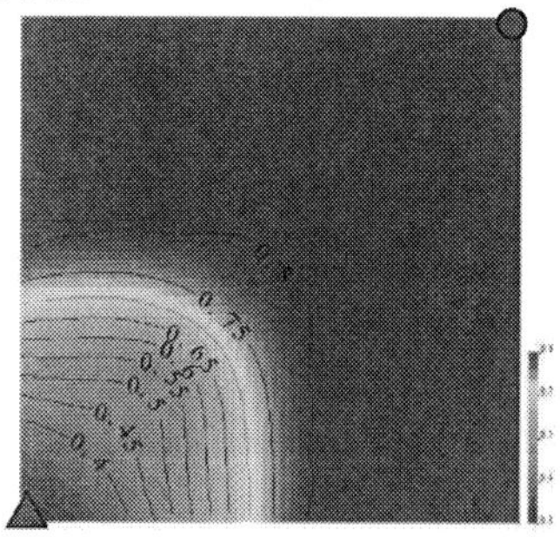

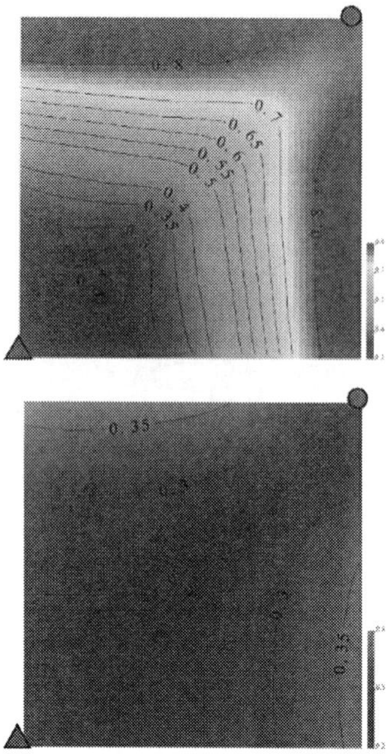

a. End of polymer flooding

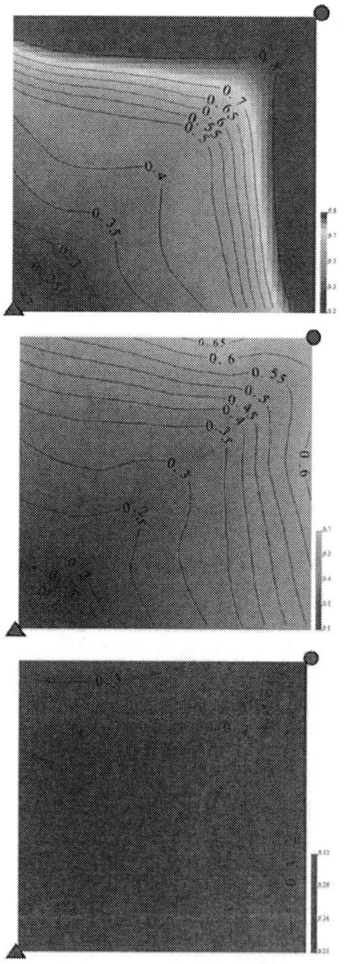

b. The end of the polytable binary drive

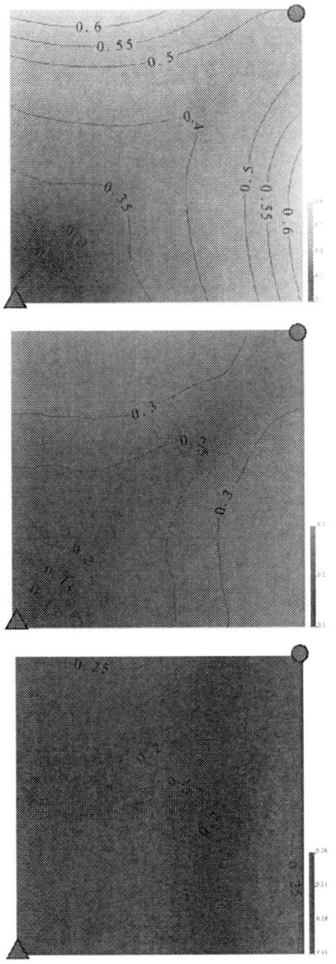

c. End of follow-up watet

Low permeability layer Medium permeability layer

High permeability layer

Figure 2. Distribution of remaining oil saturation in each layer

As can be seen from Figure 2, the remaining oil saturation of each layer gradually decreases with the progress of displacement, especially the middle permeability layer. Further analysis shows that the saturation of the high permeability layer does not change much, while the saturation of the middle and low permeability layer decreases rapidly. It shows that the high concentration polymer/surface binary can further block the hyperpermeability layer on the basis of the earlier polymer flooding. At the same time, the system enters the medium and low permeability layer to improve the emulsification and oil washing effect of the remaining oil in the affected area. In the subsequent water stage, the sealing effect of the dual system on the high permeability layer still exists, and the displacement water diffuses in the middle and low permeability layer, further reducing the remaining oil saturation in this area.

Table 8 shows the changes of oil saturation and sweep coefficient in different permeable layers.

Table 8. Variation of sweep coefficients at different horizons

Each small layer	Oil saturation, %			Coefficient of sweep, %		
	Polymer flooding	Two element drive	Drop	Polymer flooding	Two element drive	Growth
300 mD	65	36	29	21.09	90.63	69.54
900 mD	56	20	36	59.38	100	40.62
3000 mD	20	15	5	100	100	0

As can be seen from Table 8, after polymer flooding to binary flooding, the sweep coefficient of the high permeability layer has reached 100%, the oil saturation has decreased by 5 percentage points, and the oil displacement mechanism mainly focuses on improving the oil displacement efficiency. The sweep coefficient of the middle permeability layer increased by 40.62 percentage points, and the oil saturation decreased by 36 percentage points. The oil displacement mechanism was mainly based on the synergistic effect of expanding sweep volume and improving oil displacement efficiency. The sweep coefficient of low permeability layer increased by 69.5 percentage points, and the oil saturation decreased by 29 percentage points. The oil displacement mechanism mainly expanded sweep volume. From the increase of sweep volume, it can be seen that the enhancement ability of low permeability layer is the greatest, and that of medium permeability layer is the least. From the point of view of oil saturation decline, the middle permeability layer has the largest decline, the low permeability layer, the high permeability layer is the smallest, which also shows that the dual flooding mainly starts the remaining oil in the middle permeability layer.

4. Conclusion

(1) The improvement range of polymer flooding is very limited, and the improvement range is all below 10%. The oil displacement efficiency can be increased by 12%~29% by binary flooding.

(2) With the increase of surfactant concentration, the oil displacement efficiency increases, and the recovery degree of large and medium pores increases obviously. When the concentration is higher than 0.25%, the increase of oil displacement efficiency slows down, indicating that the concentration has reached the maximum oil washing efficiency, and further improving the oil displacement efficiency needs to expand the coordination of sweep volume.

(3) In the binary composite flooding stage, the contribution rate of recovery degree of middle pores is the largest, which is more than 40% or even 49%, while the contribution rate of small pores and large pores is relatively low. It can be seen that poly/surface binary composite flooding mainly acts on the remaining oil in the middle pores, and also has a certain effect on the small pores and large pores.

(4) In the large physical simulation experiment, the oil displacement mechanism of the low permeability layer is mainly based on expanding the swept volume, the middle permeability layer is mainly based on the synergistic effect of expanding the swept volume and improving the oil displacement efficiency, and the dual displacement mainly starts the remaining oil of the middle permeability layer.

Acknowledgments

This work was financially supported by Research on New High Temperature and High Salinity Chemical Flooding Technology of CNPC fund.

References

1. Zhang Zhuo, Wang Zhengxin, Xue Guoqin, et al. Heterogeneous compound flooding technology for medium-high permeability consolidated reservoir after polymer flooding[J]. Xinjiang petroleum geology,2021,42(04):475-479.
2. Yu Meng, Tei Leilei, Li Xiang, et al. Experiments on adaptability of polymer flooding for medium-high permeability reservoir in Bohai Oilfield[J].Reservoir evaluation and development,2020,10(06):40-45.
3. Zhang Xiaoqin. Experiment and application of staged chemical flooding and production of class III reservoirs[J].Petroleum geology and recovery efficiency,2022,29(03):137-145.
4. Fu Lipei, Liao Kaili, He Yanfeng, et al. Study on polymer / surfactant compound flooding technology[J]. Journal of Changzhou University (Natural science edition),2020,32(06):54-59.
5. Guo Zhiqiang. Optimization of injection parameters for polymer / surfactant binary flooding in high water cut oilfields[J].Liaoning chemical industry, 2020,49(07):814-816.
6. Liu Zheyu, Li Yiqiang, Leng Runxi, et al. Effects of pore structure on surfactant polymer flooding-based enhanced oil recovery in conglomerate reservoirs[J]. Petroleum exploration and development, 2020, 47(01):129-139.
7. Shi Bowen. Formula optimization and performance evaluation of combination flooding system for fault block Guan 109-1 in Dagang Oilfield[D]. China university of geosciences(Beijing),2021。
8. Ren Jianfang. Pilot study and application of dual flooding in Guan 109-1 fault block in Zaoyuan Oilfield[D]. Southwest Petroleum University,2018.
9. Cao Weijia. Experimental study on the optimal fluidity of chemical compound flooding in high condensed reservoir of Kongnan area[D].Northeast petroleum university,2017.
10. Fan Meng. Enhanced oil recovery mechanism of polymer flooding in high permeability ordinary heavy oil reservoir[D].Northease petroleum University,2022.
11. LIU Zheyu, WEI Qiufan, ZHANG Jing, et al.Effect of modes of pore structure on pore-scale residual oil displacement in polymer/surfactant flooding[J]. Fault-Block Oil&Gas Field, 2019, 26(3):337-341。
12. Zhang Zhentao, Jiang Hanqiao. Displacement characteristics of oil-water transition zones through nuclear magnetic resonance[J].Petroleum geology and engineering, 2019,33(04):54-57.
13. Sun Chen, Research on conglomerate reservoir remaining oil micro starting mechanism[D].China university of petroleum (Beijing).
14. Cheng Z L, Ning Z F, et al. New insights into spontaneous imbibition in tight oil sandstones with NMR[J]. Journal of petroleum science and engineering, 2019,179:455-464.
15. Zhou Desheng, Li Ming, Shi Yuhan, et al. Sensitivity analysis of imbibition stability time in tight sandstone reservoir[J]. Special oil and gas reservoirs,2018,25(2):125-129.
16. Shen Rui, Qin Jianhua, Xiong wei, et al. Study on pore structure and fluid mobility of shale oil in Jimsar Lucaogou Formation[J]. Journal of central south university (Science and Technology), 2022, 53(9):3368-3386.
17. Zhang Juan, Deng Bo, Zhang Haoyi, et al. Experimental study on static and dynamic imbibition and oil increasing mechanism of tight sand-stone[J]. Science technology and engineering, 2022, 22(24):10526-10533.
18. Jin Jun, Liu Weizhou, Wang Ziqiang, et al. Quantitative characterization of shale pore size distribution based on nuclear magnetic resonance T2 spectrum[J]. Science technology and engineering, 2022,22(16):6448-6455.

https://doi.org/10.1051/e3sconf/202337501042

Effect of perforation shear on viscosity of polymer solution

Xi Yan [1,2], Xiaoyan Wang [1,2], Wei Wang [1,2], Zeyu Lin [1,2], Guangyu Yuan [1,2], Nan Zhan [1,2], Song Han [3], Xidong Cai [4], Xing Wu [1,2], Yanbin Liang [1,2]

[1] Oil Production Technology Institute, PetroChina Dagang Oilfield, Tianjin China
[2] Key Laboratory of Nano-chemistry, CNPC, Tianjin, China
[3] The Fifth Oil Production Plant, PetroChina Dagang Oilfield, Tianjin, China
[4] Exploration and Development Research Institute, Tuha Oilfield, CNPC, Xinjiang, China

Abstract: Polymer flooding is a tertiary oil recovery technology that is very suitable for the characteristics of China's reservoirs. However, due to the fast flow rate of polymer solution in near-well zone, the shear effect of perforation blasthole and the compacted zone results in serious loss of polymer viscosity. In this paper, the polymer used in Dagang Oilfield is studied by simulation experiment through the shearing process of perforating holes, and the influence of different perforating parameters on polymer viscosity loss is analyzed, so as to provide theoretical basis for the optimization design of perforating technology in field test. The experimental results show that, the shear effect of perforation blasthole on polymer is not obvious, and the viscosity retention rate of polymer solution is greater than 96%. The size and shape of perforation blasthole have no effect on viscosity loss of polymer solution. The shear effect of compacted zone on polymer is obvious, and the viscosity retention rate of polymer solution is lower than 64% for the target block. The viscosity loss of polymer solution increases with flow rate at compacted zone, and the decrease of permeability can increase viscosity loss of polymer solution. The higher the polymer concentration is, the stronger the shear resistance is, while the higher the molecular weight is, the weaker the shear resistance is.It is suggested that perforation gun and perforation method with deep perforation depth and low compaction degree be chosen to reduce the flow rate at compacted zone and viscosity loss of polymer solution.

Key words: Perforation blasthole; compacted zone; shear effect; polymer viscosity loss; perforation parameter

1. Introduction

The industrial application of polymer flooding in China has been more than 20 years, and it has been proved to be a tertiary oil recovery technology suitable for the characteristics of oil reservoirs in China[1-4]. After polymer solution is injected into reservoir, effects relating to increasing the viscosity of water phase and polymer retention cause the decrease of permeability to improve the oil-water mobility ratio and adjust the water injection profile, which can further expand the swept volume and improve the crude oil recovery[5-6]. Therefore, after polymer solution is injected into the reservoir, the retention of solution viscosity has a great impact on polymer flooding effect[7-10].

The polymer solution is subject to continuous shearing during preparation and injection, resulting in the loss of solution viscosity. Researchers from Clausthal University of Technology in Germany pointed out that polymer for EOR will suffer mechanical shear in the whole process from ground to reservoir, such as preparation (agitator), transportation (pipeline components, pumps, valves), injection (casing holes) and flowing through porous media[11-12]. According to statistics, the total viscosity loss of polymer solution from the injection well to the detection well can reach as high as 66.7%. While the flow velocity near the well is faster than other process, and the viscosity loss caused by perforation and borehole shearing can be more than 28.8%, which reduces the ability of polymer to establish flow resistance and improve swept volume in the formation[13-14]. Therefore, this paper mainly focuses on the polymer used in the medium to high permeability and medium temperature reservoirs of Dagang Oilfield. The research about polymer shearing process through perforating blastholes is carried out to analyzes the impact of different perforating parameters on polymer viscosity, and further provides a theoretical theory for the optimization design of perforating technology in field development.

 https://doi.org/10.1051/e3sconf/202337501042

2. Physical Model Design of Perforating Hole Structure

The near well zone in reservoir is an area where the fluid regime is complex, such as linear flow, radial flow, spherical flow. Meanwhile, the distribution of flowing medium is also complicate in this area, including wire wound screen, gravel packing layer, perforating hole and compaction zone, formation pore. According to the principle of similarity, the same or similar medium can be gotten by transforming the complex spherical flow and radial flow into simple one-dimensional linear flow. The shearing degree of polymer flowing near the well depends on the flowing medium and rate. Therefore, the design principles of the physical model in this paper are: first, the flow rate of polymer in the model is the same as velocity in the actual reservoir. Secondly, the medium in the model is the same or similar to the actual medium. Through investigation, it is found that the main perforation factors affecting the polymer effect are: perforation density, aperture, hole depth, phase angle, hole shape, compaction degree of perforation compaction zone, compaction thickness of perforation compaction zone, pollution zone depth. Based on the above principles and influencing factors, this paper has established a physical model of perforating blasthole structure, as shown in Figure 1, which divides the perforation blasthole structure into blasthole part (Zone I) and compaction zone (Zone II).
For the simulation of borehole shearing, indoor simulation models of perforating blastholes with different apertures, shapes and chamfers are designed, as shown in Figure 2 which is a schematic diagram of a physical model of the blasthole. The hole shape is round with 45 ° chamfer, and the aperture is 13mm. The effects of perforation diameter, shape (shape and chamfer), and density on polymer solution shear were studied. For the simulation of shearing in compaction zone, cores with different porosity and permeability are selected, and the core length is designed to simulate the thickness of compaction zone equivalently. The core porosity and permeability are equivalent to characterize the average porosity and permeability of compaction zone to study the shear effect of compaction zone on polymer.

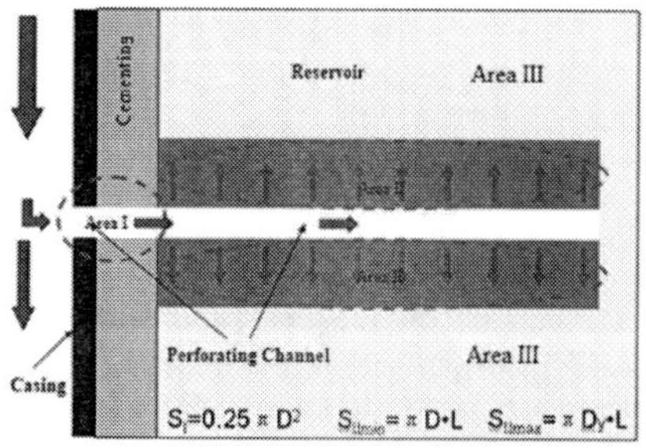

Figure 1 Schematic diagram of single hole injection fluid flow area

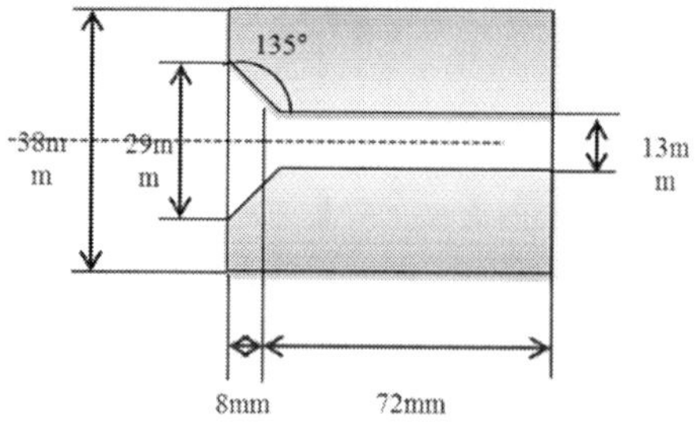

Figure 2 Simulated perforation model

(13 mm diameter, circular 45° chamfering)

2.1 Experiment Process of Shear Degradation of Polymer through Perforation

Based on the similarity criterion that the flow velocity at the perforation is consistent, the experiment flow chart of the hole shear is designed as shown in Figure 3. The specific experimental methods and steps are: 1) Preparation polymer solutions with different molecular weights and concentrations and measure the viscosity. 2) At 53 °C, the polymer solution is stably passed through the perforation-simulated device to measure the viscosity of the solution after shearing; 3) Calculate the viscosity loss of polymer solution before and after shearing.

2.2 Experimental Process of Polymer Degradation due to Shear in Compacted Zone

In this paper, the process of shear simulation experiment in perforated compaction zone is established as shown in Figure 4. Cores with different parameters are selected to simulate compaction zones with different compaction degrees, which are characterized by the permeability after compaction. Specific experimental methods and steps are as following: 1) Prepare polymer solutions with different molecular weights and concentrations, and measure the viscosity of the solution; 2) The polymer solution is stably passed through the simulated compaction belt device at 53 °C to measure the viscosity of the solution after shearing; 3) Calculate the viscosity loss of polymer solution before and after shearing.

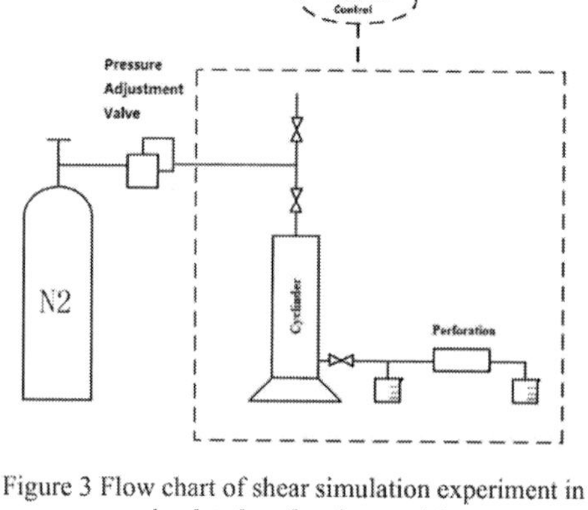

Figure 3 Flow chart of shear simulation experiment in simulated perforation model

 https://doi.org/10.1051/e3sconf/202337501042

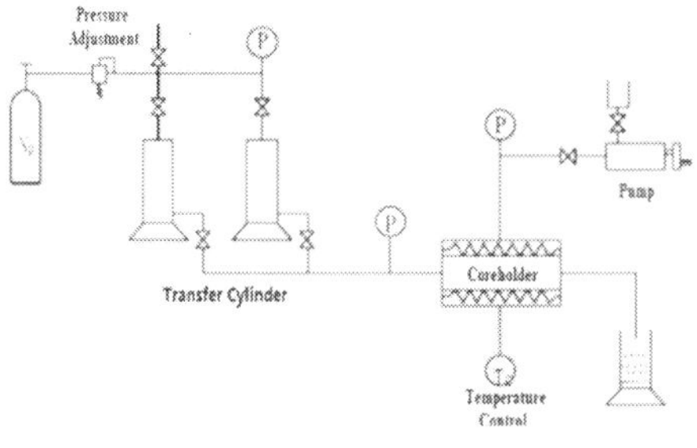

Figure 4 Flow chart of simulating shear test in compacted zone

3. Results and Discussion

3.1 Shearing Passing through Perforation

The experiments were carried out under the conditions of different pore diameters (1mm, 2mm, 6mm, 13mm), different pore shapes (round, square, round - 45 ° chamfer), different molecular weights (25 million, 30 million) and different solubility (1500mg/L, 2000mg/L) in order to analysis these factors' impact on polymer viscosity.

3.1.1 Perforation Diameter

The polymer solution with a concentration of 1500mg/L prepared by 25 million molecular weight polymer was used to conduct shear experiments under different round and square perforation shape. The experimental results are shown in Figure 5-6. Under the two perforation diameters, the relationship between viscosity loss and flow rate is basically the same. The greater the flow rate, the greater the viscosity loss caused by the cutting of perforation. Under the same flow rate, the viscosity loss caused by different perforation diameters is similar. Therefore, the perforation diameter has little effect on the shear of polymer solution. For example, in Gangxi polymer injection test area, the velocity near perforation is about 5500m/d, and the viscosity loss of corresponding polymer solution is below 4%.

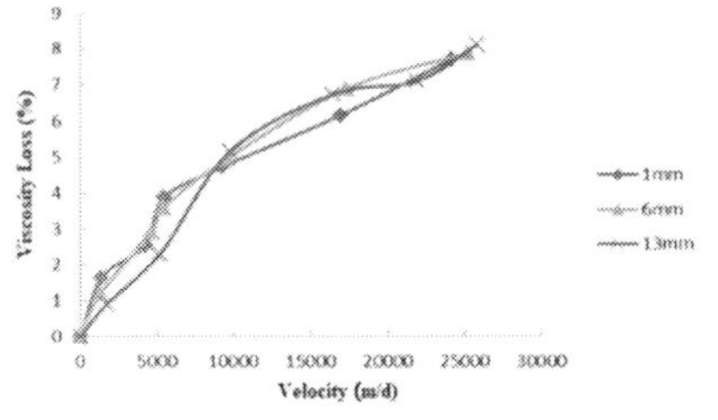

Figure 5 The relationship between velocity and viscosity loss under different diameter (circular)

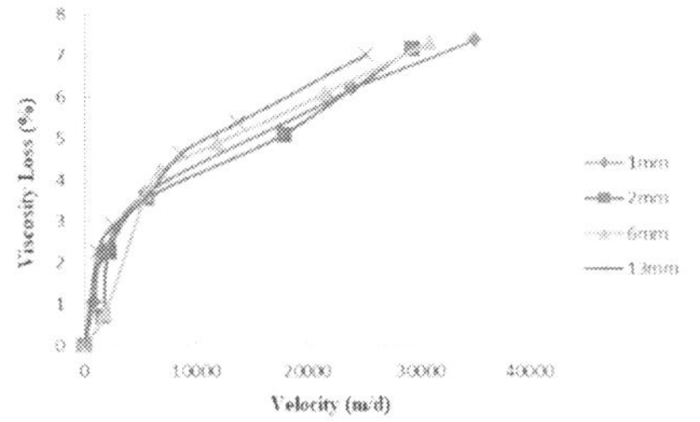

Figure 6 The relationship between velocity and viscosity loss under different diameter (square)

3.1.2 Perforation Shape

The polymer solution with a concentration of 1500mg/L prepared with a 25 million molecular weight polymer was used to conduct the shear test under the condition of no chamfer in 1mm (round and square) perforation with different shapes. The results are shown in Figure 7. At the same flow rate, the viscosity loss of polymer solution is basically not affected by different shapes of perforations. With the increase of flow rate, the viscosity loss of both types of perforations increases, and the influence of round perforations on the viscosity loss of polymer solution is slightly lower than that of square perforations.

Figure 8 shows the change of viscosity loss when comparing the round perforation (an aperture of 13mm) with or without chamfer. It can be seen that at the same flow rate, the shearing effect of the chamfered hole on the polymer solution is significantly reduced, and the existence of chamfer can reduce the viscosity loss by more than 20%.

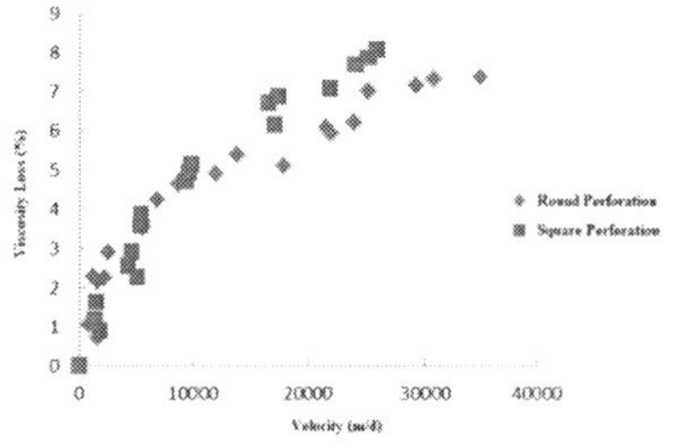

Figure 7 The relationship between velocity and viscosity loss under different shaped perforations

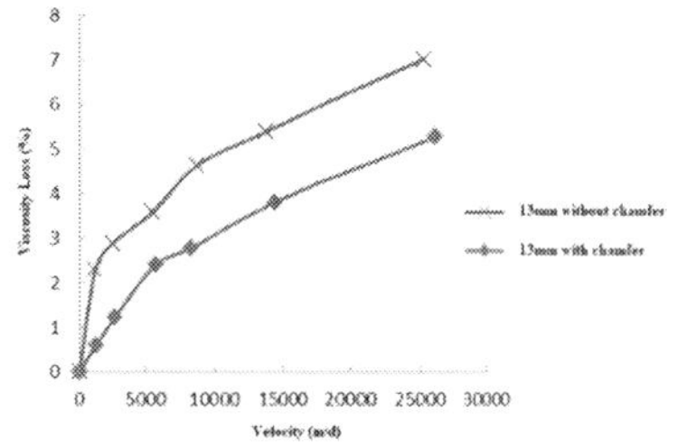

Figure 8 The relationship between velocity and viscosity loss under different chamfered perforations (circular)

 https://doi.org/10.1051/e3sconf/202337501042

3.1.3 Concentration and Molecular Weight of Polymer

Figure 9 shows the shear test results of polymer solutions with different concentrations (1500mg/L, 2000mg/L) prepared with a molecular weight of 25 million under a round 13mm diameter perforation. At the same flow rate, 2000mg/L polymer solution has a better shear resistance, and the viscosity loss of 2000mg/L polymer solution is obviously lower than 1500mg/L polymer solution. The higher the concentration of polymer solution is, the more tightly the molecular strands are intertwined, the less the damage of shear on the winding point is, and the stronger the shear resistance is.

Figure 10 shows the shear test results of 1500mg/L polymer solution prepared by 25 million molecular weight polymer and 30 million molecular weight polymers respectively under 13mm hole diameter round perforation. At the same flow rate, the viscosity loss of polymer solution with low molecular weight is obviously lower, and the maximum reduction can reach 30%, which illustrates that the shear resistance is better. The larger the molecular weight of the polymer, the more uneven the distribution of the long and short links of its molecular chain. When subjected to shearing, the molecular chain is stretched, and the distance between the long and short links increases. Therefore, the intertwining points between the links are greatly reduced, and the probability of the long links being sheared increases. The high molecular weight polymer is severely damaged by shearing, and its shear resistance is poor.

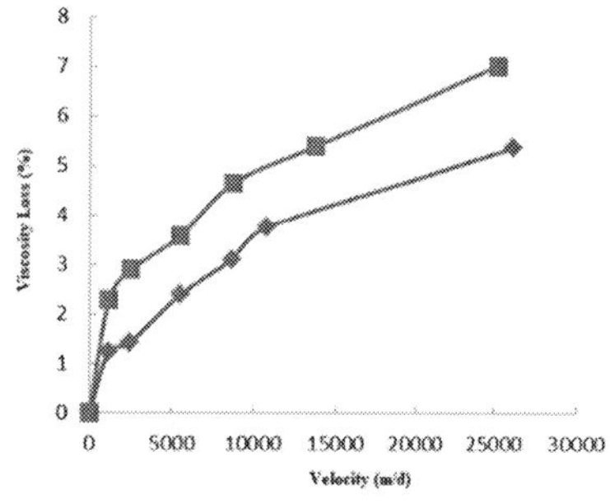

Figure 9 The relationship between velocity and viscosity loss under different polymer concentrations (Molecular weight=25 million)

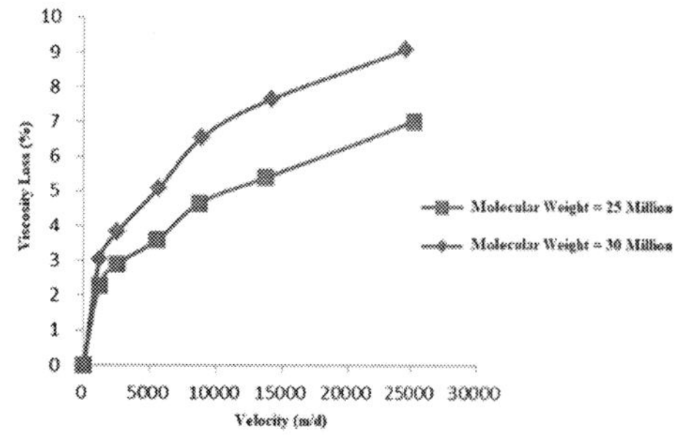

Figure 10 The relationship between velocity and viscosity loss under different molecular weight (Polymer concentration=1500 mg/L)

3.2 Shearing in Compaction Zone

Different lengths and permeability are selected. By changing the velocity, shear experiments in compaction zone under different molecular weight (25 million, 30 million) and concentration (1500mg/L, 2000mg/L) are conducted to analyze the influence of different factors on polymer viscosity. The concept of pore velocity is introduced:

$$V=\frac{Q}{\pi\times\phi\times D\times Sd\times PD\times T}$$

Which, Q—Injection velocity; V—Real velocity in pore structure; ϕ—Porosity; D—Perforation diameter; Sd—Shot density; Pd—Perforation depth; T—Thickness of perforated well section

3.2.1 Compaction Degree

A polymer solution with a molecular weight of 25 million was used to prepare a concentration of 1500 mg/L. When the compaction thickness was 1.25 cm, shear tests were conducted at different compaction degree (169 mD, 178 mD, 330 mD, 537 mD, 891 mD). The viscosity loss were shown in Figure 11. With the increase of flow rate, the viscosity loss of polymer increases gradually. When the flow rate reaches a certain extent, the viscosity loss tends to be gentle (the highest viscosity loss reaches 64%); At the same pore velocity, different compaction degrees are more sensitive. The greater the compaction degree (the smaller the permeability), the greater the impact of the compaction zone on the viscosity loss of polymer solution.

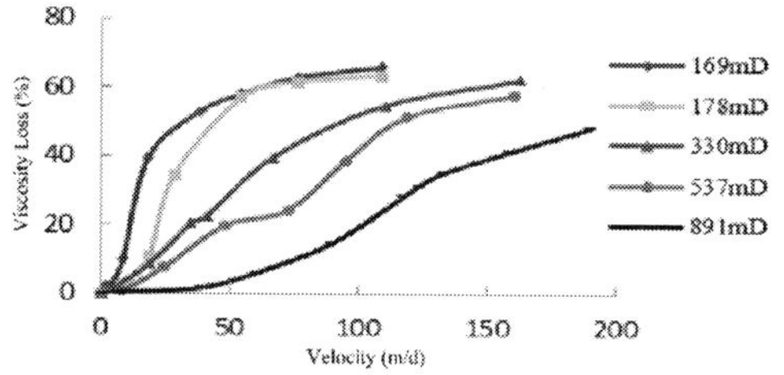

Figure 11 The relationship between velocity and viscosity loss under different permeability

3.2.2 Compacted Thickness

A polymer solution with a molecular weight of 25 million is used to prepare a concentration of 1500 mg/L. When the compaction degree is 169 mD, shear tests are carried out under the compaction bands with different compaction thicknesses (1.10cm, 1.25 cm, 1.55 cm). The viscosity changes are shown in Figure 12. With the increase of velocity, the viscosity loss in compacted zone with different compaction thickness increases. At the same compaction degree and flow rate, the greater the thickness of the compaction zone, the longer the migration distance in the porous medium, and the greater the shear effect, so the greater the impact on the viscosity loss of the polymer solution.

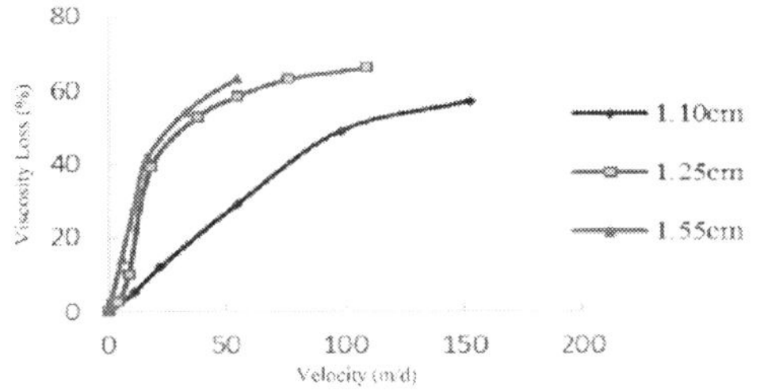

Figure 12 The relationship between velocity and viscosity loss under different thickness of the compacted zone

3.2.3 *Polymer Concentration*

The Gangxi polymer solution with different concentrations (1500mg/L, 2000mg/L) was used to conduct shear experiments under the compaction with thickness of 1.25cm (as shown in Figure 13). The results showed that under the same shear conditions, the higher the concentration of polymer solution, the lower its viscosity loss and enhanced its shear resistance.

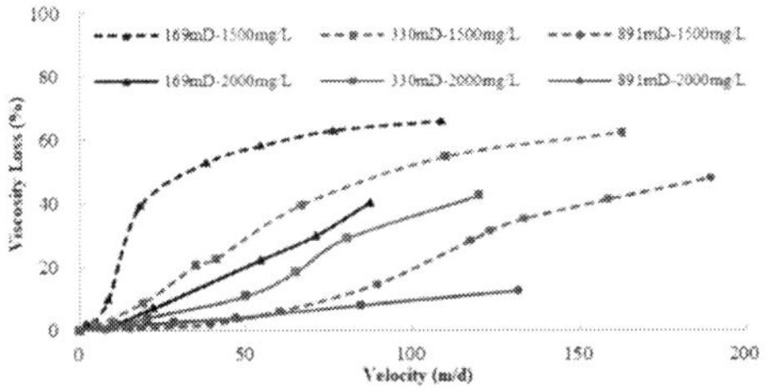

Figure 13 The relationship between velocity and viscosity loss of polymer with different concentration after sheared by compacted zone

3.2.4 *Molecular Weight*

Comparing the 2000mg/L solution prepared by 25 million and 30 million polymer, the shear results of the 1.25cm thick compaction zone are shown in Figure 14. Under the same pore flow rate and compaction conditions, the viscosity loss rate of 25 million polymer solution is obviously lower than that of 30 million polymers. The smaller the polymer molecular weight, the better its shear resistance.

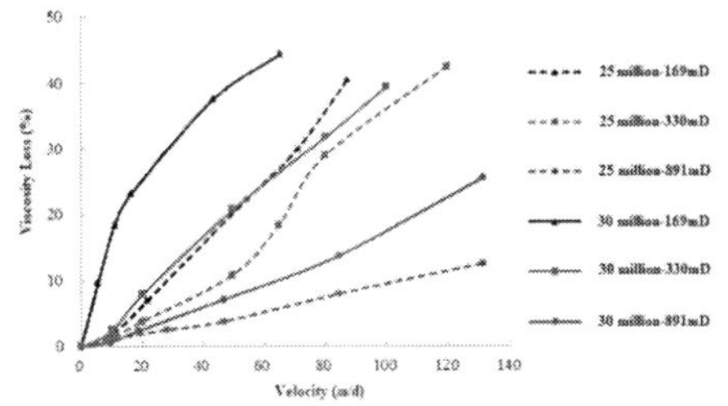

Figure 14 The relationship between velocity and viscosity loss of polymer solutions (made of different molecular weights, 2000 mg/L)

4. Conclusion

(1) The effect of perforation on polymer shear is not obvious (viscosity retention is greater than 96%); The bigger the velocity of the blast hole is, the greater the viscosity loss of polymer solution is. The size and shape of perforation have little effect on the viscosity loss;

(2) The compaction zone has obvious shear effect on polymer. With the increase of velocity, the degree of compaction (the lower the permeability) and the compaction thickness, the viscosity loss rate of polymer solution increases rapidly. Therefore, final viscosity retention rate of polymer solution tends to be gentle (under the condition of pore flow velocity in the target block, the minimum viscosity retention rate is only 36%).

(3) The higher the polymer concentration, the stronger the shear resistance. While the higher the polymer molecular weight, the weaker the shear resistance;

(4) When designing the perforation process of injection wells in the polymer injection test area, the perforating cartridge and perforation mode with large perforation depth and low compaction degree should be selected as far as possible to reduce the pore flow rate at the compaction zone and reduce the viscosity loss of polymer solution caused by the compaction zone.

References

1. Wang Demin. Cheng Jiecheng. Wu Junzheng. Wang Gang.Application of polymer flooding technology in Daqing Oilfield[J]. Acta Petrolei Sinica. 2005.26(1):74-78.
2. Gao Yan. Wang Qingping. Zhou Zongming. Zhang Aimei. Yao Huili. Effect of polymer injection rate on EOR[J]. Petroleum Geology And Recovery Efficiency.2001.8(2):64-66
3. Han Ming. Bai Baojun. Niu Lujie. Zhang Yanfu. Gao Yujun. Performance evaluation of polymer flooding in fault block Jin 16 [J]. Oil & Gas Recovery Techinology.1999(3):1-6
4. Luo Pingya. Li Huabin. Zheng Yan. Liao Guangzhi. Yang Zhenyu. EOR by ASP of Hydrophobic Associated Polymer-Alkali-surfactant in Daqing Oil Field[J]. Petroleum Geology & Oilfield Development In Daqing.2001. 20(6):1-4
5. Hill H J, Brew J R, Claridge E L, et al. The Behavior of Polymers in Porous Media[C]// SPE Improved Oil Recovery Symposium. Society of Petroleum Engineers, 1974.
6. Martin F D. Laboratory investigations in the Use of Polymers in Low Permeability Reservoirs[J]. 1974 .
7. Martin F D. Mechanical Degradation of Polyacrylamide Solutions in Core Plugs From Several Carbonate Reservoirs[J]. Spe Formation Evaluation, 1986, 1(2):139-150.
8. Maerker J M, Maerker J M. Shear Degradation of Partially Hydrolyzed Polyacrylamide Solutions[J]. Society of Petroleum Engineers Journal, 1975, 15(4):311-322.

9. LuXiangguo. Yan Wenhua. Song Helong. Gao Zhenhuan. An Experimental Study On Shear Degradation Of HPAM Solutions Flowing Through Porous Pedia[J]. Oilfield Chemistry.1995(4):375-378
10. Shao Zhenbo. Zhou Jisheng. Sun Gang. Niu Jingang. Liu Fengqi. Han Zhaorang.Studies on Mechanical Degradation of Partially Hydrolized Polyacrylamide in Course of Polymer Flooding: Changes in Relative Molecular Mass, Viscosity and Related Parameters[J]. Oilfield Chemistry.2005.22(1):72-77
11. An Yongsheng. Liu Wenli. Qi Xiangwen. Influence of perforation parameter on productivity of horizontal well[J]. Fault-Block Oil And Gas Field.2011.18(4):520-523
12. Chen Huijun. Wu Liqiang. Zhang Jun. Wei Nanning. Hu Yuzhong. Investigation into the viscosity loss of polymer solutions through different diameter nozzles[J]. Journal Of Daqing Petroleum Institute. 1999.23(3):224
13. Smith J E. Closing the Lab-Field Gap: A Look at Near-Wellbore Flow Regimes and Performance of 57 Field Projects[J]. 1994.
14. OSTERLOH.W. T. E. J. Polymer transport and rheological properties for polymer flooding in the North Sea Captain field[J]. Oil Field.1998.

Analysis on influencing factors of gully control in Loess Plateau

Na Lei [1,2,3,4,*], Panpan Zhang [1,2,3,4]

[1]Technology Innovation Center for Land Engineering and Human Settlements, Shaanxi Land Engineering Construction Group Co., Ltd and Xi'an Jiaotong University, China
[2]Shaanxi Provincial Land Engineering Construction Group Co., Ltd., Xi'an, China
[3]Institute of Land Engineering and Technology, Shaanxi Provincial Land Engineering Construction Group Co., Ltd., Xi'an, China
[4]Key Laboratory of Degraded and Unused Land Consolidation Engineering, the Ministry of Land and Resources, Xi'an, China
[5]Shaanxi Provincial Land Consolidation Engineering Technology Research Center Xi'an, China

Abstract. There are about 270,000 gullies over 500 meters in length on the Loess Plateau, and the average soil water content in the gullies is nearly three times that of the slope surface of the Loess Plateau. In the crisscrossed gullies on the Loess Plateau, there are very rich reserve resources of cultivated land and potential for increasing grain production. How to manage channel ecologically and make full use of channel resources is a problem that needs to be paid attention to. This study puts forward five factors affecting gully governance, and pays more attention to the systematization, integrality and coordination of protection, governance and utilization, which can provide reference standards for more comprehensive and accurate implementation of gully governance and land construction projects, and provide technical support for promoting economic development, ecological environment improvement and new rural construction in western China and old revolutionary base areas. It is conducive to ecological protection and high-quality development of the Yellow River Basin.

Key words: Channel management; Water resources; Soil erosion; Land degradation; Geological disaster.

1. Introduction

The Loess Plateau is crisscross with gullies and gullies, and the natural ecological environment is fragile. It is one of the most serious soil and water loss areas in the upper and middle reaches of the Yellow River [1-2]. The loess has the characteristics of light texture, loose soil, strong permeability, easy disintegration, fast consolidation and dehydration [3-4]. The people in the Loess Plateau region have made full use of the loess characteristics to create an effective measure of soil and water conservation in the practice of fighting against soil erosion for a long time -- silting dam, which has a history of hundreds of years. It can not only intercept sediment, preserve soil and water, but also silt and cultivate farmland and increase grain production [5-6]. Since the 1960s, gully damming has been popularized, and some achievements have been made in soil and water conservation and increasing cultivated land.

However, due to the lack of technical level and understanding of the occurrence law of water resources, gully governance still has some problems. Only by clarifying the factors affecting gully governance, Combining soil and water conservation, disaster prevention and reduction, agricultural production development, arable land increase, cultivated land quality improvement, and construction of high standard farmland in gully, realizing coordinated promotion and common development of gully control and land reclamation, and combining economic development, ecological environment construction and social development, can meet the needs of land consolidation in gully such a special landform and realize regional sustainable development. Only by comprehensively controlling gullies and slopes on the Loess Plateau, making full use of scientific protection, and realizing reasonable design and construction of gullies on the Loess Plateau, such as organic soil reconstruction and water resources regulation and utilization, can we scientifically control gullies with a clear target [7].

2. Uneven distribution of water resources

2.1 Spatial and temporal characteristics of water resources distribution

Water resources in the hilly and gully region of the Loess Plateau are generally poor, but some gullies are very rich, showing great spatial variability. At the same time, the precipitation in the area is uneven in distribution and changes between the years, 70% of the precipitation is mainly concentrated in July to September, often resulting

* Corresponding author: 619648133@qq.com

in drought no water available, flood season flood disaster, the shortage of effective water resources, agricultural production mainly depends on natural precipitation, a rain-fed agriculture. In order to get rid of these limiting factors, it is urgent to carry out engineering measures such as dam building, flood drainage channels, etc., to trap rainwater, flood control and disaster reduction, and to benefit and eliminate harm.

The regional difference of precipitation is: annual precipitation decreases from southeast to northwest. The annual average precipitation in some parts of the southern Qinling Mountains can reach more than 900 mm, while the annual average precipitation in northern Inner Mongolia is only 100 mm. The 400 mm precipitation contour crosses from northeast to southwest to north-central. In the south of the line, precipitation increases gradually, and it is subhumid and humid area, which is the main agricultural area. The rainfall in this region is mainly concentrated from June to September, and the precipitation accounts for more than 60% of the annual precipitation. In addition, the abundance and blight ratio of wet and dry years is very large, which makes drought and flood disasters occur frequently in this area, especially drought, almost nine droughts in ten years, but the scope and degree of disaster are different.

2.2 Extreme Rainfall

From the perspective of precipitation characteristics in gully region of Yan 'an Loess Plateau, there are two extremes: most of the rainfall days do not produce runoff, and the runoff rainfall is more than 10 mm. The annual runoff rainfall in this region is 200-250 mm, and it is mainly concentrated in rainstorm period, thus forming two extremes of river runoff or surge or close to cut-off. As a result, drought and flood disasters occurred frequently in the loess hilly region, and the agricultural harvest could be reduced by 2-3%. There is a general lack of large reservoirs that can regulate and store runoff in the loess hilly region, and the regional water surface area is very small, mainly concentrated along the river bay. There are only some small tangerines in other towns and townships, with water storage capacity below 10,000 m^3. Due to the lack of matching water conservancy projects, surface water resources are drained away in flood season, and people and animals have difficulties in drinking water in dry season. Therefore, attention should be paid to the problem of time mismatch between water supply and demand. It is necessary to strengthen the saving of water resources in rainy season for use in dry season. In addition, in some gullies, such as Yangwan Gully and Jiulongquan Gully in Baota District, there are rich groundwater resources. Coupled with the regulating effect of small watershed climate, the water resources in this region are relatively rich, and waterlogging disasters are more frequent, which is easy to produce soil flow. Therefore, it is urgent to reasonably regulate the local rich water resources, solve waterlogging disasters and realize the efficient utilization of water resources.

3. Serious land degradation and soil erosion

3.1 Land degradation

The problem of soil nutrient imbalance in the Loess Plateau has always been prominent. The application amount of organic fertilizer is low, and the excessive use of nitrogen and phosphorus fertilizer damages the soil structure, aggravates the phenomenon of soil compaction, shallower surface and poor water and fertilizer retention performance. In addition, frequent tillage also resulted in the reduction of soil organic matter and the deterioration of local soil conditions. Taking Yan 'an City as an example, the problem of soil and water loss is very prominent in the Loess Plateau region. The average sediment discharge in Yan 'an River basin for many years is more than 300 million tons. In Yan 'an, about 11 percent of the land is affected by soil erosion, and the content of soil organic matter decreases at a rate of 2 % every year.

3.2 Soil and Water Loss

In the hinterland of the hilly region of the Loess Plateau, the loess is thick and loose in structure, which is one of the most serious areas of soil and water loss in the middle reaches of the Yellow River. Although the project of returning farmland to forest has slowed down soil and water loss significantly in recent years, the problem of soil and water loss is still serious in this area, especially the short period of rainstorm is easy to form runoff flood peak, which seriously affects the safety of the gully dam. To control soil and water loss, it is necessary to treat ditches and slopes at the same time, construct auxiliary stormwater storage facilities such as stormwater storage, dam, pond and cellar, and build terraces, DAMS or gully on gentle slopes and gully, so as to achieve the comprehensive goals of soil and water conservation, agricultural development and farmers' income increase. The severe soil and water loss in the hilly and gully region of the Loess Plateau has caused the deterioration of regional ecological environment, and the land resources and soil quality have become short and barren due to soil and water loss. The soil organic matter content in the Loess Plateau region is low and the grain yield is mostly less than 1%, the land productivity is low, and the soil is very unstable. In the implementation of ditch treatment and land construction project, some areas in order to one-sided pursuit of cultivated land area, open terraces on steep slopes, mechanical excavation of slope, resulting in serious destruction of slope vegetation, stability deterioration, destruction of farmland to forest (grass) results, causing new soil and water loss.

4. Frequent geological disasters

As one of the most vulnerable ecological areas in China, the unique geological geomorphology, meteorology, hydrology and soil vegetation in the loess Gully region lead to unique natural disasters. The region is riddled with gullies, serious soil erosion, scarce groundwater resources, fragile ecological environment, frequent occurrence of geological disasters such as landslides and collapses, which have caused huge losses to society and people's lives and property, and seriously restricted the sustainable development of regional economy and society. Up to 2003, geological disasters in the loess gully region of northern Shaanxi have caused more than 1,000 casualties, damaged 30,000 kilns and houses, and caused a direct economic loss of more than 300 million yuan. The main types of geological disasters in this area are landslide, collapse, unstable slope, ground collapse, debris flow, ground fracture, ground subsidence, loess collapsion and soil erosion, and the most prone ones are landslide, collapse and debris flow. The frequent occurrence of geological disasters in gully region of Loess Plateau makes gully regulation urgent and difficult at the same time.

5. Weak agricultural infrastructure

5.1 Current situation of agricultural infrastructure

Backward supporting facilities of farmland and poor agricultural production conditions are the main bottleneck factors restricting the sustainable development of agriculture and rural areas at present. This situation is not conducive to the intensive management of land, nor to the increase of farmers' production income. At present, there are widespread prominent problems in agricultural land in gully area of Loess Plateau, such as serious water damage of gully dam, disrepair of fields and roads, inadequate irrigation and drainage facilities, and inadequate ecological and environmental protection measures for gully and gully slopes. As a result, most of the dam land, once known as "life saving land", is difficult to cultivate, land utilization rate is not high, crop yield is low and unstable, and farmers' income cannot be guaranteed. In Yan 'an City, the sown area of crops in 2010 was 261,400 hm^2, while the cultivated land area at the end of 2010 was 361,300 hm^2. Nearly 30% of the cultivated land could not be effectively used, and the utilization rate and yield rate of cultivated land were low. In 2010, the grain output in Yan 'an was only 17.4 kg/ hm^2, much lower than the national average of 24.2kg/hm^2. However, the dam land is mainly the silt and alluvial soil washed down by the surface soil of the slope, which has rich organic matter and high soil fertility. It can be built into a high and stable yield field, and has greater agricultural production potential. According to the survey, the yield of maize in Yan 'an dam land is 33.3-40 kg/ hm^2, and some high yield dam land can reach more than 50 kg/ hm^2.

5.2 Ways of agricultural infrastructure construction

How to solve the restricted factors such as water resource shortage, serious soil erosion and weak farmland infrastructure in land utilization in the loess Plateau gully region, strengthen farmland infrastructure construction and consolidate the foundation of agricultural development has become the key to the sustainable development of modern agriculture in this region. Therefore, we will carry out major projects to improve ditches and land, repair and renovate washed away or abandoned DAMS and rivers and roads, improve infrastructure such as fields, DAMS, canals, roads and forests, implement measures to improve soil improvement and fertility cultivation, improve infrastructure such as field irrigation and drainage ditches and mechanized Wells, water conservation, small rainwater harvesting and storage, mechanized roads, Bridges and culverts, and farmland and forest networks, and improve agricultural production conditions. Improving the utilization rate of land is a necessary measure to construct the national modern agriculture demonstration area in Yan 'an City.

6. Backward rural economic development

6.1 Rural economic development

At present, our country has entered an important period of urban and rural transformation and coordinated regional development. The implementation of the major land renovation project is an important platform of cracking the urban-rural dual structure and promoting industrialization, urbanization, agricultural modernization and the overall development of the city and countryside. For a long time, the loess Plateau gully region has been deeply affected by regional factors such as inconvenient road traffic, sterility of market development and fragile ecological environment, resulting in slow economic and social development. Yanchuan County, Yanchang County and Yichuan County of Yan 'an City are still national-level poor counties, and Wuqi County, Ansai County and Zichang County have just taken off their poverty titles. Although the gross domestic product of Yan 'an exceeded 100 billion yuan in 2011, the per capita exceeded 5200 yuan, and the total economic aggregate jumped to the fifth place in the province, the driving force of economic growth mainly came from the exploitation of coal and oil and gas resources. In the three industrial structures, the industry dominated, the industrial structure was unreasonable, and the regional development was also unbalanced. The vast rural areas were still poor and backward. Lack of characteristic industry and rural economic growth point, rural hollowing problem prominent, urban and rural overall planning and regional coordinated development of the task is quite arduous.

6.2 Transformation of rural economic development mode

Over the past 30 years and more of reform and opening up, industrialization and urbanization have played a limited role in driving rural areas, and people in old areas still live in poverty or low income. It is imperative to comprehensively improve rural production, living conditions and ecological environment. In recent years, the characteristics of urban and rural development transformation in the Loess Plateau region are increasingly apparent, and the employment structure, income structure and consumption structure of urban and rural residents are undergoing obvious changes. In

particular, the large-scale project of returning farmland to forest has brought about "two changes", that is, the focus of cultivated land protection has changed from slope to channel, and agricultural production has changed from wide planting and low harvest to intensive and efficient.
At present, the development of urban and rural areas in the Loess Plateau generally presents a new situation of "three increases and three decreases", that is, the increase of urban population, the increase of grain yield per unit area and the increase of farmers' income, while the decrease of rural population, the decrease of arable land and the decrease of soil and water loss caused by the conversion of farmland to forest. This transformation situation has created favorable conditions for the construction of basic farmland, the moderate scale management of land and the development of modern agriculture in this area. Therefore, in view of the urgent needs of urban and rural overall development and regional coordinated development, through further promoting the major ditch treatment and land construction projects, greatly improve the rural production conditions and farmers' living environment; promote hollow village renovation, new rural construction and modern agricultural development.

7. Summary and outlook

In the 1980s, the comprehensive management model with small watershed as the unit was popularized, silt DAMS were constructed, terraced fields and reverse sloping fields were constructed, and water conservation ecological agriculture was developed, and remarkable progress and achievements were made. In the 1990s, the country carried out the western development strategy and the ecological project construction of returning farmland to forest, so the traditional comprehensive construction mode of silt-dam is no longer suitable. Yan 'an's ecological construction led by the conversion of farmland to forest is in the forefront of the country. Since the implementation of the project, the total amount of farmland to forest has been 600 thousand hm^2, accounting for 2.5% of the country and 27% of the province. The forest coverage rate has reached 45.4%, and the regional ecological environment quality has changed significantly. The deterioration of the ecological environment in the past has been effectively checked, and the mountains and lands have initially realized a historic transformation from yellow to green. However, in order to stabilize the results of returning farmland to forest, farmers must first have enough grain allotment. In order to cooperate with the policy of returning farmland to forest, the mode of "gully improvement" is explored. Based on the actual situation of gully regions on the Loess Plateau, gully treatment and land construction are promoted according to local conditions, and comprehensive land improvement measures such as gully filling, dam restoration and construction, slope modification and so on are implemented to improve the design level and construction standards. It has achieved remarkable results in increasing the area of cultivated land, improving the quality of cultivated land, reducing soil erosion and improving the ecological environment by storing water resources, preventing flood and removing waterlogging, and optimizing the allocation of soil and water resources.
Therefore, on the basis of summarizing the influencing factors of gully, it is of great significance to further accelerate the comprehensive improvement of gully, innovate the land improvement mode of gully treatment and construction of high standard basic farmland for consolidating the conversion of farmland to forest, protecting and improving the ecological environment, effectively improving people's livelihood and consolidating the rural foundation, and realizing the coordinated development of man and nature in the Loess Plateau area of northern Shaanxi.

Acknowledgments

This work was financially supported by Technology Innovation Center for Land Engineering and Human Settlements, Shaanxi Land Engineering Construction Group Co., Ltd and Xi'an Jiaotong University (2021WHZ0092).

References

1. Z. S. Li, L. Yang, G. L. Wang. Current situation, problems and countermeasures of soil and water loss control in Loess Plateau. Acta ecologica sinica, 39(2019)7398-7409.
2. C. H. Hu, X. M. Zhang. Soil and water loss control and water and sediment change of Yellow River in Loess Plateau. Water conservancy and hydropower technology, 51(2020)1-11.
3. B. P. Zhang, Z. X. Hua, X. X. Cheng. A review on the mechanism of loess erosion and its disaster effects. Chinese Journal of Geological Hazards and Control, 32(2021)41-52.
4. J. M. Sun. Loess deposits interact with the earth's sphere. Quaternary Research, 40(2020)1-7.
5. X. Y. Liu, Y. F. Gao, S. B. Ma. Sediment reduction effect and timeliness of silt dam in Loess Plateau. Journal of Hydraulic Engineering, 49(2018)145-155.
6. H. Bao, D. X. Wang, T. Zhang. Some thoughts on the construction and management of silt dam in Loess Plateau in the new period. Soil and water conservation in China, 2(2020)23-26.
7. Y. S. Xie, Z. P. Li, J. J. Wang. Hierarchical structure and evolution of soil and water loss control model in Loess Plateau. Journal of Soil and Water Conservation, 25(2011)211-214.

https://doi.org/10.1051/e3sconf/202337501044

Capacity Analysis and Understanding of Pilot Test Well in Gas Storage A

Weiyan Li

Daqing Oilfield Gas Production Branch (Gas Storage Branch), Daqing, 163000, Heilongjiang, China

Abstract. There is a relatively simple gas reservoir structure in gas storage A, with good trap sealing, large cap rock thickness, relatively inactive fault activity, good overall reservoir connectivity, large storage capacity, water invasion occurs in only a few wells in the low part of the structure, so there are the geological conditions for building gas storage reservoir. This paper analyzed the capacity and maximum recovery capacity of horizontal wells in this block, and the pressure influence range during the recovery process of high-speed pilot production through the pilot test of reservoir construction; moreover, it got the following understanding: the horizontal wells with good development effect, high recovery capacity, and the pressure influence range in this block are mainly within 400m from the well distance; there are differences in the gas production capacity of different types of reservoirs, and the capacity contribution of Class I reservoirs is large, while the capacity contribution of Class II and III reservoirs is small, together these believe that Shengping gas storage has the capacity conditions for building reservoir.

Keywords: gas storage A; pilot test; capacity analysis; high-speed pilot production; pressure influence range; capacity contribution of different reservoirs.

1. Introduction

Block A has been developed since 2007, the development technology policy has always upheld the principle of "overall consideration and classified governance", implements the development method of "low drainage and high control", drains gas wells in low parts and produces gas with water, while produces gas with water control gas wells in high parts, it has been implemented over the years with better results. However, objective conditions such as the development of bottom water and non-homogeneous volcanic rocks in the block have brought challenges to the conversion to gas storage, and the geological understanding and dynamic features obtained from the development are not enough to meet the needs of reservoir construction.

The geological and dynamic understanding is obtained through the deployment of two pilot test horizontal wells, which provides important data support for the reservoir construction. It mainly includes two aspects: first, drilling cap rock of pilot wellbore and coring reservoirs, conducting connected processing and interpretation, core test and supporting comprehensive research of 200 square kilometers of seismic data to meet the geological evaluation requirements of reservoir construction; second, carrying out gas production tests, gas-water interface monitoring, gas production profile test, full gas reservoir shut-in well pressure test and other pilot test tests, analyzing the capacity of injection-production horizontal wells, the maximum production capacity at high-speed pilot production, the pressure influence range during pilot production, and the capacity distribution features of different types of reservoirs, etc.

2. Capacity Analysis of Pilot Test Well

2.1 Capacity Verification

Taking A Ping 1 well as an example: in the first gas testing, capacity test was conducted for six working systems, in total, the test gas pressure changes were small in the six production working systems (see Table.1-1); for working system 2, 4, 5, 6, multi-point regression binomial method calculate the open flow capacity $145\times10^4m^3/d$; multi-point regression exponential method calculate the open flow capacity $166\times10^4m^3/d$.

Table.1-1 gas testing and pilot production data table of well A Ping 1

category	production parameter	working system					
		1	2	3	4	5	6
gas testing	output ($10^4m^3/d$)	15	22	30	34	42	52
	wellhead pressure(MPa)	21.4	21.36	21.16	20.51	19.92	19.27
pilot production	output($10^4 m^3/d$)	42	55	70	100		
	wellhead pressure(MPa)	20.61	19.33	18.01	15.22		

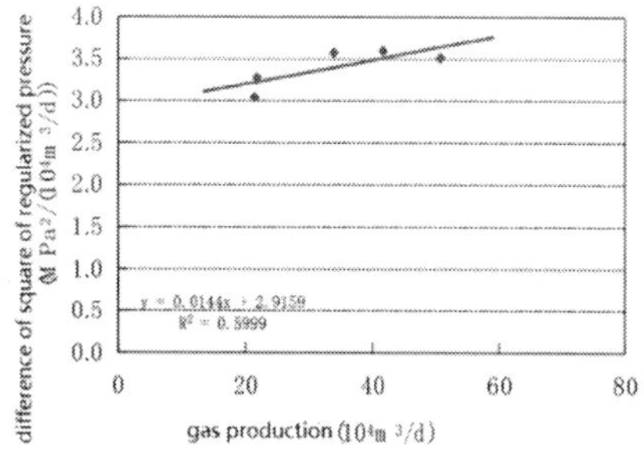

Fig.1-1 multi-point regression binomial curve (gas testing)

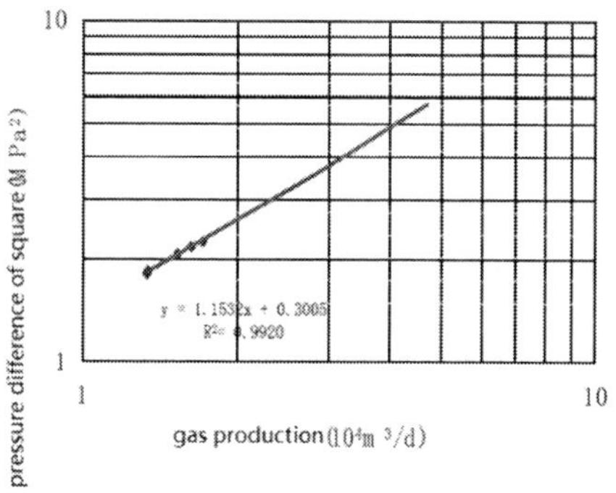

Fig.1-2 multi-point regression exponential curve (gas testing)

The multipoint regression binomial and exponential equations are as follows, respectively.

$$P_r^2 - P_{wf}^2 = 2.9159Q + 0.0144Q^2$$

$$Q = 0.5488\left(P_r^2 - P_{wf}^2\right)^{0.8672}$$

In the second gas testing, the capacity test is carried out with four working systems, in total, there is significant change in the gas testing pressure of the four output working systems (see Tables 1-2). The single-point correction binomial method is used to calculate the open flow capacity and binomial capacity equation of each working system, respectively, obtain the open flow capacity between 199~244×$10^4m^3/d$. can be seen that: the capacity calculated using working system 1 is high, and the capacity calculated using working system 2, 3 and 4 is very close, concentrated in 199~207×$10^4m^3/d$, 204.09×10^4m^3 /d on average.

Table.1-2 capacity equation and open flow capacity with single-point correction binomial method (pilot production phase)

work system number	output ($10^4m^3/d$)	wellhead pressure (MPa)	inflow model of extraction process	open flow capacity($10^4m^3/d$)
1	42	20.61	$P_r^2 - P_{wf}^2 = 2.079968Q +$	244.56
2	55	19.33	$P_r^2 - P_{wf}^2 = 2.480916Q +$	207.44
3	70	18.01	$P_r^2 - P_{wf}^2 = 2.580711Q +$	198.90
4	100	15.22	$P_r^2 - P_{wf}^2 = 2.445118Q +$	205.93

Similarly, five working systems are tested for well A Ping 2, in total: the gas testing pressure variation of the five capacity working systems is also small (see Table.1-3). The single-point correction binomial method is used to calculate the open flow capacity and binomial capacity equation of three working systems in the pilot production phase, obtain the open flow capacity between 243.53~260×$10^4m^3/d$, as shown in Table.1-4.

Table 1-3 data table of gas testing and pilot production of well A Ping 2

production parameter	working system				
	1	2	3	4	5
output($10^4m^3/d$)	10	20	33	40	53
Wellhead pressure(MPa)	21.51	21.22	21.11	20.81	20.30

Table.1-4 capacity equation and open flow capacity of single-point correction binomial method of well A Ping 2 (pilot production phase)

work system number	output ($10^4 m^3/d$)	wellhead pressure (MPa)	inflow model of production process	open flow capacity ($10^4m^3/d$)
1	33	21.11	$P_r^2 - P_{wf}^2 = 1.990026Q +$	260.73
2	40	20.81	$P_r^2 - P_{wf}^2 = 2.141329Q +$	243.53
3	53	20.30	$P_r^2 - P_{wf}^2 = 1.950351Q +$	263.96

2.2 Determination of Maximum Gas Production Capacity

According to the gas testing and pilot production capacity evaluation of well A Ping 1, the inflow and outflow curves of well A Ping 1 well are built, and the maximum production capacity is 120×$10^4m^3/d$ at the current formation pressure when the wellhead pressure is 12MPa.

https://doi.org/10.1051/e3sconf/202337501044

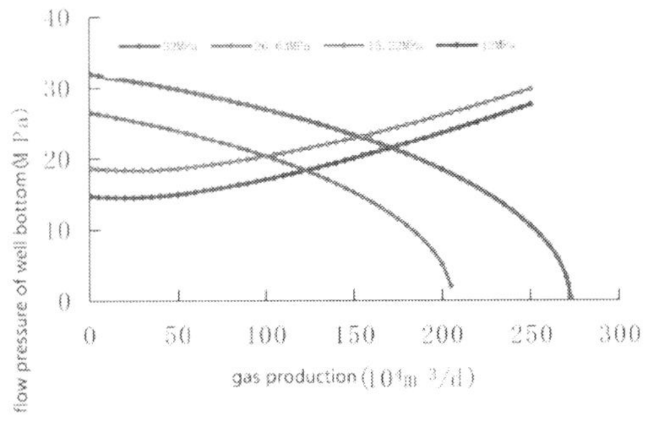

Fig.1-3 maximum production capacity curve of well A Ping 1

According to capacity evaluation of the gas testing and pilot production of well A Ping 2, the inflow and outflow curves of well A Ping 2 are built, and the maximum production capacity of current formation pressure is $130\times10^4m^3/d$ when the wellhead pressure is 12MPa.

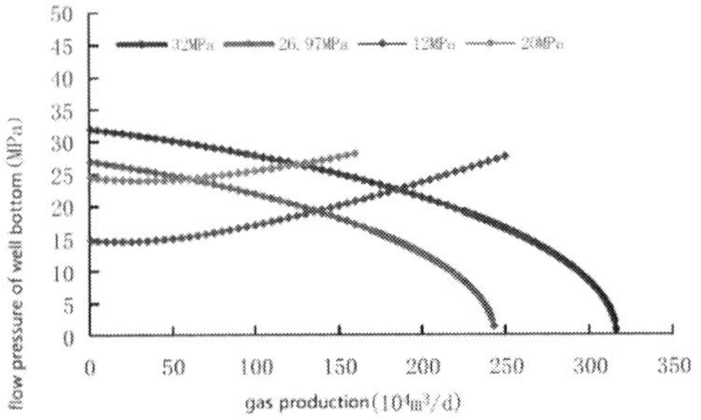

Fig.1-4 maximum production capacity curve of well A Ping 2 well

According to the above capacity analysis, we can see that the increase effect of pilot test horizontal well is obvious, the average daily gas production capacity reaches one million cubic meters, which is favorable for reservoir construction.

3. Pilot Production Evaluation of Pilot Test Well

3.1 Analysis of High-Speed Pilot Production

The pilot test wells of A Ping1 and Ping2 were put into production from August 2020 to the end of April 2022, and the pilot test wells had strong stable production capacity during the pilot production, as shown in Fig.2-1 and 2-2 below.

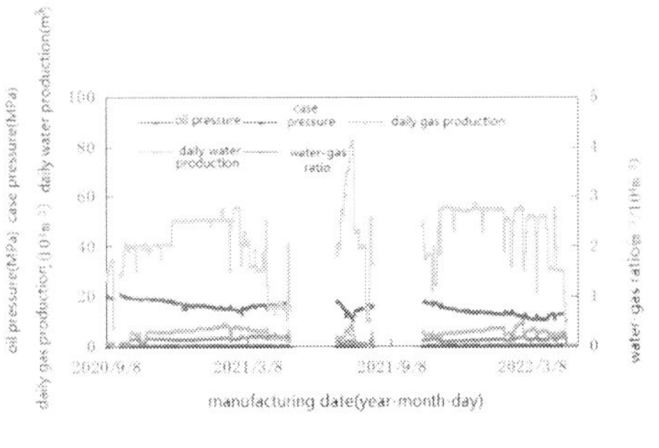

Fig.2-1 pilot production curve of well A Ping 1

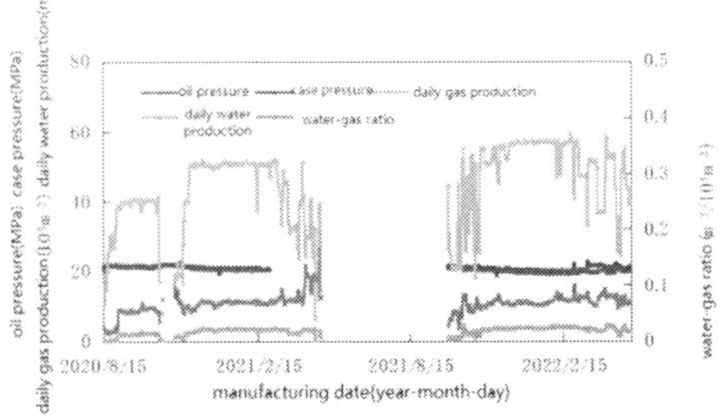

Fig.2-2 pilot production curve of well A Ping 2

The high-speed gas production test was carried out in well A Ping 1 and well Ping 1 in 2021. The production capacity of pilot production of well A Ping 2: the production capacity gradually decreased over time during the pilot production, it generally met the exponential decreasing law; during the medium-speed gas production ($40\text{-}50\times10^4m^3/d$), the production capacity was between $100\sim74\times10^4m^3/d$ during the 120-day gas production cycle, decreased by 0.3% every day; during the high-speed gas production ($50\sim83\times10^4m^3/d$), the production capacity was between $90\sim80\times10^4m^3/d$, during 15-day gas production, decreased by 1.1% every day (Fig.2-3).

The production capacity of pilot production of well A Ping: the production capacity gradually decreases over time during pilot production, and generally meets the exponential decreasing law; in the medium speed gas production stage ($40\sim50\times10^4m^3/d$), the production capacity is between $160\sim140\times10^4m^3/d$, decreased by 0.1% every day in 120 days of gas production cycle; the stable production capacity is stronger in comparison with (Fig.2-4).

According to the analysis of the high-speed pilot production, we can see that the increase effect of pilot test horizontal well is obvious, the gas production capacity of horizontal well reaches one million cubic meters, and has the capacity conditions for building storage.

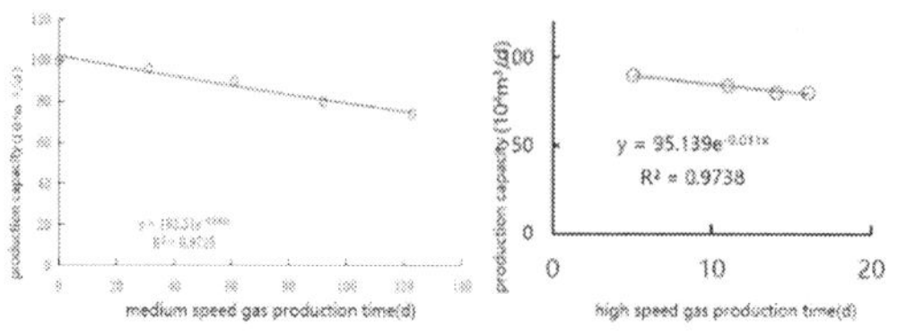

Fig.2-3 evaluation of continuous production capacity of medium-speed and high-speed pilot production in well A Ping 1

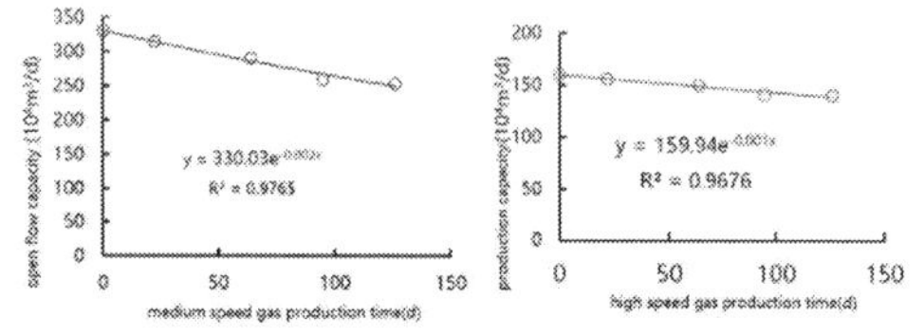

Fig.2-4 evaluation of continuous production capacity of medium-speedand high-speed pilot production in well A Ping 2

3.2 Influence Range of High-Speed Gas Production

The central pressure of the gas layer of the static pressure test of well A Ping 1 is 23.52MPa, and the flow pressure and flow temperature reflect no liquid accumulation in the wellbore above 2570m; the central pressure of the gas layer of the static monitoring of well A Ping 2 is 25.95MPa, and the flow pressure and flow temperature reflect no liquid accumulation in the wellbore above 2400m.

The interpretation results of the old wells in the block suggest that: the influence range of Class I wells is 148 to 447 m, the influence range of Class II wells is 114 to 435 m. The interpretation results of the pilot test wells suggest that: the influence range of well A Ping 1 is 398m, the permeability is 1.028md; the influence range of well A Ping 2 is 420m, the permeability is 8.64md. The influence range of high-speed gas production is generally less than 400 m. (Fig.2-5 to Fig.2-6).

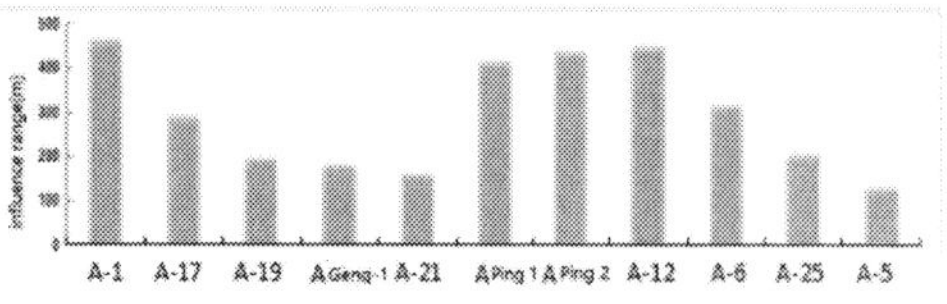

Fig.2-5 distribution of interpretation results of the gas production influence range of old wells and pilot test wells

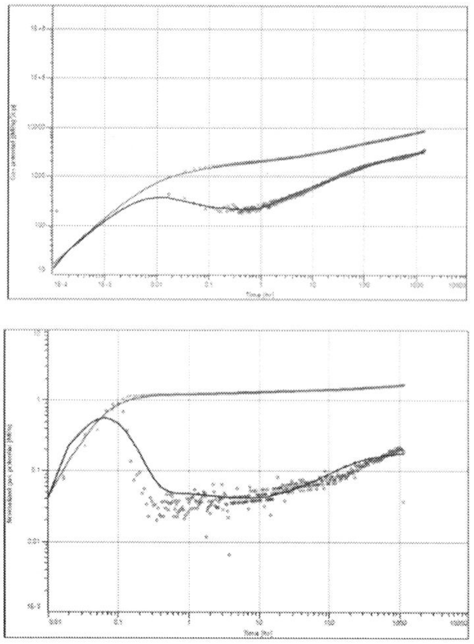

Fig.2-6 double logarithmic curve of pressure recovery of well A Ping 1 and A Ping 2 from April to June 2021

The pressure influence range test of well A Ping 1 suggests that: well A-7 with less than 400m away from adjacent wells is affected, while well A-6 and well A-Ping 1 with more than 400m are not affected. The pressure influence range test of well A Ping 2 suggests that: well A-17, A Geng 2 with less than 400m away from adjacent wells are affected, well A-21 and A-25 beyond this distance are not affected.

4. Capacity Distribution of Different Types of Reservoirs

The gas production section test requires monitoring the production section of long gas production layer, i.e., the perforation section of well is relatively long, and two wells meet the conditions after screening. The depth of perforation section of well A-17 is from 2910m to 2940m the length is 30m, the depth of perforation section of well A Geng-1 is from 2928.4m to 2990.8m, the length is 62.4m, both have condition for testing the gas production section in pilot test.

The gas production section test of well A-17 suggests that: 2938-2940m is the main production layer of the well, production contribution reaches 95%; compare the reservoir type with production contribution: the Class I high-quality reservoir of the well contributes more to the production capacity, as shown in Fig.3-1.

The gas production section test of well A Geng-1 suggests that: the production contribution 2 Class I high-quality section reaches 88%, compare reservoir type and production contribution: Class I high quality reservoir of the well contributes more to the production capacity, as shown in Fig.3-2.

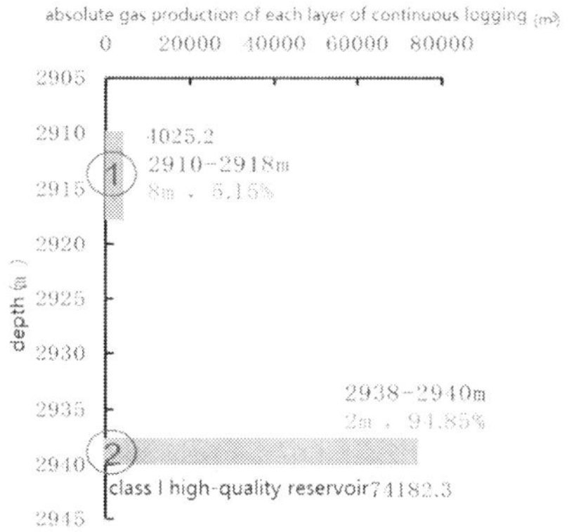

Fig.3-1 gas production section test of well A-17

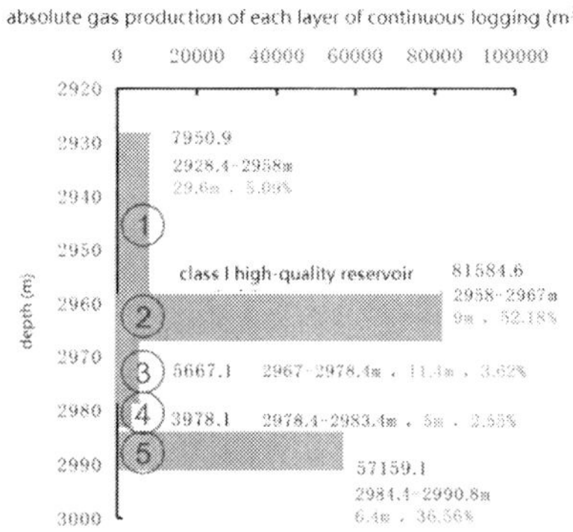

Fig.3-2 gas production section test of well A Geng-1

5. Conclusion

(1) It can be seen that the production capacity of horizontal well is high by analyzing the pilot test horizontal well capacity and maximum production capacity, the increase effect of high-speed gas production is obvious, the gas production capacity is up to one million square meters, and it has the capacity conditions for building reservoir.
(2) The influence range of different types of reservoir gas wells in the block is different, they are generally within 450m, the influence range of high-speed gas production pressure of pilot test wells is mainly within 400m, the overall connectivity of the block is relatively good, which provides favorable conditions for the balanced injection and production of gas storage and efficient operation.
(3) The features of gas production capacity of different types of reservoirs are made clear, Class I reservoirs have good physical properties and large capacity contribution, and Class II and III reservoirs have smaller capacity contribution.

References

1. Shu Ping, Gao Tao. Injection and Production Capacities for the Volcanic Gas Reservoir Converted into Underground Gas Storage. Petroleum Geology & Oilfield Development in Daqing, 2015, 34(6), P48-53.
2. Li Zonglin. Analysis of Injection-Production Capacity of Single Wells in Wen23 Gas Storage. Chemical Engineering Design Communications, 2020, 4, P21-25.
3. Xu Hongcheng, Zhu Weiping. A Method for Designing the Pilot Test Scheme of Underground Gas Storages Rebuilt from Complex Gas Reservoirs. Natural Gas Industry, 2020, Vol. 6, No. P33-37.
4. Wang Haiyan. Optimization of the Injection-Production Capacity of the Gas Storage Well Considering the Influences of the Multiple Factors. Petroleum Geology & Oilfield Development in Daqing, 2019, 38(3), P54-58.
5. Huang Jianhua, Wei Xiaofeng. Study on the Reevaluation of Preliminary Work of Shuang-6 Gas Storage. World Petroleum Industry, 2020, P49-53.

 https://doi.org/10.1051/e3sconf/202337501045

Gas distributor for ultra-large air blow-out bromine extraction plant

Shujing Chai, Wenyan Zhang, Xiaoyu Yu, Xiaocui Hao, Tao Li, Zejiang Wang and Wei Liu*

Institute of Seawater Desalination and Multipurpose Utilization, MNR (Tianjin), Tianjin 300192, China

Abstract. The initial distribution of gas and liquid is one of the key factors that determine the operation efficiency of packed towers. The development of a gas distributor for ultra-large air blow-out bromine extraction plant can improve operation efficiency, save energy and reduce consumption. In order to develop the gas distributor for ultra-large air blow-out bromine extraction plant, the guidelines for evaluating gas distributors are clarified in this paper, and six kinds of commonly gas distributor are compared in detail. Considering gas distribution inhomogeneity, liquid foam entrainment rates and distributor pressure drop, the double tangential circulation type gas distributor is the first choice in the ultra-large air blow-out method bromine extraction plant.

key word: Bromine; Gas distributor; Air blow-out method

1. Introduction

Bromine is an important fine chemical raw material. At present, there are more than 4500 bromides that can be synthesized artificially, among which more than 1800 are commonly used. Bromine flame retardants account for more than 40% of the consumption of organic flame retardants, and the commonly used ones include TBBPA, DBDPO, HBCDD, etc. [1]. High efficiency and low toxicity pesticides containing bromine are used to kill a variety of microorganisms, and the commonly used ones include bromomethane, bromopropylate, DBDMH, etc. [2-3]. As medical intermediates, more than 80 brominated drugs can be produced, among which ambroxol, ciprofloxacin, cefapirin and bronopol are common [4]. In the petroleum industry, dibromoethane is used as an antiknock agent for petroleum fuels and calcium bromide is used as a solid filling fluid for petroleum drilling. In the rubber industry, butyl bromide rubber is gradually replacing ordinary butyl rubber. According to statistics, the average annual growth rate of apparent consumption of bromine from 2007 to 2019 was about 2.57%, and the consumption of bromine exceeded 200,000 tons in 2019, with a year-on-year growth of 7.08%.

China's bromine industrial technology was developed from the underground brine in Laizhou Bay in the 1980s. About 90% of the domestic factories are uses the "Air blow-out method". The production process is that the brine input into the pipeline by the brine pump, the acid and chlorine gas are added to the pump outlet pipeline, so the bromine ion oxidizes to free bromine. Acidizing oxidation solution further reacted by the oxidation tower to be sent to the blow-out tower, and sprayed from the top of the tower. The blower blows air from the bottom of the tower to complete the gas-liquid exchange on the surface of the packing. The bromine in the brine is desorbed and blow-out, and the waste liquid is discharged from the bottom of the tower. The bromine-containing air is discharged from the top of the tower, mixed with sulfur dioxide gas and fresh water spray added in the top air passage, and absorbed downstream, and formed small droplets of hydrobromic acid, which are enriched by the absorption tower and the trap to form bromine content. The higher hydrobromic acid completion liquid is passed through chlorine gas oxidation, steam distillation, condensation, washing and separation to obtain the finished industrial bromine. The simplified process flow of bromine by acid method is shown in Figure 1. [5]

* Corresponding author: liuwei@isdmu.com.cn

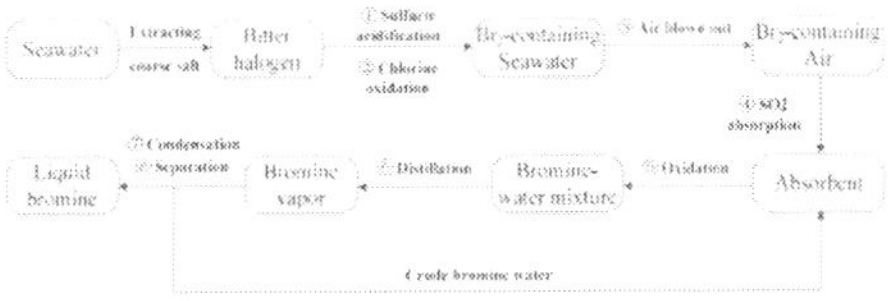

Figure 1. Bromine extraction process by air blowing.

The blow-out tower is the main equipment for the analysis and enrichment of "Air blow-out method", and is the core of the whole process. The blow-out towers in China are mainly 6.5 m in diameter, and there are some small towers of about 5 m. According to statistics, the 6.5 m diameter blow-out tower whose rated flow capacity is usually not greater than 1260 m^3/h, and the 5 m diameter blow-out tower whose rated flow capacity is about 700 m^3/h. Compared with other industries, the single tower rated flow capacity is small and energy consumption is high. The packed tower separation technology is a relatively mature process, drawing on the experience of other industries can completely increase the single tower processing capacity, so as to play the scale effect, to reduce energy consumption, improve the yield and capacity of the purpose.

For an excellent packed tower, the performance of the packing is a very important factor, but the matching tower internals (especially liquid distributor and gas distributor) also play a vital role in the performance of the packing. If there is no excellent performance of the tower interior to match, the performance of the packing will not be fully played. Especially for the large diameter and shallow bed packed tower, the initial distribution of gas and liquid is very important and often becomes the key factor for the successful application of packed column. Poor gas-liquid distribution can reduce the performance of the packing by 50%-70%.

2. Guidelines for evaluating gas distributors

In recent years, the research on packed towers has continued to make breakthroughs, and the maximum application diameter of regular packed towers in modern chemical production can reach 14-20 m. Gas distributor is an important gas initial distribution device in the packed tower, the performance of which directly affects the efficiency of the packed tower and product quality. Gas phase bias flow will cause uneven local gas-liquid distribution, reducing the separation efficiency; too much liquid entrained in the updraft will also affect product quality and yield, the poor structure of the feed distributor, will also affect the whole tower pressure drop. With the development of large void rate, low pressure drops, new high-efficiency packing and large diameter, shallow bed packed towers, the initial gas flow distribution in packed towers is receiving more and more attention.

After referring to a number of reports [6-7], we believe that a good gas distributor should have the following requirements at the same time:

- Good homogeneous distribution performance;
- Low flow resistance;
- Occupy a small space in the tower;
- Effectively prevent gas-liquid entrainment;
- Easy installation and maintenance;
- Low Cost

In the above requirements, good distribution performance and low resistance is the key indicator, the rest of the requirements in different occasions have different focus. Such as occupying a small space in the tower, in the old tower renovation is likely to be the first condition that must be met, while in the new tower design is not very important.

3. Commonly used forms of gas distributors and their performance

Domestic and foreign researchers pay more and more attention to the development of gas distributor, developed a wide range of styles, excellent performance gas distributor. In the industrial production of the common distributor are porous straight tube type, straight tube baffle type, single tangential circulation type, tangential horn type, double column blade type, double tangential circulation type, etc.

3.1 Porous straight tube type

The porous straight tube type is currently the most commonly used form of refinery pressure reduction tower, the inlet pipe extends to the center of the tower, the tube below the opening of a long hole, the gas from the hole ejected by the bottom of the tower rebound upward. At this time, the tower wall at the higher gas velocity, airflow along the mouth of the tube down the spray, in the feed tube above the formation of vortex, flow resistance is very large.

3.2 Straight tube baffle type

Straight pipe baffle type is an improvement of the porous straight pipe type, in order to reduce the impact and in the straight pipe with downward curved baffle, reducing the fluid flow resistance, gas velocity distribution and porous straight pipe similar, but if the gas-liquid two-phase feed, the liquid falls more at the entrance.

3.3 Single tangential circulation type

Single tangential annular flow inlet pipe tangential into the tower, and set the deflector, the gas-liquid fluid into the annular flow channel, in turn by the arc-shaped blade guide flow down, meet the bottom of the tower after the air flow back up into the sleeve. The center of the gas velocity is higher, the liquid is centrifugal force along the tower wall flow down, liquid foam entrainment is zero, the distributor resistance is small, while the inlet tube in the high-speed two-phase flow resistance is larger.

3.4 Tangential horn type

Tangential horn type inlet pipe tangential into the tower, the mouth of the pipe has a downward sloping horn-shaped guide hood, under the action of the guide hood downward circling flow, the gas mixture at high speed tangential into the gradually expanding horn, into the tower along the tower wall downward rotation to the bottom of the tower and then folded upward movement circling to the bottom of the tower and upward. When the gas flow contains liquid droplets, under the action of centrifugal force, the liquid droplets are separated and the mist entrainment is almost zero. The resistance is low. However, it is necessary to design a suitable angle, otherwise it will cause vibration.

3.5 Double column blade type

Double column blade type gas distributor, fluid radial into the tower, the inlet on both sides of the two guide arc vane, the top and bottom are closed, the airflow along the two vane left and right separate, rushing to the tower wall and folded upward. Because the double-row vane gas distributor tower occupies a small space, gas distribution is uniform, the flow resistance is small, in recent years the application is very widespread.

3.6 Double tangential circulation type

Double tangential circulating gas distributor is similar to single tangential circulating gas distributor, the material enters the tower in radial direction and is divided into two strands by the guide plate, which enter the annular channel along the inner cylinder and are guided to the bottom of the tower by the curved blades in turn, and folded upward. The distributor resistance is smaller, the liquid foam entrainment is smaller, the airflow distribution is more uniform.

3.7 Gas distributor performance comparison

Pan et al. [8] had systematically studied the performance of several gas distributors in terms of gas flow, gas distribution inhomogeneity M, liquid foam entrainment rates e_L and distributor pressure drop ΔP. The performance comparison of various distributors is shown in Table 1.

Table 1. Performance comparison of gas distributors.

	Porous straight tube type	Straight tube baffle type	Tangential horn type	Single tangential circulation type	Double tangential circulation type	Double column blade type
M	2.0	2.0	1.97	0.52	0.37	1.8
e_L	5.3	1.3	0	0	0.1	0.6
ΔP (Pa)	2740	843	10	49	15	30

4. Conclusion

Ultra-large air blow-out tower is of great significance to reduce production energy consumption and achieve economies of scale. As an important part of tower interior, gas distributor is indispensable. Considering the process flow and material properties of bromine extraction by air blow-out method, the support structure in the tower and the differences in the distribution inhomogeneity, pressure drop and liquid foam entrainment rates, the double tangential circulation type gas distributor is the first choice in the ultra-large air blow-out method bromine extraction device.

In the future, we will use the numerical simulation method to simulate the double tangential circulation gas distributor. A series of structural indexes such as the distance between the tower wall and the inner cylinder, the height of the inner cylinder, the height of the baffle plate, the number of baffle plates, and the opening condition of the roof were optimized by taking the distribution inhomogeneity and pressure drop as indexes.

Acknowledgments

This work was financially supported by Key R&D Program in Shandong Province (2019JZZY010807), National Key Research and Development Program of China (2022YFC290400503), and Special basic research fund for central public welfare institute of China (R-JBYWF-2021-T01).

References

1. A. Papachlimitzou, J.L. Barber, S. Losada, et al. A review of the analysis of novel brominated flame retardants, J. Chromatogr. A. 2012, 1219: 15-28.
2. R.H. Thompson, A review of the properties and usage of methyl bromide as a fumigant, J. Stored Prod. Res., 1966, 1(4): 353-376.
3. F. Sabuzi, G. Pomarico, B. Floris, et al. Sustainable bromination of organic compounds: A critical review, Coord. Chem. Rev., 2019, 385: 100-136.

https://doi.org/10.1051/e3sconf/202337501045

4. A. Jitareanu, G. Tataringa, A.M. Zbancioc, et al. Bromination-A versatile tool for drugs optimization, Med. Surg. J., 2018, 122(3): 614-626.
5. Y. Zhang, S.J. Chai, D. Song, et al. Development status and future development trend of bromine, Earth Environ. Sci., 2019, 300:032018.
6. Y.R. Dong. Modern Packed Tower Technology (III) Gas Distributor and Other Inner parts of Packed tower, Chem. Prod. Tech., 1996, 3(4):8.
7. C.Q. Hu, W.M. Song, X.B. Feng, et al. Structure optimization design and analysis of Double Tangential Circulation Gas Distributor, Petro. Equip., 2022, 51(1):8.
8. G.C. Pan, Q.F. Guo, Research on the gas distributor of packed tower, Oil Ref. Des., 1995,25 (2) : 28-32

The Vapor-Liquid Equilibrium of Bromine-Brine System at Atmospheric Pressure

Jianting Qian [1], Zhaoling Yu[1], Xincheng Yang[1], Xiaolong Zhang[1], Paisheng Xu[1], Jundong Liu[1], Wei Liu[2], Shujing Chai[2], Wenyan Zhang[2], Xiaocui Hao[2], Yuqi Wang[2], Qi Zhang[2] and Xiping Huang[2*]

[1] ShangDong HaiWang Chem. Co., LTD, Shandong, China
[2] Institute of Seawater Desalination and Multipurpose Utilization, MNR (Tianjin), Tianjin 300192, China

Abstract. The air blow-out method of extracting bromine from seawater or brine is widely used at home and abroad. The key parameter in this unit operation is the vapor-liquid equilibrium data (T, x, y) for bromine-brine system, which were obtained with the Bubble Equilibrium equipment at 10 and 20°C in this paper. The experimental equilibrium data were carried out by using Margules Equation for liquid phase and ideal equation for gas phase, and the nonlinear least square method was used to regress the parameters of Margules model for the system. It showed good consistency for the calculated data from Margules model and the experimental data.

key word: Bromine; air blow-out method; vapor-liquid equilibrium (VLE).

1. Introduction

Bromine is the basic chemical raw material for the production of flame retardants, fire extinguishing agents and photosensitive materials. It is widely used in the pharmaceutical, pesticide, fuel and oil industries, which is one of the important branches of the marine chemical industry. Extracting bromine by the air blow-out method from (concentrated) seawater or salt brine is the most widely used production process at home and abroad [3]. The main steps include acidification, oxidation, blowing out, absorption, distillation and condensation [4]: when the bromine ion is oxidized to bromine molecular by chlorine gas, it is blown out from the (concentrated) seawater or salt brine by air, and then reacts with sulfur dioxide or alkali to become the concentrated solution with high concentration of bromine ion. the bromine molecular is blown out by chlorine gas oxidation or acidizing from the concentrated solution with high concentration of bromine ion by steam distillation process. Finally, the bromine is separated from the liquid phase and condensed to obtain product [1,2].

The main unit operation equipment of the air blow-out method is the reaction tower, such as the blowing-out tower, absorption tower, and distillation tower. The design of the blowing tower is critical [5]. However, the design is mainly based on the previous data or empirical correlation so far, and few rigorous calculations are carried out in theory. This is mainly because the vapor-liquid equilibrium data of bromine-brine systems are very few. In the existing literature, the saturated vapor pressure of bromine at a specific temperature under atmospheric pressure, Henry coefficient of bromine in pure water and saturated vapor pressure data of brine in certain salt fields can be found or calculated by using the Antony Formula, although the relative volatility can also be estimated, and then the phase equilibrium and mass transfer behavior in the tower can be studied. However, due to the different kinds of raw materials, the chemical compositions are also different, so not only the saturated vapor pressure data of brine cannot be generalized, resulting in a narrow application range of this method, but also the interference of other ions, resulting in the dissolution behavior of bromine often deviates from Henry's Law. Gergert V R pointed out that due to the interference of chloride ions, the dissolution behavior of bromine in dilute sodium chloride solution does not strictly follow Henry's Law. Ostapenko L F studied the distribution coefficient of bromine between vapor and liquid phases at 15 and 35 °C, and found that when the molar fraction of bromine in liquid phase is less than 2.8×10^{-6}, its phase equilibrium behavior also deviates from Henry's law. This study shows that the distribution coefficient of bromine is not only affected by the concentration of bromine in the liquid phase, but also increases with the increase of temperature and decreases with the increase of salt concentration. The above studies are earlier, and the existing data are far from meeting the needs. Therefore, it is necessary to measure the vapor-liquid equilibrium data of specific raw materials through experiments, and then design and calculate the blow-out tower in combination with the actual process operation, which plays an important guiding role in improving the bromine blowing rate and reducing the consumption of raw materials [3,4].

* Corresponding author: huangxiping@163.com

Vapor-liquid equilibrium data mainly includes temperature, pressure, vapor phase and liquid phase composition of the system. The methods of vapor-liquid phase equilibrium determination are divided into direct determination and indirect determination. The former refers to the direct determination of the temperature, pressure and composition of liquid and vapor phases at equilibrium state through experiments. This method mainly includes distillation method, flow method, static method, quasi-static method and circulation method. The key of this method is the design of equilibrium kettle. However, in actual measurement, the gas phase composition is not easy to measure accurately, especially when the component contains heavy components. In most practical situations, T, P and x are measured, and then y is calculated indirectly. The relationship between P and x can be measured at constant temperature, and the relationship between T and x can also be measured at constant pressure. In addition, many studies also pointed out that the parameters of the activity coefficient equation can be directly determined by measuring the infinite dilution activity coefficient, so as to calculate the gas-liquid equilibrium data of the whole concentration range of the system. The main methods include boiling point meter method, gas chromatography, gas chromatography and inert gas saturation strip method. In addition, in special cases, thermodynamic estimation can also be carried out by looking up physical property data in the manual [5,6].

In this paper, the vapor-liquid equilibrium behavior of bromine - brine system is studied by using the bubbling entrainment equilibrium kettle method to directly determine the concentration of bromine in the liquid and vapor phases. The method has simple equipment and high accuracy. In view of practical application, the vapor-liquid equilibrium state of the actual raw material system containing dozens to hundreds of ppm of bromine at 10 and 20 °C at atmospheric pressure is mainly studied [7].

2. Experiment

2.1 Experimental principle

Considering that in practical application, the content of bromine in the raw material solution is very low, only tens to hundreds of ppm. It is a dilute solution, and its composition does not cover the full concentration range. Therefore, at this time, it is not appropriate to use the pure component standard state to describe the thermodynamic properties, but should use Henry's law based on the fugacity of the infinite dilution solution standard state. Henry's constant can be strictly defined by the following process of finding the limit

$$H_{2,1} = \lim_{x_2 \to 0}(y_2 \phi_2^V P / x_2) \quad (1)$$

Where: $H_{2,1}$: Henry constant of solute 2 in solvent 1,

ϕ_2^V: Vapor phase fugacity coefficient of solute 2 in solvent 1,

x_2、y_2: Molar fraction of component 2 in the liquid phase and gas phase respectively,

P: Total system pressure

2.2 Experimental process

In this experiment, an improved bubble entrainment flow equilibrium kettle was used to measure the vapor-liquid equilibrium data. It mainly consists of two balance kettles: the first is a pre-saturation kettle, in which the brine which is exactly the same as the main saturation kettle but does not contain bromine, while the second balance kettle is placed with brine solution containing different concentrations of bromine. Each balance kettle has a precision thermometer to measure the temperature of the solution directly. The whole set of balance kettle device is in the thermostatic water tank, which is externally connected to the atmosphere to keep the pressure constant. When the temperature is stable, blow the air in with constant flow rate. The air bubbles continuously through the sand core filter at the bottom of the bottle. The bromine molecular is dissolved into the air bubble and blown out after a long contact time with the solution. Then it enters the washing bottle containing potassium iodide solution for absorption, and finally enters the mass air flow meter to measure the air volume. The equipment is shown in Figure 1. The same equipment was also used in our previous work.

The main function of the pre-saturation kettle is to take out the saturated steam of the solvent from it, so that the solvent in the main balance kettle will not be vaporized, which can ensure the constant composition of the solvent in the main balance kettle; The long contact time and the tiny bubbles produced by the sand core filter nozzle ensure the full contact between gas and liquid; And use a drying filter to absorb the moisture carried out in the gas; Excessive potassium iodide solution ensures that bromine carried out in the air can be completely absorbed.

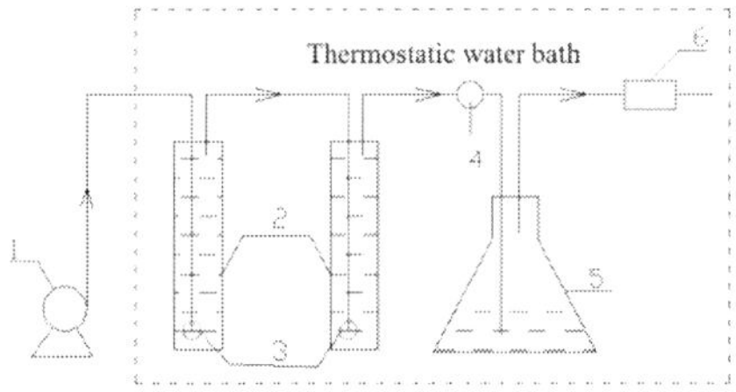

Figure 1. Equipment

1. Air pump 2. Vapor-liquid balance kettle 3. Sand core filter nozzle 4. Drying filter 5. KI conical flask 6. Gas mass flowmeter

3. Experimental data and analysis

3.1 Data results

Figure 2 (a) and (b) show the relationship between the partial pressure of bromine in the vapor phase and the molar fraction of bromine in the liquid phase in the bromine-medium brine system at 10 and 20 °C respectively. It can be seen that in dilute solution, the relationship between the two is not linear but quadratic function, which deviates from Henry's law. This is because the liquid phase in the system is a non-ideal solution, and the sulfate radical, chloride ion and other interfering impurities in the raw liquid brine affect the phase equilibrium behavior of bromine; On the other hand, it can also be found from the fitting formula that the quadratic term has little influence. If this term can be ignored, it is a linear relationship, which conforms to Henry's law. According to the fitting formula in the figure and the above theoretical analysis, the Henry constant values of the bromine-brine system are16.92 and 23.05 atm at 10 and 20 °C, respectively. The trend of this data just reflects the trend that the volatility of bromine in the brine increases with the increase of temperature.

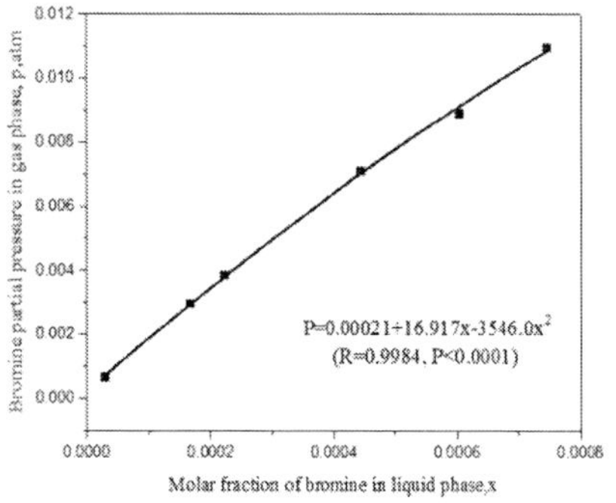

(a)

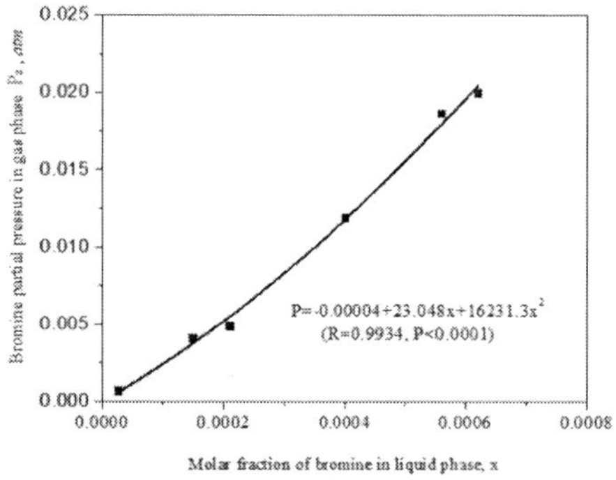

(b)

Figure 2. The Vapor-Liquid Equilibrium of bromine-brine (a) 10°C; (b) 20°C

3.2 Activity coefficient equation correlation

The thermodynamic treatment methods of vapor-liquid equilibrium include activity coefficient method and equation of state method, both of which have advantages and disadvantages, and should be selected according to different situations. Generally, the equation of state method is particularly beneficial in the calculation of high-pressure vapor-liquid equilibrium, and the activity coefficient method is often used in the low and medium pressure range. The commonly used activity coefficient equations include Whole-type equation based on normal solution, Margules equation, Van Laar equation, Wilson equation based on local composition, NTRL equation and UNIQUAC equation model. Generally speaking, for engineering applications, the first three equations can also meet the accuracy requirements, and the calculation is simple. In this paper, we use the cubic Margles equation to correlate the relationship between the activity coefficient equation and the composition. The general formula of the equation model in the binary system is as follows:

$$\ln \gamma_1 = (A+3B)x_2^2 - 4Bx_2^3 \quad (2)$$

$$\ln \gamma_2 = (A-3B)x_1^2 + 4Bx_1^3 \quad (3)$$

Among the above equation, 1 and 2 represent two substances respectively, A and B are binary parameters, which need to be solved by nonlinear least square method according to the experimental data. Considering that the experimental system is a dilute solution system and bromine is a solute, it is only necessary to correlate the activity coefficient of the solute.

The correlation results of the Margules equation at 10 °Ccan be seen in Table 1. The Margules equation is as follows:

$$\ln \gamma_1 = -424.065 \times (1-x_1)^2 + 424.162 \times (1-x_1)^3 \quad (4)$$

where: r_1 is the activity coefficient of bromine in the liquid phase, and x_1 is the mole fraction of bromine in the liquid phase.

Table 1. Correlation Results of Margules Equation at 10°C

$x_{1,exp}$	$P_{1,exp}, at$	γ_1, ca	$P_{1,cal}, at$	$\Delta P, at$	$\|\Delta P / P_{1,e}$
3.0049E-5	0.00066	1.2997	0.0007	0.00004	0.0606
1.6720E-4	0.00297	1.0511	0.0029	-0.00007	0.0236
2.2165E-4	0.00386	1.0305	0.0038	-0.00006	0.0155
4.4451E-4	0.00710	0.9451	0.0070	-0.00010	0.0140
6.0234E-4	0.00891	0.8743	0.0091	0.00019	0.0213
7.4531E-4	0.01096	0.8701	0.0108	-0.00016	0.0146

From the table 1, we can see that except for the first data, the error of other data is between 1.46-2.36%, which has a very good correlation.

Similarly, at 20 °C, the correlation results of Margules equation are shown in Table 2, and its equation is as follows

$$\ln \gamma_1 = 572.421 \times (1-x_1)^2 - 572.435 \times (1-x_1)^3 \quad (5)$$

https://doi.org/10.1051/e3sconf/202337501046

Table 2. Correlation Results of Margules Equation at 20°C

$x_{1,\exp}$	$P_{1,\exp}, at$	γ_1, ca	$P_{1,cal}, at$	$\Delta P, at.$	$\Delta P / P_{1,\exp}$
2.7432E-5	0.00067	1.0691	0.0006	-0.00007	0.1058
1.5210E-4	0.00410	1.1695	0.0038	-0.00026	0.0631
2.1036E-4	0.00488	1.0683	0.0055	0.00035	0.0669
4.0376E-4	0.01186	1.2744	0.0119	0.00005	0.0043
5.6020E-4	0.01864	1.4436	0.0180	-0.00067	0.0362
6.2011E-4	0.01994	1.3950	0.0205	0.00055	0.0277

From the table 2, we can see that except for the first data, the error of other data is between 0.43-6.69%.

4. Conclusion

In this paper, the vapor-liquid equilibrium data of bromine-brine system at 10 °C and 20 °C were measured by bubble entrainment equilibrium kettle, and their Henry constants were 16.92 and 23.05 atm, respectively. In this paper, the Margules activity coefficient equation model is used to correlate the experimental data. The results are satisfactory, which fully meets the requirements of engineering design. The experimental data can be applied to the design and calculation of chemical separation process.

Acknowledgments

This work was financially supported by Key R&D Program in Shandong Province (2019JZZY010807), National Key Research and Development Program of China (2022YFC290400503) and Special basic research fund for central public welfare institute of China (R-JBYWF-2021-T01).

References

1. L. C. Stewart, Commercial Extraction of Bromine from Sea Water. Ind. Eng. Chem., 26 (1934) 361-369.
2. A. M. E. Hamouz, R. Mann, Chemical Reaction Engineering Analysis of the Blow-out Process for Bromine Manufacture from Seawater. Ind. Eng. Chem. Res., 46 (2007) 3008-3015.
3. B. E. Poling, J. M. Prausnita, et al. The Properties of Gases and Liquids. 5th ed., McGraw-Hill Education Co. and Chemical Industry Press, 2006.
4. J. C. Eerol, J. C.Masson, Accurate measure of activity coefficient at infinite dilution by inert gas stripping and gas chromatography. Ind. Eng. Chem. Process Des. Dev., 16 (1997) 139-144.
5. K. Fukuchi, K. Miyoshi, et al. Measurement and correlation of infinite dilution activity coefficients of alkanol or ether in aqueous solution. Fluid Phase Equil., 194-197(2002) 937-945.
6. D. W. Green. Perry's Chemical Engineers' Handbook. 7th ed., New York, 1997.
7. R. H. Cai, W. Liu, Y. Yao, Y. S. Zhang, Atmospheric gas-liquid phase equilibrium of bromine-concentrated seawater system, Chem. Ind. Eng., 27 (2010) 199-204.

Preparation of high-temperature gels for enhanced oil recovery using methyl etherized melamine-formaldehyde resin

Hongbin Guo [1], Jijiang Ge [1] and Jiayi Li [2]

[1] China University of Petroleum (East China), Qingdao 266580, China
[2] Sinopec Shengli Oilfield Dongxin Oil Production Plant, Dongying 257000, China.

Abstract. Polymer gels are the most commonly used materials in the petroleum industry for reservoir conformance control and resolving excessive water production problems. In this study, high-temperature gels were prepared with HPAM or AM/AMPS as the gel-forming agent, methyl etherified melamine-formaldehyde resin (MEMFR) as the crosslinker, and nano-silica as the stabilizer, and their properties were evaluated. The results showed that HPAM-MEMFR gels were unstable at 110°C in a brine of 20665 mg/L. AM/AMPS-MEMFR gels are stable at 110°C, with gelation time ranging from 15 h to 92 h and storage modulus ranging from 3.6 Pa to 50 Pa. AM/AMPS-MEMFR gel has less than 10% dehydration after aging at 110°C for 90 days, almost no shrinkage in volume, and a significant increase in strength, making it suitable as a gel material for enhanced oil recovery.

Keywords: High-temperature gel; gel evaluation; conformance control; enhanced oil recovery.

1. Introduction

Due to the existence of fractures and high permeability areas in the formation, oil fields developed by water injection always have the problem of excessive water production [1]. The injection of plugging agents from oil recovery wells or water injection wells to seal high permeability layers or fractures is an important measure to improve the effectiveness of water injection development. A key to this technology is the development of high-performance plugging agents [2,3]. Gels are one of the most commonly used plugging agents and are formed by the reaction of a polymer and a crosslinker [4]. Gels used at high temperatures are generally prepared from organic crosslinkers. These crosslinkers include polyethyleneimine (PEI), water-soluble phenolic resins (WSPR), and combined systems of hydroquinone (HQ) and hexamethylenetetramine (HMTA).

PEI gel is friendly to the environment but has a high price, which is mainly used in offshore oil fields at present [5]. Although WSPR crosslinker is inexpensive, its application is more restricted, with disadvantages such as short storage time (less than 3 months) and high residual formaldehyde content in the product (greater than 5%). HQ and HMTA can react at high temperature to form a phenolic resin intermediate, which can play a crosslinking role similar to that of WSPR. These crosslinkers are currently used in high-temperature formations [6], but there are problems such as increasing HQ prices and restrictions on the purchase, transportation, and use of HMTA because it is a hazardous chemical.

In addition to the above-mentioned crosslinkers, methyl-etherified melamine-formaldehyde resin (MEMFR) can also crosslink with polymers containing hydroxyl and amide groups to produce gels. MEMFR is often used as an adhesive for rubber, wood, and paper in the industry, with formaldehyde content below 1% and a storage time up to one year. It was used as a gel crosslinker, but the application in the oil field did not attract attention. In this study, high-temperature gels were prepared using partially hydrolyzed polyacrylamide (HPAM) and acrylamide/2-acrylamide-2-methylpropanesulfonic acid (AM/AMPS) crosslinked with MEMFR, respectively. The performance of the gel was evaluated in terms of gel formation time, strength, and stability in a brine of 20 665 mg/L salinity at 110°C. The prepared gel can be used as an alternative system for reservoir conformance control.

2. Experimental

2.1 Materials

The HPAM used in the experiment was purchased from the product of Anhui Jucheng Fine Chemical Co., Ltd. with a hydrolysis degree of 12-14% and a relative molecular weight of 1.4×107-1.5×107 g/mol. The AM-AMPS was supplied by Qingdao Qucheng Technology Co., Ltd. with an AMPS content of 10% and a relative molecular weight of 6×106-8×106 g/mol. The crosslinker MEMFR for gel preparation was provided by Shandong Baomao Biochemical Co. In addition, nano-silica was added to the gel as a stabilizer with a particle size of 10-20 nm, which was obtained from Qingdao Jinfangge Technology Co.

The gelants in this study were prepared using synthetic brine with a salinity of 20665 mg/L. The ionic composition of the brine is shown in Table 1. Synthetic brine was prepared from potassium chloride, sodium chloride, sodium bicarbonate, anhydrous sodium sulfate, magnesium chloride hexahydrate, and calcium chloride anhydrous, supplied by Sinopharm Chemical Reagent Co.

Table 1 Composition of synthetic brine

Ions (mg/L)						TDS (mg/L)
Na^+/K^+	Ca^{2+}	Mg^{2+}	SO_4^{2-}	HCO_3^-	Cl^-	
6359	632	196	68.5	140.5	13269	20665

2.2 Methods.

2.2.1 *Preparation of gels*

Synthetic saline, MEMFR, and nano-silica were weighed according to the composition of the gelant and mixed well by stirring them. Then, HPAM or AM/AMPS were slowly added to the mixture while stirring. Next, the mixture was stirred at 120 r/min for 2 hours to obtain a homogeneous gelant. Finally, the gelant was dispensed into ampoules using a syringe and sealed using an alcohol torch for performance evaluation.

2.2.2 *Evaluation of gels*

The gels were evaluated in terms of gelation time, strength, and stability. The gelation time of the gels was determined visually using Syndansk's gel strength code [7] as a standard. The time required for the gel to reach the F level of strength at 110°C was recorded as the gelation time.

The strength of the gel was characterized by strength code and storage modulus. The energy storage modulus of the gels was measured by an Anton paar MCR92 rheometer equipped with a parallel plate system at 25°C. The test was conducted in amplitude sweep mode with a measurement gap of 1 mm, a fixed oscillation frequency of 10 rad/s, and an amplitude range of 0.1% to 1000%.

The stability of the gels was characterized by the dehydration ratio of the gels after aging for different times at 110°C. The dehydration rate was defined as the ratio of the mass of water removed from the gel after aging to the mass of the initial gelant.

3. Results and Discussion

3.1 HPAM-MEMFR gel

HPAM is the most commonly used water-soluble polymer in the oilfield. Therefore, HPAM was first used as a gel-forming agent crosslinked with MEMFR to prepare the gel. The methyl ether group in MEMDR will hydrolyze under acidic conditions to form hydroxymethyl, as shown in Fig. 1. These hydroxymethyl groups can crosslink with the amide groups on the polymer to form a gel, and the crosslinking reaction is shown in Fig. 2. We made up 16 gel formulations by orthogonal design, in which polymer concentration range of 0.4%-1.0%, cross-linker concentration range of 0.2%-0.8%, and stabilizer concentration fixed at 1%, and their gelation time and strength are shown in Table 2.

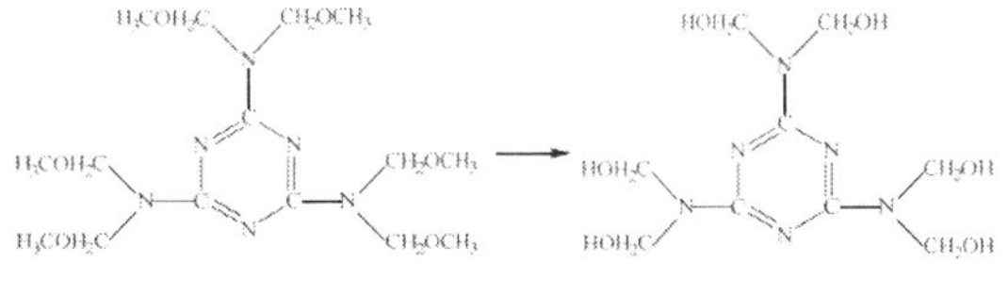

Fig. 1 Hydrolysis of methyl ether groups.

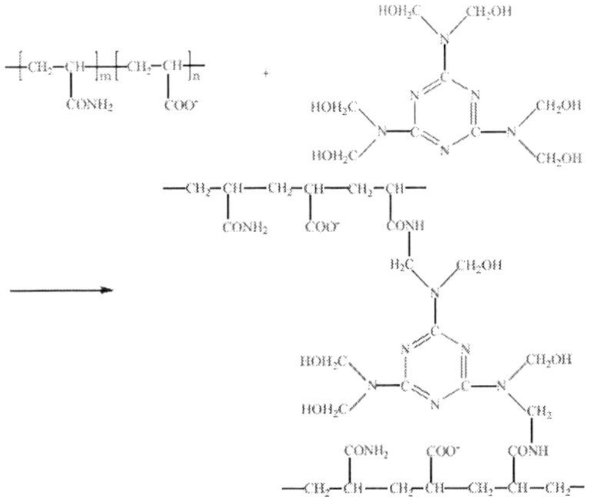

Fig. 2. Crosslinking reaction of HPAM with MEMFR.

Table 2. Composition, gelation time, and strength of gels.

No.	w(HPAM)/%	w(MEMFR)/%	w(SiO2)/%	Gelation time/h	Strength code
1	0.4	0.2	1	86h-E	E
2	0.4	0.4	1	50h	G
3	0.4	0.6	1	47h	G
4	0.4	0.8	1	41h	G
5	0.6	0.2	1	50h	G
6	0.6	0.4	1	47h	G
7	0.6	0.6	1	23h	H
8	0.6	0.8	1	18h	H
9	0.8	0.2	1	25h	G
10	0.8	0.4	1	10h	H
11	0.8	0.6	1	7h	H
12	0.8	0.8	1	6h	H(I)
13	1.0	0.2	1	23h	G
14	1.0	0.4	1	9h	H
15	1.0	0.6	1	6h	H(I)
16	1.0	0.8	1	5h	H(I)

It can be seen that the reaction of HPAM with MEMFR can form gels with high strength. The gelation

 https://doi.org/10.1051/e3sconf/202337501047

time of the gels at 110°C was in the range of 5-50 h, and the gel strength was between G-I level. However, these gels were not stable at high temperatures, and most of them showed significant syneresis when aged at 110°C up to 8 days, as shown in Fig. 3.

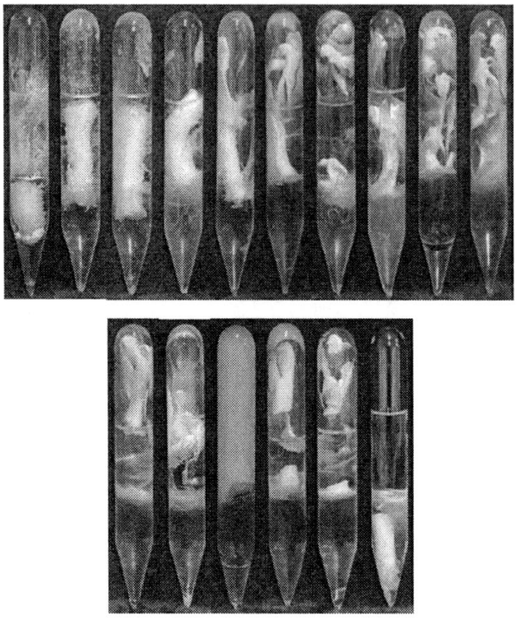

1 2 3 4 5 6 7 8 9 10 11 12 13 14 15 16

Fig. 3. The state of HPAM-MEMFR gel after aging at 110°C for 8 days.

3.2 AM/AMPS-MEMFR gel

Considering the instability of HPAM-MEMFR gels, the gels were considered to be prepared by crosslinking AM/AMPS with MEMFR for better thermal stability. Sixteen gels were prepared following the gel composition in Table 2 after replacing HPAM with AM/AMPS. The gelation time, strength, and stability of these gels were investigated.

3.2.1 Gelation time

The gelation times of gels formed by crosslinking AM/AMPS with MEMFR are given in Figure 4 and compared with the gelation times of HPAM-MEMFR gels.

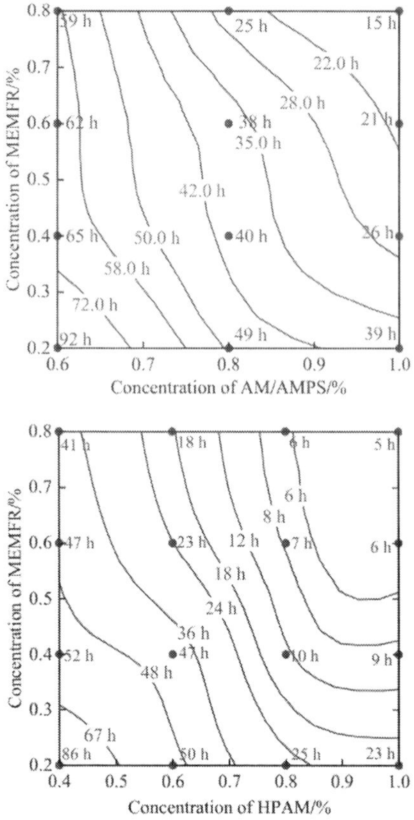

(a) AM/AMPS-MEMFR (b) HPAM-MEMFR

Fig. 4. Gel formation time of AM/AMPS-MEMFR gels and HPAM-MEMFR gels.

By comparison, although the molecular weight of AM-AMPS is slightly lower than that of HPAM, the gel formation of AM/AMPS is significantly slower than that of HPAM for the same polymer and cross-linker concentrations, which is related to the electrostatic repulsion of the cross-linker and polymer in water. The sulfonic acids on AM/AMPS molecules are strongly ionized groups, and their charged nature is less affected by electrolytes in water than the carboxylic acids on HPAM molecules. Therefore, the electrostatic repulsion between AM/AMPS and MEMFR is higher than that of HPAM, which hinders the crosslinking reaction between the crosslinker and the polymer.

3.2.2 Gel strength

The storage moduli of the gels prepared from AM/AMPS and MEMFR were determined after aging at 110°C for 5 and 90 days, respectively, as shown in Fig. 5.

The storage modulus of the gels aged for 5 days ranged from 3.6 Pa to 50.0 Pa. Except for the gels with polymer concentrations of 0.6% and 0.8% and MEMFR concentrations of 0.2%, the storage modulus of all the gels could be more than 10 Pa, which belongs to strong gels. The gel strength gradually increased with the increase of aging time. The gel storage modulus after aging at 110°C for 90 days ranged from 42.5 Pa to 645.4 Pa, showing a substantial increase in strength. This is because the primary amide on the polymer will hydrolyze under high-temperature conditions, and the carboxylate generated by hydrolysis will chelate with Ca^{2+} and Mg^{2+} in the brine, enhancing the crosslink density of the gel.

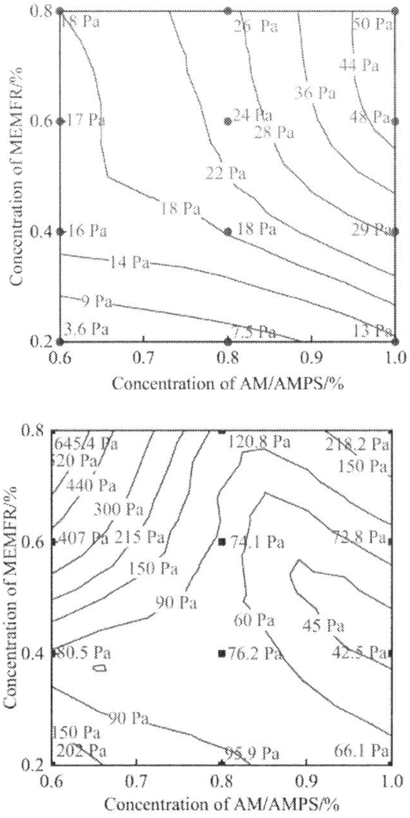

(a) Aging for 5 days (b) Aging for 90 days

Fig. 5. Storage moduli of AM/AMPS-MEMRF gels aged at 110°C for different times.

3.2.3 Gel stability

To evaluate the gel stability formed by AM/AMPS and MEMFR, their dehydration ratios after aging at 110°C for 15 and 90 days were measured, as shown in Fig. 6.

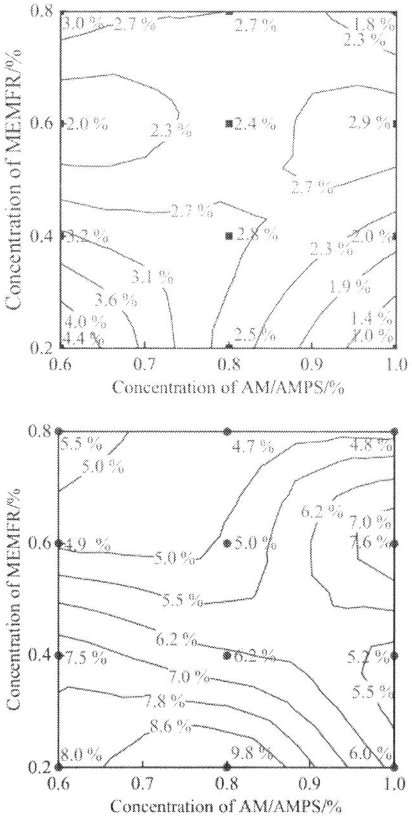

(a) Aging for 15 days (b) Aging for 90 days

Fig. 6. Dehydration ratios of AM/AMPS-MEMRF gels aged at 110°C for different times.

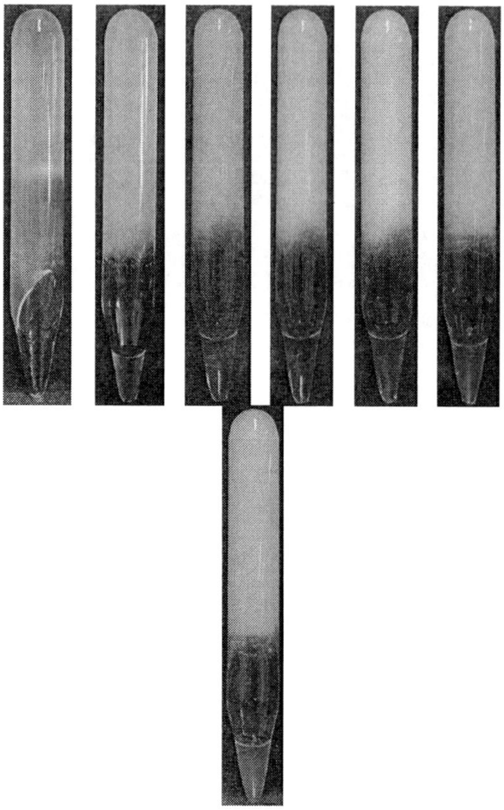

5d 15d 30d 45d 60d 75d 90d

Fig. 7. The state of gel aged at 110°C for different times.

The dehydration rate of the gel formed by AM/AMPS and MEMFR was less than 5% after aging at 110°C for 15 days and less than 10% after aging for 90 days. Their stability was significantly better than the gels formed by HPAM and MEMFR. Figure 7 shows the states of the gels composed of 0.6% AM/AMPS, 0.2% MEMFR, and 1.0% SiO2 aged at 110°C for different times. The overall volume of the gel remained unchanged and the strength increased, although small dehydration occurred. Therefore, we believe that the stability of AM/AMPS-MEMFR gels is sufficient to support their application for enhanced oil recovery.

4. Summary

Based on the above results and discussions, the following conclusions were obtained:

(1) Gels can be prepared by using HPAM as the gel-forming agent and MEMFR as the cross-linking agent. However, HPAM-MEMFR gels are unstable under high-temperature conditions and exhibit severe syneresis after aging at 110°C for 8 days.

(2) High-temperature stable gels can be prepared by crosslinking AM/AMPS with MEMFR. AM/AMPS-MEMFR gels have a gelation time range of 15 h-92 h, a storage modulus range of 3.6 Pa-50 Pa, and a dehydration ratio of less than 10% after 90 days of aging at 110 °C.

(3) AM/AMPS-MEMFR gels showed almost no volume shrinkage and a significant increase in gel strength after aging at 110°C, making them suitable as a gel system for enhanced oil recovery.

References

1. Zhu, Daoyi, Baojun Bai, and Jirui Hou. Polymer gel systems for water management in high-temperature petroleum reservoirs: a chemical review. Energy & fuels 31.12 (2017): 13063-13087.
2. Hou X., Li L., Li B., et al. Research on water control technology for flow potential adjustment in seam-hole type reservoirs. Xinjiang Oil and Gas,16.01(2020): 65-71. (in Chinese)
3. C. Norman; B. Turner; J. L. Romero, et al. A review of over 100 polymer gel injection well conformance treatments in Argentina and Venezuela: Design, Field implementation, and Evaluation. International Oil Conference and Exhibition in Mexico. Cancun, Mexico, August 2006.
4. Bai, Baojun, Jia Zhou, and Mingfei Yin. A comprehensive review of polyacrylamide polymer gels for conformance control. Petroleum exploration and development 42.4 (2015): 525-532.
5. Vasquez, Julio, and Yoann Santin. Organically crosslinked polymer sealant for near-wellbore applications and casing integrity issues: successful wellbore interventions. SPE North Africa Technical Conference and Exhibition. Cairo, Egypt, September 2015.
6. José Luis Juárez; María Rosario Rodriguez; Jesús Montes, et al. Conformance gel design for high-temperature reservoirs. SPE Europec. Virtual, December 2020.
7. Sydansk, Robert D. A newly developed chromium (III) gel technology. SPE Reservoir Engineering 5.03 (1990): 346-352.

Research on foaming agents for gas flooding in medium-temperature and high-salinity clastic reservoirs

Kexin Li, Jijiang Ge, Longjie Li, Jiasu Li and Pengju Chu

School of China University of Petroleum (East China), Qingdao 266400, China.

Abstract. Anionic surfactants are commonly used as foaming agents in foam-enhanced oil recovery, but their performance is seriously affected by high temperature and high salinity environment. However, there are not many studies on the adsorption pattern and salinity resistance performance of anionic surfactants on solid surfaces. This study evaluated the foaming performance of several anionic surfactants suitable for high-temperature, high-salinity salinity reservoirs. It was found that α-olefin sulfonate (AOS) showed good foaming performance under high temperature and high salt condition. However, the solubility of the foaming agent was low in brine with a salinity of 11×104 mg·L-1. Therefore, a co-solvent (ABS), which is a strong hydrophilic alkyl benzene sulfonate, was chosen to be compounded with AOS in this study. In this study, a foaming agent with excellent foaming performance and solubility at a temperature of 90°C and a salinity of 11×104 mg·L-1 was constructed. The adsorption of the foaming agent was less than 0.3mg/g on the surface of quartz sand, and its foaming rate and foam decay half-life after three adsorptions maintained more than 85% of the original performance. The results of the study can guide the selection of foaming agents for gas injection and mobility control in medium temperature and high salinity clastic reservoirs.

Keywords: Gas injection; EOR; foaming agent; α-olefin sulfonate; high salinity reservoir

1. Introduction

A clastic reservoir in western China, where the reservoir temperature is between 70-90°C and the reservoir water salt content is between 7-11×10^4 mg·L^{-1}, has adopted a development approach of fluid injection followed by gas injection. The key to implementing this development approach is the development of a temperature and salinity resistant foaming agent. At present, there are two main types of foamers with good temperature and salinity resistance. One type is the foaming agent based on amphoteric surfactants. These surfactants are excellent in solubility and foaming ability, but they exhibit high adsorption on the surface of sandstone [1]. The second category is anionic surfactants. AOS is a more commonly used temperature and salinity resistant foaming agent. The famous foreign foaming agent Chaser CD 1045[2] is mainly composed of AOS, but experimental evaluation showed that AOS was insoluble in water with a salinity of nearly 11×10^4 mg·L^{-1}. In this paper, the problem of AOS dissolution in high salinity water was solved by the addition of ABS. A salinity-resistant foaming agent based on AOS was constructed.

2. Experimental section

2.1 Materials and apparatus

The industrial surfactants used in this study were carboxymethyl polyoxyethylene alkyl ether (AEC-7, AEC-9, where the number indicates the number of oxyethylene links), sodium fatty alcohol polyoxyethylene ether sulfate (AES0810, where the number of oxyethylene links is about 4), α-olefin sulfonate (AOS) and strong hydrophilic alkyl benzene sulfonate (ABS). In addition, quartz sand with a particle size of 80-100 mesh was used in this paper. The instruments used in the study were a Waring mixer purchased from Qingdao Shengji Instrument Systems Co. and an SP-754 UV-Vis spectrophotometer purchased from Shanghai Spectrum Instruments Co.

The simulated formation water used in the research showed a salinity of 11×10^4 mg·L^{-1} and was formulated according to Table 1.

Table 1 Ion compositions in simulated formation water

Ionic composition/(mg·L^{-1})					Total salinity/ (mg·L^{-1})
Na^+	Ca^{2+}	Mg^{2+}	HCO_3^-	Cl^-	111901.5
36649.2	5636.3	759.4	91.8	68764.8	

2.2 Experimental methods

2.2.1 Bulk foam test of foaming agents

The experiments used simulated formation water to prepare the foaming agent solution. The experiment first preheated 100 mL of the solution to 90°C, then poured it into the steel cup of the Waring tissue masher, then used the masher to stir at 3000 r/min for 1 min, and finally quickly poured the formed dispersion system into a measuring cylinder placed in a 90°C water bath. In addition, the experiment recorded the changes in foam volume and drainage liquid volume with time in real-time. The foaming rate ψ (equal to [V/100] ×100) indicates the foaming capacity of the foaming agent, and $t_{1/2}$ indicates the stability of the foam.

2.2.2 Solubility evaluation of foaming agents

A spectrophotometer was used to test the absorbance of the standard solution at 680 nm, and the standard curve was plotted based on the absorbance. The surfactant solutions were prepared using simulated formation water, and the solutions were treated at 30°C and 90°C and the absorbance of the solutions was measured. The turbidity of the solutions was found on the standard curve. The solution's preparation method and determination procedure can be referred to as GB13200-91[3].

2.2.3 Determination of the adsorption of foaming agents

The quartz sand was acid washed firstly, then washed and dried. The surfactant solution was prepared using simulated formation water. First, 18g of the prepared solution and 6g of quartz sand were drawn into an ampoule and sealed with an alcohol torch. The sealed ampoules were placed in a constant temperature oven at 90°C for heat treatment, and then the ampoules were removed periodically, and finally, the solutions were carefully removed after cooling and evaluated for content determination and foaming performance.

A two-phase titration method was used for the determination of surfactant content. The indicator used in the two-phase titration method is an imported acidic mixed indicator, and its preparation method and titration procedure can be referred to as GB5173-1995[4].

The sample content is calculated according to equation 1.

$$X = \frac{C \times V \times M_r}{m \times \frac{25}{1000} \times 10} \quad (1)$$

Where X is the content of anionic active substance of the sample (%), C is the actual concentration of cetyltrimethylammonium bromide standard solution (mol/L), V is the volume of cetyltrimethylammonium bromide standard solution consumed in titration(mL), M_r is the molar mass of the sample (g/mol), m is the mass of the sample(g).

The sample adsorption is calculated according to equation 2.

$$A = \frac{(w_1 - w_2) \times m_l}{m_s} \quad (2)$$

Where A is the sample adsorption (mg/g), w_1 w_2 is the content of the anionic active substance in the sample before and after adsorption (%), m_l is the mass of surfactant solution added to the ampoule(g), m_s is the mass of quartz sand added to the ampoule(g).

3. Results and Discussion

3.1 Screening of anionic foaming agents

The current foaming agents with good solubility in high salinity water are mainly anionic and betaine types. Betaine surfactantshave good foaming performance, but they exhibit high adsorption on the surface of sandstone [5]. Therefore, in this paper, the foaming performance of anionic foaming agents was mainly evaluated. The evaluation results are shown in Fig. 1.

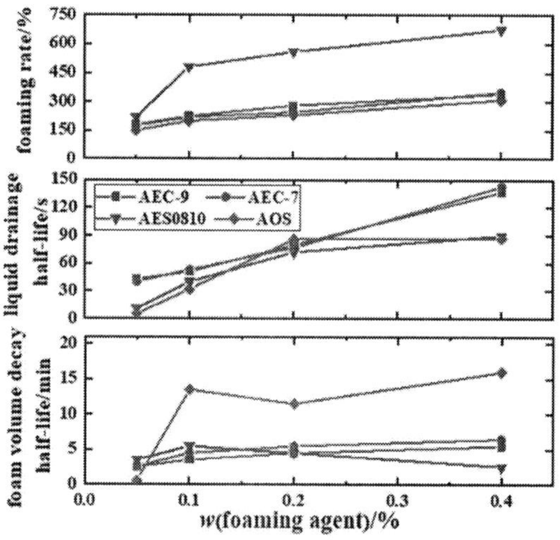

Fig. 1 Evaluation of the foaming performance of different anionic surfactants

It can be seen from Fig. 1 that the foaming rate and half-life of AEC are poor. In addition, AES0810 shows a high foaming rate but a short half-life, while AOS is poor in solubility but exhibits a high foaming rate and half-life. The carbon chain length of AES0810 is mainly 8, and that of AOS is mainly 16-18. Generally speaking, surfactants with short carbon chains are capable of rapid diffusion to the gas-liquid interface. They can easily generate foam.

https://doi.org/10.1051/e3sconf/202337501048

However, their carbon chain interactions are very weak, so they are poorly stable in foaming. In contrast, surfactants with long carbon chains exhibit slow diffusion rates and strong foam stabilization behavior.

The poor solubility of AOS in simulated brine is due to the formation of reversed micelle between AOS and Ca^{2+} (See Fig.2).

3.2 Evaluation of AOS and co-solvent composite system

ABS can form mixed micelles with AOS agents, (See Fig.2) which effectively resists the electrostatic effect of calcium and magnesium ions on the hydrophilic groups in the micelles. Thus the surfactant solubility in simulated formation water is improved.

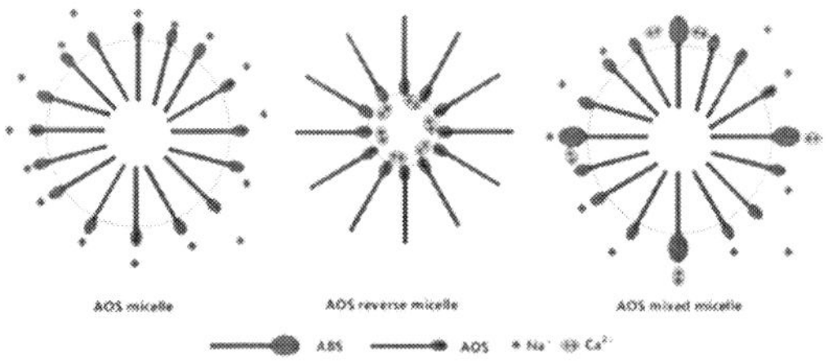

Fig. 2 Micelle and reverse micelle formed by AOS and mixed micelle formed by AOS and ABS

In this paper, AOS and ABS were mixed in a 5:1 ratio to form a system called FM51. FM11, FM31, FM51, and FM61 were composed similarly. The turbidity of the different solutions is shown in Fig. 3. From Fig. 3, it can be seen that the composite system is well soluble at 90°C when the ratio of AOS and ABS is less than 5:1. The dissolution states of AOS and FM51 at different temperatures are shown in Fig. 4.

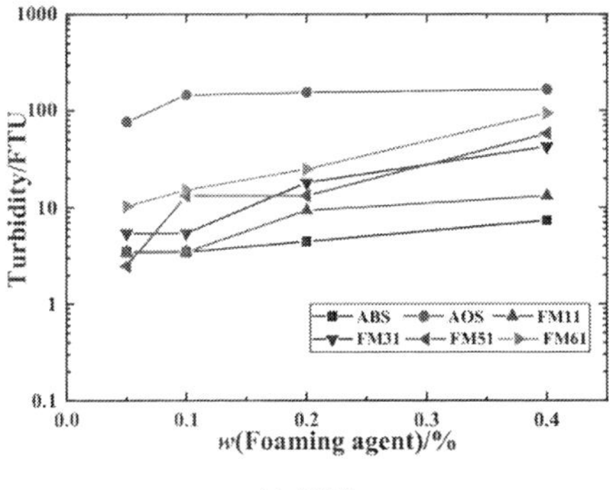

(a) 30°C

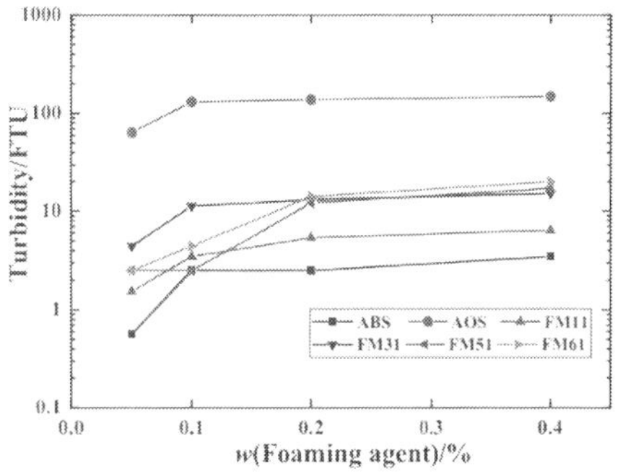

(b) 90°C

Fig. 3 Turbidity of AOS, ABS, and composite systems

0.05%AOS 0.1%AOS 0.2%AOS 0.4%AOS 0.05%FM51 0.1%FM51 0.2%FM51 0.4%FM51

(a) 30°C

0.05%AOS 0.1%AOS 0.2%AOS 0.4%AOS 0.05%FM51 0.1%FM51 0.2%FM51 0.4%FM51

(b) 90°C

Fig. 4 Dissolution state of AOS and FM51 at different temperature conditions

The foaming performance of the solution was further measured in this paper, and the results are shown in Fig. 5. It can be seen that high ABS content weakens the foaming performance of the system. so FM51 was selected as the foaming agent in the subsequent study.

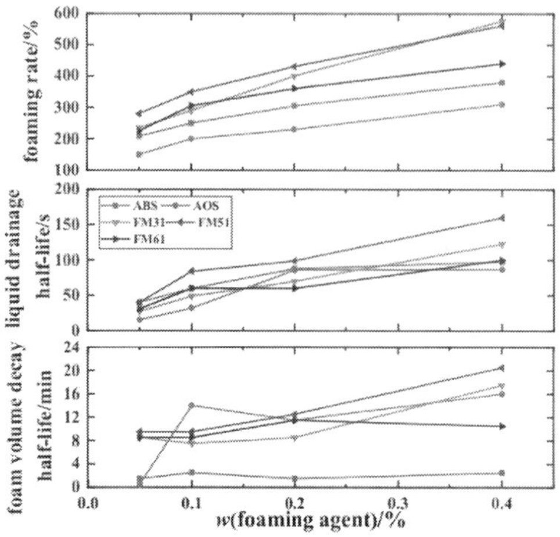

Fig. 5 Foaming performance of different composite systems

3.3 FM51 adsorption performance measurement

Surfactants with strong adsorption have higher losses in the formation process. Zhou Jimin suggested that the static adsorption of the foaming agent should be at least 1mg/g or less [6]. In this paper, the adsorption of FM51 on the surface of quartz sand was measured. The results are shown in Fig. 6.

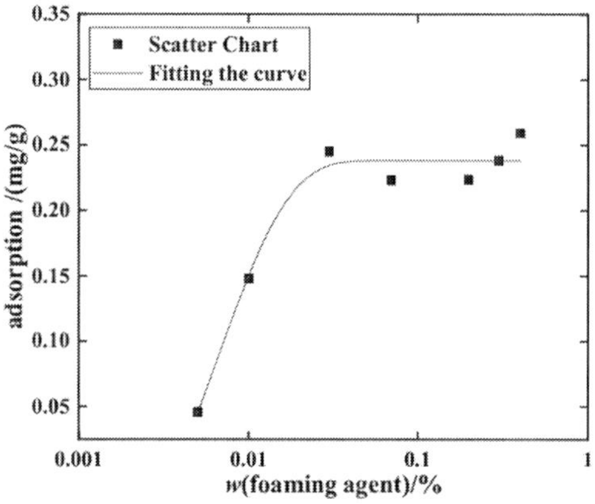

Fig. 6. Adsorption of foaming agent on the surface of quartz sand

It can be seen from Figure 6 that the static adsorption of FM51 is less than 0.3mg/g. FM51 is a complex anionic surfactant, which is negatively charged after ionization in water. It repels each other with the quartz minerals which are negatively charged overall. Therefore, the adsorption of FM51 is low.

3.4 Evaluation of the foaming performance of FM51 after adsorption

Evaluation of foaming performance after adsorption is an important part of the screening of foaming agents [7]. In this paper, the foaming performance of FM51 before and after adsorption was compared. The results are shown in Fig. 7. Although the foaming performance of FM51 decreases after three adsorptions, it can maintain more than 85% of the original performance.

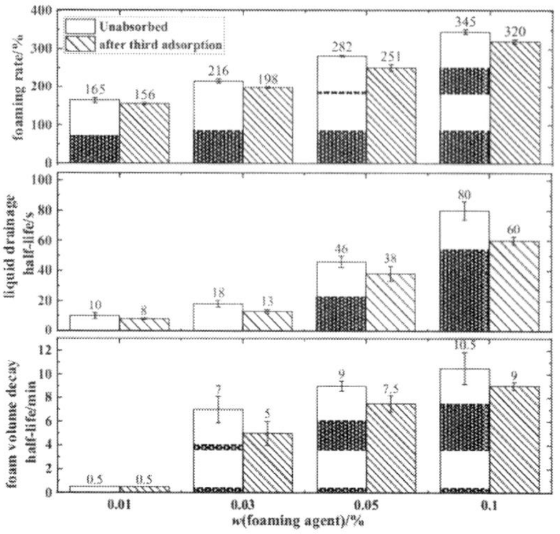

Fig. 7 Foaming performance of FM51 before and after adsorption

4. Summary

(1) At the temperature of 90°C and salinity of 11×10^4 $mg\cdot L^{-1}$, the foaming performance and half-life of AEC are worse, the foaming rate of AES0810 is higher and the half-life is shorter, while the solubility of AOS is worse, but it shows higher foaming rate and longer half-life.

(2) The compounding of AOS and ABS can improve the dissolution state of AOS, but the high content of ABS will affect the foaming performance of the system, so the system FM51 is screened as the foaming agent.

(3) The adsorption of FM51 is low, the adsorption is less than 0.3 mg/g. FM51 still keeps good foaming performance after third adsorption, and basically can keep more than 85% of the original performance.

References

1. Mannhardt K, Novosad J J, Jha K N N. Adsorption of foam-forming surfactants in berea sandstone. Journal of Canadian Petroleum Technolog. Vol. 33 (1994) No. 2, p. 34-43.
2. Prajapati R R, Bhagwat S S. Effect of Foam Boosters on Krafft Temperature. Journal of Chemical & Engineering Data. Vol. 57 (2012) No. 3, p. 869–874.
3. National Technical Supervision Agency. China Patent GB 13200-91(2000).
4. National Technical Supervision Agency. China Patent GB5173-1995(1995).
5. Liu Yunfei, Liu Qiao, Diao Guangzhi, et al. Sensitivity analysis and optimization of nitrogen-assisted steam drive injection and extraction parameters for heavy reservoirs. Xinjiang Oil and Gas. Vol. 17 (2021) No. 1, p. 55-59.

6. Zhou J, Srivastava M, Hahn R, et al. Evaluation of an Amphoteric Surfactant for CO2 Foam Applications: A Comparative Study. SPE Improved Oil Recovery Conference. USA, 2020, p. 50-53.
7. Scamehorn J F, Schechter R S, Wade W H. Adsorption of surfactants on mineral oxide surfaces from aqueous solutions: II: Binary mixtures of anionic surfactants. Journal of Colloid and Interface Science. Vol. 85(1982) No. 2, p. 479-493.

Development of high strength polyvinyl alcohol gel for water shutoff in carbonate reservoirs

Jin Chen, Jijiang Ge, Wenhui Wang, Pengfei Chen and Weixiong Chen

School of Petroleum Engineering, China University of Petroleum (East China), Qingdao 266000, China

Abstract. To provide a high-strength, non-resin organic plugging agent for water shutoff in carbonate fracture and cave reservoirs, a high-strength gel was prepared with polyvinyl alcohol 1788 as gelant and p-phthalaldehyde as crosslinker in this study. By selecting latent acids rather than strong acid such as hydrochloric acid and sulfuric acid as catalysts, and by adding clay and sulphonated phenolic resin as stabilisers to the gelation, the drawbacks of short gelation time of polyvinyl alcohol and high dehydration rate of the formed gel soaking in brine were overcame. The evaluation results showed that the gellant composed of 7wt% polyvinyl alcohol, 4wt% clay, 0.3wt% aldehyde crosslinker, 0.25wt% latent acid, 1.5wt% sulfonated phenolic resin, and 0.3wt% hexamethylenetetramine had a gelation time of 2 h at 130 °C. The storage modulus of the gel was up to 2000 Pa, and the dehydration rate of the gel after soaking in synthetic brine for 30 days was less than 15%. Furthermore, the breakthrough pressure gradient in a slim tube with a diameter of 3mm reached 15 MPa / m, which indicated that the gel is a temperature and salinity resistant plugging agent with strong plugging ability.

Keywords: Water Plugging; Crosslinked polyvinyl alcohol; latent acid

1. Introduction

The Taher carbonate reservoir is a bottom water-rich reservoir [1,2]. The rising bottom water causes excessive water production with oilfield development [3]. Due to the high formation temperature (above 110 °C) and high salinity of formation water (Salt content 22×10^4 mg/L, calcium and magnesium ion content 2×10^4 mg/L), bottom water treatment is difficult. Recently, great progress has been reported about temperature-resistant and salinity-resistant gel composed of AM-AMPS copolymer with a molecular weight over 6 million [4,5]. Polymer gel is injected into formation first and then the formation is sealed with water glass gel to prevent the polymer gel from displacing . This treatment has a better water plugging effect. However, field trials found that after two rounds of water plugging, the well fluid production dropped significantly. It is supposed that the high concentration of water glass gel caused permanent damage to the formation. So a non-resin organic high-strength plugging system is required to replace the high-strength water glass gel. In this paper, a high-strength polymer gel with good stability in high-temperature and high-salinity environments was developed.

2. Experimental section

2.1 Materials

The materials used in the experiment include Polyvinyl alcohol 1788 (industrial product, Chengdu Huamao Co., Ltd.), para-phthalaldehyde (Shanghai Aladdin Reagent Co., Ltd.), clay (Dinglong Mining), sulfonated phenolic resin (industrial product, Davey Petroleum Technology Co., Ltd.), latent acid catalyst (industrial product, Qingdao Qucheng Technology Co., Ltd.).

2.2 Methods

The storage modulus of the gel was measured using a MCR-92 rheometer and the dehydration rate of the gel was calculated according to equation (1).

$$A = \frac{m_0 - m_i}{m_0} \times 100\% \qquad (1)$$

Where: A is the dehydration rate of gel after high temperature heat treatment; m0 is the initial mass of gel; mi is the mass of different time periods of heat treatment.

3. Results and Discussion

3.1 Development of high-strength polyvinyl alcohol gel

3.1.1 Screening of cross-linkers

Aldehydes that can undergo acetal reactions with polyvinyl alcohol include formaldehyde, glyoxal, glutaraldehyde, terephthalaldehyde, and o-phthalaldehyde. Formaldehyde is highly toxic, and glyoxal and glutaraldehyde are easily volatile, and o-terephthalaldehyde and terephthalaldehyde have low volatility. Considering the low toxicity and slightly lower cost of terephthalaldehyde, this paper selected terephthalaldehyde as the cross-linking agent.

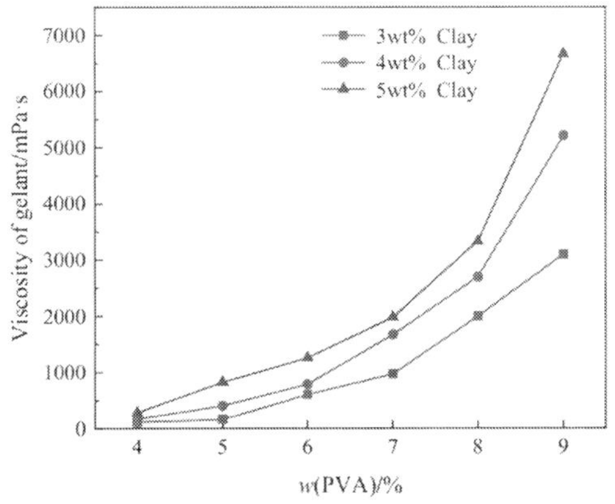

Fig. 1 Effect of polyvinyl alcohol and clay content on viscosity of gum-forming solution

Since the viscosity of the gellant increases with the polyvinyl alcohol content (as shown in Fig. 1). The viscosity of the gellant with 8wt% polyvinyl alcohol (also containing 0.2wt% crosslinker and 4wt% clay) up to 2696 mPa·s, which is close to the upper limit of the pumped fluid, so it is recommended that the amount of polyvinyl alcohol in the gellant is 4wt% to 8wt%. For terephthalaldehyde, the strength of the gel also increases with the amount of terephthalaldehyde. However, the over cross-linking reaction can reduce the hydrophilicity of the gel, and the dehydration rate of the gel in synthetic brine increases with terephthalaldehyde, so the amount of cross-linking agent is recommended to be 0.2wt% to 0.5wt%.

3.1.2 Screening of catalysts

To ensure the successful injection of plugging agent into the formation, the gellant should have a long gelation time. The above experiment used 0.2wt% mass fraction of hydrochloric acid as the catalyst, and the gelation time is shorter under high temperature, thus it is tricky to meet the construction injection requirements. To reduce the cross-linking rate of polyvinyl alcohol, latent acid is considered as the catalyst. Latent acid can be hydrolyzed in water to produce H^+, thus delaying the acetal reaction. The pH values of different concentrations of Latent acid and hydrochloric acid was shown in Fig. 2.

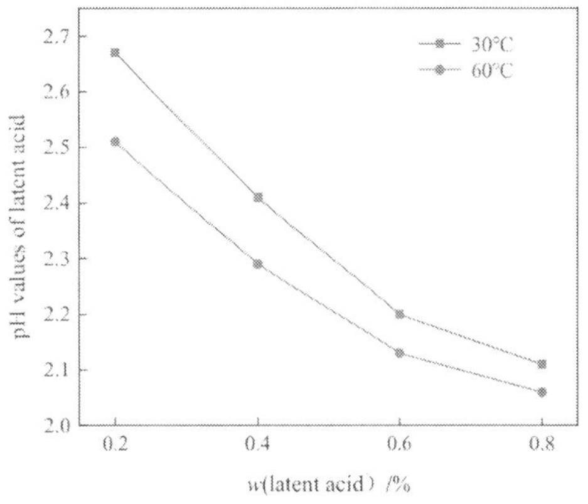

(a) pH values of latent acids

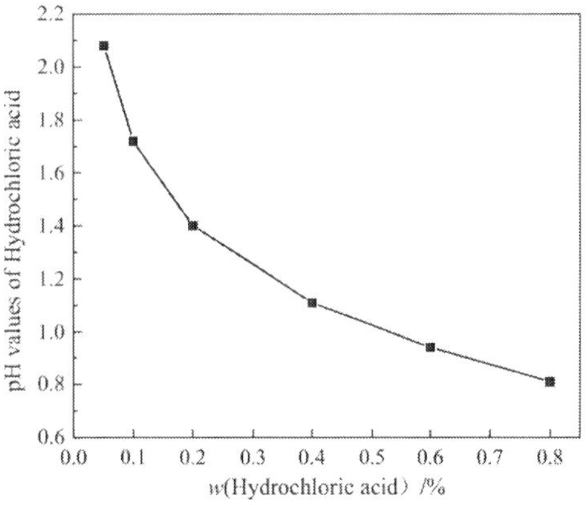

(b) pH values of hydrochloric acid

Fig. 2 pH values of different acid solutions

The gellant was formulated according to Table 1. The gelation time of the gel containing 4wt% clay, 7wt% polyvinyl alcohol, 0.3wt% crosslinker, and 0.2wt%-0.25wt% Latent acid of the catalyst can reach more than 2 h at 130°C, which can meet the requirement of near-well injection (more than 60 min).

Table 1 Formulation table for different catalyst contents

w(Clay) /%	w(PVA) /%	w(Crosslinker)/%	w(Latent acid) /%	Gelation time/h
4	7	0.3	0.20	4.0
4	7	0.3	0.25	2.5
4	7	0.3	0.30	1.5
4	7	0.3	0.35	1.0

https://doi.org/10.1051/e3sconf/202337501049

3.1.3 Screening of stabilizers

It is proposed to improve the stability of cross-linked polyvinyl alcohol by adding clay and introducing salt-resistant groups.

(1) Addition of clay. Clay is a material commonly used in hydrogels to improve the strength and water holding capacity. The cross-linked polyvinyl alcohols with different clay loading were prepared by fixing the amount of cross-linker at 0.3wt% and catalyst at 0.2wt%. The dehydration rates of these cross-linked polyvinyl alcohols aged in situ at 130 °C and immersed in synthetic brine with different aging times were shown in Fig. 3.

It can be seen that the dehydration rate of the cross-linked polyvinyl alcohol formed by adding clay to the gellant is significantly lower. The higher the content of clay and polyvinyl alcohol in the gellant, the lower the dehydration rate of cross-linked polyvinyl alcohol gel formed. Since the high viscosity (see Fig. 1) and the difficulty of pumping when the content of clay and polyvinyl alcohol in the gellant is too high, the amount of clay used is finally determined to be 4wt%.

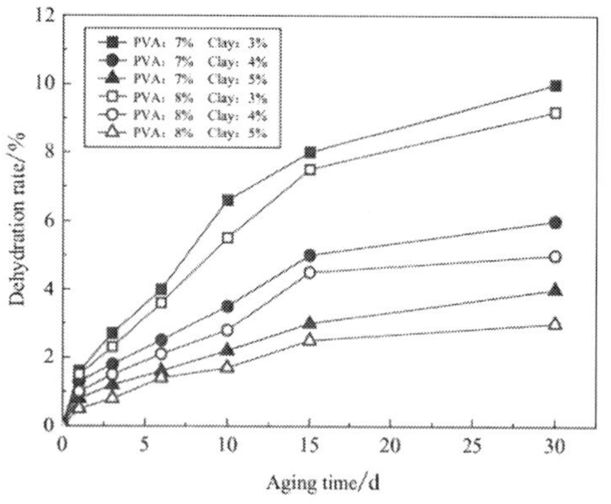

(a) Aging in ampoules

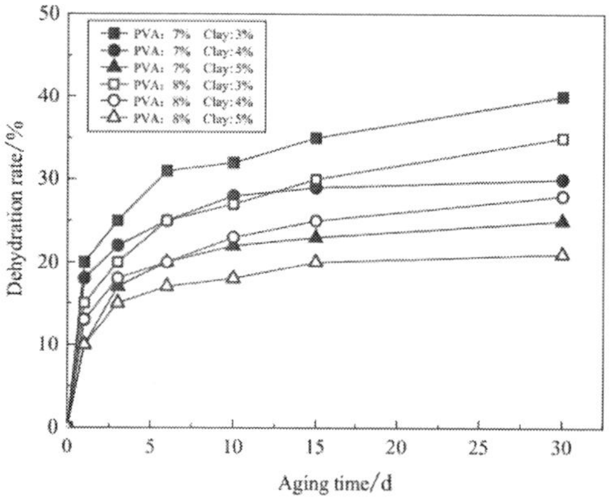

(b) Aging in simulated formation brine

Fig. 3 Effect of clay on the stability of polyvinyl alcohol jelly

(2) Addition of sulfonated phenolic resin and hexamethylenetetramine. Whereas the good salt resistance of the sulfonated phenolic resin, the temperature and salt resistance of the compound cross-linked polyvinyl alcohol is improved by adding sulfonated phenolic resin and hexamethylene tetramine.

Under the condition of the 6wt% polyvinyl alcohol, 4wt%clay, 0.3wt%cross-linker, 0.2wt%catalyst, less than 2wt% sulfonated phenolic resin loading and sulfonated phenolic resin to hexamethylenetetramine in a 5:1 mass ratio, The stability of the prepared cross-linked polymer in synthetic brine was shown in Table 2.

Table 2 Stability of cross-linked polymers immersed in simulated formation water with different sulfonated phenolic resins, hexamethylenetetramine and catalyst contents

Serial number	*w* (Hexamethylenetetramine) /%	*w* (Sulfonated phenolic resins) /%	Dehydration rate after 10 days of aging/ %	Dehydration rate after 30 days of aging/ %
1	0.1	0.5	15	20
2	0.2	1.0	12	18
3	0.3	1.5	9	13
4	0.4	2.0	8	11

As can be seen from Table 2, the dehydration rate of the produced particles under high temperature and high salt conditions decreased significantly with the increase of sulfonated phenolic resin and hexamethylenetetramine, indicating that the introduction of sulfonated phenolic resin and hexamethylenetetramine improved the temperature and salt resistance of the cross-linked polymers. After the cross-linked polymer gel particles with 1.5wt% or 2wt% sulfonated phenolic resin loading in synthetic water aged for 30 days, the dehydration rate of the two particles was close. Considering the production cost, the mass fraction of sulfonated phenolic resin could be determined as 1.5wt% and the mass fraction of hexamethylenetetramine as 0.3wt%.

3.2 Evaluation of high-strength polyvinyl alcohol gel

3.2.1 Evaluation of high-temperature stability

Firstly, 100 g synthetic formation brine was added to the homemade high-temperature tank. Secondly, 50 g gellant was slowly injected. Then, the tank was sealed and de-oxygenation with nitrogen. Lastly, placed the sealed tank in an oven at 130 °C for aging. The dehydration rate and storage modulus of the gel were measured at different heat treatment times to evaluate the long-term stability of the gel under high temperature and high salt conditions. The results are shown in Figure 4 and Fig. 5.

1d 3d 6d 10d 30d

Fig. 4 Diagram of the state of gel at different times of heat treatment

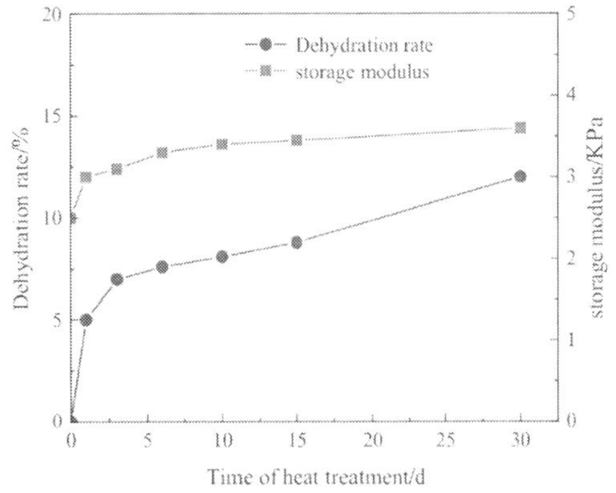

Fig. 5 Effect of heat treatment time on the dehydration rate and storage modulus of gel

The gel is formed in the upper area of the tank and the gel can still keep its shape intact after 30 days under high temperature and high salinity conditions. After 30 days of heat treatment, the storage modulus of the cross-linked polyvinyl alcohol is over 2000 Pa (much greater than the storage modulus of general strong gel) and dehydration rate is less than 15%, exihibiting as a stable high strength gel.

3.2.2 Sealing ability test of high strength cross-linked polyvinyl alcohol

In this paper, we used a slim steel tube with a diameter of 3 mm to simulate fractures and then compared the plugging capacity of two gels which composed of 7wt% polyvinyl alcohol, 4wt% clay, 0.3wt% aldehyde cross-linker, 0.25wt% latent acid, and 1.5wt% sulfonated phenolic resin, 0.3wt% hexamethylenetetramine and 1.0wt% AM-AMPS, 0.2wt% hydroquinone, 0.2wt% hexamethylenetetramine respectively. The former was formulated with tap water and the latter was formulated with synthetic brine. Both of two were aging at 130 °C for 5 days, and then the breakthrough pressure was measured at room temperature with a water injection rate of 64 m/d, and the results are shown in Fig. 6.

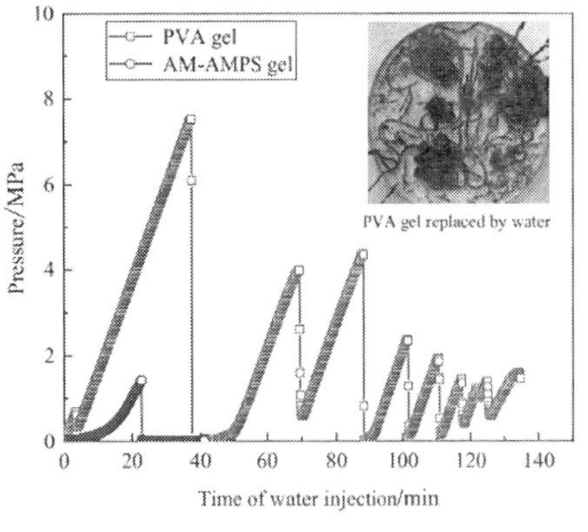

Fig. 6 Breakthrough pressure of AM-AMPS gel and PVA gel in a fine tube

It can be seen that the breakthrough pressure of cross-linked polyvinyl alcohol can reach 15 MPa/m, which is far higher than the general temperature and salt resistant gel. In addition, after the cross-linked polyvinyl alcohol was breached, the polymer in the slim tube can also re-form a blockage, resulting in repeated fluctuations in injection pressure over time.

4. Conclusion

Using latent acid as a catalyst can substantially prolong the gelation time of polyvinyl alcohol and terephthalaldehyde; Adding a certain amount of clay and sulfonated phenolic resin when preparing cross-linked polyvinyl alcohol can improve the stability of the gel system in synthetic brine; The gel composed of 7wt% polyvinyl alcohol, 4wt% clay, 0.3wt% aldehyde cross-linker, 0.25wt% latent acid, 1.5wt% sulfonated phenolic resin and 0.3wt% hexamethylenetetramine has a gelation time of 2 h at 130 °C, and a storage modulus of 2000 Pa, a dehydration rate less than 15% after 30 days of immersion in synthetic brine, and a breakthrough pressure gradient in the simulated slim tube of 3 mm diameter up to 15 MPa/ m.

References

1. Liu Lili, Chen Xiaofan, Cui Ligong, Analysis of the main factors influencing the water content variation in the slit-hole carbonate reservoir of Tahe oilfield, J. Xinjiang Oil and Gas, 2008, 4(1):3.
2. Ding YZ, Liu Q, Zhao S, Analysis of water injection effect of northern seam-hole type reservoir unit in the tenth district of Tahe oilfield, J. Xinjiang Oil and Gas, 2017, 13(1):4.
3. Xiao Chunyan, Li Wei, Xiao Shuping, Mechanisms and water uplift in marginal bottom water reservoirs, J. Fracture Block Oil and Gas Field, 2009(6):3.
4. Guo Hongbin, Ge Jijiang, Xu Ya, Preparation and Mechanism of Stability for High-Temperature and High-Salinity Gels, J. Society of Petroleum Engineers, 2022, SPE-209823-PA.
5. Tian-Ci Zhang, Ji-Jiang Ge, Hao Wu, Effect of AMPS(2-acrylamido-2-methylpropane sulfonic acid) content on the properties of polymer gels, J. Journal of Petroleum Science, 2022, 19(2):697-706.

Study on the influencing factors of productivity under pressure drive development in low permeability reservoir

Chuanzhi Cui, Guobiao Wang and Shuiqingshan Lu*

College of Petroleum Engineering, China University of Petroleum (East China), Qingdao, China

Abstract: The productivity of low-permeability reservoirs is affected by multiple factors. After the development of water injection by pressure drive, it will also be affected by factors such as construction, resulting in large differences in the productivity of different oil wells. According to reservoir engineering knowledge, the factors affecting the productivity of pressure drive water injection development are preliminarily determined. The conceptual model of pressure drive water injection development of low permeability reservoir is established by numerical simulation method. Combined with orthogonal design, the main control factors of pressure drive water injection development productivity are determined. The main control factors include injection production well spacing, water injection volume of pressure drive measures, water content before pressure drive, porosity, and quantitative analysis of the main control factors. Through correct analysis and evaluation of these factors, it provides guiding significance for the productivity prediction of pressure drive water injection development in low permeability reservoirs and the optimization of pressure drive water injection scheme.

Keywords: Low permeability reservoir; Pressure drive; Numerical simulation; Influencing factor

1. Introduction

The development technology of pressure drive water injection is a technology to realize the combination of 'fracturing-infiltration-flooding' to improve the recovery rate. The main principle is to form cracks through fracturing and quickly send the oil displacement agent to the enrichment site of remaining oil through cracks. The oil displacement agent is quickly filled into the pores, reducing the contact time and contact distance between the chemical agent and the formation and improving the oil displacement efficiency. At the same time, by injecting a large amount of oil displacement fluid, the formation energy can be replenished in a short time, the effect of energy storage before pressure can be achieved, the formation pressure can be improved and maintained, and the stable production period can be prolonged.

At present, pressure drive water injection development technology has been applied on a large scale in low permeability reservoirs. With respect to the pressure drive water injection development technology, current scholars' research directions mainly focus on the design optimization of pressure drive water injection volume and the dynamic propagation of fractures. Xu [1] used reservoir engineering methods to analyze the pressure distribution characteristics of the injection end and production end and defined the pressure drive development mechanism of low permeability reservoirs. A water injection design method for a pressure drive well is presented. Huang [2] optimized the pressure drive mechanism, the pressure drive well selection, pressure drive injection volume, injection displacement, and filling time based on field practice. Wu [3] studied the fracture propagation law of pressure drive water injection technology and optimized the water injection volume and water injection pressure difference combined with numerical simulation. Considering that the productivity of oil wells after the pressure drive is very different, and the productivity after the pressure drive is affected by multiple factors such as fluid, reservoir, well pattern, and construction [4-5], there is no mature productivity prediction method at present. There is a lack of research on the influencing factors of productivity under pressure drive development in low permeability reservoirs. Based on the productivity formula of vertically fractured wells in low permeability reservoirs and from the perspective of the integration of injection well and production well, the influencing factors related to pressure drive development are screened comprehensively. The conceptual model of pressure drive for low permeability reservoir development is established based on the actual model of pressure drive development provided by the field and the field data. The influencing factors of productivity under pressure drive development are considered in various aspects [6]. An orthogonal test determined the main controlling factors of productivity under pressure drive, and then the main controlling factors were quantitatively analyzed.

* Corresponding author: 709662647@qq.com

2. Screening of productivity factors in pressure drive development

As a new type of enhanced oil recovery technology, there is currently no relatively accurate oil well productivity formula. Through consulting the data and according to the productivity formula (1) of vertical fracturing well in low permeability reservoirs [7]:

$$Q=\frac{2\pi Kh(p_e-p_w)}{\mu B\left[\xi_e+\frac{\pi h_d}{2x_f}\left(\frac{K}{K_d}-1\right)\right]}\times\left[1-\frac{2\lambda x_f\sinh\xi_e+\pi h_d(\lambda_d-\lambda)}{\pi(p_e-p_w)}\right] \quad (1)$$

In the Formula: x_f ——Fracture half-length, m

h_d ——Thickness of contamination zone around the fracture, m

K ——Permeability of reservoir before contamination, μm²

K_d ——Contaminant zone permeability, μm²

λ ——Start-up pressure gradient of the reservoir before pollution, MPa/m

λ_d ——Contaminant zone starting pressure gradient, MPa/m

h ——Thickness of reservoir, m

μ ——Fluid viscosity, mPa·s

B ——Volumetric factor

p_e ——Formation pressure, MPa

p_w ——Bottom hole pressure, MPa

According to the productivity formula of vertically fractured wells in low permeability reservoirs, the productivity after pressure drive is related to fracture properties[8], permeability, fluid viscosity, reservoir thickness, porosity, and bottom hole pressure. The pressure drive water injection, well spacing, and pressure drive timing (water cut before pressure drive) under pressure drive development also greatly influence the productivity after the pressure drive. Finally, the influencing factors related to pressure drive water development are as follows: water cut before pressure drive, well spacing, water injection, permeability, oil viscosity, effective thickness, and porosity. Through field data observation, due to the mutual interference among various influencing factors, there is no strong correlation with productivity after pressure drive, and there is no obvious linear relationship or functional relationship. Numerical simulation software will be used for specific analysis later.

3. Numerical simulation analysis of Influencing factors of pressure drive productivity

3.1 Conceptual model of pressure drive development

According to the actual model of Block A provided by the mine and the corresponding actual production data, a pressure drive well group in the model is selected, and the water injection well is subjected to 20 days of pressure drive water injection. The total water injection volume is 8000 m³ (after pressure drive water injection, the production well is effective, and the production is improved). Due to the short-term high-pressure water injection, the pressure near the water injection well is high, and the fracture is generated and gradually expanded. In the actual block model, the pressure drive characteristics are characterized by adding cracks between the injection well and the production well[9] (the pressure drive water injection time is 20 days, which is divided into 10-time steps, and the software restart function is used to simulate the gradual expansion of cracks). Finally, the crack area with a length of 200 m, a width of 30 m, and a height throughout the whole model is gradually formed. As shown in Figure 1, different pressure drive water injection forms different cracks, and different cracks correspond to different conductivity. The conductivity of the crack is affected by changing the permeability of the crack, which corresponds to different pressure drive water injection.

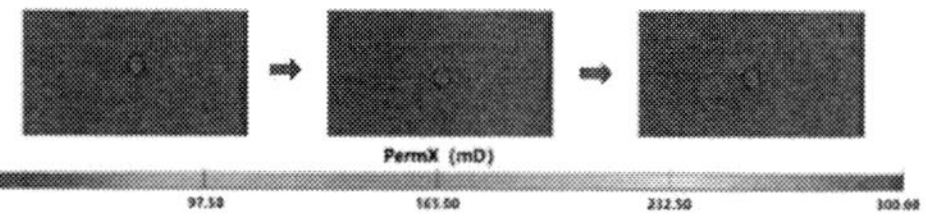

Fig 1. Gradual extension of fractures around the injection well

After the pressure drive measures are taken, a period of soaking is carried out, and constant pressure development is carried out. The production data of the model and the actual data of the block are fitted by line (changing the permeability of the added cracks). After the fitting is completed (the error between the actual production data and the model production data is within 5 %), it is obtained in the actual block model: when the pressure drive water injection is 8000 m³, the fracture permeability is 300 mD. If the injection volume increases or decreases, the fracture permeability will also increase or decrease proportionally[10]. This result is applied to the following conceptual model of pressure drive water injection development.

Based on the actual model of Block A, a conceptual model of pressure drive development is established. The same reservoir depth, initial formation pressure, initial oil saturation, PVT parameters, and relative permeability curves are used. The number of model grids is 41 * 41 * 10. The size of a single grid in the X and Y directions is 10 m, the size of the grid in the Z direction is 4 m, the buried depth of the reservoir top surface is 2900 m, the initial pressure is 29 Mpa, and the initial oil saturation is 0.6. The model is a homogeneous model with four

https://doi.org/10.1051/e3sconf/202337501050

production wells (W1, W2, W3, W4) and one injection well (Z1), which is arranged in a five-point well pattern. The model is shown in Figure 2.

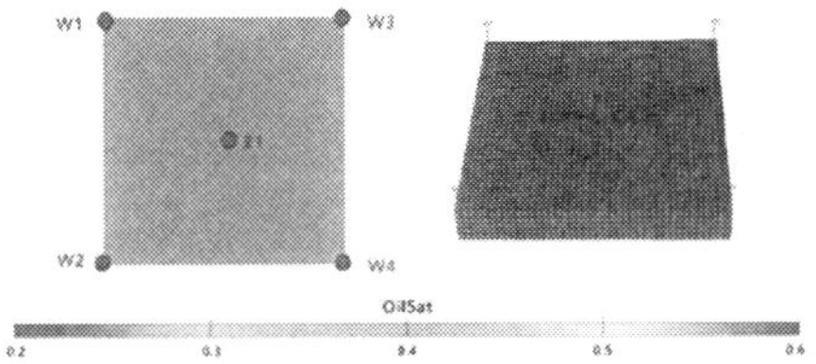

Fig 2. Initial oil saturation of model

3.2 A multi-factor study on influencing factors of pressure drive water injection development productivity

After the model is established, the injection-production balance production is carried out first. When the water cut reaches the specified size, the pressure drive is carried out, the production well is closed, the water injection well is formed by short-term high-pressure water injection, the water injection well is closed after the water injection, and the 40-day soaking is carried out. Then the production well is subjected to constant pressure production. The average daily oil production within one month after the pressure drive and the average daily oil production within three months are used as evaluation indexes.

Using orthogonal test design[11], water cut before pressure drive, injection-production well spacing, water injection volume, permeability, crude oil viscosity, effective thickness, and porosity were selected as influencing factors. According to the variation range and data concentration of the influencing factors of the field data, the reasonable range of each factor is determined, and three horizontal values are taken within the reasonable range. The selected orthogonal table is L18 (3^7). The orthogonal test factor level table is shown in Table 1 , a total of 18 tests must be done, and the test plan is prepared. The test is carried out according to the plan, and the test results (average daily oil production within one month after the pressure drive and the average daily oil production within three months) are recorded. The orthogonal test software is used for variance analysis of the test results.

Table 1. Orthogonal test factor level table

	Water cut before pressure drive (%)	Injection-production well spacing (m)	Water injection volume (m^3)	Permeability (md)	Crude oil viscosity (cp)	Effective thickness (m)	Porosity (%)
Level 1	20	150	4000	15	1	5	5
Level 2	40	300	8000	30	2	10	10
Level 3	60	450	12000	45	3	15	15

Analysis of variance shows[12] that F is more significant than that, and the larger the F-ratio is, the more significant it is. If the F-ratio is greater than the F critical value, that item is significant. The test variance table is shown in Table 2 and Table 3.

Table 2. Daily oil production test variance table within one month after pressure drive

	Water cut before pressure drive (%)	Injection-production well spacing (m)	Water injection volume (m^3)	Permeability (md)	Crude oil viscosity (cp)	Effective thickness (m)	Porosity (%)
F-ratio	271.95	494.47	261.67	33.23	1.00	48.07	187.49
F critical value	19	19	19	19	19	19	19
Significance	2	1	3	6	7	5	4

The daily oil production within one month after the pressure drive is taken as the evaluation index of the test scheme, as shown in Table 2. According to the F ratio, the influence on the development capacity of the pressure drive is ranked as follows: injection-production well spacing > water content before pressure drive > water injection volume of pressure drive measures > porosity > effective thickness > permeability > crude oil viscosity. The comparison of the F-ratio and F critical value shows that the injection-production well spacing, water content before pressure drive, water injection volume of pressure drive measures, porosity, effective thickness, and permeability have a significant influence on the productivity of pressure drive development.

Table 3. Daily oil production test variance table within three months after pressure drive

	Water cut before pressure drive (%)	Injection-production well spacing (m)	Water injection volume (m³)	Permeability (md)	Crude oil viscosity (cp)	Effective thickness (m)	Porosity (%)
F-ratio	29.84	50.25	43.68	1.00	1.24	7.32	19.99
F critical value	19	19	19	19	19	19	19
Significance	3	1	2	7	6	5	4

The daily oil production within three months after the pressure drive is taken as the evaluation index of the test scheme, as shown in Table 3. According to the F ratio, the influence on the productivity of pressure drive development is ranked as follows: injection-production well spacing > water injection volume of pressure drive measures > water content before pressure drive > porosity > effective thickness > crude oil viscosity > permeability. The comparison of the F-ratio and F critical value shows that the injection-production well spacing, water injection volume of pressure drive measures, water content before pressure drive, and porosity significantly influence the productivity of pressure drive development.

3.3 Single factor study on influencing factors of pressure drive water injection development productivity

Through the previous multi-factor analysis, the main controlling factors of pressure drive water injection development capacity are injection-production well spacing, water injection volume of pressure drive measures, water content before pressure drive, and porosity. According to the design of the pressure drive scheme, based on the data given by the mine site, within a reasonable range, the injection-production well spacing, water injection volume of pressure drive measures, water content before pressure drive, and porosity take 10 horizontal values each. The influencing factors are input into the conceptual model of pressure drive water injection development. The daily oil production within three months after the pressure drive is analyzed as the evaluation index. The pressure drive capacity results corresponding to different influencing factors are recorded to generate scatter plot analysis, the scatter plot results are shown in Figure 3.

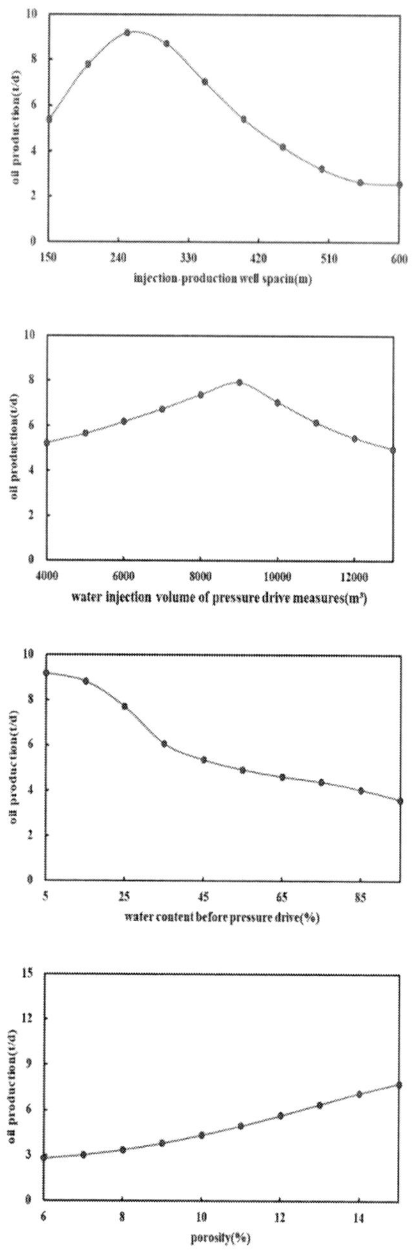

Fig 3. Quantitative analysis chart of main control factors

Injection-production well spacing: when the injection-production well spacing is less than 240 m, and the productivity after the pressure drive increases with the increase of injection-production well spacing. When the injection-production well spacing is greater than 240 m, the productivity after the pressure drive decreases with the increase of injection-production well spacing.

The water injection volume of the pressure drive measures when the water injection volume is less than 9000 m³, and the productivity after the pressure drive increases with the increase of water injection volume. When the water injection is greater than 9000 m³, the production capacity after the pressure drive decreases with the increase of water injection.

Water content before pressure drive: with the increase of water content before the pressure drive, the relative permeability of crude oil decreases, and the relative permeability of water increases, resulting in a gradual decrease in productivity after the pressure drive.

Porosity: As the porosity increases, the reserves between the well groups increase, and the productivity after the pressure drive gradually increases.

4. Conclusion

The productivity formula of a vertically fractured well in a low permeability reservoir is used to screen out the influencing factors related to pressure drive, including water cut before pressure drive, injection-production well spacing, water injection rate, permeability, crude oil viscosity, effective thickness, and porosity. Numerical simulation is used to establish a conceptual model of pressure drive water injection development in line with the actual block. The main controlling factors of productivity after a low permeability pressure drive are determined by orthogonal design, including injection-production well spacing, water injection volume of pressure drive measures, water content before pressure drive, and porosity.

Through quantitative analysis of the main control factors, it is concluded that too low injection-production well spacing will increase the water content of the production well after the pressure drive and decrease productivity after the pressure drive. Too high will lead to a decrease in the effect after pressure drive, inability to effectively displace, and reducing productivity. Therefore, the injection-production well spacing should be set between 200-300 m; the low water injection rate of pressure drive measures will lead to poor pressure drive effect, and the formation energy will not be effectively supplemented. If it is too high, water channeling will occur, resulting in an increase in water cuts in production wells. Both of them will reduce productivity after the pressure drive. The appropriate injection volume should be set according to different well groups, and the injection volume should be between 8000-10000 m^3. The higher the water cut before the pressure drive, the lower the recoverable reserves of the well group. The pressure drive water injection development should be carried out in the low permeability reservoir as soon as possible. The larger the porosity, the larger the reserves between the well groups, and the higher the productivity after the pressure drive. Through the analysis of the main control factors, this paper can provide the basis for the optimization of the field pressure drive scheme and the basis for the prediction of the production capacity after the subsequent pressure drive.

References

1. Xu Yonghui, Meng Yang, Shi Shige. Design method of water injection rate for pressure drive in low permeability reservoir J. Inner Mongolia Petrochemical Corporation, 2021, 47 (11): 118-121.
2. Huang Yue, Jin Zhirong, Qiao Chunguo, Zhang Huali, Ma Wei. Practice of pressure drive water injection technology in low permeability reservoirs of small fault blocks in Jiangsu Oilfield J. Petrochemical application, 2022,41 (06): 48-51.
3. Wu Xiaoyun, Li Xiaowen, Liu Fangfang. Mechanism and application of pressure drive development technology in deep ultra-low permeability reservoirs J.Inner Mongolia Petrochemical, 2021,47 (10) : 66-69.
4. Zhao Junping, Liang Ruobing, Yin Lijuan. Comparative analysis of three methods for predicting recoverable reserves of oil and gas fields J. Journal of Northeast Petroleum University, 2007, 31 (6): 65-67.
5. Wang Jingwei, Ji Bingyu, Zhang Wenbiao. Uncertainty analysis of oilfield recoverable reserves based on multiple factors J. Offshore Oil and Gas of China, 2016, 28 (4): 76-82.
6. Fan Chao, Li Sanshan, Li Lu, Wang Yanan. Low permeability reservoir pressure drive water injection development technology research J. Petrochemical application, 2022, 41 (01): 37-40.
7. Li Longlong, Yao Jun, Wu Minglu. Productivity calculation of low permeability vertical wells considering fracturing fluid pollution J. Daqing Petroleum Geology and Development, 2015,34 (01): 79-82.
8. Sun Xiang.Study on the effect of fracturing fluid filtration on oil increase and precipitation D.Northeast Petroleum University, 2018.
9. Zhang Yifei, Yang Yong, Sun Zhigang, Yu Chunlei, Sun Qiang, Bei Junping. Physical simulation and quantitative characterization of fractures in low permeability reservoirs J.Oil and gas geology and recovery, 2022,29 (04) : 143-149.
10. Cha Luming, Feng Qihong, Wang Sen, Xu Shigan, Liu Gaowen, Huang Wenhuan. Based on the virtual element method and damage model, the numerical simulation method of pressure drive water injection J / OL.Computational physics : 1-13 [2023-02-07].
11. Du Lan, Xie Jinzhong, Zhang Wei, Lai Qiuxiang, Chen Liang, Xia Jie, Chen Sheng. Study on the influencing factors of green bamboo container seedlings based on orthogonal test design J.World Bamboo and Rattan Newsletter, 2019,17 (02) : 12-17.
12. Xu Yadan, Qiu Yingjie, Shen Guoqing. Research on the influencing factors of automobile carbon dioxide emissions J. Agricultural equipment and vehicle engineering, 2014,52 (08) : 9-12.

 https://doi.org/10.1051/e3sconf/202337501051

Hot-in-place recycling engineering application of low penetration aged asphalt

Ning Shi [1,*], Hongyu Xiao [2], Jiang Xiao [3], Qi Liu [4]

[1]School of Civil and Architectural Engineering, Nanchang Institute of Technology, Nanchang, China
[2]Ganzhou Kangda Highway Co., Ltd, Ganzhou, China
[3]Jiangxi Ruixun Highway Co., Ltd, Ganzhou, China
[4] Jiangxi Transportation Institute Co., Ltd, Nanchang, China

Abstract. Hot-in-place recycling of asphalt pavement is one of the important regeneration technologies of pavement engineering, which apply to the pavements with fewer diseases, excellent driving quality, slight rutting, and good pavement structural strength. However, the surface layer of some road sections was aging, the penetration index of aged asphalt at 25°C is lower than 20(0.1mm), which does not meet the recommended requirements of the criterion in China. In 2019, 3 sections of test projects were laid on the expressway, each section was 1 km, for confirming the feasibility of using hot-in-place recycling with asphalt penetration less than 20 (0.1mm) through physical engineering. Experimental results confirmed that Marshall stability, residual stability, dynamic stability and water seepage of the hot-in-place recycled mixture optimized by the recycled pavement material with a penetration of aged asphalt of 15 can meet the specification requirements, Pavement Condition Index (PCI), International Roughness Index (IRI), and Rut Depth (RD) of the test pavements are in good conditions.

Keywords: Asphalt pavement; hot-in-place recycling; the penetration index; pavement condition index; International Roughness Index; Rut Depth.

1. Introduction

Hot-in-place recycling of asphalt pavement is one of the important regeneration technologies of pavement engineering. This kind of technology first heats and softens the original asphalt pavement to facilitate loosening or milling to a certain depth, and then fully mixes the old pavement material with new asphalt mixture and regeneration agent. Mixing, and finally paving and rolling equipment to complete the asphalt pavement regeneration, 100% of the original asphalt pavement can be recycled on site, and can effectively deal with rutting, bagging, oil flooding, poor flatness, water seepage, polishing, pumping, Diseases such as loosening and peeling.

The hot-in-place recycling process of asphalt pavement has many advantages: maximum utilization of resources; reduction of vehicle load; strong functionality of restored pavement; little impact on traffic; fast and convenient process, etc.

The hot-in-place recycling has little contribution to the structural strength of the pavement, and its pavement effect and service life are significantly affected by the structural strength of the original pavement. In china, for damaged road sections, the "Technical Specifications for Highway Asphalt Pavement Recycling" (JTGT5521-2019) [1] recommends that when hot-in-place recycling is used, the technical condition of the pavement should satisfy that the structural strength index of the pavement is not less than 80, and the penetration of aged asphalt within the regeneration range is not less than 20 (0.1mm), the aged asphalt content is not less than 3.8%.

In the past 10 years, more than 3,000 kilometers of expressways have been built and opened to traffic in Jiangxi Province[2]. The typical characteristics of these roads are that the daily traffic volume of the road is not large, the strength of the pavement structure is good, and only the friction performance of the road surface is reduced, polished, loose and peeling. Such as surface functional damage, the use of hot-in-place recycling process can effectively improve the pavement function and repair surface diseases. However, before designing the construction scheme of hot-in-place recycling, after investigating the actual pavement usage conditions and analyzing the material properties, we found that the surface layer of some road sections was aging due to factors such as asphalt aging during the construction stage and long-term exposure to the sun and rain during the service stage. The penetration index of asphalt at 25°C is lower than 20(0.1mm). Since this index does not meet the recommended requirements of the code, the application of hot-in-place recycling is limited.

* Corresponding author: 36517774@qq.com

The asphalt pavement of an expressway in southern China has fewer diseases, excellent driving quality, slight rutting, and good pavement structural strength. In the design stage of the pavement maintenance project of the "13th Five-Year Plan" in 2019, the asphalt pavement plan for some road sections was adopted. Geothermal regeneration construction plan, but according to the requirements, samples were taken from the original asphalt pavement, and the aged asphalt was extracted and distilled. After conducting the penetration test, it was found that the asphalt penetration of the taken samples was less than 20 (0.1mm), which did not meet the technical requirements. However, considering the short construction period of this project, the focus on improving the road service level, and the small overall maintenance scale, the hot-in-place recycling is still a better technical transformation process. In response to this problem, considering the recommended index of "the penetration index at 25°C is not less than 20 (0.1mm)" in the specification, in 2019, 3 sections of test projects were laid on the expressway, each section was 1 km, in order to demonstrate the feasibility of using hot-in-place recycling for road sections with asphalt penetration less than 20 (0.1mm) through physical engineering.

In this paper, the data collected from the 3 test projects over the past three years are summarized and evaluated.

2. Methodology

2.1 Materials

2.1.1 Asphalt

The type and grade of new asphalt selected for hot-in-place recycling should be consistent with the asphalt used for the original road surface, and asphalt with a higher penetration degree can also be used.

The on-site hot-in-place recycling project uses the same grade of SBS modified asphalt as the original pavement, and the performance indicators of the asphalt meet the requirements of the specification[3]. The test results are shown in Table 1.

Table 1. Properties of the newly mixed SBS asphalt tested

Index		Unit	Qualified range	Result	Judgment
Penetration（25°C、100g、5s）		0.1mm	40~60	52	Yes
Extensibility（5cm/min、5°C）		cm	≥20	30	Yes
Softening point（global approach）		°C	≥60	78.0	Yes
segregation (Difference of softening point between 48H)		°C	≤2.5	1.5	Yes
After RTFOT	Weight change	%	≤±1.0	-0.209	Yes
	Penetration ratio（25°C）	%	≥65	67	Yes
	Extensibility（5°C）	cm	≥15	16	Yes

2.1.2 Aggregate

The lithology of the aggregate used in this project is the same as that of the original road surface. The fine aggregate is machine-made sand whose parent material is limestone, and the filler is mineral powder ground from limestone, which is dry and clean and can flow freely from the mineral powder bin. The performance testing indicators of each grade of aggregate meet the requirements of the corresponding specifications.

2.1.3 Recycled pavement material

The recycled pavement materials used in the laboratory are sampled from the asphalt pavement of the selected test section of the expressway. The samples need to be dried at a temperature of 105°C before use to fully drain the water. The aged asphalt and sieved gradation compositions in recycled pavement materials are shown in Table 2-Table 3, respectively.

Table 2. Properties of the recycled asphalt tested

Index	Qualified range	Result
Penetration (25°C、100g、5s)	≥20	16
Asphalt content (%)	≥3.8	4.4

Table 3. Sieving Test Results of Recycled Pavement Materials

Sieve aperture (mm)	Mass percentage passing through the following square hole sieve (%)									
	16.0	13.2	9.5	4.75	2.36	1.18	0.6	0.3	0.15	0.075
1#	100	99.1	91.8	74.1	45.5	32.5	23.5	14.6	7.8	5.2
note	1# is the mineral material gradation after extraction of recycled pavement materials									

2.1.4 Asphalt rejuvenator

Considering the aging degree of the pavement asphalt in this project, we choose two kinds of asphalt rejuvenators for comparison, namely asphalt rejuvenator A produced in China and asphalt rejuvenator B imported from the Netherlands.

The test method is to add regenerated agent according to 3% and 5% of the amount of aged asphalt from indoor standard tests[4], and stir at 145°C and 600rpm for 30 minutes to measure the penetration of recycled asphalt to see if it meets the requirements, and therefore to determine the type and amount of regenerant, the test results are shown in Table 4 below.

Table 4. Penetration test results of recycled asphalt

Sample	Dosage	Result	Qualified range
A1-1	3%	25	40-60
A1-1	5%	35	40-60
B2-1	3%	30	40-60
B2-1	5%	45	40-60
Note	Asphalt aging to a penetration of 15 in test		

 https://doi.org/10.1051/e3sconf/202337501051

It can be seen from Table 4 that the effect of asphalt rejuvenator B is better than that of A, and the two dosages of A cannot meet the project requirements. According to the analysis of the project team personnel, it may be due to the limitation of the regulations that the asphalt rejuvenator produced by company A has no consideration for using aged bitumen with a penetration below 20. Therefore, a 5% dosage of B asphalt rejuvenator was used to carry out the next asphalt mixture test.
The asphalt rejuvenator is PrePhalt FBK imported from the Netherlands. Through the determination of the dosage test, it has good compatibility with the aging asphalt of this project and certain anti-aging performance, it also can ensure the durability of the recycled asphalt and asphalt mixture. It has been verified by experiments that the flash point of the asphalt regenerant is above 220°C.

2.1.5 *Newly added asphalt mixture*

For the newly added mixture used for remixing, according to the gradation of the recycled asphalt mixture (RAP) mineral material and the performance requirements of the recycled pavement, determine its gradation composition, and test the performance of the recycled mixture; the newly added asphalt mixture When its performance does not meet the relevant provisions of JTG F40[3], if the quality of recycled asphalt mixture meets the relevant provisions of JTG F40, it can be used.
In order to determine the newly added asphalt mixture gradation and asphalt ratio index, in addition to the road performance requirements of recycled asphalt mixture, construction operability, transportation, storage and other factors should also be considered.

2.2 Proportion Design of Recycled Asphalt Mixture

Before the mix design of recycled asphalt mixture, firstly, we evaluated the technical performance of original asphalt pavement and recycled asphalt pavement materials.
In the mix design of recycled asphalt mixture, secondly, we consider the mix ratio and disease characteristics of the original pavement asphalt mixtures, and carry out the mix ratio design of the recycled mixture in a targeted manner. Therefore, the type of recycled asphalt mixture is selected to be consistent with that of the original pavement asphalt mixture, as shown in Figure 1.
The mix design of recycled asphalt mixture also takes into account the actual operating conditions of the recycling plant, as well as the characteristics of the construction organization and the convenience of construction. It is required that production, mixing, spreading and compaction are easy to carry out.

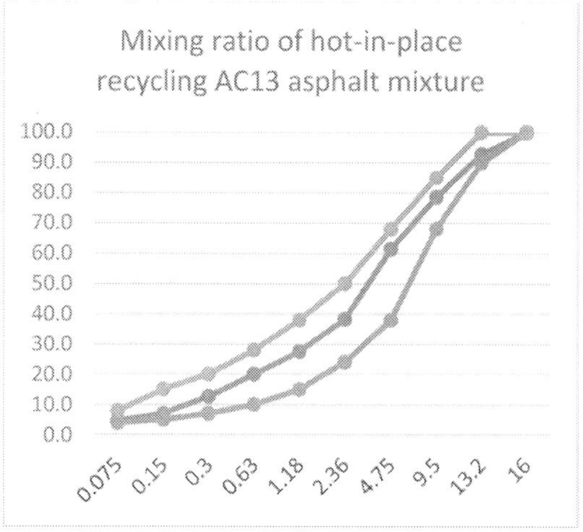

Fig. 1. Mineral grading curve of hot-in-place recycling AC-13C asphalt mixture

2.3 Asphalt mixture tests and pavement technical condition inspection

In order to ensure the quality of the recycled mixture, during the construction of the test section, we tested the road performance indicators such as Marshall stability, residual stability, dynamic stability and water seepage of the rutted plate of the recycled mixture.
After the completion of the paving of the test section, every year we examined pavement condition index (PCI), International Roughness Index (IRI), and Rut Depth (RD)to reflect the technical condition of the pavement.

3. Results and discussion

3.1 The road performance indicators

Based on experimental verification, when the hot-in-place recycling AC-13C asphalt mixture has an optimal asphalt ratio of 5.0%, all technical indicators of the asphalt mixture meet the requirements of JTG F40-2004 [3]("Technical Specifications for Construction of Highway Asphalt Pavement "), as shown in Table 5.
The results also show that the performance of the recycled asphalt mixture with the selected type and amount of asphalt rejuvenator can meet the requirements of the Chinese industry norms and the actual requirements of the project.
Obviously, the technical indicators such as Marshall stability, residual stability, dynamic stability and water seepage of the hot-in-place recycled mixture optimized by the recycled pavement material with a penetration of aged asphalt of 15 meet the specification requirements.

Table 5. Test results of recycled asphalt mixture

No.	Index	Unit	Qualified range	Result	Judgment
1	Marshall stability	kN	≥8	12.36	Yes
2	Flow value	mm	1.5-4.0	1.75	Yes
3	Asphalt saturation	%	65-75	69.1	Yes
4	Gross volume density	g/cm³	/	2.361	/
5	Theoretical maximum density	g/cm³	/	2.473	/
6	Porosity	%	4-6	4.5	Yes
7	VMA	%	≥14.5	14.5	Yes
8	Residual stability	%	≥85	85.3	Yes
9	Dynamic stability	number of times /mm	DS≥2800	3320	Yes
10	Water seepage	ml / min	/	0	Yes

3.2 The technical condition of the test pavements

Before the construction, the design unit entrusted the inspection unit to conduct a full range of inspections on the technical performance of the expressway asphalt pavement (indices such as Pavement Condition Index, International Roughness Index and Rut Depth).

In 2019, we selected 3 road sections of 1 km each (A, B, and C respectively) to carry out the test road verification work. In order to test and observe the technical status of the test road surface, we have carried out the same inspection work for two consecutive years from 2020 to 2021. The detected indicators include PCI, IRI and RD.The same equipment, spanning 3 years, 3 tests, Figure 2 to Figure 4 are the test results.

It can be seen from Figure 2 that after the hot-in-place regeneration, one year later, the PCI condition of the three sections of asphalt pavement improved significantly, with an average increase of 26.5%, and the best result was 100%. After two years of vehicle loading, the average PCI of the test sections was 92%, which is still better than the situation before hot-in-place regeneration in the field, and the average increase of PCI after two years was 17.8%.

It can be seen from Figure 3 that the average International Roughness Index of the three test sections in 2019 was 0.8 m/km, which was in an excellent technical state. After the hot-in-place regeneration, the average level of International Roughness Index is increased to 1.2 m/km, and the maximum value of section A is 1.5 m/km. We all know that the smaller the IRI, the smoother the road surface. Obviously, the construction of hot-in-place regeneration reduces the smoothness of the road surface. Even so, the smoothness of the road surface in 2019 is very good and rare, and the smoothness of the road surface after construction in 2020 still meets the requirements of Chinese standards[5]. With road traffic running smoothly, the average IRI of the three test roads returns to 1.1 m/km in 2021.

It can be seen from Figure 4 that there were basically no road ruts on the three test roads in 2019, with an average Rut Depth of 4.8mm. This state also corresponded to the excellent road smoothness mentioned above. In 2020, after the hot-in-place regeneration construction treatment, the average rut depth of the test roads reduced to 3.3mm. Two years later, in 2021, the rut depth of the test roads still maintain an excellent level of 4.0mm on average.

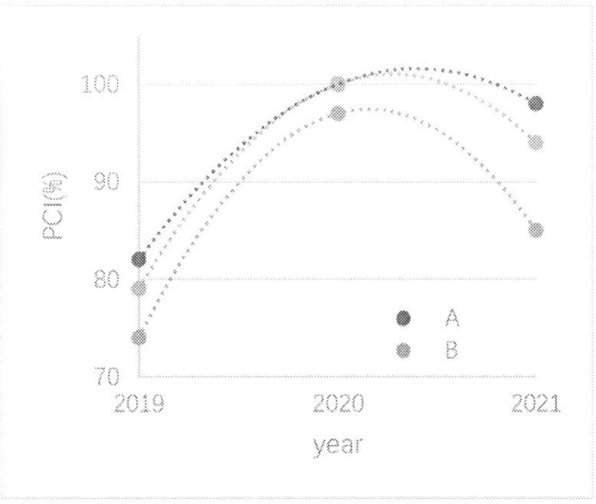

Fig. 2. Pavement condition index of the test road

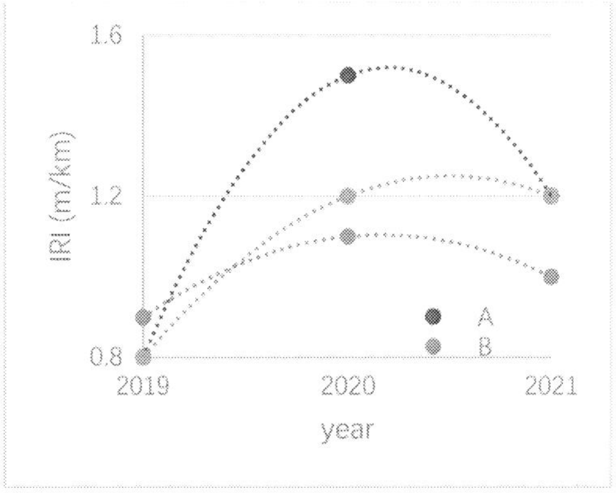

Fig. 3. International Roughness Index of the test road

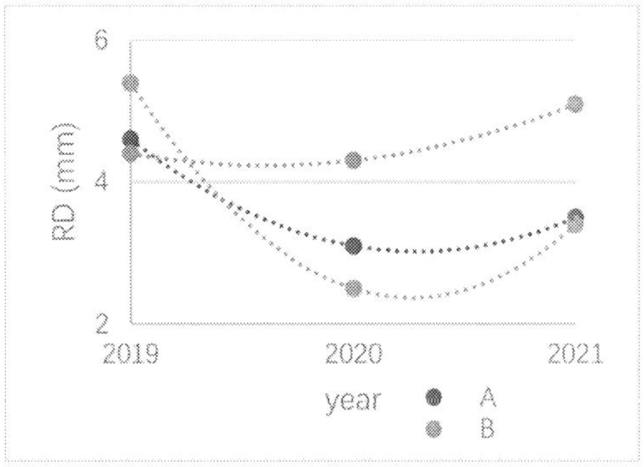

Fig. 4. Rut Depth of the test road

 https://doi.org/10.1051/e3sconf/202337501051

4. Conclusions

A large number of asphalt pavements with sufficient pavement structural strength and excellent pavement technical condition were found to have a penetration index of aged asphalt that could not meet the specification requirements after the hot-in-place regeneration process in the maintenance project to improve the service function level of the pavement. We selected three highway sections with aging asphalt penetration of 16(0.1mm) to carry out experimental research. The conclusions are as follows:
(1) The technical indicators such as Marshall stability, residual stability, dynamic stability and water seepage of the hot-in-place recycled mixture optimized by the recycled pavement material with a penetration of aged asphalt of 15 can meet the specification requirements.
(2) After the hot-in-place regeneration, the PCI condition of asphalt pavement improve significantly. After two years of vehicle loading, the average PCI of the test sections is still better than the situation before hot-in-place regeneration.
(3) The construction of hot-in-place regeneration may cut down the smoothness of the road surface, but the International Roughness Index can meet the requirements of Chinese standards.
(4) The hot-in-place regeneration can heal road ruts and maintain the Rut Depth of the pavement in a smaller state.

Acknowledgments

This work was financially supported by Science and Technology Research Project Fund of Jiangxi Provincial Department of Education (GJJ190978)and Science and Technology Research Project Fund of Jiangxi Provincial Department of Transportation (2022H0011).

References

1. JTGT5521-2019, Technical Specifications for Highway Asphalt Pavement Recycling[S]. Beijing, Ministry of Transport of the People's Republic of China,2019.
2. Jiangxi Provincial Department of Transportation, Jiangxi Traffic Yearbook-2021[M], 2021.
3. JTG F40-2004. Technical specification for construction of highway asphalt pavement[s]. Beijing, Ministry of Transport of the People's Republic of China,2004.
4. JTG E20-2011.Standard Test of Bitumen and Bituminous Mixtures for Highway[s]. Beijing, Ministry of Transport of the People's Republic of China,2011.
5. JTG F80/1-2017 Inspection and Evaluation Quality Standards for Highway Engineering Section1: Civil Engineering. Beijing, Ministry of Transport of the People's Republic of China,2017.

Study on the method of identifying fractures by ant tracking technique

Juan Zhang [1], Ke Liu [1,*], Xianbo Han [2]

[1] School of Earth Sciences, Northeast Petroleum University, Daqing, Heilongjiang, China
[2] School of Electrical Engineering, Northeast Petroleum University, Daqing, Heilongjiang, China

Abstract. Fracture is an effective reservoir space and important seepage channel for low permeability sandstone reservoirs, which is very important for reservoir exploration and development. Conventional fracture prediction methods have high cost, slow operation speed and low resolution, which limits the promotion of fracture identification methods. Therefore, this paper proposes a fracture prediction method based on ant tracking. This method is based on post-stack seismic data. Firstly, the seismic data is denoised and the fault continuity is enhanced by using structural smoothing and variance body attributes. On this basis, the active ant tracking is used to enhance the fault boundary of the data body. Then, the data body is subjected to multiple ant tracking calculations by optimizing parameters, and finally the suitable parameter values are determined to realize the prediction of natural fractures. The results show that the technology is used to predict natural fractures, and the prediction results have high resolution and reasonable fracture distribution.

Keywords. Post-stack seismic data; ant tracking; variance body; fracture prediction.

1. Introduction

The identification and prediction of reservoir fractures is an important task for the exploration and development of fractured reservoirs, which is of great significance for the adjustment of development plans and the improvement of oil recovery. At present, the identification and prediction methods of reservoir fractures at home and abroad mainly include geological methods, logging methods, seismic methods, structural mechanics methods, dynamic analysis methods[1] and multi-information fusion methods[2]. Geological method[3] refers to the method of directly observing and describing fractures through field outcrops and cores. The fracture parameters obtained by this method are accurate, but the lateral resolution of fractures is low. The logging method can indirectly reflect the development of fractures near the well trajectory through the change of physical characteristics of fractured reservoirs[4,5], but the high cost of imaging logging, limited data and multiple solutions limit the wide application of this method. Seismic methods can detect fractures through the changes of propagation time, velocity, amplitude and frequency of seismic waves in anisotropic media in different azimuths[6], such as coherence attribute analysis technology[7,8], multi-scale edge fracture detection[9] and ant tracking[10]. Ant technology is a very popular post-stack fracture prediction technology in recent years. Compared with conventional post-stack fracture prediction technology, this method has the advantages of short running time and high prediction accuracy, and has good application effect in many areas[11-13]. The application of ant tracking technology in fracture detection in X area is discussed.

2. Basic Principle of Ant Tracking Technology

Ant technology, also known as automatic fracture tracking technology, is a seismic attribute based on ant colony algorithm developed by Schlumberger. The ant colony algorithm was first proposed by Dorigo M et al.[14] as a stochastic optimization algorithm. The algorithm follows the principle that ants use pheromones that attract ants to convey information between their nests and food sources to find the shortest path. Set a large number of such ' ants ' in the seismic data volume and let each ' ant ' move forward along the possible fault plane and release pheromones to track the fault plane through the path of the ant 's forward movement. The essence is to track according to the difference between amplitude and phase in the seismic data, that is, the characteristics of discontinuity in the data volume, move forward along possible fault planes and fractures, and search for fault traces until they are fully characterized.

3. Technical process

3.1 Seismic data preprocessing

Before the ant tracking, the seismic data volume of the same set of SEGY format is first analyzed, and its thinning processing is converted into a sparse data volume with small memory space and clearer plane structure. Therefore, for the study area, the selection of sparse data volume for subsequent ant tracking construction can greatly improve the operation speed, and effectively save the time of parameter test, comparison and section and slice retrieval. After the thinning process, we first perform structural smoothing on the thinned data volume. Compared with the original seismic profile before structural smoothing, the seismic profile after structural smoothing has clearer in-phase axis dislocation and more vertical faults (Figure 1, the red arrow in the figure indicates the in-phase axis dislocation area). The volume noise of seismic data is suppressed and the continuity of seismic events is enhanced after structural smoothing. At the same time, due to the smoothing effect of the fault boundary, the corresponding seismic signal becomes enhanced and smoothed.

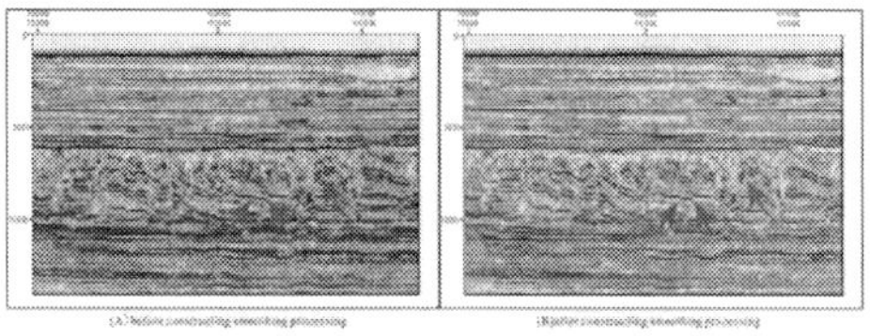

Fig. 1 Seismic section comparison before and after structural smoothing, the red arrow in the figure indicates the in-phase axis dislocation area

3.2 Variance system

The basis of ant body is to detect the discontinuity of seismic data. The common effective attribute bodies are variance body and chaos body. For the seismic data in this area, the variance attribute characterizes the fault better, so it is selected as the basic attribute of ant tracking. The variance attributes are extracted from the preprocessed data volume, and the variance body (Figure 2) is obtained. It can be seen that the quality of the variance body is obviously improved, the local fault relationship is clear, and the interference of noise and non-fault factors is effectively weakened.

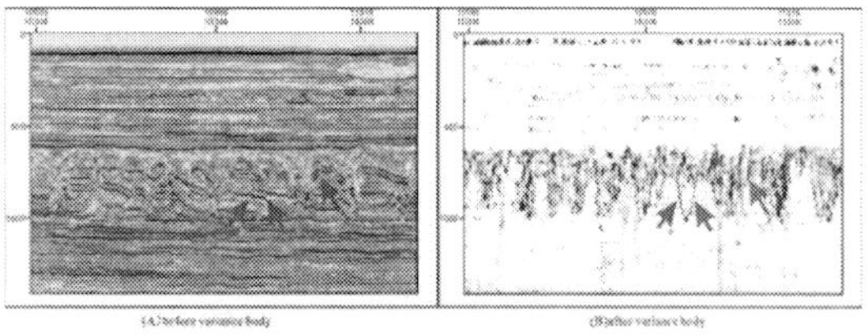

Fig. 2 Comparison of seismic sections before and after variance body, the red arrow in the figure indicates the in-phase axis dislocation area

3.3 Ant body tracking

1. Key parameter

Ant tracking properties controlled by the tracking parameters, the algorithm mainly includes 6 parameters, parameters meaning and characteristics are shown in Table 1.

Table. 1 Parameter Setting and Meaning of Ant Tracking Method

parameter	meaning	characteristic
Initial ant boundary	define the initial distribution range of ants	the smaller the parameter, the easier it is to identify small scale faults
tracking deviation	set the deviation range of ant searchable direction	the larger the parameter, the better the identification of bending fault
step size in search	step length of each ant search	the larger the parameter, the greater the search capability, but some details may be ignored
Illegal step	allow search beyond defined step size	the larger the parameter, the more continuous
Legal step	number of legal steps that must be included in the search path	the smaller the parameter, the more limited and discontinuous the fracture search
Search threshold value	percentage of illegal steps allowed during tracking	the larger the parameter, the stronger the search capability

Tracking parameter effect comparison

Firstly, the "passive" ant tracking method is selected to track the above variance body for the first time. The passive ant tracking algorithm is "lazy ant", which tends to track extremely strong signals and abandon weak signals, so it helps to suppress noise and reflect the trend of large faults (Fig.3). The specific parameters are : initial boundary is 7, tracking deviation is 2, step size in search is 3, Illegal step is 1, Legal step is 3, Search threshold value is 5%. Thus an edge - enhanced ant tracking attribute data body is obtained.

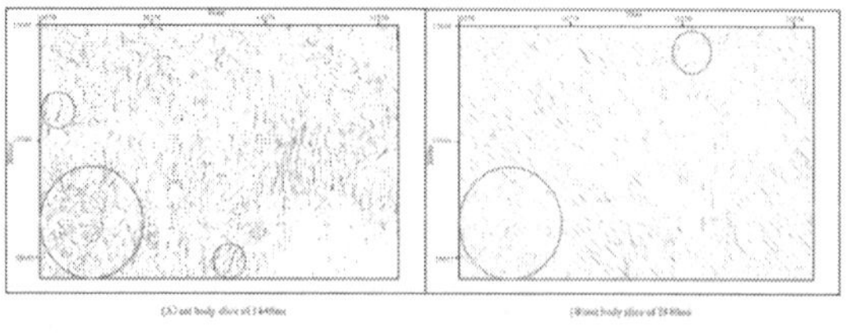

Fig. 3 First ant tracking, red and blue areas depict contrast areas for fractures.

A second ant tracking is performed based on the first ant tracking data body described above. Petrel was used to conduct multiple tests on the secondary ant tracking parameters to determine the optimal parameter values suitable for the study area. The ant parameters in Fig. 4 are: initial boundary is 5, tracking deviation is 3, step size in search is 3, Illegal step is 3, Legal step is 1, Search threshold value is 10%. Compared with the first ant body, the parameter ant body is more detailed in fracture characterization and effectively reduces noise interference. Among them, Figure 4 (a) in the red circle range and 5 (b) in the red circle range compared to Figure 3 (a) (b) fracture characterization more clearly, but the parameter to identify more discontinuous points, to maximize the prediction of the location of discontinuous points, used to predict the fracture resolution is too low, the fracture is more fragmented.

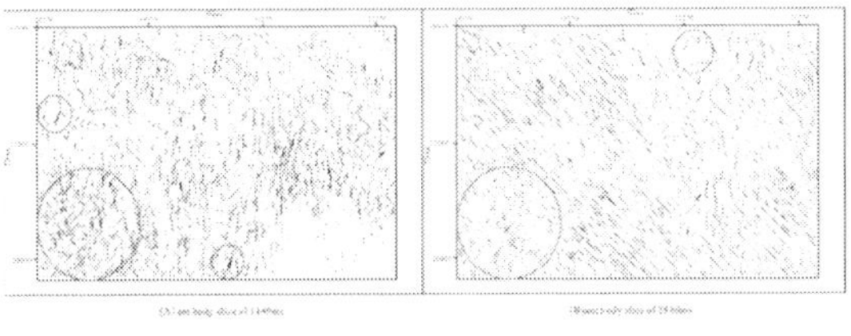

Fig. 4 First ant tracking, the tracking parameter is 5333110,red and blue areas depict contrast areas for fractures.

In order to obtain high-definition ants, we continue to optimize the extraction parameters of ants. The parameters of ants in Figure 5 are as follows: initial boundary is 4, tracking deviation is 2, step size in search is 3, Illegal step is 1, Legal step is 3, Search threshold value is 5%. Compared with the ant body with the parameter of 5333110, the ant body with this parameter is more accurate in identifying fractures, but the micro is not detailed enough and the resolution between fractures is not clear.

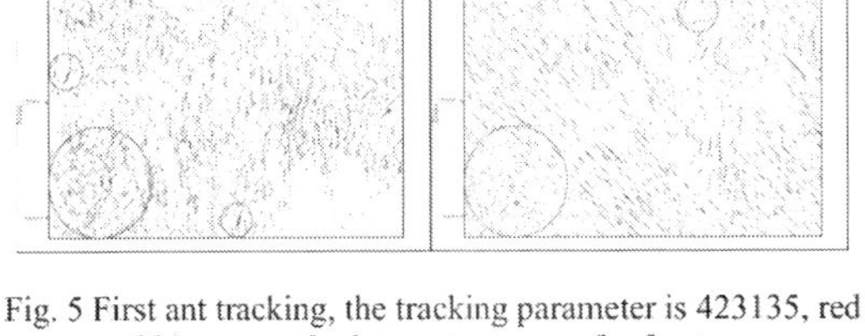

Fig. 5 First ant tracking, the tracking parameter is 423135, red and blue areas depict contrast areas for fractures.

In order to improve the accuracy of local microscopic prediction, initial boundary is 5, tracking deviation is 2, step size in search is 3, Illegal step is 1, Legal step is 3, Search threshold value is 5%. The horizontal slices of the ant body for the parameters are observed and analyzed, as shown in Fig.6. It can be seen that the parameters describe the fractures in detail and can see the direction of a single fracture.

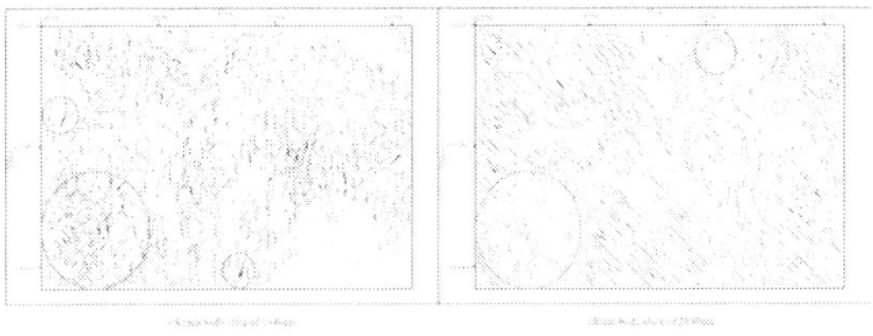

Fig. 6 First ant tracking, the tracking parameter is 523135, red and blue areas depict contrast areas for fractures.

4. Application effect analysis

The fracture characteristics of three structural zones A, B and C in X layer of the study area are statistically analyzed. Among them, A represents the diapir development zone, B represents the diapir eaffected zone, and C represents the diapir slope zone. The statistical results are shown in Fig.8, the east of the diapir is dominated by NW-trending fractures, and the fracture density tends to decrease away from the diapir area, which is consistent with the geological law.

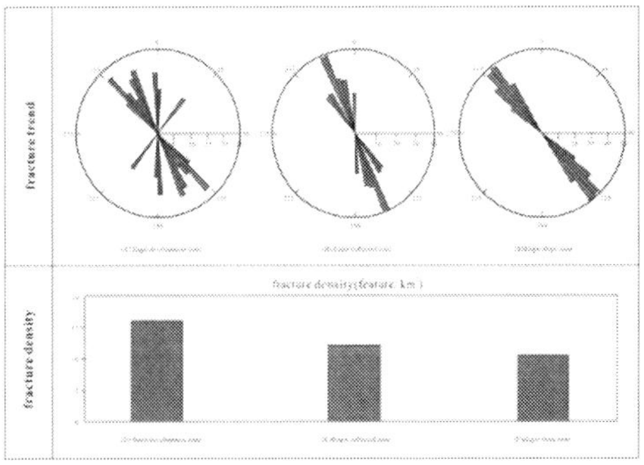

Fig. 7 Fracture statistical diagram

5. Summary and conclusions

Ant tracking data volume can show the development of small structures more intuitively and quickly. Through the study of fracture prediction in a certain area, the following conclusions can be drawn :

(1) Ant tracking algorithm can eliminate the subjective factors of interpreters, making the interpretation results of small faults more objective.

(2) Ant tracking is an advanced technology of post-stack fracture prediction based on ant colony algorithm. Before ant tracking, it is necessary to filter the original seismic data to improve the signal-to-noise ratio of seismic data and provide high-quality data for post-stack attribute extraction.

(3) Through the statistics of fracture development characteristics, the reliability of fracture prediction results is confirmed, which provides an important basis for the next oil and gas prediction.

References

1. Lei Chen,Jianjun Xu,Shuang Chen,Hui Yang,Linhu Liu. Power Quality Disturbance Signal Denoising Based on Overcomplete Representation. IEEJ TRANSACTIONS ON ELECTRICAL AND ELECTRONIC ENGINEERING.2022; 17: 544–555
2. Xu Jianjun, Huang Lida, Yan Limei, Yi Na. Insulator Self-Explosion Defect Detection Based on Hierarchical Multi-Task Deep Learning. Transactions of China Electrotechnical Society, 2021,36(07):1407-1415.
3. Limei,LIU Yongqiang,XU Jianjun,et al.Broken string diagnosis of composite insulator based on Grabcut segmentation and filler area discrimination.Power System Protection and Control,2021,49(22):114-119
4. Yi, Q. Wang, L. Yan, et al., A multi-stage game model for the false data injection attack from attacker's perspective. Sustainable Energy Grids & Networks 28 (2021).
5. Na Yi,Jianjun Xu,Limei Yan,Lin Huang. Task Optimization and Scheduling of Distributed Cyber-physical System Based on Improved Ant Colony Algorithm. Future Generation Computer Systems, 109(Aug. 2020),134-148.
6. Yang Zhao, Jianjun Xu, Jingchun Wu. A New Method for Bad Data Identification of Oilfield Power System Based on Enhanced Gravitational Search-Fuzzy C-Means Algorithm. IEEE Transactions on Industrial Informatics. VOL. 15, NO. 11, NOVEMBER 2019 5963-5970
7. Wang Dan, Jia Yuewei, Wei Shuijian, et al. Comprehensive prediction of post-stack fractures in the fourth member of Xujiahe Formation in Xinchang [J]. Geophysical and geochemical exploration, 2014,38 (5):1038 ~ 1044.
8. Dorigo M, Maniezzo V. Colorni A. Ant system : by acolony of robots optimization [J].IEEE Transactions on SystemsMan & Cybernetics Part B Cybernetics A Publication of the IEEE Systems Man & Cybernetics Society, 1996, 26 (1) : 29-41.

Study on the influence of different key parameters on EOR in LB block

Qunyi Wang [1], Yongbin Bi [1, 2], Yang Zhang [1], Tongfeng Cao [1], Dong Liu [1]

[1] Jidong Oilfield Company, PETROCHINA, Tangshan, China
[2] Northeast Petroleum University, Daqing, China

Abstract. The high temperature reservoir is in the high water cut stage after long-term development, and the inefficiency or invalid circulation caused by injected water is aggravated. With a long development history, the well pattern has been adjusted many times, and the remaining oil distribution is highly dispersed. Based on the high-temperature reservoir conditions of the target block, the influence analysis of different key parameters on EOR effect of LB- block was carried out. The gas flooding and chemical flooding potential were evaluated by numerical simulation method, and the EOR effect of different oil flooding systems was defined. The gas flooding and binary flooding numerical models were established according to the well pattern, reservoir and fluid physical parameters. The development effect is predicted under different formation dip, reservoir permeability, oil viscosity, oil saturation, effective thickness and reservoir temperature. The simulation results show that the key factors of EOR in gas flooding and binary flooding are formation dip.

Keywords: reservoir permeability; reservoir thickness; viscosity; formation dip; EOR.

1. Introduction

After long-term water injection development, the dominant seepage channels in the middle Liuzan area are developed, and the development characteristics are high and low at present, and the stable production situation is severe. In order to further improve the reservoir development effect, it is urgent to carry out EOR program research on the basis of fine geological re-understanding research. But there are three problems at present: first, because of the strong heterogeneity in plane, inter-layer and inter-layer, the shape of sand body is complex, the inheritance of longitudinal sand body is poor, the superposition relation is complex, and the fine reservoir description is difficult; Second, the reservoir is buried deep, the temperature is high, the high temperature resistance EOR drive medium optimization is difficult, third, the target block is complex fault block multi-layer heterogeneous sandstone reservoir, EOR program design and optimization is difficult[1-5].

①The injection and production well pattern has a high control degree of reserves (91%), a high control degree of water flooding reserves (77.6%), a high utilization degree of water flooding reserves (65.4%), a high recovery degree of geological reserves (32.1%), a high comprehensive water cut (96.5%), a low recovery rate of geological reserves (0.22%), and an unsatisfactory development effect.

② Conventional water flooding has little potential to improve the development effect. Therefore, it is urgent to change the development mode. However, high reservoir temperature restricts the effect of tertiary oil recovery.

2. State of the art

Wu Wei[6] et al. set up an evaluation method for emulsifying performance and emulsifying viscosity of emulsion surfactant to further improve the recovery factor of ultra-high temperature and medium-low permeability reservoirs, and studied the structural-activity relationship of emulsion surfactant. It provides technical support for greatly enhancing oil recovery in ultra-high temperature and low permeability reservoirs.

Shi Jing[7] et al. developed a new dual compound oil drive system to meet the development needs of high temperature and high salt heavy oil reservoirs in Sheng li Oilfield. The successful pilot test provides a reference for similar reservoirs to greatly improve the recovery factor research.

For high temperature and high salt reservoirs, Yuan Chengdong[8] et al. proposed micro-scale dispersing gel and surfactant flooding technology in view of reservoir conditions and water injection development. It provides a new technology and method for enhancing oil recovery in high temperature and high salt reservoir and has practical significance for the efficient development of high temperature and high salt reservoir. It has important theoretical significance to enrich the oil displacement mechanism of surfactant and the migration and plugging mechanism of disperse adhesives in porous media.

Tao Jiping[9] et al. developed a new nitrogen foam flooding system with temperature and salt resistance and low interfacial tension for high temperature and high salt reservoirs, which can only effectively control the high permeability bands, but also better emulsify the crude oil, peel and carry the crude oil for recovery, and adapt to a wider range of formation conditions.

Aiming at low saturation, high viscosity and high temperature reservoir, Han Qi[10] et al. co-polymerized various functional monomers with AM and synthesized a new high temperature resistant polymer GL-100 under appropriate reaction conditions. The research experience of injection system is provided for the study of enhanced oil recovery technology by chemical flooding in high temperature reservoir.

According to Zhou Yumeng[11] et al., the contradiction between reservoir layers is prominent, the intra - layer interference is gradually intensified, the heterogeneity is enhanced, the water absorption profile is not uniform, the injection water line is rapidly advancing along the crack of high permeability large pore channel, and the water channeling phenomenon is serious. The composite ion gel profile control agent and the slow expansion flexible viscoelastic particle profile control agent were developed to effectively solve the phenomenon of injected water advancing rapidly along the high permeability large pore, and the effect of oil increase and precipitation was obvious, which played an important guiding role in the comprehensive management of deep reservoir profile control in this block.

In order to meet the requirements of high reliability and stability in oilfield fracturing development and high performance of profile control agent, Wenjin Zhao [12]et al. adopted layer-by-layer assembly technology, microfluid technology, ion gelation method and reverse microemulsion polymerization method to prepare a series of composite gel profile control agent with temperature sensing function and intelligent control of gel water plugging.

Liu Xiaomin[13] et al. systematically screened and evaluated the surfactant foam flooding system in view of the problem of surfactant foam being exposed to oil and the conditions of high temperature and high salt reservoirs in Daqing oil reservoir. After that, nanoparticle were introduced to improve the foam performance, providing guidance for finding a system that can simultaneously obtain ultra-low oil/water inter-facial tension and excellent foam performance.

Weijia Cao[14] et al. Aiming at the problem of the large distance between oil and water in high-temperature oil reservoirs, the electrode gel was used as the displacement agent and evaluated. The research results provided theoretical and technical support for the development of inorganic gel and its application in Wells.

For surfactant flooding in high-temperature and high-salt oil reservoirs, Yuan-Li Li[15]ect. Study adopted sulfobetaine and sulfonate surfactant derived from 1, 3-dialkyl glycerol ether as the surfactant complex system. This formula has good resistance to the adsorption of quartz sand, and reduces interface tension, making it easier for crude oil to form water-in-oil emulsion.

Fenglan Zhao[16] et al. Feasibility of using organic base instead of inorganic base for ASP flooding in heterogeneous high temperature reservoir. The effect of alkali on formation was evaluated in laboratory environment, and the mechanism of organic alkali ASP flooding was studied, which provided guidance for high temperature heterogeneous reservoir composite flooding system[16].

3. Result Analysis and Discussion

Numerical Simulation Methods Gas Flooding and Chemical flooding potential Evaluation In order to determine the EOR effect of different oil flooding systems under different key parameters, the project team established the ideal numerical simulation model of gas flooding and binary flooding according to the well pattern, reservoir and fluid physical parameters of the fault block in the north of Liu90. This model is used to study the influence of key parameters on EOR of gas flooding and binary flooding.

1、Potential evaluation of gas flooding and chemical flooding by numerical simulation method.

The numerical simulation model of typical well group is used to predict the potential of enhanced oil recovery efficiency by chemical flooding and gas flooding, the results are shown in the figure 1 and figure 2 below.

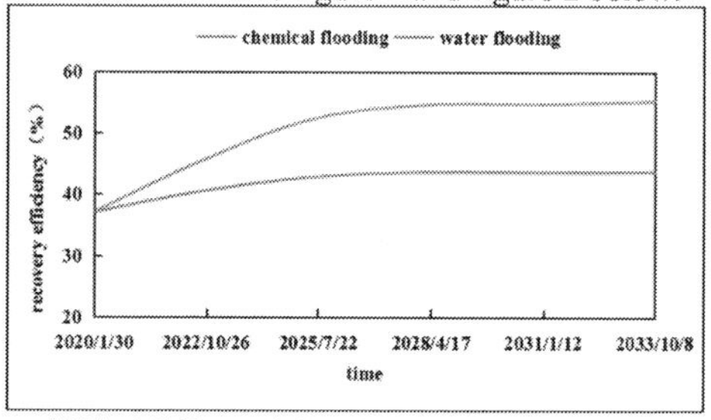

FIG. 1 Effect of chemical flooding on EOR

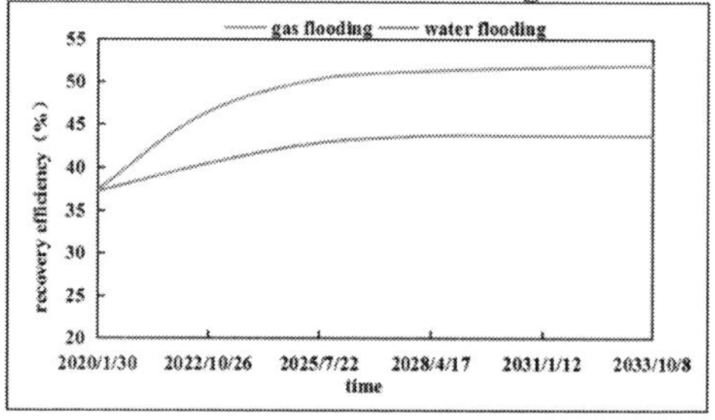

FIG. 2 EOR effect of gas flooding

The results show that compared with water flooding, chemical flooding increases oil recovery efficiency to 12.33% when rate of water content reaches 98%, compared with water flooding, gas flooding increases oil recovery efficiency to 8.31% when rate of water content reaches 98%. So the application of gas flooding and water flooding in complex fault block reservoirs can have a good effect of enhancing oil recovery efficiency.

Table 1 EOR effects of chemical flooding and gas flooding

Medium of displacement	Chemical drive	Gas drive	Water drive
Ultimate recovery factor (%)	55.32	51.30	42.99
Increased oil recovery value (%)	12.33	8.31	-

In order to determine the effect of different oil displacement systems on enhanced oil recovery efficiency under different key parameters,the project team has established the ideal numerical simulation model of gas flooding and binary flooding according to the well pattern form, reservoir and fluid physical parameters of Liu90 north fault block. The ideal model is used to study the influence of key parameters on the enhanced oil recovery efficiency effect of gas flooding and binary flooding. In the study, the effects of gas flooding and binary flooding under the conditions of different formation dip, reservoir permeability,crude oil viscosity, oil saturation, effective thickness and reservoir temperature are predicted respectively,,and thus clarify the influence of key parameters on gas flooding and binary flooding effect.

The effect of strati-graphic dip on enhanced oil recovery efficiency by gas flooding and binary flooding.

The oil reservoir in Liuzan Middle Area belongs to a typical complex fault block oil reservoir, each fault block is affected by the ground stress in the process of sedimentation. While forming fault block, it will cause a certain inclination and strati-graphic dip.Influenced by the strati-graphic dip, the three-phase flow of oil, gas and water in the reservoir will be affected by the strati-graphic dip, Three phases generate differentiation under the effect of gravitational differentiation, so as to affect the development effect of the reservoir. It is great significant for the efficient development of complex fault block reservoirs in Liuzan central area to clarify the effect of formation dip on the effects of different enhanced oil recovery efficiency methods.

In this study, ideal models are used to predict the effect of enhanced oil recovery when the strati-graphic dips are 5 °, 15 ° and 30 ° respectively. Through the enhanced oil recovery effect of binary flooding and gas flooding under different strati-graphic dip, the influence of strati-graphic dip on enhanced oil recovery effect of binary flooding and gas flooding is clarified. See Fig.3 for enhanced oil recovery effect of binary flooding and gas flooding under different strati-graphic dip.

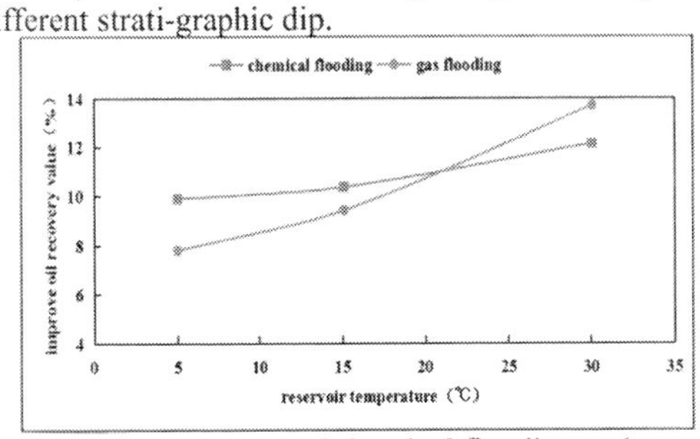

FIG. 3 EOR effects of chemical flooding and gas flooding under different dip angles

It can be seen from the enhanced oil recovery effect of binary flooding and gas flooding under different strati-graphic dip that the enhanced oil recovery of chemical flooding and gas flooding increases with the increase of strati-graphic dip , When the strati-graphic dip increases from 5 ° to 30 °, the enhanced oil recovery value of binary flooding increases from 9.89% to 12.15%, and that of gas flooding increases from 7.89% to 13.45%, It shows that when the strati-graphic dip is increasing, the increase of enhanced oil recovery by gas flooding is greater than binary flooding, the influence of strati-graphic dip on the development effect of gas flooding is greater than that of enhanced oil recovery by binary flooding. Therefore, the greater the strati-graphic dip in the actual reservoir, the better the effect of enhanced oil recovery by binary flooding and gas flooding . At the same time, it shows that the strati-graphic dip is the key factor affecting the enhanced oil recovery by gas flooding.

The effect of reservoir permeability on enhanced oil recovery by gas flooding and binary flooding.

Permeability is an important parameter to evaluate the physical property of the reservoir. The value will affect the characteristics of oil and water flow and enrichment in the reservoir.If the permeability is too low, the residual oil saturation in the reservoir will be too high, and the movable oil saturation will decrease, thus affecting the crude oil production of production wells. In the study, the effects of binary flooding and gas flooding on enhanced oil recovery are predicted under the conditions of 50, 100, 150 and 300mD permeability respectively, so as to clarify the influence of reservoir permeability on the effects of binary flooding and gas flooding on enhanced oil recovery. See Fig.4 for enhanced oil recovery effect of binary flooding and gas flooding under different permeability conditions.

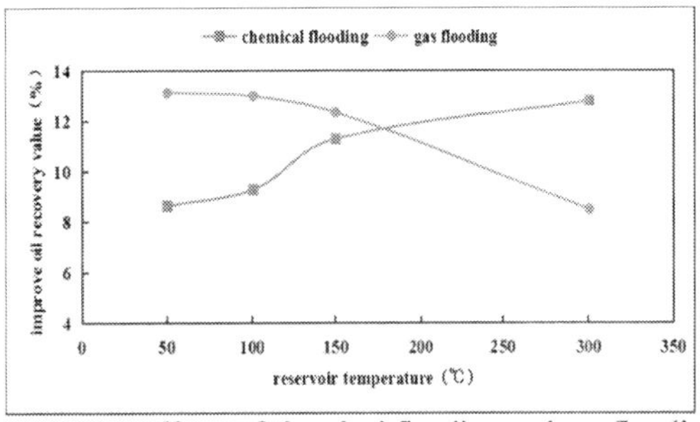

FIG. 4 EOR effects of chemical flooding and gas flooding under different permeability conditions

From the numerical simulation results of binary flooding and gas flooding under different reservoir permeability conditions, it can be seen that with the increase of permeability, the recovery efficiency of binary flooding increases, while that of gas flooding decreases gradually. When the permeability increases from 50mD to 300mD, the enhanced oil recovery value of binary flooding increases from 8.71% to 12.83%, and that of gas flooding decreases from 13.17% to 8.58%. This is because in the process of permeability increase, the adsorption loss of surfactant and polymer and the viscosity loss of polymer decrease during the injection of surfactant and polymer into the reservoir, so the effect of enhanced oil recovery by binary flooding increases with the increase of reservoir permeability; When the permeability is too high, the gas in the gas flooding tends to burst along the high permeability formation, thus forming a gas channeling channel between injection and production wells, thus reducing the effect of gas flooding on enhancing oil recovery. Therefore, it can be seen from the above analysis that permeability is one of the key factors affecting the effect of binary flooding and gas flooding to enhance oil recovery.

(3) Effect of crude oil viscosity on the effect of enhanced oil recovery by gas flooding and binary flooding. The viscosity of crude oil will affect the mobility ratio of displacement medium to displaced fluid in the process of gas flooding and binary flooding.If the mobility ratio is too large, it will lead to the injected binary system or gas channeling too early, thus greatly reducing gas flooding and binary flooding to improve oil recovery.In this study, the effects of enhanced oil recovery of gas flooding and binary flooding are predicted when the viscosity of crude oil is 3.6,10,20.The effect of enhanced oil recovery of binary flooding and gas flooding under different crude oil viscosity is shown in figure 5 .

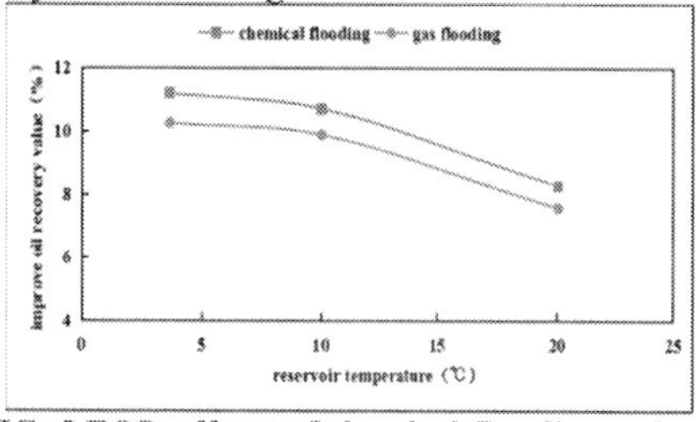

FIG. 5 EOR effects of chemical flooding and gas flooding under different crude oil viscosity

From the effect of enhanced oil recovery of binary flooding and gas flooding under different crude oil viscosity conditions, it can be seen that the effect of enhanced oil recovery of binary flooding and gas flooding decreases with the increase of crude oil viscosity. When the crude oil viscosity increases from 3.6 to 20, the recovery efficiency of binary flooding decreases from 11.25% to 8.28%, and the recovery efficiency of gas flooding decreases from 10.35% to 7.65%.It shows that in the process of increasing the viscosity of crude oil, the effect of improving mobility ratio of binary flooding becomes worse, and the effect of adverse mobility ratio of gas flooding increases, so the recovery efficiency of both binary flooding and gas flooding decreases with the increase of crude oil viscosity.

(4) Effect of oil saturation on the effect of enhanced oil recovery by gas flooding and binary flooding.

The level of oil saturation directly affects the crude oil reserves in a certain area of the reservoir, which is affected by the residual oil saturation. if the oil saturation in a certain area of the reservoir is too low, the oil well production is affected because the oil phase relative permeability is too low, thus affecting the recovery efficiency of gas flooding and binary flooding. This study predicts the effect of enhanced oil recovery of gas flooding and binary flooding under the conditions of 30%,40% and 60% oil saturation respectively, so as to clarify the effect of oil saturation on the effect of enhanced oil recovery of gas flooding and binary flooding. The prediction effect of enhanced oil recovery by gas flooding and binary flooding under different oil saturation conditions is shown in figure 6.

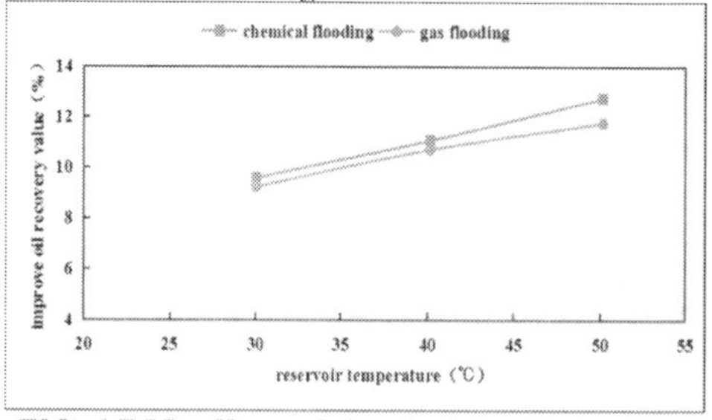

FIG. 6 EOR effects of chemical flooding and gas flooding under different oil saturation conditions

From the chart of predicting the effect of enhanced oil recovery of gas flooding and binary flooding under different oil saturation conditions, it can be seen that the recovery efficiency of gas flooding and binary flooding increases with the increase of the initial oil saturation. When the oil saturation increases from 30% to 50%, the recovery efficiency of binary flooding increases from 9.57% to 12.65%, and the recovery efficiency of gas flooding increases from 9.31% to 11.85%. It shows where the higher the remaining oil saturation after water flooding, the higher the effect of gas flooding and binary flooding to improve oil recovery, At the same time, oil saturation is also an important parameter affecting gas flooding and binary flooding, but it has little variability due to the distribution of initial oil saturation in the reservoir. therefore, in the actual production process, this parameter has little influence on the enhanced oil recovery of gas flooding and binary flooding.

Influence of reservoir thickness on gas flooding and binary flooding for enhanced oil recovery.

Reservoir thickness affects the distribution of oil, water and gas in flooding process.The larger the reservoir thickness is, the more obvious the gravity differentiation of gas, water and oil will be, which will significantly affect the inflow and output dynamics at the production well end.In this study,the effect of enhanced oil recovery gas flooding and binary flooding were predicted under the conditions of oil reservoir thickness of 0.5m, 0.7m, 1.0m and 2.0m, so as to clarify the influence of oil reservoir thickness on the effect of enhanced oil recovery gas flooding and binary flooding.The chart of the effect of enhanced oil recovery gas flooding and binary flooding under different reservoir thicknesses are shown in Figure 7.

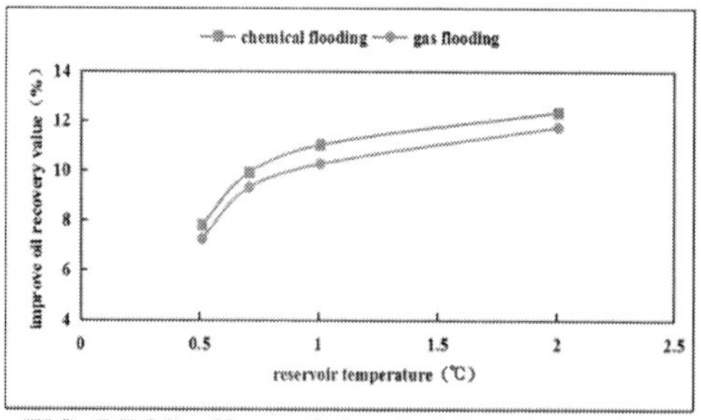

FIG. 7 EOR effects of chemical flooding and gas flooding under different reservoir thicknesses

It can be seen from the chart of the effect of enhanced oil recovery of gas flooding and binary flooding under different reservoir thicknesses,with the increase of reservoir thickness, the effect of enhanced oil recovery of chemical flooding increases first and then flattens out, but the effect of enhanced oil recovery of gas flooding increases first and then slightly decreases.When the reservoir thickness increases from 0.5m to 2.0m, the effect of enhanced oil recovery of binary flooding increases from 7.84% to 12.35%, and the effect of enhanced oil recovery of gas flooding increases from 7.32% to 11.83%.It indicates that the effect of enhanced oil recovery of gas flooding and binary flooding will increase with the increase of reservoir thickness. When the reservoir thickness is too large, a large amount of residual oil will be formed in the reservoir due to binary, injected gas and gravity differentiation of three phases of crude oil, which will affect the effect of enhanced oil recovery of gas flooding and chemical flooding.

Influence of reservoir temperature on gas flooding and binary flooding for enhanced oil recovery.

Reservoir temperature will affect the physical properties of crude oil and injected fluids in the reservoir, which will affect the effect of enhanced oil recovery of gas flooding and Two element drive.In this study, the effect of enhanced oil recovery gas flooding and binary flooding were predicted respectively at reservoir temperatures of 90, 100 and 110, so as to clarify the influence of reservoir temperature on the effect of enhanced oil recovery gas flooding and binary flooding.The chart of the effect of enhanced oil recovery gas flooding and binary flooding under different reservoir temperature conditions is shown in Figure 8.

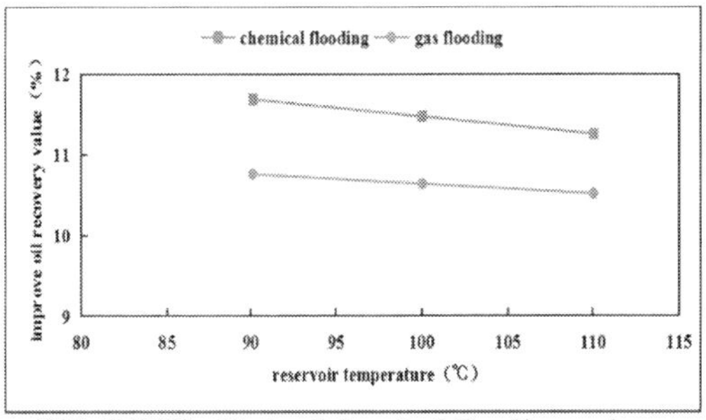

FIG. 8 EOR effects of chemical flooding and gas flooding at different reservoir temperatures

It can be seen from the chart of the effect of enhanced oil recovery of gas flooding and binary flooding under different reservoir temperature,With the increase of reservoir temperature, the effect of enhanced oil recovery gas flooding and binary flooding decreases.When the reservoir temperature increased from 90 to 110, the recovery efficiency of binary flooding decreases from 11.68% to 11.25%,, and the recovery efficiency of gas flooding decreases from 10.78% to 10.35%.However, it can be seen from the effect of enhanced oil recovery that the change of reservoir temperature has little influence on the effect of enhanced oil recovery gas flooding and binary flooding.

Based on the above analysis results, it can be concluded that the key factors for enhanced oil recovery with gas and binary flooding are formation dip Angle, reservoir permeability, crude oil viscosity and reservoir thickness.The influence degree of formation dip > reservoir permeability > reservoir thickness > crude oil viscosity.

4. Conclusion

(1) The EOR potential of chemical flooding and gas flooding was predicted by using typical well group numerical simulation model. The EOR potential of chemical flooding and gas flooding could be increased by 12.33% and 8.31% respectively on the basis of water flooding. Both gas flooding and water flooding can enhance oil recovery in high temperature fault block reservoirs.

(2)Based on the reservoir conditions of the target block, the paper analyzes the influencing factors of formation dip, reservoir permeability, oil viscosity, oil saturation, reservoir thickness and reservoir temperature, and identifies the key factors of EOR in gas flooding and dual flooding as formation dip, reservoir permeability, oil viscosity and reservoir thickness respectively. The influence degree is the formation dip > Reservoir permeability > Reservoir thickness > Viscosity of crude oil.

(3) Based on the numerical simulation method, the sensitive factors were analyzed, and the results showed that the permeability limit was:Chemical flooding & GT. 150 mD, gas flooding & Lt. 150 mD; single layer thickness limit: both chemical flooding and gas flooding should be greater than 0.7 m; reservoir dip limit: both chemical flooding and gas flooding should be greater than 15° ; The viscosity of crude oil should be less than 10 mPa·s, and the oil saturation should be more than 30% .

References

1. Kong Zhaoko, Shuanghe oilfield high temperature reservoir chemical oil displacement agent synthesis and field application. Henan province, Sinopec Henan oilfield branch, 2016-08-29.
2. Cui, Zhang Liang, Ren Shaoran, Zhang Yang. Geochemical characteristics and burial efficiency during CO flooding and sealing in reservoirs [J] . Journal of China University of Petroleum Science, 2017,41(06) : 123-131.
3. Zhang Shaodong, study on microbial flooding technology in high water cut reservoir. Petroleum engineering technology research institute, Sinopec Shengli Oil Field branch, Shandong province, 2013-05-09.
4. Han Jie. Study on polymer flooding and composite flooding technology in high temperature reservoir [D] . China University of Geosciences (Beijing) , 2010.
5. TAO Lei. Pilot study on Enhanced Oil recovery by AMPS polymer flooding in high temperature reservoir [J]. Journal of Oil and Gas Technology, 2011,33(09):141-144 170.

6. Liu Xiaomin. Study on foam system for oil displacement [D] . Kangnam University, 2022. DOI: 10.27169/d. CNKI. Gwqgu. 2022.000095.
7. Weijia Cao,Kun Xie,Bao Cao,Xiangguo Lu,Zhongyuan Tian. Inorganic gel enhanced oil recovery in high temperature reservoir[J]. Journal of Petroleum Science and Engineering,2021,196.
8. Zhao Wenjin. Study on preparation and properties of multi-component copolymer temperature-sensitive gel profile control agent [D] . Lanzhou University of Technology, 2020. DOI: 10.27206/d. CNKI. GGSGU. 2020.000830.
9. Yuan-li Li,Zhao Chen,Xiao-min Liu,Ling Wang, Zheng-gang Cui. 1,3-Dialkyl glyceryl ethers derivatives as surfactant for enhanced oil recovery in high salinity and high temperature reservoir conditions[J]. Colloids and Surfaces A: Physicochemical and Engineering Aspects, 2020, 589(C).
10. Shi Jing, Cao Xulong, Wang Hongyan, Guo Shufeng, Xia Xiran. Complex flooding technology for high temperature and high salinity heavy oil reservoirs in Shengli Oil Field [J] . Special Oil and gas reservoir, 2018, 25(04) : 129-133.
11. Zhou Yumeng. Study on deep profile control technology of high temperature reservoir in block 437 of Shuanghe oilfield [D] . Xi'an Shiyou University, 2018.
12. Tao Jiapin. Study on nitrogen foam flooding system with low interfacial tension in high temperature and high salinity reservoir [d] . China University of Petroleum (East China) , 2018. DOI: 10.27644/d. CNKI. GSYDU.2018.000809.
13. Wu Wei. EOR technology of emulsion Surfactant flooding in ultra-high temperature medium-low permeability reservoirs [J] . Oil and gas geology and recovery, 2018,25(02) : 72-76 + 82. DOI: 10.13673/J. CNKI. CN37-1359/TE. 2018.02.011.
14. Han Qi. Laboratory experimental study on high temperature polymer flooding in block GAO30[d] . China University of Petroleum (East China) , 2017. DOI: 10.27644/d. CNKI. GSYDU.2017.000272.
15. Yuan Chengdong. Study on dispersing gel-Surfactant flooding mechanism in high temperature and high salinity reservoir [d] . Southwest Petroleum University, 2016.
16. Fenglan Zhao,Yunfei Ma,Jirui Hou,Jinxing Tang, Donghai Xie. Feasibility and mechanism of compound flooding of high-temperature reservoirs using organic alkali[J]. Journal of Petroleum Science and Engineering, 2015,135.

Fine dissection of meandering river sand bodies under dense well nets

Ye Liu, Hong Lu, Yun Hu

College of Earth Sciences, Northeast Petroleum University, Daqing, China

Abstract: The main oil layer of Grape Flower in X Oilfield has entered the late stage of high-extra-high water cut, and the remaining oil distribution is uneven, so as to ensure the development effect of the purpose block. Based on the theory of fine sedimentology, the sedimentation characteristics of $PI1_1$ and $PI2_1^2$ deposition time units were studied. Combined with the sedimentary and logging data of the meandering river delta, the logging facies model was classified in detail and established. Eleven microfacies of 2 subfacies in the delta and the delta shunt plain were identified. $PI1_1$ and $PI2_1^2$ deposition microphase layout maps were drawn. According to the sedimentary microfacies and conventional flooded logging interpretation data, four flooding distribution histograms were plotted to obtain the residual oil distribution law. The research results provide theoretical support for the later development adjustment.

Keyword: Deposition time unit; logging phase model; sedimentary microfacies; flooding characteristics; Residual oil distribution.

1. Introduction

Since the early stage of development,XB Development Zone has been explored and developed for more than 60 years, and has experienced the layout of the basic well network, the first encryption adjustment, the second encryption adjustment, the third encryption adjustment, and the polymer flooding [1-6]. The main oil layer of the grape flower has entered the late stage of high-high water cut, with low recovery and very little production, and after long-term water injection, most of the remaining oil is stored in the sand body at different scales. XB Development Zone is a new type of anti-salt polyflooding promotion block, and the degree of sand body cutting and stacking is different, and the internal connection degree is different, resulting in large differences in the residual oil in different sand bodies.

he main reservoir PI1 and PI2 oil layer sand bodies in Q area are mainly dominated by the delta front margin and the delta diverted plain[7], and under the guidance of the principle of sequence stratigraphy, the PI oil formation in the northern part of the Changyuan X oilfield in Daqing was finely divided and compared with the sedimentary time units, and the isochronous stratigraphic lattice of the target layer was established[8]. The logging facies model was established by identifying the logging data under the guidance of the sedimentary model [9]. The flooding characteristics of different sedimentary microfacies were used to further illustrate the important influence of physical properties on the remaining oil.

2. Overview of the study area

The central test area of Q area is located in the northern part of Daqing Changyuan X oilfield in the Songliao Basin [10]. Three rows from Q6 District to the north, and one row to Q8 District to the south. The study area is an anticline structure, which has the characteristics of a gentle structure and basically symmetrical wings. The target layer is the oil layer of grape flower in the first section of the Yaojia Formation. The PI1 reservoir is a delta intra-delta leading edge deposit, and the PI2 reservoir is a delta shunt plain sediment. The PI1 reservoir group and PI2 reservoir group in the test area were divided into $PI1_1$, $PI1_2$, $PI2_1$ and PI22 deposition units, and the $PI2_1$ deposition units were subdivided into $PI2_1^1$ and $PI2_1^2$ deposition units. Take $PI1_2$ and $PI2_1^2$ deposition time units as examples.

3. Sedimentary facies features

3.1 Well logging phase establishment

According to the differences in sedimentary environments in the study area, logging curves such as GR, SP, RMG/RMN were selected for comprehensive analysis and judgment, and 10 logging facies models of the leading edge subfacies in the delta and 11 subfacies modes of the delta shunt plain were established in the study area according to the different energy units (Figure 1).

https://doi.org/10.1051/e3sconf/202337501054

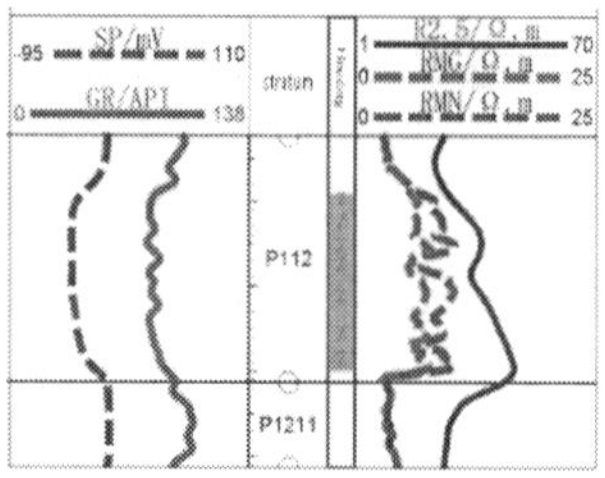

a.Underwater main river channel Q-Q-933 ($PI1_2$)

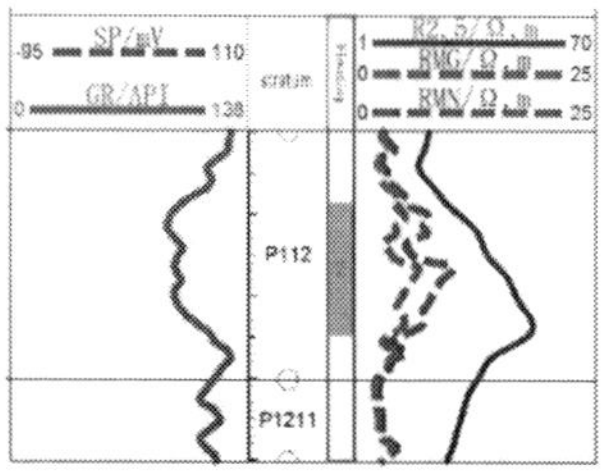

b.Underwater first-class river channel Q-Q-626($PI1_2$)

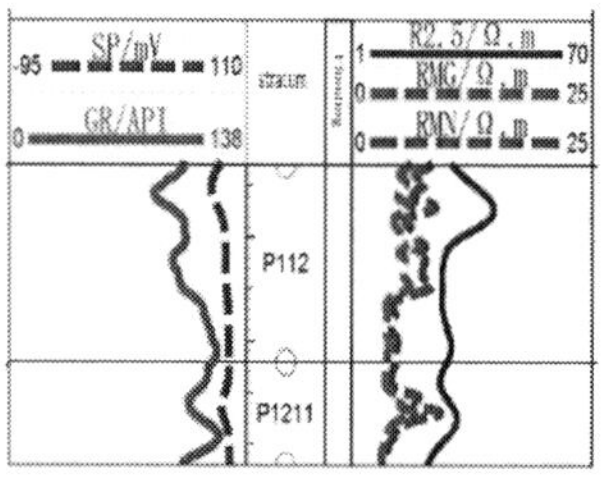

c.Underwater secondary river channel Q-Q-624($PI1_2$)

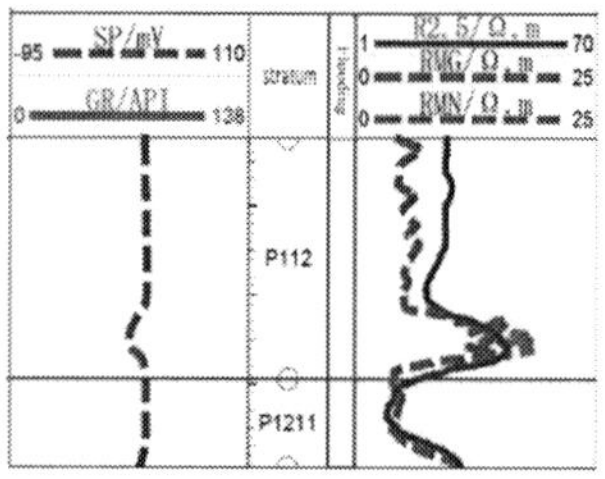

d.Underwater three-level river channel Q-Q-26($PI1_2$)

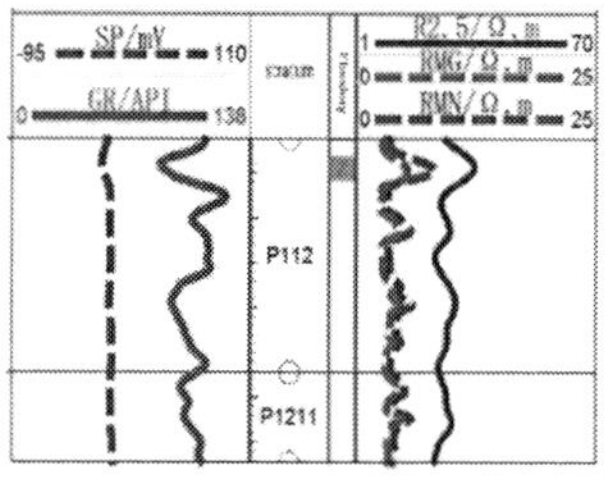

e.Underwater natural dike Q-Q-932($PI1_2$)

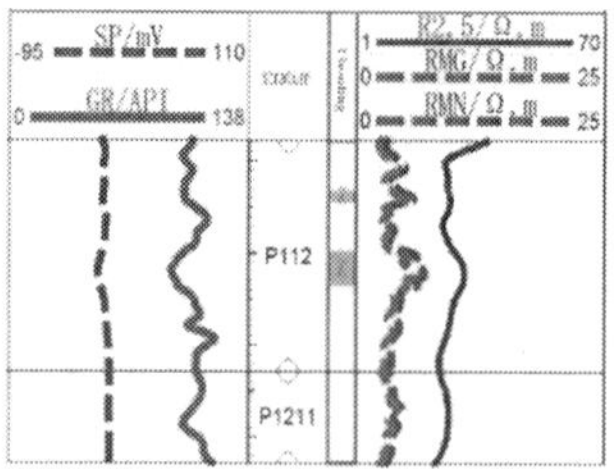

f.Underwater breaking fan Q-Q-930 ($PI1_2$)

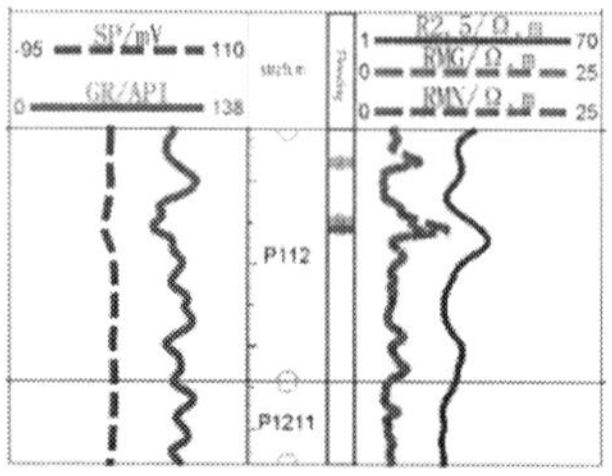

g.Thin layer sand body Q-Q-728($PI1_2$)

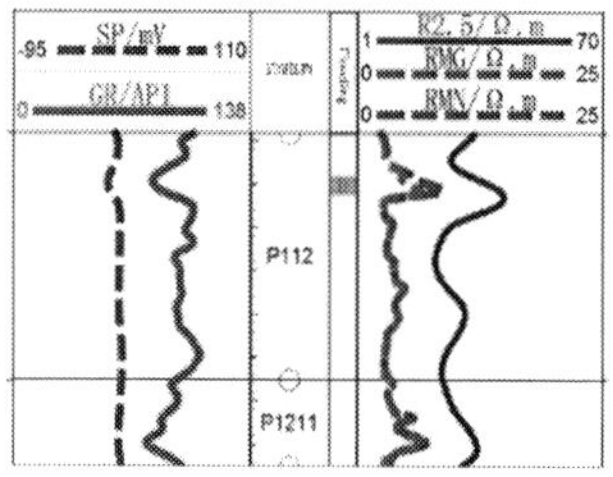

h.Thin layer sand constant Q-Q-729($PI1_2$)

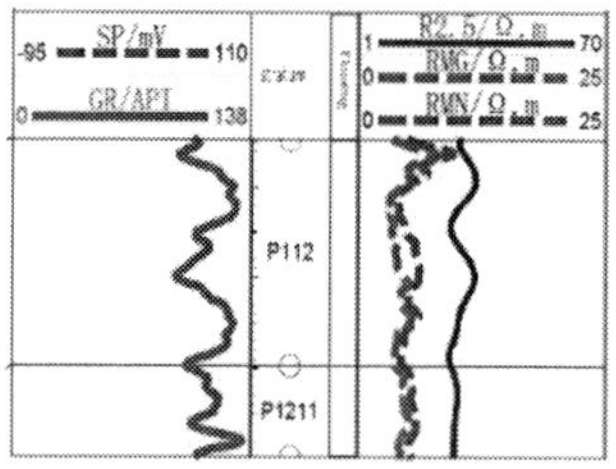

i.Thin layer sand edge Q-Q626($PI1_2$)

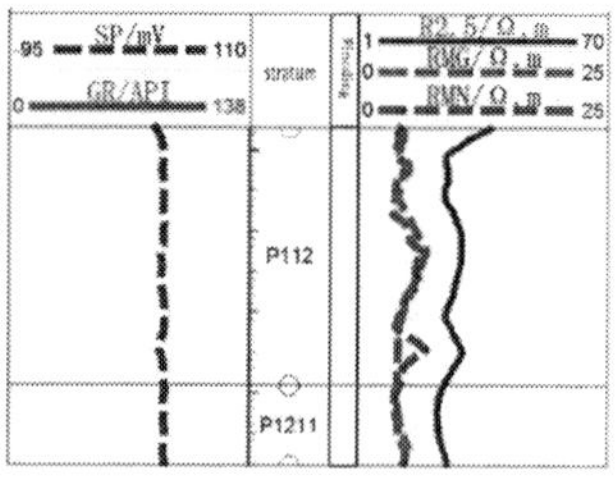

j.Underwater diversion bay Q-Q-123($PI1_2$)

 https://doi.org/10.1051/e3sconf/202337501054

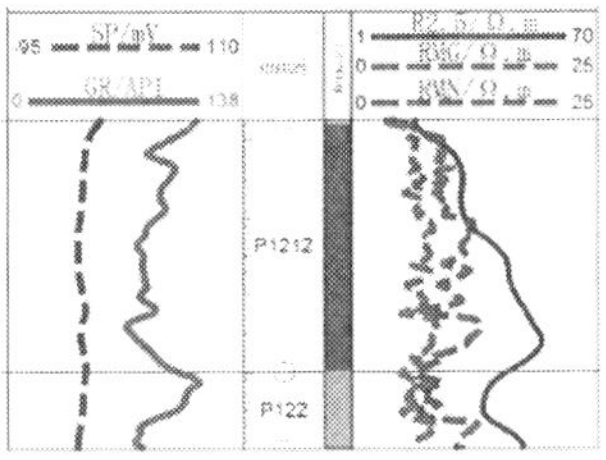

k.Main river channel Q-Q-Q740($PI2_1^2$)

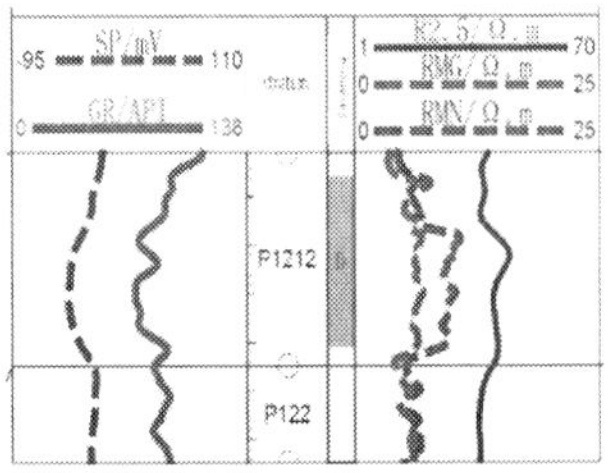

l.I-type river channel Q-Q-Q945($PI2_1^2$)

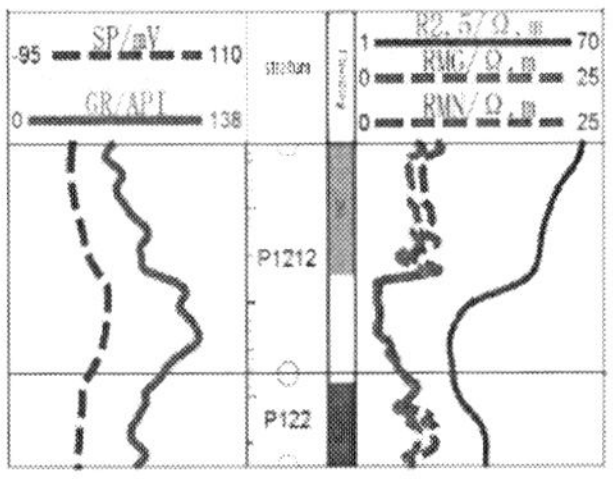

m.П-type river channel Q-Q-Q946($PI2_1^2$)

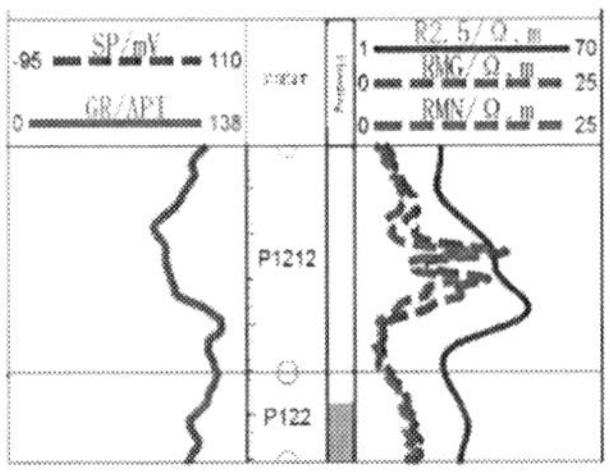

n.III-type river channel Q-Q-635($PI2_1^2$)

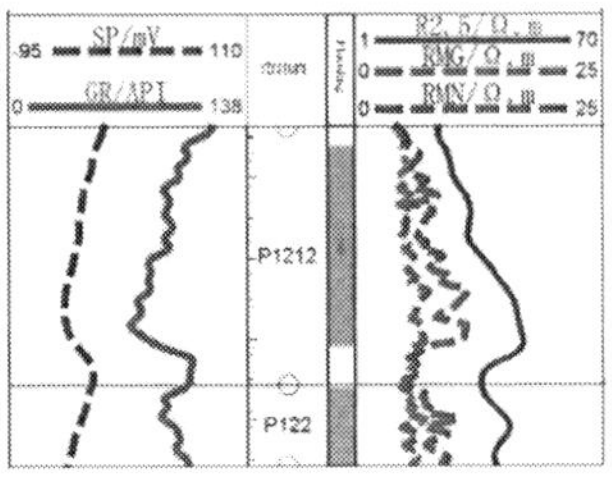

o.Abandoned river channel Q-Q-739($PI2_1^2$)

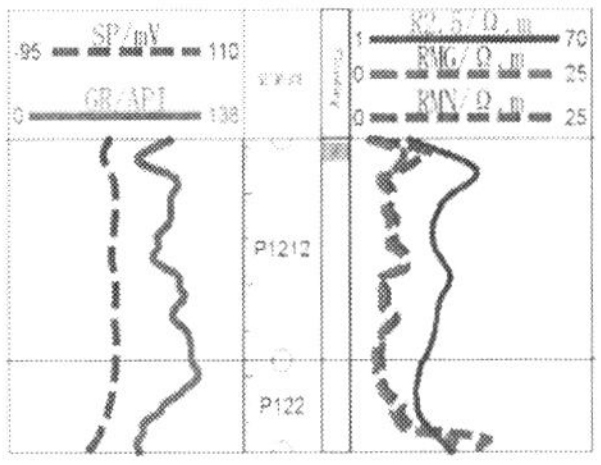

p.Natural dike Q-Q-Q635($PI2_1^2$)

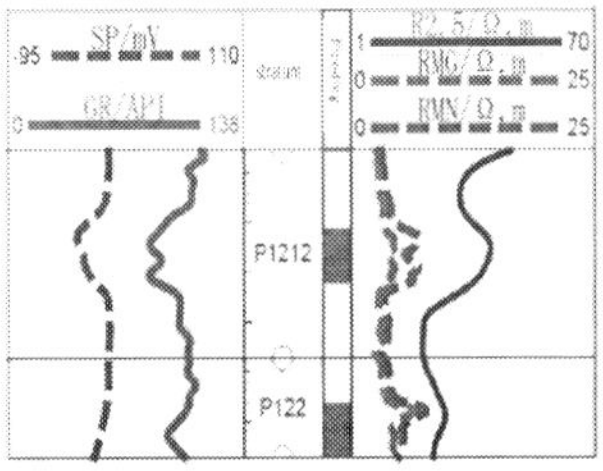

q.Break the fan Q-Q-Q943($PI2_1^2$)

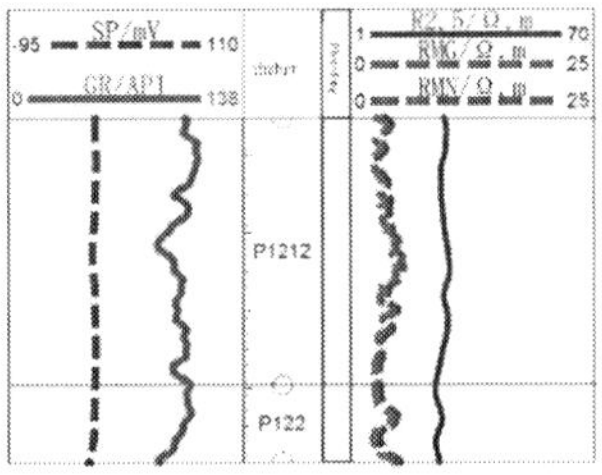

r.Overflow sand body Q-Q-951($PI2_1^2$)

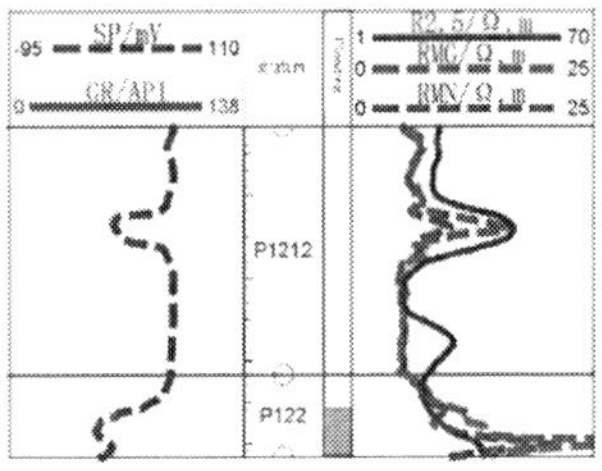

s.Overflow sand constant Q-Q-136($PI2_1^2$)

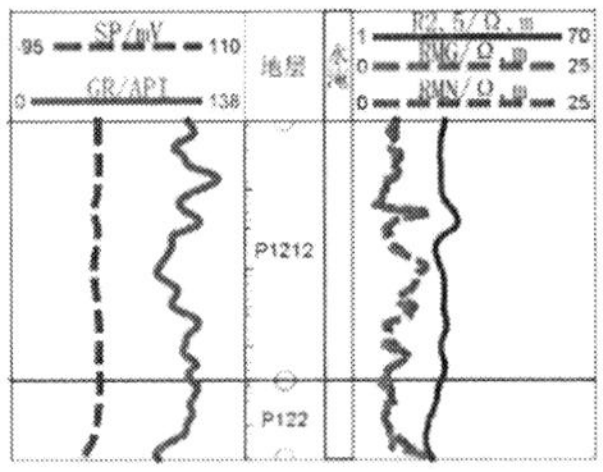

t.Overflow sand edge Q-Q-745($PI2_1^2$)

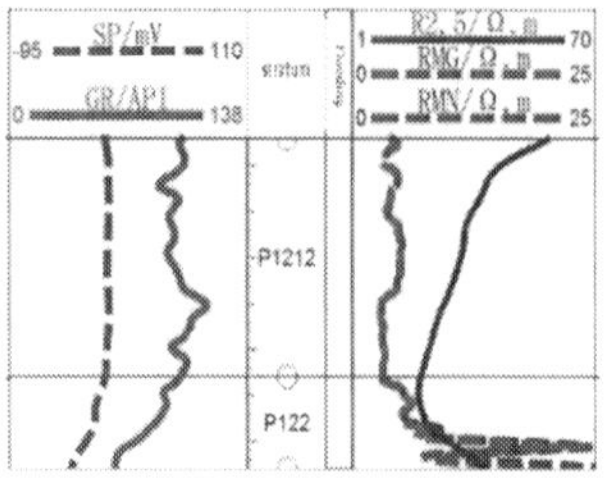

u.Shunt mud Q-Q-Q745(PI2_1^2)

Fig .1 Sedimentary microfacies logging model

3.2 Intra-delta leading microfacies

(1)Underwater diversion channel: it is mainly divided into main river channel, type I channel, Π type channel and III type river channel. The main river channel as a whole is high-extremely high amplitude and amplitude difference. The logging curves such as GR, SP, RMG/RMN are thick box type. The overall channel of type I is high to medium amplitude, and the logging curves such as GR, SP, RMG/RMN are thick box type. The overall channel of Π type is high to medium amplitude, the GR and SP logging curves are bell-shaped, and the RMG/RMN logging curve is box-shaped. The Ш-type river channel is high to medium amplitude as a whole, and the whole bell shape.
(2)Underwater natural dike: the underwater extension of onshore natural dike, GR, RMN/RMG and other logging curves are seriously toothed, and the lithology is fine sand and silt.
(3)Underwater Breaking Fan: Located on the side of an underwater diversion channel, it has characteristics similar to that of an underwater natural dike, with a coarser grain size than an underwater natural dike.
(4)Mat-shaped sand: It is not limited by river channels and develops in a large area, and the degree of development is greater than that of overflow sand.
(5)Underwater diversion bay: A relatively concave bay area between underwater diversion channels. GR, RMN/RMG and other logging curves are seriously dentified, and the lithology is mainly mud.

3.3 Delta shunt plain microfacies

(1)Diversion river: GR, SP, RMG/RMN and other logging curves are selected for analysis, and according to the shape and thickness of the logging curve, the diversion river channel is divided into main river, primary river, secondary river channel and tertiary river. In general, the main river channel has a high-very high amplitude and amplitude difference; The logging curves such as GR, SP, RMG/RMN are thick box type. The overall first-class river course is high-medium amplitude. The secondary rivers were generally high to medium amplitude, the GR and SP logging curves were bell-shaped, and the RMG/RMN logging curves were box-shaped. The overall height of the tertiary channel is high-medium amplitude, and the logging curve is bell-shaped.
(2)Abandoned river channel: SP and GR show a bell-shaped positive rhythm, the thickness is thin, the upper part is dominated by sand and mud interlayer, and the bottom is generally high.
(3)Onshore natural embankment: the amplitude of the logging curves such as GR, RMN/RMG is low, the peak of the GR and SP logging curves is visible locally, and the sand and mud interlayer on the logging curve. The lithology is dominated by silt and silty clay.
(4)Breaking fan: The logging curves such as GR, SP, RMG/RMN are funnel-shaped, and the thickness of the breaking fan is thinner than that of other sand bodies. The grain size is slightly thicker than that of natural, and the lithology is mainly fine sandstone and siltstone.
(5)Sludge between splits: GR, SP and other logging curves are approximate to mudstone baselines. The main lithology is mud with small amounts of silt and fine sand.
(6)Overflow sand: the logging curve is finger-shaped, and the vertical thickness is thin.

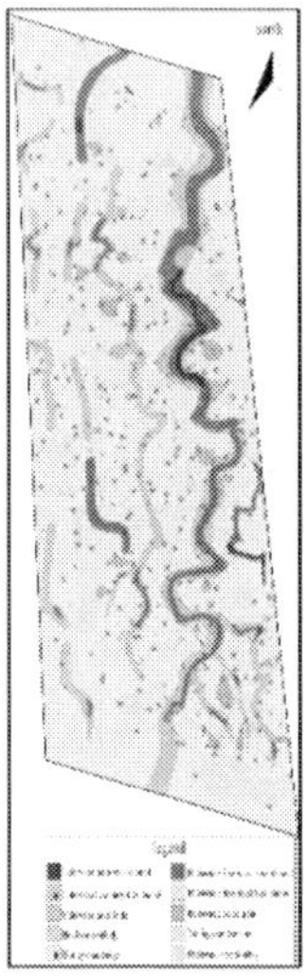

a.PI1_2 unit deposition microfacies

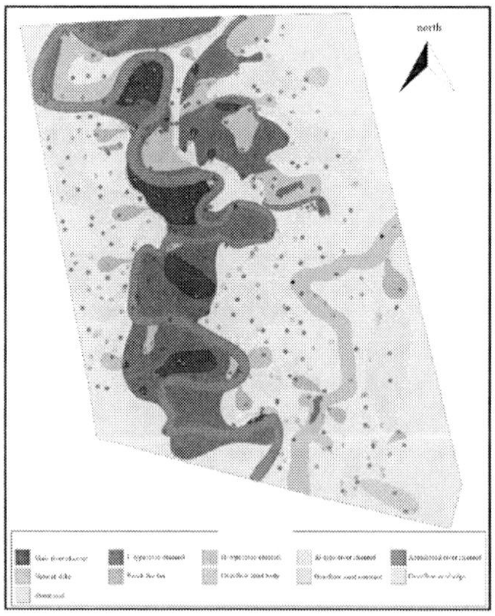

b.PI2_1^2 unit deposition microfacies

Fig.2 Layout of the microfacies plane of each small layer sedimentation

4. Residual oil distribution characteristics

4.1 Effect of different sand bodies on the remaining oil

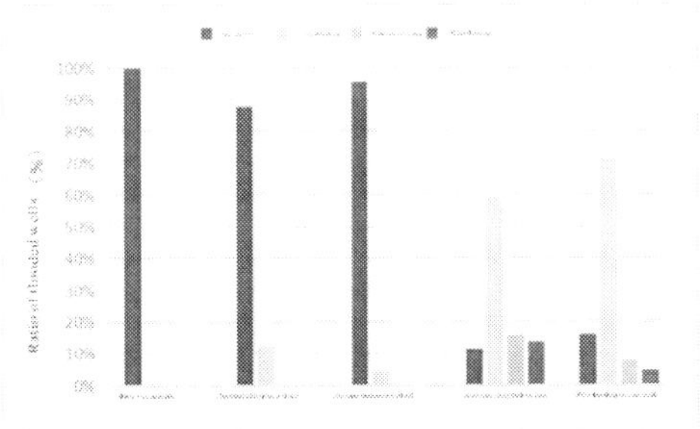

a.Histogram of the distribution of flooded wells in different batches of thin-bed sand network

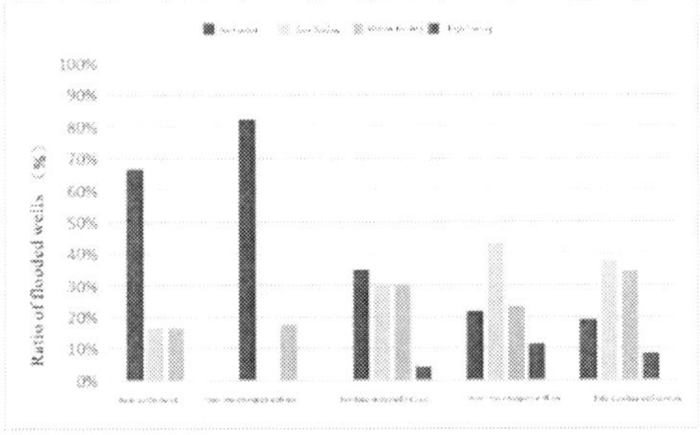

b.Histogram of the distribution of the ratio of flooded wells in different batches of well networks in underwater diversion channels

Fig.3 Histogram of flooding distribution of different microfacies of $PI1_2$

In summary, from the basic well network to the poly-flooded well network, the number of flooded wells in the underwater diversion channel increased, the proportion of unflooded wells in thin-bed sand gradually decreased, the proportion of low flooded wells gradually increased, and the ratio of medium and high flooded wells showed a decreasing trend. As a result, the residual oil stored in thin layer sand is higher and the residual oil stored in the underwater diversion channel is low.

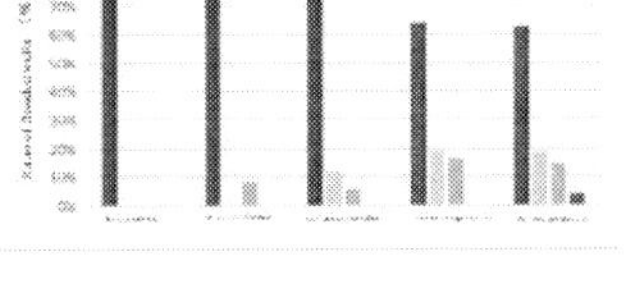

a.Histogram of the distribution of flooded wells in different batches of overflow sand

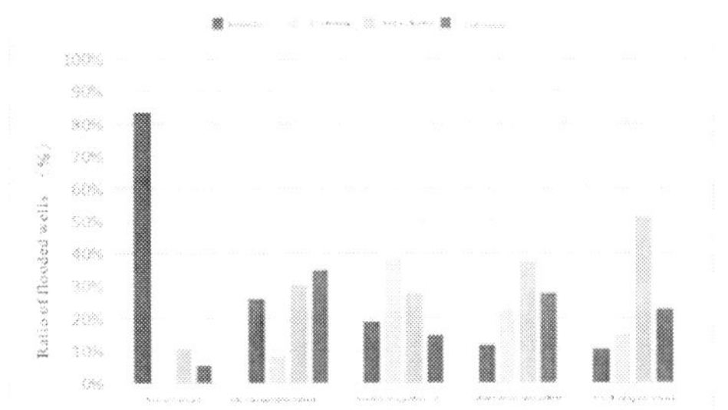

b.Histogram of the distribution of the ratio of flooded wells in different batches of well networks in the diversion channels

Fig.4 Histogram of flooding distribution of different microfacies of $PI2_1^2$

The different types of energy phases, flooded wells and flooded well ratios at different stages in $PI2_1^2$ deposition time unit in the target area were studied. The results show that in the basic dense well network, the ratio of unflooded wells in the diversion channel is the highest, and in other batches of well networks, the ratio of unflooded wells gradually decreases, and finally drops to 10.79%. The ratio of the number of wells in medium water is high, and the ratio of the number of wells can reach 11.11%, and the ratio of high flooding is 5.56%. In the diversion channel, the ratio of well number from the basic well network to the low flood in the secondary well network increased to 38.30%, and the low flood ratio from the tertiary well network to the poly flood well network decreased, and the well number ratio decreased to 15.11%. For the ratio of the number of wells flooded in medium water, the ratio of well number can reach 30.43% from the basic well network to the primary infill well network, and then decreases to 27.66% by the secondary infill well network, and the maximum ratio increases to 51.08% from the tertiary infill well network to the poly flood well network. In the basic well network, primary infill well network, secondary infill well network, tertiary well network and poly flood well network, high flooding phenomenon appeared, and showed a trend of first increasing and then decreasing and then rising and decreasing [10].

5. Conclusion

(1)The sedimentation of Q area of X oilfield is the leading edge subfacies in the delta and the delta divergence plain subfacies, and 11 microfacies are mainly developed, namely underwater diversion channel, underwater natural dike, underwater break fan, mat-shaped sand, and underwater diversion bay. Diversion of river channels, abandoned river channels, onshore natural embankments, breaking fans, diversion of mud, overflow sand.

(2)On the plane, the size and geometry of the sand body have a close influence on the distribution of residual oil, and the reservoir has good physical properties, which is prone to flooding and low residual oil content. The reservoir has poor physical properties and is not easy to flood, or the flooding condition is not good, and the residual oil content is high.

References

1. SHI Chengfang,WU Xiaohui. Development mode and evolution trend of La, Sa and Xing oilfields[J].Petroleum Geology & Development in Daqing, 2019,38(05):45-50.
2. Wang Yao. Adjustment and effect analysis of well network in the west layer of the first and second rows of the north[J].Journal of Petrochemical Universities, 2019,32(02):59-65.
3. SONG Guofen,WANG Chunli,LU Xiangguo. Oil-displacement agent and reservoir rock interaction and its influence on seepage characteristics[J].Petroleum Geology & Development in Daqing,2017,36(02):95-101.
4. LIU Xia,ZHANG Ye. Study on auxiliary design and analysis system of adjusting well drilling geology[J].Fault-Block Oil & Gas Field, 2015, 22(06):825-828.
5. Ma Zhonghui. Analysis of water injection effect of shortening test period in a block[J].Science and Technology and Enterprise,2015(13):155.
6. YANG Fengbo. Research and application of comprehensive potential mining technology in ultra-high water cut period of Lamadian Oilfield[D]. Daqing Petroleum University, 2006.
7. XU Xiao. Study on sedimentary facies and diagenetic facies of Buyeo oil layer in the Fuxin uplift belt in southern Songliao Basin[D].China University of Petroleum(East China),2016.
8. ZHANG Jingjun,LIU Chengzhi, JIANG Guochao, FAN Tongsheng. High-resolution sequence stratigraphy of PI-SII reservoir in Xingnan area[J]. Science Technology and Engineering, 2011, 11 (14):3182-3187+3191.
9. LENG Xingbo. Study on sedimentary microfacies of Zhao261 block of Tooutai Oilfield[J].Inner Mongolia Petrochemical Industry,2010,36(08):223-224.
10. LI Zhandong, ZHANG Lishuang, LI Li, LIANG Shun, SHI Hao, TIAN Mi, WANG Yang, ZHANG Shuxin. Journal of Liaoning Shihua University, 2017,37(04):34-38+43.

 https://doi.org/10.1051/e3sconf/202337501055

Application of multiphase flow fiber optic fluid production profile testing technology in W1 well

Fuji Yu

Daqing Oilfield Liability Company No.8 Oil Production Plant, Daqing, Heilongjiang Province, China

Abstract. In recent years, the number of completions of tight oil horizontal wells in oil field X has increased year by year. Horizontal well fracturing technology has been widely used in the development of tight oil reservoirs in oil field X. However, there are still some technical difficulties in the dynamic monitoring of horizontal wells. Monitoring is mainly for inclined vertical wells, and there are few dynamic monitoring methods and means for horizontal wells. Therefore, the production of each cluster after fracturing in horizontal wells has not been accurately monitored. To obtain the output of each cluster after fracturing in horizontal wells, analyze and evaluate fracturing effects, and guide the preparation of fracturing plans, it is necessary to conduct fiber-optic fluid production profile testing and research. Optical fiber monitoring technology for horizontal wells has developed particularly rapidly in recent years. At present, a relatively mature distributed optical fiber temperature DTS supporting technology has been formed. It is widely used in the monitoring of unconventional reservoirs in some large oil fields abroad, meeting the actual monitoring needs, and has achieved good results. better application effect. This paper expounds the basic principle and well selection principle of horizontal well optical fiber testing, and introduces the knowledge obtained in practical application. It is of great significance to accelerate the application of horizontal well optical fiber monitoring technology in the development of tight oil reservoirs in X oil field, improve the dynamic monitoring of horizontal wells and evaluate the fracturing effect.

Keywords: Output of each cluster of horizontal wells; Fiber Monitoring; Liquid production profile test.

1. Test principle of multiphase flow fiber optic liquid production profile

1.1 Optical fiber monitoring principle

Fiber optic monitoring (DTS) in horizontal wells uses the Raman effect for temperature measurements. A laser sends a pulse of light along the fiber. Some light is reflected back as backscatter, which can be analyzed to separate the backscattered light from the incident pulse. , and filtered into discrete wavelength signals, the speed of light remains constant, and backscattered light measurements per meter of fiber can be obtained.

In Raman scattering, the energy transfer between scattering molecules and photons is related to temperature. When the scattered light appears at a frequency lower than that of the incident light, the scattered photons release energy to the medium in the form of an extra phonon in the interaction. The resulting new spectral signature is called the Stokes band. In the opposite case, scattering increases the frequency and energy is transferred from the medium to the light. This is called anti-Stokes scattering. The anti-Stokes band is more sensitive to temperature than the Stokes band. Both The ratio of the signals is often used for temperature estimation. The Stokes signal with a longer wavelength has a very weak temperature sensitivity, and the backscattered light of the anti-Stokes signal with a shorter wavelength has a strong sensitivity to temperature. The ratio of the anti-Stokes to Stokes signal is proportional to the temperature of the scattering medium. Based on the Raman scattering principle, the scattered light intensity is measured to obtain the temperature distributed along the fiber. The temperature point is located by measuring the time when the backscattered light returns to the starting end. Similar to the radar echo technology, the distributed temperature measurement of the fiber is realized.

* Corresponding author: yufuji@petrochina.com.cn

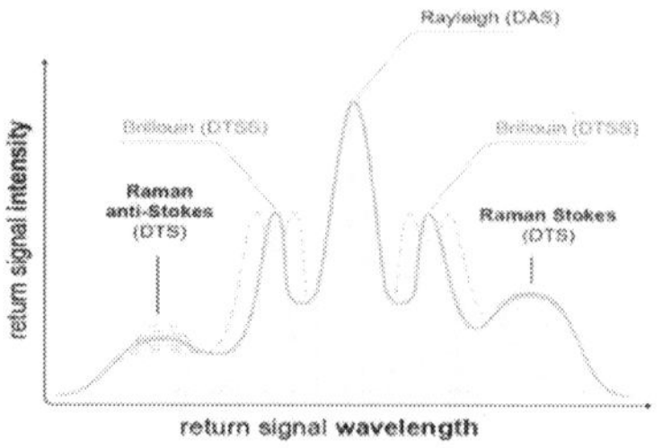

Fig. 1. Inverse discrete spectroscopy

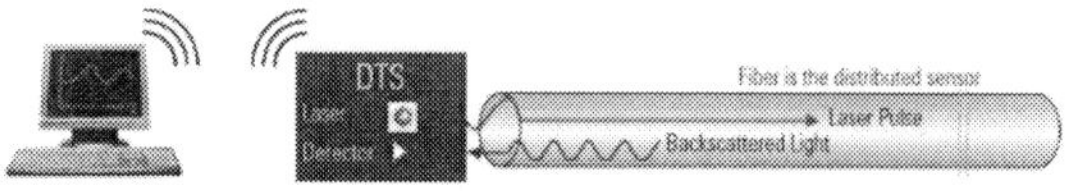

Fig. 2. Schematic diagram of horizontal well fiber test.

2. Application in tight oil horizontal wells

2.1 Test well selection

In order to measure the production of each cluster, comprehensively analyze and evaluate the fracturing effect, Well W1 with a longer horizontal section was selected for testing. Well W1 is located in block Z. The block is located in the east wing of the downdip part of the F nose structure. It is located on the west side of the F nose structure in the Q Sag. It develops fluvial-delta facies deposits. microphase. Main layer FI7 sublayer and dessert layer FI4 sublayer. The coiled tubing fiber optic technology is used to monitor the liquid production profile. By formulating different supporting test systems, the liquid production capacity of a single section (cluster) of horizontal wells under different working systems is monitored to ensure the effective identification of the contribution rate of oil production and water production of each section (cluster). The liquid production capacity of each segment (cluster) was qualitatively analyzed. After the construction is completed, the liquid production capacity of each section (cluster) is quantitatively analyzed, and the oil production and water production status of the subsection (cluster), and the production contribution rate of the section (cluster) are given.

The target layer of Well W1, layer FI7, has a drilling depth of 4670m and a horizontal section of 2472m. 2096m of sandstone were encountered, the sandstone drilling rate was 84.79%, 1974m of oil-bearing sandstone was encountered, and the oil-bearing sandstone was drilled with a drilling rate of 79.85%, including 1637m of oil immersion and 337m of oil spots. The thickness of the target sandstone is 5.0m, the effective thickness is 4.4m, the porosity is 11.7%, and the permeability is 1.2mD. According to the temperature data of 23 surrounding layers, the temperature of Y oil layer is 86.0℃-93.3℃, and the average geothermal gradient is 4.84℃/100m. The formation pressure gradient of Y oil layer in the well area is 0.9~1.08MPa/100m, with an average of 0.97MPa/100m, which is a normal pressure system reservoir.

The first section of the well is perforated by tubing, the second to twenty-seventh sections of cable transmission composite bridge plugs; the casing multi-section and multi-cluster volume fracturing 27 sections and 163 clusters; the average cluster spacing of infill is 15m; Volume 2200m^3, proppant 220-240m^3; Paragraph 10-13, liquid volume 1500m^3, proppant 175m^3; Paragraph 14, volume 1753m^3, proppant 210m^3; Paragraph 17, volume 1044m^3, proppant 105m^3; Paragraph 15-16,18-27 section liquid volume 2000m^3, proppant 240-280m^3; total liquid volume 53468m^3 proppant 6421.8m^3.

2.2 Coiled tubing fiber optic data acquisition

2.2.1 Monitoring feasibility analysis

In order to successfully complete this optical fiber test task, the factors that may lead to logging failure are analyzed as follows:

1) Whether the design and production system can be achieved through the control of the nozzle.

2) The casing-to-coiled tubing cannot reach the target layer due to fracturing.

3) Coiled tubing self-locking caused by bottom hole pressure and wellbore trajectory.

4) During the test, the coiled tubing was stuck due to sand production in the well. Based on the above analysis, the well can be adjusted to design the test production system and meet the test requirements. There is no self-locking during the running of the coiled tubing, the wellbore is in good condition, and there is no risk of encountering resistance caused by casing changes, and the coiled tubing can be run into the test target layer. There is no sand production during single well drainage and adjacent well drainage, and it is judged that there is no sand burial risk during the test.

2.2.2 Monitoring main equipment

Table 1.Horizontal well fiber optic test equipment statistics table

Classification	Heading	Models and Specifications	Amount
Collection equipment	DTS modulation equipment	AP Sensing, DTS N4415A, N4386B 1550nm、1064nm	3piece
	DTS Analysis software	DTS configurator ISP Analysis	3set
	Auxiliary equipment	Optical fiber fusion splicer Optical cable connection protection Connect the fiber optic cable	1set
Ground equipment	Coiled tubing fiber optic equipment	Coiled Tubing Vehicle Coiled Tubing Fiber Optic Coiled Tubing Injection Head	1set
Downhole tools	Storage pressure gauge	Storage pressure gauge	2set

2.2.3 Fiber monitoring situation

After the fiber optic coiled tubing enters the well to the bottom, it is circulated to monitor the ground temperature baseline, and the monitoring time is 6 hours. Carry out the first production system, the monitoring time is 12 hours, and obtain production data to meet the data analysis requirements; carry out the second production system monitoring, the monitoring time is 12 hours, and obtain production data to meet the data analysis requirements; carry out the third production system monitoring, monitoring After 12 hours, the production data was obtained to meet the data analysis requirements; the monitoring was completed; the fiber optic coiled tubing was pulled out of the wellbore; the site was sorted out, and all construction operations ended.

In this well logging, shut-in, 6.35mm nozzle, 7.94mm nozzle, and 9.53mm nozzle production were carried out to monitor the output under the three nozzle production systems. Combining all the data to carry out calibration analysis, quantitatively explain the output of each level under the three production systems. The DTS test method is used to carry out continuous monitoring of the whole wellbore, to complete the interpretation of distributed temperature data, to carry out the correction and interpretation of gas production profiles in all directions and from multiple angles; to carry out comparative analysis of geological engineering parameters.

2.3 Coiled tubing fiber optic test results interpretation

The 6.35mm nozzle has a production rate of about $1.0m^3/h$. Figure 3 shows the production contribution percentage of each perforation section in the model (from the heel end to the toe end of the horizontal well, the perforation sections are S27-S14 in sequence). It can be seen that S25 -S27 section has a low contribution to the liquid production, and each section towards the toe end has a relatively high contribution to the liquid production. The specific liquid production contribution percentage of each section is shown in the table.

The 7.94mm nozzle, the production is about $1.26m^3/h$ Figure 4 shows the production contribution percentage of each perforation section in the model (from the heel to the toe of the horizontal well, the perforation sections are S27-S14 in sequence). It can be seen that the liquid production contribution of S25-S27 sections is low, and the liquid production contribution of each section towards the toe end is relatively high. The specific liquid production contribution percentage of each section is shown in the table.

The 9.53mm nozzle, the output is about $1.48m^3/h$. Figure 5 shows the production contribution percentage of each perforation section in the model (from the heel to the toe of the horizontal well, the perforation sections are S27-S14 in sequence). It can be seen that the liquid production contribution of S25-S27 sections is low, and the liquid production contribution of each section towards the toe end is relatively high. The specific liquid production contribution percentage of each section is shown in the table.

It can be seen from the inversion results that under different nozzles, the production contribution ratio of each perforation section is roughly the same. The contribution of S25-S27 stage liquid production is relatively low, and the S15-S16 and S18-S21 stage liquid production contribution is relatively high. The interpretation results of the fluid production contribution of Well W1 are shown in the figure. Under different nozzles, the fluid production contribution of each section is roughly the same. The contribution of S15-S21 is higher, while the contribution of S25-S27 is lower.Under different nozzles, the production contribution ratio of each perforation cluster is roughly the same. 83-88, 93-94, 97-124, 136-137 clusters contributed relatively high liquid production.

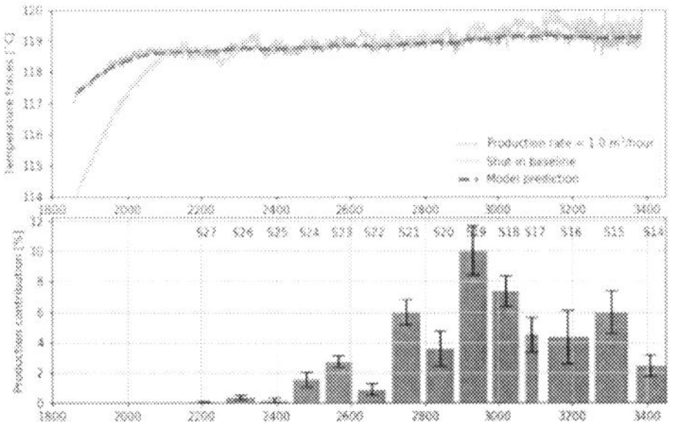

Fig.3.contribution to various sections under the 6.35mm nozzle

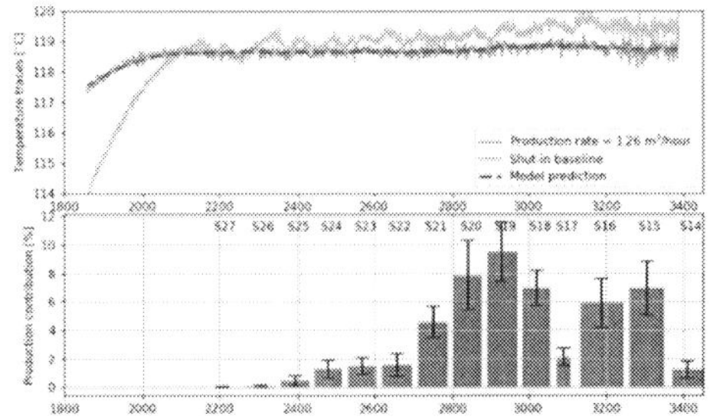

Fig.4.contribution to various sections under the 7.94mm nozzle

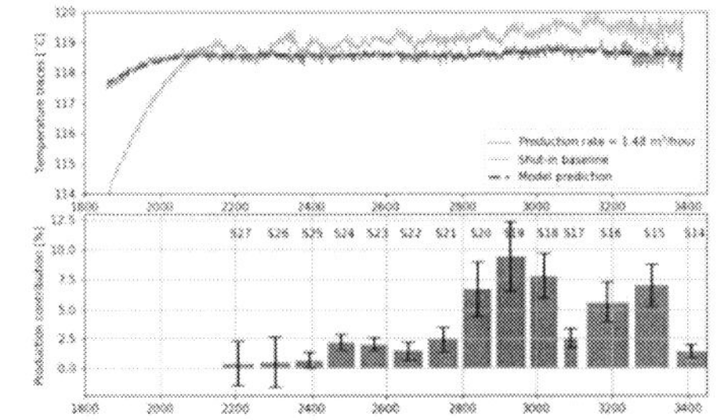

Fig.5. contribution to various sections under the 9.53mm nozzle

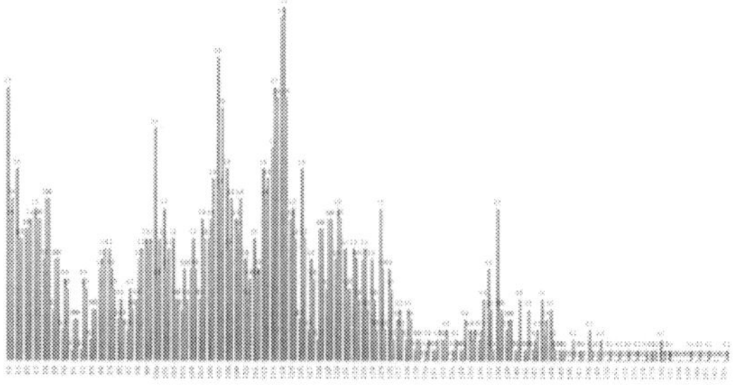

Fig.6.Contribution of each cluster fluid production under different nozzles

6.35mm nozz 7.94mm nozz 9.53mm nozz

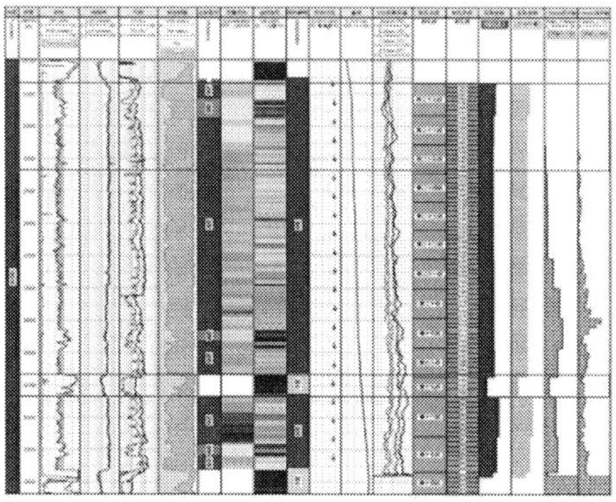

Fig.7.Production liquid contribution

2.4 Comparative analysis of test results and drilling data

Section 1-4 of Well W1 has relatively low natural gamma values and relatively high quantitative fluorescence values. The formation sand bodies are well developed and have good oil content (physical properties). The fiber test results show that they have a high contribution to productivity. The fiber monitoring proves that this section has It is the main oil-producing interval of horizontal wells.

The natural gamma value of the fifth section of Well W1 is relatively low, and the formation sand bodies are developed. The quantitative fluorescence value is lower than that of the 1-4 section, and the oil content (physical property) is worse than that of the 1-4 section. The test results show that the productivity contribution is slightly lower, and the well section can be selected as a re-fracturing in the later stage.

The sixth section of Well W1, the natural gamma value is relatively high, the quantitative fluorescence value is relatively low, and the oil-bearing formation is relatively poor. The test results show that there is basically no contribution to productivity. The later fracturing plan can consider no fracturing or increase in this type of reservoir. Intercluster fracturing.

The seventeenth section of Well W1 has high shale content and has not encountered oil-bearing sandstone. The fracturing process adopts the penetrating fracturing process. Under different nozzles, the fluid production contribution of the S17 section is equivalent to that of the 16th and 19th sections that encountered oil-bearing sandstone. The effect of well penetration fracturing is obvious and effective.

3. Conclusion and understanding

Optical fiber monitoring can continuously record the temperature data of the whole well section. Compared with conventional tracer testing, it can effectively reflect the production of each cluster of oil and water.

Since the starting pressure gradient of the oil phase is higher than that of the water phase, through the comparative analysis of the oil-water ratio in the flowback under different nozzles, amplifying the pressure difference can effectively drive the oil production and reduce the water content.

The well section with low contribution rate of optical fiber monitoring is an important potential interval in the later stage. The horizontal well penetration section has a certain fracturing effect. In the later stage, when the geological understanding is sufficient, the well section can be preferably implemented for the penetration fracture.

References

1. Xu Bangcai. Application test of coiled tubing optical fiber gas production profile testing technology [J]. Journal of Jianghan Petroleum Staff University, 2016, 2901: 26-29.
2. Wang Weijia. Coiled tubing optical fiber logging technology and its application in shale gas wells [J]. Petroleum Drilling and Production Technology, 2016, 3802: 206-209.
3. Feng Xiaowei, Zhao Yi, Yang Peng, Zhou Jincheng. Application of distributed optical fiber temperature monitoring technology in production profile interpretation of fracturing horizontal wells [J]. Reservoir Evaluation and Development, 2021, 1104: 542-549.
4. Liu Jiancheng, Duan Yinlu, Wen Wen, Zheng Xiaomin, Pei Yang, Du Xu. Application of Coiled Tubing Optical Fiber Logging Technology in Gas Production Profile [J]. Logging Technology, 2021, 4503: 253-259.
5. Luo Hongwen, Li Haitao, An Shujie, Xin Ye, Li Lei, Zou Shunliang, Li Ying, Liu Chang. Analysis of factors affecting the temperature profile of fracturing horizontal wells in tight gas reservoirs [J]. Special Oil and Gas Reservoirs, 2021, 2804: 150-157 .
6. Liu Weiming, Li Haitao, Wang Yongqing, Luo Hongwen, Shao Changchun. Experiment on temperature distribution characteristics of horizontal wells in gas reservoirs based on DTS test [J]. Fault block oil and gas fields, 2020, 2702: 228-232.
7. Dai Yuexiang. Logging technology and application analysis of distributed optical fiber well temperature and fluid production profile in horizontal wells [A]. Xi'an Petroleum University, China University of

Petroleum (East China), Shaanxi Petroleum Society. Proceedings of the 2021 International Conference on Oil and Gas Field Exploration and Development (Volume 1) [C]. Xi'an Petroleum University, China University of Petroleum (East China), Shaanxi Petroleum Society:, 2021:9.

8. Zhang Ruiduo, Duan Yonggang, Cai Junjun. A new method for predicting horizontal wellbore temperature profile [J]. Complex Oil and Gas Reservoirs, 2017, 1003: 39-43.
9. Long Hua. Research and application of horizontal well fluid production profile testing technology [J]. Chemical Management, 2014, 30: 51-52.
10. Zhu Shiyan. Theoretical Research on Interpretation of Horizontal Well Production Profile Based on Distributed Optical Fiber Temperature Test [D]. Southwest Petroleum University, 2016.
11. Li Haitao, Luo Hongwen, Xiang Yuxing, An Shujie, Li Ying, Jiang Beibei, Xie Bin, Xin Ye. Application status and prospect of DTS/DAS technology in fracturing monitoring of horizontal wells [J]. Xinjiang Petroleum Natural Gas, 2021, 1704: 62-73.
12. Li Yahui. Interpretation model and realization of horizontal well production profile in bottom water gas reservoirs based on DTS data [D]. Southwest Petroleum University, 2018.
13. Yan Zhenghe. A new method for monitoring horizontal well production profile [J]. Oil and Gas Well Testing, 2022, 3102: 49-56.

E3S Web of Conferences 375, 01056 (2023)
ESAT 2023
https://doi.org/10.1051/e3sconf/202337501056

Research Progress on Stability Analysis Methods of Granite Residual Soil Slope

Zian Wang [1], Chao Zheng [1], Le Zhang [1], Kangwei Xiong [1], Zhanlin Li [2]

[1] Guangzhou Environmental Services Co., LTD., Guangzhou 510000, China
[2] School of Civil and Transportation Engineering, Guangdong University of Technology, Guangzhou, Guangdong 510006, China

Abstract: Based on the instability and failure of granite residual soil slope, this paper will introduce the stability of granite residual soil slope from theoretical research, indoor and outdoor simulation tests, and numerical analysis. Based on the review of previous work and the latest research results, the stability analysis of granite residual soil slope is discussed, and the main influencing factors of granite residual soil slope instability are summarized. The main factors are: its own disintegration property; The degree of microcracks in its internal structure.

Key words: Granite residual soil; Slope instability; Theoretical research; Indoor and outdoor simulation test; Numerical analysis; influence factor.

1. Introduction

Since the 20th century, with the rapid development of urban construction, in order to ensure the highway safety, ensure the stability of embankment or cutting slope, save money, and reduce the land area as much as possible, artificial slopes formed by construction are more and more common. Among them, in Fujian, Guangdong and other coastal areas, granite residual soil is widely distributed, which accounts for more than 40% of the total slope. Through investigation, it is found that granite and residual soil account for about 35% of the total area of granite residual soil slope in Guangdong and Fujian respectively. In Guangxi, Hunan and Jiangxi, granite residual soil blocks account for about 15% of the total land area of the country [1], and the granite residual soil slopes built thereby are also widely distributed. Due to the disintegration of granite residual soil, granite residual soil slope has also become a potential factor endangering urban safety. According to research, more than 60% of countries in the world have suffered disasters in granite residual soil engineering [2]. Fig.1 shows the geological disasters of unsaturated granite residual soil in Guangdong Province and its adjacent areas, which have caused serious losses to people's lives and property. Most of the disasters are caused by subgrade collapse and slope instability.

According to statistics, there are many reasons for slope instability and failure, which can be divided into external factors and internal factors. The internal factors are mainly the lithological characteristics of the stratum and the internal structural characteristics of the slope. The external causes are mainly precipitation and earthquake, and the current main causes are human activities. If the slope is not properly treated and treated in time, life and property will be lost once the slope is unstable.

Fig. 1 Geological hazards of unsaturated granite residual soil [2]

Granite eluvial soil slope is a special slope. Its main component, "granite eluvial soil", is the debris left in place by granite under physical and chemical weathering. It contains kaolin, montmorillonite and other mineral components. The cementation structure formed by a large amount of free silicon oxide, iron and aluminum makes it have a strong structure, which is characterized by high heterogeneity, obvious anisotropy, easy to collapse under unloading, and strong water absorption, Special soil easily softened in case of water [3]. The grain grading curve of granite residual soil is also very special, with the grading characteristics of "big at both ends, fine in the middle". The coarse particles that make up the granite residual soil are wrapped by the free oxide of soluble salts, and the residual connection between some original rock mineral grains is retained between the coarse particles, forming a soil skeleton with a certain strength. However, the medium, fine sand and silt missing from the grading curve are not enough to fill the soil skeleton, which makes the granite residual soil have special structure. This structure is the main reason why granite residual soil becomes a problem soil. The special soil structure and particle size distribution of the residual soil make it exhibit the abnormal phenomenon of high compressibility and high shear strength. The problems of granite residual soil deserve attention, and most of them are unsaturated. Under saturated and unsaturated conditions, the soil properties of granite residual soil are significantly different, and the engineering properties are more complex. In view of this, domestic and foreign scholars have carried out a lot of research on the stability analysis of granite residual soil slope, mainly focusing on theoretical research, indoor and outdoor simulation tests and numerical analysis. This paper intends to introduce the basic physical characteristics of granite residual soil and the stability analysis of the slope through the research on the stability of granite residual soil slope, and point out the direction for the stability analysis of granite residual soil slope in the future.

2. Analysis and research method of granite residual soil slope stability

2.1 Theoretical research

For the theoretical study of granite residual soil, foreign scholars have made active exploration: Sammori et al. [4] took Richaxd equation as the seepage control equation, analyzed the transient seepage on the slope according to Galerkin finite element method, and analyzed the sensitivity of factors affecting the stability of the slope using the simplified Bishop method. Finally, they proposed that soil permeability, soil layer thickness and slope shape have a significant relationship with the stability of the slope. C. W. W. Ng et al. [5,6] simulated and analyzed the transient seepage and stability changes of common slopes in Hong Kong under different rainfall conditions and initial boundary conditions with the finite element method, and proposed that the slip phenomenon was caused by the reduction of the matrix suction of the surface soil mass of the slope caused by rainwater infiltration, while the rainfall intensity and early rainfall had an impact on the slope stability.

Deng et al. [7] used a slope stability calculation method based on saturated unsaturated percolation theory, saturated unsaturated shear strength theory with softening effect, and considered the changes of substrate suction with softening effect, and found that under the same rainfall conditions, the effect of long-time minor rainfall on the increase of internal water level lines in a high diking slope with granitic residue soils was significantly greater than that of Short-time Heavy Rainfall on the increase of internal water level lines in the slope; Chan-KeeKim [8] studied the influence of different initial water content on the strength of granite residual soil, and found that the influence of initial water content and matrix suction on the internal friction angle can be ignored, while the cohesion increases with the increase of matrix suction; Saffari [9] conducted a constant rate unsaturated multi-stage consolidated drained triaxial shear test on granite residual soil in Sumatran, Malaysia, and fitted the shear strength parameters of saturated and unsaturated residual soil with mathematical formulas. The results can reflect the influence of matrix suction and net confining pressure on strength. In-MoLee [10], Zhang [11], Shen [12] respectively analyzed the strength of granite residual soil from different conditions such as vertical stress, initial dry density, initial saturation, initial suction, and found that the soil is a granular material with complex composition, whose particle composition reflects the weathering process of the soil and is closely related to the structure, mechanical properties and hydraulic properties.

Granite residual soil is a kind of soil widely distributed in South China. Whether the granite residual soil slope is under construction, completed or in the maintenance stage, its soil mass is mostly formed by the physical and chemical differentiation of granite. The soil will have a certain strength without disturbance and water infiltration. However, once affected by external factors, its meso structure will be destroyed, its strength will drop, and it can be crushed under the pressure of people. Therefore, through theoretical research, it provides theoretical support for long-term protection, design and construction, risk control and prevention of granite residual soil slope.

2.2 Indoor and outdoor simulation test

In natural state, granite residual soil has low compressibility, high bearing capacity and high shear strength, but its mechanical properties are significantly reduced due to its softening and disintegration characteristics after encountering water. Therefore, domestic and foreign scholars have discussed the physical and mechanical properties of granite residual soil through indoor tests. Wu [13], Ma [14], Frank C.Townsend [15] studied the physical and mechanical properties of granite residual soil through a series of indoor tests and in-situ tests; Chainchye E. Wang [16] learned the deformation characteristics of granite residual soil residual soil through consolidation test, triaxial compression, resonant column and torsional shear test.

Zhao [17] studied the anti-scouring ability of different grass species and soil fixation structures under different rainfall intensity and different slope conditions through the combination of different grass species and soil fixation structures in the field test area; Xiao [18] analyzed the slope erosion characteristics of different slopes, soil and plant protection measures under the conditions of rainstorm and runoff erosion through the highway slope simulation test, and studied the mechanism of rill formation in the slope; Lu [19] understood the scouring mechanism of loess slope through field investigation, and carried out indoor simulation test to study and analyze the action mechanism of grass mud protection mode, intermittent rainfall and slope ratio on slope scouring; Lin [20] learned through indoor tests that high-intensity rainfall is easy to induce sliding failure of soil slope, and long-time low-intensity rainfall is easy to induce sliding failure of soil slope and the scale of landslide is relatively large, and proposed that rainfall intensity and accumulated rainfall should be considered as the parameters for rainfall early warning benchmark; Yang [21] analyzed the impact of rainfall conditions, soil structure strength, pavement runoff and slope shape on the erosion of loess slope based on indoor model test and field investigation data, and discussed the erosion law and protection countermeasures of loess slope.

Granite residual soil is a kind of "special" soil, which is very easy to induce geological disasters. The key characteristics of granite residual soil are obtained through indoor and outdoor simulation tests to understand the influence of different factors on the stability of granite residual soil slope, and the key factors causing disasters of granite residual soil slope are investigated and studied to prevent such soil slope disasters. At the same time, the mechanical properties of granite residual soil are affected by indoor test disturbance and other factors, and its strength and deformation parameters are often difficult to reflect its true characteristics. The excellent engineering properties of granite residual soil have not been reasonably applied. This situation wastes both construction funds and foundation bearing capacity resources, so there is a large research space for comprehensive indoor and outdoor simulation tests on the properties of granite residual soil.

2.3 Numerical analysis

At present, the numerical analysis method is one of the stability analysis methods of granite residual soil slope widely used by scholars at home and abroad. Due to the difficulty in sampling and large disturbance of granite residual soil, scholars tend to adopt numerical analysis when studying the impact of rainfall on the stability of granite residual soil slope.

Yang [22] used MIDAS/GTS finite element software to simulate the working conditions of different rainfall intensity and rainfall time conditions, and analyzed the coupling of rainfall infiltration flow field and stress field, and discussed its role in the stability of residual slope soil slope; Deng [23] established a finite element model to analyze the steady or transient seepage field of the eluvial soil cutting slope, obtained the safety factor and matrix suction of the slope under rainfall conditions by using the limit equilibrium method, and analyzed the shape of the landslide mass and its formation reasons at different times under rainfall conditions; Chen [24] used the Plaxis finite element program to analyze the deformation characteristics of the slope, indicating that the existence of the weak interlayer in the granite eluvial soil not only increased the overall deformation of the slope, but also determined the position of the sliding surface when the slope was unstable and damaged; Qiu [25] analyzed the relationship between the collapse and rainfall in combination with the field survey data of the collapse disaster of the high fill residual soil embankment, and analyzed the mechanism of the collapse phenomenon of the high fill residual soil embankment using the finite element seepage software Seep/W and the limit equilibrium slope stability software Slope/w; Fang [26] studied the stress strain state and stability of the granite residual soil slope after the excavation limit by using the numerical method FLAC3D software in combination with the strength reduction method, and analyzed the stability of the granite residual soil slope by using the limit equilibrium method Slope/W software to judge the stability of the granite residual soil slope.

After the long-term development of theory and practice, the stability analysis of granite residual soil slope is becoming more and more perfect. At the initial stage, simple soil mechanics was introduced into the slope calculation, and gradually developed from two-dimensional calculation to three-dimensional calculation, from the limit equilibrium method to the stage of multi method and multi theory. After entering the 21st century, numerical simulation analysis methods are more and more used, and multi-disciplinary intersection is more and more valued. Artificial intelligence method, neural grid method and other analysis methods developed in recent years have enriched and diversified modern slope analysis methods [27-31]. At present, various analysis methods make the stability calculation of granite residual soil slope more reliable, which is a key development direction in the future.

3. Factors and forms affecting the stability of granite residual soil slope

According to the principles and methods of soil mechanics [32], the reasons for the instability of granite residual soil slope can be divided into two categories: the force generated during the construction of the project has destroyed the original equilibrium state in the soil, such as the excavation of cutting or foundation pit and the filling of embankment, which makes the side slope surface have stacking materials, vehicle loads, etc; The negative effect of reducing the soil structure strength due to the change of the external environment leads to the instability and damage of the slope, such as long-term high-intensity rainfall erosion, vibration caused by construction near the slope, and the effect of seismic force. Therefore, the influence of soil seepage, slope shape, rainfall conditions, sliding surface and other factors on the stability of granite residual soil slope can be understood by using finite element, reliability analysis, fuzzy theory and other methods. For granite residual soil slope, its slope instability is mainly affected by two factors: its own disintegration property; The degree of microcracks in its internal structure. If the soil is relatively homogeneous, the degree of microcracks is light, and most of them are sliding failures; On the contrary, it belongs to collapse.

3.1 Slip failure

There are two main types of failure: softening slip; Slip tension crack. The former is mostly located in low-lying places, with shallow buried depth of groundwater, and the overlying soil layer belongs to alluvium. After excavation, groundwater will drain towards the slope. Once the surface of the slope is soaked, it is easy to collapse, and the soil will soften and collapse, so that the stress release of the soil will be accelerated, the micro fissures in the soil will expand, groundwater will infiltrate, and argillization expansion reaction will occur in the fissures, resulting in weak strength of the structural surface. At the same time, the small particles in the fissures will be washed away by water, and the pores will continue to expand, eventually leading to softening and sliding. The slip tension crack of slopes are mostly located in low hills, and the groundwater level is relatively stable in the residual soil. When excavating, once the groundwater level is exceeded, the soil is easy to collapse, as is the case with softening and sliding, and tension cracks are formed on the upper part.

The groundwater plays a leading role in the sliding failure. In the case of softened sliding, the water level can be lowered first, and anchor retaining walls can be used for support. If there is no drainage, anti-sliding piles should be selected. In the case of slip tension crack, shotcrete and anchor support can be selected for the upper part, and the lower part needs to be drained first, and then anchor bolts or soil nails are selected for support.

3.2 Collapse failure

Some hillsides are located in high terrain, and the groundwater is buried deep. Generally, the excavation cannot reach the groundwater level. The overlying soil layer is mostly slope deposit with low water content, which has good mechanical properties and high shear strength. At the beginning of excavation, the slope is relatively stable. However, over a long period of time, cracks expand, surface water seeps, and the strength of the slope weakens, so the anti-sliding friction resistance of each soil mass decreases. Under the action of gravity, wedge collapse is easily caused. Some slopes are easy to form mud ditches and soil caves under the scouring of rainwater, and finally collapse.

At the beginning, the collapse phenomenon is mostly local collapse, and then the area continues to expand. The key to control the collapse is to control the local collapse. In case of wedge collapse, it can be supported by anchor bolt walls, and grass can be planted on the slope. In case of scouring and collapse, surface water shall be drained in time, and turf shall be planted on the slope.

4. Strength theory of granite residual soil unsaturated soil

The behavior of unsaturated soil is more complex than that of saturated soil. In addition to solid, liquid and gas phases, the contraction membrane (i.e. water air interface) is considered as an independent fourth phase. Since the 1940s, Bishop [33], Fredlund [34], Shen [35], Miao [36], Yin [37] and Zhao [38] have successively put forward many shear strength theories of unsaturated soil.

4.1 Single variable effective stress

Fredlund and Morgenstern [34] applied the formula of effective stress of saturated soil to the theoretical study of unsaturated soil, and proposed to express the effective stress of unsaturated soil by controlling a single value effective stress or stress state variable related to soil parameters. Later, scholars found that the description of the stress state of soil should be separated from the nature of the soil itself, and the imperfections of the univariate effective stress were obvious. For example, Bishop's univariate formula has neither theoretical support nor experimental verification. "χ" in Eq. (1) is a parameter reflecting the influence of unsaturated soil gas phase, but it has no clear physical significance and has been rarely mentioned so far.

$$\sigma' = (\sigma - u_a) + \chi(u_a - u_w) \quad (1)$$

Where: σ' is the total stress; $(\sigma - u_a)$ is the net stress (the difference between the principal stress and the air pressure); $(u_a - u_w)$ is the matrix suction (the difference between air pressure and pore water pressure).

4.2 Two variable effective stress

For the research on mechanical properties of unsaturated soil, Bishop effective stress principle can be used for calculation, but many scholars slowly found that Bishop effective stress and deformation characteristics of unsaturated soil are not a single corresponding relationship. Scholars changed their previous inherent concepts, and began to use two stress variables to describe the stress state of unsaturated soil, which is called the dual variable theory. The stress state of unsaturated soil is described by single value stress variable. Two of the three state variables $(\sigma - u_a)$, $(u_a - u_w)$ and $(\sigma - u_w)$ can be combined in pairs to obtain these two independent stress state variables. Fredlund and Rahardio [39] think that $(\sigma - u_a)$ and $(u_a - u_w)$ are most suitable for engineering practice. They bring the bivariate effective stress proposed by Rahardio into the Mohr-Coulomb strength failure criterion, and obtain the bivariate unsaturated soil failure criterion as shown in Eq. (2):

$$\tau_f = c' + (\sigma - u_a)\tan\varphi' + (u_a - u_w)\tan\varphi^b \quad (2)$$

Lamborn [40] also proposed the shear strength expression with $(\sigma - u_a)$ and $(u_a - u_w)$ as double stress variables, as shown in Eq (3):

$$\tau_f = c' + (\sigma - u_a)\tan\varphi' + (u_a - u_w)\theta_w \tan\varphi \quad (3)$$

Where: θ_W is the volume water content of the sample; τf is the shear strength; $tan\varphi'$ and $tan\varphi^b$ are parameters related to internal friction angle.

The fitting formula proposed by Abramento and Carvallo [42] according to the test results is shown in Eq. (4):

$$\tau_f = c' + (\sigma - u_a)\tan\varphi' + \alpha(u_a - u_w{}^{\beta}) \quad (4)$$

Where: α and β are fitting parameters.

4.3 Other forms of shear strength

With the deepening of unsaturated soil strength, researchers began to consider deepening and revising the original formula. Richards [42] considered the component of solute suction in matrix suction, as shown in Eq. (5):

$$c' = \sigma - u_a + X_m + X_S(h_s + u_a) \quad (5)$$

Where, X_m is the effective stress parameter considering the matrix suction; X_s is the effective stress parameter considering the solute suction, and h_s is the solute suction. Aitchison [43] considered both solute suction and soil parameters related to stress path, and proposed a modified effective stress formula similar to Richard's.
Chinese scholars also put forward many shear strength formulas related to suction. Shen [33], from the perspective of nonlinear relationship between matrix suction and unsaturated soil, established a hyperbolic relationship between generalized suction u_s and strength as shown in Eq. (6):

$$\tau_f = c' + (\sigma - u_a)\tan\varphi' + \frac{u_s}{1+du_s}\tan\varphi' \quad (6)$$

Where, d is the fitting parameter.

5. Conclusion and expectation

The process and influencing factors of granite residual soil slope instability and failure are very complex. This paper introduces the research methods of granite residual soil slope stability, and summarizes the failure forms of granite residual soil slope. The following conclusions are drawn:
(1) The basic physical properties of granite residual soil are analyzed through theoretical research; The key characteristics of granite residual soil are obtained through indoor and outdoor simulation tests, and the effects of slope, rainfall intensity, rainfall time, protection methods and other factors on soil slope erosion under rainfall are discussed; The stability of granite residual soil slope is judged by numerical analysis, and the position of sliding surface is determined when the slope fails.
(2) The main reasons for the instability of granite residual soil slope are as follows: the force generated during the construction of seepage works destroyed the original equilibrium state in the soil; The negative effect of decreasing the soil structure strength due to the change of the external environment leads to the instability and failure of the slope. For granite residual soil slope, its slope instability is mainly affected by two factors: its own disintegration property; The degree of microcracks in its internal structure.
(3) The strength theory of granite residual unsaturated soil includes single variable effective stress, double variable effective stress and other forms of shear strength.
Through the research, the following points are found and put forward: (1) The slope stability analysis theory has been deeply and systematically studied on the basis of previous efforts, and the quantitative analysis results of slope stability have been very close to the actual engineering situation. However, the research on granite residual soil slope is few, especially the analysis and calculation research on soil slope after support is less than the theoretical analysis on soil slope stability. The effect of slope support and the design parameters introduced by support have a lot of research space; (2) The granite residual soil slope has a certain shear strength. What kind of materials can be added to enhance the strength of the slope and reduce the hazard of landslide; (3) Some scholars adopt outdoor simulation test, but outdoor simulation test is time-consuming and laborious; Some scholars adopt the research method of combining indoor simulation test with software analysis, but the basic properties of indoor remolded residual soil are greatly different from that of on-site granite residual soil. The former is obviously weaker than the latter. How to eliminate the difference between the two remains to be studied.
In a word, it is hoped that this study can provide some theoretical basis and guidance for the future stability engineering of granite residual soil slope.

References

1. Editorial Board of Chinese Physical Geography, Chinese Academy of Sciences. 1980, physical geography and geomorphology of China. Beijing: Science Press, 152 -160.
2. Liu Yiao. Study on Unsaturated Mechanical Characteristics of Granite Residual Soil in Shenzhen Area and its Application of Constitutive Model[D]. Jilin University, 2021. (In Chinese)
3. ZHOU xiaowen, LIU pan, HU liming, et al. An experimental study of shear yield characteristics of structured granite residual soil. Rock and Soil Mechanics, 2015(S2): 157-163. (In Chinese)
4. Sammori T, Tsuboyama Y Parametric study on slope stability with numerical simulation in consideration of seepage process [M].In:Proc 6th Int. Symp. On Landside.Bell(ed)Bailema, Rotterdam, 1991:539-544.
5. Ng C W W, Shi Q. A study on stability analysis of shallow layer slope due to raining permeation[J]. Computer and Geotachnics, 1998, 22(1): 1-28.
6. Ng C W W, Wang B, Tung Y K. Effect of rain infiltration on the stability of slopes[J]. Canadian Geotechnical Joural, 2001, 12(5): 1049-1062. (In Chinese)
7. DENG Xi, HE Zhongming, FU Hongyuan, ZENG Ling. Impact of Rainfall on Slope Stability of High Embankment with Granite Residual Soil[J]. Mining and Metallurgical Engineering, 2016, 36(04): 11-15. (In Chinese)
8. Chan-Kee Kim. Behavior of unsaturated weathered residual granite soil with initialwater contents[J]. Engineering Geology. 2010:1-10. (in Chinese)
9. Pooya Saffari. Shear Strength of Unsaturated Malaysia Granitic Residual Soil[J]. Journal of Testing and Evaluation. 2019, 47(1): 640-653.
10. In-Mo Lee. Effect of stress state on the unsaturated shear strength of a weathered granite[J]. Can Geotech. J. 2005, 42: 624-631.
11. ZHANG Long, CHEN Zhenghan, ZHOU Fengxi, et al. Test verification of stress variables for unsaturated soils[J]. Chinese Journal of Geotechnical Engineering, 2017, 39(2): 380-384. (In Chinese)

12. SHEN Qiuhua. Experimental study on influencing factors of unsaturated shear strength index of granite residual soil[J]. Journal of Mapping and Exploration. 2020, 95-96. (In Chinese)
13. Wu Nengsen. A Study on Characteristics and Some Engineering Problems of Granite Residual Soil with Structural Nature[D]. Nanjing Forestry University, 2005. (In Chinese)
14. MA Haiyi, LU Zude. Investigation of physico-mechanical properties for highly weathered granite of nuclear power station in coastal site of western Guangdong province [J]. Rock and Soil Mechanics, 2012, 33(02): 361-366+374. (In Chinese)
15. Frank C. Townsend. Geotechnical Characteristics of Residual Soil[J]. Construction and Building Materials, Journal of Geotechnical Engineering, 19885, 111(1): 77-94.
16. Chainchye E. Wang, Roy H. Borden. Deformation Characteristics of Piedmont Residual Soils[J]. Journal of Geotechnical Engineering, 1996, 122(10): 822-830.
17. ZHAO Minghua, JIANG Desong, CHEN Changfu, et al. Research on On-the-Spot and Indoor Washing-Out Test of Ecology Protection of Rocky Slopes[J]. Journal of Hunan University (Natural Sciences), 2044, 31(5): 77-81. (In Chinese)
18. XIAO Peiqing, SHI Xuejian, CHEN Jiangnan, et al. Experimental Study on Protecting Speedway Slope Under Rainfall and Flow Scouring [J]. Bulletin of Soil and Water Conservation, 2004, 24(01): 16-18. (In Chinese)
19. Lu Hao. Study on the loess highway slope erosion mechanism and experimental adaptability of grass-mud prtection[D]. Chang'an University, 2014. (In Chinese)
20. LIN Hungchou, YU Yuzhen, LI Guangxin, et al. INFLUENCE OF RAINFALL CHARACTERISTICS ON SOIL SLOPE FAILURE [J]. Chinese Journal of Geotechnical Engineering, 2009, 28(01): 198-204. (In Chinese)
21. Yang Guofeng. Study on Erosion Damage and Prevention Countermeasures of Highway Slope in Loess Area of Northern Shaanxi [D]. Chang'an University, 2004. (In Chinese)
22. YANG Jiao, WANG Yu, LEI Fuhong, et al. Research on the Residual Soil Slope Stability Simulation under Rainfall Infiltration[J] . Safety and Environmental Engineering, 2012, 19(03): 5-10. (In Chinese)
23. DENG Tongfa, GUI Yong, LUO Sihai, et al. Study on Slope Stability of Granite Residual Soil Cutting Excavation with Rainfall[J]. Journal of Earth Sciences and Environment, 2012, 34(04): 88-94. (In Chinese)
24. CHEN Wei, JIAN Wenbin, DONG Yansong, et al. Stability Study of a Granite Residual Soil Slope with Weak Interlayers[J]. Journal of Water Resources and Architectural, 2014, 12(06): 107-111. (In Chinese)
25. QIU Luyang, LIU Shuchuan, LI Dayong. A case study of failure mechanism of residual soil filled embankment[J]. Rock and Soil Mechanics, 2007, 28(10): 2161-2166. (In Chinese)
26. Fang Wenkai. Study on Grantie Residual Soil High Slope Stability and Support Scheme[D]. JiLin University, 2020. (In Chinese)
27. Revilla J, Castillo E. The calculus of variations applied to stability of slopes[J]. Géotechnique, 1977, 27(1):1-11.
28. Cho S E. Infltration analysis to evaluate the surficial stability of two-layered slopes considering railfall characteristic[J]. Engineering Geology, 2009, 105: 32-43.
29. Chugh A K. A method for locating critical slip surfaces in slope stability analysis. [J]. Canadian Geotechnical Journal, 2001, 39(3): 765-770.
30. LI Shouju, LIU Yingxi, HE Xiang, et al. Global Search Algorithm of Minimum Safety Factor for Slope Stability Analysis Base on Annealing Simulation[J]. Chinese Journal of Geotechnical Engineering, 2003, 22(2):236-240. (In Chinese)
31. CAO Ping, ZHANG Ke, WANG Yixian, et al. Mixed Search Algorithm of Critical Slip Surface of Complex Slope[J]. Chinese Journal of Geotechnical Engineering, 2010, 29(4): 814-821. (In Chinese)
32. LI Jingpei, LIANG Fayun, ZHAO Chunfeng. Soil Mechanics [M]. Beijing: Higher Education Press, 2008.
33. Bishop A W, Blight G E. Some aspects of effective stress in saturated and partly saturated soils [J]. Geo-technique, 1963, 13(3): 177-197.
34. Fredlund D G, Morgenstern N R.Stress state variables for unsaturated soils [J]. Journal of Geo-technical and Geo-environmental Engineering, 1977, 103(5): 447-466.
35. SHEN Zhujiang. Unified deformation theory of generalized suction and unsaturated soils [J]. Chinese Journal of Geotechnical Engineering, 1996(02): 1-9. (In Chinese)
36. MIAO Linchang,Yin Zongze. Shear strength of unsaturated soils [J]. Rock and Soil Mechanics, 1999(03): 1-6. (In Chinese)
37. YIN Zongze, ZHOU Jian, ZHAO Zhonghui, et al. Constitutive Relations and Deformation Calculation for Unsaturated Soils[J]. Chinese Journal of Geotechnical Engineering, 2006, (02): 137-146. (In Chinese)
38. ZHAO Chenggang, CAI Guoqing. Principle of Generalized Effective Stress for Unsaturated Soils[J]. Rock and Soil Mechanics, 2009,30(11): 3232-3236. (In Chinese)
39. D.G. Fredlund and H.Rahardjo. Soil Mechanics Principles for Highway Engineering in Regions, Transportation Res.Record 1137. pp.1-11, 1987.
40. L. Lam, D.G Fredlund and S. L. Barbour, Transient Seepage Model for Saturated-Unsaturated Soil

Systems: A Geotech-nical Engineering Approach, Can Geot. Jou , 1988, 24(4): 565-580.

41. Abramento M, Whittle A J. Shear-Lag Analysis of Planar Soil Reinforcement inPlane-Strain Compression[J]. Journal of Engineering Mechanics, 1993, 119(2): 270-291.
42. B. G. Richards, The Significance of Moisture Flow and Equilibria in Unsaturated soils in Relation to the Design of Engineering Structures Built on Shallow Foundations in Australia, presented at the Symp on Pemeability and Capillary, Amer.soc. Testing Materials, Atlantic City, NJ, 1966.
43. Aitchison G D. Twenty-five years of application of soil survey principles in the practice of foundation engineering[J]. Geoderma, 1973, 10(1-2): 99-112.

Preparation of Quaternary Ammonium Gemini Surfactant and Its Application in Insulator Cleaning

Xinyuan Wan [1], Xiaojian Xia [1], Yiyang Chen [1], Deyuan Lin [1], Yicheng Hong [1], Wenzhe Zhang [2,3], Linxi Hou [2]

[1] Fujian Power Co., Ltd. Electric Power Research Institute, Fuzhou, Fujian Province, 350007, China
[2] College of Petrochemical Engineering, Fuzhou University, Fuzhou, Fujian Province, 350108, China
[3] College of Petrochemical Engineering, Fuzhou University, No.2, Xueyuan Road, New District, Fuzhou, China

Abstract. In this paper, quaternary ammonium salt type Gemini surfactant G12 was designed and synthesized, and its structure was characterized by infrared and nuclear magnetic resonance. The surface tension, permeability, emulsification ability and cleaning of insulator surface contamination in the composite system were tested, and the test results showed that it had low critical micelle concentration (cmc) and good emulsification performance. Due to the presence of rigid groups, the permeability is average. When quaternary ammonium Gemini surfactant G12 is mixed with permeant JFC-E, and is combined with auxiliary and organic solvent, the cleaning rate of insulators can reach more than 90%.

Key words: Gemini surfactant, insulator cleaning4, mixture, emulsibility, permeability.

1. Introduction

Insulators are important components in the circuit transportation system. Due to long-term exposure to outdoor environment, the surface area of insulators is severely polluted, which may lead to potential safety hazards[1]. These hygroscopic and conductive filth particles such as solid, liquid and gas filth particles adhered to the insulator surface greatly reduce the insulation level of the insulator under the joint action of fog, dew, rain, snow, frost and other weather conditions. As a result, "pollution flashover" malfunction will occur even under the operating voltage, resulting in power outage[2,3]. This can cause serious economic loss to Chinese electric power system[4].

Studies have shown that the main components of insulator pollution are soluble NaCl, CaSO4, CaO, insoluble oil, shellac and SiO2[5,6,7,8,9]. By studying the existing methods of cleaning electric equipment with chemical cleaning agent, it is found that there are still problems of weak cleaning ability and unsatisfactory cleaning effect[10,11,12].

Surfactants are an important component of water-based cleaning agents[13]. Due to the existence of intermolecular repulsion, traditional surfactants tend to have large molecular spacing when they are aligned on the surface of water, which has limited effect on the reduction of surface tension. Gemini surfactant is a new type of surfactant containing two hydrophilic head groups and two hydrophobic tail chains[14]. The strong force of the covalent bond pulls into the distance between the two hydrophilic groups, which makes them more tightly arranged at the liquid level and has better performance[15,16,17,18].

2. Experimental conditions and methods

2.1 Materials and Instruments

Phthalic anhydride, epichlorohydrin, ethylene glycol, dipropylene glycol butyl ether and other reagents were commercially available analytical pure; Surfactant iso-octanol polyoxyethylene ether (JFC-E), BASF (China) Co., LTD. Sodium dodecyl diphenyl ether disulfonate (2A1), DOW Inc.

SFT-D1 surface tension instrument, Beijing Hake Experimental Instrument Factory; HARKE-SPCAX1 Contact Angle Tester, Beijing Hake Experimental Instrument Factory; Nicolet iS50 infrared spectrometer, Thermo Fisher; AVANCE III 500 NMR instrument, Bruker, Switzerland.

2.2 Experimental methods

2.2.1 Synthesis of phthalate monoesters

6.21g ethylene glycol (0.10 mole) and 37.03g phthalic anhydride (0.25 mole) were added to a 250 mL round-bottom flask, followed by 0.2 mL concentrated sulfuric acid as catalyst, and 100 mL acetone as solvent. It was refluxed at 65℃ for 4h. Acetone recrystallized three times to obtain phthalate monoester (81% yield).

* Corresponding author: 936917551@qq.com.

https://doi.org/10.1051/e3sconf/202337502001

2.2.2 Synthesis of phthalate monoester - chloropropane

2.94g phthalate monoester (0.10 mol) and 1.85g epichlorohydrin (0.20 mol) were added to a 250 mL round-bottom flask, and then 100 mL toluene was added as solvent. It was heated to 105℃ for reflux reaction for 6h, and then dried to obtain phthalate monoester chloro-propane (83% yield).

2.2.3 Synthesis process of quaternary ammonium Gemini surfactant G12

0.10 mol phthalate monoester chloro-propane and 0.20 mol dodecyl dimethyl tert-amine were added to a 250 ml round-bottom flask, and the reaction was carried out at 65℃ for 4h to obtain a yellow thick liquid, which was recrystallized with acetone to obtain G12 (78% yield).

2.3 Characterization and testing methods

2.3.1 Infrared absorption spectrum

KBr tablet was used for the test sample, the test resolution was 4 cm^{-1}, the number of scans was 16, and the scan range was 400-4000 cm^{-1}.

2.3.2 Nuclear magnetic resonance hydrogen spectrum

Deuterated chloroform was used as the solvent and the reference was used for calibration. The test frequency was 400 MHz and the temperature was 298.15K.

2.3.3 Surface tension test

A series of aqueous surfactant solutions with different concentrations were prepared and the surface tension was measured by ring method at 25℃. The corresponding γ-lg$C$ curve was plotted, and then the critical micelle concentration (C_{cmc}) and the corresponding surface tension (γ_{cmc}) were obtained. Parallel test three times, take the average as the measurement result.

2.3.4 Penetration capacity test

Using the canvas settling method, the standard canvas was placed on the surface of 500 ml of 0.5% surfactant aqueous solution, and the settling time of the canvas from the surface to the bottom of the beaker was recorded for three consecutive times, and the average value was taken.

2.3.5 Emulsifying ability test

Using the phase separation method. 20 ml 25 g/L aqueous surfactant was put into a 100 mL stopper cylinder, and 20 mL liquid paraffin was added into the cylinder. Then the cylinder was placed in a water bath at 34℃ for 5 minutes. After 10 times of vigorous shaking, it was left to stand, and the time needed to separate 10 ml of aqueous phase was recorded. The operation was repeated for 5 times, and the average value was taken.

2.3.6 Cleaning ability test

Silica, kaolin, 30# transformer oil and sodium chloride were configured into pollution standards according to the mass ratio of 1:10:88:1, weighing 100 mm×100 mm tiles and recording as m1. 3g filth was coated on the surface of the tiles, dried at 60℃ for 12h, and weighed and recorded as m2. The tiles were soaked in the cleanser for 20 minutes, removed and dried for 2 hours, and the weight was recorded as m3. Cleaning rate = (m2-m1-m3) / (m2-m1) ×100%.

3. Results and discussion

3.1 Structural characterization

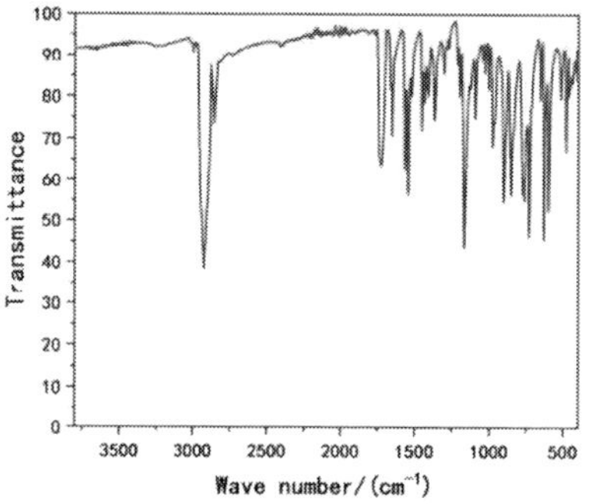

Figure 1. The FTIR spectrum of G12.

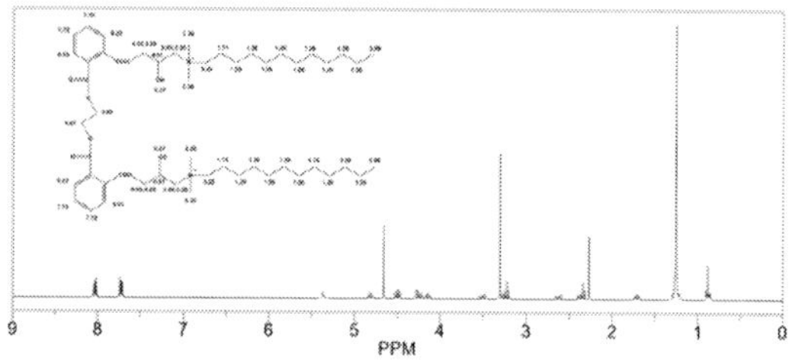

Figure 2. The ^{1}H NMR spectrum of G12.

The FTIR and ^{1}H NMR of G12 are shown in Figure 1 and Figure 2 respectively. The characteristic peak of alkane chain appears near 2922 cm^{-1}2853 cm^{-1}; The stretching vibration peak of C=O bond in ester group appears at 1724 cm^{-1}; C-N bond stretching vibration peak appeared near 1467 cm^{-1}; The characteristic absorption peak of benzene ring appears near 1581 cm^{-1} and 1567 cm^{-1}.

G12(400 MHz, $COCl_3$, TMS): 0.89(t, 6H, $(CH_2)_{13}CH_3$), 3.38(t, 12H,N-CH_3), 1.26(m, 32H, CH_2-CH_2-CH_2-$(CH_2)_{10}$-CH_3), 3.50(t, 12H,N-CH_3), 3.24(d, 4H, N-CH_2-$(H_2)_2$ CH_3), 1.29(d, 4H, CH_2-CH_2-CH_2-$(CH_2)_{10}$-CH_3), 3.50~3.52(m, 4H, N-CH_2-CH), 5.37(m, 2H, OH), 1.71(d, 4H, N-CH_2-CH_2-$(CH_2)_{11}$-CH_3), 4.25~4.50(m, 4H, OH-CH-CH_2-O), 3.22(d, 4H, N-CH_2-(CH_2)12-CH_3), 4.81(m,

2H, 2(CH$_2$)-CH-OH), 7.72 ~ 8.03(d, 8h, C_6H_4), 4.66(d, 4H, O-CH$_2$-CH$_2$-O).
According to the results of FTIR and ^{1}H NMR, it can be determined that the product structure is consistent with the target product.

3.2 Surface tension test

The surface tension of surfactant aqueous solutions with different concentrations at 25 °C was measured by the ring method and the results are summarized in Figure 3. It can be seen from Figure 43 that the C_{cmc} of G12 is 1.6 × 10^{-4} mol/L, γ_{cmc} is 30.3 mN/m. The reason for the surface tension can reduce to 30.3 mN/m is mainly because the rigid group contained in the molecular structure can pull into the space between the two hydrophilic head groups in the form of covalent bond, making the surfactant molecules more closely arranged on the liquid surface.

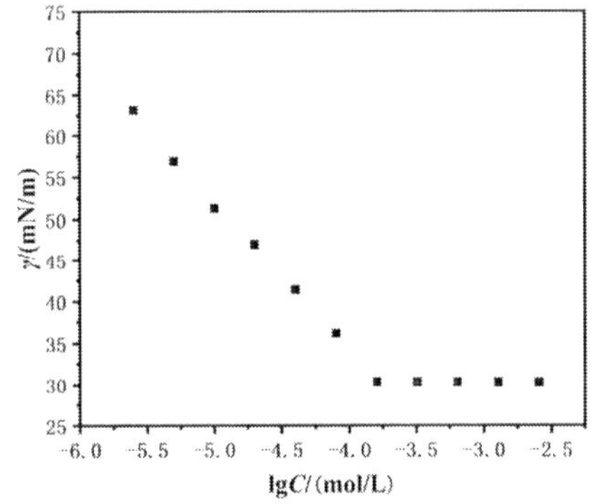

Figure 3. The Surface tension test result of G12.

3.3 Permeability test

The wettability and permeability of G12 and JFC-E were tested by canvas sedimentation method, and the results are summarized in Table 1. It can be seen that the wetting permeability of G12 is general, mainly because the steric hindrance of rigid group is large, which hinders the penetration of surfactant. Therefore, in actual use, G12 needs to be used combine with others to achieve better application effect by using the synergistic effect.

Table 1. Canvas contact time of G12 and JFC-E.

Sample	Test 1	Test 2	Test 3
G12	137"47	135"39	138"51
JFC-E	20"56	22"30	23"19

3.4 Emulsification capacity test

The emulsification ability of G12 and 2A1 were tested by phase separation method, and the experimental results are shown in Table 2. It can be seen that the emulsification performance of G12 is stronger than 2A1, so it is feasible to use G12 in the cleaning field.

Table 2. Time for 10 mL aqueous phase separation of G12 and 2A1.

Sample	Test 1	Test 2	Test 3	Test 4	Test 5
G12	35"09	34"29	35"77	34"26	35"49
2A1	29"28	29"59	29"06	30"44	29"86

3.5 Cleaning capacity test

G12 was applied to the cleaning agent system for insulator surface cleaning, and three cleaning agents were prepared respectively. The composition of detergent is shown in Table 3. The dried porcelain chips were cleaned in three kinds of cleaning agents for a period of time, and the Cleaning efficiency results are shown in Table 4. It shows that the detergent mixed with two surfactants has the best cleaning effect.
We mixed the dirt directly with the cleaning solution, stirred it and placed it for 1 min to observe the dispersion of the system. The stability of No. 1 detergent lotion added with G12 as emulsifier alone is even better than that of No. 3 detergent added with G12 and JFC-E composite system. The main reason is that G12 has better emulsification effect and can better disperse the dirt in the cleaning agent. The reason for the better cleaning effect of the composite system on the porcelain chip is that JFC-E has well permeability, which can play a synergistic effect when used with G12. JFC-E penetrates into the interface of insoluble dirt such as porcelain chips, silica, kaolin, etc. to peel off the dirt. G12 quickly wraps up these dirties to form lotion, which is dispersed in the cleaning liquid system to achieve better cleaning effect.

Table 3. The formulas of different cleaning agents.

Detergent 1		Detergent 2		Detergent 3	
Component	Content	Component	Content	Component	Content
cholamine	4	cholamine	4	cholamine	4
PEG	4	PEG	4	PEG	4
inhibiter	1	inhibiter	1	inhibiter	1
G12	0.5	G12	0	G12	0.5
JFC-E	0	JFC-E	0.5	JFC-E	0.5
water	90.5	water	90.5	water	90

Table 4. Cleaning Results of different formulas.

Sample	Detergent 1	Detergent 2	Detergent 3
Cleaning efficiency(%)	83.56	72.88	95.32

4. Conclusion

The quaternary ammonium salt gemini surfactant was designed and synthesized, and its structure was characterized by IR and NMR. The surfactant has low cmc concentration and can reduce the surface tension of water to 30.3 mN/m. The surfactant mixed with auxiliaries, organic solvents, penetrants, etc. was used to prepared detergent which is used for insulator pollution cleaning. The detergent can achieve good cleaning effect after mixing with penetrant.

Acknowledgements

This work was supported by the State Grid Technology Project under Grant No. 521304200028.

References

1. Gu Z. Pollution flashover and prevention of transmission lines [J]. Public electricity, 2004(6)38-39.
2. Zhan Q, Ruan J, Zhong L. Natural fouling characteristics of overhead line insulators in the Pearl River Delta region [J]. High voltage technology, 2010,36(11):2662-2667 .
3. Shu L, Ran Q, Jiang X, et al. Comparison of AC flashover characteristics of artificial pollution of porcelain and glass insulators [J]. High voltage technology, 2007,33(3):9-13 .
4. Wen X, Lan L, Jiang R. Statistical Characteristics and Risk Assessment of Insulator Pollution Flashover [J]. High Voltage Technology, 2011,37(1):1-7 .
5. SungnamC, Chulbaep. Analysis of contaminants adhered on polymeric insulator used in coastal and industrial areas. [C] // Electrical insulation and Dielectric Phenomena .Daejeon korea: [s.n.], 2008.
6. Li H, Ye X, Liu G, et al. Source analysis of natural pollution chemical composition of insulators along transmission lines in Guangzhou [J]. High voltage technology, 2011, 37(8):1937-1943.
7. Ren S, Yang D. Analysis of anion composition in natural pollution of high-voltage insulators [J]. Journal of Beijing Institute of Petrochemical Technology, 2009, 17(2):11-14.
8. Song Y, Liu G, Li H. Comparative analysis of natural pollution components of insulators in highway area and highway farmland area [J]. High voltage apparatus, 2010, 46(5):22-25 .
9. Feng B, Chang Y, Liu X, et al. Study on natural pollution composition of high-voltage insulators [J]. Hunan Electric Power, 2007, 27(1):9-10 .
10. Zhu Zhengguo, Zhong Jianling. Analysis of Substation Insulator Cleaning Test Data [J]. High Voltage Apparatus, 2008, 44(5):472-473 .
11. Jiang K. Electrically charged chemical cleaning of power equipment [J]. High voltage technology, 2001, 27(3):53-54 .
12. Rong X. Feasibility Study on Live Cleaning and Maintenance Technology for Antipollution Flashover of High Voltage Circuit [J]. Cleaning World, 2005, 21(9):13-17 .
13. Studies on Micellarization Properties, Interfacial Activity and Chemical Biology of Quaternary Ammonium Gemini Surfactants [D]. Zhejiang: Department of Chemistry, Zhejiang University, 2016.
14. Menger F M, Keiper J S. Gemini Surfactants [J]. Angew Chem Int Ed Engl, 2000, 39(11): 1906-1920.
15. Dan L, Fang W, Wang H, et al. Gold/Oil Nanofluids Stabilized by a Gemini Surfactant and Their Catalytic Property[J]. Industrial & Engineering Chemistry Research, 2013, 52(24): 8109-8113.
16. Zhang S, Jing Y, Wu J, et al. Micellization Parameters of Six Gemini Quaternary Ammonium Surfactants from Measurements of Conductivity and Surface Tension[J]. Journal of Chemical & Engineering Data, 2014, 59(9): 2891-2900.
17. Kumar D, Neo K, Rub M. Effect of Alkanediyl-α,ω-Type Cationic Dimeric (Gemini) Surfactants on the Reaction Rate of Ninhydrin with [Cu(II)-Gly-Tyr]+ Complex[J]. Journal of Surfactants & Detergents, 2016, 19(1): 101-109.
18. More U, Kumari P, Vaid Z, et al. Interaction Between Ionic Liquids and Gemini Surfactant: A Detailed Investigation into the Role of Ionic Liquids in Modifying Properties of Aqueous Gemini Surfactant[J]. Journal of Surfactants & Detergents, 2016, 19(1): 75-89.

Development prospect of heat supply and peak shaving of nuclear power units in Shandong power grid

Lingkai Zhu [1], Wei Zheng[1], Ziwei Zhong [1], Zhiqiang Gong [1], Junshan Guo [1]

State Grid Shandong Electric Power Research Institute, Jinan, China

Abstract. Under the current national "dual carbon" economic development goal, nuclear energy, as a clean, low-carbon, safe and efficient form of energy, has the advantage of natural heat. Central heating is an important area that consumes coal and generates carbon emissions. Therefore, the development of nuclear heating is conducive to improving the energy utilization structure and upgrading the industry. Based on the current situation of the nuclear motor assembly machine in Shandong Province and the power supply composition structure of the provincial power grid, this paper has carried out the relevant technical analysis and prospects for the peak shaving of nuclear heating.

1. Introduction

In October 2021, the State Council issued the Action Plan for Carbon Peak by 2030, which proposed to "actively, safely and orderly develop nuclear power" and "actively and steadily carry out demonstration of nuclear heating"[1]. This is the second time since the government work report in March 2021 first proposed "actively and orderly develop nuclear power on the premise of ensuring safety", and the central government mentioned "actively, safely and orderly develop nuclear power", which means that nuclear power may usher in a new round of development opportunities.

By the end of 2016, the heating area of urban and rural buildings in northern China was about 20.6 billion square meters, the heating coal used was about 400 million tons of standard coal[2]. Nuclear heating is a safe, simple, mature and reliable technology. If it can be successfully used for urban heating and popularized, it will help alleviate the haze problem caused by coal heating[3]. As of June 2022, China has 54 nuclear power generating units in operation, 23 nuclear power generating units under construction, and the number of nuclear power generating units under operation and construction is the second in the world. It is estimated that by 2025, the installed capacity of nuclear power operation in China will reach about 70 million kilowatts[4].

Shandong Province is a traditional large heating province. After the "double carbon" goal was proposed, the task of reducing coal combustion in Shandong Province was severe. A large number of small thermoelectric and small boilers were shut down, and large direct dispatching public thermoelectric units were used to replace heat supply. The heating area of directly transferred public thermal power has increased year by year with a compound growth rate of 10%. It will be 341.84 million square meters in 2019, 379.33 million square meters in 2020, and 413.29 million square meters in 2021. At the same time, the proportion of heat supply transformation has increased year by year. By the end of 2021, the number of directly transferred public thermal power medium heat motor units has reached 144 [5], the installed capacity is 51.985 million kilowatts, and the installed proportion is close to 90%. There are only 13 pure condensing units in the province, The number and capacity of direct transfer public thermal power heating transformation have reached the limit, so vigorously developing nuclear power heating can effectively supplement the heating gap caused by the shutdown of small units, while meeting the development requirements of clean and low-carbon.

Shandong Province is a typical receiving power grid. In 2021, the maximum load of incoming electricity from outside the province will reach 21.47 million kilowatts, accounting for one fourth of the maximum load of power consumption in the whole society. The total installed capacity of wind power and photovoltaic has exceeded 40 million kilowatts. The power supply composition of Shandong power grid is becoming increasingly complex. External power does not participate in peak shaving, and the uncertainty of wind and solar grid connection is characterized by "reverse peak shaving". In the case of low pumped storage capacity in the province, the peak shaving of the power grid mainly depends on thermal power, This has also brought about two major problems. First, the coal price has remained high in the past two years, the thermal power is in a state of substantial loss, and the willingness to take the initiative to generate electricity has fallen to the freezing point. The situation of power supply guarantee is severe; Second, the direct transfer of public thermal power heating has increased year by year, and the peak shaving capacity affected by heating has decreased significantly in the heating season.

* Corresponding author: zhulingkai@woyoxin.com

https://doi.org/10.1051/e3sconf/202337502002

The contradiction between heating and peak shaving is prominent.

2. Status quo of nuclear heating peak shaving in Shandong Province

At present, the nuclear power installed capacity in Shandong Province is two 1250MW imported AP1000 third-generation nuclear power units in Haiyang Nuclear Power Station Phase I of Shandong Nuclear Power Co., Ltd. According to the Action Plan for Shandong Nuclear Energy Development and Construction Project issued by Shandong Provincial Energy Administration, by 2025, the installed capacity of nuclear power in operation and under construction in the province will be more than 13 million kilowatts, including 5.7 million kilowatts in operation and 30 million square meters of nuclear heating area; In 2030, the installed capacity of nuclear power in operation and under construction in the province will be more than 30 million kilowatts, including 13 million kilowatts in operation and installation, and the nuclear heating area will be 200 million square meters. During the 14th Five Year Plan period, nuclear power in Shandong Province will usher in a period of rapid development, and the impact of nuclear power on people's livelihood heating and grid peak shaving will gradually increase[6].

In terms of nuclear power heating, the low parameter steam extraction of nuclear power unit steam turbine is used for urban heating in nearby areas, realizing the energy ladder utilization of nuclear power, improving the thermal efficiency of nuclear power units, further reducing the carbon emissions and soot emissions in this area, and improving the contribution of nuclear power to regional environmental protection. Since 2019, Haiyang Nuclear Power has begun to undertake part of the people's livelihood heating tasks in Haiyang urban area. In Phase I, part of the plant auxiliary steam is extracted for external heating, with the maximum extraction capacity of 50t/h. On November 15, 2021, Shandong Haiyang Nuclear Power Heating Phase II Project will be put into operation. High exhaust steam extraction of Unit 1 will be adopted for external heating, with the maximum steam extraction volume of 346t/h. The heating area of Haiyang Nuclear Power will reach 5.2 million square meters, covering the urban area of Haiyang. Haiyang Nuclear Power Unit 1 will replace 12 coal-fired boilers, becoming the largest cogeneration unit in the world. On April 1, 2022, the "Nuanhe No.1" Haiyang Nuclear Energy Heating Project exceeded the heating task in the first heating season, providing 2 million GJ of clean heat and nearly 5 million square meters of heating area. Compared with coal-fired heating, nuclear heating saved 180,000 tons of raw coal consumption, reduced the amount of heat released to the environment decreased by 1.5 million GJ[7].

In terms of nuclear power peak shaving, the current load adjustment of nuclear power units is realized by regulating the output power of nuclear reactors. In the operation of PWR nuclear power plant, the control rod displacement and boron solution concentration are generally used as control variables to achieve the reactor load tracking control, which has a great impact on the life and operation safety of nuclear power units. Therefore, based on safety, economy and other considerations, Haiyang Nuclear Power does not participate in the peak shaving of the power grid at present, but operates with basic load. However, with the increasing peak valley difference of Shandong power grid and the pressure of new energy consumption, the demand for peak shaving of the power grid is increasing, and nuclear power peak shaving is gradually put on the agenda.

3. Development of Nuclear Power Heating and Peak shaving Technology at Home and Abroad

3.1 Research status and development trend abroad

In terms of nuclear heating, Bento et al. conducted safety evaluation on SECURE low-temperature nuclear heating stations developed in Sweden and Finland in 1978[2]. Rämä and others put forward a concept of flexible nuclear cogeneration system. Nuclear power plants operating according to this concept can use steam at different pressure levels to adjust the heat and electricity produced. In terms of nuclear power cogeneration heating, in 1964, the first nuclear power heating project in the world was put into use in the coastal area of the Swedish city of Ågesta. The main function of the nuclear power plant is to provide central heating and produce a small amount of electricity for Farsta, a suburb of Stockholm. In 1969, Miller of the Oak Ridge National Laboratory of the US Department of Energy proposed a nuclear power cogeneration system[8]. This system can provide most of the heat energy for the heating system or air conditioning system of nearby cities by using the heat from the back pressure steam turbine unit of the nuclear power plant or the secondary circuit of the nuclear power plant, and reduce the waste of the heat energy of the nuclear power plant.

In terms of nuclear power peak shaving, Westinghouse PWR nuclear power unit has achieved the record of more than 600 daily load tracking operations in four consecutive refueling cycles. The pressurized water reactor units in France are capable of peak regulation from the design point of view. During the first 65% life cycle of the fuel, the minimum output can be reduced to 27% of the rated power. During the first 65% - 90% life cycle of the fuel, the adjustable range of the units decreases linearly. The minimum technical output gradually increases. At 90% - 100% life cycle, the units do not have the capacity to regulate. Because nuclear power units account for a large proportion, France frequently participates in peak regulation of the power grid.

3.2 Research status and development trend in China

At present, China has the third generation nuclear power technology of "Hualong No.1" and "Guohe No.1". In terms of nuclear heating, China has developed "Yanlong", NHR-200I, NHR-200II and other types of low-temperature heating reactors. Xu Jiming briefly described the status and role of nuclear energy in the current energy development of the world and China, including the principles and technical route recommendations that should be followed for the development of nuclear heating[9]. Zhou Zhengdao et al. took the AP1000 third generation million nuclear power unit as an example to compare and analyze the two heat supply schemes, such as HP cylinder steam extraction and reheat hot section steam extraction, and introduced the impact of each steam extraction scheme on the unit's economy and safety[10]. Taking the AP1000 nuclear power generation unit as an example, Wu Xinzhuang and others proposed to use the absorption heat pump to extract the heat of the nuclear island equipment cooling water and the conventional island closed cooling water system as the first stage heating of the heat network, recover the waste heat of equipment operation, improve the unit economy, and then provide hot water after further heating through the heat network heater[11]. Low temperature heating nuclear reactors have not been popularized and applied in China due to factors such as economy and residents' sensitivity, and cogeneration of nuclear power units has been commercially applied in Haiyang Nuclear Power Plant.

In terms of nuclear power peak shaving, the control rod displacement and boron solution concentration are generally used as control quantities to achieve the reactor load tracking control, which can achieve "12-3-6-3" daily load tracking, that is, 12 h full output at peak load, 3 h linear load reduction at night, 6 h operation on low power platform (generally 50% rated power), and 3 h linear load increase at morning to full output. However, the above regulation modes have adverse effects on the operation safety and economy of the unit. Therefore, domestic scholars have proposed joint peak shaving and auxiliary peak shaving by comprehensive utilization of nuclear energy, and Zhao Jie has proposed the mode of pumped storage combined with nuclear power peak shaving, which can meet the peak shaving demand of the power grid and ensure that nuclear power has basic load[12]. Gao Xiaotian investigated the status quo of hydrogen energy preparation, storage, transportation and utilization, explored the hydrogen energy storage technology route suitable for nuclear power peak shaving, and analyzed the economic benefits of different hydrogen energy utilization technology routes by taking the existing operation of a nuclear power plant as an example[13]. Yang Zhenshuai proposed a peak shaving system and operation method of nuclear power plant based on compressed air energy storage, which can realize full gradient peak shaving of nuclear power plant under the premise of ensuring the safe operation of nuclear power generator set at rated load[14]. Lin Yi proposed a nuclear power peak shaving system and method based on seawater desalination technology, which can meet the requirements of nuclear power generator units to participate in peak shaving by means of hydropower generation, and extend the operation time of nuclear power with base load[15].

4. Summary and outlook

At present, there are two main methods for nuclear power peak shaving. One is to participate in peak shaving by adjusting the power of nuclear reactor, and the other is to cooperate with hydrogen production, energy storage, seawater desalination and other comprehensive utilization methods to participate in peak shaving. For nuclear power units that are responsible for heat supply, extracting some steam from the turbine side for heat supply will reduce the output of the unit. If "thermoelectric decoupling" can be achieved, the output reduction caused by heat supply will become a peak shaving resource instead, The unit output can be changed by adjusting the steam extraction volume without regulating the output power of the nuclear reactor to participate in the peak shaving of the power grid. Taking Haiyang Nuclear Power as an example, at present, the high pressure exhaust steam extraction capacity of Haiyang Nuclear Power for external heating is 300t/h, which affects the output of the unit by about 50000 kilowatts. If Haiyang Nuclear Power realizes "thermoelectric decoupling" by adding heat storage tanks, electric boilers, etc., the unit can flexibly operate within the load range of 1.2-125 million kilowatts to participate in peak shaving of the power grid. Therefore, the exploration of a new mode of nuclear power generator units participating in power grid peak shaving through "thermoelectric decoupling" has avoided the potential safety hazard of traditional peak shaving by adjusting nuclear reactor power, which is conducive to promoting the participation of provincial nuclear power generator units in peak shaving, and it is urgent to carry out research. Relying on the mature thermal power heating transformation technology route, it is of great significance to carry out the research on the heating transformation and economic evaluation of nuclear power units, to promote the substitution of nuclear power for thermal power heating, and to help achieve the goal of "double carbon".

Acknowledgments

This work was supported by the scientific and technological project of State Grid Shandong Electric Power Research Institute "Research on Thermoelectric Cooperative Regulation Technology of Nuclear Power Units" (ZY-2022-10).

References

1. Xinling Wang. Actively Responding to Climate Change and Promoting Clean and Low Carbon Transformation - Interpretation of the Action Plan for Carbon Peak by 2030[J]. China Electric Power Industry, 2021,(12):22-25.
2. Yantui Li, Yunsheng Bai, Shaoyang Han, etc. Analysis on the current situation and trend of nuclear heating development [J]. China Nuclear Science and Technology Progress Report, 2019, (6): 129-134.

3. Yanan Lun. Discussion on heating application of low temperature heating reactor [J]. Green Building Materials, 2018, (12): 186-187.
4. Jing Gao. China's number of nuclear power units under construction in operation is the second in the world [N]. Economic Information Daily, 2022-8-15 (7).
5. Jinxu Lao, Wei Zheng, Zhiqiang Gong, etc. Research on heating and peak shaving performance of different types of thermoelectric units in Shandong Power Grid [J]. Shandong Electric Power Technology, 2022,49 (293): 34-38.
6. Fang Wu. Holding up the leading role of nuclear energy to promote green transformation [J] Shandong State owned Assets, 2022, (06): 56-60.
7. Su Mu, Huizhong Liu. Shandong Nuclear Power, Breakthrough [N]. Yantai Daily, 2022-7-14(8).
8. Solomykov Aleksandr. Research on Sino Russian Comparison of Nuclear Heating and Optimization of Basic Heat Load [D] Dalian: Dalian University of Technology,2020.
9. Jiming Xu. My opinion on the technical route of nuclear energy development in China [J]. Nuclear Physics Review, 1997,14 (1): 59-61.
10. Zhengdao Zhou, Weiwei Bao, Yingwu Qin, etc. Frequency conversion transformation and economic analysis of 600 MW unit electric feed water pump [J]. Zhejiang Electric Power, 2020, 39 (6): 63-67.
11. Xinzhuang Wu, Shuan Xia. Economic analysis of steam extraction heating for AP1000 nuclear power units [J]. Turbine Technology, 2020, 62 (6): 475-477.
12. Jie Zhao, Dichen Liu, Qingsheng Lei, etc. Research on Nuclear Power Units Participating in Grid Peak shaving and Joint Operation with Pumped Storage Power Plants [J]. Journal of Electrical Engineering of China, 2011, 31 (7): 1-6.
13. Xiaotian Gao, Kexin Zheng, Chunrong Cai, etc. Research on the economy of hydrogen storage for nuclear power peak shaving [J]. Southern Energy Construction 2021,8 (4): 1-8.
14. Zhenshuai Yang, Huanran Wang, Ruixiong Li, etc. Internal combustion engine supercharging compressed air energy storage cold heat and power cogeneration system [J]. Energy Storage Science and Technology, 2020,9 (6): 1917-1925.
15. Yi Lin, Xi'an Pan, Zhangsui Lin, etc. Overview of peak shaving technology in power system with high proportion of nuclear power [J]. Modern Power, 2020,37 (1): 51-58.

 https://doi.org/10.1051/e3sconf/202337502003

Research on Technical supervision of Steam turbine in nuclear power plant

Yongqiang Che, Yisen Chen

State Grid Shandong Electric Power Research Institute, Jinan 250003, China

Abstract. Based on the technical characteristics of nuclear power turbine, the technical differences between nuclear power turbine and thermal power turbine are analyzed. Based on the premise of nuclear safety and combined with the characteristics of nuclear power plants, the feasible supervision mechanism of conventional island turbine and the daily management responsibilities of nuclear power supervision are studied. Extensive research and collection of nuclear power related data, including technical specifications, operation procedures, maintenance procedures, etc., through the analysis and collation of these data, the technical supervision content of nuclear power conventional island turbine is studied, highlighting the differences with thermal power turbine technical supervision.

Keywords: Nuclear power turbine; technical supervision; nuclear safety.

1. Introduction

In recent years, with the proposal of the dual-carbon goal, nuclear power, as a kind of clean energy and can meet the dual-carbon requirements, has attracted more and more attention, and domestic nuclear power has ushered in a period of rapid development. Nuclear power, in essence, belongs to the power industry, and should carry out technical supervision according to the traditional management mode of the power industry[1]. Technical supervision has made a great contribution to the safe operation of power grid and unit, and has become a common and always existing supervision and management method in the electric power industry. Another attribute of nuclear power is that it is nuclear-related and has a bearing on public safety, so special emphasis is placed on nuclear safety[2]. For nuclear safety, there is already a complete and independent supervision system, but the main focus of nuclear safety supervision is on the safety equipment on the nuclear island, and less attention is paid to the supervision of the system and equipment on the conventional island, such as the turbine body, auxiliary equipment, generators and so on. Generally speaking, the current domestic nuclear power plants have been able to basically meet the requirements of the technical supervision of the power industry from the technical point of view, but the lack of the construction of the technical supervision system of the power industry[3].

At present, some steam turbine technical supervision work has been carried out on the conventional island of nuclear power plants in China. However, due to the technical supervision standards that can be directly referred to in the implementation process, relatively complete technical supervision system standards of thermal power generation industry can only be referred to in large numbers[4-5]. Since the operation parameters of conventional nuclear power island are lower than that of thermal power generation, the system, equipment and structure of conventional nuclear power island are also different from that of thermal power generation, so the technical standards of thermal power generation often do not meet the technical supervision requirements of conventional nuclear power island. In terms of the supervision of the conventional island of nuclear power, if a set of conforming supervision system of electric power technology can be established according to the characteristics of nuclear power technology, it will be a supplement to the existing nuclear safety supervision, which is both necessary and feasible from the technical and implementation levels. Therefore, it is a very necessary task to study the technical supervision mechanism and supervision content of conventional island turbine in nuclear power plant.

Foreign institutions related to nuclear power operation generally focus on technical management, such as the International Atomic Energy Agency, the World Nuclear Operators Association, the Owners' Federation, the United States Nuclear Power Operation Institute and many other well-known foreign organizations have issued a series of standard guidelines in terms of technical management practices, and in terms of system and equipment supervision. Such as performance test, design and manufacturing, nuclear island core, equipment maintenance, medium - and long-term projects and other aspects of the formation of a relatively perfect technical management system[6]. The form of foreign nuclear power technology management is mainly implemented through technical support, its core point is to support the safe operation of the unit, it is directly supervised health condition; Performance testing is the direct monitoring of the availability of the plant's systems and equipment; Medium - and long-term project supervision is used to solve the major problems existing in the unit[7].

The main implementation unit of technical supervision of domestic nuclear power plants is the Institute of Electrical Science under each nuclear island group, which is responsible for technical service and technical supervision at the same time. It has a good knowledge of its nuclear power plants, so it is relatively easy to carry out the work. However, it should be realized that there are few researches on the technical supervision of nuclear power in China. At present, domestic nuclear power plants can basically meet the requirements of technical supervision of the electric power industry from the technical point of view, but there is a lack of construction of technical supervision system of the electric power industry. At present, some steam turbine technical supervision work has been carried out on the conventional island of nuclear power plants in China. However, due to the technical supervision standards that can be directly referred to in the implementation process, relatively complete technical supervision system standards of thermal power generation industry can only be referred to in large numbers[8].

2. Technical characteristics of steam turbine in nuclear power plant

2.1 Low main steam parameters

Common nuclear power reactors include pressurized water reactor, boiling water reactor, heavy water reactor and air-cooled reactor, etc. The main steam pressure of conventional island turbine is generally saturated steam or slightly superheated steam, and the pressure is generally between 5 and 7MPa[9]. The steam of the steam turbine in the boiling water reactor nuclear power plant is in contact with the nuclear reactor, while the steam working medium of other types of steam turbine is not in contact with the reactor, so there is basically no problem of non-radioactive contamination. The operation personnel of the nuclear power turbine do not need additional special protection. New types of nuclear reactors emerging in recent years, such as improved gas-cooled reactors, molten salt proliferation reactors, fast neutron reactors, etc., are similar to traditional thermal power units. Nuclear islands can provide superheated steam above 500℃ [10]. Therefore, subcritical or supercritical high-temperature reheat turbines can be selected for the selection of conventional island turbines, and the types and technical requirements of the turbines used are basically the same as those of conventional thermal power plants.

2.2 Large main steam flow

The obvious difference between nuclear power turbine and thermal power turbine is that the steam flow of nuclear power turbine is very large. Take the AP1000 unit of Haiyang Nuclear Power Plant as an example, the main steam flow is up to 6000t/h. This is mainly due to the low initial parameters and low steam pressure of nuclear power turbine, which makes the ideal enthalpy drop of nuclear power turbine much lower than that of thermal power plant. Therefore, under the same unit output conditions, the steam mass flow required by nuclear power turbine is about 2 times that of thermal power turbine. In addition, the steam pressure of nuclear power turbine is smaller than that of thermal power turbine, and the steam density is also smaller, resulting in the specific steam capacity of nuclear power turbine is larger than that of thermal power turbine. Therefore, the volume flow of main steam of nuclear power turbine is about 3-5 times of that of thermal power turbine.

The volume flow of nuclear steam makes the whole structure of the turbine very large. The centrifugal force of the steam turbine is directly proportional to the square of the rotating speed. For a steam turbine with a full speed of 3000r/min, the length of its final stage blade cannot be too long due to the restrictions of centrifugal force and material strength, and the volume flow of nuclear steam is large, so it is necessary to increase the number of low pressure cylinders. However, the maximum volume flow of steam at the exhaust outlet of the low pressure cylinder mainly depends on the length of the final stage blade. In order to avoid the excessive number of low-pressure cylinders, which may lead to too long turbine shaft, most turbines in large-capacity nuclear power plants adopt the half-speed turbine with a speed of 1500r/min, which can greatly reduce the centrifugal force and avoid the restrictions emphasized by materials. Moreover, the length of the final stage blade of the low-pressure cylinder can be increased, thus increasing the volume flow of the exhaust port. In addition, the low pressure cylinder adopts the double flow structure.

2.3 High steam humidity

The steam humidity of nuclear power turbine is generally relatively high, which means that the main steam is basically saturated or with small superheat, resulting in the flow part of the turbine is basically in the wet steam working state, which is very different from that of thermal power unit. For the low-pressure cylinder of the nuclear power turbine, under such wet steam working conditions, the humidity of the steam at the exhaust outlet of the low-pressure cylinder will reach about 25% if no dehumidification measures are taken, thus the safe and reliable operation of the final stage blade cannot be guaranteed. In addition, high humidity also brings many problems to the safe and economic operation of the steam turbine. The water content of the steam is large, and the water droplets may erode and corrodes the structure of the steam turbine, thus affecting their life and performance. There is also the problem of reduced efficiency in the steam turbine, which is mainly because the accelerated water droplets consume more energy than the accelerated steam flow, and the water droplets impact on the blade to produce a certain braking effect, which brings additional losses. The overspeed problem of steam turbine in nuclear

https://doi.org/10.1051/e3sconf/202337502003

power plant is much more prominent than that of steam turbine in thermal power plant. Therefore, special measures need to be taken. This is mainly due to the high water content. Therefore, in the design of wet steam turbine in nuclear power plant, the separation of water and the prevention of water erosion should be emphasized.

2.4 Half-speed machine

With the continuous improvement of the single capacity of nuclear power turbine, the new projects are basically above 1000MW. At present, the largest single capacity of nuclear power plant in China is 1750MW. The volume flow of nuclear steam makes the whole structure of the turbine very large. The centrifugal force of the steam turbine during operation is proportional to the square of the rotating speed. Therefore, limited by blade materials and centrifugal force, the length of the final stage blade of the full-speed turbine cannot meet the requirements of the flow area. However, the centrifugal force can be significantly reduced by reducing the turbine speed, such as 1500r/min, so longer final stage blade can be selected.

In addition, under the same output, the steam flow of nuclear power turbine is more than 2 times that of ultra-supercritical thermal power unit, and the number of low pressure cylinders is also more than that of thermal power unit, generally 3-4 low pressure cylinders. Moreover, due to the limitation of the parameters of the steam generator of nuclear reactor, the design of nuclear power turbine should include base load.

2.5 Reheater for soda separation

If no additional measures are taken, the high humidity of the high-pressure cylinder exhaust steam of nuclear power turbine will result in high humidity of the steam in the low-pressure cylinder. In order to effectively control the steam humidity, it is necessary to add a soda separator. The soda separation reheater has also become an indispensable equipment for nuclear power turbine, which is also an obvious difference from the thermal power turbine. The soda separation reheater consists of two parts: soda separator and reheater. In thermal power units, in order to ensure that the inlet steam of the low pressure cylinder retains a certain superheat, and at the same time, the steam humidity of the final blade of the low pressure cylinder can be controlled at about 10%, the exhaust steam of the high pressure cylinder is generally introduced into the boiler reheater to heat again, and then into the low pressure cylinder to do work, so as to improve the steam superheat and reduce the exhaust steam humidity.

However, in nuclear power units, since there is only superheater without reheater on nuclear island, the steam generator in nuclear island is equivalent to the superheater of thermal power boiler. The steam provided by nuclear island is generally saturated steam with pressure of 5-7MPa and humidity below 0.25%, and the main steam itself has no superheat. When the saturated main steam enters the high-pressure cylinder of the steam turbine to expand and do work, when the steam pressure drops to about 1.0-1.5MPa, the humidity of the steam has increased to 10%-13%. If the steam with such high humidity enters the low-pressure cylinder to continue to expand and do work, then the exhaust humidity of the final stage blade of the low-pressure cylinder will reach 20%-25%. This will significantly reduce the efficiency of the steam turbine and cause serious water erosion in the flow part of the steam turbine. Serious water erosion may cause the unit to fail to work normally.

Therefore, the steam separation reheater must be added in the nuclear power unit to control the humidity of the exhaust steam at the last stage of the low pressure cylinder. The reheater is usually composed of two parts: Soda separator and reheater, after the high-pressure cylinder exhaust steam is introduced into the soda separation reheater, the soda separator first separates the wet steam, dehydrates the water in the wet steam, and then enters the reheater to reheat the wet steam through the high-pressure cylinder extraction or main steam, so that the inlet steam of the low pressure cylinder has a superheat of 70-80℃. So that the low pressure cylinder last stage exhaust steam humidity can be controlled within the safety line of 10%.

3. Technical Management

Different from thermal power overhaul cycle, the overhaul (preventive maintenance) cycle of nuclear power turbine is determined by the refueling overhaul cycle (C) of nuclear power plant, which is generally 18 months. For each refueling overhaul cycle, only part of the steam turbine equipment is serviced (preventive maintenance). The regular island overhaul cycle is currently 4 refueling overhaul cycles. For example, in Haiyang Nuclear Power Plant, the overhaul of HY101 completed the maintenance of high-pressure cylinders and No.1 low-pressure cylinders, and the maintenance of HY201 completed the maintenance of high-pressure cylinders.

The technical departments of traditional thermal power plants are generally three departments: Production and Technology Department, Maintenance Department and Operation Department, while the technical departments related to nuclear power plants have system equipment department, technical support department, operation Department, overhaul department, mechanical maintenance department, Engineering Department and design management Department, etc. Each department has section-level departments under it. The division of labor is overlapping and repeated, but the management levels and responsibilities are different. Because the nuclear power industry has a high level of safety requirements, there is a big difference between it and thermal power plants in terms of management system and professional setting. First of all, there is a big difference between the establishment of nuclear power plants and thermal power plants. The technical departments of traditional thermal power plants are generally three departments: Production and Technology Department, Maintenance Department and Operation Department, while the technical departments related to nuclear power plants have system equipment department, technical support department, operation Department, overhaul department, mechanical maintenance department, Engineering Department and design management Department, etc. Each department has section-level departments under it. The division of labor is overlapping and repeated, but the management levels and responsibilities are different. Secondly, the management of steam turbine equipment and systems is three-dimensional. Different departments monitor, test and analyze the health status, performance and safety of equipment from different perspectives to make a comprehensive evaluation. Third, the management of steam turbine equipment and systems is fragmented. The division of labor of technicians is meticulous, and each of them is specially responsible for the management of some equipment and systems. The work within their management scope can be detailed and in-depth, but they lack a comprehensive understanding of relevant equipment and systems. Finally, the division of labor between different departments to deal with the same problem is clear and clear. Problem raising, cause analysis and treatment measures, implementation and supervision, acceptance and other links belong to different departments. After the completion of one link, it will be reviewed and entered into the next link. Online closed-loop management is implemented.

4. Supervision mechanism and management responsibilities

Conventional island turbine technical supervision of nuclear power plant covers the whole process supervision and management of turbine body, auxiliary equipment and related systems, such as selection, design, manufacturing, installation, commissioning, operation and maintenance, overhaul, performance test, life and aging management, etc. The technical supervision of conventional island steam turbines in nuclear power plants takes safety and quality as the center, takes technical standards as the basis, takes testing and inspection as the main means, and combines the application of new technologies, new processes, new materials and new equipment (hereinafter referred to as the "Four new") to carry out the work dynamically.

The technical contents of the technical supervision of the conventional island steam turbine of nuclear power plant should be specified and specific requirements should be put forward according to the technical characteristics, structure, parameters of the steam turbine of nuclear power plant and the influence of the steam turbine failure on the nuclear island (nuclear safety). Nuclear power plants and relevant units (nuclear power groups, Electric power Research Institutes, etc.) shall, in accordance with the requirements of DL/T 1051 and with reference to relevant standards, formulate implementation rules and carry out technical supervision of steam turbines on regular island of nuclear power plants.

An organizational system of technical supervision under the responsibility of the leaders in charge of nuclear power plants shall be established, the working system and process of technical supervision shall be perfected, and the corresponding detailed rules of supervision shall be formulated. A technical supervision network should be established at different levels to clarify the responsibilities and powers of technical supervision organs and personnel at different levels. Technical supervision should be carried out according to the characteristics of nuclear power plant management, and requirements should be put forward for the traceability, integrity and standardization of quality documents.

A supervision work plan shall be formulated, which shall include work requirements, work contents, responsible units, time nodes, etc., and a summary of supervision work shall be issued every six months or a year. Technical supervision meetings should be organized regularly to coordinate and solve specific problems in supervision work. The technical supervisory personnel shall check the new information in time, correctly evaluate the effectiveness and accuracy of the relevant technical standards of the equipment or facilities under supervision according to the newly promulgated national and industrial standards and the technical requirements of the equipment, and rectify the problems found in time. Technical supervision personnel should have the corresponding professional ability and skills, and each unit should organize regular special training work.

The technical supervision shall take every quarter (year) as a cycle to summarize the supervision work. The content of the summary report shall at least include the completion of the main supervision work, the highlights of the work, the difficulties of the work, the results of the work, the problems found, the cause analysis, the treatment process and the improvement measures, and the next supervision work plan. The technical supervision shall timely inspect the discovered problems or the problems of other units. If the problems are found, the technical supervision shall issue an early warning or alarm sheet, and follow up the implementation of the rectification. The problems found in the process of technical supervision shall be included or supplemented into the next annual supervision work plan, and the implementation of the rectification shall be supervised in the process of implementation.

5. Conclusion

This paper studies the technical characteristics of the conventional island turbine in nuclear power plant and analyzes the differences between it and the conventional thermal turbine. Based on the premise of nuclear safety and combined with the characteristics of nuclear power plants, the feasible supervision mechanism of conventional island steam turbine and the daily management responsibilities of nuclear power supervision are studied. Through extensive research and collection of nuclear power related data, including technical specifications, operation procedures, maintenance procedures, etc., through the analysis and collation of these data, this paper studies the technical supervision content of nuclear power conventional island turbine, and highlights the differences from thermal power turbine technical supervision. According to the life flow of nuclear power turbine, this project provides the technical supervision contents and requirements covering the whole process supervision and management, such as selection, design, construction, installation, commissioning, operation and maintenance, overhaul, performance test, life and aging management, etc.

The technical supervision of conventional island turbine in nuclear power plant is the whole process of turbine construction and production, including design review, equipment selection, supervision and acceptance, installation and commissioning, trial production, operation, maintenance and technical transformation. This project is a basic theoretical research, the research results are directly applied to the whole field above, applicable to the proposed nuclear power plant, under construction, in operation, has a very wide range of application prospects. Through the research of this project, a good foundation has been laid for the technical supervision of conventional island turbine in nuclear power plant.

Acknowledgments

This work was financially supported by ZY-2022-18 fund.

References

1. WANG Xin. Status Quo and development of nuclear power technology Supervision and management [J]. Science and Technology Vision,2013(31):337+352.
2. Sun Yecong, Liang Xueyuan, Liu Shangyuan, Deng Ruiyuan, Niu Jingjuan, Wang Xun, Meng De. Current situation of national standards related to nuclear safety and suggestions for further work [J]. Nuclear Standards Metrology and Quality, 2022(02):10-14.
3. Mao Haiyan, Jiang Shizhao. Analysis of Components of Technology management System [J]. China Electric Power Enterprise Management,2022(27):60-61.
4. HUANG Wenyao. Research on the value of Technical Supervision in Electric Power Safety Production [J]. China New Technology and New Products,2017(22):146-147.
5. Guo Junwen, Establishment and Application of Power Technology Supervision and management System in Power generation Group. Xi 'an Thermal Engineering Research Institute Co., LTD., Shaanxi Province,2017-11-14.
6. Du Aiguo, Wei Guojun, Guan Yufeng, Li Youyi, Yun Yun. Exploration of a new Model of group Reactor Management in the field of nuclear power technical Support [J]. China Electric Power Education,2020(S1):370.
7. ZUO Junru. Development Strategy of Foreign nuclear power Enterprises [J]. China Nuclear Power,2017,10(02):286-287.
8. Zeng Yapeng, Liu Huiming. Technology Supervision and Intelligent Management of Steam Turbine Supervision in a Thermal Power Plant [J]. Jiangxi Electric Power,2021,45(02):38-40.
9. Chen Ying. The present status of nuclear power steam turbine [J]. Journal of thermal turbine, 2006 (02) : 104-107.
10. Li Yong, Liu Xiansheng, Gu Meimei, He Xiaozhong. New 210 MW grade high temperature gas cooled reactor nuclear power steam turbine structure features [J]. Journal of thermal turbine, 2021, 50 (03) : 187-191.

https://doi.org/10.1051/e3sconf/202337502004

Application of Lidar in Comparison of Wind Speed and Wind Direction Meters in Wind Power Field

Hua Wang, Jinshan Wang, Wenqiang Tian, Zhiqiang Bao, Bao Wang, Yanxu Gou

Inner Mongolia Datang International Zhuozi New Energy Co., Ltd Ulanchap 012300, China

Abstract. Wind speed and wind direction indicator in wind turbine is a very important component in the whole power generation system. Inaccurate or incorrect sensor signals will not only make the wind turbine inefficient, but also affect the safe operation of the wind turbine and the life of other components. After the warranty period, after the wind power plant takes over the wind power operation and maintenance, the maintenance demand of the above sensors will be greatly increased. In this project, numerical simulation is used to evaluate the status of wind speed and direction indicator, so as to automatically capture data, compare data, and generate correction coefficients.

Keywords. Wind Speed; Wind Direction; Comparison; Indicator; Wind Turbine

1. Background

Wind turbines rely on dozens of micro sensors to monitor their operating conditions, and monitor the structure, internal sensitive locations and system components of wind turbines at any time to ensure efficient and safe operation. Among them, wind speed and direction indicator, vibration, speed, temperature, inclination and other sensors are very important components in the entire power generation system. Inaccurate or incorrect sensor signals will not only make the fan inefficient, but also affect the safe operation of the fan and the life of other components [1].

In addition, the fan manufacturer is mainly responsible for the calibration of the fan sensor, which is mainly carried out by sampling inspection [2]. The purpose is to meet the requirements of the contract terms, not to consider the safe and efficient operation of the unit. With the expiration of the warranty period and the improvement of the importance attached to the above work by major power generation groups, the demand for fan sensor maintenance is increasing in a blowout manner [3]. However, the current situation of wind power generation enterprises is that there are few operation and maintenance personnel, and the professional coverage of personnel is incomplete, which makes it difficult to achieve a comprehensive assessment of wind turbine sensors [4].

2. Necessity of technical transformation

Wind speed and wind direction indicator, vibration, temperature, inclination and displacement sensor in wind turbine are very important components in the whole power generation system [5]. Inaccurate or incorrect sensor signals will not only make the fan inefficient, but also affect the safe operation of the fan and the life of other components. After the warranty period, after the wind power plant takes over the wind power operation and maintenance, the maintenance demand of the above sensors will be greatly increased. However, the current situation of wind power generation enterprises is that there are few operation and maintenance personnel and the personnel coverage is incomplete.

For the wind speed and direction indicator, Doppler lidar technology is used in this project to achieve on-site and real-time calibration without dismantling the wind speed and direction indicator. At the same time, combined with numerical simulation technology, the deviation of the wind speed and direction indicator itself and the data deviation affected by the impeller are quantified, and finally the wind speed and direction indicator is calibrated in situ, and the unit power generation efficiency is improved by correcting the data deviation of the wind speed and direction indicator.

3. Main problems solved through technical transformation

This project will realize the transformation from the traditional sensor inspection mode to the judgment of the real-time working state of the sensor. With the goal of the ultimate safe and efficient production of the object, it will realize the real object-oriented measurement and make the measurement serve the object.

The purpose of this technology is to develop a sensor status assessment technology to meet the actual needs of the wind power generation industry, change the sensor status assessment work from labor-intensive to intelligent, reduce manual operation, reduce the difficulty of implementation, improve operability, solve the contradiction between the high requirements for status supervision in existing wind power generation and the difficulties in actual implementation, and fill the technical gap for the wind power generation industry.

This project is mainly aimed at wind power plants or power generation groups that are about to or have expired their warranty period. When the rights and responsibilities of the wind turbine are transferred after the warranty period, the wind power plant or the power generation group will pay more attention to the safety of the operating state of the unit, and will carry out the annual sensor calibration in accordance with relevant regulations. Therefore, there is an urgent technical demand.

4. Scheme demonstration

This technology is based on Doppler lidar technology, combined with numerical simulation model analysis, to achieve the in-situ static and dynamic calibration of wind direction and anemometer of wind turbine, and evaluate the real working state of wind speed and anemometer of unit. In addition, the mobile multi in one comparison sensor is arranged in the fan room to realize the sensor comparison status evaluation without disassembling the original sensor.

Doppler lidar is used to retrieve the radial wind speed by detecting the backscattered signal of aerosol particles and the Doppler frequency shift of the local oscillator of the system. It has the characteristics of high spatial and temporal resolution and high measurement accuracy. It is one of the most effective means to measure the fan performance curve at present, and has been applied to the wind measurement in the early stage of the project, the fan curve test and the post evaluation of wind farms at home and abroad.

At present, this technology has not been used in the field of wind turbine anemometer measurement. Due to installation error, mechanical wear and the influence of wake, the error of traditional anemometer will be caused. However, it is impossible to judge whether the deviation is caused by equipment failure or blade wake simply by comparing the lidar measurement results with the unit data.

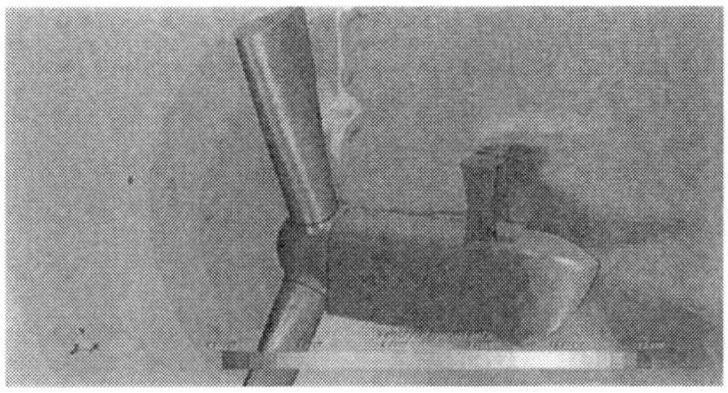

Fig. 1 Numerical simulation of the influence of impeller on anemometer

This technology uses numerical simulation modeling to calculate the influence of the wake on the anemometer. The input conditions include wind speed, wind direction, wind density, etc., and forms a matrix database to coordinate with the laser radar to call data.

Project scale and scheme:

Study on Quantifying Interference Factors of Anemometer and Wind Direction Meter by Numerical Simulation

It is impossible to accurately evaluate the status of the anemometer and wind direction indicator simply by comparing the Doppler lidar equipment with the anemometer and wind direction indicator. It is necessary to further clarify other interference factors, including the impact of impeller rotation under the operating state and the impact of impeller structure on the anemometer and wind direction indicator under the Y-type shutdown state, to analyze the initial yaw value of the wind direction indicator and optimize the unit operation.

The influence of the impeller on the anemometer can be calculated through numerical simulation. Taking the model building of 37.5m blade model as an example (Fig. 4), due to the interference of impeller, the wind speed data measured at the anemometer position in the engine room fluctuates periodically at 11m/s wind speed, as shown in Fig. 2, the anemometer results are always small, the detected value fluctuates between 8.5-10m/s, and there is 10-23% deviation from the incoming 11m/s wind speed.

Fig. 2 Modeling of 37.5m blade model (contour map with vorticity=0.8)

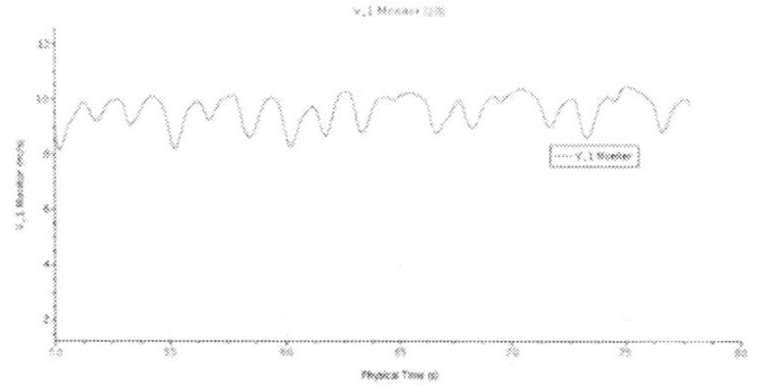

Fig. 3 Wind Speed Results of Existing Anemometers (11m/s)

As shown in Figure 3, when the theoretical wind vane should be at 0 °, it is interfered by the impeller, and the wind vane also has a certain angle deviation. Through big data analysis, it is usually interfered by the impeller, and the deviation of the wind vane is about 4 ° - 10 °.

The research content of this part belongs to theoretical calculation research, which is an important means to establish the impeller interference database through calculation, and 0 is an important reference basis for comparison between Doppler lidar measured data and wind speed and wind direction indicator.

Carry out research on the calibration method of Doppler lidar, focusing on the layout mode, scanning mode, dynamic screening process and static screening process of lidar; The corresponding software operating system is matched and developed. A screening method for the status of anemometers and wind direction indicators is developed for fans in operation; Develop the screening method of anemometer and wind direction indicator status under the Y-type shutdown status, and form a complete dynamic+static down wind speed and wind direction indicator status screening process.

Doppler lidar calibration technology is currently based on the International Electrotechnical Commission standard IEC61400-12-1 (2017). The equipment accessories used in this technology have been rigorously tested in Germany, and the measured data are compared with the data measured by the Class I wind cup anemometer in the international IEC standard. The correlation between them at any level is up to 99%, and the verification result is excellent. In this technology, the accuracy of numerical simulation wind speed can reach 0.05m/s, the accuracy of wind direction measurement can reach 0.1 °, the accuracy of lidar wind speed measurement can reach 0.1m/s, and the accuracy of wind direction measurement can reach 1 °, which fully meet the technical requirements of this project.

The calibration work will use the dynamic and static two-stage screening method to evaluate the two states of the fans in the wind farm:

In the first stage, dynamic screening: the Doppler lidar is arranged within the range of (0-50) meters on the ground at the front end of the wind direction of the wind turbine to measure the wind speed data at the installation height of the wind vane of the wind turbine; According to the wind speed data measured by Doppler lidar in real time, the correction coefficient of the wind vane of the wind turbine disturbed by the impeller under the non-stop state is called;

5. Conclusion

Compare the corrected wind speed data with the data measured by the wind vane of the wind turbine; Set the screening deviation limit, and enter the second stage when the deviation between the corrected wind speed data and the data measured by the wind vane of the wind turbine exceeds the screening deviation limit; When the deviation between the corrected wind speed data and the data measured by the wind vane of the wind turbine does not exceed the screening deviation limit, the recorded data deviation is used for the initial correction setting of the wind vane system; The laser radar is arranged at the nearest position at the front end of the fan to measure the wind speed and direction data at the installation height of the anemometer, and call the matrix database data to participate in the comparison according to the real-time wind changes. So far, the first phase of status screening has been completed.

The second stage, static screening: conduct shutdown test for wind turbines whose data deviation in the second stage exceeds the screening deviation limit. Shut down the fan blade in Y shape, place the Doppler lidar within (0-50) meters of the front ground in the wind direction of the wind turbine, and measure the wind speed data at the installation height of the wind vane of the wind turbine; According to the real-time wind speed data measured by Doppler lidar, the correction coefficient of the wind vane of the wind turbine disturbed by the impeller under the shutdown state is called; Compare the corrected wind speed data with the data measured by the wind vane of the wind turbine; Set the screening deviation limit. When the deviation between the corrected wind speed data and the data measured by the wind vane of the wind turbine exceeds the screening deviation limit, it is determined that the wind vane of the wind turbine under test is faulty and needs to be repaired or replaced.

The research content of this part is experimental research, which is an important research content of Doppler lidar application in wind speed and wind direction indicator status assessment, and an important technical means for verification of related methods.

References

1. Liu Yang, Fu Yang, Huang Ling-ling et al. Reborn and upgrading: Optimum repowering planning for offshore wind farms[J] Energy Reports, 2022, 8
2. Rostami Saeed, Rostamzadeh Hadi, Fatehi Rouhollah A new wind turbine driven trigeneration system applicable for humid and windy areas, working with various nanofluids[J] Journal of Cleaner Production, 2021, 296
3. Xu Fang Yuan, Tang Rui Xin, Xu Si Bin et al. Neural network-based photovoltaic generation capacity prediction system with benefit-oriented modification[J] Energy, 2021, 223
4. Kumar Pravin, Singh Rajesh Kr, Paul Justin et al. Analyzing challenges for sustainable supply chain of electric vehicle batteries using a hybrid approach of Delphi and Best-Worst Method[J] Resources, Conservation & Recycling, 2021, 175
5. Sebastiani Alessandro, Castellani Francesco, Crasto Giorgio et al. Data analysis and simulation of the Lillgrund wind farm[J] Wind Energy, 2020, 24(6)

https://doi.org/10.1051/e3sconf/202337502005

Application of Mobile Vibration Online Comparison Sensor in Wind Power Field

Hua Wang, Jinshan Wang, Wenqiang Tian, Zhiqiang Bao, Bao Wang, Yanxu Gou

Inner Mongolia Datang International Zhuozi New Energy Co., Ltd Ulanchap 012300, China

Abstract. MEMS sensor is a new type of sensor manufactured by micro electronics and micro machining technology. Compared with traditional sensors, it has the characteristics of small size, light weight, low cost, low power consumption, high reliability, suitable for batch production, easy integration and intelligent realization. This project adopts a multi in one sensor based on MEMS principle. The sensor is a wireless passive three-axis sensor that integrates acceleration, speed, displacement, temperature, inclination and other parameters. It is very convenient to install and disassemble, and can realize the cycle monitoring of the whole wind field.

Keywords: MEMS sensor; Mobile Vibration; Online Comparison Sensor; Wind Power.

1. Background

The main vibration sources in the wind turbine include the main shaft, gearbox, generator, engine room, tower, etc., which are all important components in the entire power generation system [1]. The occurrence of vibration will not only make the fan inefficient, but also affect the safe operation of the fan and the life of other components [2]. At present, all types of vibration sensors are equipped on the newly put into operation wind turbine generator units, but the status assessment of the sensors is still blank. After installation, they will not be disassembled or submitted for inspection, and no assessment status scheme has been provided [3].

2. Main basis

Wind power equipment has been operating under the action of complex alternating loads such as gusts all day for a long time, and the harsh service environment has seriously affected the operation safety and maintenance guarantee of wind power equipment [4]. The alternating load acts on the wind turbine blades and is transmitted through the bearings, shafts, gears, generators and other components in the drive chain, making the drive chain prone to failure during service. At present, the monitoring system widely equipped on wind power equipment is SCADA system, which can monitor the operating status of wind power equipment in terms of current, voltage, grid connection and other conditions, and has the functions of alarm and report; However, the state parameters monitored by the system are limited, mainly including current, voltage, power and other signals, and the vibration monitoring and fault diagnosis functions for key components are still lacking [5].

Foreign countries, especially western developed countries, have developed condition monitoring equipment and analysis software for wind power equipment for a long time. Among the condition monitoring equipment for the drive chain of wind power equipment, the more famous developers and their monitoring systems include: the wind turbine remote monitoring and vibration system VIBROWEB XP and the condition monitoring software OMNITREND of Germany PRUFTECHNIK; ADAPT of Bently Nevada Corporation Condition monitoring system of wind power generation equipment drive chain; Winergy AG, a wind electronics company of Flender, Germany, has built a full set of wind power operation and maintenance plans. The service department can install a status monitoring system to detect the operation of each transmission component of the wind power generation equipment in real time and online, and provide maintenance and repair services for the transmission system; New Zealand Commtest Company has developed vboine and other systems to diagnose the fault of the transmission chain of wind power generation equipment.

Although the domestic vibration monitoring technology started late, the development of the domestic monitoring system has entered a rapid development stage driven by the huge domestic market demand for remote operation and maintenance of wind power. Anhui Rongzhi Rixin Technology Co., Ltd. has specially developed an online monitoring and fault diagnosis system WindCMS for wind power equipment, which can monitor various operating data during operation and automatically store them. Beijing Tangzhi Technology Development Co., Ltd. has developed the online fault diagnosis system JK10460

for wind turbines, focusing on automatic real-time diagnosis. XJ Electric Co., Ltd. has developed an online status monitoring system FJK-530 for wind turbine generators, which can identify and evaluate the potential faults of wind turbine equipment through vibration signals, and evaluate the service life. Domestic wind power equipment OEMs have also developed health management systems for the later operation and maintenance of wind power, to improve the utilization and intelligence of wind turbines. Xinjiang Goldwind Technology Co., Ltd. has developed the intelligent health management system SPHM, which has realized intelligent fault diagnosis, classified conventional faults, difficult faults and "difficult and miscellaneous diseases", formed standard fault work orders and standard processing results, thus achieving the effect of intelligent operation and maintenance deployment. For fault diagnosis of wind turbine generator units, the fault diagnosis system developed by Guodian United Power Technology Co., Ltd. can quickly determine the type of fault of wind turbine generator units, quickly and accurately extract the main faults from a large number of relevant fault information, thereby improving the efficiency of unit fault troubleshooting and reducing manual troubleshooting time. The intelligent fault diagnosis and early warning protection of wind power equipment can reduce the cost and increase the efficiency of wind power operation and maintenance.
It can be seen that the monitoring system based on vibration, inclination, temperature, displacement and other parameters is relatively well developed, but the evaluation of the sensor status in the above system is still blank. This technology aims to achieve the sensor status evaluation in the monitoring system through technical means to ensure the quality of analysis data.

3. Basic information of the original system or equipment:

With the vigorous development of wind power, the sensors arranged on wind turbines are also being upgraded and improved synchronously. At present, most of the newly built units are equipped with vibration, inclination and other sensors. However, the state evaluation method for sensors is still traditional, mostly offline, that is, the sensors in use are removed from the installation position and sent to the metrological department for verification. However, unlike thermal power units, once the wind turbine sensor is installed, it is almost difficult to disassemble it for inspection. In addition, many factors such as installation, disassembly, transportation and comprehensive error are ignored in the inspection. Only the state of the sensor under laboratory conditions is evaluated, not the true state under working conditions.

4. Scheme demonstration:

Develop a set of in-situ comparison methods and algorithms for wind turbine generator vibration, inclination, temperature and displacement sensors, analyze the possible vibration interference factors in the position of the wind turbine generator room, form a vibration sensor evaluation system, conduct a comprehensive evaluation of the sensor body, transmission system, etc., and analyze the cause of error. By arranging multiple in one sensors at the main vibration source, the vibration sensor body and data transmission system are evaluated separately by comparison.
This part focuses on the basic research of data analysis algorithm.
Based on the accumulation of sensor data arranged within three months, the alarm range is analyzed. This section analyzes the monitoring data of the main bearing rear end cover, gearbox low speed end, gearbox high speed end, generator front shaft end cover, generator rear shaft end cover, and generator bottom frame to determine the boundary value of the vibration evaluation area.

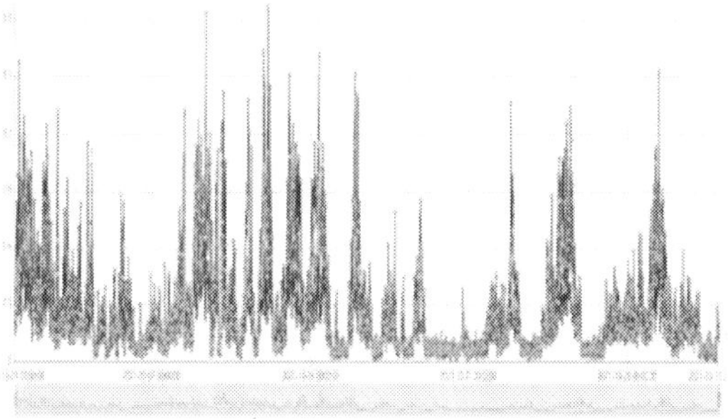

Fig. 1 Acceleration data sampling of main bearing rear end cover

 https://doi.org/10.1051/e3sconf/202337502005

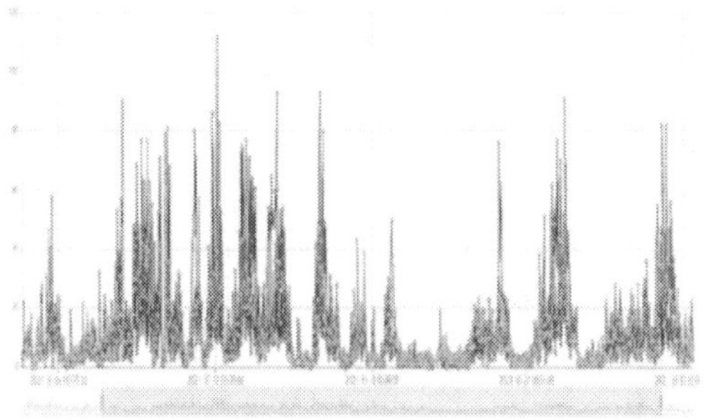

Fig. 2 Speed data sampling of main bearing rear end cover

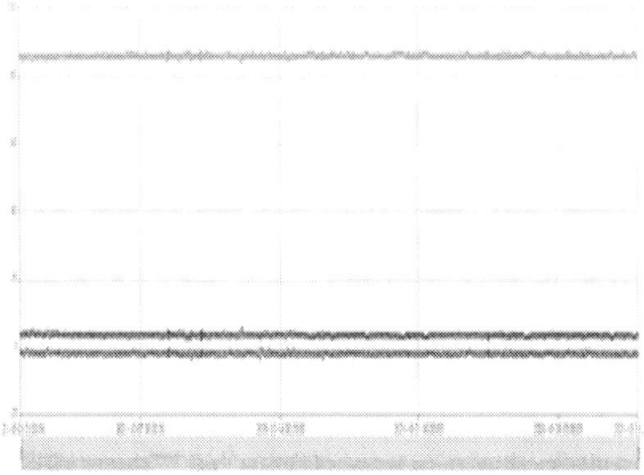

Fig. 3 Main bearing rear end cap displacement data sampling

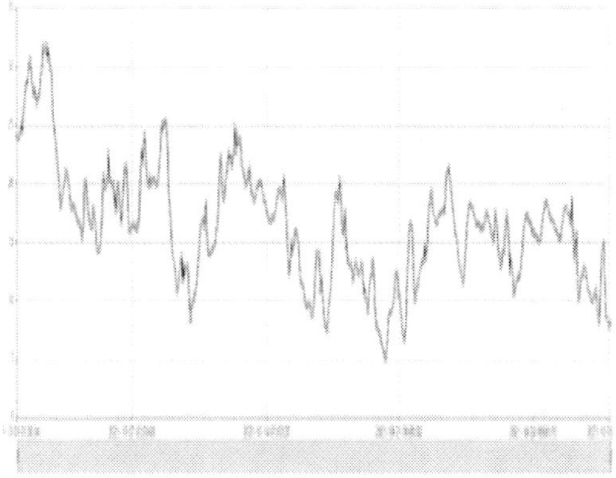

Fig. 4 Main bearing rear end cap inclination data sampling

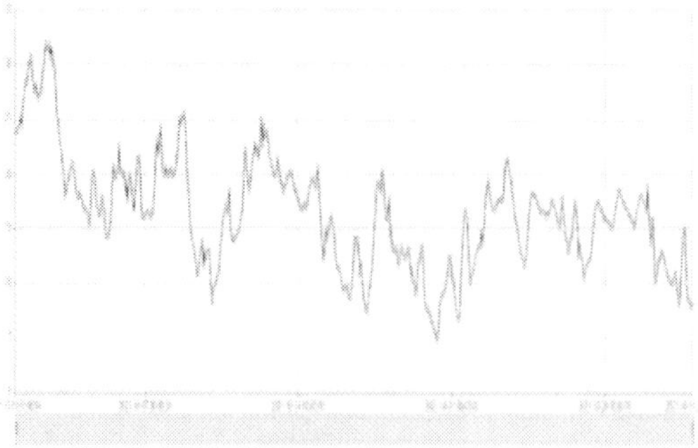

Fig. 5 Temperature data sampling of rear end cover of main bearing

5. Conclusion

The main bearing position mainly focuses on the axial and radial acceleration and speed parameters, where the root mean square of acceleration triaxial data is 0.09-0.137m/s2, and the root mean square of velocity triaxial data is 0.24-0.48 mm/s. Displacement parameters are helpful to compare acceleration and velocity parameters; The inclination parameter is relatively stable; In addition, temperature monitoring shall be combined with outdoor temperature to determine whether there is abnormality.

https://doi.org/10.1051/e3sconf/202337502005

References

1. Xiaolin Ge, Quan Chen, Yang Fu et al. Optimization of maintenance scheduling for offshore wind turbines considering the wake effect of arbitrary wind direction[J] Electric Power Systems Research, 2020, 184(C)
2. Yizhong Chen, Jing Li, Li He Tradeoffs in cost competitiveness and emission reduction within microgrid sustainable development considering price-based demand response[J] Science of the Total Environment, 2020, 703(C)
3. Pan Keda, Chen Zhaohua, Lai Chun Sing et al. An unsupervised data-driven approach for behind-the-meter photovoltaic power generation disaggregation[J] Applied Energy, 2022, 309
4. Ying Deng, Rashid Asif, Sheng Liu He et al. Research on power increase adaptive control strategy based on 5 MW wind turbine[J] Energy Reports, 2021, 7(S1)
5. Jiang Hongli, Ma Shiying, Song Ruihua et al. Analysis on subsynchronous oscillation stability for large scale offshore wind integration[J] Energy Reports, 2020, 6(S9)

https://doi.org/10.1051/e3sconf/202337502006

Study on Amplification Effect of Wind Turbine Blade Stopping Position on Vortex Induced Vibration of Tower

Ziyue Wang *, Shoulei Jiao, Guowei Zhang, Xiaoxu Chen

CRRC Shandong Wind Power Co., Ltd., Jinan 250000, China

Abstract. This project takes XX unit as the analysis object to study the first-order vibration wind speed of wind turbine under shutdown vibration. The vortex-induced vibration response is calculated and the displacement under resonant wind speed is calculated by fluid-solid coupling method. The above analysis techniques and methods are forward-looking and advanced in vibration characteristics of domestic wind turbines. On the basis of fluid dynamics analysis, the suppression measures are studied emphatically. Through the optimization and comparison of different inhibition schemes, the effect of inhibition measures was quantitatively analyzed. The analysis shows that the vibration characteristics of wind turbine with high flexible tower has its own particularity. The first order vibration probability and vibration displacement can be effectively reduced by pneumatic method and external measures.

Keywords: Wind Turbine; vibration characteristics; first-order vibration; fluid-solid.

1. Research Background

Vortex-induced vibration (VIV) is a complex air-elastic coupling phenomenon[1]. The wind passes around the non-fluid section of the structure to form vortexes, and the periodic shedding of vortexes will arouse the structure's vibration perpendicular to the direction of incoming flow[2-3]. In the case of specific wind speed, when the frequency of vortex shedding coincides with or is close to the natural frequency of the structure, resonance occurs, and the vibration amplitude is large, so it will cause greater fatigue damage to the structure[4-5].

As for the research on VIV, the current research methods are mainly based on calculation evaluation and field test verification based on standards. However, due to the complexity of the problem and the fact that the phenomenon of VIV intensification caused by blades is difficult to be calculated by engineering algorithms, it is impossible to provide guidance for the test work, and the analysis of this content in the industry is not sufficient, so it is necessary to conduct in-depth research. From the perspective of mechanism research and engineering application, this project plans to use CFD simulation calculation method to study the vortex-induced vibration risk of unit during and after hoisting, and verify the consistency of simulation results through field tests.

2. Vortex-excited vibration background

The vortex generation of vortex induced vibration can be divided into 2S, 2P, 2T and 2C. In the form of 2S vortex, two vortexes with little difference in intensity fall off in each vibration period, and the two vortexes rotate in opposite directions. The well-known Karman vortex street is 2S form. In Figure 2-6, (a), 2S form trailing vortexes occur during vortex-induced vibration. In the form of 2P vortex, two vortex pairs with different strengths occur on both sides of the cylinder in opposite directions, as shown in Figure 1 (b) and (c).

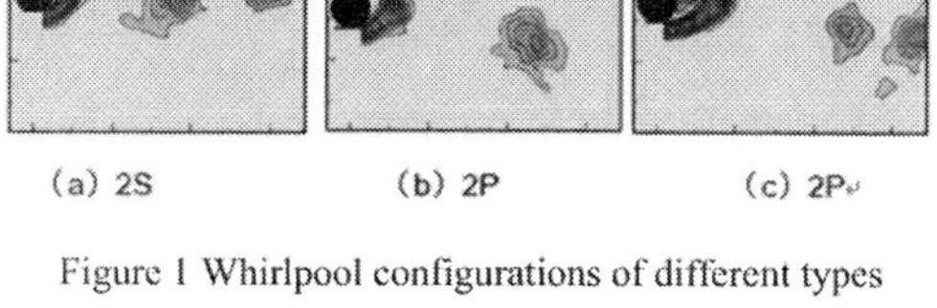

(a) 2S (b) 2P (c) 2P

Figure 1 Whirlpool configurations of different types

* Corresponding author: wangziyue.sd@crrcgc.cc

The vortex form of 2T is formed every half cycle, three vortices occur on the cylinder, two rotating in the same direction and the other opposite. And the intensity of the vortex in the same direction is obviously greater than that of the single vortex. The analysis shows that the type of 2T vortex is affected by the mass ratio and the degree of freedom. With the increase of the mass ratio in the case of two degrees of freedom, the 2T vortex type gradually transforms to 2P. Type 2C vortex mode means that one vortex pair occurs on each side of the cylinder, with the same internal rotation and opposite rotation of the two vortex pairs, as shown in Figure 2.

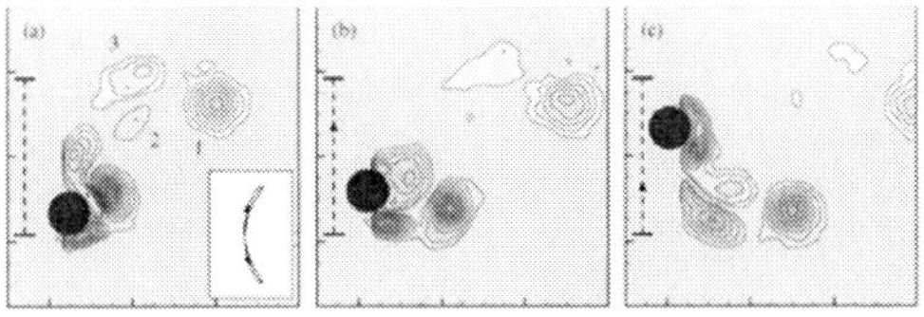

(a) 2T

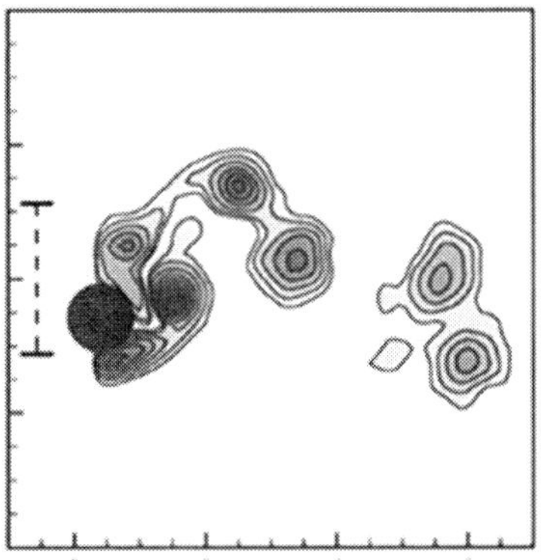

(b) 2C

Figure 2 Different types of whirlpools

A follow-up study found the P+S vortex pattern. In the form of P+S vortex, one vortex pair and one counter-rotating vortex occur respectively on both sides of the cylinder, as shown in Figure3. At first, this vortex form was only found in forced vibration, and it was believed that it could not appear in self-excited vibration. In recent years, many scholars have discovered the P+S vortex form in self-excited vibration studies, and supplemented the production principle. The formation of P+S vortex only occurs in the Reynolds number range of 300 to 500, and is not found at higher and lower Reynolds numbers. So far, the research on the morphology of P+S vortex is only superficial, and further research is needed.

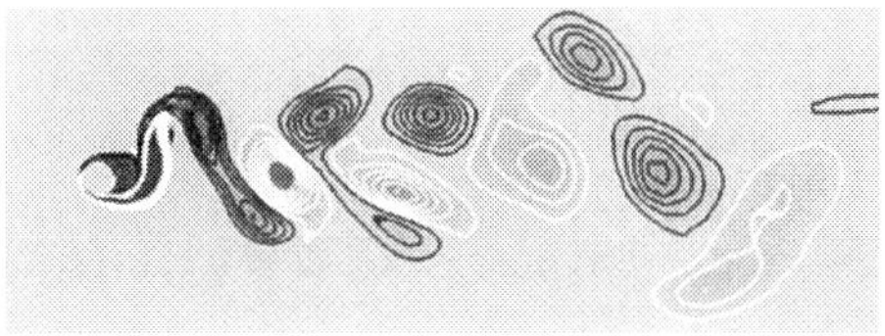

Figure 3 P+S vortex

3. Pitch Angle variation _ flow condition

According to the results of two-dimensional calculation, the calculation model of the whole machine was established. The model stops unchanged and the pitch angles are 90 degrees, 0 degrees, 15 degrees, 30 degrees and 45 degrees respectively.

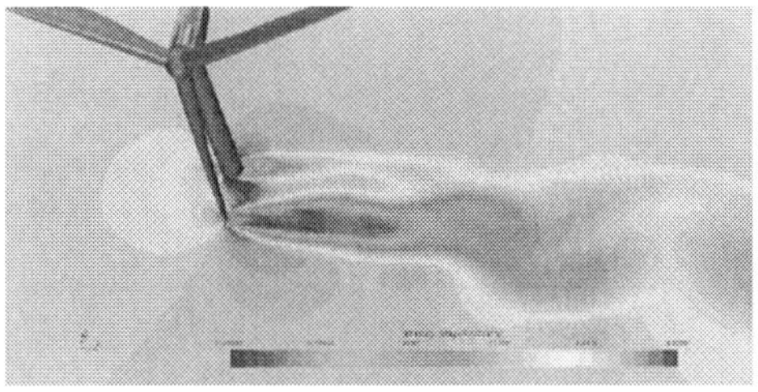

Figure 4 Contrast of different pitch angles at 120m height (90 degrees)

From the cloud image calculation results of the whole machine, it is consistent with the 2D results. Due to the influence of double column flow around the tower cylinder shedding vortex is greatly affected by the blade. This effect is eliminated when the distance between the blade and the tower barrel is greater than 2D, and when the distance between the blade and the tower barrel is less than 100m.

4. Parking impact

Figure 5 Impact of 30 ° on wake flow

Figure 5 shows the change of wake flow at 30 degrees at the terminal. From the perspective of wake impact, the front-end blade has an important effect on the wake of the tower barrel. The vortex shedding of the tower barrel is intensified by the wake action of the front blade. More intense double column shedding was formed. This result is consistent with misplacement shedding in double column shedding.

 https://doi.org/10.1051/e3sconf/202337502006

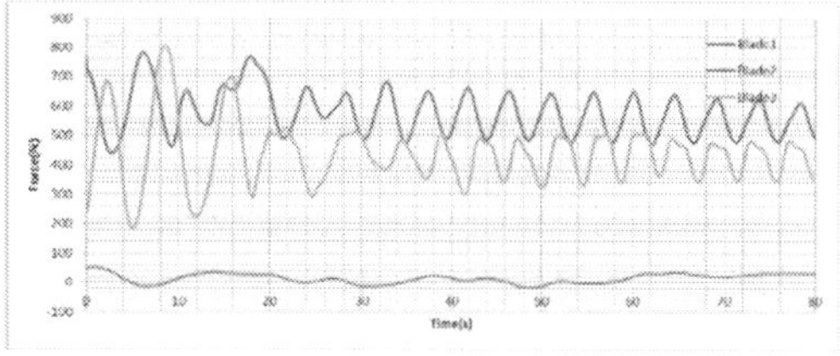

Figure 6 Mechanical changes at 30 degrees in the terminal

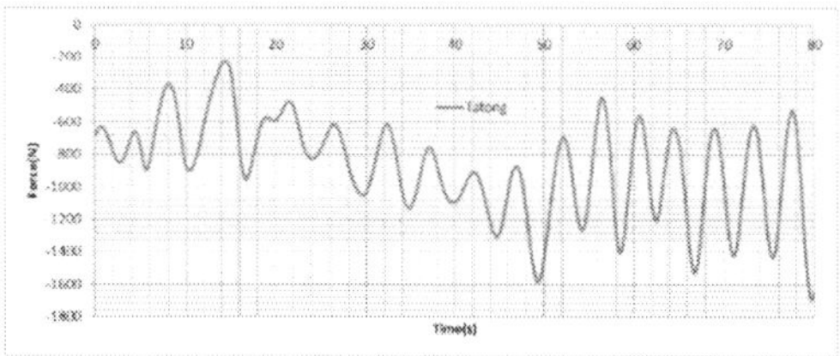

Figure 7 Force change of 30 ° tower barrel at the parking position

At 30 degrees, the wind is blowing upstream from the side. Both the blades and the tower barrel form their own strong periodic loads. The vibration of the tower barrel is changed into the superimposed vibration of the tower barrel and the blade. When the standing position is 30 degrees, the two upstream blades are in vibration state, and this superposition effect will significantly magnify the vibration amplitude of the tower barrel.

5. Stop with pitch Angle

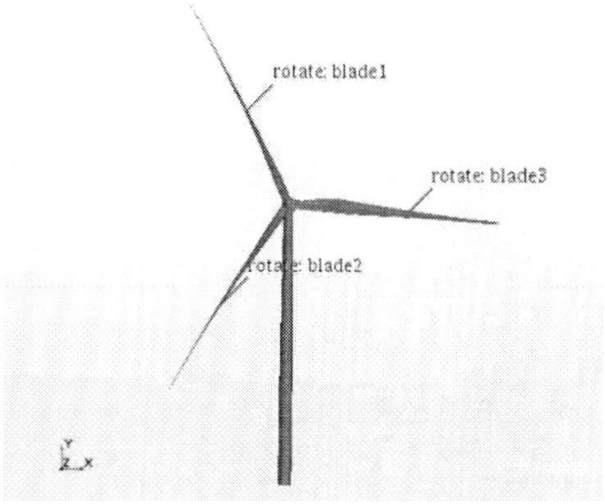

Figure 8 Single blade pitch Angle 0 °

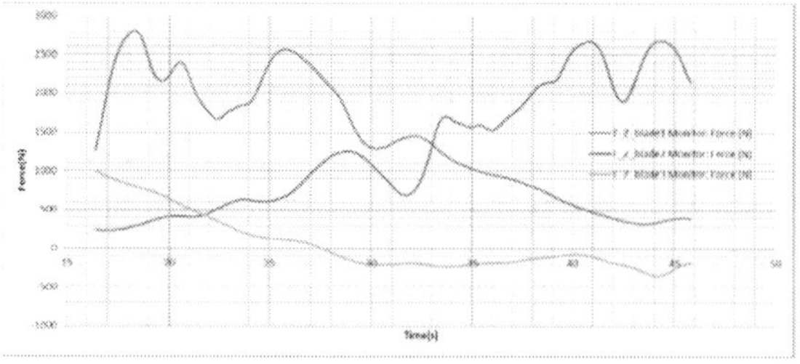

Figure 9 Forces on blades in the Z direction

It is calculated that the wind wheel rotates at a low speed (1rpm) at a low wind speed. The pitch Angle of blade3 is 0 degrees, and that of blade1 and 2 is 90 degrees. FIG. 9 shows the stress changes of three blades in the Z direction. Under the combined action of low speed rotation and pitch Angle, the blade's periodic force in the Z direction is eliminated. The overall z-direction amplitude will be greatly reduced.

6. Summary

In this chapter, the influence of the standing position of the blade on the vibration characteristics is calculated. Some conclusions are drawn as follows:

1. Due to the influence of double column flow around the tower cylinder shedding vortex is greatly affected by the blade. This effect is eliminated when the distance between the blade and the tower barrel is greater than 2D, and when the distance between the blade and the tower barrel is less than 100m.
2. The superimposed force of the blade on the tower barrel changes with the change of blade pitch Angle. According to the two-dimensional and three-dimensional integrated results, the column with pitch Angle <45 degrees is eliminated by blade influence. When the pitch Angle is less than 60 degrees, the vibration state of the cylinder is greatly reduced. At the same time, the aerodynamic force of the blade changes under the influence of pitch Angle.

3, the parking position has a blade in the upstream of the tower barrel, such as deflection 30 degrees. The upstream blade also has an independent vibration shedding vortex, which is superimposed with the vibration of the tower barrel.

4. To sum up, the adjustment of blade pitch Angle can change the vibration characteristics of blade and tower barrel. Pitch Angle should be at least 60 degrees or less.
5. After comprehensive consideration, the expected effect is achieved through the effect of parking position and pitch Angle.

References

1. Health Effects Related to Wind Turbine Sound, Including Low-Frequency Sound and Infrasound[J]. Irene Kamp;Frits Berg. Acoustics Australia. 2018(1)
2. Studies on Infrasound Noise Generated by Operation of Low-Power Wind Turbine[J]. Pierzga R.;Boczar T.;Wotzka D.;Zmarzły D.. Acta Physica Polonica A. 2013(3)

3. Wen Xiaoqiang, Xu Ziang Wind turbine fault diagnosis based on ReliefF-PCA and DNN[J] Expert Systems With Applications, 2021, 178
4. Guo Sheng, Yang Tao, Hua Haochen et al. Coupling fault diagnosis of wind turbine gearbox based on multitask parallel convolutional neural networks with overall information[J] Renewable Energy, 2021, 178
5. Karimpour Mostafa, Schmid Robert, Tan Ying Exact output regulation for wind turbine active power control[J] Control Engineering Practice, 2021, 114

Effect of dam removal of small hydropower station on migration and distribution of biogenic elements: Progress and prospects

Cai Li [1,2,*], Linjie Guan[1,2], Kun Yang [1,2], Li Zhang[1,2], Qingsong Fan [1,2]

[1] Changjiang Spatial Information Technology Engineering Co., Ltd., Wuhan 430010, China
[2] Hydro information Perception and Big Data Engineering Technology Research Center of Hubei Province, Wuhan 430010, China

Abstract: The small hydropower stations have reached a new balance after years of operation. The ecological balance of the river would been broken by the dam removal and the reservoir area above the dam would disappear. Consequently, a series of changes occurred in water depth, velocity of flow, the concentration of organic and inorganic nutrients, distribution and form of organic matter, aquatic organism, etc. It is unclear what would happen for the hydrodynamic conditions, sediment characteristics and aquatic organism community structure of the reservoir area and the lower reaches after the dam removal. Therefore, it is urgent to study the impact of dam removal on the transportation and redistribution of organic debris and the potential ecological effects with benthos as indicators in the river where the small hydropower station is located.

Keywords: biogenic elements; organic detritus; distribution and morphology; diversion-type small hydropower station; dam removal

1. Introduction

According to statistics, there were 24100 small hydropower stations in the Yangtze River Economic Belt of China by the end of 2017. In a specific historical period, these small hydropower stations have played a huge role in energy utilization, Replacing firewood with electricity, energy conservation and emission reduction. Due to poor management and operation, some small hydropower stations have led to the cutoff of some small and medium-sized rivers, which has affected the transmission of river material and energy and the community structure of aquatic organisms to a certain extent. This furtherly has had a certain impact on the river ecological environment of the river. In China, most of small hydropower stations were built in the 1960s and 1970s, and most of them have been in operation more than 30 years. Consequently, the physical and chemical characteristics and the aquatic biological community of the river located by the small hydropower station have reached a new balance through natural selection and succession.

In December 2018, the Opinion on the Classification and Rectification of Small Hydropower Stations in the Yangtze River Economic Belt have been jointly issued by the Ministry of Water Resources and three other departments of China. Relevant departments of provinces and cities in the Yangtze River Economic Belt should carry out assessment, classification, rectification and management for these small hydropower stations, according to the issued opinion. The small stations are required to be divided into three categories: immediate demolition stations, timed shutdown stations and guided rectification stations. For the immediate demolition stations, the dam and ancillary facilities should be partially or completely removed in principle, those of which with comprehensive benefits such as flood control, irrigation and water supply, the dam or diversion canal can not be removed, but complete discharge and monitoring facilities for ecological flow are ensured. For the timed shutdown stations, it is required to complete reconstruction of discharge facilities and installation of monitoring equipment for ecological flow and connect to the provincial ecological flow monitoring platform. The ecological balance of the river would been broken by the dam removal and the reservoir area above the dam would disappear. The upstream water habitat of the dam would be restored from a still water reservoir to a flowing river channel. Consequently, a series of changes occurred in water depth, velocity of flow, the concentration of organic and inorganic nutrients, distribution and form of organic matter, aquatic organism, etc. What would happen for the hydrodynamic conditions, sediment characteristics and aquatic organism community structure of the reservoir area and the lower reaches after the dam removal? Positive or negative? Single or cumulative? How would the change of tributary water environment affect the main stream? These are not clear at present and need for further research.

* Corresponding author: 476530436@qq.com

2. Influence of dam removal on distribution and morphology of biogenic elements

According to the River Continuum Concept (Vannote et al., 1980), from the source to the open lower reaches the river is a continuous, flowing, unique and complete ecosystem, the exogenous and endogenous nutrients of which are characteristic of continuous changes. The transport of biogenic elements is of great significance to the river ecosystem and has always been a hotspot in the study of the watershed geochemical material cycle. The main forms of biogenic elements from different sources (including internal and external sources) in reservoirs were shown in Table 1 (Deng et al., 2018). The biogenic elements in rivers are transported, floated and transformed under the combined physical, chemical and biological effects of climate, landform, animals, plants, microorganisms, etc., taking water and sand as carriers (Han et al., 2014; Maavara et al., 2014; Marx et al., 2017). Biogenic substances, as important environmental elements and the foundation of river ecosystem, physical sedimentation, have different chemical dissolution, mineralization and bio-absorption processes in the reservoir due to their different properties and functions, thus further leading to changes in retention efficiency and chemometric characteristics of biogenic elements. The cyclic process of reservoir biogenic elements was shown in Figure 1 (Deng et al., 2018).

Table 1 The main forms of biogenic elements in reservoirs (Deng et al., 2018)

Elements	Main forms
C	POC (Particle Organic Carbon), PIC (Particle Inorganic Carbon), DOC (Dissolved Organic Carbon), DIC (Dissolved Inorganic Carbon)
N	PON (Particle Organic Nitrogen), DON (Dissolved Organic Nitrogen), DIN (Dissolved Inorganic Nitrogen, including NO_x-N and NH_4-N)
P	TDP (Total Dissolved Phosphorus), POP (Particle Organic Phosphorus), EP (Exchangeable Phosphorus), UPP (Unreactive Particle Phosphorus)
Si	DSi (Dissolved Silicon), PSi (Particle Silicon), SSi (Sediment Silico), BSi (Biogenic Silicon)

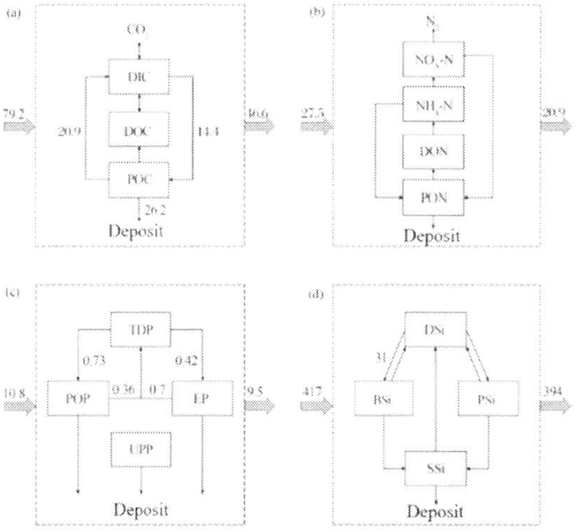

Fig. 1 Biogenic elements cyclic processes in reservoirs (The wide arrows refer to the inflow and outflow fluxes and the dotted box refers to the cycle process of each element in the reservoir. The unit is Tg/a) (Deng et al., 2018)

The classification and rectification of small hydropower stations, particularly these with removing barrage or blocking the water intake would significantly change the hydrological regime characteristics (including flow, water level, water depth, flow velocity and wet perimeter, etc.) of the original river channel. Consequently, the reservoir reach upstream of the dam would change from still water habitat to flowing water habitat, and the water reach under the dam would change from a water reducing or cutoff reach to a rapid flow river reach, with significant changes in water depth and flow velocity. At the same time, a large number of sediments intercepted and accumulated in the upstream reservoir area of the barrage for many years including organic detritus and other biogenic substances would be transported to the downstream river section, with the removal of water retaining structures or the increase of discharge flow. The change of sediment type and composition of different affected river sections might further affect benthos, plankton, fish and other aquatic biological communities and their habitats. At present, there is little research and attention on the impact of dam demolition of diversion run-of-river small power station on organic debris and its ecological effects in river ecosystems in China.

3. Influence of dam removal on transport and decomposition of organic debris

Limited by the development stage of small hydropower and the research area and research objects, there are relatively few studies on the impact of dam demolition on the ecological environment in China at present. Some studies qualitatively describe the ecological impact of dam demolition, lacking of on-site observation and systematic quantitative research. In United States largest number of dams have been demolished and some relevant researches on the impact by dam demolition have been done. The evolution of upstream and downstream rivers after dam demolition and the impact of dam demolition on plants in the bank zone have been studied (Doyle et al., 2003; Kim et al., 2015).

Studies have shown that suspended particulate organic detritus and sediment organic detritus in most rivers, especially large rivers, are mainly exogenous (Tesi et al., 2007), and part of coarse particulate organic detritus (CPOM) are derived from plant leaves (Devol and Hedges, 2001). On the whole, river organic matter is also mainly from external sources. For some small watersheds with little human disturbance, the source of river organic matter may mainly be the photosynthesis of algae in the water body. The source of suspended particulate organic carbon and nitrogen of rivers and lakes in some plain areas have been studied by domestic researchers using stable isotope technology (Li et al., 2011; Wang et al., 2014).

The research shows that all the organic matter in the river have been transported in the form of coarse particles at the beginning. The output of coarse suspended solids is characteristic of obvious seasonal trend and randomness and most of them occurred in flood seasons (Ni et al., 2013). It was shown that different nutrient sources possess greater influence on the coarse sediment organic debris than that on the fine organic debris in rivers (Tant et al., 2013). The decomposition of organic detritus in rivers includes two main processes: rapid weight loss in the initial stage due to leaching and decomposition due to invertebrate feeding and mechanical damage. The leaching stage is mainly the abiotic loss of soluble components, which largely depends on the activities of decomposers in the decomposition stage (Xiong and Nilsson, 1997). The biological factors affecting the decomposition rate of organic detritus are mainly related to the activities of invertebrates and microorganisms in rivers. The abiotic factors include temperature and hydro-chemical characteristics. The effect of nitrogen in lakes dominated by submerged plants on organic matters from the sediment has been studied by using fluorescence spectroscopy (Yao et al, 2015).

4. Influence of dam building on form and distribution of organic debris

There were few studies on the impact of dam removal of small hydropower stations on the organic debris in the sediment. Considering that dam removal and dam construction are two relative processes, comparative analysis can be carried out in some aspects. The influence on form and distribution of organic debris under the dam construction were demonstrated. The main impact of dam construction on rivers is that the discharge of downstream reach reduced obviously, thus reducing the area of aquatic habitats, the continuity (Anderson et al., 2006; James et al., 2008) and the assimilation capacity (Liu, et al. 2005) relative to the original river. Although a small low dam can still continue to affect the transport (Nislow et al, 2011) and utilization efficiency of the river organic matter (von Schiller et al., 2011).

Whether low or high dams are formed by hydropower stations, they can intercept organic matter (Flores et al., 2011; Sanchez-Vadal et al., 2013), thus reducing the input of organic matter in the downstream. To some extent, the reduction of water flow in the downstream reach caused by the dam barrier would promote the accumulation of organic matter (Brookshire and Dwire 2003; Dewson et al., 2007). Organic matter was distributed and balanced under the influence of these two opposite actions with the regional specificity. Gessner and Chauvet (2002) suggested that the changes of organic matter accumulation rule and other biological and abiotic factors may lead to the reduction of the decomposition rate of organic matter in water bodies. Other studies showed that the water intake of diversion run-of-river small power stations has significantly changed the storage and decomposition of coarse organic debris in rivers in winter and the number of predators of macroinvertebrate benthos is also significantly reduced (Arroita et al., 2015). At present, the impact of dam construction on organic debris was mainly concentrated on dissolved and suspended organic matter under the impact of large-scale hydraulic projects. The research on organic debris of sediment in mountain rivers under the dam removal of small diversion hydropower stations is almost blank.

5. Research prospects and suggestions

To sum up, the current research is mainly focused on suspended or dissolved organic matter in water bodies in plain areas, and the research on organic debris in sediment is mainly focused on the relationship with benthos and other organisms. The foreign reports are also focused on the impact of damming on organic debris in sediment and there is almost no research on the organic detritus in the sediment of rivers in mountainous areas under the dam demolition of small diversion hydropower stations. In the future, it is suggested to carry out researches from the following aspects: the distribution rule of organic debris in the sediment of the river reach affected by the diversion type small hydropower station; the characteristics and mechanism of the influence of dam removal on the transport flux and distribution of organic detritus in the sediment; the relationship between organic detritus of sediment and hydraulic characteristics such as water velocity and depth; ecological effects of changes in organic detritus of sediment with benthos as indicator organisms.

Acknowledgments

This work was financially supported by Exploratory pilot Project of Digital Twin Danjiangkou Reservoir of Middle-Line South-to-North Water-Transfer Project (Grant No. GXTC-C-22870043).

References

1. Anderson E P, Freeman M C, Pringle C M. Ecological consequences of hydropower development in Central America: impacts of small dams and water diversion on neotropical stream fish

 https://doi.org/10.1051/e3sconf/202337502007

assemblages [J]. River research and applications, 2006, 22(4): 397-411.

2. Arroita M, Aristi I, Díez J, et al. Impact of water abstraction on storage and breakdown of coarse organic matter in mountain streams [J]. Science of the Total Environment, 2015, 503: 233-240.
3. Brookshire E N J, Dwire K A. Controls on patterns of coarse organic particle retention in headwater streams [J]. Journal of the North American Benthological Society, 2003, 22(1): 17-34.
4. Devol A H, Hedges J I. Organic matter and nutrients in the mainstem Amazon River [J]. The biogeochemistry of the Amazon basin, 2001, 15: 275-306.
5. Dewson Z S, James A B W, Death R G. Stream ecosystem functioning under reduced flow conditions [J]. Ecological Applications, 2007, 17(6): 1797-1808.
6. Doyle M W, Stanley E H, Selle A R, et al. Predicting the depth of erosion in reservoirs following dam removal using bank stability analysis [J]. International Journal of Sediment Research, 2003, 18(2): 115-121 .
7. Flores L, Larranaga A, Diez J, et al. Experimental wood addition in streams: effects on organic matter storage and breakdown [J]. Freshwater Biology, 2011, 56(10): 2156-2167.
8. Gessner M O, Chauvet E. A case for using litter breakdown to assess functional stream integrity [J]. Ecological applications, 2002, 12(2): 498-510.
9. Han H, Lu X, Burger D F, et al. Nitrogen dynamics at the sediment- water interface in a tropical reservoir [J]. Ecological Engineering, 2014, 73:146-153.
10. James A B W, Dewson Z S, Death R G. The effect of experimental flow reductions on macroinvertebrate drift in natural and streamside channels [J]. River research and Applications, 2008, 24(1): 22-35.
11. Kim S, Toda Y, Tsujimoto T. Geomorphological and riparian vegetation responses following a low-head dam removal: a study based on literature review [J]. International Journal of River Basin Management, 2015, 13(3): 315-324.
12. Li H Y, Wang C, Ji H B, et al. Stable carbon and nitrogen isotope composition of particulate organic matter in water source of Beijing [J]. Acta Scientiae Circumstantiae, 2011, 31 (12): 2663-2671.
13. Liu W C, Liu S Y, Hsu M H, et al. Water quality modeling to determine minimum instream flow for fish survival in tidal rivers [J]. Journal of Environmental Management, 2005, 76(4): 293-308.
14. Maavara T, Diurr H H, Cappellen P Van. Worldwide retention of nutrient silicon by river damming: From sparse data set to global estimate [J]. Global Biogeochemical Cycles, 2014, 28(8): 842-855.
15. Marx A, Dusek J, Jankovec J, et al. A review of CO2 and associated carbon dynamics in headwater streams: A global perspective [J]. Reviews of Geophysics, 2017, 55(2): 560-585.
16. Ni Z K, Wang S R, Zhao H C, et al. The sources of organic carbon and nitrogen of suspended particulate matter in inflow river of Erhai Lake [J]. Research of Environmental Sciences, 2013, 26 (3): 287-293.
17. Nislow K H, Hudy M, Letcher B H, et al. Variation in local abundance and species richness of stream fishes in relation to dispersal barriers: implications for management and conservation [J]. Freshwater Biology, 2011, 56: 2135-2144.
18. Sanchez-Vidal A, Higueras M, Martí E, et al. Riverine transport of terrestrial organic matter to the North Catalan margin, NW Mediterranean Sea [J]. Progress in oceanography, 2013, 118: 71-80.
19. Tant C J, Rosemond A D, First M R. Stream nutrient enrichment has a greater effect on coarse than on fine benthic organic matter [J]. Freshwater Science, 2013, 32(4): 1111-1121.
20. Tesi T, Miserocchi S, Goni M A, et al. Organic matter origin and distribution in suspended particulate materials and surficial sediments from the western Adriatic Sea (Italy) [J]. Estuarine, Coastal and Shelf Science, 2007, 73(3-4): 431-446.
21. Vannote R L, Minshall G W, Cummins K W, et al. The river continuum concept [J]. Canadian journal of fisheries and aquatic sciences, 1980, 37(1): 130-137.
22. Von Schiller D, Acuña V, Graeber D, et al. Contraction, fragmentation and expansion dynamics determine nutrient availability in a Mediterranean forest stream [J]. Aquatic Sciences, 2011, 73(4): 485-497.
23. Wang M L, Zhang D L, Lai J P, et al. Distribution and sources of stable organic carbon and nitrogen isotopes in suspended particulate organic matter of Poyang Lake [J]. China Environmental Science, 2014, 34(9): 2342-2350.
24. Xiong S, Nilsson C. Dynamics of leaf litter accumulation and its effects on riparian vegetation: a review [J]. The Botanical Review, 1997, 63(3): 240-264.
25. Yao X, Wang S, Jiao L, et al. Resolving the influence of nitrogen abundances on sediment organic matter in macrophyte-dominated lakes, using fluorescence spectroscopy [J]. Journal of Environmental Sciences, 2015, 27: 197-206.

A Novel Doping Layer Strategy to Realize High Efficiency Layer-by-Layer Organic Solar Cells

Xinrui Li, Xiaoyang Du, Silu Tao *

School of Optoelectronic Science and Engineering, University of Electronic Science and Technology of China (UESTC), Chengdu 610054, China

Abstract: Molecular doping is an effective mean to achieve high performance organic solar cells (OSCs). However, the introduction of dopants aggravates the problem of morphological complexity in bulk heterojunction (BHJ) OSCs, and the choice of solvent after mixing with the donor and acceptor is greatly limited. Here, we innovatively propose a solution to solve the above problems by inserting a dopant layer between the donor layer and the acceptor layer to construct OSCs with a stacked structure through layer-by-layer (LbL) spin-coating solution method. Compared with the control devices (16.95%), the performance of PM6/PABA/BO-4Cl devices with the addition of the dopant layer 4-Aminobenzoic acid (PABA) was significantly enhanced, achieving an efficiency of 17.46%. Morphological characterization and charge analysis showed that the performance improvement was attribute to the film morphology optimization by the dopant located at the D/A interface, while effectively increasing the exciton dissociation rate and charge mobility of OSCs. Thus, our work demonstrates that the doping layer strategy coupled with sequential solution deposition is an effective way to construct efficient devices and is a promising alternative to BHJ OSCs.

Key words: Organic solar cells, Dopant layer, Sequential solution deposition, Morphology.

1. Introduction

In recent years, the problem of environmental pollution has become more and more serious, and solar cells using renewable energy sources are an effective approach to alleviate the above problem. Among them, organic solar cells (OSCs) have attracted much attention due to their advantages such as simple fabrication process, low cost, the ability to prepare flexible devices and their potential sustainability[1-3]. Thanks to the efforts of researchers, representative polymer donors such as PM6 and D18 coupled with high-performance non-fullerene acceptor materials such as Y6 and its derivatives have been able to achieve OSCs with efficiency exceeding 18%[4-6]. However, the continued improvement of the efficiency of OSCs remains the primary goal of researchers. It is well known that the common strategies to obtain high-performance devices are as follows: developing novel materials[7, 8], designing new device structures[9, 10], and using dopants[11, 12]. The former is time-consuming and expensive. The complex fabrication process of new device structures is not conducive to future commercial applications. In contrast, the use of dopant strategy is a simple and more feasible way to improve device performance.

In the early period, conventional OSCs were prepared with bulk heterojunction (BHJ) structure, where donor and acceptor were mixed homogeneously and spin-coated to form[13-15]. Therefore, the introduction of dopant in BHJ devices requires that donor, acceptor, and the dopant can be dissolved in the same solvent, which makes the choice of solvent very limited. At the same time, the introduction of dopants tends to increase the complexity of BHJ devices morphology. By contrast, sequential solution deposition structure by spin coating of donor and acceptor layer individually to fabricate Layer-by-layer (LbL) OSCs has become a promising alternative to BHJ structure[16-20]. LbL devices with the stacked structure plays a significant role in the vertical morphology, the fine tuning of the morphology and the direct charge transport. Moreover, one of its unique advantages is that it is less dependent on the ratio of donor and acceptor materials as well as the choice of solvent. Inspired by the sequential spin-coating solution method, we innovatively applied dopants into LbL OSCs that introducing the dopant layer between the donor layer and the acceptor layer.

In this paper, we used 4-aminobenzoic acid (PABA) and 3, 4-Diaminobenzoicacid (DABA) as the dopant layer for the preparation of high-performance OSCs with stacked structure. Among them, PM6/PABA/BO-4CL devices exhibited the better results. Compared with the control devices (16.95%), the PM6/PABA/BO-4CL devices with the introduction of the dopant layer PABA achieved a maximum PCE of 17.46%, a VOC of 0.835 V, a JSC of 26.77 mA cm-2, and an impact factor of 78.02%. Moreover, the solvents applied in OSCs were eco-friendly

* Corresponding author: silutao@uestc.edu.cn

solvents that the donor, dopant and acceptor were dissolved in o-xylene, methanol, and o-xylene, respectively. As a result, the devices we prepared were eco-green energy devices. It is worth noting that methanol and o-xylene are orthogonal solvents, so the introduction of dopant layer can also protect the donor layer from being affected when spin-coating the acceptor solution. Charge dynamic characterization and PL spectroscopy demonstrated that the introduction of the dopant layer effectively enhanced the exciton dissociation and charge collection rates as well as accelerated the charge transfer (CT) process between donor and acceptor. AFM and TEM tests indicated that interfacial doping was an effective method to optimize the film morphology. Therefore, this work demonstrates that the insertion of a suitable doping layer between the donor and acceptor layers is an effective way to fabricate efficient green LbL OSCs.

2. Experimental Section

All devices in this paper adopted a conventional structure that Indium Tin Oxide (ITO)/PEDOT: PSS/Active Layer/PDINN/Ag. The glass substrates with ITO were sonicated in ethanol, acetone and ethanol solvent for 30mins respectively. After sonication, the substrates were blown dry with nitrogen and then placed into UV ozone treatment for 20mins. PEDOT: PSS was uniformly dropped onto the surface of the above treated ITO glass substrate and spin coated at 4000 rpm for 30 s, followed by annealing at 150°C for 15 mins in the air atmosphere. The substrates coated with PEDOT: PSS film were transferred to a glove box, and then the donor solution, dopant, and acceptor solution were spin-coated sequentially to form an active layer based on LbL structure, followed by thermal annealing at 100 °C for 10 min. Among them, donor PM6 was dissolved in O-XY solvent at a concentration of 10 mg/ml. The dopants PABA or DABA was dissolved in methanol solvent at a concentration of 0.5 mg/ml. Acceptor BO-4Cl was dissolved in O-XY solvent at a concentration of 8 mg/ml and 0.5% v/v DIO was added. After that, PDINN solution dissolved in methanol with a concentration of 1 mg/mL is spin-coated onto the active layer at 3000 rpm for 30 seconds to form the electron transport layer. Finally, 100 nm silver (Ag) electrode was deposited by vacuum vapor deposition under vacuum conditions of 5×10^{-4} Pa.

3. Results and Discussion

In this work, we chose PM6/ BO-4Cl as the main system and added PABA or DABA as the interfacial dopant between the donor layer and the acceptor layer to fabricate green LbL OSCs. The molecular structure of the materials and the flow chart of the active layer are shown in Figure 1. The normalized absorption spectra of pure PM6, PM6/PABA, PM6/DABA and pure BO-4Cl films are given in Figure 2a. Among them, the absorption of PM6 was mainly in the short-wave region of 500-650 nm, with the main absorption peaks as well as the shoulder peaks located at 575 nm and 615 nm, respectively. And the absorption of BO-4Cl was in the region of 700-900 nm. The complete complementarity of photon absorption by PM6 and BO-4Cl facilitates the preparation of high-performance devices. It is noteworthy that the relative intensity between the main peak and the shoulder peak of the films changed after spin-coating the dopant layer on PM6, which indicated that the introduction of the dopant layer affected the crystallization of the donor films. Furthermore, the absorption spectra of PM6/BO-4Cl, PM6/PABA/BO-4Cl and PM6/DABA/BO-4Cl films are detailed in Figure 2b. It can be seen that the absorption intensity of the films after inserting the dopant layer was significantly higher than that of the control film. This revealed that the addition of the dopant layer between the donor layer and the acceptor layer optimized the aggregation of the polymer PM6 with better molecular alignment, which increased the photon capture and enabled the devices to obtain higher J_{SC} values.

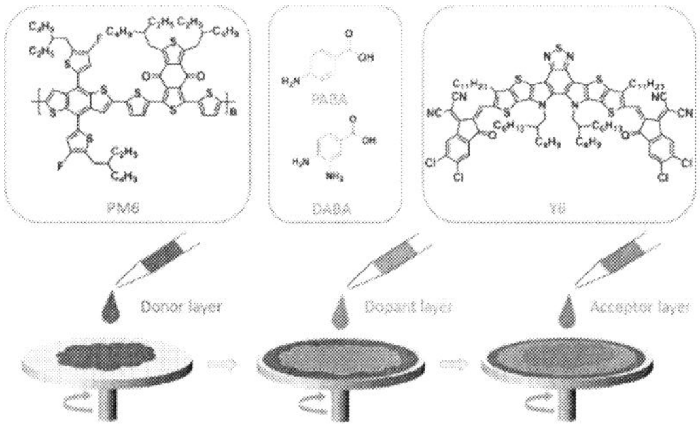

Figure 1. Molecular structure of PM6, Y6, the dopants PABA and DABA as well as the flow chart of the active layer.

The PL emission spectra of pure PM6 and LbL films with or without the dopant layer are visible in Figure 2c, where PM6 had a strong emission peak at 600-800 nm. After spin coating BO-4Cl on PM6 film by sequential solution deposition method, the emission peak was quenched. Further, the quenched situation was more effective after introducing the interfacial doping layer between the donor layer and the acceptor layer. Compared to PM6/BO-4Cl films (94.37%), the quenched efficiency of PM6/PABA/BO-4Cl and PM6/DABA/BO-4Cl were 97.98% and 97.32%, respectively. This indicated that the insertion of the doping layer allowed the excitons generated in active layer to dissociate into free charges more efficient.

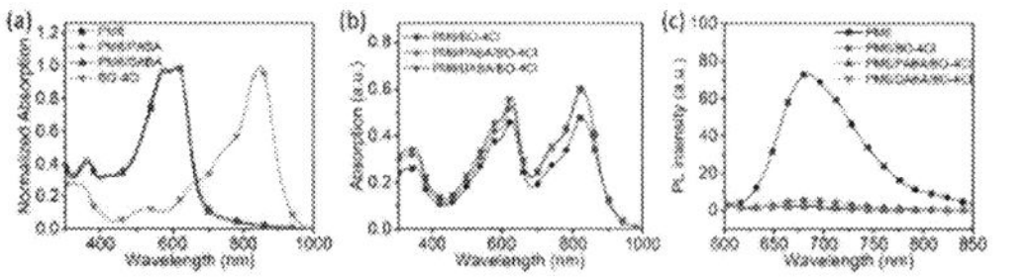

Figure 2. (a) Absorption spectra of pure PM6, PM6/PABA, PM6/DABA and pure BO-4Cl film. (b) Absorption spectra of PM6/BO-4Cl, PM6/PABA/BO-4Cl and PM6/DABA/BO-4Cl LbL films. (c) PL spectra of pure PM6 film, PM6/BO-4Cl, PM6/PABA/BO-4Cl and PM6/DABA/BO-4Cl films.

We fabricated OSCs with the conventional structure. Figure 3a provides the J-V characteristic curves of the devices based on PM6/BO-4Cl system with or without the

https://doi.org/10.1051/e3sconf/202337502008

doping layer, and the detailed parameters are listed in Table 1. It can be found that the optimal PCE of the PM6/BO-4Cl devices was 16.95%, where V_{OC} was 0.834 V, J_{SC} was 26.40 mA cm^{-2} and FF is 76.87%. With the introduction of the doping layer PABA or DABA, the devices exhibited a certain degree of improvement in J_{SC}, FF and PCE. Among them, the PM6/PABA/BO-4Cl device had the best performance with PCE of 17.46%, where V_{OC} was 0.835 V, J_{SC} was 26.77 mA cm^{-2} and FF was 78.02%. From the above, it implied that the doping layer contributed to the performance enhancement of LbL devices. This is probably due to the active layer morphology was improved as well as charge transport and collection was enhanced after the addition of PABA or DABA.

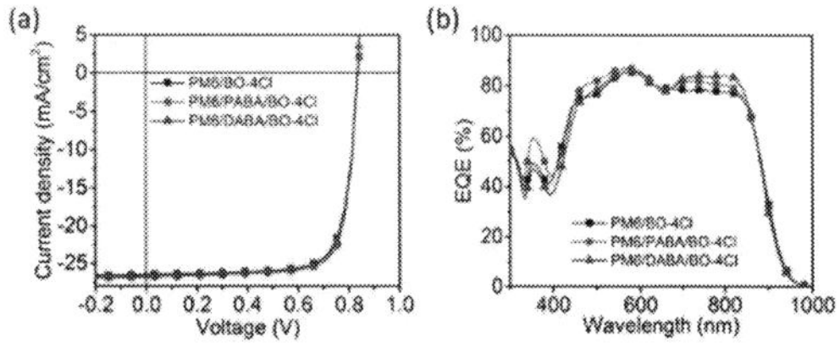

Figure 3. (a) J–V curves and (b) EQE spectra of PM6/BO-4Cl, PM6/PABA/BO-4Cl and PM6/DABA/BO-4Cl devices.

Table 1. The relevant detailed performance parameters of OSCs.

Devices	V_{OC} [V]	J_{SC}[mA cm^{-2}]	FF [%]	PCE (ave) [%]
PM6/BO-4Cl	0.834	26.40	76.87	16.95 (16.80)
PM6/PABA/BO-4Cl	0.835	26.77	78.02	17.46 (17.34)
PM6/DABA/BO-4Cl	0.833	26.54	77.70	17.17 (17.04)

The EQE spectra of LbL OSCs are displayed in Figure 3b. In the range of 450-850 nm, the EQE intensity of the control OSCs (PM6/BO-4Cl) was around 80%. When the doped layer was inserted, the EQE intensity was enhanced in the range of 300-900 nm. And in the range of 500-800 nm, the EQE intensity reached more than 80%. The increase of EQE intensity suggested that the introduction of PADA or DABA between the donor layer and the acceptor layer improved the photoelectric conversion efficiency, which was consistent with the trend of J_{SC}. And comparing the calculated J_{SC} from the EQE integration with the value obtained from the J-V curve, the error was within 5%.

The carrier mobility of the LbL devices with or without the doped layer were further measured using space charge-limited current (SCLC), as detailed in Table 2 and Figure 4a-b. Compared to the PM6/BO-4Cl-based devices (3.74×10^{-4}cm^{2}V^{-1}s^{-1}), the hole mobility (μ_h) of PM6/PABA/BO-4Cl and PM6/DABA/BO-4Cl devices were enhanced to 4.34×10^{-4}cm^{2}V^{-1}s^{-1} and 4.10×10^{-4}cm^{2}V^{-1}s^{-1}. Similarly, the introduction of the doping layer had a positive effect on the increase of electron mobility (μ_e). μ_e of PM6/BO-4Cl, PM6/PABA/BO-4Cl and PM6/DABA/BO-4Cl devices were 2.96×10^{-4}cm^{2}V^{-1}s^{-1}, 3.66×10^{-4}cm^{2}V^{-1}s^{-1} and 3.44×10^{-4}cm^{2}V^{-1}s^{-1}, respectively. After calculation, the $\mu h/\mu e$ value of PM6/PABA/BO-4Cl device was closer to 1 among the above devices, indicating that the device had a more balanced charge transport, which was consistent with the FF results obtained from the tests.

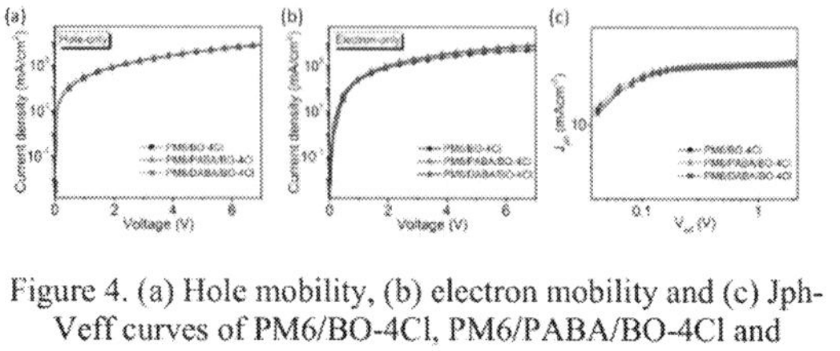

Figure 4. (a) Hole mobility, (b) electron mobility and (c) Jph-Veff curves of PM6/BO-4Cl, PM6/PABA/BO-4Cl and PM6/DABA/BO-4Cl devices.

Table S2. Summary of charge mobilities of PM6/BO-4Cl, PM6/PABA/BO-4Cl and PM6/DABA/BO-4Cl devices.

Devices	μ_h ($\times10^{-4}$cm^{2}V^{-1}s^{-1})	μ_e ($\times10^{-4}$cm^{2}V^{-1}s^{-1})	μ_h/μ_e
PM6/BO-4Cl	3.74	2.96	1.26
PM6/PABA/BO-4Cl	4.34	3.66	1.18
PM6/DABA/BO-4Cl	4.10	3.44	1.19

To further illustrate the effect of the doping layer on the charge dynamics of LbL devices, the relationship curves between J_{ph} and V_{eff} were observed, as shown in Figure 4c. Theoretically, when the value of V_{eff} is large enough (typically, $V_{eff} = 2.0$ V), the electrode can fully extract the photogenerated carriers, at which time J_{ph} is close to the saturation value of the current (J_{sat}). And the J_{power}/J_{sat} value can be used to express the charge collection efficiency (η_{coll}), where J_{power} is the corresponding current density value at the maximum power. [92, 93]. After calculation, the η_{coll} values of PM6/BO-4Cl, PM6/PABA/BO-4Cl and PM6/DABA/BO-4Cl devices were 87.4%, 88.7% and 88.0%, respectively. It can be seen that LbL devices containing the doping layer, especially for PM6/PABA/BO-4Cl OSCs, had higher η_{coll} values compared with the control devices, indicating that the introduction of the doping layer promoted exciton dissociation, facilitated charge transport and collection.

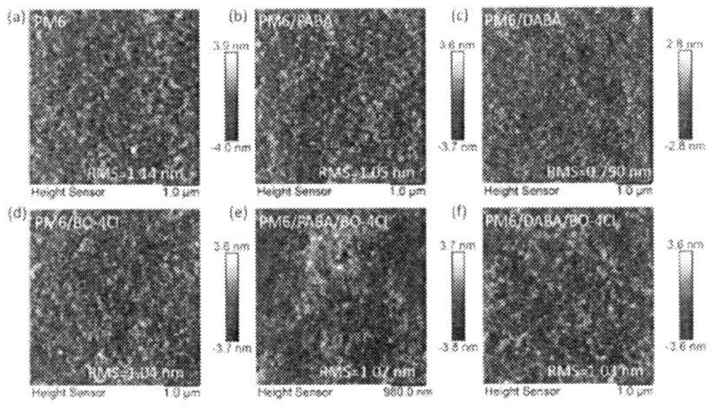

Figure 5. AFM images of (a) pure PM6 film, (b) PM6/PABA film, (c) PM6/DABA film. AFM images of (d) PM6/BO-4Cl, (e) PM6/PABA/BO-4Cl and (f) PM6/DABA/BO-4Cl films.

In order to investigate the effect of the introduction of the doping layer on the active layer morphology, we observed the change of the film surface morphology by AFM, as shown in Figure 5a-c. It can be seen that the root mean square roughness (RMS) of the pure PM6 film was 1.14 nm. After spin-coating the doping layer PABA or DABA on PM6 film, the value of RMS was 1.05 nm and 0.79 nm, respectively. The introduction of the doping layer resulted in a decrease in the surface roughness of the film, indicating that PABA or DABA affected PM6 film, making the film surface smoother and preventing excessive penetration of acceptor, resulting in a P-I-N morphology favorable for charge transport. AFM images of LbL films are given in Figure 5d-f. The RMS values of the PM6/BO-4Cl, PM6/PABA/BO-4Cl and PM6/DABA/BO-4Cl films were 1.04 nm, 1.07 nm, and 1.03 nm, respectively. After adding the doped layer, the surface roughness of the films remained almost unchanged. It is attribute to the presence of the dopant layer between the donor layer and the acceptor layer, which mainly affects the interface and has less influence on the surface morphology of the overall active layer.

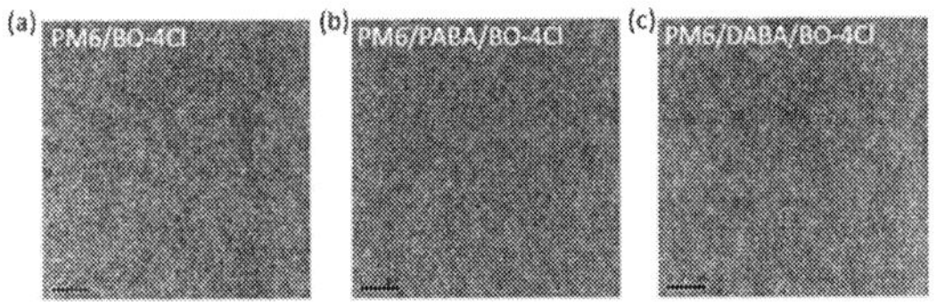

Figure 6. TEM images of (a) PM6/BO-4Cl, (b) PM6/PABA/BO-4Cl and (c) PM6/DABA/BO-4Cl films.

Figures 6a-c represent the corresponding TEM images of the LBL films described above. It can be observed that the phase separation between donor and acceptor in PM6/BO-4Cl film was relatively pronounced. In contrast, the LbL films after the introduction of the dopant layer, both PM6/PABA/BO-4Cl and PM6/DABA/BO-4Cl films showed a fine bicontinuous interpenetrating network. In particular, the distribution of donor and acceptor was more uniform in the PM6/PABA/BO-4Cl films. It could be attributed to the presence of the dopant at the interface, which given the active layer film a better microscopic morphology and thus improving devices efficiency.

4. Conclusion

In this work, we have prepared efficient green LbL devices by inserting the dopant layer between the donor and acceptor layer. For control LbL OSCs based on PM6/BO-4Cl system, the PCE of 16.95% was obtained, while the efficiency of LbL devices constructed after the introduction of doping layer PABA or DABA were higher than that of the control devices. Among them, the PM6/PABA/BO-4Cl devices obtained the optimized PCE of 17.46%, which is one of the highest values for the efficiency of green LbL devices. The exciton dissociation and charge collection rates calculated from the J_{ph} -V_{eff} curves indicated that in LbL devices containing the doped layer, excitons dissociate more fully into free charges and charges are more easily collected by the corresponding electrodes, leading to devices with higher FF and J_{SC}. Morphological characterization illustrated that interfacial doping between the donor layer and the acceptor layer contributed to improved material distribution and optimized film morphology. Therefore, the simplicity of interfacial doping and its effectiveness in improving OSCs performance make this approach promising for future applications of green LbL OSCs.

Acknowledgments

The work was supported by the National Natural Science Foundation of China (NSFC Grant Nos. 62075029, 52130304, 62105055), the Sichuan Provincial Regional Innovation Cooperation Project (2022YFQ0078). The China Postdoctoral Science Foundation (2020TQ0058 and 2021M7006). The Fundamental Research Funds for the Central Universities (Program No. ZYGX2021J017).

References

1. Li Yaowen, Xu Guiying, Cui Chaohua, Li Yongfang. Flexible and Semitransparent Organic Solar Cells[J]. Advanced Energy Materials, 2018, 8(7): 1701791.
2. Dong Pei, Rodrigues Marco-Tulio F., Zhang Jing, Borges Raquel S., Kalaga Kaushik, Reddy Arava L. M., Silva Glaura G., Ajayan Pulickel M., Lou Jun. A flexible solar cell/supercapacitor integrated energy device[J]. Nano Energy, 2017, 42: 181-186.
3. He Chengliang, Pan Youwen, Ouyang Yanni, Shen Qing, Gao Yuan, Yan Kangrong, Fang Jin, Chen Yiyao, Ma Chang-Qi, Min Jie, Zhang Chunfeng, Zuo Lijian, Chen Hongzheng. Manipulating the D:A interfacial energetics and intermolecular packing for 19.2% efficiency organic photovoltaics[J]. Energy & Environmental Science, 2022, 15(6): 2537-2544.
4. Qin Jianqiang, Zhang Lixiu, Zuo Chuantian, Xiao Zuo, Yuan Yongbo, Yang Shangfeng, Hao Feng, Cheng Ming, Sun Kuan, Bao Qinye, Bin Zhengyang, Jin Zhiwen, Ding Liming. A chlorinated copolymer donor demonstrates a 18.13% power conversion efficiency[J]. Journal of Semiconductors, 2021, 42(1): 010501.

https://doi.org/10.1051/e3sconf/202337502008

5. Ma X., Zeng A., Gao J., Hu Z., Xu C., Son J. H., Jeong S. Y., Zhang C., Li M., Wang K., Yan H., Ma Z., Wang Y., Woo H. Y., Zhang F. Approaching 18% efficiency of ternary organic photovoltaics with wide bandgap polymer donor and well compatible Y6 : Y6-1O as acceptor[J]. Natl Sci Rev, 2021, 8(8): nwaa305.
6. Sun R., Wu Y., Yang X., Gao Y., Chen Z., Li K., Qiao J., Wang T., Guo J., Liu C., Hao X., Zhu H., Min J. Single-Junction Organic Solar Cells with 19.17% Efficiency Enabled by Introducing One Asymmetric Guest Acceptor[J]. Advanced Materials, 2022, 34(26): e2110147.
7. Lin Y., Wang J., Zhang Z. G., Bai H., Li Y., Zhu D., Zhan X. An electron acceptor challenging fullerenes for efficient polymer solar cells[J]. Advanced Materials, 2015, 27(7): 1170-1174.
8. Li Huan, Zhao Yifan, Fang Jin, Zhu Xiangwei, Xia Benzheng, Lu Kun, Wang Zhen, Zhang Jianqi, Guo Xuefeng, Wei Zhixiang. Improve the Performance of the All-Small-Molecule Nonfullerene Organic Solar Cells through Enhancing the Crystallinity of Acceptors[J]. Advanced Energy Materials, 2018: 1702377.
9. Zheng Zhong, Wang Jianqiu, Bi Pengqing, Ren Junzhen, Wang Yafei, Yang Yi, Liu Xiaoyu, Zhang Shaoqing, Hou Jianhui. Tandem Organic Solar Cell with 20.2% Efficiency[J]. Joule, 2022, 6(1): 171-184.
10. Firdaus Yuliar, He Qiao, Lin Yuanbao, Nugroho Ferry Anggoro Ardy, Le Corre Vincent M., Yengel Emre, Balawi Ahmed H., Seitkhan Akmaral, Laquai Frédéric, Langhammer Christoph, Liu Feng, Heeney Martin, Anthopoulos Thomas D. Novel wide-bandgap non-fullerene acceptors for efficient tandem organic solar cells[J]. Journal of Materials Chemistry A, 2020, 8(3): 1164-1175.
11. Sprau Christian, Kattenbusch Jens, Li Yonghe, Müller Erich, Gerthsen Dagmar, Berger Rüdiger, Michels Jasper J., Colsmann Alexander. Revisiting Solvent Additives for the Fabrication of Polymer:Fullerene Solar Cells: Exploring a Series of Benzaldehydes[J]. Solar RRL, 2021, 5(9): 2100238.
12. Tremolet De Villers Bertrand J., O'hara Kathryn A., Ostrowski David P., Biddle Perry H., Shaheen Sean E., Chabinyc Michael L., Olson Dana C., Kopidakis Nikos. Removal of Residual Diiodooctane Improves Photostability of High-Performance Organic Solar Cell Polymers[J]. Chemistry of Materials, 2016, 28(3): 876-884.
13. Xia Tian, Cai Yunhao, Fu Huiting, Sun Yanming. Optimal bulk-heterojunction morphology enabled by fibril network strategy for high-performance organic solar cells[J]. Science China Chemistry, 2019, 62(6): 662-668.
14. Hexemer A., Bras W., Glossinger J., Schaible E., Gann E., Kirian R., Macdowell A., Church M., Rude B., Padmore H. A SAXS/WAXS/GISAXS Beamline with Multilayer Monochromator [M]. XIV INTERNATIONAL CONFERENCE ON SMALL-ANGLE SCATTERING (SAS09). 2010.
15. Wen Zhen-Chuan, Yin Hang, Hao Xiao-Tao. Recent progress of PM6:Y6-based high efficiency organic solar cells[J]. Surfaces and Interfaces, 2021, 23: 100921.
16. Sun Rui, Guo Jing, Sun Chenkai, Wang Tao, Luo Zhenghui, Zhang Zhuohan, Jiao Xuechen, Tang Weihua, Yang Chuluo, Li Yongfang, Min Jie. A universal layer-by-layer solution-processing approach for efficient non-fullerene organic solar cells[J]. Energy & Environmental Science, 2019, 12(1): 384-395.
17. Song Y., Zhang K., Dong S., Xia R., Huang F., Cao Y. Semitransparent Organic Solar Cells Enabled by a Sequentially Deposited Bilayer Structure[J]. ACS Appl Mater Interfaces, 2020, 12(16): 18473-18481.
18. Li Bangbang, Zhang Xuanyu, Wu Ziang, Yang Jie, Liu Bin, Liao Qiaogan, Wang Junwei, Feng Kui, Chen Rui, Woo Han Young, Ye Fei, Niu Li, Guo Xugang, Sun Huiliang. Over 16% efficiency all-polymer solar cells by sequential deposition[J]. Science China Chemistry, 2022, 65(6): 1157-1163.
19. Kang H., Zhang X., Xu X., Li Y., Li S., Cheng Q., Huang L., Jing Y., Zhou H., Ma Z., Zhang Y. Strongly Reduced Non-Radiative Voltage Losses in Organic Solar Cells Prepared with Sequential Film Deposition[J]. J Phys Chem Lett, 2021, 12(43): 10663-10670.
20. Xu Xiaopeng, Yu Liyang, Meng Huifeng, Dai Liming, Yan He, Li Ruipeng, Peng Qiang. Polymer Solar Cells with 18.74% Efficiency: From Bulk Heterojunction to Interdigitated Bulk Heterojunction[J]. Advanced Functional Materials, 2021, 32(4): 2108797.

Unit Commitment Considering the Different Frequency supporting capability of Synchronous Generators

Jiaqi Zhang*, Guoyi Xu, Beibei Li, and Lei Han

State Key Laboratory of Alternate Electrical Power System with Renewable Energy Sources (North China Electric Power University), Beijing 102206, China

Abstract. With the high penetration of renewable power in the power system, the inertia level and frequency regulation capability are reduced, the system frequency safety are facing a major challenge. Considering the different frequency support capability of the synchronous generators, the effect of adjustment coefficient of different units on the minimum frequency has been analyzed based on the sensitivity analysis. A multi-objective frequency constraint unit commitment model considering the sensitivity of adjustment coefficient is proposed, and simulation results show that the proposed model increases the minimum frequency value, improves the frequency stability of the power system.

Keywords. unit commitment, frequency response, adjustment coefficient, frequency constraint

1 Introduction

With the high penetration of renewable power in the power system, the proportion of synchronous units decreases, resulting in the reduction of the system inertia level and frequency regulation capacity[1-3], the frequency security is facing major challenges[4].

The traditional unit commitment requires the turbines to reserve a certain proportion of active power to provide frequency regulation[5]. But when the disturbance increases, further increase of reserve capacity will reduce the economic benefits of system operation[6]. Therefore, it is necessary to optimize the unit commitment model to ensure the system frequency safety.

In order to solve the problem of frequency stability, many scholars add frequency constraint to the unit commitment model, considering the frequency regulation methods of wind power, solar power and other renewable power[7-9]. On the basis of average system frequency modle(ASF model), the reference[10] considers the virtual inertia control and primary frequency regulation of wind turbine, constructs the unit commitment model considering the minimum frequency constraint. The references[11-12] consider the dynamic constraints of the maximum rate of change of frequency (RoCoF), maximum frequency deviation and steady-state frequency deviation in the whole process. However, at present, most of renewable energy units still use maximum power point tracking to control their output, the frequency regulation capability is limited. Most of the frequency regulation requirements are still supported by synchronous generators. In addition, the proportion of synchronous generators will decline in the future, how to make full use of the limited synchronous generators resources to adjust the frequency fluctuations is of great significance. However, the existing literature on unit commitment lacks consideration of the difference in frequency regulation capability of synchronous units.

In this paper, a multi-objective frequency constraint unit commitment model considering the sensitivity of synchronous generator adjustment coefficient is proposed. The effect of adjustment coefficient of different units on the maximum frequency deviation has been analyzed based on the sensitivity analysis. Then, the sensitivity factor is added to the unit commitment model to adjust the starting sequence of the units, so as to maximize the frequency regulation ability of each synchronous generator and improve the system frequency response ability. NSGA-II is used to solve the model. Finally, a 10-machine 39-bus system with wind turbines is utilized to validate the proposed model.

2 Sensitivity analysis of adjustment coefficient

After the disturbance occurs, the characteristic quantities used to describe the frequency change include: RoCoF $d\Delta f/dt$, maximum frequency deviation Δf_{max} or minimum frequency f_{nadir}. Among them, the maximum frequency deviation Δf_{max} can reflect the ultimate limit state of the frequency response process[13].

Compared with inertia, dead zone and other parameters, the adjustment coefficient R has a greater impact on the maximum frequency deviation[14]. Therefore, the maximum frequency deviation of the system after disturbance is further analyzed from the angle of the adjustment coefficient.

* Corresponding author: letizia1698@163.com

For a multi-machine system, due to the difference of disturbance location, unit inertia and frequency response characteristics, the contribution of different generators to system frequency regulation is different[15]. So in the unit commitment problem, when the load demand increases at the following moment, and a new unit is required to start up, the frequency deviation after disturbance will be different when selecting different units to start up. This difference is difficult to be clearly described. Therefore, the introduction of the concept of sensitivity of adjustment coefficient can effectively reflect the results of the comprehensive impact of various factors[16].

Through the simulation of the system, it is convenient to obtain the frequency change process of the system after different types of disturbances. First, obtain the maximum frequency deviation value, and then we can calculate the sensitivity of different generator nodes.

Suppose there are n generator nodes, and the sensitivity S_{Ri} of the ith generator can be expressed as:

$$S_{Ri} = \frac{\left|\Delta f_{\max}(R_0 + \Delta R_i) - \Delta f_{\max}(R_0)\right| / \Delta f_{\max}(R_0)}{\Delta R_i / R_0} \quad (1)$$

Where: S_{Ri} is the sensitivity of the adjustment coefficient of the ith generator, indicating the relative change of the maximum frequency deviation caused by the adjustment of the adjustment coefficient, which can reflect the ability of generator to support the system frequency; $i=1,2,\ldots,n$; $\Delta f_{max}(R_0)$ is the maximum frequency deviation, ΔR_i is the perturbation of the adjustment coefficient of the ith generator, $\Delta f_{max}(R_0+\Delta R_i)$ is the maximum frequency deviation after the adjustment of the adjustment coefficient of the ith generator.

3 Unit commitment model considering sensitivity and frequency constraints

3.1. Objective function

Traditional unit commitment model usually aims at the lowest generation cost. On this basis, the sensitivity of the adjustment coefficient is taken as the second objective function to form a multi-objective problem:

$$\begin{cases} \min f_1 = \sum_{i=1}^{n}\sum_{t=1}^{T}\left[\left(a_i + b_i P_{it} + c_i P_{it}^2\right) + u_{it}\left(1 - u_{i,t-1}\right)B_i\right] \\ \min f_2 = \sum_{i=1}^{n}\sum_{t=1}^{T} u_{it}\frac{1}{S_{Ri}} \end{cases} \quad (2)$$

Where: f_1 is the power generation cost; n is the total number of generators; T is the total scheduling period; P_{it} is the output of the ith generator at time t; a_i, b_i and c_i are the generation cost coefficients of the ith generator; u_{it} is the operation status of the ith generator at time t, 1 indicates startup and 0 indicates shutdown; B_i is the start-up cost of the ith generator; f_2 is the reciprocal of the sum of sensitivities, since it is desired to start the units with high sensitivity first.

3.2. Constraints of traditional unit commitment

The constraints of the traditional unit commitment problem are as follows.

(1)Power Balance Constraints:

$$\sum_{i=1}^{n} u_{it}P_{it} + P_{wt} = P_{Lt} \quad (3)$$

Where: P_{wt} is the output of wind turbine at time t; P_{Lt} is the total load demand at time t.

(2) Generation Limits:

$$P_{i,\min} \le P_{it} \le P_{i,\max} \quad (4)$$

Where: $P_{i,\min}$ and $P_{i,\max}$ are the lower and upper limits on generator active power of the ith generator.

(3) System Reserve Constraints:

$$\sum_{i=1}^{n} u_{it}(P_{i,\max} - P_{it}) \ge \lambda P_{Lt} \quad (5)$$

Where: λ is the reserve factor, taken as 0.05.

(4) Ramp Limits:

$$-D_{i,\text{down}} \le P_{it} - P_{i,t-1} \le D_{i,\text{up}} \quad (6)$$

Where, $D_{i,\text{down}}$ and $D_{i,\text{up}}$ are the maximum limit values of the output drop and rise of the ith generator.

(5) Minimum Startup and Shutdown Time Limits:

$$X_{Sit} \ge T_{Si} \ 、 X_{Oit} \ge T_{Oi} \quad (7)$$

Where: X_{Sit} and X_{Oit} respectively represent the number of hours that the ith generator has been offline or online; T_{Si} and T_{Oi} respectively represent the minimum hours that the ith generator should be offline or online.

3.3. Frequency Limit Constraints

The ASF model of multi-machine system is established as shown in Fig.1:

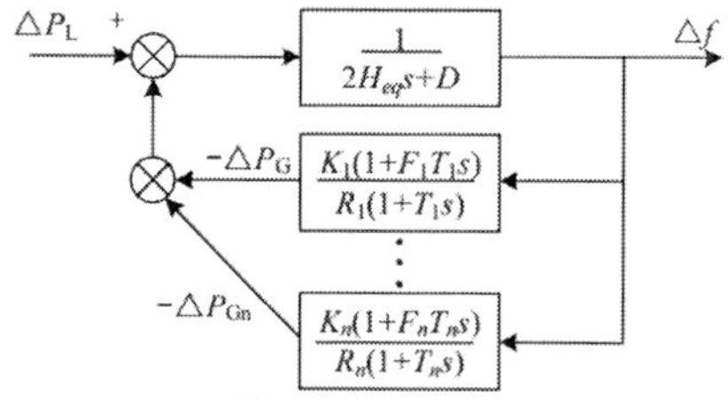

Fig. 1 ASF model

It can be seen from Fig.1:

$$\Delta P_L - \sum_{i=1}^{n}\frac{K_i(1 + F_i T_i s)}{R_i(1 + T_i s)}\Delta f = (2H_{eq}s + D)\Delta f \quad (8)$$

Where, ΔP_L represents power disturbance; K_i is the mechanical power gain factor of the ith generator; F_i is the power factor from HP turbine of the ith generator; T_i is the governor time constant of the ith generator; R_i is the static adjustment coefficient of the ith generator; D is the load damping factor; H_{eq} represents the equivalent inertia of the system.

Replace all T_i in equation (8) with the unified Teq[10]:

$$\Delta f = \frac{\Delta P_L(1 + T_{eq}s)}{(2H_{eq}s + D)(1 + T_{eq}s) + \sum_{i=1}^{n}\frac{K_i}{R_i}(1 + FT_{eq}s)} \quad (9)$$

Get the frequency domain expression of frequency

deviation, then the time domain expression of Δf can be obtained by inverse Laplace transform. Take the derivative of it, make $d\Delta f(t) / dt=0$, we can get the time t_m when the frequency reaches the lowest point, and the maximum frequency deviation(p.u.):

$$\Delta f_{max} = -\frac{\Delta P_L}{D+R_{eq}}[1+e^{-\xi\omega_n t_m}\sqrt{\frac{1+T_{eq}^2\omega_n^2-2T_{eq}\xi\omega_n}{1-\xi^2}}\cos(\omega_n\sqrt{1-\xi^2}t_m-\theta)] \quad (10)$$

$$t_m = \frac{1}{\omega_n\sqrt{1-\xi^2}}tg^{-1}\frac{T_{eq}\omega_n\sqrt{1-\xi^2}}{T_{eq}\omega_n\xi-1} \quad (11)$$

$$\theta = tg^{-1}\frac{\xi - T_{eq}\omega_n}{\sqrt{1-\xi^2}} \quad (12)$$

Where: ω_n is the natural frequency; ξ is the damping ratio:

$$\begin{cases} \omega_n = \sqrt{\frac{D+R_{eq}}{2H_{eq}T_{eq}}} \text{、} \xi = \frac{2H_{eq}+(D+F_{eq})T_{eq}}{2\sqrt{2H_{eq}T_{eq}(D+R_{eq})}} \\ R_{eq} = \sum_{i=1}^{n}\frac{K_i}{R_i} \text{、} F_{eq} = \sum_{i=1}^{n}\frac{K_i}{R_i}F_i \end{cases} \quad (13)$$

The frequency constraints can be constructed as:

$$f_{c1} \geq f_N + f_N\Delta f_{max} \geq f_{c2} \quad (14)$$

where, f_{c1} and f_{c2} are the upper and lower limits of frequency, which are generally set as the setting values of the system's high frequency machine cutting and low frequency load shedding.

3.4. Optimization steps considering sensitivity and frequency constraints

The steps of unit commitment considering the sensitivity of adjustment coefficient and frequency constraints proposed in this paper are as follows:

Step 1: Based on the system's grid structure, load forecast, etc, prepare the expected contingency set.

Step 2: Conduct offline simulation for the disturbances in the contingency set, calculate the sensitivity of each synchronous generator according to formula (1), get the sensitivity order, form the expected contingency sensitivity information set.

Step 3: Choose the accident with the most serious impact on the system, and search the corresponding sensitivity order information in the contingency set.

Step 4: Add the sensitivity information into the unit commitment as an objective function, establish a multi-objective unit commitment optimization model. Set the starting sequence of the synchronous units, give priority to the units with high sensitivity of the adjustment coefficient. NSGA-II is used to solve the problem.

4 Simulation calculation and analysis

Based on the IEEE 10-machine 39-bus system with wind turbines, the unit commitment model proposed in this paper is verified by using MATLAB and PSASP simulation software. The parameters of synchronous genetators are referred to [17], and the load data and wind power prediction data are referred to [18],, D=2.

4.1 Calculation of the sensitivity of adjustment coefficient

According to the calculation steps proposed in subsection 3.4, taking the sudden increase of load disturbance as an example, calculate the minimum frequency when different load node increased 150MW, and the results are shown in Table 1.

Table 1. Minimum frequency when different load increased 150MW

Load node	f_{nadir}/Hz	Load node	f_{nadir}/Hz	Load node	f_{nadir}/Hz
L3	49.6763	L18	49.6757	L27	49.6748
L4	49.6765	L20	49.6672	L28	49.6635
L7	49.6775	L21	49.6754	L29	49.658
L8	49.6771	L23	49.6751	L31	49.6588
L12	49.6825	L24	49.6754	L39	49.6590
L15	49.6762	L25	49.6641		
L16	49.6746	L26	49.6737		

According to the results in Table 1, select L29 as the disturbance node, which has the biggest frequency deviation under the same disturbance. Set load L29 increase 150MW, and calculate the sensitivity according to Equation (1). The results are shown in Table 2:

Table 2. Sensitivity of units when load29 increased 150MW

	S/%		S/%		S/%
G30	3.8012	G34	2.3392	G37	3.5088
G31	3.6550	G35	3.3626	G38	4.5322
G32	3.5088	G36	3.0702	G39	4.5322
G33	3.2164				

It can be obtained that under the L29 increase of 150 MW, the sequence of the sensitivity is:

G38,G39>G30>G31>G32,G37>G35>G33,G36>G34

4.2 Simulation analysis of unit commitment

Set the minimum frequency limit of the frequency constraint to 49Hz [19]. Set load L29 increase, and the power gap is 10% of the total load at that time. Compare the unit commitment results of these two cases:

Case 1: consider frequency constraints

Case 2: consider synchronous units' sensitivity sequence of adjustment coefficient and frequency constraints

4.2.1 Startup and shutdown status

NSGA-II algorithm is used to solve the model, and the startup and shutdown states of the two cases are shown in Figure 2 and Figure 3.

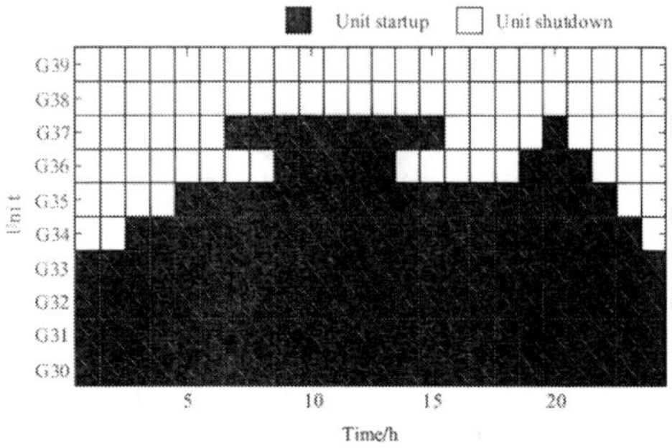

Fig. 2. Units' on-off conditions without considering sensitivity

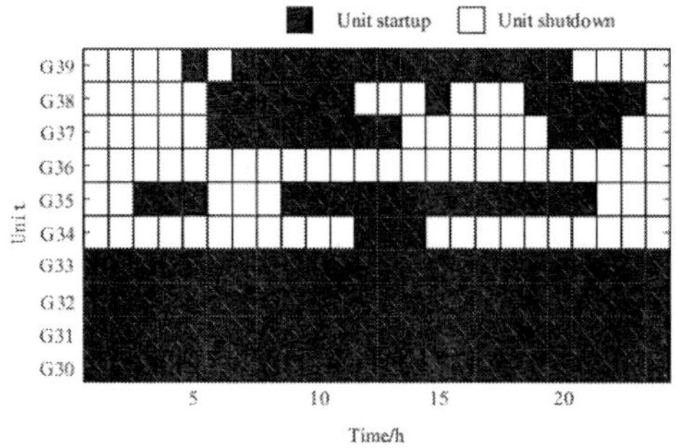

Fig. 3. Units' on-off conditions considering sensitivity

It can be seen from Figure 2 and Figure 3 that the total number of online units at each time under the two cases is the same, but the specific online units are different. In case 2, the units with higher sensitivity such as G39 and G38 are given priority to start up, and the total number of online hours for the units with lower sensitivity such as G34 and G36 are reduced.

4.2.2 Minimum frequency value

According to the unit commitment results obtained from the two cases, the simulation is carried out in PSASP, and the minimum frequency values at 24 times are compared, as shown in Figure 4.

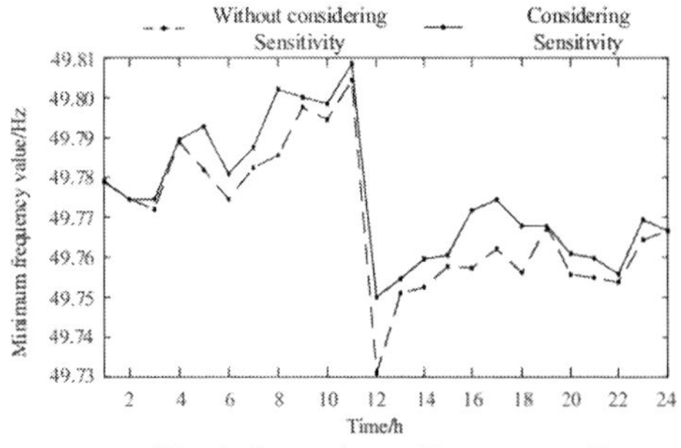

Fig. 4. Comparison of frequency nadir

We can see from Figure 4, after considering the sensitivity on the basis of frequency constraints, the minimum frequency has been improved, because case 2 allows the units with higher sensitivity to be started first, which providing better frequency support capability.

5 Conclusion

In order to deal with the system frequency problem caused by the high penetration of renewable power in the power system, this paper proposes a unit commitment model considering the sensitivity of synchronous generator adjustment coefficient. The main conclusions are as follows:

The sensitivity method is used to analyze the influence of different units' adjustment coefficient on the minimum frequency value, so as to characterize the frequency support capability of different units. The units with high sensitivity are more conducive to improving the system frequency stability. Add the sensitivity information into the unit commitment model on the basis of considering frequency constraints. Use the multi-objective optimization algorithm NSGA-II to solve the problem. The calculation and simulation results show that the minimum frequency value after disturbance can be further improved after considering sensitivity.

References

1. Z.Wang, L.L.Li, Z.H.Li, PSPC. J. 49, 5-73.(2021)
2. D.D.Li, Q.Liu,B.Xu, PSPC. J. 49, 24-33.(2021)
3. S.C.Li,B.J.Tian, H.Z.Li, PSPC. J. 50, 60-71.(2022)
4. C.Yan, Y.Tang, J.Dai, PCMP. J. 6, 1-13.(2021)
5. M.Malekpour, M.Zare, R. Azizipanah-Abarghooee. IET GENER TRANSM DIS. J. 14. (2020)
6. H.Li, Z.Lu, Y.Qiao, IEEE PES General Meeting 2020. C. (2020)
7. J.J.Rong, M.Zhou, G.Y.Li, Power System Technology. J. 46, 1335-1346. (2022)
8. J.X.OuYang, Y.F.Yuan, M.Y.Li, Power System Technology. J. 45, 2192-2201. (2021)
9. B.R.Cai, L.Yang, W.Huang, PSPC. J. 49, 169-177. (2021)
10. G.W.Cai, C.Zhong, G.Wu, AEPS. J. 45, 134-142. (2021)
11. X.L.Ge, Y.Liu, Y.Fu, Proceedings of the CSEE. J. 41, 4043-4058. (2021)
12. S.F.Guo, W.Li, S.S.Hu, Electric Power Engineering Technology. J. 39, 61-67. (2020)
13. X.F.Fan, X.Zhong, X.Q.Chang, PSPC. J. 43, 55-60. (2015)
14. F.Wang, H.F.Li, G.Y.Xu, PSPC. J. 48, 1-8. (2020)
15. F.Wang, G.Y.Xu, C.Shao, Electrical Measurement & Instrumentation. J. 1-8. (2022)
16. L.D.Zhao, X.C.Xing, Electrical Measurement & Instrumentation. J. 55, 31-35. (2018)

 https://doi.org/10.1051/e3sconf/202337502009

17. W.Ongsakul, N.Petcharaks, IEEE T POWER SYST. J. 19, 620-628. (2004)
18. B.Wang, D.Y.Yang, G.W.Cai, Power System Technology. J. 44, 2513-2519. (2020)
19. X.J.Pan, M.S.Fan, W.C.Zhang, ELECTRIC POWER. J. 53, 76-82. (2020)

Integrated Photo - rechargeable Batteries: Photoactive Nanomaterials and Opportunities

Liwen You

Faculty of Mathematical and Physical Sciences, University College London, London, United Kingdom

Abstract: The demand for fossil fuels has been increasing over the last few decades but will one day be depleted and researchers are now using biomass to alleviate the fuel crisis. This paper concentrates on a range of current devices with intrinsic solar energy collection, conversion and storage properties, different classes of cells as well as their areas of application and recent research advances. Nanomaterials, meanwhile, are key to making significant progress in the study of photovoltaic electrodes for solar rechargeable batteries, and this paper describes seven currently commonly used semiconductor and nanomaterials. This not only alleviates the severe environmental pollution and greenhouse effect caused by fossil fuels, but also makes a significant contribution to the sustainability of human existence.

Keywords: Photo-rechargeable battery, Nanomaterial, Photoactive.

1. Introduction

For the past few decades, fossil fuels have been used to facilitate daily human life. But as the world's population grows, so does the demand for fossil fuels, which are a non-renewable resource that will one day be exhausted. Research data suggests that in the future mankind can rely on fossil fuels to diversify its energy sources, converting non-renewable energy sources into renewable ones such as photovoltaic (PV), hydroelectric, wind, thermal and biomass. Researchers are now harnessing solar energy and applying it to solar energy storage cells, thus alleviating the fossil fuel crisis.

These renewable energies can be used in providing a greener and safer energy source for mankind and, most importantly, they are sustainable, which means that they will never run out. On the other hand, as the burning of fossil fuels brings about severe environmental pollution and the greenhouse effect, the consumption of renewable energy sources can largely reduce the ecological problems caused by fossil fuels, which makes a great contribution to the sustainability of human existence.

2. Solar-charged electrochemical energy storage: state of the art

2.1 Converting and storing solar energy

Researchers have now developed a range of devices with intrinsic solar energy collection, conversion and storage properties [1]. In practice, however, due to fluctuating conditions, solar cells need to be combined with storage devices to balance energy supply and demand [2]. Experiments have shown that the weight and volumetric energy density of a battery are key parameters for its use in real-life devices. Therefore there are currently several methods of calculating specific energy (weight) and energy density (volume) applicable to the different stages of battery development: (i) materials exploration, (ii) electrode design, and (iii) cell-level engineering. These calculations help to establish a fair and robust method to compare energy metrics [3-4].

2.2 Photovoltaic cells combined with rechargeable batteries

Photovoltaics (PV) is a renewable energy source that converts sunlight into electricity through the photovoltaic effect and has become one of the most widely used energy sources due to its high efficiency and green nature [5-7]. By 2020, the cost of producing photovoltaic energy has fallen to less than $0.05 per kilowatt hour, compared to $0.05 per kilowatt hour for fossil fuels and $0.03 per kilowatt hour for coal [7]. As photovoltaic technology is increasingly researched and developed, the future of solar energy will be cheaper and more capable of meeting human needs than fossil energy.

Current researchers have classified photovoltaic cells into five different types based on raw materials and operating principles [8], namely silicon solar cells, thin-film solar cells, dye-sensitised solar cells, organic solar cells and chalcogenide solar cells [9,10]. Silicon-based solar cells dominate due to their high efficiency of approximately 25% [11,12]. The second generation of photovoltaic cells, known as solar thin film cells, is made up of multiple layers of photovoltaic materials and although it is cheap to produce, it is inefficient and contains toxic substances, which can lead to environmental pollution and ecological damage [13]. For the third generation of photovoltaic cells (including organic solar cells, dye-sensitised solar cells and chalcogenide solar cells), although they have shown their superior performance on a research scale, they nevertheless still lag behind other generations of solar cells due to limitations in stability, efficiency, reproducibility and scalability.

2.3 Photo-charge or photo-assisted charge

Photo-charge cells enable an effective combination of solar energy collection and energy conversion/storage functions, offering a potential solution for the large-scale utilisation of unlimited and cost-effective solar energy and alleviating the limitations of conventional energy storage devices. Solar energy can not only be stored directly during the charging process, enabling various photo-assisted/ photo-charged rechargeable batteries, but also in certain devices to improve the discharge performance through the photovoltaic effect. Several integration/combination strategies have been developed by researchers for direct conversion and/or solar energy storage[14,15]. Depending on the electrode number used, the different device structures can be classified as 1) two-electrode, 2) three-electrode, and 3) T-type systems.

2.4 Working principle of the photo-assisted rechargeable metal batteries

In general, rechargeable batteries work by using visible light to induce charge storage and release on demand. Photoelectrochemical (PEC) systems provide a method of converting light energy into electricity or chemical fuel through an electrochemical reaction. The key to a photoelectrochemical cell is the electrolyte used and the semiconductor photoelectrode where the redox chemistry occurs [16]. In contrast, the PEC system of semiconductor photoelectrodes works on the principle of charge separation occurring upon excitation by photon energy and the formation of a region of space charge at the semiconductor-electrolyte interface. The key to self-charging lies in the ability of the material itself to generate photoelectrons and thus excite charge transfer in the dark on demand [17].

2.5 Photo-assisted rechargeable lithium metal batteries

Although graphite has proved to be the best and most reliable substance used to date for making cathodes, it holds a limited number of ions. Researchers have been hoping to replace graphite with lithium metal foil, which can hold more ions, but typically lithium metal foil reacts adversely with the electrolyte, which can lead to overheating and even combustion. Lithium metal batteries (LMBs) are batteries that use lithium metal as the negative electrode, and the cathode material that goes with it can be oxygen, monolithic sulphur, metal oxides and other substances. Lithium metal batteries were primary batteries when they were first proposed in 1912, but after a century of development, the technology is now well suited to the dissolution and deposition of lithium metal during the charging and discharging process, i.e. secondary batteries. Recent researchers have used nanostructured halloysite nanotubes (NHNT) to create high-performance polyvinyl alcohol composite diaphragms (OPVA/NHNT diaphragms), which can effectively improve the electrochemical performance and safety of lithium metal batteries while reducing the overall production cost of lithium metal batteries.

2.6 Photo-assisted rechargeable Li-ion intercalative electrodes

Since Sony commercialised the first lithium-ion battery with $LiCoO_2$(LCO)-graphite insertion chemistry in 1991, there have been many developments of new electrode materials [20-22]. Current researchers have used intercalation techniques to modify the "shuttle" effect and multiplicity performance of lithium batteries.

Tong et al. explored the application of different hydrated metal ion intercalated layered vanadium-based electrode materials in battery cathode materials by preparing LVO nanosheets synthesized by hydrothermal method after wet mixing of $LiOH$-H_2O and V_2O_5, and investigated their electrochemical performance and energy storage mechanism. This novel design will have important theoretical and practical implications[23].

2.7 Photo-assisted rechargeable lithium-ion batteries

The most commonly used batteries today are based on metal anodes, of which lithium, zinc and sodium are the most representative examples of anode materials. Of these, lithium-based materials are most commonly used as solar rechargeable batteries[24-27]. The main advantage of using lithium is its large theoretical energy density and, in addition, due to its small size, lithium ions can be easily inserted into other nanomaterials without changing their structure[28,29].

Designing lithium-ion batteries (LIBs) is key to portable devices and mobile electrical devices, which currently exist mainly in areas such as electric vehicles [30-32]. The fast charging and self-powering of batteries is therefore an important but challenging key technology. With the depletion of fossil fuels and global energy depletion, people are forced to look for new and renewable sources of energy, but without efficient energy storage the full use of energy cannot be achieved, so modern devices are placing additional demands on existing LIB technology to meet the fast-paced society and emerging needs, especially in terms of power density and charging rates [33-37].

Wang et al. designed a lithium-ion battery with photoelectric charging (photo LIB), using LiV2O5 as the photocathode, which can be light-assisted fast charging and photo-assisted [38]. The device achieves the highest full-spectrum light energy conversion efficiency to date of 9 % in pure light charging mode, demonstrating an efficient self-powered mode. The results show that reversible vanadium charge transfer and Li+ insertion in V_2O_5 under light conditions can improve the performance.

This study provides new avenues for integrating energy conversion and storage into practical applications.

2.8 Photo-assisted rechargeable lithium-oxygen batteries

Lithium-oxygen batteries have attracted a lot of attention in the electric vehicle sector due to their ultra-high theoretical energy density (~3560 $Whkg^{-1}$), which far exceeds that of conventional lithium batteries (~387 $Whkg^{-1}$). The overall discharge/charge reaction of Li-ox batteries (with a theoretical potential of 2.96 V) involves the reversible formation and decomposition of $Li\text{-}O_2$ [39]. However, the high charge overpotential severely limits further cell development due to the insolubility and poor conductivity of $Li\text{-}O_2$. The current researchers can effectively solve this problem by integrating light through the photovoltaic effect, efficiently achieving the conversion and storage of the anode energy. Recent research has shown that the use of photocathodes in combination with lithium-oxygen cells can effectively reduce the charge overpotential by absorbing solar energy, thereby increasing the energy efficiency of the cells.

2.9 Photo-assisted rechargeable lithium-sulfur batteries

Compared to the currently popular lithium-ion batteries, lithium-sulfur (Li-S) batteries are considered to be one of the most promising candidates for portable electric devices and electric vehicles as an emerging energy storage device due to their ultra-high energy density (2600 $Whkg^{-1}$), high safety and low cost[40]. Zhou and coworkers proposed a lithium-sulphur hybrid cell consisting of a lithium anode, an organic electrolyte and a Pt/CdS-containing cell[41]. By absorbing solar energy, the photoexcited pores of Pt/CdS can oxidise the discharge products, thus allowing a reproducible photo-charging process without external electrical input. However, the reaction produces by-products that cannot be used secondarily, and this continuous consumption and relatively slow charging and discharging efficiency has led to the slow replacement of such cells.

However, the commercialisation of lithium-sulphur batteries has been hampered to some extent by a number of problems of its own. The main problems of lithium-sulphur batteries are twofold: 1) the lithium polysulphides (LiPSs) formed during the discharge process tend to dissolve into the electrolyte and are deposited across the semi-permeable membrane onto the surface of the negative lithium metal, which not only leads to capacity degradation but also hinders the further charging and discharging process; 2) the insulating properties of the active substance sulphur and the discharge products lead to poor multiplier performance. Researchers are now using calcium titanite cells to improve on lithium-sulphur batteries, and these hybrid cells exhibit good photo-charging capabilities.

2.10 Photo-assisted rechargeable Lithium-Iodine2 batteries

Among a variety of batteries, lithium-iodine (Li-I) batteries, which usually consist of a lithium anode and an iodine-based calcium solvate (I^-/I^{3-}), have a high theoretical potential and capacity (211 $mAhg^{-1}$) and are a promising type of large-scale energy storage battery. In 2015, Yu et al. reported the first example of a photosensitive Li-I cell with a three-electrode structure using dye-sensitized titanium dioxide photoelectrodes[42]. To meet the practical needs of light-assisted charging cells, researchers have investigated the application of new carbon materials in lithium-ion battery anode materials and fuel cell cathode oxygen reduction electrocatalytic materials, and explored excellent photoelectric electrodes with low cost, good visible light absorption and good electrolyte stability.

2.11 Photo-assisted rechargeable metal zinc batteries

Not only to solve the problem of energy shortage and consumption, but also to eliminate the main obstacle to modern electronic product design - the flexible battery, researchers have recently developed rechargeable zinc ion batteries (ZIBs) in addition to the continuous improvement of lithium batteries, which have the advantage that zinc metal anodes have a better stability during cycling [43,44]. As a result, zinc metal is easier to use as an anode material, thus simplifying the design of the battery [45,46].

This type of battery has a zinc anode, a metal oxide cathode and a solid polymer electrolyte. The cathode and anode are located on either side of the cell and the zinc ions flow from the anode to the cathode in the electrolyte to initiate a chemical reaction that continuously generates electricity. Because zinc is not very active in the environment and can be as small as a few hundred microns (the broad band of two hairs), these ultra-miniature batteries could be used in digital smart tags, for example to check the freshness of food. the Imprint team has produced a lithium battery that could be used in flexible sensors, especially for devices that are worn on the human body or even implanted inside the body, making zinc batteries a safer option.

Buddha uses photo-assisted zinc ion cells (hv-ZIBs), where the electrodes consist of layers of grown zinc oxide and molybdenum disulfide [47]. These are able to collect solar energy and store it in the corresponding material, while at the same time alleviating the need for solar cells or power converters. Experimental results demonstrate a capacity retention of up to 82% after 200 cycles, a photocathode with a light-to-charge conversion efficiency of 1.8% and a capacity increase of up to 38.8% in the presence of light.

2.12 Photo-assisted rechargeable metal sodium batteries

Sodium ion batteries are more suitable as large-scale energy storage devices, which have three advantages: 1) sodium has better safety properties as an energy storage material compared to lithium; 2) sodium is abundant in the earth's reserves, with 2.4% of crustal species of metallic sodium; and 3) sodium is cheap. For the first time, researchers have now embedded a titanium dioxide (TiO_2) photoelectrode into the positive electrode of a new sodium ion battery to achieve efficient conversion and storage of solar energy. The light-assisted rechargeable sodium ion battery utilizes an aqueous Na_2SO_4 anode electrolyte and NaI cathode electrolyte as the active substances for the negative and positive electrodes respectively, and the TiO_2 photoelectrode is embedded in the positive electrolyte as the solar energy conversion and storage primitive. When the battery is charged under simulated solar illumination, the TiO_2 photoelectrode is excited by light to produce electrons and holes, and the discharge process of the battery is similar to that of a conventional sodium ion battery. The ultra-low charging potential plateau is significantly lower than the discharging potential plateau, resulting in an energy conversion efficiency of 190%, equivalent to a saving of nearly 90% of the input power. The research results provide a new idea for the development of low-cost, high-safety performance photoelectric conversion and storage devices, while promoting the practical application of photo-assisted batteries.

2.13 Pioneering works on bifunctional electrodes

Photovoltaic cell (PE) is one of the key factors in the development of high performance and is often the limit to overall efficiency. The choice of materials and design of structures is therefore crucial for efficient conversion and storage of solar energy[48]. The PE can consist of a simple semiconductor or a photovoltaic cell. In a photovoltaic cell, the PE is part of the energy conversion unit and works independently of the storage system. In an integrated two-electrode photovoltaic cell, the PE is a bifunctional photovoltaic cell because it has to fulfil two purposes: to obtain energy from the sun/sunlight; and to store the obtained energy by chemical means. The storage of energy by means of a redox reaction/ion transfer electrolyte requires efficient charge transfer between the PE and the electrolyte, which are closely related to the nanostructure of the material[49,50]. Researchers can now more easily tune the band gap and electronic band structure of nanomaterials by changing their size and doping [51].

3. Photo-induced mechanisms

3.1 Basic concepts in photo-batteries

A photovoltaic cell is a semiconductor element that generates an electric potential in the presence of light. It is a component that generates an electric potential in response to the irradiation of light. It is generally used for photoelectric conversion, photoelectric detection and light energy utilization. There are many types of photovoltaic cells, commonly used are selenium photovoltaic cells, silicon photovoltaic cells and thallium sulphide and silver sulphide photovoltaic cells.

Photovoltaic power generation is a technology that uses the photovoltaic effect of semiconductors to convert light energy directly into electricity. The key component of this technology is the solar cell. Solar cells are connected in series and then encapsulated and protected to form a large area solar cell module, which is then combined with power controllers and other components to form a photovoltaic power generation device. The advantages of photovoltaic power generation are fewer restrictions, safety and reliability, no noise and low pollution.

3.2 Dependence on the electrode architecture

3.2.1 Influence of the electrode thickness

Lithium-ion batteries (LIBs) have become the most attractive source of power in today's pure or hybrid electric vehicles, but the fact remains that there is still range anxiety, insufficient fast charging capability and a range of safety issues. Reducing the proportion of materials that are not electrochemically active is a good way to increase the energy density of a battery. For example, the electrode composition can be changed by reducing the proportion of active binders and conductive agents, which can be disguised as an increase in electrochemically active material content. Studies have shown that at low multipliers there is little difference in capacity between electrodes and little difference in voltage plateau. When the electrode thickness is greater, the voltage plateau increases and decreases slightly during charging and discharging, which indicates greater polarisation for thicker electrodes.

3.2.2 Influence of the porous network

Combining electrodes, diaphragms, collectors and even graphite cathodes of similar area capacity, the researchers found that reducing porosity results in higher battery capacity at lower magnifications, however, as charge and discharge magnifications increase, electrodes with higher porosity exhibit superior performance due to the faster transfer kinetics and mass transfer characteristics of higher porosity electrodes at higher magnifications. Therefore, while reducing the contact resistance and effective lithium ion diffusion rate in the electrolyte, reducing the porosity of the electrode will increase the specific resistance and charge transfer resistance of the cell.

3.3 Role of the electrolyte

Electrolyte chemistry plays an important role due to the high reactivity of the electrode-electrolyte interface and the corresponding layered oxides to high states of charge (SOC). The electrolyte composition can therefore be adapted to promote a more stable cycle[52]. By comparison, it was found that the electrolyte generally uses about 7 mol/l of KOH solution (there is also a certain amount of NaOH instead of KOH), but of course a small amount of other components such as LiOH are added to the electrolyte, but some impurities such as carbonates, chlorides, sulphides, etc. are required.

4. Photoactive nanomaterials working as bifunctional photoelectrodes

4.1 Dye-sensitized photoelectrodes

Gratzel Cells introduced the third generation of solar cells in 1988, known as dye-sensitised solar cells (DSSCs).DSSCs are photoelectrochemical solar cells consisting of a glass substrate, a transparent conductor, a semiconductor material, a dye, an electrolyte and a cathode in a five component structure [53,54]. In the past two decades, academic research on dye-sensitised solar cells (DSSCs) has shown tremendous progress. Researchers have indicated that dye-sensitised solar cells hold promise as one of the alternative options to replace conventional silicon-based photovoltaics. The DSSC works in four basic steps: photon absorption, electron injection, carrier transport and current collection. The advantage of DSSC is that it can achieve better performance through low material cost and simple fabrication [55,56]. However, DSSC glass substrates have limitations due to their stiffness, high mass and expensive nature [57,58].

Since dyes are primarily responsible for the absorption of light in the system, the researchers wanted a dye whose absorption covered a wide range of wavelengths in the visible region and even extended into the near infrared (NIR) part. Equally important, given the lifetime of the device, the dye should be thermally stable and have strong interfacial binding to the metal oxide [59,60]. Xu et al. designed a composite formed by TiO2/N719 dye/Cu2S to be used as a photocathode in a solar rechargeable Li-S cell [61]. By depositing cuprous sulphide on the dye, a bifunctional PE was created that could collect light and store it.

4.2 Transition metal oxide-based semiconductors

Transition metal oxide cathodes tend to have intrinsically low conductivity, which can lead to inhomogeneous charge distribution in the electrode [62]. To mitigate the potential charge inhomogeneity and electrode composition, coating methods using conductive polymers or carbonaceous substances have been intensively investigated. Layered oxide materials remain the current cathode of choice for lithium-ion batteries, particularly in the automotive sector, where a number of low cobalt and high nickel compositions are ready for commercialisation. In order to mitigate the effects of the harmful degradation processes of nickel-rich compositions, research has shown that strategies to introduce electrochemical activity, stabilise the entry of cations into the structure and apply stable surface coatings have proven to be successful in extending cycle life.

The following are currently being studied by researchers: 1) titanium dioxide, a well-known n-type semiconductor with high stability, biocompatibility and high electron mobility [63]; 2) ferric oxide, an n-type semiconductor with magnetic properties. α-Fe_2O_3 is one of the most abundant metal oxides on Earth and it is thermodynamically more stable than other iron oxides [64]. It is chemically stable over a wide pH range, has a low cost and a relatively narrow band gap (2.1-2.3 eV), which facilitates light absorption in the visible spectral range; 3) tungsten trioxide is an n-type semiconductor with a band gap of 2.6-3.0 eV [65]. Tungsten trioxide is non-toxic and has good chemical and photochemical stability [66]. Tungsten trioxide nanoparticles have a high specific surface area and high electron mobility; 4) vanadium oxide, vanadium is considered to be a very abundant element in the earth's crust [67]. It has a crystal structure with different oxygen coordination sites, the most common crystal structures being vanadium dioxide and vanadium pentoxide [68]. These two materials have been used as cathode materials for lithium-ion batteries and are promising materials for photovoltaic cells with high specific capacity, energy density and good photocatalytic properties [69-71]; 5) molybdenum oxide, which, like vanadium oxide, is an n-type semiconductor. Molybdenum oxides also have light absorption in the visible range of the spectrum, and molybdenum trioxide has attracted a lot of attention in the last decades due to its non-toxicity and excellent properties in the fields of photovoltaics, energy storage, gas sensing and catalysis [72-75]; 6) cobalt oxide, which, like many metal transition oxides, can be used in photocatalysis, including secondary lithium-ion batteries [76,77]. Co_3O_4 is a p-type semiconductor with a direct band gap of 1.4-1.8 eV and an indirect band gap of 2.2 eV.

4.3 Chalcogenide-based nanomaterials

Metal-sulphur compounds are often referred to as MX_2, where X stands for elements of the VI A group (X: S, Se and Te) and M is a transition metal. MX_2 is a two-dimensional material with a large interlayer distance [78]. Due to the large interlayer spacing, interlayer van der Waals interactions are weak and metal ions can be inserted into their structure [79]. As a result, they have been used as electrodes in lithium-ion and sodium-ion batteries [80,81].

Cadmium sulphide (CdS) is the most commonly used sulphur group compound. It is an n-type II-VI semiconductor that is widely used as a photocatalyst due to its light-absorbing ability in the visible light range. It has a low cost and high electrical conductivity. In order to improve the stability and performance of cadmium sulphide, Li et al. proposed the introduction of Pt as a co-catalyst for lithium-sulphur PBATs [82]. Experimental results showed that the S_2- ions produced during discharge were oxidised to polysulphides and electrons were transferred to the nanoparticles, which would reduce the H^+ in the electrolyte and produce a valuable fuel, H_2.

4.4 Elemental-based nanomaterials

Graphite-like nanomaterials for Li-CO2 batteries. Hybrid photocathodes of silicon carbide grown on reduced graphene oxide (SiC/rGO), where the silicon carbide is synthesized in situ on the reduced graphene oxide surface [83]. Both the two-dimensional nanosheet reduced graphene oxide and the Si-OH bonding on the surface of the silicon carbide favoured CO2 adsorption.2021, C. Jia et al. proposed 3 nm sheet-thick super-arch size silicone oxide (2.05 nm) as a photoactive material with energy storage properties, which provided impressive discharge/charge voltage retention (98 and 93%, respectively) [84].

4.5 Perovskite-based nanomaterials

Chalcogenide materials have the advantages of tunable band gap, high absorption coefficient, long exciton diffusion length, good carrier mobility and low exciton binding energy [85-87]. The power conversion efficiency (PCE) of chalcogenide solar cells has now risen rapidly to 25.5% [88]. However, point defects, luminescence or heating effects of the material lead to losses of energy and PCE of chalcogenide solar cells below the radiation limit defined by SQ theory [89].

Factors that affect the PCE of chalcogenide solar cells include the composition of the chalcogenide material, the charge transport material of the transport layer, and interfacial defects. These defects can cause or accelerate the degradation of chalcogenide, leading to non-radiative compounding, which affects the performance and stability of chalcogenide solar cells [90].

Current researchers have introduced several additives to improve their surface morphology and crystallinity, resulting in highly purified and smooth surface chalcogenide layers. yang et al. introduced ammonium benzene sulfonate (ABS) as a ligand molecule to reduce the defect density, which effectively retarded the crystallisation process and produced high-quality, stable chalcogenide films [91].

At present, the preparation of highly efficient, high quality and stable chalcogenide films remains the main challenge in achieving a wide range of PSC applications. At the same time, a number of mechanisms still need to be further investigated by a combination of experimental, computational and simulation means.

4.6 Organic-based photoactive materials

The main advantage of using organic-based materials in batteries is the ability to tune their physicochemical, optical and electrical properties in a simple synthesis step. Organic-based nanomaterials open up a large number of possible materials. There are examples of small organic molecules, polymers, metal organic frameworks (MOFs) and covalent organic frameworks (COFs) that have been used in rechargeable batteries [92-95].

Zhang et al. published a photocathode in a fuel-free photochemical cell (PEC) containing titanium dioxide for the water oxidation reaction [96]. This cell produces both water and oxygen, rather than carbon dioxide, which would benefit the environment.

4.7 Carbon nitride-based nanomaterials

Graphitic carbon nitride (g-C3N4) is a metal-free semiconductor that was first reported by Wang et al. in 2009 and has attracted significant attention since then [97]. Graphitized carbon nitride has the advantages of low cost, environmental friendliness, medium band gap (~2.7 eV), high redox capability, high surface activity and stable photo- and physicochemical properties [98-99]. It consists of a tri-triazine ring as a high nitrogen content building block and abundant triangular nanopores that can provide a large number of active sites for ion adsorption and redox reactions [100].

Liu et al. reported the first light-assisted rechargeable battery using g-C3N4 as a bifunctional PE [101]. Zhu et al. reported that using g-C3N4 to increase the number of active sites on its surface can help facilitate photocatalytic reactions [102]. More importantly, it can extend the light absorption range, and the experimental demonstration of the dual-electrode system yielded an energy efficiency of 92.5% after 50 cycles.

5. Challenges and outlook

The development of photoactive cells, especially integrated two-electrode configurations, is still in its infancy and presents many problems and challenges. So far, the maximum overall efficiency of a two-electrode photovoltaic cell is 9%. How to achieve an efficient and cost effective use of solar energy remains the focus of current research. Researchers are constantly improving the performance of cells by means of optimising relevant parameters such as energy conversion/storage efficiency, long-term durability, energy and power density.

However, exploring low-cost, advanced and compatible photovoltaic electrode materials while optimising the cell assembly is key to obtaining high overall efficiency, high capacity and energy density. Therefore, the development of electrolytes that are also photostable and compatible is also critical to overall stability. At the same time, nanomaterials are the key to significant progress in research on photovoltaics for solar rechargeable cells. Future generations of photovoltaic cells with high energy and photovoltaic efficiency will certainly rely on nanostructured materials.

6. Conclusions

This paper concentrates on a range of current devices with intrinsic solar energy collection, conversion and storage properties, the different classes of cells as well as their application areas and recent research advances. At the same time, nanomaterials are key to making significant progress in the study of photovoltaic electrodes for solar rechargeable cells, and this paper presents seven currently commonly used semiconductor and nanomaterials. Although the fabrication of a low-cost, stable and efficient photovoltaic cell remains a challenge, the projected growth in this field will eventually lead to a viable commercial solution. The rational use of a renewable energy source, solar energy, not only alleviates the fossil fuel crisis, but also the severe environmental pollution and greenhouse effect caused by fossil fuels, which contributes significantly to the sustainable development of human existence.

References

1. Li, Qi, Li, Na., Ishida, et al. Saving electric energy by integrating a photoelectrode into a Li-ion battery [J]. Journal of Materials Chemistry, A. Materials for energy and sustainability,2015,3(42).
2. Boruah Buddha Deka, Wen Bo, De Volder Michael. Light Rechargeable Lithium-Ion Batteries Using V2O5 Cathodes.[J]. Nano letters,2021,21(8):3527-3532.
3. Richard Schmuch, Ralf Wagner, Gerhard Hörpel, et al. Performance and cost of materials for lithium-based rechargeable automotive batteries [J]. Nature Energy, 2018,3(4): 267-278.
4. Shuru Chen, Chaojiang Niu, Hongkyung Lee, et al. Critical Parameters for Evaluating Coin Cells and Pouch Cells of Rechargeable Li-Metal Batteries [J]. Joule, 2019, 3(4): 1094-1105.
5. Samaneh Mozaffari, Mohammad Reza Nateghi, Mahmood Borhani Zarandi. An overview of the Challenges in the commercialization of dye sensitized solar cells [J]. Renewable and Sustainable Energy Reviews, 2016,71.
6. P. Chelvanathan, S.A. Shahahmadi, F. Arith, K. Sobayel, et al. Effects of RF magnetron sputtering deposition process parameters on the properties of molybdenum thin films [J]. Thin Solid Films, 2017, 638, 213–219.
7. A. Nizamuddin, F. Arith, I.J. Rong, M. Zaimi, et al. Investigation of copper(I)thiocyanate (CuSCN) as a hole transporting layer for perovskite solar cells application, J. Adv. Res. Fluid Mech. Therm. Sci. 2021,78, 153–159.
8. José Antonio Luceño-Sánchez, Ana María Díez-Pascual, Rafael Peña Capilla. Materials for Photovoltaics: State of Art and Recent Developments [J]. International Journal of Molecular Sciences, 2019, 20(4).
9. N.S. Noorasid, F. Arith, A.N.M. Mustafa, et al. Numerical analysis of ultrathin TiO2 photoanode layer of dye sensitized solar cell by using SCAPS-1D [J]. Proceedings of the IEEE Regional Symposium on Micro and Nanoelectronics, 2021, 96–99.
10. O.V. Aliyaselvam, F. Arith, M.K. Nor, et al. Solution processed of solid state HTL of CuSCN layer at low annealing temperature for emerging solar cell [J]. Int. J. Renew. Energy Res, 2021.
11. Utpal Gangopadhyay, Sukhendu Jana, Sayan Das. State of Art of Solar Photovoltaic Technology [J]. Conference Papers in Energy, 2013.
12. N.K.A. Hamed, M.K. Ahmad, N.S.T. Urus, et al. Performance comparison between silicon solar panel and dye-sensitized solar panel in Malaysia [J]. AIP Conf. Proc. 2017.
13. M.S. Jamal, M.S. Bashar, A.K.M. Hasan, et al. Fabrication techniques and morphological analysis of perovskite absorber layer for high-efficiency perovskite solar cell: A review [J]. Renewable and Sustainable Energy Reviews,2018,98.
14. Ehsanul Kabir, Pawan Kumar, Sandeep Kumar, et al. Solar energy: Potential and future prospects [J]. Renewable and Sustainable Energy Reviews, 2018,82:894–900.
15. Zhang Kai, Gao Ke, Xia Ruoxi, et al. High-Performance Polymer Tandem Solar Cells Employing a New n-Type Conjugated Polymer as an Interconnecting Layer [J]. Advanced materials (Deerfield Beach, Fla.), 2016,28(24):4817-4823.
16. P. Chelvanathan, S.A. Shahahmadi, F. Arith, K. Sobayel, et al. Effects of RF magnetron sputtering deposition process parameters on the properties of molybdenum thin films [J]. Thin Solid Films,2017,638, 213–219.
17. Sumit NAGAR, Kamal SHARMA, A. K. PANDEY. Effect of graphene and its derivatives on thermo-mechanical properties of phase change materials and its applications: a comprehensive review [J]. Frontiers in Energy, 2022.
18. Li Jingyi, Wang Zhenyu, Zhou Zhiwei, et al. Cathode-electrolyte integrating strategy enabling solid-state lithium metal battery with enhanced cycle stability [J]. Journal of Power Sources,2022,544.
19. Zhang Wendi, Fan Qianxiao, Zhang Dongmei, et al. Dynamic charge modulate lithium uniform plating functional composite anode for dendrite-free lithium metal batteries [J]. Nano Energy,2022,102.
20. Boruah, B. D.; Wen, B.; Volder, M. D. Light rechargeable lithium-ion batteries using V2O5 cathodes [J]. Nano Lett. 2021, 21, 3527−3532.
21. V. Etacheri, R. Marom, R. Elazari. Challenges in the development of advanced Li-ion batteries: a review [J]. Energy Environ. Sci. 2011, 3243-3262.
22. N.S. Choi, Z.H. Chen, S.A. Freunberger, et al. Challenges facing lithium batteries and electrical double-layer capacitors [J]. Angew. Chem. 2012, 9994-10024.
23. Boruah, B. D.; Wen, B.; Volder, M. D. Light rechargeable lithium-ion batteries using V2O5 cathodes [J]. Nano Lett. 2021, 21, 3527−3532.
24. N. F. Yan, X. P. Gao. Photo-assisted rechargeable metal batteries for energy conversion and storage [J]. Energy Environ. Mater. 2021.
25. A. Paolella, A. Vijh, A. Guerfi, et al. Review—Li-ion photo-batteries: challenges and opportunities [J]. J. Electrochem. Soc. 2020.
26. D. Schmidt, M. D. Hager, U. S. Schubert. Photorechargeable electric energy storage systems [J]. Adv. Energy Mater.2016.
27. Y. Wu, C. Li, Z. Tian. Solar-driven integrated energy systems: state of the art and challenges [J]. Power Sources. 2020.
28. Doron Aurbach, Bryan D. McCloskey, Linda F. Nazar. Advances in understanding mechanisms

underpinning lithium – air batteries [J]. Nature Energy,2016,1(9).

29. F. Jiao, P. G. Bruce. Mesoporous crystalline β-MnO2 — a reversible positive electrode for rechargeable lithium batteries [J]. Adv. Mater. 2007.
30. Jang Wook Choi, Doron Aurbach. Promise and reality of post-lithium-ion batteries with high energy densities [J]. Nature Reviews Materials,2016,1(4).
31. Li Matthew, Lu Jun, Chen Zhongwei, et al. 30 Years of Lithium-Ion Batteries [J]. Advanced materials (Deerfield Beach, Fla.),2018,30(33).
32. Tarascon J M, Armand M. Issues and challenges facing rechargeable lithium batteries [J]. Nature, 2001, 414: 359–367.
33. Tang Yuxin, Zhang Yanyan, Li Wenlong, et al. Rational material design for ultrafast rechargeable lithium-ion batteries [J]. Chemical Society reviews, 2015,44(17): 5926–5940.
34. Jian Duan, Xuan Tang, Haifeng Dai, et al. Building Safe Lithium-Ion Batteries for Electric Vehicles: A Review [J]. Electrochemical Energy Reviews, 2020, 3(411).:1–42.
35. Lee Wontae, Muhammad Shoaib, Sergey Chernov, et al. Advances in the Cathode Materials for Lithium Rechargeable Batteries [J]. Angewandte Chemie, 2020, 59(7):2578–2605.
36. Yong Lu, Jun Chen. Prospects of organic electrode materials for practical lithium batteries [J]. Nature Reviews Chemistry,2020,4(3): 127–142.
37. Yayuan Liu, Yangying Zhu, Yi Cui. Challenges and opportunities towards fast-charging battery materials [J]. Nature Energy, 2019,4:540–550.
38. Wang Jie, Wang Yan, Zhu Chaofeng, et al. Photoinduced Rechargeable Lithium-Ion Battery [J]. ACS applied materials & interfaces, 2022.14, 4071–4078.
39. K. Amine, H. Tukamoto, H. Yasuda. A New Three-Volt SpinelLi1 + x Mn1.5Ni0.5 O 4 for Secondary Lithium Batteries [J]. Journal of The Electrochemical Society,2019,143(5).
40. H. J. Peng, S. Urbonaite, C. Villevieille. Consequences of Electrolyte Degradation for the Electrochemical Performance of Li1+x(NiaCobMn1 [J]. Journal of The Electrochemical Society, 2015, 162(13).
41. Qi Li, Yang Liu, Shaohua Guo. Solar energy storage in the rechargeable batteries [J]. Nano Today,2017,16, 46–60.
42. M. Yu, W. D. McCulloch, D. R. Beauchamp, Z. Huang, X. Ren, Y. Wu, J. Am. Chem. Soc. 2015, 137, 8332–8335.
43. Deka Boruah, Angus Mathieson, M De Volder. Photo-rechargeable Zinc-ion Batteries [J]. Energy & Environmental Science,2020,13(8): 2414–2421.
44. Boruah, B. D.; Mathieson, A.; Volder, M. D. Vanadium Dioxide Cathodes for High-Rate Photo-Rechargeable Zinc-Ion Batteries [J]. Adv. Energy Mater. 2021, 11, 2100115.
45. Boya Tang, Lutong Shan, Shuquan Liang, et al. Issues and opportunities facing aqueous zinc-ion batteries [J]. Energy & Environmental Science, 2019,12, 3288–3304.
46. Appel, J. H.; Li, D. O.; Podlevsky, J. D.; et al. Low Cytotoxicity and Genotoxicity of TwoDimensional MoS2 and WS2. ACS Biomater [J]. Sci. Eng. 2016, 2, 361–367.
47. Deka Boruah, Buddha ; Wen, Bo ; De Volder, Michael. Molybdenum Disulfide – Zinc Oxide Photocathodes for Photo-Rechargeable ZincIon Batteries [J]. Advanced energy materials. 2021.
48. Qi Li, Yang Liu, Shaohua Guo. Solar energy storage in the rechargeable batteries [J]. Nano Today,2017,16, 46–60.
49. Z. Wang, H. C. Chiu, A. Paolella. Lithium photo-intercalation of cdssensitized WO3 anode for energy storage and photoelectrochromic applications [J]. ChemSusChem, 2019, 12, 2220-2230.
50. Zhai Tianyou, Liu Haimei, Li Huiqiao, et al. Centimeter-long V2O5 nanowires: from synthesis to field-emission, electrochemical, electrical transport, and photoconductive properties [J]. Advanced materials (Deerfield Beach, Fla.),2010,22(23): 2547–2552.
51. Josny Joy, Jinu Mathew, Soney C. George. Nanomaterials for photoelectrochemical water splitting – review [J]. International Journal of Hydrogen Energy, 2018,43(10):4804–4817.
52. M.D. Levi, K. Gamolsky, D. Aurbach. Determination of the Li ion chemical diffusion coefficient for the topotactic solid-state reactions occurring via a two-phase or single-phase solid solution pathway [J]. Journal of Electroanalytical Chemistry,1999,477(1).
53. Sharma Khushboo, Sharma Vinay, Sharma S S. Dye-Sensitized Solar Cells: Fundamentals and Current Status[J]. Nanoscale research letters,2018,13(1).
54. Carmen Cavallo, Francesco Di Pascasio, Alessandro Latini. Nanostructured Semiconductor Materials for Dye-Sensitized Solar Cells [J]. Journal of Nanomaterials, 2017.
55. Rahul Kumar, Veena Sahajwalla, Parag Bhargava. Fabrication of a counter electrode for dye-sensitized solar cells (DSSCs) using a carbon material produced with the organic ligand 2-methyl-8-hydroxyquinolinol (Mq) [J]. Nanoscale Advances, 2019.
56. C. Wang, X. Zhang, D. Cao, H. Yin, X. Li, P. Cheng, B. Mi, Z. Gao, W. Deng, In situ preparation of hierarchically structured dual-layer TiO2 films by E-spray method for efficient dye-sensitized solar cells [J]. Org. Electron. 2017, 135–141.
57. Bo Wang, Lei L. Kerr. Dye sensitized solar cells on paper substrates[J]. Solar Energy Materials and Solar Cells, 2011,95(8).

58. A R Yugis, R F Mansa, C S Sipaut. Review on Metallic and Plastic Flexible Dye Sensitized Solar Cell [J]. IOP Conference Series: Materials Science and Engineering,2015,78(1).
59. Md. K. Nazeeruddin, Etienne Baranoff, Michael Grätzel. Dye-sensitized solar cells: A brief overview[J]. Solar Energy,2011,85(6):1172–1178.
60. Brian E. Hardin, Henry J. Snaith, Michael D. McGehee. The renaissance of dye-sensitized solar cells [J]. Nature Photonics,2012,6(3):162–169.
61. C. Xu, X. Zhang, L. Duan, et al. A photoassisted rechargeable battery: synergy, compatibility, and stability of a TiO2/dye/Cu2S bifunctional composite electrode [J]. Nanoscale, 2020.
62. G. B. Less, J. H. Seo, S. Han. Micro [J]. Journal of The Electrochemical Society,2012,159(6).
63. Zhenguo Yang, Daiwon Choi, Sebastien Kerisit, et al. Nanostructures and lithium electrochemical reactivity of lithium titanites and titanium oxides: A review [J]. Journal of Power Sources, 2009,192(2):588–598.
64. Ludovico Macera, Giuliana Taglieri, Valeria Daniele, et al. Nano-Sized Fe(III) Oxide Particles Starting from an Innovative and Eco-Friendly Synthesis Method[J]. Nanomaterials, 2020,10(2):323–342.
65. Yao Yu, Sang Dandan, Susu Duan, et al. Excellent optoelectronic applications and electrical transport behavior of the n-WO3 nanostructures/p-diamond heterojunction: A new perspective[J]. Nanotechnology, 2021, 32(33).
66. J. C. Murillo-Sierra, A. Hernández-Ramírez, L. Hinojosa-Reyes, et al. A review on the development of visible light-responsive WO3-based photocatalysts for environmental applications [J]. Chem. Eng. J. Adv. 2021.
67. Liu Minsu, Su Bin, Tang Yue, et al. Recent Advances in Nanostructured Vanadium Oxides and Composites for Energy Conversion [J]. Advanced Energy Materials,2017,7(23).
68. Lee Shinbuhm, Ivanov Ilia N, Keum Jong K, et al. Epitaxial stabilization and phase instability of VO2 polymorphs [J]. Scientific reports,2016,6(1).
69. Mattelaer Felix, Geryl Kobe, Rampelberg Geert, et al. Amorphous and Crystalline Vanadium Oxides as High-Energy and High-Power Cathodes for Three-Dimensional Thin-Film Lithium Ion Batteries [J]. ACS applied materials & interfaces, 2017, 9(15). 13121–13131.
70. Jing Wang, Shengzhi Yao, Weiqing Lin, et al. Improving the electrochemical properties of high-voltage lithium nickel manganese oxide by surface coating with vanadium oxides for lithium ion batteries [J]. Journal of Power Sources, 2015, 280.114–124.
71. Hongbin Zhao, Lanying Pan, Siyi Xing, et al. Vanadium oxides–reduced graphene oxide composite for lithium-ion batteries and supercapacitors with improved electrochemical performance [J]. Journal of Power Sources, 2013,222.21–31.
72. E. Avigad, L. Etgar. Studying the effect of MoO3 in holeconductor-free perovskite solar cells [J]. ACS Energy Lett. 2018,2240–2245.
73. Y. Zhang, P. Chen, Q. Wang, et al. High-capacity and kinetically accelerated lithium storage in MoO3 enabled by oxygen vacancies and heterostructure [J].Adv. Energy Mater. 2021.
74. Nanotechnology - Nanoflakes; Findings in Nanoflakes Reported from University of Brescia (Gold Functionalized Moo3 Nano Flakes for Gas Sensing Applications) [J]. Nanotechnology Weekly, 2019.
75. Yuhua Zhu, Yuan Yao, Zhu Luo, et al. Nanostructured MoO3 for Efficient Energy and Environmental Catalysis [J]. Molecules, 2019, 25(1):18–44.
76. J. S. Lee, M. S. Jo, R. Saroha, et al. Hierarchically welldeveloped porous graphene nanofibers comprising N-doped graphitic C-coated cobalt oxide hollow nanospheres as anodes for high-rate Li-ion batteries[J]. Small, 2020.
77. Kai Wang, Jiajia Wan, Yuxuan Xiang, et al. Recent advances and historical developments of high voltage lithium cobalt oxide materials for rechargeable Li-ion batteries [J]. Journal of Power Sources,2020,460(C).
78. J. Theerthagiri, K. Karuppasamy, G. Durai, et al. Recent advances in metal chalcogenides (MX; X = S, Se) nanostructures for electrochemical supercapacitor applications: a brief review [J]. Nanomaterials, 2018.
79. Deng Wentao, Chen Jun, Yang Li, et al. Solid Solution Metal Chalcogenides for Sodium-Ion Batteries: The Recent Advances as Anodes [J]. Small (Weinheim an der Bergstrasse, Germany), 2021,17(35).
80. Yeonwoong Jung, Yu Zhou, Judy J. Cha. Intercalation in two-dimensional transition metal chalcogenides [J]. Inorganic Chemistry Frontiers, 2016,3(4):452–463.
81. Y. Zhang, L. Zhang, T. a. Lv, et al. Twodimensional transition metal chalcogenides for alkali metal ions storage [J]. ChemSusChem, 2020.
82. Na Li, Yarong Wang, Daiming Tang, et al. Integrating a Photocatalyst into a Hybrid Lithium–Sulfur Battery for Direct Storage of Solar Energy [J]. Angewandte Chemie,2015,127(32):9271–9274.
83. Kojima Akihiro, Teshima Kenjiro, Shirai Yasuo, et al. Organometal halide perovskites as visible-light sensitizers for photovoltaic cells [J]. Journal of the American Chemical Society,2009,131(17):6050 – 6051.
84. Wehrenfennig Christian, Eperon Giles E, Johnston Michael B, et al. High charge carrier mobilities and lifetimes in organolead trihalide perovskites [J]. Advanced materials (Deerfield Beach, Fla.), 2014,26(10):1584–1589.

https://doi.org/10.1051/e3sconf/202337502010

85. Z. Li, M. L. Li, X. X. Wang, et al. In situ fabricated photo-electro-catalytic hybrid cathode for lightassisted lithium–CO2 batteries [J]. Mater. Chem. 2020.

86. C. Jia, F. Zhang, L. She, et al. Ultra-large sized siloxene nanosheets as bifunctional photocatalyst for a Li-O2 battery with superior round-trip efficiency and extra-long durability [J]. Angew. Chem. Int. Ed, 2021.

87. Henry J. Snaith. Perovskites: The Emergence of a New Era for Low-Cost, High-Efficiency Solar Cells [J]. J. Phys. Chem. Lett. 2013,4(21):3623–3630.

88. Green Martin, Dunlop Ewan, HohlEbinger Jochen, et al. Solar cell efficiency tables (version 57) [J]. Progress in Photovoltaics: Research and Applications,2020,29(1).

89. Sarritzu Valerio, Sestu Nicola, Marongiu Daniela, et al. Optical determination of Shockley-Read-Hall and interface recombination currents in hybrid perovskites [J]. Scientific reports,2017,7(1):1–10.

90. Chen Bo, Rudd Peter N, Yang Shuang, et al. Imperfections and their passivation in halide perovskite solar cells [J]. Chemical Society reviews,2019,48(14):3842–3867.

91. Yi Yang, Huirong Peng, Cheng Liu, et al. Bi-functional additive engineering for high-performance perovskite solar cells with reduced trap density [J]. Journal of Materials Chemistry A,2019, 6450–6458.

92. Chen Yuan, Wang Chengliang. Designing High Performance Organic Batteries [J]. Accounts of chemical research,2020,53(11):2636–2647.

93. Yang Xu, Min Zhou, Yong Lei. Organic materials for rechargeable sodium-ion batteries [J]. Materials Today,2018,21(1):60–78.

94. J. Xie, Q. Zhang. Recent progress in aqueous monovalention batteries with organic materials as promising electrodes [J]. Mater. Today Energy, 2020.

95. Xie Jian, Zhang Qichun. Recent Progress in Multivalent Metal (Mg, Zn, Ca, and Al) and Metal-Ion Rechargeable Batteries with Organic Materials as Promising Electrodes [J]. Small (Weinheim an der Bergstrasse, Germany),2019,15(15).

96. Zhang Bingqing, He Lihue, Yao Tingting, et al. Simultaneous Photoelectrocatalytic Water Oxidation and Oxygen Reduction for Solar Electricity Production in Alkaline Solution[J]. ChemSusChem, 2019,12(5):1026–1032.

97. Wang Xinchen, Maeda Kazuhiko, Thomas Arne, et al. A metal-free polymeric photocatalyst for hydrogen production from water under visible light [J]. Nature materials,2009,8(1):76–80.

98. Q. Zhang, X. Liu, M. Chaker, et al. Advancing graphitic carbon nitride-based photocatalysts toward broadband solar energy harvesting [J]. ACS Mater. Lett. 2021.

99. Nanotechnology - Photocatalytics; Investigators at King Abdul-Aziz University Describe Findings in Photocatalytics (g-C3N4-Based Heterostructured Photocatalysts) [J]. Nanotechnology Weekly,2018.

100. Y. Yoon, M. Lee, S. K. Kim, et al. A strategy for synthesis of carbon nitride induced chemically doped 2D MXene for highperformance supercapacitor electrodes [J]. Adv. Energy Mater. 2018.

101. Y. Liu, N. Li, S. Wu, et al. Reducing the charging voltage of a Li– O2 battery to 1.9 V by incorporating a photocatalyst [J]. Energy Environ. 2015.

102. Z. Zhu, Y. Ni, Q. Lv, et al. Surface plasmon mediates the visible light–responsive lithium–oxygen battery with Au nanoparticles on defective carbon nitride [J]. Proc. Natl. Acad. Sci. Unit. States Am.2021.

https://doi.org/10.1051/e3sconf/202337502011

Utilizing CO_2 from refinery for methanol and electricity co-production: system assessment

Qian Ma [1,*], Xiaodan Yu [1], Song Xia [2], Xu Zhang [1], Anqi Liu [1], Bo Yuan [1], Xiaojun Chen [3]

[1] Research Institute of Safety & Environment Technology, China National Petroleum Corporation, Beijing 102206, China
[2] Daqing Oilfield water supply Company, Daqing 163000, China
[3] Chinese Academy of Environmental Planning, Beijing 100012, China

Abstract. As one of the major energy-consuming and high-emission industries, the refineries account for 10% of global industrial carbon emissions, of which about 25% is emitted by fluid catalytic cracking (FCC) process. Therefore, CO_2 capture from FCC flue gas create the possibility for low-carbon refineries. We developed a novel CO_2 utilization process to methanol production modeling using Aspen Plus. Meanwhile, organic Rankine Cycle (ORC) power generation technology was coupled to recover the waste heat of the system. A five-million-ton FCC unit in China was selected as a case study. We conducted some analysis for the process, the results show that the developed system boosts the energy efficiency of the FCC unit by 2.8%. The annual capacity of the waste heat power generation unit is 4.8 GWh, with a thermal efficiency of 5.9%.

1. Introduction

Globally, annual carbon dioxide (CO_2) emissions have increased sharply during past few decades.The increasing greenhouse gas emissions exacerbate climate change and leads to various extreme weather events[1].Thus, the consensus on CO_2 reduction has prompted most countries to make commitments and put them into action.China also put forward the strategic goal of achieving carbon peak by 2030 and carbon neutrality by 2060 in 2020.

The chemical industry is one of the industries with high energy consumption and high emissions. Refineries, as a major contributor to the chemical industry, occupy 10% of the global industrial carbon emissions [2]. Although the energy structure is gradually transforming, driven by the rapid economic development , the carbon emissions caused by energy use are still rising , increasing by about 7% by 2040[3]. In the face of "double carbon target", refineries must assume corresponding responsibility to contribute to China's transition toward a lower carbon future. Many refineries' units emit CO_2, Given the FCC process has a high CO_2 emissions, accounting for 20-35% of the total carbon emissions of the refinery[4].Therefore, capturing CO_2 from FCC flue gas will be an important step to reduce the total CO_2 emissions of refineries. Meanwhile, the dry gas of refinery is rich in hydrogen, methane, nitrogen, ethylene, ethane, etc., and the production of dry gas in FCC is the largest[5-6]. But at present, the dry gas in domestic refineries is not well used that used as fuel gas or torch. Therefore, the utilization of dry gas in refinery promises an optimistic outlook for decarbonization.

In this study, we designed a methanol production system that uses the CO_2 recovered from the flue gas of FCC unit and the CH_4 and H_2 separated from dry gas as feedstock. A 5-million-tons FCC unit in China was selected for case study. Fig.1 shows the conceptual design of the hybrid system, including carbon capture, methane dry reforming, methanol synthesis, methanol distillation, and ORC units. To be specific, the CO_2 in the FCC flue gas was captured by carbon capture unit using monoethanolamine (MEA) , followed by syngas generation. Then the reaction of syngas with hydrogen via methanol synthesis unit to produce crude methanol, which sent to distillation unit to purify. After the optimization of HEN, the low-grade waste heat derived from methanol distillation unit was utilized by ORC unit for electricity generation.

Methanol, as a chemical intermediate, has attracted much attention. China is the world's largest producer and consumer of methanol. Based on the essential situation, the domestic methanol production is mainly from coal-to-methanol, which is characterized of high energy consumption and high carbon emission intensity [7]. It is not conducive to the low-carbon development of China's economy. Natural gas is the most ideal raw material for methanol synthesis. However, at the restrictions of region, it doesn't work in China based on the current energy structure that rich in coal, insufficient oil and limited natural gas supply.

Meanwhile, the generated low-grade waste heat was also recovered and utilized by coupling with Organic Rankine Cycle (ORC) power generation technology. The ORC is a state-of-the-art technology that combines the main process system with power generation system [8]. Although energy use has been optimized under the heat exchanger network , some low-grade heat has still not been well utilized. Thus, we deployed ORC system to recover the waste heat from the methanol distillation unit, converting the low-grade waste heat into electricity.

* Corresponding author: ma-qian@cnpc.com.cn

Fig.1. Aspen model diagram of the FCC co-production

2. Methodology

2.1 Overall energy efficiency calculation

According to the First Law of Thermodynamics, the overall energy efficiency of the methanol production system is calculated as Eq 1.

$$\eta = \frac{\sum E_o}{\sum E_i} \times 100\% \qquad (1)$$

where, η denotes the energy efficiency of the methanol production system, Eo denotes the energy output of the system , and Ei denotes all the input energy of the system.

2.2 Sustainability evaluation

The pros and cons of the new process system are affected by various factors. In order to evaluate the new process more comprehensively, we assessed the sustainability of the CTM from a multi-dimensional perspective. In general, the four main aspects of technology, environment, economy and society contained 12 sub-indicators are considered respectively. To evaluate and compare sustainability of different alternatives more reasonably and scientifically, a further processing of these indicators is adopted by use of normalization method, according to eq. (12). In addition, the setting of some sub-indicators such as technical maturity refers to a previous study of Huang[9].

$$x_{ij} = \frac{x_{ij} - worst(x_i)}{best(x_i) - worst(x_i)} \qquad (2)$$

where xij is the indicator j for process i ; best(xj) is the assumed best case of indicator j; worst(xj) is the assumed worst case of indicator j; xij, which varies in the range [0,1], is the normalized indicator j for process i.

3. Results and discussion

3.1 Energy efficiency of the CO_2–dry gas to methanol production system

Based on the Aspen simulation tool and theoretical calculations, the input and output energy of the CO2–dry gas to methanol production system are summarized in Table 1. It can be seen that the system's energy input mainly derives from the feedstock, fuel gas, and electricity, with a total energy input of 36815 TJ, while the total energy output was estimated to be 28958 TJ . According to Eq (1), the overall energy efficiency of the CO_2–dry gas to methanol production system was 78.6%.

Table 1 Input and output energy of the"CO2-dry gas" methanol production system

Input		Export	
Material	Energy (TJ)	Material	Energy (TJ)
methane	18900	Methanol	28933
H_2	14.8	3.5Mpa Steam	7.47
Fule gas	10400	Electricity	17.3
Electricity	7500		
Sum	36815	28958	
Overall energy efficiency		78. 6%	

Fig. 2 compares the energy efficiency of the FCC process, the CO_2–dry gas to methanol production process (CTM), and the FCC combined with the CO_2–dry gas to methanol process (FCC+ CTM). As for the FCC unit, its energy efficiency is relatively lower than CTM process (78.6%), which is only 57.9%. With the co-production of the process, the energy efficiency of FCC increased from 57.9% to 60.7%, prompting a 2.8% energy efficiency improvement.It can be seen that through carbon recovery and recycling can effectively improve the energy utilization efficiency of the original FCC unit.

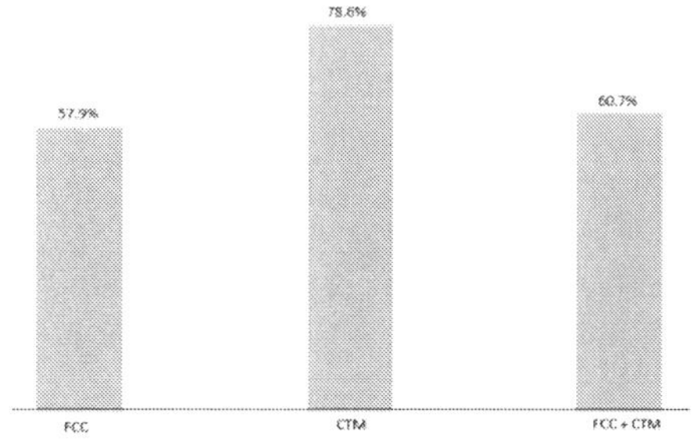

Fig. 2 Energy efficiency comparison of process integration

3.2 Electricity generation analysis based on ORC

There is a large quantity of low-grade heat in the methanol production system, and this waste heat is difficult to use for heat exchange. However, the ORC process enables efficient utilization of these energies for electricity generation. Waste heat mainly derives from the cooling process of the methanol product. Considering the impacts of the working fluid temperature and expansion ratio on the ORC process, we selected R290 as the working fluid for ORC power generation. For the designed system with an annual production capacity of 1.45 million tons of methanol, the net output power was 4798 MW, and the annual power generation capacity was 4.798M kWh. The thermal efficiency of the ORC unit was calculated to be 5.9%.

3.3 Sustainability analysis

To evaluate the CTM process more conveniently and intuitively, sustainability evaluation was developed that enable to identified advantages and drawbacks of new process compared with original unit (FCC) and traditional technology. The specific setting values of different indicators are listed in Table 2 and the results (Fig.7) show that CTM brings more environmental benefits in comparison to FCC and traditional process. However, it also has some weak points, such as low energy security. Besides, the technology is still immature and needs further study and improvement. In short, because of its good carbon utilization and high internal recovery rate, and it also shows outstanding advantages in energy efficiency. Therefore, It is a step closer to the realization of urban low-carbon refineries.

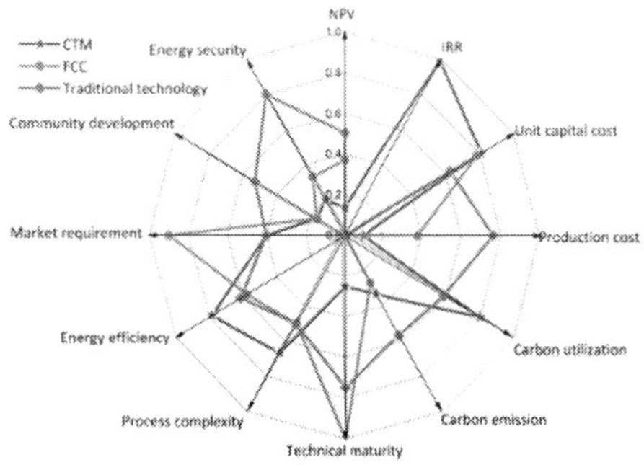

Fig.3 Sustainability comparison of CTM, FCC and Traditional

Table 2 Comparison of Indicators for CTM, FCC and Traditional technology

Indicator (subindicator)	CTM	FCC [12,13]	Traditional technology [10,11,16]	Reference value	
				Best	Worst
Economic					
NPV (RMB/B Yuan)	12.74	17.33	20.1	30	10[14]
IRR (%)	22.6	14.9	17.15	22.65	17.15[11]
Unit capital cost (RMB/M Yuan)	17.16	17.8	21.76	12.4	37.4[11]
Production cost (RMB/Yuan)	1690	1511	1323	1204	1700 [11]
Technical					
Technical maturity	0.250	1.00	0.750	1	0[14]
Process complexity (%)	20	40	40	30	50[11]
Energy efficiency (%)	0.787	0.507	0.614	100	0[14]
Environmental					
Carbon utilization (t/t production)	0.75	0.00	0.55	0.93	0[14]
Carbon emission (t/t production)	0.333	0.27	0.566	0	2.99[14]
Social					
Market requirement (%)	40	90	40	100	0[11]
Community development (staff/1000t)	5	5	16	30.3	0[15]
Energy security (%)	20	33	8	100	0[15]

4. Conclusion

Implementation of the new process, not only can energy efficiency of original FCC unit be improved, but the GHG can also be utilized. In other words, it has obvious net carbon emission reduction potential for refinery. The energy efficiency analysis showed that the energy efficiencies of the original FCC process, the CO_2–dry gas methanol production process in the basic case, and the integrated FCC process with the CO_2–dry gas methanol production process were 57.9%, 78.6% and 60.7%, respectively, illustrating a 2.8% energy efficiency improvement, which was attributed to the process integration. Moreover, reuse or make full use of methane in dry gas is a new thinking of promote production while reducing emissions, which CH4 and CO2 are the main GHG. The design ingeniously saves raw materials and turn "waste" into a benefit. Also, it is necessary to apply the new system in other carbon capture and utilization industries that can achieve a beneficial effect on GHG abatement for future sustainable energy systems.

To completely understand the sustainable performance of the CO_2–dry gas to methanol production system, other environmental footprints (such as the water and environmental footprints) should be further quantified, which are directions for future studies.

Acknowledgments

This work was supported by National Key R&D Program of China No. 2022YFC3702900. Thanks are also due the China National Petroleum Corporation (CNPC) Research Institute of Safety & Environment Technology for access to the database.

References

1. Frame D J, Wehner M F, Noy I, et al. The economic costs of Hurricane Harvey attributable to climate change [J]. Climatic Change, 2020, 160(2): 271-281.

2. Xiaobing Zhang, Jing Xu. Optimal policies for climate change: A joint consideration of CO2 and methane. Applied Energy 2018; 211: 1021–1029
3. BP. BP Energy Outlook 2021edition; 2021; https://www.bp.com. [accessed 06 June 2021]
4. Biswas J, Maxwell I E. Recent process-and catalyst-related developments in fluid catalytic cracking. Applied Catalysis 1990; 63(1): 197-258.
5. Güleç F, Meredith W, Sun C G, et al. A novel approach to CO2 capture in fluid catalytic cracking—chemical looping combustion. Fuel 2019; 244: 140-150.
6. Wei M, Qian F, Du W, et al. Study on the integration of fluid catalytic cracking unit in refinery with solvent-based carbon capture through process simulation. Fuel, 2018;219: 364-374
7. Li S H,Sun X J, Liu L L, Du J.A full process optimization of methanol production integrated with co-generation based on the co-gasification of biomass and coal.Energy 2023;267:126566
8. Cao Y, Mihardjo L W W, Dahari M, et al. Waste heat from a biomass fueled gas turbine for power generation via an ORC or compressor inlet cooling via an absorption refrigeration cycle: A thermoeconomic comparison[J]. Applied Thermal Engineering, 2021, 182: 116117.
9. Huang, Y, Yi, Q, Wei GQ et al. Energy use, greenhouse gases emission and cost effectiveness of an integrated high- and low-temperature Fisher-Tropsch synthesis plant from a lifecycle viewpoint. Applied Energy 2018; 228:1009-1019.
10. Zhou H, Yang S, Xiao H, et al. Modeling and techno-economic analysis of shale-to-liquid and coal-to-liquid fuels processes. Energy 2016; 109: 201-210.
11. Yu Han. Construction and comprehensive evaluation of industrial chain for carbon emission and utilization of coal to olefin; China University of Petroleum (Beijing); 2018. (in Chinese)
12. Delin Kong. Contrast of Processing Programmes of C4 Fractions from Fluid Catalytic Cracking Unit. Contemporary Chemical Industry 2003; 2: 103-105. (in Chinese)
13. Qianqian Song. The construction of carbon industrial chain of refineries and low-carbon refineries in China; China University of Petroleum (Beijing); 2014. (in Chinese)
14. Omar Y. Abdelaziza, Wafaa M. Hosny, Mamdouh A. Gadallac, et al. Novel process technologies for conversion of carbon dioxide from industrial flue gas streams into methanol. Journal of CO2 Utilization 2017; 21: 52–63.
15. Siyu Yang, Li Xiao, Andrzej Kraslawski, et al. Sustainability Assessment of the Coal/Biomass to Fischer–Tropsch Fuel Processes. ACS Sustain.Chem.Eng. 2014; 2: 80–87.
16. Qin Z, Zhai G, Wu X, et al. Carbon footprint evaluation of coal-to-methanol chain with the hierarchical attribution management and life cycle assessment. Energy conversion and management 2016; 124: 168-17.

Utilizing carbon dioxide from refinery for methanol and electricity co-production: system design and economic assessment

Qian Ma [1,*], Yuxi Wang [1], Xu Zhang [1], Xiaodan Yu [1], Anqi Liu[1], Bo Yuan [1], Xiaojun Chen[3]

[1] Research Institute of Safety & Environment Technology, China National Petroleum Corporation, Beijing 102206, China
[2] Daqing Oilfield water supply Company, Daqing 163000, China
[3] Chinese Academy of Environmental Planning, Beijing 100012, China

Abstract. Mitigating greenhouse gas (GHG) emissions to curb climate change has become a consensus among international community. Refining is one of the major industries with high energy consumption and high emissions, which is responsible for 4–10% of global carbon dioxide (CO_2) emissions and approximately 25% is generated by fluid catalytic cracking (FCC) units. The flue gas discharged from FCC units has a high CO_2 content, presenting potential for methanol and electricity co-production production when the methane in the dry gas is considered. To unlock this green chance for enterprises, we designed a methanol production system that uses the CO_2 recovered from the flue gas of fluid catalytic cracking (FCC) unit in a refinery and the CH4 separated from dry gas as feed stock (hydrogen to be added from an internal hydrogen production unit of the refinery). We analyzed economic feasibility of the process, the results show that the developed system is economically feasible for annual methanol yields of 1.0–2.5 Mt and the internal rate of return increases by 8.3%.

1. Introduction

As the world's largest emitter of CO_2, China has made a commitment to become carbon neutral by 2060 and to begin cutting its CO_2 emissions within the coming decades. China's ambitious plan to meet the goal of carbon neutrality requires a set of diversified emission reduction strategy, which has become a long-term strategy to ensure the achievement of China's carbon neutral goal[1]. With the rapid urbanization and economic development of national economy, environmental problems caused by carbon emissions have attracted more and more attention. Refining and chemical industry is one of the main industries represents a significant source of China's GHG emissions[2]. In the face of increasingly severe environmental protection requirements, bear the corresponding responsibility for emission reduction and develop strategies to work toward the deep decarbonization of the refineries are critical to meet the goal of carbon neutrality.

Studies on low-carbon refinery development are not rare. Johansson et al. [3] concluded that fuel substitution helps to reduce short-term CO_2 emissions in a refinery. Zhang improved refinery energy efficiency by recovering low temperature waste heat through Organic Rankine Cycle power generation[4]. Thus, previous studies primarily focused on improving the energy efficiency of refineries, whereas endeavors to direct GHG emission reduction via CO_2 capture and utilization are usually lacking.

In addition to improve refinery energy efficiency, CO_2 is a valuable C1 raw material. It offers a good opportunity for refineries to improve decarbonization while to improve CO_2's economic value through chemical conversion[5]. Fernández-Dacosta C used Aspen to simulate CO_2 capture from a hydrogen production unit to co-produce dimethyl-ether (DME) and polyols, providing refineries with an approach to mitigate CO_2 emissions [6].Through carbon capture technology to recover CO_2 from flue gas, and then realizing its utilization by chemical reaction promises an optimistic outlook for low-carbon refineries. In addition to environmental benefits, converting CO_2 into high added value chemicals can also bring considerable economic benefits[7]. At present, Chinese petrochemical enterprises are facing the dual pressure of increasing oil-gas supply and the reduction of CO_2 emissions. How to balance this dilemma, this is the significance of this study.

We developed a novel CO_2 utilization process in refinery using Aspen Plus. Specifically,we designed a process turning "waste" into wealth in refinery that uses the CO_2 recovered from the flue gas of an FCC unit and the CH_4 separated from dry gas as feedstock for producing methanol, Fig. 1 displays the conceptual design of this study.

* Corresponding author: ma-qian@cnpc.com.cn

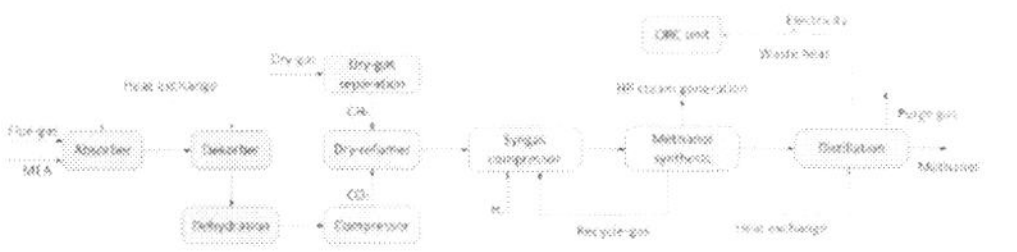

Fig.1. System design of the FCC co-production

2. Methodology of economic evaluation

In this study, several key indicators, such as the total capital cost (TCI), the total product cost of methanol (TPC), the accumulative net present value (NPV) and internal rate of return (IRR) were calculated for economic evaluation due to their wide application in economics [8,9,10]. TCI mainly includes fixed asset investments and fluid capital. Peters proposed that the fixed asset investments of general chemical plants account for approximately 85% of the total project investment. He also proposed that there is an exponential relationship between the cost of similar chemical process equipment and scale rather than a simple linear relationship. In this paper, the scaling factor method is used to estimate the purchased equipment cost [11], which is expressed as

$$C_B = C_A \times \left(\frac{S_B}{S_A}\right)^n \quad , \quad (1)$$

where C_A and C_B are the equipment cost of the reference scale and the required scale to be estimated, respectively; S_A and S_B denote the scale of A and B, respectively; and n is the equipment scale factor. The TCI is the sum of all purchased equipment investment and fixed investment (instrument and control, pipeline, installation costs, etc.), and its calculation equation is expressed as

where I is the total purchased equipment investment, F_i is the ratio of the other purchased equipment cost to the total equipment investment, and i is the components of investment.

TPC is all the costs generated in the process of producing products, which is divided into two parts: operation cost and general expense. It is determined by Eq. (3).

$$T_{TPC} = C_{RM} + C_U + C_{O\&M} + C_{P\&R} + C_D + C_{T\&I} + C_{PO} + C_{GE} \quad (3)$$

where C_{RM}, C_U, $C_{O\&M}$, $C_{P\&R}$, C_D, $C_{T\&I}$, C_{PO} and C_{GE} denote raw materials, utilities, operation and maintenance costs, patents and royalties, depreciation, taxes and insurance, plant overhead and general expenses, respectively. Among these, the depreciation cost adopts the average life method with an assumption of a 20-year plant life and 5% salvage value.

NPV and IRR are also significant criteria used to assess the feasibility of a system. The specific method for NPV is given by

$$TCI = I \times (1 + \sum F_i) \quad (2)$$

$$NPV = f(t,n) = \sum_{t=0}^{t} \frac{CI - CO}{[1+i]^t} \quad , \quad (4)$$

$$\sum_{t=0}^{t} \frac{NPV}{[1+IRR]^t} = 0 \quad , \quad (5)$$

where CI and CO denotes the inflow cash and outflow of cash, t is the cash flow year, i is the discount rate and n is the total number of years.

3. Results and discussion

Fig. 3 displays the IRR, TCI and constitution of costs with methanol production ranging from 0.35 Mt/yr to 3.5 Mt/yr. From the point of view of IRR, it is economically infeasible when the capacity is less than 0.7Mt/yr, which is lower than the value specified in the industry standard of China (12%). As can be seen, the value of IRR increases with increasing methanol capacity, which is similar to the effect of plant scale on capital investment. Compared with the traditional process, the IRR increases by 8.25 at the production of 1.8 Mt/yr [12]. With production changes in the scope of 3.5–2.45 Mt/yr, the TCI and constituent costs increase exponentially with increasing production. It is obvious that production of 2.45Mt/yr is a threshold, and increasing production beyond this point leads to significant increases in TCI. This bottleneck implies that there is a requirement for more processing units and power consumption to meet the higher capacities of methanol. The unit investment (TCI divided by production) decreases dramatically when the production is no more than 1.5 million tons. In general, the TCI for fixed asset investments is dominant, followed by working capital and equipment investment.

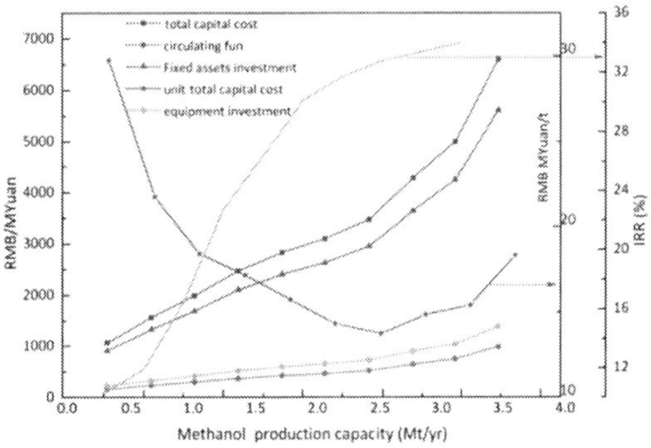

Fig. 3 Variations in total capital and constituent costs with changing methanol production

Fig. 4a shows the production cost distribution with different levels of methanol production. It is obvious that TPC increases were gradually caused by increasing production. Moreover, the raw material costs as well as O&M costs and utilities occupy the greatest portion of the TPC, which means the TPC seems to be highly affected by the prices of raw materials, O&M and utilities costs, which is consistent with the sensitivity analysis of methanol production cost. To obtain a better understanding of the economic performance, the unit product cost of methanol was evaluated (see Fig. 4b). Obviously, when the yield is within 145 t, the unit product cost drops conspicuously. Then the downward trend flattens until it reached a minimum of 1690 Yuan/t at 2.45 Mt/yr, after which the trend goes up for methanol production over 245 t. This indicates that the optimal methanol production is approximately 125–325 t from the perspective of production cost. Fig. 5 shows the comparison of accumulative NPV under different methanol production levels. As shown in the enlarged part

 https://doi.org/10.1051/e3sconf/202337502012

of the red circle, the longest payback period is 7 years, while the capacity is 0.35 Mt/yr. When the capacity is in the range 1.05–3.15 Mt/yr, the payback period is 4 years.

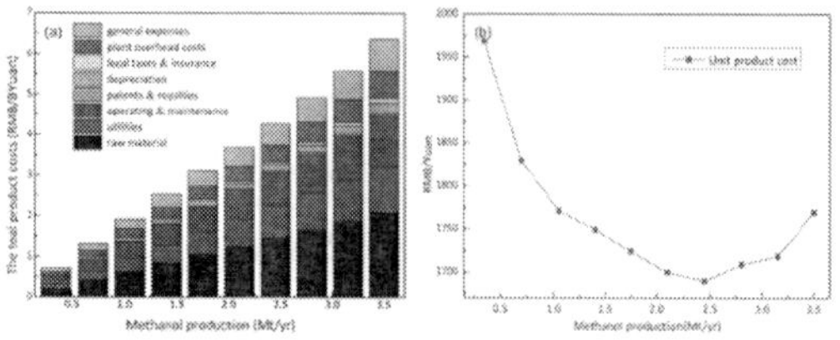

Fig.4 Variations in the TPC and unit product cost at different methanol production (Mt/yr)

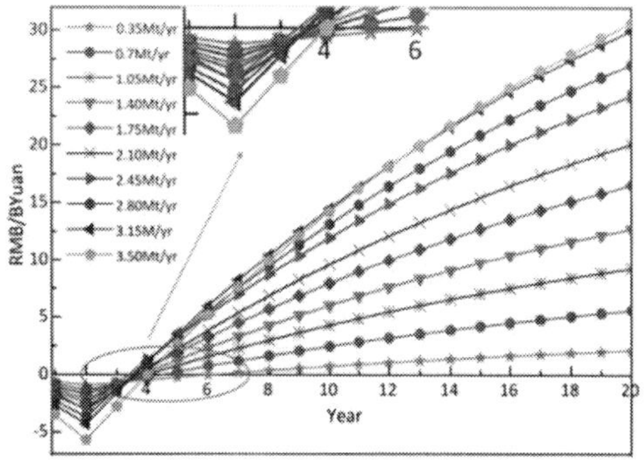

Fig. 5 Cumulative cash flow of the CO2–dry gas methanol production process

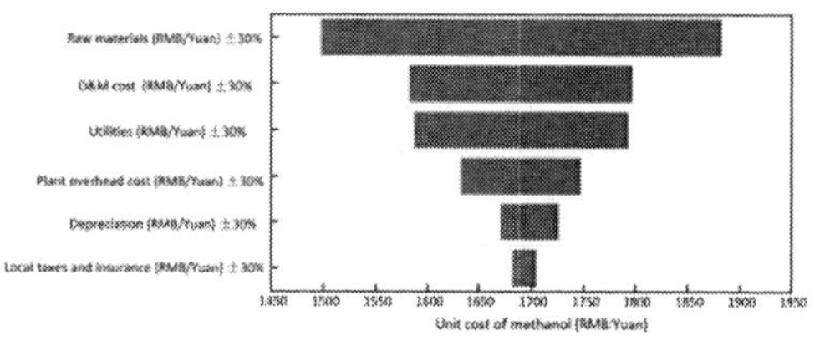

Fig.6 Sensitivity analysis of the most influential parameters on unit production cost of methanol

Finally, a sensitivity analysis was conducted to evaluate the effects of uncertainties of some important factors. It can be seen from Fig. 6 that the raw materials, O&M and utilities costs are the three main factors that affect the unit of methanol cost. Among these, raw materials comprise the principal sensitive factor, while the O&M and utilities costs are less so. When the price of raw materials fluctuates by ±30%, the variation range of unit product cost can reach 11.4%, followed by O&M and utilities costs with variations of 6.4% and 6.1%, respectively. Finally, reasonable changes in depreciation and local taxes and insurance have little impact on the methanol product cost.

Considering the above analysis results of TPI, TPC and NPV, when the production scale of methanol is within approximately 100–250 million, the economic performance of the CTM process is better.

4. Conclusion

In this study, we proposed a promising way to reduce greenhouse gas emissions in the petrochemical field by recovering the CO_2 in the flue gas of the FCC unit and the methane in dry gas to unlock a solution for green refinery development. Specifically, a CO_2–dry gas to methanol production process was modeled based on Aspen Plus. The techno-economic and environmental performance of the system was analyzed. The results show that the proposed process could not only demonstrate excellent performance in mitigating carbon emissions in refineries but also make it possible to make good use of the methane in dry gas compared to the traditional methods. With the aid of economic analysis, the economic performance of the CTM process is better considering the annual production capacity within 100–300 million and compared with the conventional technique of methanol production, IRR increases by 8.3%. In addition, sensitivity analysis shows that the raw materials, O&M and utilities costs are the three main factors that affect the unit of methanol cost.

Acknowledgments

This work was supported by National Key R&D Program of China No. 2022YFC3702900. Thanks are also due the China National Petroleum Corporation (CNPC) Research Institute of Safety & Environment Technology for access to the database.

References

1. Liu S Y, Jiang Y C, Yu S Y et al.Electric power supply structure transformation model of China for peaking carbon dioxide emissions and achieving carbon neutrality.Energy Reports 2022;8:541-548
2. Gabriela N S,Pedro R R, Alexandre S. Renewable hydrogen production to deal with wind power surpluses and mitigate carbon dioxide emissions from oil refineries.Applied Energy 2022;311:118631
3. Johansson D, Rootzén J, Berntsson T, et al. Assessment of strategies for CO2 abatement in the European petroleum refining industry. Energy 2012;42(1): 375-386.
4. Zhang MQ. Organic Rankine Cycle power generation using low temperature waste heat in refinery and its benefit analysis. Petroleum processing and petrochemicals 2020; 51(12): 92-95.
5. Pérez-Fortes, Mar, et al. Methanol synthesis using captured CO2 as raw material: Techno-economic and environmental assessment. Applied Energy 2016; 161:718-732.
6. Fernández-Dacosta C, Stojcheva V, Ramirez A. Closing carbon cycles: Evaluating the performance of multi-product CO2 utilisation and storage configurations in a refinery. Journal of CO2 Utilization 2018; 23: 128-142.

7. Rafiee, Ahmad, et al. Trends in CO2 conversion and utilization: A review from process systems perspective. Journal of environmental chemical engineering 2018; 65: 5771-5794
8. Zhou H, Yang S, Xiao H, et al. Modeling and techno-economic analysis of shale-to-liquid and coal-to-liquid fuels processes. Energy 2016; 109: 201-210.
9. Kontou V,Grimekis D,Braimakis K.Techno-economic assessment of dimethyl carbonate production based on carbon capture and utilization and power-to-fuel technology.Renewable and Sustainable Energy Reviews 2022;157:112006.
10. Zhang HF, Desideri U.Techno-economic optimization of power-to-methanol with co-electrolysis of CO2 and H2O in solid-oxide electrolyzers.Energy 2020;199:117498.
11. Kobe K. A. Plant design and economics for chemical engineers (Peters, Max S.). McGraw-Hill; 1991.
12. Li J Y, Ma X X, Liu H et al. Life cycle assessment and economic analysis of methanol production from coke oven gas compared with coal and natural gas routes.Journal of Cleaner Production 2018; 185:299-308.

 https://doi.org/10.1051/e3sconf/202337502013

Calculation model of non-nuclear steam rush in steam turbine of nuclear power unit Construction and application

Junfeng Liu, Zhen Kang, Miaomiao Wang, Shima Chao, Jingbin Xi, Zhijun Wu, Wei Wang and Zhenlu Zhang

[1] Xi'an Thermal Power Research Institute Co.,Ltd, Xi'an 710000, China
[2] Huaneng Nuclear Energy Technology Research Institute co., ltd, Shang Hai 200126, China
[3] Jiangsu Nuclear Power Co., Ltd, Lian Yungang 222042, China.

Abstract. According to the technical characteristics of steam turbines and their thermodynamic systems of various types of nuclear power units, a calculation software suitable for evaluating the feasibility of non-nuclear steam rushing of steam turbines of various types of nuclear power units is developed The software can accurately evaluate and check the non-nuclear steam rushing parameters by comparing the steam intake and heat consumption required by the steam turbine with the effective steam quantity and heat storage provided by the primary loop, and solve the quantitative evaluation problem of the feasibility of non-nuclear steam rushing for steam turbines of various types of nuclear power units.

Keywords: Nuclear power unit; Turbine rushing; Non-nuclear vapor; Thermodynamic model.

1. Introduction

At present, the installed capacity and technical level of nuclear power units in China have developed rapidly Because the third and fourth generation advanced nuclear power units under construction have adopted a large number of new technologies, new processes and new equipment, the construction period of nuclear island lags behind that of conventional island, and there is still enough time to carry out the first impulse test of turbo-generator unit before the first critical test of reactor Therefore, it is necessary to use non-nuclear steam for the first rotary test of turbo-generator set to find out and deal with the problems existing in turbo-generator set as soon as possible, so as to realize the seamless connection between nuclear island and conventional island commissioning test and shorten the construction period to the maximum extent[1-6].

Regarding the evaluation of the feasibility of non-nuclear steam run-in, literature [1] proves that it uses non-nuclear steam through the theoretical analysis and calculation of the comprehensive heat balance between the nuclear steam supply system and the steam turbine generator system before and after run-in of the AP1000 nuclear power steam turbine generator It is feasible to conduct the first run-through test of steam; literature [2] started from the first law of thermodynamics, established the calculation formula of non-nuclear run-in balance, and clarified the feasibility of non-nuclear steam run-in of turbogenerator units, and conducted the test in Unit 1 of Fuqing Nuclear Power Plant Successfully implemented; literature [3] based on the technical characteristics of the pebble bed modular high-temperature gas-cooled reactor, carried out comprehensive heat balance calculations for the two methods of heat storage in the primary circuit and steam supply from the auxiliary boiler, and obtained a feasible non-nuclear steam flushing scheme . Due to the large differences in the steam turbines of various types of nuclear power units, pressurized water reactors and fast reactor nuclear power units generally use half-speed steam turbine generators, and high-temperature gas-cooled reactors use full-speed steam turbine generators. The calculation methods in the existing literature are all limited to the selection of flushing steam parameters considering the characteristics of their own units, and there is no calculation model suitable for comprehensively evaluating the feasibility of non-nuclear steam flushing of various types of nuclear power units. This paper intends to establish a quantitative evaluation calculation model applicable to the feasibility of non-nuclear steam flushing of steam turbines of various types of nuclear power. units.

2. Thermodynamic Model

To realize non-nuclear steam flushing of nuclear power unit, two necessary conditions must be met: (1) there is enough steam intake to make the turbo-generator set rise from zero speed (turning speed has little influence) to rated speed; (2) There is enough heat storage of non-nuclear steam to meet the heat consumption of turbo-generator set during empty load test under impulse and rated speed This research results establish the thermal balance model diagram of turbo-generator set as shown in Figure 1 below.

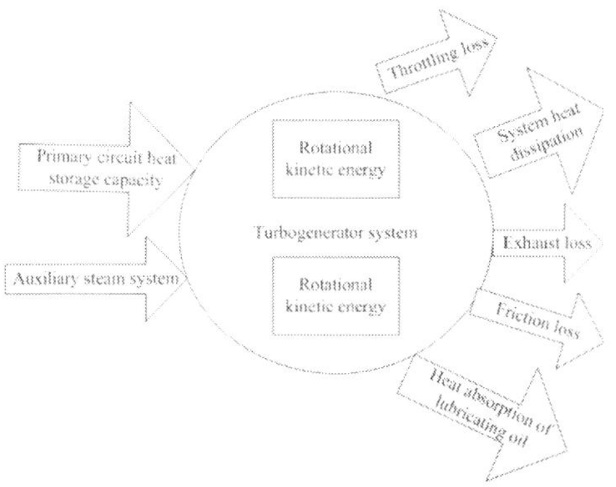

Fig. 1 Heat balance model diagram of turbo-generator set

2.1 Flushing Steam Intake

According to the pressure and temperature given by the steam turbine manufacturer, the inlet steam pressure and temperature under rated working conditions and the rated flow rate, the steam flow rate during the rushing period can be calculated by Friuger's formula:

$$\frac{q}{q_0} = \sqrt{\frac{T_0}{T}}\sqrt{\frac{P_1^2 - P_2^2}{P_{1,0}^2 - P_{2,0}^2}} \quad (1)$$

Where: q is the main steam flow rate, kg/s; T is the grade group temperature, ℃; P is the stage group pressure, MPa;; Subscript 0 indicates the known standard working condition (rated working condition) parameter, subscript 1 indicates the parameter before the stage group and subscript 2 indicates the parameter after the stage group.

During the process of steam turbine rising from zero speed to rated speed, the relationship between inlet steam flow and time is approximately linear When t ≥ t 0, the steam quantity required for steam turbine rushing is:

$$m=\frac{1}{2}\times q\times t_0+q\times(t-t_0)=q\times(t-\frac{t_0}{2}) \quad (2)$$

Where: m is the amount of steam, kg; t is the total time s of non-nuclear steam rush and test, s; t_0 is duration from zero speed to rated speed, s.

2.2 Heat Consumption of Steam Turbine

From the heat balance model diagram of turbo-generator set in Fig. 1, it can be seen that the heat consumed during the period when the turbo-generator set is driven up to the rated speed by the rotational kinetic energy provided by the primary loop heat storage during the rush period includes heat absorption system of steam turbine metal body, heat dissipation, throttling, exhaust steam and friction loss, heat absorption of lubricating oil and steam supply heat of shaft seal The heat of each part is calculated separately below .

According to the kinetic energy theorem, the rotational kinetic energy of the rotor that should be input after the rotor of the turbo-generator set reaches the rated speed is as follows:

$$Q_1 = \frac{1}{2}\times GD^2\times\omega^2 = \frac{1}{2}\times GD^2\times(\frac{2\pi n}{60})^2 \quad (3)$$

Where: GD^2 is the moment of inertia of the entire turbine generator shafting, kg·m²; ω is the angular velocity, rad/s; n is the rotational speed, r/min.

Metal components such as rotor stator blades and cylinder diaphragm will absorb part of steam heat during steam turbine rushing. The total heat absorption of metal is:

$$Q_2 = c_1 m_1(T_2 - T_1) \quad (4)$$

Where: c_1 is the specific heat capacity of metal, J/(kg °C); m_1 is metal mass, kg; T_1 and T_2 is the initial temperature and equilibrium temperature of metal,℃.

Most of the steam inlet modes of steam turbines in nuclear power plants adopt throttling regulation. The throttling loss mainly comes from the steam inlet loss of steam turbine regulating valves under different overlapping degrees. The calculation formula is as follows:

$$Q_3 = \sum_{i=1}^{n} q_i(H_{1,i} - H_{2,i})\ t \quad (5)$$

Where: q_i is the steam inlet flow rate of nozzles between stages of each regulating valve ,kg/h, H is the steam enthalpy value , kJ/kg; Subscript 1 indicates the parameters before the steam inlet nozzle and subscript 2 indicates the parameters after the steam inlet nozzle; i is the number of nozzles from 1 to n.

The heat dissipation of turbo-generator system mainly comes from the heat dissipation of thermal insulation materials laid on the cylinder surface to the environment, which can be obtained by Fourier heat conduction law:

$$Q_4 = A\frac{\lambda}{\delta}(T_3 - T_4)t \quad (6)$$

Where: λ is the thermal conductivity ,W/(m • K); A is heat exchange area, m2 ; δ is the thickness m of thermal insulation material, m ; T_3 and T_4 are the temperature of the inner and outer sides of the insulation material , °C.

Exhaust steam loss is caused by friction swirl in the process of exhaust steam flow from the last stage rotor blade and then into the condenser through the exhaust pipe The calculation formula of pressure loss of exhaust pipe is:

$$\Delta p = \frac{q_1^2 v_1}{A_1^2}\left[(\zeta_{ex} - 1)\frac{1}{f^2}\right] \quad (7)$$

According to the concept of thermal equivalent, the heat loss of exhaust steam can be obtained as follows:

$$Q_5 = \Delta pV = \frac{q_1^2 v_1}{A_1^2}\left[(\zeta_{ex} - 1)\frac{1}{f^2}\right]V \quad (8)$$

Where: q_1 is the final exhaust mass flow rate, kg/s; v_1 is the specific volume of steam after the last rotor blade, m³/kg; A_1 is the inlet area of exhaust pipe,m²; ζ_{ex} is the

https://doi.org/10.1051/e3sconf/202337502013

loss coefficient of exhaust pipe; f is the diffuser of exhaust pipe, that is the ratio of inlet and outlet areas of exhaust pipe; V is exhaust pipe volume, m^3

The friction loss in the process of steam turbine rushing comes from the bearing part and the generator part, and the friction loss is as follows:

$$Q_6 = \int_0^t T_r \frac{d\omega}{dt} t \tag{9}$$

The friction torque of each radial bearing is:

$$T_r = \frac{4\mu F_r D_r}{2\pi} \tag{10}$$

Radial loads are:

$$F_r = k_r m_2 g \tag{11}$$

Where: T_r is the friction resistance moment of radial bearing; μ is the friction factor of bearing contact surface; F_r is the radial load N acting on the radial bearing, N; D_r is the diameter m at the contact surface of radial bearing, m; Kr is the radial load coefficient of the rotor, and the value is 0.5; m_2 is the rotor mass,kg; g is the acceleration of gravity and the value is 9.8 m/s2

Heat absorption of steam turbine lubricating oil is:

$$Q_7 = c_2 m_2 (T_5 - T_6) \tag{12}$$

Where: c_2 is the specific heat capacity of lubricating oil, J/(kg ° C); m_2 is the mass of lubricating oil ,kg; T_5 and T_6 are the return temperature and inlet temperature of lubricating oil,°C

Turbine shaft seal steam supply from auxiliary steam system shaft seal steam supply high pressure cylinder and low pressure cylinder shaft seal and finally flow into steam turbine generator set exhaust steam The heat provided is:

$$Q_8 = q_2 t\ (H_1 - H_2) + q_3 t\ (H_3 - H_2) \tag{13}$$

Where: q_2 and q_3 are the inlet steam flow rates of high pressure cylinder and low pressure cylinder , kg/s; H_1, H_2 and H_3 are the inlet and exhaust enthalpy of high pressure cylinder and inlet enthalpy of low pressure cylinder, kJ/kg.

Based on the above, the total heat consumption of steam turbine can be obtained:

$$Q = Q_1 + Q_2 + Q_3 + Q_4 + Q_5 + Q_6 + Q_7 - Q_8 \tag{14}$$

2.3 Primary circuit heat storage

Only when the effective steam quantity and heat storage quantity provided by the primary loop are greater than the steam intake quantity and heat consumption required by the steam turbine for impulse rotation can the demand of impulse rotation and empty load test be met The main systems for heat exchange in the primary loop of PWR nuclear power unit during impulse rotation include main pump coolant regulator and its electric heating system steam generator pressure vessel core and chemical and volumetric control system The main systems for heat exchange in the primary circuit of HTGR unit include main helium blower helium coolant steam generator pressure vessel pebble bed core and internals The main systems for heat exchange in the primary circuit of fast neutron reactor unit include sodium pump in the primary circuit, sodium coolant in the secondary circuit, sodium pump, buffer tank, steam generator and superheater Through the system design data, the effective heat supply provided by the primary circuit of the above types of nuclear power units is checked respectively, and the effective steam supply in the impact test time is calculated $Q_{Thermal\ storage} > Q$, $m_{Steam} > m$,The primary heat storage generated by the primary power source (main pump or main fan) can meet the requirements of the non-nuclear impulse test of the steam turbine of the nuclear power unit When any condition exists, the auxiliary steam should provide additional heat to raise the feed water temperature of the secondary loop or pass the auxiliary steam into the steam generator for heat exchange to increase the heat storage capacity of the primary loop, so that the heat storage capacity of the primary loop can finally meet the requirements of the non-nuclear steam rush test.

3. Computing Software Development

3.1 Software Development Process

The software is edited by using the MODULE module of matlab programming language The data input module, steam turbine inlet flow calculation module, steam turbine heat consumption calculation module, primary loop heat storage calculation module, impulse parameter checking and selection module and post-processing module are constructed in C/C + + programming language environment The relationship between the modules and the software calculation flow are shown in Figures 2.

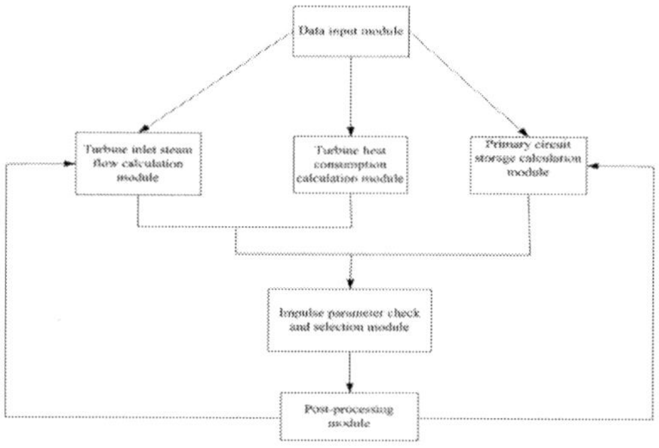

Fig. 2 Software Module Composition

3.2 Software function module

The data input module includes technical parameters of various types of nuclear power plant steam turbines such as pressurized water reactors, fast reactors, and high-temperature gas-cooled reactors. It has the functions of query, addition, modification, and deletion, which is convenient for the next step of fast and accurate thermodynamic calculations. It also has software expansion functions. The steam turbine inlet steam flow calculation module and the steam turbine heat consumption calculation module respectively calculate the rush rotation by the software calculation process shown in Fig. Inlet steam flow and heat consumption required for the test. The primary circuit thermal storage module includes three core sub-modules of pressurized water reactor, fast reactor and high-temperature gas-cooled reactor. heat storage. The check and selection module of the running parameters compares the intake steam flow and heat consumption obtained by the above modules with the heat storage of the primary circuit. If the heat storage of the primary circuit does not meet the running demand, start the auxiliary steam supply steam source check calculation program , and get the selection of parameters that meet the conditions of the reversal, and then re-check the calculation until the optimal thermodynamic parameters of the reversal are selected. The post-processing module includes a data summary, printing module and curve drawing module, which can realize data collection and online printing, and convert the result data into intuitive graphics to express through ORIGIN, TECPLOT and other drawing software.

4. Example validation

According to the technical standard given by the steam turbine manufacturer, the main equipment test parameters of primary and secondary circuits before and after non-nuclear steam flushing are shown in Table 1 below.

Table 1 Non-nuclear steam impulse test parameters of primary and secondary circuit main equipment

Status parameters	Initial value of test	End value of test
Pressurizer level/m	0.96	-3.94
Pressurizer pressure/MPa	15.5	15.3
Pressurizer power/kW	1440	0
Pressurizer temperature/°C	291.4	255
Main pump power/kW	5734	0
Steam generator water level/m	0.18	-0.76
Steam generator pressure /MPa	7.6	4.3
Steam generator temperature /°C	291.4	255

According to the above technical parameters of the primary circuit and steam turbine of the nuclear power unit, the calculation results of the thermodynamic model of the software are analyzed and compared with the actual impact results as shown in Table 2 below

Table 2 Calculation and test values of non-nuclear steam impact

parameters	Model calculation value	Test value
Test time/min	70	68.75
Primary circuit provides heat storage /kJ	2.09×10^8	2.2×10^8
Effective steam volume/t	85.21	82.15
Steam turbine impulse heat consumtion/kJ	1.9×10^8	1.81×10^8
Steam turbine impulse starting steam intake /t	80.26	80.18

From the calculation results in Table 2, it can be seen that the deviation rate between the model calculation value and the actual value is in the range of 1% ~ 5%. Considering that there is a certain heat loss in the primary loop system during the actual ramping process and there is a deviation between the actual heat consumption and the designed heat consumption value of the steam turbine, the deviation rate is within the allowable range of calculation error

5. Conclusion

In this paper, C/C++ language is used to compile the non-nuclear steam rushing calculation software of the steam turbine of the nuclear power unit, and the calculation software of the non-nuclear steam rushing and heat consumption of the steam turbine of the nuclear power unit is obtained. The software consists of a data input module, a steam turbine inlet steam flow calculation module, It consists of six basic modules: steam turbine heat consumption calculation module, primary circuit heat storage calculation module, flush parameter check and selection module, and post-processing module. Through engineering examples, it is verified that the computing software structure design is appropriate, the architecture is stable, the scalability and reusability are good, the reliability and accuracy meet the requirements, and it has certain engineering application value.

References

1. Chen Biyun, Li Xintong, Li Bin, etc. Feasibility Analysis and Calculation on Initial Startup of an AP1000 Nuclear Power Unit with Non-nuclear Steam [J]. Journal of Chinese Society of Power Enigineering 2014, 34 (11): 915-919, 920.
2. Xiao Bo, He Liu . Feasibility Analysis and Optimization for Turbo-Generator Rushing with Non-Nuclear Steam in Fuqing Nuclear Power Plant [J]. Nuclear Power Engineering 2018, 39 (3) : 122-127.

 https://doi.org/10.1051/e3sconf/202337502013

3. LIU Junfeng, CHEN Zhigang, GAO Jinghui, etc. Feasibility Analysis on Impulse Starting of HTR-PM Nuclear Power Unit with Non-Nuclear Steam [J]. Thermal Turbine 2018, 47 (4) : 293-295, 308.
4. Bao Xudong. Startup Practice of Non-nuclear Steam of 1 089 MW Nuclear Turbine Generating Unit [J]. Zhejiang Electric Power 2016, 35 (6) : 57-60.
5. TANG Mei-yu,HE Zhao-hui.Initial Startup of Qinshan II 650MW Nuclear Unit with Non-nuclear Steam[J]. Turbine Technology2005, 47 (2) : 142-144, 146.
6. Zhang Yabo, Zhang Donghui. The first rushing of CEFR turbine [J]. Chinese Journal of Nuclear Sciences and Engineering 2010, 30 (1) : 19-24.

An improved multi-objective gray target model for project resource allocation selection Research

Yaqiang Xu *

CITIC Treated Water Into River Engineering Investment Co., Ltd., Wuhan 430202, China

Abstract: In response to the traditional method of using a single subjective or objective assignment, AHP (hierarchical analysis) subjective weights and CRITIC (hierarchical relevance of importance criteria method) objective weights are coupled with the Lagrange multiplier method to obtain combined weights, and the multi-objective gray target decision model is used to establish event set, countermeasure set, and situation set to obtain the resource ranking that affects the project resource deployment problem, which provides a basis for the project participants to make decisions and plan for the project. Taking the municipal road project as an example, a judgment matrix is established. The results show that the model constructed in this paper is relatively consistent with the actual situation.

Keywords: resource allocation; AHP; CRITIC; Lagrange multiplier method; gray target decision

1. Introduction

At present, the international community has formed a consensus on green transformation and taken practical actions. Efficient resource utilization has been a major issue that the construction of global ecological civilization has been striving to crack. Although the earth is rich in resources and the total amount of resources is relatively large, due to the rough development since the industrial revolution, resources such as land, energy, and water have formed a hard constraint and tight balance for development, and resource consumption has lit up the red light. In recent years, experts and scholars have devoted a lot of attention to the issue of resource deployment, and the areas involved include aviation[1] , railroads[2] , rail transportation[3] , industrial workshops[4], and many other aspects.

Resource deployment is the rationalization of resource utilization, and the key to achieving optimization is to improve the influencing factors closely related to the desired project goals, such as project time management to make the project more efficient by optimizing the project schedule, cost management is to control the project budget by optimizing the consumption of resources required for the project, and quality management is to ensure the production quality by optimizing the production activities.[5] Quality management is the optimization of production activities to ensure the quality of production.

In engineering project management, the deployment of resources is often intricate and critical, and to analyze key resource elements such as labor, materials, and machinery, Cao et al.[6] To improve the assembly line management of assembly buildings, the defects of the existing production process are analyzed and an assembly line integrated supermarket is proposed to optimize the raw material scheduling method. Wu et al.[7] studied the tower crane planning problem of construction projects and proposed a Spatio-temporal planning model of tower cranes for construction projects based on simulated annealing, saving 31.10% of the total cost on average. For the engineering multi-resource balancing problem, Wang et al.[8] To solve the problem that the traditional multi-mode resource-constrained project scheduling problem (MRCPSP) model is difficult to characterize multiple relationships between process duration, cost, and resource requirements simultaneously, an MRCPSP model is proposed and the arithmetic case is solved using constrained planning (CP). Wang et al.[9] To efficiently and stably solve the multi-resource balancing problem in the process of construction project management, an optimization algorithm based on subset simulation is proposed, which has a large improvement in the stability of obtaining the optimal solution compared with the currently widely used genetic algorithm. To conduct a comprehensive resource scheduling analysis for project clusters, Feng et al.[10] borrowed the single project schedule optimization method under the condition of total resource constraint, constructed the project group schedule optimization model under the resource constraint, and determined the contract items that can be resource output and input to finally achieve the purpose of project group schedule optimization. jiang et al.[11] conducted risk evaluation and prediction of resource conflict problem of hydropower project group from the perspective of construction enterprises, estimated the

* Corresponding author: xuyq@citic.com

weights of each evaluation index by entropy weight method, established homogeneous inverse risk evaluation and prediction model for static risk evaluation by using the five-element coefficient of association, and predicted risk development trend dynamically by using the partial coefficient of association.

From the above-related studies, scholars have made more research results on project resource deployment, but few of them have explored the overall deployment of project resources from a global perspective, and only a single consideration of qualitative or quantitative calculation has been made in calculating the weights of each index, thus resulting in the omission of some information. In this paper, through expert scoring and qualitative analysis of hierarchical analysis (AHP), weights are combined and calculated by the Lagrange multiplier method coupled with CRITIC (Criteria Importance Though Intercrieria Correlation), and then a multi-objective weighted gray target decision model is established to rank the importance of resource deployment problems of construction projects. Finally, the applicability and robustness of the model are verified by example projects. The findings are helpful for all parties involved in the project to identify the important and difficult points of the resource allocation problem before the project implementation and to plan the resources well in advance to ensure the smooth implementation of the project.

2. Conventional gray target decision model construction

Grey target decision making (GSDM) is proposed by Prof. Ju-Long Deng[12] It is roughly classified into rectangular, spherical, and ellipsoidal, and is mainly based on the set of evaluated sequences to construct grey targets. It is widely used to solve decision-making problems under multi-objective and uncertain conditions. It determines the bull's eye by unifying the objective data of each indicator, judging the benefit, cost, intermediate, and interval indicators, determining the bull's eye, and calculating the bull's eye distance of each solution by using the Euclidean space with the bull's eye as the reference point and ranking the bull's eye distance to determine the optimal solution.The model calculation steps are specified as follows.

STEP1: Determine the scope of the study event set, response set, and situation set.The set of events in a study area is the set of events in the study area, and is denoted as, where $A=\{a_1, a_2, \cdots, a_n\}$ $a_i(i=1,2, \cdots, n)$ is an individual event in the study area. The corresponding set of all possible responses is called the response set, and is denoted as, where $B=\{b_1, b_2, \cdots, b_m\}$ $b_i(i=1,2, \cdots, m)$ is an individual response within the study area. The Cartesian product of the event set $A=\{a_1, a_2, \cdots, a_n\}$ and the response set $B=\{b_1, b_2, \cdots, b_m\}$ is the set of situations in the study area, denoted as s_{ij} . $u_{ij}=(u_{ij}^{(1)}, u_{ij}^{(2)}, \cdots, u_{ij}^{(s)})\in S^s$ is the effect vector of the situation . s_{ij} $u_{ij}^{(k)}(k=1,2, \cdots, s)$ is the effective value of the situation s_{ij} under the k objective.

STEP2: Qualitative ranking of the situations under each indicator.Qualitative analysis of the effect vector $u_{ij}=(u_{ij}^{(1)}, u_{ij}^{(2)}, \cdots, u_{ij}^{(s)})\in S^s$ of the situation s_{ij} for each effect value $u_{ij}^{(k)}$ under the target k $(k=1,2, \cdots, s)$. The values are assigned according to the expert scoring method, marking the optimal effect value as 1 and the next best value as 2. The effect values are ranked according to this increasing value law.

STEP3: Determine the optimal effect vector and build the gray target model according to the optimal effect obtained from the ranking. For the solution decision problem with s targets, the optimal effect vector is $r_0=(r_0^{(1)}, r_0^{(2)}, \cdots, r_0^{(s)})$ after ranking, where: $r_0^{(k)}$ is the optimal effect value of the effect vector u_{ij} under the target . k Where the spherical target space is formed based on each effect vector and the optimal effect vector r_0 is called s dimensional spherical gray target which R is

$$R=\left\{\left(r^{(1)}, r^{(2)}, \cdots, r^{(s)}\right)|(r^{(1)}-r_0^{(1)})^2+(r^{(2)}-r_0^{(2)})^2+\cdots+(r^{(s)}-r_0^{(s)})^2 \le R^2\right\} \quad (1)$$

Where: r_0 is the bull's-eye; $r^{(i)}$ is each effect vector.STEP4: Calculate the bull's-eye distance.For the program $r_1=(r_1^{(1)}, r_1^{(2)}, \cdots, r_1^{(s)})\in R$, the bull's-eye distance is

$$|r_1-r_0|=\sqrt{(r^{(1)}-r_0^{(1)})^2+(r^{(2)}-r_0^{(2)})^2+\cdots+(r^{(s)}-r_0^{(s)})^2} \quad (2)$$

STEP5: Sort the solutions according to the bullseye distance size.Calculate the bull's-eye distance for each situation effect vector based on the gray target and evaluate the decision options. For different situations s_{ij} , the different effect vectors corresponding to s_{hj} are $u_{ij}=(u_{ij}^{(1)}, u_{ij}^{(2)}, \cdots, u_{ij}^{(s)})$ and $u_{hj}=(u_{hj}^{(1)}, u_{hj}^{(2)}, \cdots, u_{hj}^{(s)})$, and if the bull's-eye distance is $|u_{ij}-r_0|\ge|u_{hj}-r_0|$, then the situation s_{hj} is said to be superior to s_{ij} , which is considered equivalent if the sign of the equation is valid.

3. AHP Method

Hierarchical analysis (AHP) is a systematic hierarchical modeling method that was first proposed by T.L. Saaty, a famous American operations researcher, in the 1970s.[13] It is a systematic hierarchical modeling method combining qualitative and quantitative approaches firstly proposed to establish a matrix by two-by-two comparison to determine the weights of each index, which has the advantages of low data volume requirement, the easy establishment of the evaluation system, low calculation workload and mature development of system tools. This method is often used in the study of multi-objective, multi-criteria, multi-factor, multi-level unstructured complex decision problems, especially strategic decision problems. The specific steps of the hierarchical analysis method (AHP) are as follows.STEP1: Build a recursive hierarchical model. Hierarchize the decision problem, construct a hierarchical structural model and decompose it into three levels: goal level, criterion level, and solution level.STEP2: Construct the judgment matrix. Adopting the approach of factor two-by-two comparison to build the matrix, b_{ij} indicates the influence of the impact factor indicators B_i and on the target B_j A, and similarly, $1/b_{ij}$ indicates the influence of B_j and B_j on the target A, to build the matrix $B=(B_{ij})_{n\times n}$, and the determination of B_{ij} uses 1~9 and it's reciprocal as the scale.

STEP3: Hierarchical single ranking and consistency test. The eigenvectors of the judgment matrix B corresponding to the largest eigenvalue $\lambda_{\max}$ are normalized to obtain the ranking weights of the relative importance of the corresponding factors at the same level as a factor at the previous level, which is the hierarchical single ranking. The consistency test of the judgment matrix needs to calculate the consistency index (CI, RI) and consistency ratio (CR), and when $CR<0.1$, the consistency of the judgment matrix is considered to be acceptable. CI The formulas for calculating RI and CR are

$$\left\{\begin{array}{l} CI=\dfrac{\lambda_{\max}-n}{n-1} \\ RI=\dfrac{\lambda'_{\max}-n}{n-1} \\ CR=\dfrac{CI}{RI} \end{array}\right\} \quad (3)$$

Equation (3): $\lambda'_{\max}$ is the average of the maximum eigenvalues of 500 sample matrices constructed using a random method and randomly drawing numbers from 1 to 9 and their reciprocals to construct positive reciprocal inverse matrices.

STEP4: Hierarchical total ranking and consistency check. The weights are synthesized based on the results of the hierarchical single ranking to obtain the ranking weights of each factor on the total goal w_{AHP}.

4. CRITIC Method

The conventional objective methods of weight calculation include the entropy method and the CRITIC method.[14] The CRITIC method is a more scientific and objective weighting method, compared with the entropy method, which only considers the information quantity of indicators and ignores the correlation between indicators, the CRITIC method determines the weights of indicators based on the comparison intensity and conflict between evaluation indicators. The larger the standard deviation of the value of the indicator, the greater the amount of information reflected, and the greater the weight; correlation refers to the correlation coefficient between two indicators, the smaller the correlation coefficient, the more similar the amount of information reflected, and the smaller the weight. The specific calculation steps are as follows.STEP1: Standardized matrix.

$$m_{ij}=\frac{x_{ij}-x_{i\text{-min}}}{x_{i\text{-max}}-x_{i\text{-min}}} \quad (4)$$

$$m_{ij}=\frac{x_{i\text{-max}}-x_{ij}}{x_{i\text{-max}}-x_{i\text{-min}}} \quad (5)$$

m_{ij} is the normalized value, x_{ij} is the original value of the variable j, $x_{i-\max}$ is the maximum value of the indicator i, and $x_{i-\min}$ is the minimum value of the indicator. i

STEP2: Calculate the standard deviation.

$$\sigma_j=\sqrt{\frac{\Sigma_{i=1}^{n}(m_{ij}-\bar{m}_j)^2}{n-1}} \quad (6)$$

$$\bar{m}_j=\frac{1}{n}\sum_{i=1}^{n}m_{ij} \quad (7)$$

$\bar{m}_j$ is the arithmetic mean of the variable. j STEP3: Construct the correlation coefficient matrix.

$$R_j=\sum_{i=1}^{n}(1-r_{ij}) \quad (8)$$

$$r_{ij}=\frac{\sum_{i,j=1}^{n}(m_i-\bar{m})(m_j-\bar{m})}{\sqrt{\sum_{i=1}^{n}(m_i-\bar{m})^2\sum_{j=1}^{n}(m_j-\bar{m})^2}} \quad (9)$$

STEP4: Calculate the amount of information.

$$c_j=\sigma_j R_j \quad (10)$$

STEP5: Normalize c_j to get the objective weight w_{CRI}.

5. Lagrange multiplier method

Since the hierarchical analysis method is a subjective assignment method, which is subject to the interference of human factors, and the CRITIC method is an objective assignment method, which tends to ignore the subjective intention of the evaluator and the actual weight relationship among the evaluation indexes. Therefore, the two methods are coupled to form a combined weighting method to determine the comprehensive weight of each evaluation index by complementing the subjective and objective. The specific calculation process of combined weighting adopts the principle of minimum information entropy and the Lagrange multiplier method.[15] The specific calculation steps are as follows.

 https://doi.org/10.1051/e3sconf/202337502014

$$\begin{cases} \min F=\sum_{j=1}^{m}\lambda_j(\ln\lambda_j-\ln w_{AHP})+\sum_{j=1}^{m}\lambda_j(\ln\lambda_j-\ln w_{CRI}) \\ s.t.\sum_{j=1}^{m}\lambda_j=1,\ \lambda_j>0,\ j=1,2,\ \cdots,\ m \end{cases} \quad (11)$$

The above optimization problem is solved according to the Lagrange multiplier method, calculated as

$$\lambda_j=\frac{\sqrt{w_{AHP}w_{CRI}}}{\sum_{j=1}^{m}\sqrt{w_{AHP}w_{CRI}}} \quad (12)$$

6. Empirical analysis

6.1 Project basic introduction and decision index setting

YB Avenue Municipal Supporting Project is located in the northern part of Jiangxia District, Wuhan City, Hubei Province, China, and is an urban secondary road in Guanggu South Health Industry Park, which mainly undertakes the service functions of freight transportation and residents' traffic travel inside and outside the region, with a planned red line width of 30 meters. The main construction content includes road engineering, bridge engineering, drainage engineering, traffic engineering, lighting engineering, electric power telecommunications, and landscape greening engineering. The project construction period is 36 months, and the total investment is 472,517,800 yuan, including the engineering cost of 264,239,700 yuan. The project adopts a PPP (Public Private Partnership) mode of construction, the implementation agency is Jiangxia District Water and Lake Bureau, the SPV company, and the construction unit is CITIC Clearwater into the River (Wuhan) Investment and Construction Co.The project is a key project of the investment infrastructure supporting the Guanggu South Health Industry Park, which is characterized by a tight construction cycle, high investment intensity, multiple cross-construction, and wide distribution of work surfaces. Therefore, the resource allocation of the project is a high concern for all parties involved in the construction. Based on the resources themselves, we believe that resource deployment will have an impact on five aspects: project financial evaluation X_1, project extension cost X_2, project cycle pressure X_3, task risk factor X_4, and the number of related tasks X_5, indicators are detailed in Table 1, where X_1、X_2、X_3 is a property of the project itself and X_4、X_5 is a property of the tasks themselves.

Table 1 Resource deployment impact indicators

Indicator Name	Properties of the indicator	Marking of indicators	Literature on Indicators
Project Financial Evaluation	Beneficial	X_1	[16]
Project Extension Costs	Cost Type	X_2	[17]
Project cycle pressure	Cost Type	X_3	[18]
Mission Risk Factor	Cost Type	X_4	[19]
Number of related tasks	Interval type	X_5	[20]

Combining the specific situation of this project, after reviewing the literature and consulting with project experts, it was concluded that the five resources Y_1, Y_2, Y_3, Y_4, and Y_5 best reflect the impact of their deployment issues on the project. Through a questionnaire survey of nine industry experts, index scoring boundaries were set, 0-1 scoring was completed, and the data were averaged to form the following scoring matrix Table 2.

Table 2 Resource deployment impact indicators

Resource Category	Project Financial Evaluation X_1	Project Extension Costs X_2	Project cycle pressure X_3	Mission Risk Factor X_4	Number of related tasks X_5
Resources Y_1	0.045	0.233	0.175	0.134	0.138
Resources Y_2	0.064	0.342	0.286	0.152	0.156
Resources Y_3	0.086	0.462	0.312	0.171	0.347
Resources Y_4	0.061	0.382	0.254	0.202	0.191
Resources Y_5	0.047	0.217	0.183	0.336	0.198
Boundaries	0.045	0.425	0.357	0.427	[0.120,0.450]

6.2 Calculation of project resource decision index weights

The AHP method and formula (3) are used to calculate weight index $X_1 \sim X_5$ w_{AHP}, the CRITIC method and formula (4)-(10) are used to calculate $X_1 \sim X_5$ weight index w_{CRI}, and finally, the Lagrange multiplier method and formula (11)-(12) are used to calculate the comprehensive weight λ_j, which forms the following judgment matrix Table 3.

Table 3 Calculation of the weights of independent and coupled methods

Indicator Name	AHP Method		CRITIC Method				AHP-CRITIC
	Eigenvector	w_{AHP}	Variability of indicators σ_j	Conflicting indicators R_j	Amount of information c_j	w_{CRI}	λ_j
Indicator	1.351	0.270	0.402	4.727	1.899	0.218	0.252

s X_1							
Indicators X_2	1.694	0.339	0.420	5.144	2.162	0.248	0.301
Indicators X_3	0.501	0.100	0.446	5.014	2.238	0.257	0.166
Indicators X_4	0.653	0.131	0.399	4.531	1.808	0.208	0.171
Indicators X_5	0.800	0.160	0.130	4.569	0.596	0.069	0.109

According to the calculation results, the weight vector obtained by the AHP method alone is w_{AHP} = (0.270,0.339,0.100,0.131,0.160), X_2 has the largest weight value, and X_3 has the smallest weight value; the weight vector obtained by the CRITIC method alone is w_{CRI} = (0.218,0.248,0.257,0.208,0.069), X_3 has the largest weight value and X_5 has the smallest weight value By coupling the two subjective and objective weight calculation methods through the Lagrange multiplier method, the combined weight of λ_j = (0.252,0.301,0.166,0.171,0.109), X_2 has the largest weight value and X_5 has the smallest weight value. Figure 3 shows that the objective and subjective biases can be better integrated through coupled calculations.

6.3 Project resource allocation decisions

According to Eqs. (1)-(2), the AHP-CRITIC combination weights are used to obtain the score of each resource based on the software, resource $Y_1 \sim Y_5$ = (0.627,0.486,0.457,0.433,0.577), formation Figure 2, i.e., $Y_1 > Y_5 > Y_2 > Y_3 > Y_4$, which results in the largest impact of the resource Y_1 's blending problem on the project, and the smallest impact of the resource Y_4 's blending problem on the project. By comparing AHP-Multi-objective Grey Target, CRITIC - Multi-objective Grey Target, and AHP-CRITIC-Multi-objective Grey Target, formation Table 4, the impact ranking of resource redeployment problem formed by multiple methods is similar, and AHP-CRITIC-Multi-objective Grey Target combines subjective judgment and objective evaluation to reflect the problem more comprehensively.

Combined with the actual progress of this project example, its resource deployment problems do conform to the conclusion of this paper in order, that the resources Y_1 have a great impact on the progress of the project, and problems such as deployment difficulties often occur during construction, and have a certain impact on the financial measurement of the project; the resources Y_5 have the second largest impact, with a large amount of resource deployment and a high-risk factor, and are prone to various chain reactions due to their deployment problems; the impact gap between the resources Y_2、 Y_3、 Y_4 is smaller, but also in The project progress affects the project progress, there are certain mutual constraints, and it is easy to form resource cross-deployment, which causes a large coordination workload.

Table 4 Decision ranking of independent and coupled methods

Resource Category	AHP		CRITIC		AHP-CRITIC	
	Score	Sort by	Score	Sort by	Score	Sort by
Resources Y_1	0.561	2	0.701	1	0.627	1
Resources Y_2	0.457	3	0.511	3	0.486	3
Resources Y_3	0.449	4	0.462	5	0.457	4
Resources Y_4	0.401	5	0.471	4	0.433	5
Resources Y_5	0.564	1	0.602	2	0.577	2

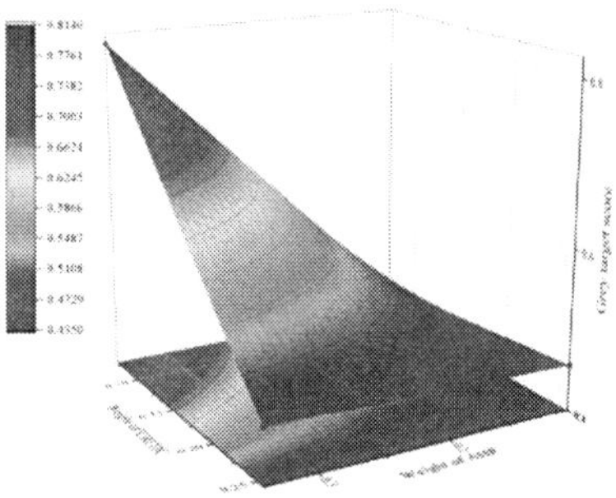

Figure 1 Weight calculation and gray target decision score

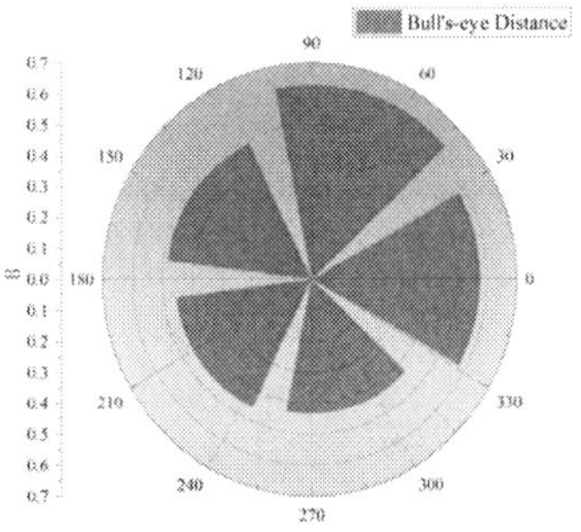

Figure 2 Bull's-eye distance

7. Conclusion and improvement

(1) Considering the qualitative and quantitative analysis of index weights and calculating the comprehensive weights by coupling the Lagrange multiplier method, a multi-objective weighted gray target decision model based on AHP-CRITIC is constructed to study the project resource deployment selection problem.

(2) The multi-objective gray target weighted decision model is validated through the example calculation of the YB Avenue project, the Euclidean spatial bull's-eye distance is calculated, the bull's-eye distances of the options are compared, the ranking of the options is derived, and the key factors affecting the resource allocation problem of the project are identified.
(3) Through an in-depth analysis of the YB Avenue project, the reasonableness and feasibility of the conclusions in the example calculations were verified in light of the actual situation of project implementation.
Meanwhile, this paper can further adapt the large-scale group decision algorithm to continuously improve the accuracy of the decision and form a decision database, which is easy to be used in various engineering projects.

References

1. Deng Q, Santos B F, Curran R. A practical dynamic programming based methodology for aircraft maintenance check scheduling optimization [J]. European journal of operational research, 2020, 281(2):256-273.
2. Heil J, Hoffmann K, Buscher U. Railway crew scheduling: models, methods, and applications [J]. European Journal of Operational Research, 2020,283(2):405-425.
3. Tang J, Xu L, Luo C, et al. Multi-disruption resilience assessment of rail transit systems with optimized commuter flows [J]. Reliability Engineering & System Safety, 2021,214:107715.
4. Lunardi W T, Birgin E G, Ronconi D P, et al. Metaheuristics for the online printing shop scheduling problem [J]. European Journal of Operational Research, 2021,293(2):419-441.
5. Guo K, Zhang L. Multi-objective optimization for improved project management: current status and future directions [J]. Automation in Construction, 2022,139:104256.
6. Cao J, Zhao P, Liu G. Optimizing the production process of modular construction using an assembly line-integrated supermarket [J]. Automation in Construction, 2022,142:104495.
7. Wu K, García De Soto B, Zhang F. Spatio-temporal planning for tower cranes in construction projects with simulated annealing [J]. Automation in construction, 2020,111:103060.
8. Wang, Heng-Wei, Lin, Jia-Rui, Zhang, Jian-Ping. A resource-constrained project scheduling problem considering productivity and process [J]. Journal of Tsinghua University (Natural Science Edition), 2020,60(03):271-277.
9. Wang J, Liu KX, Zhang XQ, et al. Multi-resource equilibrium optimization algorithm for construction projects based on subset simulation[J]. Journal of Hunan University (Natural Science Edition), 2021,48(07):168-176.
10. Feng Jingchun, Li Xueming, Feng Hui, et al. Research on project group duration optimization model under resource constraints[J]. Science and Technology Management Research, 2019,39(11):219-225.
11. Jiang X, Hu YW, Li W, et al. Evaluation and prediction of resource conflict risk in hydropower project clusters from the perspective of construction companies[J]. South-North Water Diversion and Water Resources Science and Technology, 2019,17(05):202-208.
12. Li Q, Liu S. F. A gray target decision model for two-stage equipment maintenance policy selection [J]. Control and Decision, 2022:1-9.
13. Bhadra D, Dhar N R, Abdus Salam M. Sensitivity analysis of the integrated AHP-TOPSIS and CRITIC-TOPSIS method for selection of the natural fiber [J]. Materials today : proceedings, 2022,56:2618-2629.
14. Peng J, Zhang J. Urban flooding risk assessment based on GIS- game theory combination weight: A case study of Zhengzhou City [J]. International journal of disaster risk reduction, 2022,77:103080.
15. Wu Jiahui, Wang Weiqing, Wang Haiyun, et al. Static voltage stability analysis of power grids based on comprehensive weight theory and Fuzzy-TOPSIS integrated evaluation[J]. Power System Protection and Control, 2018,46(23):77-85.
16. Feng Jingchun, Liu Qin, Chen Rundong, et al. Optimizing the allocation of A-supplied non-commodity resources for project clusters based on synthetic event technology[J]. Journal of Systems Management, 2022:1-12.
17. Zhang JG, Li J, Cheng NO. A two-stage monitoring and warning mechanism for critical chain multi-projects based on integrated priority[J]. Journal of Management Engineering, 2021,35(06):226-233.
18. Zhang JG, Li K. Research on critical chain schedule monitoring method based on project resource segmentation [J]. Management Review, 2021,33(06):224-231.
19. Li Qian, Wei Jie-Lin, Liu Feng-Tao. Construction and application of risk contagion network of engineering project cluster based on complex system[J]. Journal of Railway Science and Engineering, 2022:1-8.
20. Su C.X., Kil J.X., Zou X., et al. A new 0-1 hybrid linear optimization model for a class of local resource-constrained project scheduling problems [J]. Chinese Management Science, 2022:1-15.

Research progress of energy equalization topology of power lithium battery pack

Yinbao Miao [1, 2], Wenhua Zhang [1, 2, *], Weihao Liu[1,2], Dongqi Kang[1,2], Shuai Wang[1,2], Zhe Chen[1,2], Jia Liu[1], Biaoxian Chen[1], Leijing Zhu[1]

[1] Nanchang Institute of Technology, Nanchang 330000, China
[2] Key Laboratory of Precision Drive and Control of Jiangxi Province, Nanchang 330096, China

Abstract. At present, with the vigorous development of electric vehicle industry, power battery, as an important power source of electric vehicle, has attracted more and more attention of researchers. Among many kinds of power batteries, lithium-ion battery has been widely studied by many experts and scholars because of its clean, pollution-free and other advantages. As an important part of battery management, battery energy equalization technology makes the energy in the battery pack flow between single batteries by building an equalization circuit, which provides a strong guarantee for the efficient output and stable operation of lithium-ion batteries. This paper summarizes the current equilibrium topology of lithium-ion battery pack by summarizing and combing various types of existing equilibrium structures. According to the development process of equilibrium technology, this paper introduces the working principles, advantages and disadvantages of passive equalization, active equalization and active passive hybrid equalization structure, and summarizes the design idea of current equilibrium circuit by systematically summarizing the characteristics of the development of equilibrium circuit structure in recent years, According to the latest research progress, grasp the future development direction. Finally, according to the current balanced technology development, this paper puts forward personal views and prospects for the future balanced technology development.

1. Introduction

In recent years, with the society vigorously promoting green travel, electric vehicles have gradually entered the public's vision. Compared with traditional fuel vehicles, electric vehicles are sought after by the public for their clean, pollution-free, environment-friendly and other characteristics. Nowadays, the electric vehicle industry is developing rapidly and has the potential to replace fuel vehicles as the mainstream of future vehicle development. As an important power source of electric vehicles, lithium-ion batteries have been widely studied by many scholars because of their advantages such as high energy density, long service life and no memory effect[1] . Usually, power battery packs are connected by multiple battery cells in series and parallel, so as to achieve the performance indicators such as output voltage and battery capacity required by electric vehicles[2] . However, considering the differences in the real environment and production process, there are differences in the performance of individual cells such as internal resistance and self discharge rate, and the actual performance of individual cells can not be consistent. With the accumulation of the operation time of the battery pack, the performance differences of battery cells will be gradually enlarged, leading to the imbalance of energy among cells [3] . Based on the "barrel effect", the capacity of the battery pack will be restricted by the low-power cells, which will easily lead to excessive charging and discharging of the battery pack, cause irreversible damage to the battery cells, and affect the overall performance and service life of the battery pack[4] . The equalization technology can solve the problem of inconsistent energy of individual batteries well. Whether it is to consume the power of cells through energy consuming elements or to transfer energy between cells through energy storage elements, the energy balance of battery packs can be achieved and the actual effect is stable and reliable. The in-depth research of battery equalization technology is of great value to the development of electric vehicles in the future.

2. Introduction to equalization method

The purpose of the energy balance technology[5] is to reduce the energy difference between the individual cells in the battery pack, maintain the battery power at a relatively balanced level, weaken the damage caused by

* Corresponding author: 2015994552@nit.edu.cn

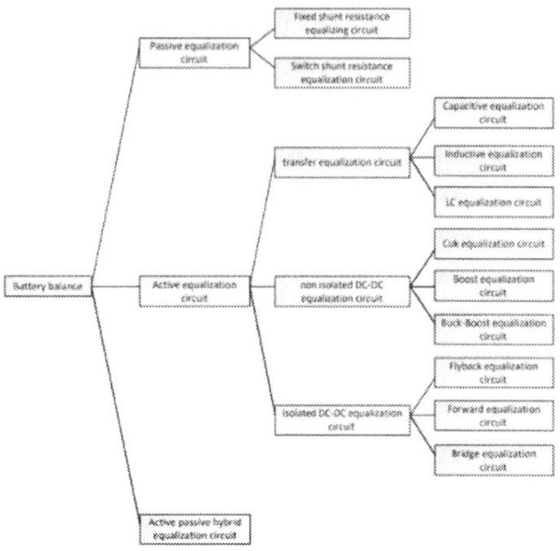

Figure 1. Classification of basic battery balancing topology

energy imbalance to the output performance of the battery pack, and extend the service life. According to the different treatment methods of energy in the equalization process, energy equalization can be divided into passive equalization and active equalization[6] . With the continuous development of equalization technology, the category of active passive hybrid equalization has emerged in recent years. The active passive hybrid equalization circuit combines the advantages of active equalization and passive equalization, and can further improve the energy equalization effect of the battery pack. The specific classification of these three types of equilibrium methods is shown in Fig.1.

3. Passive equalization circuit

The passive equalization method uses the method of parallel resistance branches at both ends of each cell to consume the energy of the cell, and releases the electric energy to the outside in the form of heat, so as to achieve the goal of consistent energy of each cell. Passive equalization circuits are simple in structure, stable, reliable and expandable, but resistance heating will shorten the service life of batteries. Nowadays, they are mostly used in low-power equalization circuits[7,8].

The reason why passive equalization can achieve uniform battery power is that the current flowing through the parallel resistance branch is in direct proportion to the terminal voltage of each cell, and the terminal voltage of the cells with different energy reserves is different, which makes the energy loss of the cells with high power consumption fast and the energy loss of the cells with low power consumption slow, so as to finally achieve uniform power within the battery pack[9] .

At present, there are two main passive equalization methods, namely, fixed shunt resistance equalization and switched shunt resistance equalization[10] . The fixed shunt resistance structure realizes power balance by directly connecting resistors in parallel at both ends of each cell. The advantages of this structure are simple structure, low cost, good equalization effect, stability and reliability; On the other hand, this structure will cause serious and uncontrollable energy loss. Excess energy will be lost in the form of heat, resulting in energy waste. In addition, the heat generated by the resistance will accelerate the aging of the battery pack, affect the output performance, and have higher requirements for the thermal management system. Compared with the fixed shunt resistor structure, the switch shunt resistor structure aims to achieve balanced power loss control to a certain extent. By adding a switch to the shunt resistor branch and controlling the on-off of the switch to achieve energy balance, the balanced structure is more flexible and the energy loss is reduced. At the same time, taking into account the advantages of simple structure, economy and reliability, it is still widely used in practical industrial applications. However, the problem of energy loss in the form of thermal energy still exists, which affects the service life of the battery pack[10] , and the equalization speed is slow. This equalization method is not suitable for the balanced use of high-power battery packs[11] .

4. Active equalization circuit

The active equalization method is to transfer energy from high-power cells to low-power cells by using energy storage elements such as capacitors and inductors. Compared with passive equalization, active equalization greatly reduces the power loss, greatly improves the energy utilization rate of the battery pack and has low heat output, reducing the pressure of the thermal management system[12] . Active equalization technology is a hot research direction of equalization technology at present.

Active equalization structure can be divided into transfer equalization structure[13] and DC-DC equalization structure according to whether DC-DC conversion circuit is used in equalization circuit. On this basis, DC-DC equalization structure can be further divided into non isolated DC-DC equalization structure[14] and isolated DC-DC equalization structure[15] according to whether electrical isolation is performed in equalization circuit.

4.1 transfer equalization structure

Transfer equalization circuit refers to the energy equalization circuit that uses energy storage elements as the medium of intermediate energy transfer. According to different energy storage components, it can be divided into capacitive equalization circuit, inductive equalization circuit and LC equalization circuit[16] . Capacitive equalization circuit[17,18] realizes the transfer of electric energy through capacitors. Since capacitive equalization realizes the transmission of electric energy based on the voltage difference between cells, the size of the voltage difference between cells will greatly affect the equalization rate of cells. As the equalization time continues, the electric voltage difference will gradually decrease, the energy transfer efficiency will decrease, and the equalization effect will become worse. Inductive equalization[19,20] uses the inductor as an energy transfer device, making the control of the equalization circuit more convenient and accurate, and reducing the loss in the energy transfer process. LC circuit equalization method[21,22] solves the problem of small voltage difference in capacitor equalization circuit by introducing LC oscillation circuit, but the switching frequency is high and the circuit control is complex.

In order to solve the problem that the voltage of each single battery in series is inconsistent in practical application, literature[23] proposed a voltage equalization circuit based on LC circuit and bidirectional switching unit. The circuit is composed of a series cell unit, a bidirectional switch unit and an LC circuit. The cell with the highest voltage is transferred to the LC circuit, and then transferred from the LC circuit to the cell with the lowest voltage. The equalization path is independent of the cell location, which effectively shortens the energy transmission path and improves the energy utilization rate. The characteristics of easy expansion and small size of the circuit further enhance its practicability.

Literature[24] proposed a battery strings equalizer circuit based on switching capacitance. In the circuit design, considering the small size and easy control of the capacitor, the capacitor is used as the carrier of energy transfer. Through multi-channel energy transmission, the energy in the battery pack flows from the highest voltage battery cell to the lowest voltage battery cell, which not only improves the equalization speed, but also shortens the energy transmission path to achieve simultaneous equalization of multiple battery cells. However, the disadvantage is that with the increase of the number of cells, considerable switching loss will be brought, and it is impossible to ensure that the equalization effect is always at a high level.

Literature[25] proposed series-parallel battery pack balancing circuit with single inductor. The series battery pack is connected in parallel with a single inductance element, and the energy flow is controlled by a control element to achieve energy storage and transfer in the inductor, so as to achieve simultaneous equalization of each battery cell. This equalization circuit structure has the characteristics of simple structure, small size, easy expansion, etc. The equalization speed is improved compared with the conventional single inductor structure, but it is still restricted by the single battery series parallel structure.

4.2 Non isolated DC-DC equalization structure

The equalization circuit of non isolated DC-DC converter relies on DC-DC circuit to realize energy equalization. According to the structure classification, it can be divided into: Cuk structure, Boost structure, Buck Boost structure, etc. The balancing method is more flexible, easy to topology, high integration, and excellent balancing performance, which is an important research direction[26] .

In order to solve the problem of unbalanced power distribution among cells in the battery pack, literature[27] proposed a charge discharge equalization circuit for parallel battery packs. During the charging process, the cell with the voltage up to the cut-off voltage is removed from the charging circuit; During discharge, Cuk circuit is used to compensate the unit with the lowest power, so as to balance the power distribution in the battery pack. This scheme realizes the rapid equalization of the battery cell with remarkable equalization effect, effectively maintains the service performance of the battery pack, maintains the charge and discharge capacity of the battery pack, and extends the service life of the battery pack.

Reference[28] proposed a reconfigurable Buck Boost charging equalization structure to improve the battery equalization speed and reduce the energy loss during the equalization process. This structure combines the characteristics of the Buck Boost circuit, which can realize the power balance between non adjacent battery cells, and the reconfigurable circuit, which has short equalization time and low energy loss, making the performance of this reconfigurable Buck Boost circuit significantly reduce the equalization time and improve the energy transfer efficiency compared with the conventional Buck Boost circuit.

To solve the problem of unbalanced energy distribution of large battery packs, literature[29] proposed an energy balanced structure with multiple structures and multiple states for a series battery system. The cell in the battery pack is divided into several units to realize the internal independent equalization of each unit. The equalization process is divided into multiple states. The Buck is constructed by controlling the on-off of the switch, and the Boost circuit completes the energy transfer between each energy unit and the external variable voltage source. In the static state, each unit is balanced simultaneously, which improves the equalization speed of the battery pack. This multiple multi state equalization structure solves the problems of slow equalization speed and low equalization efficiency of large battery packs, and greatly improves the equalization speed and service life.

4.3 Isolated DC-DC equalization structure

The isolated DC-DC equalization structure is mainly based on the isolated DC-DC conversion circuit structure. The energy transfer is realized through the transformer structure, and the input and output terminals are electrically isolated. Common isolated DC-DC equalization circuits are mainly divided into flyback circuit, forward circuit and bridge circuit[27] . Isolated DC-DC equalization circuit is widely concerned because of its fast equalization speed, safety and reliability, and high stability[30] . At present, flyback structure is a kind of equalization structure that is widely used in various isolated DC-DC equalization structures.

Reference[31] proposed a lithium battery pack equalization circuit based on flyback converter. This circuit design connects the corresponding battery cell with other battery cells through the main and secondary coils of the transformer, and the energy flows through the primary and secondary side coils to achieve the purpose of equalization. The flyback transformer itself has the characteristics of electrical isolation. Each equalization module is relatively independent, and each module can be equalized at the same time. Compared with the traditional equalization circuit, the equalization speed and efficiency of this circuit are effectively improved.

In order to solve the problem of single equalization mode, literature[32] proposed a multi state equalization structure based on three winding transformer, which designed multiple equalization circuits integrated in the same circuit for the possible energy imbalance of the battery pack. By controlling the on-off of the switch, the battery pack equalization mode is adjusted in real time, and the battery pack is always in the best energy balance state, so as to achieve the optimization of the output performance of the battery pack. This design scheme has short equalization path and can realize bidirectional transfer of energy between objects. The equalization effect and fast equalization speed are improved, further improving the output performance and service life of the battery pack.

5. Active passive hybrid equalization structure

Active passive hybrid equalization structure is a new type of energy equalization structure which combines the advantages of active equalization and passive equalization. The active passive hybrid equalization structure combines the advantages of passive equalization, such as simple structure, strong scalability, stability and reliability, and the advantages of active equalization, such as high equalization rate and low energy loss. In the circuit design process, balance the equalization rate and energy utilization rate to further optimize the energy equalization effect of the circuit and improve the equalization efficiency of the actual circuit.

In order to solve the problem of slow equalization speed and low equalization efficiency of the battery pack, literature[33] proposed an active passive hybrid equalization circuit. The circuit is divided into resistance module and LC oscillation circuit module. The LC oscillation circuit is used to ensure the equalization speed of the battery, and the resistance passive equalization circuit is used to delay the capacity growth rate at the initial and final stages of charging and discharging, so as to quickly estimate the battery charging and discharging state.

Reference[34] proposed a hybrid equalization circuit, which combines the switch shunt resistor structure and the isolated DC-DC equalization structure to passively equalize the extremely high power cells, quickly release the excessive power of the cells. When the power dispersion of each cell is moderate, the active equalization structure transfers the power from the whole to the low power cells to achieve energy equalization. These two circuit designs give consideration to the equalization efficiency and speed, reduce the requirements of the circuit on the equalization strategy, and maintain the battery performance and service life.

Based on the above introduction to the equalization structure of the battery pack, the advantages and disadvantages of the basic equalization structure are analyzed, as shown in Table 1.

Table 1. Comparison of advantages and disadvantages of basic equalization structure

energy equalization circuit	advantage	disadvantage
Fixed shunt resistor equalization circuit	Simple structure, low cost, good scalability, stability and reliability	Large energy loss, low balancing efficiency and high heat generation are unfavorable to the battery pack
switched shunt resistance equalization circuit	Simple structure, low cost, reduce energy loss	The energy utilization rate is low, and the heat emission is unfavorable to the battery pack
transfer equalization circuit	Simple structure, easy to control, high energy utilization and good scalability	Slow equalizing speed, many control switches
non isolated DC-DC equalization circuit	High equalization efficiency and short energy transmission path	Complex control and large number of control switches
isolated DC-DC equalization circuit	High equalizing speed and efficiency, and bidirectional energy flow	High cost, large circuit volume, not conducive to expansion
Active passive hybrid equalization circuit	High equalizing speed and efficiency	Increased energy loss and calorific value of the circuit

6. Conclusion

At present, the electric vehicle industry is developing rapidly. As the main power source of electric vehicles, lithium-ion power batteries are the focus of current research in terms of capacity expansion and performance optimization. While continuously pursuing the high performance of the battery pack, it is still necessary to comprehensively consider the stability and safety of the battery pack. The battery energy equalization technology can improve the safety and stability of the battery pack to a certain extent, improve the energy utilization rate, reduce the generation of heat, improve the overall battery life, slow down the aging speed, and improve the stability and safety of the battery pack.

Energy equalization technology, as the main means to solve the uneven power distribution of power battery pack, is of great value to the development of China's electric vehicle industry. This paper combs the development process of power battery energy balance technology at the current stage from two perspectives. First of all, horizontally sort out the existing types of balanced structure and the current latest achievements; Secondly, it introduces the research and design direction of each equilibrium type vertically, summarizes the current research progress, and grasps the future development direction.

With the constant introduction of various equilibrium structures, from passive equilibrium, active equilibrium to active passive hybrid equilibrium, the structure tends to be relatively complex, while the equilibrium effect, rate and energy utilization rate are constantly improving. Compared with the power grid energy storage equalization technology, the requirements of the on-board power battery for the equalization circuit tend to be small in size, high in energy utilization, fast in equalization speed, etc., in order to achieve a longer mileage. The development of equilibrium technology has changed from passive equilibrium, active equilibrium to active passive hybrid equilibrium, from the initial requirement of stability and reliability to the design concept of low loss and high energy utilization, and then to the current equilibrium rate, energy utilization, and refined and simplified structure, which reflects that under the guidance of the market, the transformation of the design concept of equilibrium technology will be the development direction in the future.

In the aspect of balanced circuit function design, in order to make the application scope more extensive, the circuit design tends to multi structure mixed, multi-mode, multi state design ideas. Compared with the single equalization method, the design idea of multi-mode and multi-state is more conducive to broaden the practical application range of the equalization circuit, but the circuit structure is more complex, which is a stage in the development process. However, how to maintain a good energy balance effect while meeting the overall lightweight of the battery pack needs to be considered in the next step.

To sum up, in the current environment of rapid development of electric vehicles, battery energy equalization technology has developed rapidly. At present, the focus of energy equalization technology is how to take into account the needs of battery packs in terms of energy utilization, equalization speed, and structure simplification, so as to further optimize and innovate the energy equalization technology.

Acknowledgements

This work was supported by the Jiangxi Provincial Education Department Project(No. GJJ211915; No. GJJ211913), the National Natural Science Foundation of China(No. 22269014;62041406) and Jiangxi science and technology support plan (No.20171BAB206040).

References

1. LU L G, LI J Q, HUA J F, OUYANG M G. A review on the key issues of the lithium-ion battery management[J]. Science & Technology Review, 2016,34(06):39-51.
2. WANG Z M, ZHOU Y S, ZOU S Z, WANG P F, LIU H P. Research on balance system of Li-ion battery module based on efficiency optimization[J]. Chinese Journal of Power Sources, 2019,43(08):1292-1295.
3. ZHAI E N, HUA J, CUI X Y, MA H F, WANG J S. Design and implementation of power battery active balance system[J]. Chinese Journal of Power Sources, 2020,44(02):249-252.
4. CARTER J,FAN Z,CAO J. Cell equalisation circuits: A review[J]. Journal of Power Sources, 2020,448(C):227489-227489.
5. Javier Gallardo-Lozano,Enrique Romero-Cadaval,M. Isabel Milanes-Montero,Miguel A. Guerrero-Martinez. Battery equalization active methods[J]. Journal of Power Sources, 2014,246:934-949.
6. DAOWD M, OMAR N, VAN DEN BOSSCHE P, et al. A review of passive and active battery balancing based on MATLAB/Simulink[J]. International Review of ElectricalEngineering, 2011,6(7):2974-2989.
7. KUTKT N H, DIVAN D M. Dynamic equalization techniques for series battery stacks[C]// International Telecommunications Energy Conference. Boston,MA,USA: IEEE Xplore, 1996.514-521.
8. LINDEMARK B. Individual cell voltage equalizers (ICE) for reliable battery performance[C]// [Proceedings] Thirteenth International Telecommunications Energy Conference - INTELEC 91.Kyoto,Japan: IEEE, 2002.196-201.
9. TAN Z F, YANG R, HE D W, SUN R L. Review of battery balance technology for electric vechicles[J]. Chinese Journal of Power Sources,2020,44(04):624-627.
10. WU C L, LIU Z X, CHENG Y Q, HU W B. Review of power lithium battery pack equalization topology[J]. Electronic Measurement Technology, 2021,44(15):7-14.
11. CAO J W, XIA B Z, ZhOU J. An active equalization method for lithium-ion batteries based on flyback transformer and variable step size generalized predictive control[J]. Energies, 2021,14(1):207-207.
12. Lee Kyung Min,Lee Sang Won,Choi Yoon Geol,Kang Bongkoo. Active balancing of li-Ion battery cells using transformer as energy carrier[J]. IEEE Transactions on Industrial Electronics, 2017,64(2):1251-1257.
13. PENG F , MEMBER S , IEEE, et al. A hierarchical ZVS battery equalizer based on bipolar CCM buck-boost units[C]// 2018 IEEE Energy Conversion Congress and Exposition (ECCE). Portland,OR,USA: IEEE, 2018.2120-2126.
14. EISEROUGI A A, MASSOUD A M, AHMED S. A unipolar /bipolar high-voltage pulse generator based on positive and negative buck - boost DC-DC converters operating in discontinuous conduction

mode[J]. IEEE Transactions on Industrial Electronics, 2017,64 (7) : 5368 - 5379.

15. LI S, MI C C, ZHANG M. A high-efficiency active battery-balancing circuit using multiwinding transformer [J]. IEEE Transactions on Industry Applications,2013,49(1) : 198 - 207.
16. LIU H R, DU C F, LI B, CHEN S L, GUO Y X. Research on high speed energy equalizer for battery pack based on mixed chopped circuit[J], Transactions of China Electrotechnical Society,2018,33(S2):472-478.
17. Shang Yunlong,Xia Bing,Lu Fei,Zhang Chenghui,Cui Naxin,Mi Chunting Chris. A Switched-Coupling-Capacitor Equalizer for Series-Connected Battery Strings[J]. IEEE Transactions on Power Electronics,2017,32(10):7694-7706.
18. Kim M Y, Kim C H, Kim J H, et al. A chain structure of switched capacitor for improved cell balancing speed of lithium-ion batteries[J]. IEEE Transactions on Industrial Electronics, 2014, 61(8): 3989-3999.
19. Li W B, KANG L Y, GUO X W, et al. Multi-objective predictive balancing control of battery packs based on predictive current[J]. Energies, 2016, 9(4): 1-12.
20. WEI Y W , LIU G T , XIONG S N , et al. Research on power equalization using a low-loss DC-DC chopper for lithium-ion batteries in electric vehicle[J]. Iop Conference, 2017, 52(1):012038-012038.
21. CHEN W, YE S S, LIU D. Cascaded battery cell equalizer for energy storage system and its selfbalancing control strategy[J]. Electrical Engineering, 2018, 19(1): 45-49.
22. LV H, LIU C Z, YIN D, et al. The design and optimize of equalization schemes for underwater power LiFePO4 battery stack[J]. Transactions of China Electrotechnical Society, 2016, 31(19): 232-238.
23. XU S G, LI K L, ZHOU G H, ZHANG X B, GAO K. Bidirectional voltage equalization circuit based on LC unit[J]. Electric Machines and Control, 2021,25(10):39-48.
24. LIANG H Z, LI N, HAN X W, HUANG Z. A battery strings equalizer circuit based on switching capacitance[J]. Power Electronics, 2021,55(08):28-35.
25. GUO X W, LIU Z, KANG L Y, HU Z G, GENG J H. Series-parallel battery pack balancing method with single inductor[J]. Electric Machines and Control, 2021,25(12):87-95.
26. CAI M Y, ZHANG E, LIN J, WANG K L, JIANG K, ZHOU M. Review on balancing topology of lithium-ion battery pack[J]. Proceedings of the CSEE, 2021,41(15):5294-5311.
27. YIN R, ZHANG Z H, LIU H R, GUO Y X, ZHANG K X. Research on charge- discharge equalizer and equalization strategy of parallel battery[J]. Chinese Journal of Power Sources, 2021,45(10):1298-1302.
28. YU Z A, XIONG Y Y. Research on charge equalization of lithium battery pack based on reconfigurable circuit[J]. Automobile Technology, 2022(06):27-32.
29. LIU H R, ZHANG K X, GUO Y X, YIN R, QIAN J. A multi-state energy equalizer and control methods for series battery system[J]. Transactions of China Electrotechnical Society, 2020,35(S2):652-660.
30. SHI F D, SONG D W, XIONG H, LIU J Z. Research progress in Li-ion battery equalization control methods[J]. Battery Bimonthly, 2019,49(03):251-254.
31. YAN C, MO Y M, XIONG W. Active equalization circuit of lithium battery pack based on flyback converter[J]. Science Technology and Engineering, 2021,21(04):1415-1421.
32. ZHENG Z, WANG X S, LI B, HUANG T, LI P K. Adaptive interleaved control equalization scheme for Li-ion battery packs based on Three-winding transformers[J/OL]. Energy Storage Science and Technology, 1-11[2022-04-20].DOI:10.19799/j.cnki.2095-4239.2022.0004.
33. WANG L J, SHAN E Z. Active pssive equalization strategy for lithium battery pack based on dynamic dual threshold[J]. Electric Machines and Control,2022,26(01):126-136.
34. MA C Y, WANG Q L, ZHANG D, ZHANG C J. Energy balance control of series connected lithium battery based on SOC[J/OL]. Journal of Power Supply, 1-8[2022-04-20].

    http://kns.cnki.net/kcms/detail/12.1420.TM.20220215.1535.010.html

Performance Test of a 660MW Thermal Power Unit Steam Turbine Based on the Background of "Double Carbon"

Ziwei Zhong[1,*], Lingkai Zhu[1], Wei Zheng[1], Zhiqiang Gong [1], Panfeng Shang [1], Junshan Guo[1], Jun Liu [1], Yanpeng Zhang[1]

State Grid Shandong Electric Power Research Institute, Jinan, China

Abstract. Under the background of "carbon peaking and carbon neutralization", China proposes a new power system with new energy as the main body. Due to the strong randomness, volatility and anti peak shaving characteristics of new energy power generation, in order to promote the consumption of new energy and ensure the power supply, coal power generation units will assume the dual tasks of deep peak shaving and power supply guarantee in the future, gradually changing from electric power supply to regulated power supply. In order to accurately grasp the thermal performance of the steam turbine after the flow passage transformation of the unit, the thermal performance assessment test of the steam turbine of the unit was carried out. Through thermal test, determine the heat rate and efficiency of high, medium and low pressure cylinders of the unit, and provide reference for future economic operation of the unit.

key word: Deep peak shaving; Thermal performance assessment test.

1. Introduction

The steam turbine of a 660MW power plant is N660-24.2/566/566 supercritical, one intermediate reheat, single shaft, three cylinder and four exhaust condensing steam turbine produced by Shanghai Turbine Works. In order to further increase capacity and reduce consumption, the power plant entrusted Shanghai Electric Group Co., Ltd. to comprehensively upgrade the flow passage of high, medium and low pressure cylinders of the steam turbine. In order to accurately grasp the thermal performance of the steam turbine after the flow passage transformation of the unit, the thermal performance assessment test of the steam turbine of the unit was carried out. Through thermal test, determine the heat rate and efficiency of high, medium and low pressure cylinders of the unit, and provide reference for future economic operation of the unit.

Table 1. Basic parameters of test unit

Rated power	660MW
Steam turbine inlet pressure	24.2MPa
Steam turbine inlet temperature	566°C
Temperature before reheat steam inlet valve	566°C
Rated back pressure	4.2kPa
Back pressure in summer	6.5kPa
Speed	3000r/min
Driving mode of feed water pump	Small machine drive
Direction of rotation	The turbine is clockwise when looking at the generator end
Heat consumption under rated working condition	7545.1kJ/kW.h(1802.1kcal/kW.h)

* Corresponding author: dkyzhongziwei@163.com

2. Performance test process of steam turbine

2.1 Test conditions

The thermal performance test of steam turbine is conducted under seven effective working conditions, namely T01 (3VWO1), T02 (3VWO2), T03 (TRL), T04 (TMCR), T05 (2VWO1), T06 (50% THA) and T07 (2VWO2). See the following table for specific test conditions and test time.

Table 2. Test conditions and test schedule

No.	Test condition	Test date	Test time
T01	3VWO1	July 23, 2022	11:30-13:30
T02	3VOW2	July 23, 2022	14:00-16:00
T03	TRL	July 24, 2022	10:15-12:15
T04	TMCR	July 24, 2022	14:35-15:35
T05	2VWO1	July 24, 2022	21:30-23:30
T06	50%THA	July 24, 2022	1:30-2:30
T07	2VWO2	July 24, 2022	18:00-20:00
T01	3VWO1	July 23, 2022	11:30-13:30
T02	3VOW2	July 23, 2022	14:00-16:00
T03	TRL	July 24, 2022	10:15-12:15
T04	TMCR	July 24, 2022	14:35-15:35

2.2 Test standards and benchmarks

The test points are arranged according to the requirements of ASME PTC 6-2004 test procedure.

2.2.1 Main measuring points include generator power, main steam pressure and temperature, high discharge pressure and temperature, reheat pressure and temperature, feedwater temperature, intermediate discharge pressure and temperature, exhaust pressure of low pressure cylinder, extraction pressure and temperature of each section, steam inlet pressure and temperature of each heater, inlet and outlet temperature and drainage temperature of each heater, main condensate flow, steam inlet flow of small turbine, etc.

2.2.2 Multiple measuring points shall be arranged for important measuring points that have a great impact on the test results, such as main steam, high pressure exhaust, reheat, intermediate exhaust, feedwater, deaerator outlet water temperature, low pressure cylinder exhaust steam pressure, etc.

2.2.3. Main flow measurement: high precision and low precision recommended by the calibrated ASME PTC6 standard β The main condensate flow is measured by the pressure tapping flow nozzle at the throat. The nozzle is installed on the condensate pipe between the outlet of No. 5 low pressure heater and deaerator, and the pressure is taken by two groups of pressure tapping holes.

2.2.4 Auxiliary flow measurement: the inlet steam flow of the small turbine is measured with a standard throttling device; The inlet and return flow of seal water of feed pump shall be measured by ultrasonic flowmeter; The desuperheating water flow of reheater takes the DCS historical data.

2.2.5 Measurement of exhaust pressure of low pressure cylinder: the exhaust pressure of low pressure cylinder is measured by cage probe, which is arranged at the throat of the interface between condenser and exhaust cylinder. One cage probe is uniformly arranged at each exhaust port, totaling 4 probes.

2.3.6. Measurement of open leakage in the system: the amount of open leakage that cannot be isolated from the thermal system shall be measured by the tester with a measuring cylinder and stopwatch using the volumetric method.

According to the requirements of each test condition, the operating personnel shall adjust the operating parameters of the unit and keep them stable. The deviation and fluctuation shall meet the requirements of the test procedures, and the regenerative system shall be put into normal operation. Adjust the valve position of high-pressure control valve according to the requirements of test conditions (load), and keep the valve position unchanged during the test.The unit operates under unit system and is isolated strictly according to the system isolation list. Make up water to the system, adjust the water level of deaerator water tank and hot well to a higher value, keep the water level of each heater normal and stable, and stop making up water. During the test, the water level of deaerator water tank and hot well changes steadily to avoid violent fluctuations.After the unit operates stably for more than half an hour and the test instrument and data acquisition system are confirmed to work normally, start the test recording. The test duration under each working condition is 1-2 hours. The data acquisition cycle of the data acquisition device is once every 30 seconds; Data in DCS shall be recorded once every 1 second; Each parameter recorded manually, with a reading cycle of 5 minutes

2.3 Test data processing and calculation

The main condensate flow is calculated by the discharge coefficient obtained through calibration, and the flow calculation formula is:

$$q_c = \frac{\pi}{4} C \cdot \varepsilon \cdot d^2 \sqrt{\frac{2\Delta p \cdot \rho_{f1}}{1-\beta^4}}$$

kg/s

In the formula:

C — Nozzle discharge coefficient;

ε — Expansion coefficient of fluid;

d — Nozzle throat diameter under actual operation state, m;

Δp — Nozzle differential pressure, Pa;

ρ_{fl} — Density of measured medium, kg/m^3;

β — The ratio of nozzle throat diameter to pipe inner diameter under actual operation condition.

Determination of steam leakage flow of shaft seal and valve stem:
The air leakage of high pressure valve rod and the air leakage of shaft seal before and after high and medium pressure are all taken as design values.Distribution of unknown leakage of the system: the unknown leakage of the system shall not exceed 0.5% of the main steam flow, and the unknown leakage shall be divided according to 50% of the steam turbine and 50% of the boiler.

$$\Delta G = G_{da} + G_{hw} - G_{ml}$$

In the formula:

ΔG — Unknown leakage of the system, t/h;

G_{da} — Equivalent flow of deaerator water tank water level change (water level drop is positive), t/h;

G_{hw} — Equivalent flow of condenser hot well water level change (water level drop is positive), t/h;

G_{ml} — Open leakage of the system, t/h.

The calculation formula of feedwater flow is as follows:

$$G_{ffw} = G_{cw} + G_{da} + G_{HP3d} + G_{DAstm} + G_{mfs} - G_{rhs}$$

In the formula:

G_{ffw} — Final feedwater flow, t/h;

G_{cw} — Main condensate flow, t/h;

G_{HP3d} — Drainage flow of 3 # high-pressure heater, t/h;

G_{DAstm} — Deaerator steam inlet flow, t/h;

G_{mfs} — Flow difference between inlet and return of sealing water of feed pump, t/h;

G_{rhs} — Desuperheating water flow of reheater, t/h.

The calculation formula of main steam flow is as follows:

$$G_{ms} = G_{ffw} - \Delta G_b$$

In the formula:

G_{ms} — Main steam flow, t/h;

ΔG_b — Boiler side leakage, t/h.

The calculation formula of cold reheat steam flow is as follows:

$$G_{crh} = G_{ms} - G_{vlHP} - G_{glHP} - \sum_{i=1}^{2} G_{exi}$$

In the formula:

G_{crh} — Cold reheat steam flow, t/h;

G_{vlHP} — Total steam leakage of high-pressure valve rod, t/h;

G_{glHP} — Total steam leakage from the front shaft seal to the intermediate pressure cylinder of the high pressure cylinder and the rear shaft seal of the high pressure cylinder, t/h;

G_{ex} — Extraction flow, t/h.

The calculation formula of cold reheat steam flow is as follows:

$$G_{hrh} = G_{crh} + G_{rhs}$$

In the formula:

G_{hrh} — Hot reheat steam flow, t/h.

The calculation formula of test heat rate is as follows:

$$HR = \frac{G_{ms} \times h_{ms} + G_{hrh} \times h_{hrh} - G_{ms} \times h_{ffw} - G_{crh} \times h_{crh} - G_{rhs} \times h_{rhs}}{P_g}$$

In the formula:

HR — Test heat rate, kJ/(kW. h);

h — Enthalpy corresponding to each flow, kJ/kg;

P_g — Generator active power, MW;

https://doi.org/10.1051/e3sconf/202337502016

3. Conclusion

The turbine output test conditions include TRL and TMCR. The main results are listed in the following table:

Table 3.Main results of unit output test conditions

parameter	unit	TRL	TMCR
Generator active power	MW	680.1870	694.2040
Pressure after regulating stage	MPa	18.0532	18.2058
Main steam pressure	MPa	24.3213	24.3236
Main steam temperature	°C	565.06	564.94
Enthalpy of main steam	kJ/kg	3391.79	3391.40
Main steam flow	t/h	1946.6843	1960.9472
Unknown leakage	t/h	7.8881	4.7068
Unknown leakage rate	%	0.41	0.24
Reheat pressure	MPa	4.2360	4.2704
Reheat temperature	°C	563.54	563.47
Reheat enthalpy	kJ/kg	3587.26	3586.78
Reheat flow	t/h	1695.4189	1708.3157
High discharge pressure	MPa	4.5895	4.6264
High discharge temperature	°C	319.79	320.57
High exhaust enthalpy	kJ/kg	2999.33	3000.41
Cold reheat flow	t/h	1642.9189	1653.2757
Intermediate discharge pressure	MPa	1.0068	1.0137
Intermediate discharge temperature	°C	348.93	348.63
Intermediate exhaust enthalpy	kJ/kg	3156.14	3155.36
Exhaust pressure of low pressure cylinder	kPa	7.2225	6.0435
Feedwater pressure	MPa	28.2007	28.2432
Feed water temperature	°C	280.38	281.05
Feedwater enthalpy	kJ/kg	1231.71	1234.99
Feedwater flow	t/h	1946.6843	1964.2828
Desuperheating water flow of reheater	t/h	52.5	55.04
Class II corrected electric power	MW	682.2337	706.7006

The test electric power of the unit turbine under TRL (nameplate output, design exhaust pressure 6.5kPa) condition is 680.1870 MW, and the electric power after Class II correction is 682.2337 MW, reaching the design value of 680 MW;

The test power of the unit turbine under TMCR (maximum continuous output) condition is 694.2040MW, and the power after Class II correction is 706.7006MW, reaching the design value of 699.9MW.

Acknowledgements

This work is supported by the science and technology project of Electric Power Research Institute of State Grid Shandong electric power company(ZY-2022-15), which studies Research on peak capacity improvement of thermal power unit based on electric heating coordination optimization under guaranteed supply demand.

References

1. Liu Xiaohong, Zhu Litong, Yang Shoumin, etc. Study on calculation method of overall performance test of thermal power plant [J].Thermal Power Generation, 2006, 35 (3): 3-10.
2. Wang Xuedong, Wang Xuetong, Yang Jianzhu, etc. Energy saving transformation and thermal performance assessment test of flow passage of Unit 2 in Laiwu Power Plant [J].Shandong Electric Power Technology, 1998 (4): 4-12.
3. Ji Ansen.Results and Analysis of Thermal Performance Test of Unit #12 in Wuhu Power Plant before Overhaul [J]. Anhui Electric Power Technical Information, 2000, 000 (002): 7-8.
4. Weng Sicheng.Thermal performance test of SAIC 300MW steam turbine (Wangting Power Plant Unit 12) [J]. Shanghai Turbine, 1983 (01): 5-20.
5. Song Guoliang, Chang Hao, Wang Baoyu, etc. Study on Numerical Calculation Method of Uncertainty of Thermal Performance Test Results of Thermal Power Plants [C].Promoting the Revolution of Energy Production and Consumption - Proceedings of the 10th Yangtze River Delta Energy Forum two thousand and thirteen.
6. Wang Jizhou Study on thermal system mechanism model and off design operating characteristics of coal-fired power generation units [D].Huazhong University of Science and Technology, 2015.

 https://doi.org/10.1051/e3sconf/202337502017

Application Research on Zero Output of Double Low Pressure Cylinder of 600MW Unit

Jun Liu, Lingkai Zhu

Shandong Electric Power Research Institute, Ji'nan 250003, China

Abstract: Under the dual carbon target, the rapid development of new energy promotes the increasing peak shaving capacity of thermal power units as basic power sources. Taking 600MW unit as an example, this paper studies the feasibility of zero output of low pressure cylinder for heat supply transformation of 600MW unit, and analyzes the peak shaving capacity of the unit by cutting off double low pressure cylinder and single low pressure cylinder.

1. Introduction

The National Development and Reform Commission and the National Energy Administration issued the Fourteenth Five Year Modern Energy System Plan (the Plan). Compared with the Fourteenth Five Year Plan for National Economic and Social Development of the People's Republic of China and the Outline of Vision Goals for 2035 previously released, there are many changes in the expression of coal power. At the same time, the Plan also puts forward specific goals for the thermal power flexibility transformation and "three transformation linkage" of coal power: strive to achieve a cumulative flexibility transformation scale of more than 200 million kilowatts of coal motor units by 2025, focus on the flexibility transformation of 300000 kilowatts and below coal power units, and research and promote the flexibility transformation of 600000 kilowatts subcritical coal motor units in areas with difficult peak shaving. By 2025, the proportion of flexibly regulated power supply will reach about 24%, and the power demand side response capacity will reach 3%~5% of the maximum power load. We will vigorously promote the "three transformation linkage" of coal power energy conservation and carbon reduction transformation, flexibility transformation and heat supply transformation, and the scale of energy conservation transformation during the "Fourteenth Five Year Plan" period will not be less than 350 million kilowatts.

2. Analysis of heat supply transformation

The Notice on Improving the Flexibility of Directly Dispatched Utility Coal fired Power Generation Units in the Province issued by the Provincial Energy Administration clearly pointed out that during the "Fourteenth Five Year Plan" period, the flexibility of coal fired power generation units should be implemented according to 20% of the capacity every year, in which the minimum technical output of pure condensing units and condensing units should reach 30% and 40% of the rated capacity respectively under the condition of stable combustion, and the new pure condensing units and condensing units should reach 30% and 40% respectively, The condensing unit with the heat electricity ratio less than 50% shall be modified according to the standard of the condensing unit.

The steam turbine of a factory is a supercritical pressure extraction condensing steam turbine, with the model of C660/543-24.2/0.45/566/566 supercritical, one intermediate reheat, single shaft, four cylinders and four exhaust, and extraction condensing. The maximum continuous output is 710MW, the rated output is 660MW, the rated heating steam extraction is 560t/h, and the maximum steam extraction is 810t/h. The unit adopts compound variable pressure operation mode, and the turbine has seven stages of non adjustable regenerative steam extraction and one stage of regulated heating steam extraction. With the continuous expansion of urban areas and the continuous development of industry, the heating steam consumption for residents and industrial steam consumption are increasing; Considering the number of users, the nature of steam consumption, the law of steam consumption, and the pipeline loss in the central heating area, the design heat load of steam in the industrial area will be 355t/h at the maximum, 296t/h at the average, and 190t/h at the minimum in 2021; In 2022, the maximum is 515t/h, the average is 446t/h, and the minimum is 350t/h; In 2023, the maximum is 655t/h, the average is 576t/h, and the minimum is 510t/h. The extraction temperature of industrial steam extraction is 260 °C - 290 °C, and the extraction pressure is 2.30MPa-2.55MPa。

According to the current heating and heating load and industrial steam extraction load, under the most unfavorable conditions in the heating season, the maximum heating capacity of the plant needs to reach 2844t/h. With the annual development of the heating and

 https://doi.org/10.1051/e3sconf/202337502017

heating load and industrial steam extraction load in the city, the long-term heating and heating area and industrial steam extraction demand show an increasing trend year by year。According to the current operation of 5 units, the maximum total steam extraction capacity can only reach 2056t/h; If there is no relevant transformation to improve the heating capacity, it is necessary to reduce the steam supply in the industrial zone to ensure the heating demand of people's livelihood during the winter heating period.

For this reason, the characteristics of typical technical schemes for flexibility transformation of heat supply units are analyzed, and the comparison of various transformation schemes is shown in Table 1.

Table 1 Comparison of various transformation schemes

technology name	Lowpressure cylinderZero output technology	Heat storage and heating	Electric boiler heating	Optical shaft and high back pressure transformation
Principle	Cut off the steam inlet of low pressure cylinder for heat supply to realize zero output heat supply of low pressure cylinder.	The storage tank system is set at the heat supply network side to weaken the time coupling degree of thermal electrical load.	An electric boiler is arranged at the heat source side, and electricity is used as the heat source to realize thermoelectric decoupling.	Through equipment (double rotors) transformation, steam turbine exhaust steam is used for heat supply.
Characteristic	Cut off the steam inlet of low pressure cylinder during operation, and the operation mode is flexible	The single tank heat storage technology of atmospherictemperature gradient layer is adopted.	Direct electric heating boiler and regenerative electric heating boiler are mostly used.	The double rotor scheme is adopted for high heating economy.
Advantage	Small investment; Good heating economy; Flexible operation mode;	Small transformation of the original system; The heating is economical.	Small transformation of the original system; Strong thermoelectric decoupling capability.	Strong heating capacity; Good heating economy.
Shortcoming	Not fully implemented Thermoelectric decoupling.	High investment; Large floorarea;Theadaptabilityto long-term low load peak shaving of power grid is poor.	High investment; The heating economy is poor.	High investment; High maintenance cost; High requirements for heat load.

Comprehensively compare the technical advantages and disadvantages of each scheme, and in combination with the unit operation status and transformation requirements of the plant, it is recommended to implement the transformation scheme with zero output technology of low pressure cylinder as the route.

3. Analysis of steam extraction capacity of unit under zero output condition

3.1 Analysis of heating and steam extraction capacity of single cut off B low-pressure cylinder

When the low pressure cylinder is put into zero output operation, the cooling steam flow of the single low pressure cylinder is considered as 30t/h, and the heat supply and steam extraction capacity of the unit is calculated under the zero output condition when the low pressure cylinder B is cut off alone and the low pressure cylinders A and B are cut off simultaneously. See Table 2 for the check results of heat supply and steam extraction capacity under zero output working condition of the unit with single removal of low pressure cylinder B.

Table 2 Heat supply and steam extraction capacity of unit under zero output condition with single removal of low pressure cylinder B

project	Company	QG100MS	QG75MS	QG50MS	QG40MS	QG30MS
Power generation	MW	468.45	357.17	251.49	206.33	157.13
Main steam flow	t/h	1905.60	1429.20	952.80	762.24	571.68
Heating extraction flow	t/h	953.23	728.50	474.79	363.91	243.40
Industrial extraction steam flow	t/h	0.00	0.00	0.00	0.00	0.00
Total heat supply	MW	686.57	535.75	350.60	269.39	177.85
Heating ratio	--	0.49	0.49	0.46	0.43	0.37
Thermoelectric ratio	--	1.47	1.50	1.39	1.31	1.13
Power generation heat rate	kJ/kWh	5400.55	5606.50	5990.01	6312.06	6915.45
Coal consumption rate for power generation	g/kWh	202.32	210.04	224.41	236.47	259.08
Heating benchmark	W/m2	45.00	45.00	45.00	45.00	45.00
heating area	W/m2	1525.72	1190.56	779.11	598.65	395.21
Peak shaving capacity	%	70.98	54.12	38.10	31.26	23.81

Remarks: QG100MS in the table refers to the zero output operation of separate low pressure cylinder B when the main steam flow is 100% rated.

Under the zero output condition of single removal of low pressure cylinder B, when the rated main steam flow is 100% (QG100MS - single cylinder removal condition), the heat and steam extraction capacity of the unit under the zero output condition of single cylinder low pressure cylinder is 953.23t/h, equivalent to 686.57MW of heat load, 468.45MW of power generation, corresponding to 0.49 of heat supply ratio, 1.47 of heat and power ratio, and 202.32g/kWh of coal consumption; Compared with the theoretical maximum heat supply capacity of 859.70t/h before the unit transformation, it has increased by 93.54t/h, and the coal consumption for power generation has decreased by 12.81g/kWh.

Under the zero output condition of single removal of low pressure cylinder B, when 75% of the rated main steam flow (QG75MS - single cylinder removal condition), the heat and steam extraction capacity of the unit under the zero output condition of single cylinder low pressure cylinder is 728.50 t/h, equivalent to 535.75 MW of heat load, about 357.17 MW of power generation, corresponding to 0.49 of heat supply ratio, 1.50 of heat and power ratio, and 210.04 g/kWh of coal consumption.

Under zero output condition of single removal of low pressure cylinder B and 50% of rated main steam flow (QG50MS single cylinder removal condition), the heat and steam extraction capacity of the unit under zero output condition of single cylinder low pressure cylinder removal is 474.79t/h, equivalent to 350.60MW of heat load, about 251.49MW of power generation, corresponding to 0.46 of heat supply ratio, 1.39 of heat and power ratio, and 224.41g/kWh of coal consumption.

Under the zero output condition of single removal of low pressure cylinder B, when the rated main steam flow is 40% (QG40MS - single cylinder removal condition), the heat and steam extraction capacity of the unit under the zero output condition of single cylinder low pressure cylinder is 363.91t/h, equivalent to 269.39MW of heat load, about 206.33MW of power generation, corresponding to 0.43 of heat supply ratio, 1.31 of heat and power ratio, and 236.47/kWh of coal consumption.

Under the zero output condition of single removal of low pressure cylinder B, when the rated main steam flow is 30% (QG30MS - single cylinder removal condition), the heat and steam extraction capacity of the unit under the zero output condition of single cylinder low pressure cylinder is 243.40t/h, which is equivalent to 177.85MW of heat load, 157.13MW of power generation, 0.37 of heat supply ratio, 1.13 of heat and power ratio, and 259.08g/kWh of coal consumption; The peak shaving capacity under this working condition reaches 23.81%.

3.2 Analysis of heating and steam extraction capacity under zero output condition when cutting off low pressure cylinders A and B

See Table 3 for the check results of heat supply and steam extraction capacity under zero output condition when the unit cuts off low pressure cylinders A and B.

Table 3 Heat supply and steam extraction capacity of the unit under zero output condition when cutting off low pressure cylinders A and B

Project	Company	QG100MS	QG75MS	QG50MS	QG40MS	QG30MS
Power generation	MW	455.62	344.78	240.26	195.52	147.39
Heating extraction flow	t/h	1048.63	822.58	567.42	455.98	334.89
Industrial extraction	t/h	0.00	0.00	0.00	0.00	0.00

Project	Company	QG100MS	QG75MS	QG50MS	QG40MS	QG30MS
steam flow						
Total heat supply	MW	755.28	604.92	418.97	337.51	244.65
Heating ratio	--	0.54	0.55	0.54	0.53	0.51
Thermoelectric ratio	--	1.66	1.75	1.74	1.73	1.66
Power generation heat rate	kJ/kWh	5010.24	5086.22	5245.71	5406.67	5740.93
Coal consumption rate for power generation	g/kWh	187.70	190.55	196.52	202.55	215.07
Heating benchmark	W/m2	45.00	45.00	45.00	45.00	45.00
heating area	W/m2	1678.39	1344.26	931.05	750.02	543.66
Peak shaving capacity	%	69.03	52.24	36.40	29.62	22.33

Remarks: QG100MS in the table refers to zero output operation of low pressure cylinders A and B at the same time when 100% of the rated main steam flow is achieved.

Under the zero output condition of simultaneously cutting off the low pressure cylinders A and B, when the rated main steam flow is 100% (QG100MS - double cylinder switching condition), the heat and steam extraction capacity of the unit under the zero output condition of double cylinder low pressure cylinder switching is 1048.63t/h, equivalent to 755.28MW of heat load, about 455.62MW of power generation, corresponding to 0.54 of heat supply ratio, 1.66 of heat and power ratio, and 187.70g/kWh of coal consumption for power generation; Compared with the theoretical maximum heat supply capacity of 859.70t/h before the unit transformation, it has increased by 188.94t/h, and the coal consumption for power generation has decreased by 27.44g/kWh.

Under the zero output condition of simultaneously cutting off the low pressure cylinders A and B, when 75% of the rated main steam flow (QG75MS - double cylinder switching condition), the heat and steam extraction capacity of the unit under the zero output condition of double cylinder low pressure cylinder switching is 822.58t/h, equivalent to 604.92MW of heat load, 344.78MW of power generation, corresponding to 0.55 of heat supply ratio, 1.75 of heat and power ratio, and 190.55g/kWh of coal consumption.

Under the zero output condition of simultaneously cutting off the low pressure cylinders A and B, when 50% of the rated main steam flow (QG50MS - double cylinder switching condition), the heat supply and steam extraction capacity of the unit under the zero output condition of double cylinder low pressure cylinders is 567.42t/h, equivalent to 418.97MW of heat supply load, about 240.26MW of power generation, corresponding to 0.54 of heat supply ratio, 1.74 of heat and power ratio, and 196.52g/kWh of coal consumption.

Under the zero output condition of simultaneously cutting off the low pressure cylinders A and B, when the rated main steam flow is 40% (QG740MS - double cylinder switching condition), the heat and steam extraction capacity of the unit under the zero output condition of double cylinder low pressure cylinder switching is 455.98t/h, equivalent to 337.51MW of heating load, about 195.52MW of power generation, corresponding to 0.53 of heat supply ratio, 1.73 of heat and power ratio, and 202.55g/kWh of coal consumption.

Under the zero output condition of simultaneously cutting off the low pressure cylinders A and B, when 30% of the rated main steam flow (QG30MS - double cylinder switching condition), the heat and steam extraction capacity of the unit under the zero output condition of double cylinder low pressure cylinder switching is 334.89t/h, equivalent to 244.65MW of heat load, 147.39MW of power generation, corresponding to 0.51 of heat supply ratio, 1.66 of heat and power ratio, and 215.07g/kWh of coal consumption; The peak shaving capacity under this working condition reaches 22.33%.

4. Comparison and analysis of two cylinder cutting methods

For two different cylinder cutting methods, three typical working conditions are selected for analysis and comparison, as shown in Table 4.

Table 4 Comparison of Typical Working Conditions of Two Cylinder Cutting Modes

Project	Company	QG100MS		QG50MS		QG30MS	
		Double cylinder cutting	Single cylinder cutting	Double cylinder cutting	Single cylinder cutting	Double cylinder cutting	Single cylinder cutting
Power generation	MW	455	468	240	251	147	157
Main steam flow	t/h	1905	1905	952	952	571	571
Heating extraction flow	t/h	1048	953	567	474	335	243
Industrial extraction steam flow	t/h	0.00	0.00	0.00	0.00	0.00	0.00
Total heat supply	MW	755	686	419	350	245	178
Heating ratio	0.00	0.54	0.49	0.54	0.46	0.51	0.37

 https://doi.org/10.1051/e3sconf/202337502017

Thermoelectric ratio	0.00	1.66	1.47	1.74	1.39	1.66	1.13
Power generation heat rate	kJ/kWh	5010	5400	5245	5990	5741	6915
Coal consumption rate for power generation	g/kWh	187	202	196	224	215	259
Heating benchmark	W/m^2	45	45	45	45	45	45
heating area	万m^2	1678	1525	931	779	543	395
Peak shaving capacity	%	69	71	36	38	22	24

Note: QG100MS in the table refers to 100% rated main steam flow.

It can be seen from the above table that under the same steam flow, the peak shaving capacity of the unit when switching double cylinders is about 1W higher than that when switching single cylinders; At the end of the load month, the higher the coal consumption is, the higher the heat rate of power generation is. The heat rate of power generation under the double cylinder switching mode and the horizontal bar switching mode are lower, and the lower the load is, the greater the difference in the heat rate of power generation is.

5. Conclusion

Under the zero output condition of single removal of low pressure cylinder B, when the rated main steam flow is 100% (QG100MS single cylinder removal condition), the heat supply and steam extraction capacity of the unit is 953.23t/h, and the coal consumption for power generation is 202.32g/kWh. Compared with the theoretical maximum heat supply capacity of 859.70t/h before the unit transformation, the heat supply capacity is increased by 93.54t/h, and the coal consumption for power generation is reduced by 12.81g/kWh; When the rated main steam flow is 30% (QG30MS single cylinder switching condition), the heat and steam extraction capacity of the unit is 243.40t/h, the generating power is about 157.13MW, and the electric peak shaving capacity reaches 23.81%. Under the zero output condition of simultaneously cutting off the low pressure cylinders A and B, when the rated main steam flow is 100% (QG100MS double cylinder cutting condition), the heat supply and steam extraction capacity of the unit is 1048.63t/h, and the coal consumption for power generation is 187.70g/kWh, which is 188.94t/h higher than the theoretical maximum heat supply capacity of 859.70t/h before the unit transformation, and the coal consumption for power generation is 27.44g/kWh lower; When the rated main steam flow is 30% (QG30MS double cylinder switching condition), the heat and steam extraction capacity of the unit is 334.89t/h, the generating power is about 147.39MW, and the electric peak shaving capacity reaches 22.33%.

The implementation of flexible heating transformation can greatly improve the peak shaving capacity and low load heating capacity of units, meet the needs of flexible transformation and heating capacity increase, and enhance the competitiveness and profitability of units in the peak shaving market of electric auxiliary services.

Acknowledgments

This work was supported by the scientific and technological project of State Grid Shandong Electric Power Research Institute "Research on Thermoelectric Cooperative Regulation Technology of Nuclear Power Units" (ZY-2022-10).

References

1. Li Qiong et al. Research and application of peak shaving capacity of subcritical 300MW heating units [C]. Proceedings of the National Symposium on flexible transformation technology of thermal power units. 2018. Beijing
2. Zheng Fei, Chen Xiaoli, et al. Research on deep peak shaving flexibility transformation technology of extraction steam heating unit [J]. Steam turbine technology. 2021.4
3. Tian Gang, Liu Lihua, et al. Low pressure cylinder removal heating transformation scheme and peak shaving performance analysis of 350MW Unit [J]. Steam turbine technology. December 2019
4. Song Wenlei, he Xiang, et al. Application Research on flexible deep peak shaving based on 300MW subcritical heating unit [J]. Power station system engineering. July 2020
5. Zhou Xingang, Yang XINGSEN, Ding Junqi, Li daobo, Cui Hui. Verification and flexibility improvement strategy of renewable energy peak shaving units in Shandong Province [J]. Shandong electric power technology, 2019,46 (01): 58-61 + 66

 https://doi.org/10.1051/e3sconf/202337502018

Research on Intelligent Regulation System of Solar Panels Driven By Low-Voltage Electric Energy

Wei Zhang, Xuefeng Lin

School of Intelligent Manufacturing and Traffic Chongqing Vocational Institute of Engineering Chongqing, China

Abstract. This paper proposes a design method for tracking solar panel light tracking control system based on microcontroller. The main structure of the system includes light intensity detection module, automatic control module, intelligent information processing module, human-computer interaction module, computer control module and interface circuit module. In this paper, the photoelectric method is used to track the position of the sun, the control process is modeled and simulated in the system. The system is optimally controlled by adding a Kalman filter to the output of the classic PID control. Actual tests show that the input voltage of the circuit is 1.2~5.5V, the maximum charging current is 1.2A, the minimum conversion efficiency of the circuit is not less than 72%, and the maximum conversion efficiency is more than 92%, and the power energy output of the solar panel can be collected and charged even under very weak lighting conditions. Solar energy utilization rate is high.

Keywords: low voltage electric energy drive; Solar panels; Intelligent adjustment; Solar cells; Electrical energy collection.

1. Introduction

With the continuous advancement of green energy construction, as well as the intensification of global energy shortage and environmental pollution and other issues, the urgent demand for solar photovoltaic power generation has been put forward, in the application design of solar photovoltaic power generation devices, the light chasing control design of solar photovoltaic power generation as an important application direction has received great attention from people, the construction of tracking solar photovoltaic panel light tracking control system, combined with the solar photovoltaic circuit lamp light chasing control design, improve the utilization rate of solar energy [1]. Through the design of solar panel automatic light tracking system, the establishment of light tracking control adaptive information acquisition model, through the automatic detection and perception of light intensity, mechanical devices and motor devices timely adjustment, to achieve automatic light tracking control of solar energy, the use of a full range of rotation tracking methods, solar photovoltaic panel control, improve the photovoltaic control panel all-round tracking and adaptive control capabilities, related tracking solar light tracking control system design method research has attracted great attention from people. This paper proposes a design method for tracking solar panel light chasing control system based on a single microcomputer, and the main framework of the system includes light intensity detection module, automatic control module, intelligent information processing module, human-computer interaction module, computer control module and interface circuit module, etc., first of all, the overall design analysis of the system, and then the hardware modular design of the tracking solar light tracking control system, and finally the simulation test analysis, to draw the effectiveness conclusion.

2. Principle and design

2.1 Scenario Overview

The solar panel charges the battery and supercapacitor after passing through the protection circuit and the charge management circuit. The output of the supercapacitor is connected to the input of the boost conversion circuit, and the output of the battery is connected to the input of the buck conversion circuit. At the same time, the supercapacitor output is connected to the voltage comparator circuit (the reference voltage is the discharge cutoff voltage set by the supercapacitor), and the output of the voltage comparator is directly connected to the enable end of the supercapacitor boost conversion circuit, and the second is connected to the enable end of the battery buck conversion circuit after passing through the inverter. The outputs of the battery buck conversion circuit and the supercapacitor boost conversion circuit are connected to the load. When the solar panel is working normally, the supercapacitor voltage is greater than the set cutoff voltage, the boost conversion circuit enables the active terminal, and the load system is supplied by the supercapacitor; When the solar panel is not working for a short time, the super capacitor is used to discharge deeply, make full use of its energy storage, as long as the super capacitor voltage is greater than the set cutoff voltage, the load system can be guaranteed to work normally; When the solar panel is not working for a long time, the supercapacitor is deeply discharged, and its voltage is less than the set cutoff voltage, the battery buck conversion circuit enables the active end, and the power supply of the load system is provided by the battery [2]. The system operation is supported by supercapacitors, taking advantage of the long-life characteristics of supercapacitors. In a few long rainy days, the battery support system works, which not only makes full use of the characteristics of large battery capacity, but also extends the service life of the battery. According to the above design scheme, the block diagram is shown in Figure 1.

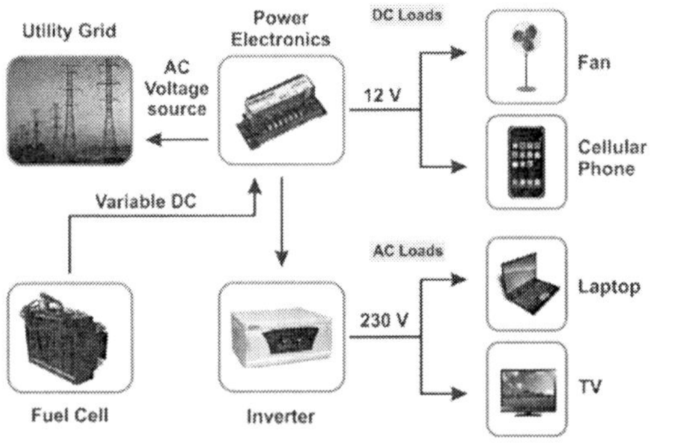

Fig. 1. Solar cell conditioning scheme

2.2 Solar charger

Solar panels are the most important products that convert light energy into electrical energy, and the biggest feature is that the output voltage and current change greatly with the change of light intensity. For example, a solar panel with a nominal 5V, 800mA output, in the case of direct sunlight, the output voltage can reach up to 6V, and in rainy weather, the outdoor test output voltage is only 2~3V. The ordinary lithium battery operates at 3.6V and the charging voltage is 4.2V. Therefore, in order to charge ordinary lithium batteries through solar panels, the simplest way is to choose a solar panel with high output voltage, and charge the lithium battery after stepping down to 4.2V, as shown in Figure 2.

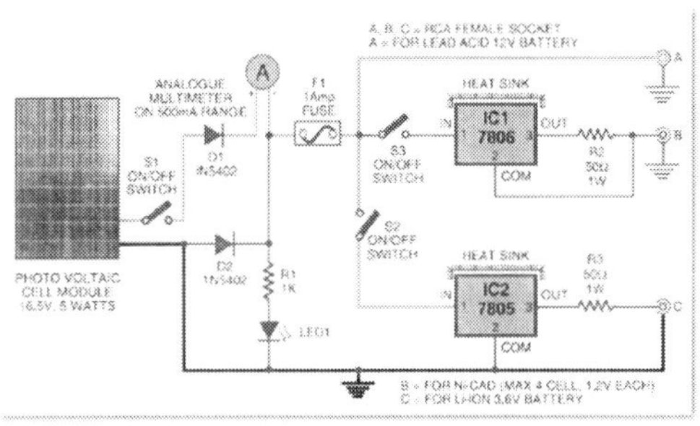

Fig. 2. Buck charging schematic

The circuit of this scheme is simple, but in actual use, when the light is less than a certain intensity, the output voltage of the solar panel is reduced, and the charger cannot be guaranteed to work properly. Therefore, the scheme has significant limitations [3]. To solve this problem, an improved approach is to first boost the output of the solar panel to stabilize to a fixed voltage, and then step down to 4.2V to charge the lithium battery as shown in Figure 3.

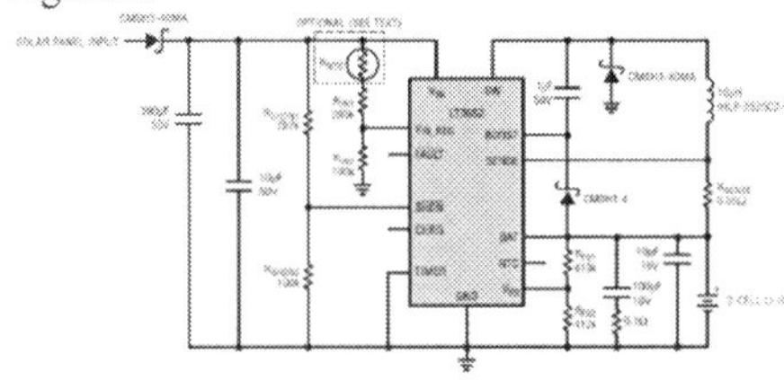

Fig. 3. Buck-buck charging schematic

Because the input voltage range of the boost circuit is generally relatively wide, it can be greatly guaranteed to charge the lithium battery in the case of a large range of light changes. The circuit structure of this scheme is complex, and the output of the solar panel is first boosted and then stepped, and the efficiency of the entire circuit is greatly reduced. At the same time, a serious problem with the scheme is that when the light is very strong, the output voltage of the solar panel may be higher than the maximum input voltage of the boost circuit [4]. If a solar panel with a lower output voltage is selected, when the light is weak, the output voltage of the solar panel will be less than the minimum input voltage of the boost circuit. The scheme uses a buck and boost automatic switching circuit, when the output voltage of the solar panel is higher than a certain value, the buck circuit is used to charge the lithium battery, and vice versa, the boost circuit is used to charge the lithium battery. Since only the first stage circuit is in operation at any one time, the efficiency of the entire circuit is very high. In order to achieve this function in this article, the core part of the circuit uses the TPS61200 intelligent power chip produced by TI. The chip has built-in boost and buck circuitry and can be automatically switched according to the relationship between input and output voltages. When the input voltage is lower than the output voltage, the chip operates in the boost mode; When the input voltage is higher than the output voltage, the chip automatically switches to the step-down mode. At the same time, the chip has a very low start-up voltage and operating voltage, and the circuit is still able to collect and charge the lithium battery from the output of the solar panel under extremely weak light.

3. System hardware and software design

3.1 Hardware design

The tracking system consists of a sensor, a controller, a driver, a stepper motor, a tracking mechanism, and a solar panel as shown in Figure 4.

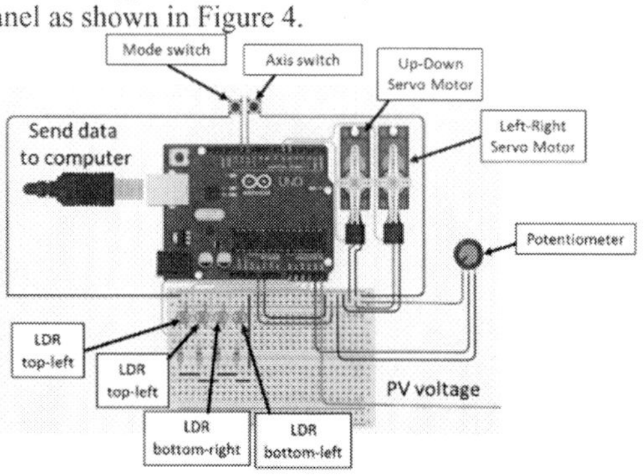

Fig. 4. The composition of the tracking system

3.1.1 Controller

Q series PLC programmable controller is a medium and large PLC series products developed by Mitsubishi from the A series PLC basis. The basic components of Q series PLC include power modules, CPU modules, substrates, I/O modules, etc. The 00JCPU of the Mitsubishi Q Series PLC is an all-in-one system for the main substrate and expansion substrate that can control multiple input and output points, making it suitable for use in small devices [5]. This system uses QX40DC input module, QY41P transistor output module, Q64AD analog-to-digital conversion module. The QX40 is used as a switching input. The QY41P is a transistor-type module that can send high-speed pulses and can be used as the input pulse signal of stepper motors. Q64AD has four AD inputs, which are set to high resolution and acquire four sensor signals and environmental signals.

3.1.2 Sensors

The light receiving component of the solar sensor is mainly composed of a four-quadrant photovoltaic detector. The four-quadrant detector is a photovoltaic detector device that arranges four photovoltaic panels with exactly the same performance according to the requirements of rectangular coordinates, and there is a "ten" shaped channel between them.

3.2 Software algorithm design

The steering gear is mainly composed of a brushless DC motor and a position control circuit, which can usually accurately control the angle of its rotation through the D) signal, which is the core of controlling the attitude of the solar panel [6]. The steering gear in the design can be wound from 0° to 180° in one plane. If you use ordinary motor modeling methods, the model will become cumbersome and difficult to analyze. Therefore, the impedance model is used in the application to model the motion process of the servo. Its motion is considered a damping motion as shown in Figure 5.

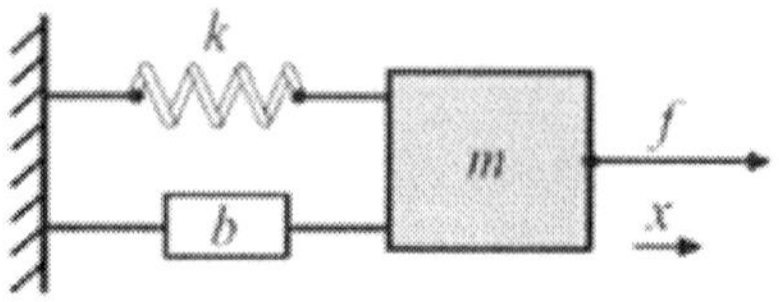

Fig. 5. Equivalent model of servo rotation

According to the impedance model, it can be obtained:

$$m\ddot{x} + b\dot{x} + kx = f \tag{1}$$

This is a typical second-order system whose transfer function can be described as:

$$G(s) = \frac{\omega_n^2}{s^2 + 2\varepsilon\omega_n s + \omega_n^2} \tag{2}$$

According to Equation 2), override the transfer function as

$$\frac{X(s)}{F(s)} = \frac{\frac{1}{M}}{s^2 + \left(\frac{B}{M}\right)s + \left(\frac{k}{M}\right)} \tag{3}$$

M is the inertial parameter and k is the stiffness parameter. M can be calculated according to the torque and rotation angle acceleration given in the servo parameter table. k The larger the stiffness, the greater the stiffness, and too small will cause a large tracking error, in order to simplify the model, and in the process of rotation, the deformation of the force arm is very small. We consider $/k = 0.1.B$ to be the damping parameter. The value of E is obtained by adjusting ε. In this design, the weight of the solar panel is much smaller than the torque of the servo, so the value of ε is taken between. During the entire control process, the difference between the two photosensitive sensors is taken as the input, and the D signal of the control steering gear is used as the output of the controller. The system operates with measurement noise and observation noise. Therefore, a Kalman filter is added to the system output.

This is a typical second-order system. Introducing the transfer function of formula (C) into a state variable can write out its equation of state and output equations respectively:

$$\dot{x}(t) = Fx(t) + Hu(t) \tag{4}$$

$$y(t) = Lx(t) \tag{5}$$

Where $u(t)$ is the output. Where the transfer function parameters are known, the value of the unknown coefficient matrix in the equation can be calculated directly by $MATLAB$. According to equations (5) and (7), and discretize them, write the equations of state and observation equations of the Kalman filter are:

$$x(k) = Fx(k-1) + H(u(k) + w(k)) \quad (6)$$

$$y_\upsilon(k) = Lx(k) + \upsilon(k) \quad (7)$$

Where $w(k)$ is the process noise and $\upsilon(k)$ is the measurement noise. They are all Gaussian white noise.

4. System detection

For ordinary voltage conversion circuits, one of the most important parameters is the conversion efficiency. For the energy collection circuit, in addition to efficiency, the main performance index is the lowest condition of the circuit start-up, that is, the lowest input voltage and current. Therefore, this article uses circuitry to test the input minimum voltage and current [7]. From the test results, it can be seen that when the input minimum voltage is 1.2V, in order to ensure normal charging, the input current is 44.2mA, that is, the output power of the solar panel is 55mW, which is very easy to achieve. As the input voltage increases, the current required gradually decreases, and the output power of the solar panel is reduced to 30mW. The test results are shown in Figure 6.

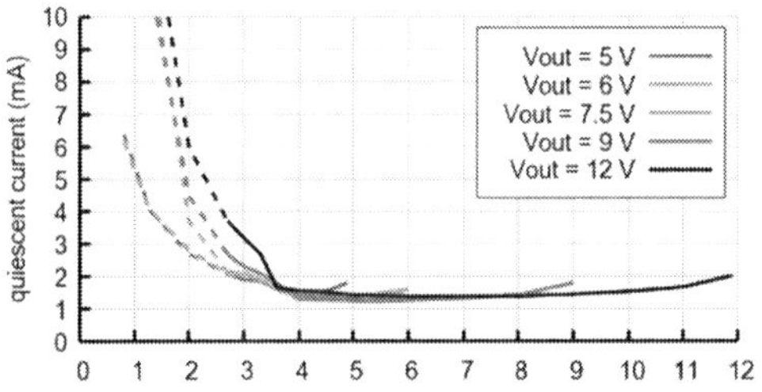

Fig. 6. The minimum start-up voltage is related to the input current, input power

5. Conclusion

By improving the original solar output characteristic measurement device, the angle between the light source and the solar panel can be adjusted, so that the influence of the light angle and the inclination angle of the solar panel on the output efficiency of the cell is studied, and the two-dimensional relationship curve of output efficiency with angle is obtained. Experiments have found that with the increase of lateral light inclination, the output power shows a tendency to increase first and then decrease. As the longitudinal inclination of the solar panel increases, the output power decreases. The improvement process of this experimental project helps to strengthen students' hands-on practical ability, and at the same time, Ife has a deep understanding of the rotation and rotation of the earth.

Acknowledgments

The work described in this paper was fully supported by the Science and Technology Research Program of Chongqing Municipal Education Commission (Grant No. KJQN202103403）.

References

1. CUI Bingyan, CHEN Peng, WU Xiaoxuan, et al. Mechanism analysis and motion simulation of adjustable solar panels. Mechanical Transmission, vol. 44, pp.79-84,June 2020.
2. JIAO Yuetan, CHEN Qin, Montaser A. M. Alzaky, et al. Study on EVA in solar panels separated by moderate heat treatment coupled with sulfuric acid carbonization. Chemical Industry, vol. 50, pp.1761-1764,July 2021.
3. WU Shuangying, GUO Haitao, XIAO Lan, et al. Experimental study on local spot phenomenon of photovoltaic panels under linear light source. Journal of Engineering Thermophysics, vol. 5, pp.2129-2133,September 2020.
4. QIAO Shuang, LIU Jihong, FU Guangsheng, et al. Research progress on lateral photovoltaic effect in structural materials of solar cells. Chinese Science Bulletin, vol. 65, pp.160-165,December 2020.
5. ZHOU Ying, MAO Li, ZHANG Yan,et al. Application of improved CNN in defect detection of solar panels. Computer Simulation, vol. 37, pp.79-82,March 2020.
6. HU Jinpeng, ZHANG Xuewu, ZHANG Zhuo. Defect detection method of outdoor solar panel based on improved DenseNet network. Optics and Optoelectronic Technology, vol. 20, pp.67-74,February 2022.
7. Wang Yuanxin, Zhang Peng, Wu Xiaojie, et al. Experimental study on wireless optical communication and energy transmission composite system based on solar panels. Acta Photonica Sinica, vol. 49, pp.100-106,May 2020.

Economic Day-ahead scheduling of SOFC and biomass-based Integrated Tri-generation Energy System using artificial bee colony algorithm

Pengxiang Zhao [1,*], Jianjun Zhu [1,2], Yanru Zhang [1,2], Qiang Wang [1,2], Xian Yang [1], Lin Cong [1]

[1] State Grid Integrated Energy Service Group Co., Ltd, Beijing, 100052, China
[2] State Energy Biological Power Generation Group Co., Ltd, Beijing, 100052, China

Abstract: In the process of operation, integrated energy systems encounter a series of complex optimization problems such as complex conversion of multiple types of energy, coupling of multiple types of equipment, and uneven distribution of multiple demands. In recent years and in the process of achieving the goal of "double carbon" in China, more and more problems of optimal scheduling of biomass integrated energy systems have become hot topics of research. In this paper, we propose an artificial bee colony algorithm-based operation and scheduling method for a biomass fuel cell (SOFC) cooling, heating, and power triple-supply system with the objective function of maximizing the economic profit of the integrated energy system (including fuel, operation, and maintenance costs), and the constraints of energy conservation, system safety, and operation state. The equipment included are biomass boiler and corresponding steam turbine, biogas digester, SOFC, gas storage tank, etc. The results of the algorithm include the economic benefits of using the artificial bee colony algorithm for the same load scenario. The operational scheduling results show that the artificial bee colony algorithm is able to maximize the profitability of the integrated energy system while reducing carbon emissions.

Key Words: Integrated energy system; optimal design; Biomass; Artificial Bee Colony algorithm

1. Introduction

As the total reserves of non-renewable energy sources such as oil, coal and natural gas are decreasing worldwide, and a series of environmental problems caused by the massive use of fossil energy sources, the search for new renewable and clean alternative energy sources is gradually becoming a concern of thermoelectricity in the world. Biomass energy refers to the conversion of biomass and other organic matter produced by green plants through photosynthesis into energy that can be used directly as fuel or converted into gaseous or liquid fuel carriers. As a kind of renewable, energy-saving and environment-friendly energy with great potential, biomass energy has a great role in optimizing the energy consumption structure, linking the energy supply tension, increasing agricultural income and improving environmental quality.

Biomass is organic matter produced by plants through photosynthesis and its transformation products. As an energy source with huge reserves and not yet fully utilized, biomass energy has the advantages of being renewable and zero-emission, which has received wide attention from all over the world [3]. The integrated energy system of biomass, electricity, heat, gas and fertilizer contains four types of energy: heat, electricity, gas and fertilizer, and is characterized by various types of loads and abundant energy supply equipment. It is necessary to take into account the system economy, safety, environmental protection and other objectives when scheduling, and use IOT technology and information technology to unify and implement the scheduling of all energy supply and storage equipment in the region, so as to achieve the effect of optimizing energy supply to regional thermoelectric loads and improving energy utilization efficiency [1,2]. In the vast areas of northern China, especially in Inner Mongolia and northeast China, there is a large amount of biomass resources to be developed, so the establishment of an integrated energy system with biomass as the core can promote the full and efficient utilization of biomass energy, which is of great significance to implement the national energy policy, achieve carbon peaking and carbon neutrality as soon as possible, improve the energy structure, and promote industrial restructuring and other goals.

Unlike the traditional power system problem, the energy optimization management problem of the integrated energy coordination based on biomass boilers and biogas digesters is more difficult: (1) there is a coupling relationship between heat, electricity and gas in the

* Corresponding author:

https://doi.org/10.1051/e3sconf/202337503001

system, all of which are affected by the biomass boiler; (2) there is a certain percentage of new energy generation such as photovoltaic power generation in the system, which is easily affected by the external environment, and the form of power generation is often quite stochastic, volatile and intermittent. Therefore, high-precision scenery resources and load prediction algorithms are needed to support the energy management of the combined heat and electrical energy system. However, the accuracy of these new energy power forecasts is difficult to guarantee, especially as the time span becomes longer, the accuracy of the forecasts becomes lower, for example, the current wind power day-ahead forecasts can reach 25% to 40% error, and for intra-day forecasts may be even greater. With the continuous development of short-term and ultra-short-term forecasting technologies based on big data, and the accumulation of historical data such as renewable energy output and heat and gas load, the forecasting accuracy and speed have been significantly improved, but it is still difficult to guarantee the efficient and economic operation of combined heat and electricity energy systems.

Traditional biomass utilization is mainly by means of direct straw combustion for power generation and anaerobic fermentation in biogas digesters for biogas production. Biomass gas (CH4-containing gas produced by biomass conversion) has a high inert component, low calorific value, and unstable composition, which makes its utilization difficult, while the development of fuel cells provides an opportunity for its large-scale utilization. As a kind of electrochemical energy conversion device, solid oxide fuel cell (SOFC) has high electrical efficiency, wide fuel adaptability, no pollution, and has advantages that other kinds of fuel cells cannot match when processing biomass gas, mainly as follows: (1) SOFC can directly use the biomass gas after pollutant purification treatment, without reforming, water gas conversion and other pre-treatments. (2) SOFC has good adaptability to changes in biomass gas composition and can be used directly as a fuel even for gases with very low calorific value[4]. Therefore, SOFC has good prospects for utilization in integrated energy systems with biomass as the core.

Good matching between various energy sources and performance of integrated energy systems cannot be achieved without rational planning. There are relatively few studies on the planning of integrated energy systems including biomass, and the literature [5] proposed a user participation willingness assessment model based on improved PMV-PPD (predicted mean vote-predicted percentage of dissatisfied) index. Based on the cloud model theory, a price-based integrated demand response model considering mixed uncertainty is established. And on this basis, the optimal dispatching strategy of the integrated energy system is proposed by considering the uncertainties of various types of energy response volume boundaries and price elasticity coefficients. In [6], an optimal scheduling strategy based on integrated demand response and master-slave game is proposed for a multi-micro-grid integrated energy system containing electrical energy interaction, and an integrated demand response model containing multiple types of loads such as transferable interruptible electrical loads, transferable non-interruptible electrical loads, flexible thermal loads and cold loads is established. In [7], a deep reinforcement learning method based on a flexible actor-judge is proposed to realize the optimization of integrated electricity-gas energy dispatching in multiple scenarios through the interaction between an intelligent body and an energy system and adaptive learning of control strategies. In the literature [8], an optimal scheduling model of integrated energy system considering integrated demand response and multi-energy storage devices is constructed with the objective function of minimizing the total operating cost of integrated energy system, and the uncertainty of renewable energy output is handled by using conditional value-at-risk approach. However, the above work does not fully consider the renewable energy access, cooling/heating/electricity multi-load demand and SOFC does not supply power to the system, i.e., the above system still depends on the power supply from the external grid.

In summary, this paper proposes a scheduling optimization model for SOFC in a biomass-core electric-thermal-gas-fertilizer multi-energy flow integrated energy system, in which a biomass direct-fired generator set, SOFC as electric-thermal coupling equipment, can both generate electricity and supply heat to the outside, and there are also gas supply equipment (digester) and energy storage equipment (gas storage tank). In this paper, we firstly establish the model of each key equipment in the system, consider the energy coupling, energy balance constraint, equipment safety constraint and operation state constraint among each energy conversion equipment, and solve the system by artificial bee colony optimization algorithm with the objective of maximizing annual profit, so as to obtain the optimal scheduling scheme before the system day. Finally, the feasibility and effectiveness of the method are verified based on the analysis of calculation cases.

2. Mathematical model of biomass system equipment

The biomass-electricity-heat-gas-fertilizer cogeneration system studied in this paper is shown in Figure 1. The power generation equipment in the system includes biomass direct-fired generator set and SOFC, and the electricity generated by the system is used to meet the electric load demand of users; the heat generated by burning straw in the biomass boiler and the waste heat generated during the operation of SOFC can be used as the heat source for external heat supply; the biogas produced by the digester is transported to the storage tank after being dried, deodorized and desulfurized by the filter, gas-water separator and purification regulator through the gas transmission pipeline. On the one hand, it can supply gas to customers, and on the other hand, it can be used as fuel for SOFC operation. The heat generated by the biomass boiler in this system can be used for the heat preservation of the digester, the fly ash can be used as the PH regulator of the digester, and the combination of digestate and fly ash can be used to produce fertilizer, which not only makes the ash and waste heat of direct-

https://doi.org/10.1051/e3sconf/202337503001

fired power generation available, but also solves the problems of heat preservation, PH regulator and efficient utilization of digestate and digestate in the process of anaerobic fermentation.

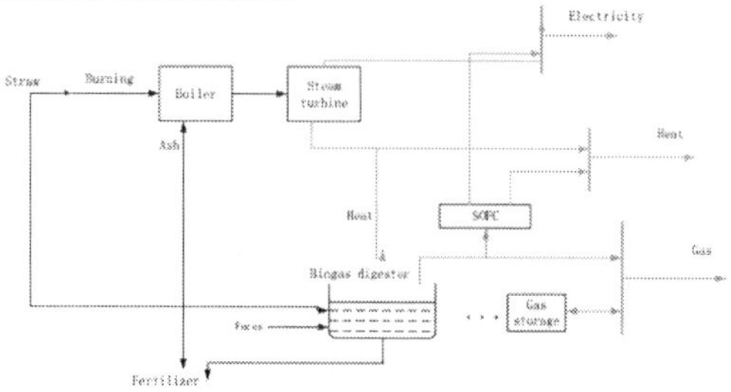

Fig.1 Flow chart of biomass electric heating and gas fertilizer polygeneration system

2.1 Biomass boilers and steam turbines

Biomass boilers can burn straw under permissible working conditions to produce heat, and their operating mode can be expressed as follows

$$Q_{BB} = \eta_h^{BB} F_{BB} \qquad (1)$$

where Q_{BB} is the thermal power output of biomass boiler, FBB the calorific value input for the fuel, η_h^{BB} the thermal efficiency of biomass boiler.

The steam turbine studied in this paper is an extracted condensing turbine, and the turbine body contains a steam extraction port, which is mainly used for external steam extraction for heat supply. For the mathematical modeling of the turbine body, the input parameters are: main steam parameters(Flow D_0 ,Pressure P_0 ,Temperature T_0)、Vapor extraction parameters(Flow D_1 , Pressure P_1 , Temperature T_1), the output parameters are: Turbine power generation Pe 、Turbine heat supply Q_h [12]。

Q_h is calculated as:

$$\begin{cases} Q_h = D_1 \cdot h_1 \\ h_1 = f\left(P_1, T_1\right) \end{cases} \qquad (2)$$

The power equation of the turbine is:

$$\begin{cases} Pe = \left[D_0\left(h_0 - h_1\right) + \left(D_0 - D_1\right)\left(h_1 - h_c\right)\right]\eta_m \eta_g \\ h_0 = f\left(P_0, T_0\right) \end{cases} \qquad (3)$$

where h_0 is the main steam enthalpy; h_c is the exhaust enthalpy; η_m and are the mechanical drive efficiency and generator efficiency, respectively, which are determined by the system structure and generator performance and can be calculated from the heat balance data.

2.2 Biogas digester

2.2.1 Calculation of gas production and biogas fertilizer yield

Biogas fermentation, also known as anaerobic digestion, is a process in which organic materials (manure, straw, etc.) are subjected to certain conditions of temperature, humidity, pH and anaerobic conditions to produce biogas through the decomposition and metabolism of anaerobic microorganisms [13]. Biogas fermentation feedstock consists mainly of water and total solids (materials other than water, mainly manure and straw). The gas production rate of raw materials is generally expressed as the gas production per unit weight of total solids, which is inconsistent for different raw materials and varies for the same raw materials due to differences in gas production rates between species. Usually, gas production can be estimated from the data in Table 1.

Table 1 Estimation of total solid content and actual gas production rate of biogas raw materials

Material type	Total solids content (%)	Gas production rate (m^3/kg·TS)
Corn stalks	80～90	0.30
Sorghum straw	80～90	0.30
Pig manure	18～20	0.35
Human manure	18～20	0.35
Chicken manure	30	0.35

Therefore, the total feedstock solids quantity and gas production can be calculated by the following equation.

$$Y_B = \sum_i^n M_i D_i \qquad (4)$$

$$V_{MG} = \sum_i^n M_i D_i \beta_i \qquad (5)$$

where Y_B is the total solids quantity of raw materials, n is the number of raw material types, M_i is the weight of the fresh material of the i-th ingredient, D_i is the total solids content of the i-th raw material, V_{MG} is biogas production, β_i is the gas production rate of the i-th feedstock.

According to the conservation of mass, the mass of the remaining digestate can be calculated, assuming that 75% of the digestate can be used as organic fertilizer, the mass of organic fertilizer produced is[13]:

$$M_{MF} = (Y_B - \rho_b \times V_b) \times 0.75 \qquad (6)$$

where M_{MF} is the yield of biomass fertilizer in one fermentation cycle; ρ_b is the biogas density and is taken as 0.7kg/m3 at room temperature.

2.2.2 Heat transfer modeling

The activity of bacterial flora during fermentation is affected by the temperature of the digestate, which in turn affects the efficiency of gas production in biogas digesters. The low ambient temperature in winter in northern areas of China leads to less or even no gas production in biogas digesters, and some digesters are even frozen[14]. In order to ensure the normal fermentation of biogas digesters, it is often necessary to take insulation measures for biogas digesters and to support the corresponding warming system[15].

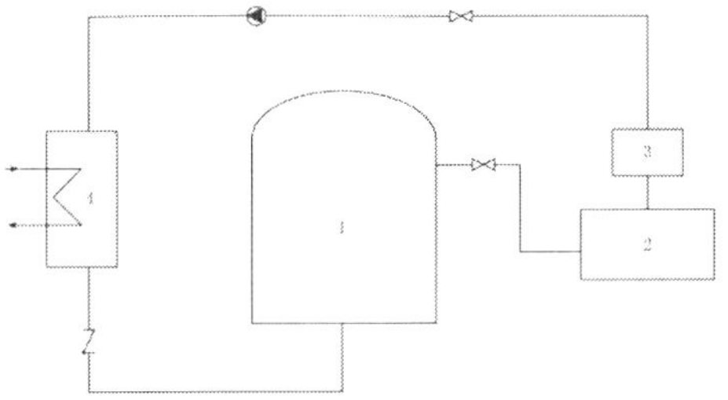

1. Biogas digester 2. Water tank 3. Filter 4. Heat exchangers
Fig.2 Brief flow chart of biogas pool warming system

The biogas digester heating system mainly consists of digester, water tank, filter, heat exchange device and heat exchange pump. The water is stored in the tank as the heat transfer medium of the system, and absorbs the heat from the steam generated by the biomass boiler in the heat exchange device, and then flows through the internal heating coil of the digester to transfer the heat to the fermentation liquid. It is assumed that the inner wall of the digester is made of enameled steel plate and the insulation material is made of polystyrene foam board. The flow of the digester heating system is shown in Figure 2.

In this paper, the heat transfer calculation model is established by referring to the method of literature[16] for heat transfer calculation. The main heat loss of this biogas digester system comes from the heat dissipation of the digester and tank, and the heat loss of the incoming and outgoing feed liquids, and the main heat gain is the heat of biological fermentation. According to the heat balance principle, the heat balance equation can be obtained as follows:

$$Q_{sr} = Q_{zq} + Q_{sx} + Q_{ly} - Q_{sw} \quad (7)$$

where Q_{sr} is the total heat loss from the digester; Q_{zq} is the amount of heat dissipated by the biogas digester; Q_{sx} is the water tank heat dissipation; Q_{ly} is the inlet and outlet liquid heat loss; Q_{sw} is biodigestion heat.

(1)Heat dissipation from biogas digester

The digester is approximated as a cylindrical shape, and only the case where the digester is in contact with air all around is considered. Based on the available studies, it is known that the reactor heat transfer model is[17]:

$$Q_{zq} = \sum_{i=1}^{n} K_i A_i \left(t_z - t_0\right) \quad (8)$$

where t_z is the temperature of the feed liquid in the digester and t_0 is the ambient temperature.

The heat dissipation of the digester comes mainly from the top, side walls and bottom parts. The top load factor is:

$$K_1 A_1 = \frac{A_1}{\frac{1}{h_n} + \frac{\delta_d}{\lambda_d} + \sum_{i=1}^{n} \frac{\delta_i}{\lambda_i} + \frac{1}{h_n}} \quad (9)$$

where A_1 is the heat transfer area at the top of the digester; h_n is the heat transfer coefficient of the internal top surface of the biogas digester; δ_d is the thickness of the biogas digester wall; λ_d is the thermal conductivity of the biogas digester wall; δ_i is the thickness of the i-th layer of insulation material; λ_i is the thermal conductivity of the i-th layer of insulation material; h_w is the thermal conductivity of the i-th layer of insulation material.

The sidewall load factor is:

$$K_2 A_2 = \frac{H_f}{\frac{1}{h_{fb} \pi d_1} + \sum_{i=1}^{n} \frac{1}{2\pi\lambda_i} \ln \frac{d_{i+1}}{d_i} + \frac{1}{h_w \pi d_{n+1}}} \quad (10)$$

where H_f is the height of the material liquid in the digester; h_{fb} is the surface heat transfer coefficient of the inner wall of the biogas digester; h_w is the heat transfer coefficient on the external surface of the digester; d_1 is the internal diameter of the digester; λ_i is the thermal conductivity of the i-th layer of insulation material; d_i is the inner diameter of the i-th layer of insulation material of the digester; d_{i+1} is the outside diameter of the i-th layer of insulation material of the digester; d_{n+1} is the outside diameter of the biogas digester.

The bottom load factor is:

$$K_3 A_3 = \frac{A_3}{\frac{1}{h_{fd}} + \frac{\delta_d}{\lambda_d} + \sum_{i=1}^{n} \frac{\delta_i}{\lambda_i} + \frac{1}{h_w}} \quad (11)$$

where A_3 is the heat transfer area at the bottom of the digester; h_{fd} is the heat transfer coefficient of the inner bottom surface of the biogas digester. The inner wall thickness and thermal conductivity of the biogas digester are: δ_d =0.01m, λ_d =2.05 $W/(m^2 \cdot k)$ .The insulation thickness and thermal conductivity are as follows δ_2 =0.2m, λ_2 =0.05 $W/(m^2 \cdot k)$. Considering the fast heat and mass transfer rate of the fluid in the biogas digester, $1/(h_{fb} \pi d_1)$ and $1/h_{fd}$ are small, Ignoring these two heat transfer coefficients on the water side.

(2)Water tank heat dissipation

The water tank is used to store the hot water required to heat the digester, and only the case where it is in contact with air all around is considered. The heat dissipation from the circumferential walls of the tank is considered as a one-dimensional steady-state heat conduction problem that varies along the radius; the heat dissipation from the top and bottom surfaces of the tank is also considered as a one-dimensional steady-state heat conduction problem through the flat walls [18]:

$$Q_{sx} = \left(\frac{l}{r_1} + \frac{A}{r_2}\right) \times \left(T_{st} - t_0\right) \quad (12)$$

where l is the height of the material in the tank; A is the sum of the area of the top and bottom of the water storage tank; r_1 is the thermal resistance of the circumferential surface thermal conductivity ; r_2 is the thermal resistance of the upper and lower bottom surface thermal conductivity; T_{st} is the water tank storage temperature.

(3)Inlet and outlet heat loss

The one-time feeding solution, after a complete fermentation cycle in the digester, will all be renewed, and the average mass flow rate will be:

$$m = \frac{M}{T} \quad (13)$$

where m is the average mass flow rate; M is the total mass of feed liquid; T is the fermentation period, where 16d is taken for calculation.

Since the heat taken away by water vapor and biogas in the feed solution is very little and is often neglected in engineering, the heat loss in and out of the feed is mainly the heat consumption required for the fermentation solution to rise from the initial temperature of the feed to the fermentation temperature, which can be derived from the heat transfer heat flow calculation formula:

$$Q_{ly} = cm\left(t_z - t_{ly}\right) \quad (14)$$

where c is the specific heat of the liquid, the liquid is approximated as water; t_{ly} is the temperature of the liquid when feeding.

(4)Anaerobic fermentation bioheat

Related studies have shown that 3% of the effective energy (16.91 kJ/kg) of the biogas fermentation stock solution is released as heat [19]. The biogas digester feed liquor will continue to generate biothermal heat during the fermentation cycle, and the average daily biothermal heat generated will be:

$$Q_{ft} = \frac{M \times 16.91 \times 3\%}{T} \quad (15)$$

According to the heat balance relationship Equation (7), the calculation results of the above sub-items are organized to obtain the total heat loss of the biogas digester . Thus the heat to be supplied to the digester by the biomass boiler is:

$$Q_h = \left(1 + \Psi\right) \cdot Q_{sr} \quad (16)$$

where ψ is the heat pipe heat loss rate, equaling 0.1%.

2.3 SOFC Model

A fuel cell is a device that converts energy stored in fuels such as hydrogen, natural gas and oxygen into electrical energy through chemical reactions without pollution and also recovers waste heat for utilization.The electro-thermal efficiency in the SOFC model varies with the output power and the empirical relationship between its chemical reaction output electrical power and thermal power can be fitted through a large number of investigations[20] as follows:

$$\frac{\eta_E}{\eta_{E,r}} = 1.042 - 0.045\frac{P_r^{FC}}{P^{FC}} \quad (17)$$

$$\frac{\eta_T}{\eta_{T,r}} = 0.2477\frac{P^{FC}}{P_r^{FC}} + 0.8326 - 0.0757\frac{P_r^{FC}}{P^{FC}} \quad (18)$$

$$H^{FC} = P^{FC}\frac{\eta_T}{\eta_E} \quad (19)$$

where P^{FC} is the output electrical power of SOFC, η_E and η_T are the electrical and thermal efficiency of SOFC respectively, r represents the rated operating condition, and the corresponding output thermal power H^{FC} can be calculated by the energy balance principle. The coupling between P^{FC}、H^{FC}、η_E and η_T increases the complexity of SOFC model operation.

2.4 Gas storage tank

When excess biogas is generated in the integrated energy system, it is partially stored in the storage tank, which is modeled as:

$$V_{t+1,zq} = V_{t,zq} + V_{t,zq,in} \cdot \eta_{zq,in} \quad (20)$$

When the gas storage tank provides biogas, its outgassing model is :

$$V_{t+1,zq} = V_{t,zq} - V_{t,zq,out} / \eta_{zq,out} \quad (21)$$

Where $V_{t+1,zq}$ is the current gas volume of the storage tank; $V_{t,zq}$ is the gas volume at the last moment of the storage tank; $V_{t,zq,in}$ is the amount of biogas entering the storage tank; $V_{t,zq,out}$ is the amount of biogas to be supplied by the storage tank; $\eta_{zq,in}$ is the storage efficiency of the storage tank, and the value of this paper is 0.9; $\eta_{zq,out}$ is the discharge efficiency of the storage tank, and the value of this paper is 0.9.

3. Run scheduling model solving based on artificial bee colony algorithm

The structure of the day-ahead economic scheduling model based on the artificial bee colony algorithm is shown in Figure 4. The assumptions of the model are as follows: the equipment output is assumed to be a continuous variable; the equipment is fault-free during the optimized operation.

3.1 Optimization models and algorithms

3.1.1 Optimization variables

For the integrated energy system studied in this paper, the capacities of biomass boilers and digesters have been determined by the project, and the amount of boiler straw, digester straw, biogas production, gas storage, boiler power generation, boiler heat supply, electricity purchase, and gas purchase are selected as optimization variables in this study.

3.1.2 Objective function

In this paper, considering the investment cost of the system and the economy of operation, the objective function is the maximum annual profit of the system, that is $\max f = Profit - Cost$ 。

$$Profit = P_{MG} + P_{MF} + P_{PG} + P_{HS} \quad (22)$$

where $Profit$ is the annual operating income of the system, including biogas income, biogas fertilizer income, power generation income, heating income.

The total cost is the sum of the annual equivalent investment cost and the annual operating cost of the equipment.

$$Cost = C_{inv} + C_{run} \quad (23)$$

The annual equivalent investment cost of equipment is the cost value of the total investment cost of the system by equal allocation to each year of the operating cycle, which includes the annual equivalent investment cost of equipment such as digesters, SOFCs, and storage tanks. The equation for calculating the annual equivalent investment cost of equipment is shown below[22]:

$$C_{inv} = C_{inv,GS} + C_{inv,SOFC} \quad (24)$$

$$C_{inv,i} = \frac{r(1+r)^n}{(1+r)^n - 1} c_{inv,i} S_{inv,i}, \quad i = GS, \text{SOFC} \quad (25)$$

where $C_{inv,GS}$、$C_{inv,SOFC}$ are the annual investment cost of gas storage tank and SOFC respectively; r is the discount rate; n is the full life cycle; $c_{inv,i}$ is the unit capacity investment cost of the i-th equipment; $S_{inv,i}$ is the installed capacity of the i-th equipment.

Annual operating costs include annual straw purchase cost C_{SC}, electricity purchase cost $C_{grid,e,t}$, gas purchase cost $C_{grid,g,t}$:

$$C_{run} = C_{sc} + \sum_{t=1}^{8760} \left(C_{grid,e,t} + C_{grid,g,t} \right) \quad (26)$$

3.1.3 Constraints

The constraints of the integrated biomass-based energy system day-ahead economic optimal dispatch model include power balance constraints, equipment capacity constraints, equipment operation constraints, etc.

The power balance constraints are calculated as $P_{swz}^t + P_{SOFC}^t + P_{Grid}^t = P_{Load}^t$, where P_{Grid}^t is the hour-by-hour power exchange value between the integrated energy system and the external grid; P_{Load}^t is the hour-by-hour load value; P_{swz}^t is the hour-by-hour power generation of the biomass boiler; P_{SOFC}^t is the hour-by-hour power generation of the SOFC.

For the thermal load, the power balance constraint is calculated as $H_{swz}^t + H_{SOFC}^t = \mathrm{H}_{Load}^t$, where H_{Load}^t is the hour-by-hour load value; H_{swz}^t is the hour-by-hour heat supply of biomass boiler; H_{SOFC}^t is the hour-by-hour heat supply of SOFC.

The equipment capacity and operating constraints within the integrated energy system are calculated as follows:

For biomass boiler power generation: $P_{swz}^{min} \le P_{swz}^t \le P_{swz}^{max}$

For biomass boiler heat supply: $H_{swz}^{min} \le H_{swz}^t \le H_{swz}^{max}$

For SOFC heat supply: $H_{SOFC}^{min} \le H_{SOFC}^t \le H_{SOFC}^{max}$

For SOFC power generation: $0 \le P_{SOFC}^t \le P_{SOFC}^{max}$

For gas storage equipment storage capacity: $0 \le Q_{cq}{}^t \le Q_{cq}^{max}$

For biogas digester gas production: $0 \le Q_{zq,i}{}^t \le Q_{zq,i}^{max}$

where P_{swz}^{min} and P_{swz}^{max} are the minimum power generation output and maximum power generation output of biomass boiler; H_{swz}^{min} and H_{swz}^{max} are the minimum heating output and maximum heating output of biomass boiler; H_{SOFC}^{min} and H_{SOFC}^{max} are the minimum heating output and maximum heating output of SOFC; P_{SOFC}^{max} is the maximum power generation capacity of SOFC; Q_{cq}^{max} is the maximum storage capacity of gas storage equipment; $Q_{zq,i}^{max}$ is the maximum gas production capacity of i-th biogas digester.

3.1.4 Optimization algorithm

In this paper, an artificial bee colony intelligent optimization algorithm is chosen to solve the optimal configuration of the integrated energy system. In the algorithm search process, the employing and following bees are responsible for finding the optimal solution, and the detecting bees are responsible for avoiding the situation of falling into the local optimum during the search process, and generating new nectar sources randomly once they fall into the local optimum, whose search process is mainly as follows:

(1) Initialization: Determine the number of populations, the maximum number of iterations, the problem dimension D and the search range of the solution, and generate D-dimensional random solutions in the search space (i=1,2,3,...,SN).

(2) Each hired bee generates a new random solution according to Equations (27) at the beginning of the search phase.

$$v_{ij} = x_{ij} + \rho_{ij} \cdot (x_{ij} - x_{kj}) \quad (27)$$

where $k \in \{1,2,3,\ldots,SN\}$, $j \in \{1,2,3,\ldots,D\}$ and $k \ne i$, random number $\rho_{ij} \in [-1,1]$.Calculate the fitness value fit_i corresponding to the new solution, and if fit_i is better than the old solution, then the new solution replaces the old solution. Conversely, the old solution is retained.

(3) After all hired bees complete the search, share the solution information with the following bees and calculate the probability value of each solution according to Equations (28).

$$p_i = \frac{fit_i}{\sum_{k=1}^{SN} fit_k} \quad (28)$$

Then, based on the principle of roulette, a random number is generated in the range of [-1,1]. If the probability value of the solution is greater than the random number, a new solution is generated according to Equations (27), and if the fitness value of the new solution A is better than the old one, the new solution replaces the old one. Conversely, the old solution is retained.

(4) After all following bees finish searching, if a solution is not updated for more than M times, it means that the solution falls into local optimum, and the solution is discarded. The detecting bees generate new solutions according to Equations (29),

https://doi.org/10.1051/e3sconf/202337503001

$$x_{ij} = x_{\min j} + rand(0,1) \cdot (x_{\max j} - x_{\min j}) \quad (29)$$

where $j \in \{1,2,3,\ldots,D\}$ .Then repeat the above search process.

(5) The maximum number of iterations is satisfied and the optimal solution is output.

For this original ABC optimization algorithm, the new food source search strategy is Equations (27), in which the base solution x_{kj} selected for the neighborhood search is a random food source and the search direction is more random. However, the random food source does not always proceed in a favorable direction when leading the swarm search, therefore, this paper adopts an improved artificial swarm optimization algorithm IGABC based on a hybrid search strategy of the best nearest neighbor and the current optimal solution, and its specific search strategy is shown below.

$$v_{ij} = \begin{cases} \text{Existing best neighbor} \\ x_{ij} + \mu_{ij} \cdot \left(x_{\text{best neighbour}} - x_{ij}\right) + \psi_{ij} \cdot \left(y_j - x_{ij}\right) \\ \text{or not} \\ x_{ij} + \varphi_{ij} \cdot \left(x_{ij} - x_{kj}\right) + \psi_{ij} \cdot \left(y_j - x_{ij}\right) \end{cases} \quad (30)$$

where μ_{ij} is a random value in [0,1].

The basic principle of the IGABC algorithm search strategy is: when the best nearest neighbor exists for this food source, the hiring bee and the following bee search for a better food source through the best nearest neighbor and the current global optimal solution; otherwise, they search for a better food source through the stochastic food source and the current global optimal solution. According to the optimization process of IGABC, a two-stage stochastic planning IGABC optimal planning platform for integrated energy systems is established as shown in Figure 4.

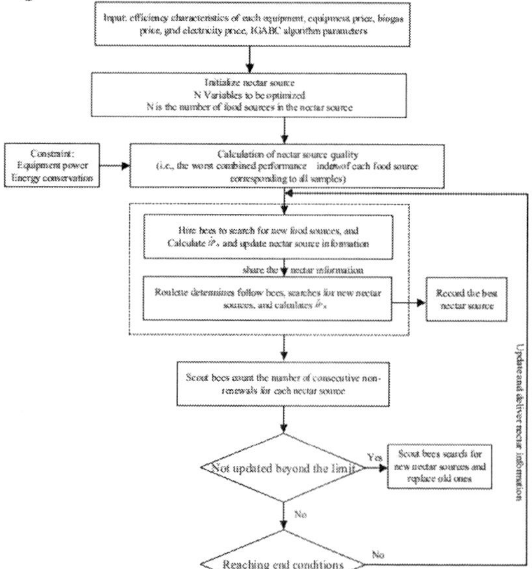

Fig.4 IGABC optimization platform for two - stage stochastic programming of integrated energy system

4. Example analysis

4.1 System Introduction

This paper is based on the example of a typical biomass-electricity-heat-gas-fertilizer co-generation system in a region of Inner Mongolia. In this integrated energy system, there are 8 biogas digesters with a diameter of 10 m and a volume of 300 m3. The raw materials entering the digesters are corn straw and pig manure, which are fermented at a high temperature of 55°C. The location of the research subject is set in an alpine region, where the equipment operates all year round. Based on the local climatic conditions, the typical days are divided into typical summer days, typical winter days and typical transitional season days based on the local historical data. The electrical load and gas load of each typical day are shown in Figure 5 and Figure 6. The gas load demand varies greatly with seasonal changes, with the highest demand in winter due to heating, the second highest in the excessive season, and the smallest in summer, with the peak of gas load occurring around 8:00 am and 10:00 pm.

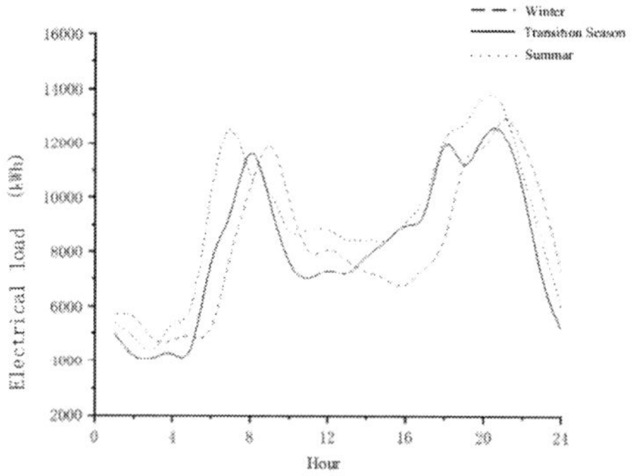

Fig.5 Typical daily electricity load

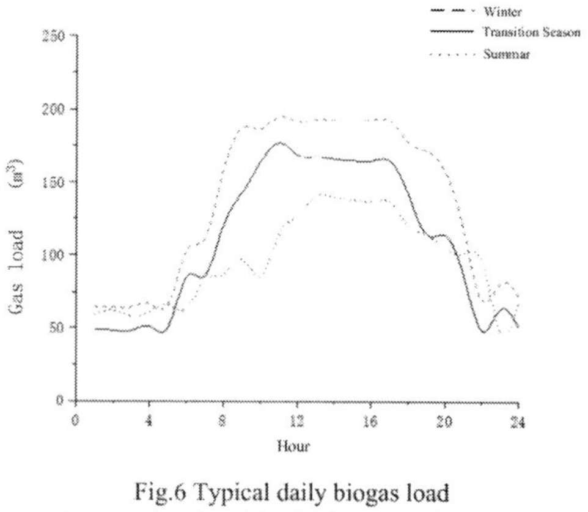

Fig.6 Typical daily biogas load

The equipment involved in the integrated energy system in this study includes biomass boilers, steam turbines, SOFC, digesters, and storage tanks. The input to the system includes straw and manure, and the output is electricity, heat, biogas, and biomass fertilizer, and some of its unit prices are shown in Table 2.

Table 2 Unit prices of some equipment and resources in the integrated energy system

 https://doi.org/10.1051/e3sconf/202337503001

Projects	Price	Source
SOFC Construction Costs	130rmb/kW	[23]
Gas Storage Tank Construction Cost	100rmb /m^3	Project Settings
Metered Heat Price	180rmb /t	Project Settings
Straw price	100rmb /t	Project Settings
Biogas fertilizer price	1000rmb /t	[13]

Since the system power generation participates in grid peaking, the electricity selling price can be obtained by referring to the peaking price in the literature[24], and the biogas price is set by the project as 1.5 rmb/m3. Electricity price and biogas price are uncertainties in the integrated energy system, and in this paper, the lower grid electricity price is set to obey a uniform distribution, and the biogas price obeys a triangular distribution, which can be obtained by Monte Carlo sampling for 100 sets of scenarios.

4.2 Analysis of optimization scheduling results

Using the above model and algorithm for calculation and analysis, Figures 7-16 give the amount of straw entering the boiler hour by hour, the amount of straw entering the digester hour by hour, the amount of gas produced by the digester hour by hour, the amount of electricity generated by the system hour by hour, the amount of heat supplied to the system hour by hour, the amount of heat purchased hour by hour, and the amount of gas purchased hour by hour, Electrical load and equipment electrical power , Heat load and equipment thermal power and Gas load and equipment gas power respectively.

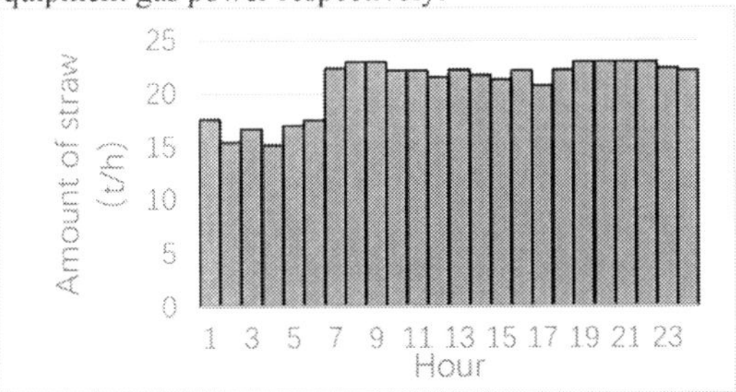

Fig.7 Amount of straw fed into the boiler

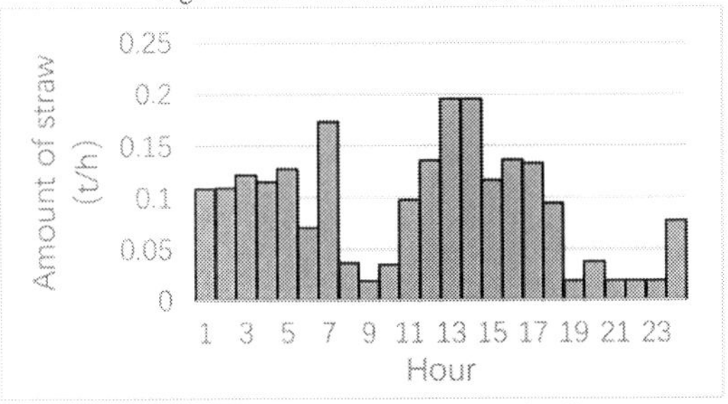

Fig.8 Amount of straw fed into the digester

From the optimization results of the previous day's operation, we can see that under uncertain conditions, the SOFC does not work for most of the time because the turbine pumping can basically meet the heat load. 9:00-10:00 time period reaches the peak heat load, and the electric load is also larger at this time, the turbine pumping is very small, and the biogas production is also smaller, so it is necessary to purchase a large amount of biogas to meet the heat load through SOFC.

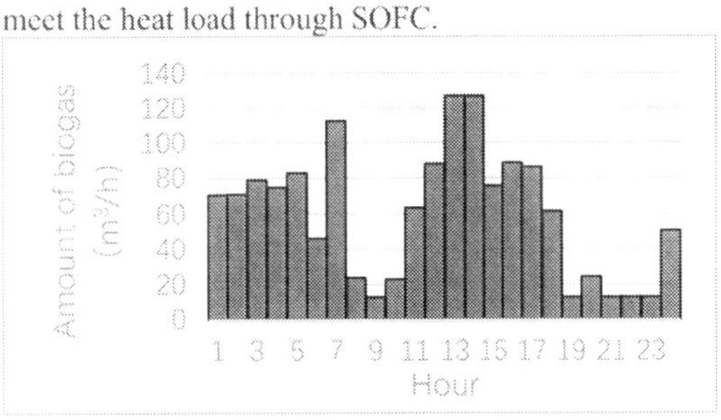

Fig.9 Amount of biogas

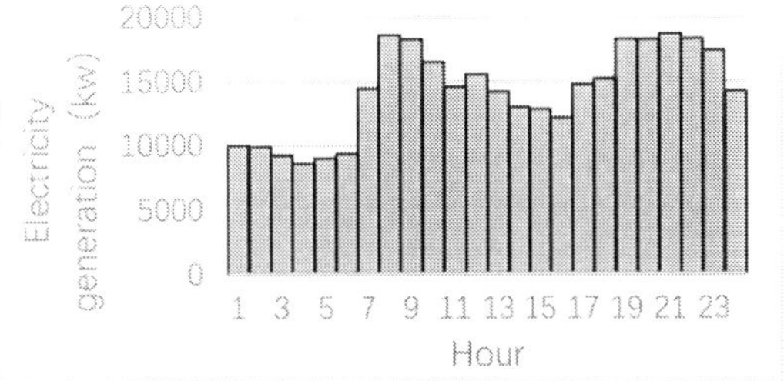

Fig.10 Electricity generation

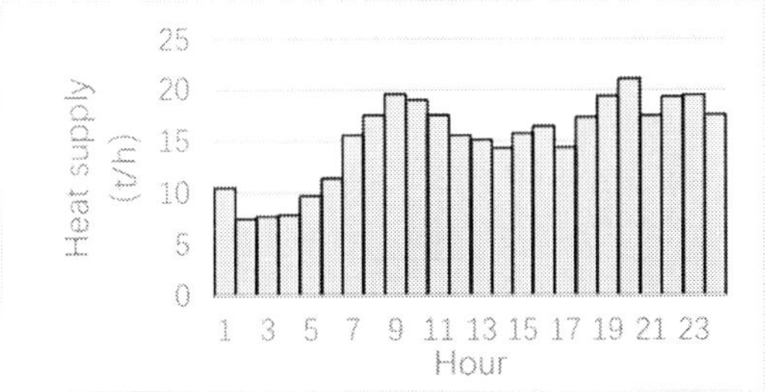

Fig.11 Heat supply

In the time period of 1:00-7:00 the electric, thermal and gas loads do not reach the peak, and the system chooses to fill the gas storage tank appropriately during this time to cope with the higher load in the subsequent moments. After 7:00 the energy storage device discharges gas to the outside, and the gas storage volume keeps decreasing until it drops to zero at 10:00. In the day-ahead dispatch, the maximum power purchase is made during the 21:00-22:00 time period, and the maximum purchase volume is made during the 9:00-10:00 time period. If the maximum power purchase is 2000kw and the maximum gas purchase volume is 100m^3/h, there are four unsatisfied points.

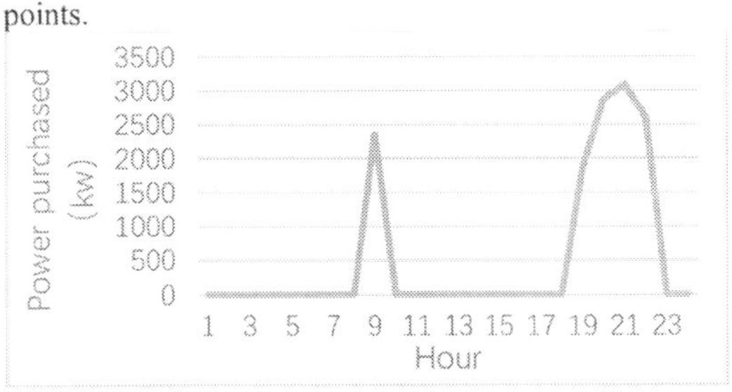

Fig.12 Power purchased

 https://doi.org/10.1051/e3sconf/202337503001

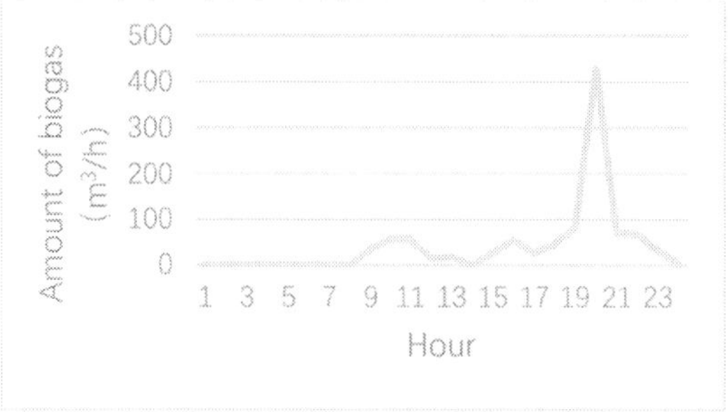

Fig.13 Amount of biogas purchased

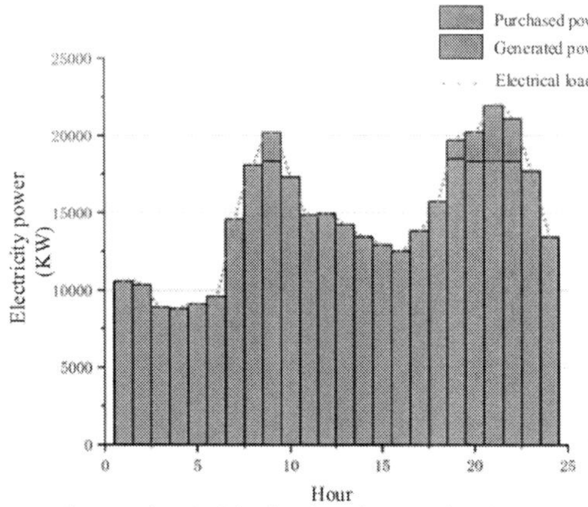

Fig.14 Electrical load and equipment electrical power

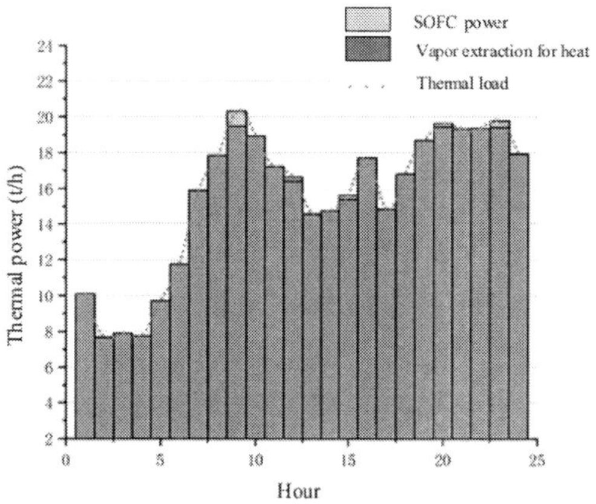

Fig.15 Heat load and equipment thermal power

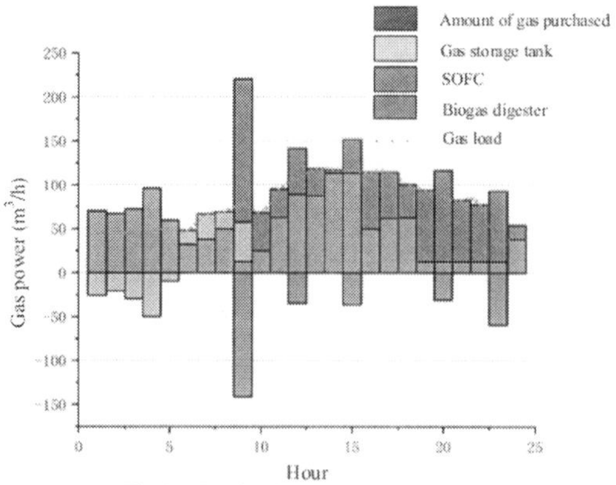

Fig.16 Gas load and equipment gas power

Under uncertain conditions, the maximum load case is larger due to the heat load, the turbine steam extraction priority heating to meet the heat load, while the amount of steam used to heat the digester is small, and the digester working at 55 °C is very little, so the amount of gas production is small, relying on gas purchase to meet the gas load. The amount of biogas production is relatively larger in the minimum load case, which is because there is more extraction steam available to heat the digester when the thermal load is low, and more straw fermented in the digester, more biogas fertilizer is produced, and the biogas fertilizer yield is larger, so the amount of straw entering the digester is larger, and the excess biogas produced is stored in the gas storage tank. The electricity, heat and gas loads are higher at the maximum load case, so the profit and carbon reduction are larger.

5. Conclusion

This paper proposes an economic scheduling optimization method based on an artificial bee colony optimization algorithm with the objective of maximizing annual profits for an integrated energy system including SOFC with biomass as the core. Based on the mathematical model of each energy device, the proposed method takes into account the energy balance constraint, equipment safety constraint, operation state constraint and coupling characteristics of energy sources, while taking into account the investment cost and operation cost of electricity and gas purchase, as well as the revenue of biogas and biogas fertilizer and power generation. The feasibility of the proposed method is verified by using the actual data of a typical biomass cogeneration system in Inner Mongolia. The results show that according to the proposed optimization method based on the swarm algorithm, the profit of the whole system can be greatly increased based on the reduction of carbon emission, which is a certain guidance for the production practice.

Acknowledgements

This work was financially supported by State Grid Corporation Science and Technology Project "Research and application of rural comprehensive energy recycling

https://doi.org/10.1051/e3sconf/202337503001

technology based on biomass-power-heat coupling" (5400-202031205A-0-0-00).

References

1. Zichi Wang,Bingyin Lei,Lingyi Yang. Multi-objective two-level optimal scheduling of regional integrated energy system considering combined demand response of electricity and heat,2021,33(7):120-127.DOI:10.19635/j.cnki.csu-epsa.000750.
2. Guo Z, Zhang R, Wang L, et al. Optimal operation of regional integrated energy system considering demand response[J]. Applied Thermal Engineering, 2021, 191: 116860..
3. Shouqiang Li. Two-stage optimal scheduling of integrated energy systems in townships with biomass energy [D]. North China Electric Power University (Beijing),2021.DOI:10.27140/d.cnki.ghbbu.2021.000360.
4. Qingshan Zhu. Prospect analysis of solid oxide fuel cells for efficient utilization of biomass gas [J]. Journal of Process Engineering, 2007(02):419-424.
5. Zhihao Yang, Hang Liu, Ming Wen, Haipeng Zhao, et al. Optimal scheduling strategy for integrated energy systems considering hybrid uncertainty in integrated demand response[J]. Power Construction, 2022, 43(09):66-76.
6. Peng Li, Difan Wu et al. Optimal scheduling strategy of multi-microgrid integrated energy system based on integrated demand response and master-slave game [J]. Chinese Journal of Electrical Engineering, 2021, 41 (4) : 1307-1321+1538 .
7. Qiao Ji, Wang Xinying, Zhang Qing, Zhang Dongxia, Pu Tianjiao. Flexible actuator-judge based
8. Electric-gas integrated energy system optimal dispatching based on deep reinforcement learning with flexible actuator-judge [J]. Chinese Journal of Electrical Engineering. 2021, 41 (3) : 819-833.
9. Chen Zhonghua, Gao Zhenyu, Chen Jiamin, Wu Xianping, Fu Xueqian, Chen Xianqing . Research on collaborative planning of integrated energy systems considering uncertainty factors [J]. Power System Protection and Control, 2021, 49(08): 32-40.
10. Rezvan A T, Gharneh N S, Gharehpetian G B. Optimization of distributed generation capacities in buildings under uncertainty in load demand[J]. Energy and Buildings, 2013, 57: 58-64.
11. Sun Wen. Modeling Simulation and Characterization of Distributed Energy System [D]. Southeast University, 2016.
12. Xiao Wei . Feasibility study of resource utilization of farmland waste in Fengxian District [D]. Shanghai Jiaotong University, 2012.
13. Liu Shumin, Han Jingyu, Yue Haiyang. Comprehensive development and utilization of biogas in cold regions of northern China [J]. Journal of Inner Mongolia Agricultural University (Natural Science Edition), 2002(04): 83-86.
14. Zhang Changai, Liu Ying, Wang Yanqin, Yuan Changbo, Yao LiResearch progress of biogas overwintering technology [J]. China Agronomy Bulletin, 2008(08): 469-472.
15. Tan Ling . Research on multiphase flow biogas digester warming technology [D]. Southwest Jiaotong University, 2016.
16. Sha Wei, Liu Jianyu, Chen Zexing, Li Wentao. Study of insulation thickness of anaerobic digester reactor for biogas in cold regions - based on limited temperature drop value and temperature drop time [J]. Agricultural Mechanization Research, 2013, 35(01): 228-231.
17. Chen Dandan. Research on the design of tiered hot water storage tank and its effect on the efficiency of solar collector [D]. Lanzhou University of Technology, 2014.
18. Xiang Yangyang, Gao Zhengyang, Li Shiping, Ou Wei, Huang Zheping, Xue Yuanyuan . Temperature control technology and economic analysis of rural biogas fermentation [J]. Anhui Agronomy Bulletin (first half of the month), 2010, 16(15): 169-171+175.
19. Sun Wen, Chen Ziwei, Zhang Yuqiong, Su Zhi-gang, Sun Li. Dynamic planning-based economic dispatching of SOFC cooling, heating and power tri-generation integrated energy system by day [J/OL]. Chinese Journal of Electrical Engineering:1-10[2022-04-10].DOI:10.13334/j.0258-8013.pcsee.211743.
20. Walker W E, Harremoës P, Rotmans J, et al. Defining uncertainty: a conceptual basis for uncertainty management in model-based decision support[J]. Integrated assessment, 2003, 4(1): 5-17.
21. Gao Yang, Xi Xiaohui, Liu Xiaohui, Wang Xiaojun. Integrated energy system planning for industrial parks considering price elasticity of biomass supply [J]. Power Construction, 2021, 42(04): 49-58.
22. Hu, Yike. Study on zero emission alternatives for small-scale coal-fired thermal power plants based on SOFC and EV battery stepped utilization [D]. Xiangtan University, 2021. DOI:10.27426/d.cnki.gxtdu. 2021.001058.
23. Liu Zhen, Guo Zhenzhu, He Liangping. Research on optimization of commercial electricity consumption based on peak tariff mechanism[J]. Journal of Chongqing University of Technology (Social Science),2017,31(04):45-50.

Study on the thermal decomposition reaction process and kinetics of SF6 and tungsten

Mengyuan Xu[1], Houhua Lv[2], Yunfang Liu[5], Ren Yang[1], Yanhua Han[1], Liuyu Xu[3], Xu Li[4], Jiuhong Li[1], Xiao Zhang[2], Yajun Jian[2*], Weiqiang Zhang[2,*]

[1]State Grid Shaanxi Electric Power Research Institute, 710054, Xi'an, China
[2]School of Chemistry and Chemical Engineering, Shaanxi Normal University, 710119, Xi'an, China
[3]Shandong Taikai High Voltage Switchgear Co.,Ltd., 271000, Tai'an, China
[4]State Grid Shaanxi Electric Power Company, 710054, Xi'an, China
[5]South China Institute of Environmental Science, Ministry of Ecology and Environment, 510655, Guangzhou, China

Abstract The thermal reaction of SF_6 with tungsten powder was investigated by thermogravimetry (TG) and differential scanning calorimetry (DSC), combined with the characterization of solid decomposition products of SF_6 and tungsten by SEM, EDS, XPS and Raman. Based on above experiments, a two-stage decomposition process of SF_6 with tungsten was proposed: the first stage is the surface vulcanization reaction $W+SF_6 \rightarrow WS_2$, mainly taken place at 600 °C, with the activation energy valued at 152.8 $kJ \cdot mol^{-1}$, according to Avrami-Erofeev equation; the second stage is the endothermic fluorination reaction $WS_2+SF_6 \rightarrow WF_6$, mainly taken place at 750 °C,, with the activation energy valued at 126.0 $kJ \cdot mol^{-1}$, according to the Avrami-Erofeev equation.

1 Introduction

SF_6 is a synthetic inert gas, colorless, tasteless, non-toxic, with excellent insulation performance, and has been applied in many important fields including gas insulated switchgear, semiconductor processing, aerospace etc [1]. On the other hand, SF_6 is a greenhouse gas, whose threat to the environment is 23900 times that of CO_2[2]. Therefore, its extensive use has attracted many researchers to understand its chemical properties. In contrast to the widely believed chemical inertia, SF_6 can react with some compounds like metal or low-valent complex under regular conditions. For example SF_6 can react with hot sodium film at 200 °C [3]. At -61 °C, SF_6 can be reduced with element Cs to CsS_2 in liquid ammonia [4]. Basta et al. found that the reaction of organotitanium and organozirconium complexes with SF_6 could react quickly at room temperature or even below room temperature [5]. Metal tungsten is a key material in electronic and electrical equipment, widely used in high voltage circuit breakers and plasma etching. Due to close contact of tungsten with SF6 under these harsh conditions, decomposition of SF6 inevitably taken place. For example, SF_6 corrodes copper-tungsten alloy and generates WF_6 under the action of arc[6]. SF_6-O_2 plasma corrodes refractory metal W to generate WF_6, WO_3, and WOF_4 [7], and Peignon et al. believed that WS_2 was an important intermediate for further oxidation or fluorination to obtain aforementioned decomposition products[8]. Fluoridation, vulcanization and oxidation occur during tungsten and SF_6 plasma etching [8-11]. At present, although researchers have basically confirmed the decomposition products of SF_6 caused by tungsten, the relevant decomposition processes are still unclear. Thermal decomposition reaction is a classical method to study the transformation of substances, which can reveal the thermodynamics and dynamics of the transformation process. However, the apparent activation energy and thermal decomposition kinetics of SF_6 reaction with tungsten powder have not been reported. Herein we adopt synchronous thermogravimetric/differential scanning calorimetric technique investigated thermal decomposition reaction of SF_6 and tungsten powder. Based on the principle of thermal analysis kinetics, the reaction temperature of W powder in SF_6 atmosphere was determined, the apparent activation energy of the reaction was calculated, and the kinetic equation of thermal decomposition reaction was obtained. Meanwhile, the decomposition product and the change of microstructure and composition at high temperature was determined by means of environmental scanning electron microscope (SEM), X-ray energy spectrum analysis (EDS), X-ray photoelectron spectrometer (XPS) and Raman spectroscopy characterization.

2 Experimental parts

2.1 The raw material

SF_6 gas purity 99.99% purchased Chengdu Kemet Special Gas Co., LTD., 4-6 μm tungsten powder.

* Corresponding author: Weiqiang Zhang: zwq@snnu.edu.cn, Yajun Jian: yajunjian@snnu.edu.cn; [+] These authors contributed equally to this work.

2.2 Thermal decomposition reaction of SF_6 and tungsten powder

The thermal decomposition behavior of SF_6 and tungsten was monitored by HCT-4 comprehensive thermal analyzer(Beijing Hengjiu Experimental Equipment Co., LTD.) to obtain the kinetic equation under following conditions: α-Al_2O_3 crucible was used as the reference, the gas flow rate was 100 mL·min^{-1}, sample dosage was 30-50 mg, the heating rate was 5, 10, 20 °C·min^{-1}, and the sample pool was α-Al_2O_3, the temperature range was 25~900°C, and the temperature is ventilated at room temperature for 60 min to ensure that the furnace is filled with SF_6 gas and then the samples are tested by TG and heat flow at different heating rates.

2.3 Preparation of thermal decomposition reaction products

Solid decomposition products were obtained under constant temperature: under the atmosphere of SF_6 , Al_2O_3 crucible was used as the reference, the flow rate was 100 mL · min^{-1}, and the furnace was ventilated at room temperature for 60 min to ensure that the furnace was filled with SF_6 gas, and the heating rate was 30 °C·min^{-1} to 615 °C for 60 min, to make sure SF_6 and tungsten powder react completely. Solid samples were collected for further characterization after the reaction.

2.4 Characterization of solid decomposition products

The morphology and size of decomposition products were characterized by Quanta 200 environmental scanning electron microscope and EDS spectroscopy, with the parameter accelerating voltage at 20 kV and accelerating current at 2 μA. The chemical composition and element binding state of the products were analyzed by Axis UltraDLD X-ray photoelectron spectroscopy (Shimadzu, Japan), and the energy reference was stained carbon C1s (Eb=284.6 eV). The chemical composition was identified by a Reflex laser Raman spectrometer with a transmittance efficiency greater than 30%, scanning range 100-4000 cm^{-1}, signal to noise ratio of silicon third-order peak was better than 22:1 and spectral resolution as visible full spectrum ≤ 1 cm^{-1}.

3 Discussion of results

3.1 Thermal decomposition reaction of SF_6 and tungsten powder

High voltage circuit breaker contacts are mainly composed of tungsten, but the research on the chemical reaction of tungsten in SF_6 atmosphere is not clear, and current research mainly focuses on the plasma etching of SF_6 and tungsten. In order to study the reaction mechanism of SF_6 and tungsten at high temperature, a thermal decomposition reaction model of SF_6 and tungsten powder was established to study the reaction process.

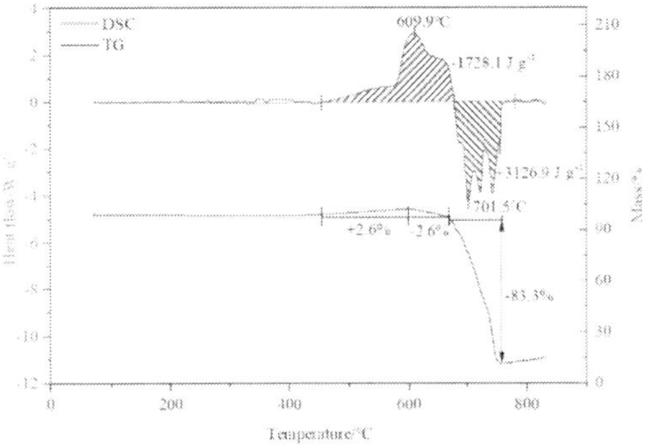

Fig.1 Thermal decomposition reaction of W powder in SF_6 atmosphere

It can be seen from Fig 1 that under SF_6 atmosphere, when heated at a heating rate of 5 °C·min^{-1}, the weight gain of the substance occurs in the range of 457-598 °C with a weight gain rate of 2.6%, and the weight loss of the substance occurs in the range of 598-678 °C with a weight loss rate of 2.6%. The weight gain rate is basically the same as the weight loss rate. The maximum exothermic peak appears on the DSC curve at 609 °C, which may be due to the sulfide reaction $W+SF_6 \rightarrow WS_2$ on the surface, and the heat release is 1728.1 J·g^{-1}. Substantial weight loss occurs at 678-754 °C, with a loss rate of 83.3%. At 702 °C, an endothermic peak appears on the DSC curve, which may be caused by surface fluoridation $WS_2+SF_6 \rightarrow WF_6$, the heat absorption is 3126.9 J·g^{-1}.

3.2 Characterization of solid products of thermal decomposition of SF_6 and tungsten powder

The morphology and elements of raw materials of tungsten powder and solid products after thermal decomposition reaction of SF_6 and tungsten powder were analyzed by environmental scanning electron microscopy (SEM) energy spectrum analyzer. The results are shown in Fig 2. Fig 2a shows the raw material of 4-6 μm tungsten powder, which is distributed in irregular polygons. Fig. 2b shows the scanning electron microscopy of solid decomposition products of 4-6 μm tungsten powder at 615 °C for 60 min. Compared with unprocessed tungsten powder, smaller particles in μm magnitude are generated on the surface, indicating that tungsten powder has undergone chemical reaction with SF_6. Fig. 2c shows the EDS spectra of 4-6 μm tungsten powder raw material, containing only W element. Fig. 2d shows the EDS diagram of tungsten powder at 615 °C for 60 min, which shows that the measured sample contains W, S and O elements, and their mass ratios are 76.91%, 17.06% and 6.03%, respectively. It is speculated that oxides and sulfides may be generated.

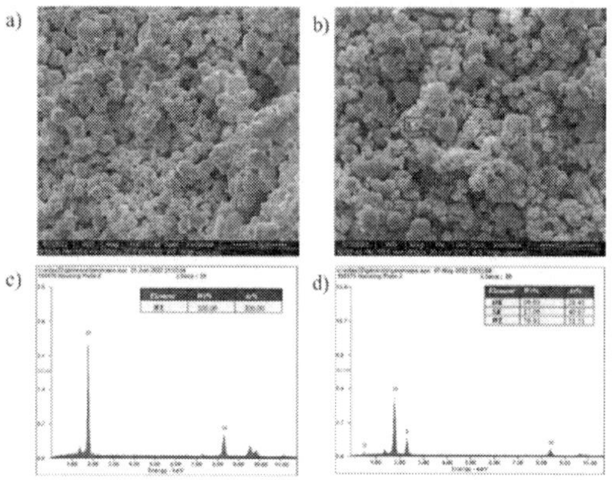

Fig.2 SEM-EDS of tungsten powder raw material, solid products after thermal decomposition reaction of SF_6 and tungsten powder

(**Fig. 2a** raw material of 4-6 μm tungsten powder; **Fig. 2b** SEM images of the solid decomposition products of 4-6 μm tungsten powder at 615 °C for 60 min. **Fig. 2c** EDS spectra of 4-6 μm tungsten powder raw materials. **Fig. 2d** EDS diagram of tungsten powder at 615 °C for 60 min.)

Fig. 3 shows the typical Raman spectra of the solid products of SF_6 reacting with W powder under 532 nm laser excitation in the frequency range of 100-800 cm^{-1}. WS_2 has 10 Raman modes, labeled as first-order modes of LA(M), LA(K), $E^1_{2g}(\Gamma)$ and $A_{1g}(\Gamma)$, second-order modes of 2LA(M) and some combined modes [12-15]. $E^1_{2g}(\Gamma)$ and $A_{1g}(\Gamma)$ are two optical phonon modes in the center of the Brillouin region. $E^1_{2g}(\Gamma)$ corresponds to the in-plane optical mode, and $A_{1g}(\Gamma)$ corresponds to the out-of-plane vibration of the sulfur atom. LA(M) is a longitudinal acoustic mode at point M, which is the collective motion of atoms in the lattice in the plane, similar to sound waves. The additional peaks correspond to the multiphonon modes combined with these first-order modes. Three peaks with good resolution can be observed in Fig. 2, which are 174.1, 350.9 and 420.1 cm^{-1} corresponding to the first modes LA(M), $E^1_{2g}(\Gamma)$ and $A_{1g}(\Gamma)$, respectively.

In order to determine the surface chemical composition and elemental valence of SF_6 reacting with tungsten powder, X-ray photoelectron spectroscopy (XPS) was performed on the solid decomposition products, and the results are shown in Fig 4. It can be seen from Fig. 4A that there are four elements: C, W, O and S after SF_6 reacts with tungsten powder. The C1s peak at 284.6 eV is derived from vacuum pollution. Fig 4b is the high-resolution map of W4f, where the characteristic peaks at 32.4 eV and 34.5 eV correspond to W4f 7/2 and W4f 5/2 of W4+ in WS_2, respectively. The characteristic peaks at 35.5 eV and 37.8 eV correspond to W4f 7/2 and W4f 5/2 of W^{6+} in WO_3, respectively. Fig 4c is the high-resolution diagram of S 2p, where the characteristic peaks at 35.5 eV and 37.8 eV correspond to W4f 7/2 and W4f 5/2 of W^{6+} in WO_3, respectively. Fig 4c is the high-resolution diagram of S2p, where the characteristic peaks

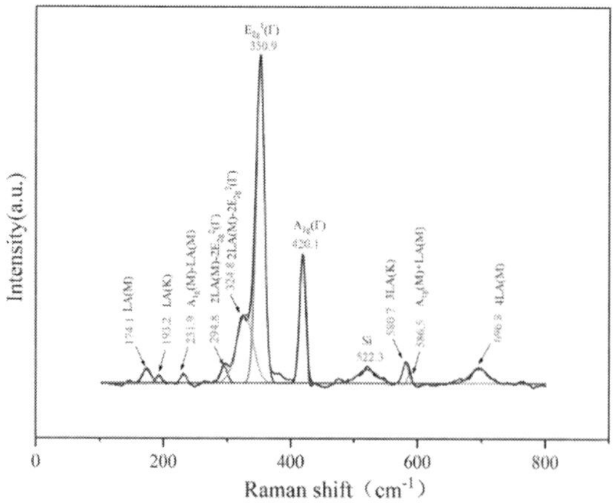

Fig. 3 Raman spectra of solid products of W powder in SF_6 atmosphere

at 162.0 eV and 163.3 eV correspond to S2p3/2 and S2p1/2 of S^{2-} in WS_2 respectively [16-18]. It is speculated that the formation of WS_2 is caused by the reaction between W and SF_6, and a small amount of WO_3 may be caused by the reaction between solid decomposition products and O_2 in the air during the cooling process, but F element is not detected, which is speculated to be caused by the volatilization of WF_6.

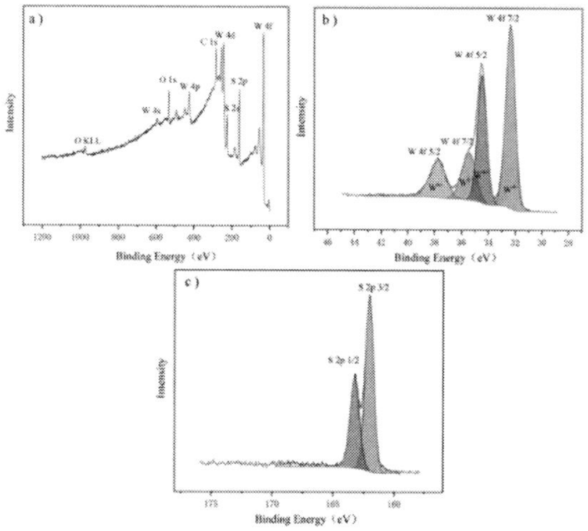

Fig. 4 XPS spectra of solid products of W powder in SF_6 atmosphere a) Full spectrum; b) W4f; c)S2p

3.3 Kinetic analysis of thermal decomposition reaction between SF_6 and tungsten powder

In order to further study the thermal decomposition reaction between SF_6 and tungsten powder and determine the most probable mechanism function F (α), the non-isothermal thermal decomposition kinetics of tungsten powder under SF_6 atmosphere was analyzed on a comprehensive thermal analyzer with different heating rates. The kinetic parameters of thermal decomposition of SF_6 with tungsten powder were calculated by using the Kissinger equation and the Flynn-Wall-Ozawa equation.

Table 1 shows the kinetic parameters of the thermal decomposition of SF_6 and tungsten powder calculated at different heating rates (β). It can be seen from Table 1 that the activation energy of the first stage of the sulfurization reaction of SF_6 and tungsten powder is 152.8 $kJ \cdot mol^{-1}$, and the activation energy of the second stage of the fluoridation reaction is 126.0 $kJ \cdot mol^{-1}$.

Ti was obtained from DSC curves of thermal decomposit ion reaction between SF_6 and tungsten powder at differen t heating rates (β), and the apparent activation energy E_α corresponding to the reaction fraction α was calculated b y Ozawa equation. α and E_α were used to draw α~T and E_α ~α curves, and the results were shown in Fig.5.

Table 1. Apparent activation energy of tungsten powder in SF_6 atmosphere at different heating rates

Stage	β / (°C·min^{-1})	T_p / °C	E_k / (kJ·mol^{-1})	lg(A_k / s^{-1})	r_k	E_o / (kJ·mol^{-1})	r_o
First stage	5	609	149.2	6.16	0.979	156.3	0.982
	10	627					
	20	665					
Mean: Eo=(149.2+156.3)/2=152.8 kJ·mol^{-1}							
Second stage	5	702	120.9	3.67	0.933	131.0	0.947
	10	717					
	20	779					
Mean: Eo=(120.9+131.0)/2=125.95 kJ·mol^{-1}							

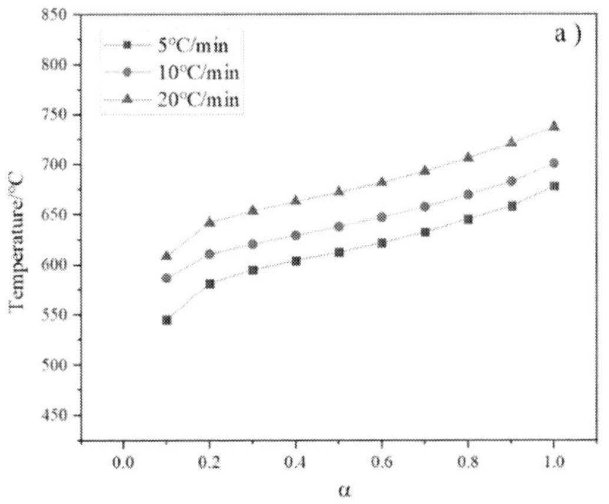

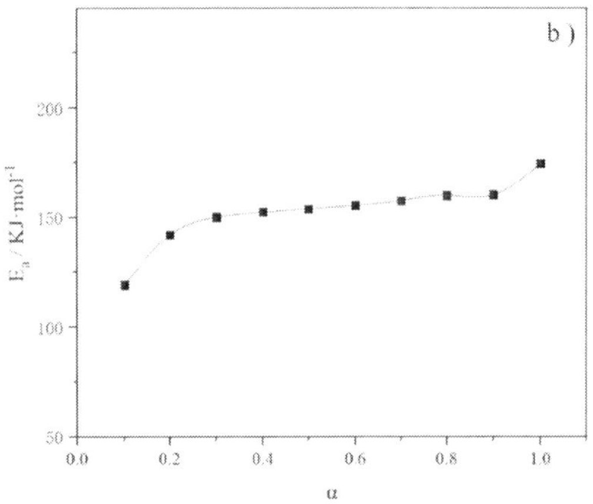

Fig. 5a) T~α curves of SF_6 and W powder at different heating rates in the first stage of thermal reaction obtained by Ozawa method

b) E~α curve of the first stage of the thermal reaction between SF_6 and W powder obtained by Ozawa method

Table 2. calculated values of kinetic parameters for the first stage of SF_6 and W powder thermal reaction

β/ (K/min)	Eq.	E/kJ·mol^{-1}	Log (A/s^{-1})	r	Q
5	General Integral	148.13	5.93	0.999	1.63e^{-4}
	Mac-Callum-Tanner	155.37	6.46	0.999	3.03e^{-5}
	Statava-Sestak	154.87	6.44	0.999	3.03e^{-4}
	Agrawal	148.13	5.93	0.999	1.63e^{-4}
10	General Integral	152.30	6.22	0.999	2.03 e^{-4}
	Mac-Callum-Tanner	159.99	6.78	0.999	3.78 e^{-5}
	Statava-Sestak	159.25	6.73	0.999	3.78 e^{-4}
	Agrawal	152.30	6.21	0.999	2.03 e^{-4}
20	General Integral	158.08	6.51	0.999	1.61 e^{-4}
	Mac-Callum-Tanner	153.501	6.78	0.999	2.89 e^{-5}
	Statava-Sestak	153.112	6.78	0.999	3.00 e^{-5}
	Agrawal	158.08	6.50	0.999	1.61 e^{-4}
	Mean	154.42	6.44		

Stage 1: The conversion rate of 0.4-0.6 interval is different heating rate β corresponding to the thermodynamic data into the equation after the non-isothermal calculation method to obtain the activation energy Mean Eo =154.42 kJ mo^{l-1}, through the Kissinger method and Ozawa method equation to calculate the activation energy Mean Eo=152.80 kJ mol^{-1}, and the activation energies are close. Table 2 shows that each G(α) and F (α) in 41 mechanism functions [19] are substituted

into the General Integral, Mac-Callum-Tanner, Statava-Sestak and Agrawal equations [20]. The kinetic parameters of thermal decomposition reaction between SF_6 and tungsten powder were determined by linear regression and logical selection. By comparison and analysis of the value of activation energy E_0, the most possible mechanism functions can be obtained as the No. 15 function. The differential formula of the mechanism function of No. 15 is $f(\alpha)=4/3[-\ln(1-\alpha)]^{1/4}$. $Ek=149.20$ kJ·mol^{-1} and $A=10^{6.16}$ s^{-1} calculated by Kissinger equation were substituted into equation $d\alpha/dt=(1/\beta)Af(\alpha)\ e^{-Ek/RT}$. The kinetic equation of the first stage thermal decomposition of SF_6 with tungsten powder is obtained as $d\alpha/dt=10^{6.16}(1/\beta)\ 4/3[-\ln(1-\alpha)]^{1/4}\exp(-1.492\times10^5/RT)$.

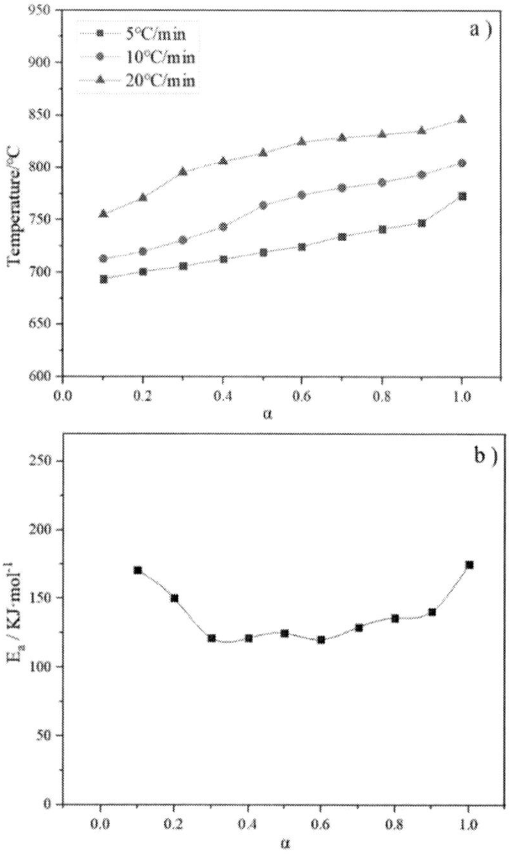

Fig. 4 a) T~α curves of SF_6 and W powder at different heating rates in the second stage of thermal reaction obtained by Ozawa method; **b)** Ozawa method obtained the E~α curve of the second phase of the thermal reaction between SF_6 and W powder

Table 4 Calculated values of kinetic parameters for the second stage of SF_6 and W powder thermal reaction

β/ (K/min)	Eq.	E/kJ·mol^{-1}	Log (A/s^{-1})	r	Q
5	The General Integral	124.57	3.65	0.991	2.01 e^{-3}
	Mac-Callum-Tanner	133.44	4.25	0.993	3.75 e^{-4}
	Statava-Sestak	134.17	4.38	0.993	3.75 e^{-4}
	Agrawal	124.57	3.64	0.991	2.01 e^{-3}
10	The General Integral	121.82	3.28	0.990	6.90 e^{-3}
	Mac-Callum-Tanner	131.25	3.90	0.992	1.30 e^{-3}
	Statava-Sestak	132.10	4.05	0.992	1.30 e^{-3}
	Agrawal	121.82	3.27	0.990	6.90e^{-3}
20	The General Integral	118.49	3.30	0.992	1.68e^{-3}
	Mac-Callum-Tanner	128.86	3.98	0.994	3.17e^{-4}
	Statava-Sestak	129.84	4.14	0.994	3.17e^{-4}
	Agrawal	118.49	3.29	0.992	1.68e^{-3}
	Mean	126.62	3.76		

The conversion rate is 0.3-0.7 in the interval of different heating rate β corresponding to the thermodynamic data into the equation after non-isothermal calculation method to obtain the activation energy Mean $Eo=126.62$ kJ·mol^{-1}, through the Kissinger method and Ozawa method equation calculated the activation energy Mean $Eo=125.95$ kJ·mol^{-1}, and the activation energies are close. Compared with the value of activation energy E_0 calculated by 41 mechanism functions, it can be found that the most possible mechanism functions are No. 12 function. The differential formula of mechanism function No. 12 is $F(\alpha)=5/2[-\ln(1-\alpha)]^{3/5}$. $E_k=120.90$ kJ·mol^{-1} and $A=10^{3.76}$ s^{-1} calculated by Kissinger equation were substituted into equation $d\alpha/dt=(1/\beta)Af(\alpha)\ e^{-Ek/RT}$. The kinetic equation of the first stage thermal decomposition of SF_6 with tungsten powder is obtained as $d\alpha/dt=10^{3.76}(1/\beta)5/2[-\ln(1-\alpha)]^{3/5}\exp(-1.209\times10^5/RT)$.

4 Conclusion

In summary, thermal decomposition reaction of tungsten powder with SF_6 was studied through synchronous thermogravimetric/differential scanning calorimetric technique. In addition, environmental scanning electron microscope (SEM), X-ray energy spectrum analysis (EDS), X-ray photoelectron spectrometer (XPS) and Raman spectroscopy characterization was carried out for solid breakdown products under the condition of high temperature, obtaining the change of microstructure, elements, chemical composition and structure. Based on the principle of thermal analysis kinetics, the reaction temperature of W powder in SF_6 atmosphere was determined, the apparent activation energy of the reaction was calculated, and the kinetic equation of thermal decomposition reaction was obtained. The main conclusions are as follows:

(1) Tungsten powder will interact with SF_6 at 600 °C, and will interact with SF_6 to form WS_2 and WO_3 at 615 °C. When the temperature reaches 750 °C, that is, before reaching the melting and boiling point, it will evaporate completely in the form of gas WF_6.

(2) The reaction between SF_6 and tungsten can be divided into two stages. The first stage is exothermic reaction, $W+SF_6 \rightarrow WS_2$, $\Delta H = -1728.1\ J \cdot g^{-1}$. The thermal decomposition kinetic equation of sulfide reaction is as follows: $d\alpha/dt=10^{6.16}(1/\beta)\ 4/3[-\ln(1-\alpha)]^{1/4} \exp(-1.492 \times 10^5/RT)$. The second stage is endothermic reaction, reaction equation: $WS_2+SF_6 \rightarrow WF_6$, $\Delta H =+3126.9\ J\cdot g^{-1}$. The kinetic equation of thermal decomposition of fluorination reaction is: $d\alpha/dt=10^{3.76}(1/\beta)5/2[-\ln(1-\alpha)]^{3/5}\exp(-1.209\times10^5/RT)$.

Acknowledgement

The authors acknowledge the support of Project Supported by Science and Technology Project of State Grid Shaanx Electric Power Co. Ltd. (Research and application of SF_6 switch latent fault based on isotope tracing and material decomposition alignment)(project no. : 5226KY22000H)

References

1. L.G. Christophorou, J.K. Olthoff, R.J. Van Brunt, IEEE Electr. Insul. Mag, 13, 20 (1997)
2. J. Zhang, J.Z. Zhou, and Q. Liu, Environ. Sci. Technol, 47, 6493(2013)
3. H. C. Cowen, F. Riding, E. Warhurst, J. Chem. Soc, 4168 (1953)
4. R. Basta, B.G. Harvey, A.M. Arif and R.D. Ernst, J. Am. Chem. Soc, 127, 11924 (2005)
5. K. Hirooka, H. Kuwahara, M. Noshiro and Y. Jitsugii, IEE Japan Trans., 95, 483 (1975)
6. W.G. Chen, L.L. Dong, et al., J. Mater., Sci. - Mater. Electron, 27, 5584, (2016)
7. A. Picard and G. Turban, Plasma Chem. Plasma Process, 5, 333 (1985).
8. M.C. Peignon, C. Cardinaud, and G. Turban, J. Appl. Phys, 70, 3314 (1991).
9. G. Turban, J.F. Coulon, Thin Solid Films, 176, 289, (1989)
10. M.C. Peignon, C. Cardinaud, G. Turban, J. Appl. Phys, 70, 3314, (1991)
11. R. Petri, D. Henry, and N. Sadeghi, J. Appl. Phys, 72, 2644, (1992)
12. N. Peimyoo, J. Shang, W. Yang, Y. Wang, C. Cong and T. Yu, Nano Res, 8, 1210, (2015)
13. A. Berkdemir, H.R. Gutierrez, A.R. Botello-Mendez, et al., Sci. Rep.-UK, 3, 1755, (2013)
14. S. Qiao, H. Yang, Z. Bai, G. Peng, X. Zhang, Proc. of the 2017 5th Int. Conf. on Mechatronics, Materials, Chemistry and Computer Engineering (ICMMCCE 2017)
15. C. Cong, J. Shang, X. Wu, et al., Adv. Opt. Mater, 2,131 (2014)
16. Tang M, Wu Y, Yang J, et al., J. Alloy. Compd, 858,157747. (2021)
17. X. Zhou, X. Yang, H. Li, et al., J. Mater. Chem, A 5, 15552 (2017)
18. Y. Yue, J. Chen, Y. Zhang, et al., ACS Appl. Mater. Interfaces, 10, 22435 (2018)
19. R.Z. Hu, Q.Z. Shi, Thermal Analysis Kinetics Beijing: Science Press (2001)
20. R.Z. Hu, Z.Q. Yang, Y.J. Liang, Thermochim. Acta, 176, 209, (1991)

Study on Calculation of Inlet Steam and Inlet Water Flow of Drain Pump of Low Pressure Heater

Lingkai Zhu [1], Wei Zheng [1], Qian Wang [1], Ziwei Zhong [1], Zhiqiang Gong [1]

[1] State Grid Shandong Electric Power Research Institute, Jinan, China

Abstract. In the traditional drainage system of low pressure heater of thermal power unit, the measuring point of drain pump is usually added to directly measure the water and steam flow of low pressure heater of drain pump with instruments. This method is time-consuming and labor-intensive, and the measurement results are affected by many environmental factors. In order to solve this problem, this paper proposes an automatic calculation method based on the principle of heat balance.

1. Introduction

In recent years, with the development of thermal power units to large capacity and high parameters, the influence of regenerative system on thermal economy of units has gradually increased[1]. The drainage system of the low pressure heater of the traditional thermal power unit is mainly of the self draining type. The drainage of the low pressure heater flows automatically from high to low and finally discharges to the condenser. The drainage water flows automatically step by step[2,3]. On the one hand, the drainage water releases heat from the lower level to drain the lower level steam inlet, reducing the effect of regenerative steam inlet; on the other hand, the drainage water is discharged to the condenser, increasing the cold source loss. Therefore, in a large capacity thermal power unit, a drainage pump is set in a certain stage or several stages of low-pressure heater system to drain the water at this stage to the outlet of the heater at this stage, improving the outlet water temperature of the heater, and squeezing the steam inlet of a higher stage of low-pressure heater, Meanwhile, the loss of cold source is reduced.

The traditional measurement method usually uses the instrument to measure directly[4,5]. When adding drain pump measuring points in the heat recovery system, appropriate flow meters should be selected according to its performance requirements, fluid characteristics, installation requirements, environmental conditions and cost factors. After adding this measuring point, there will be a certain impact on the flow, and it needs to spend manpower and material resources to maintain it. Therefore, it is time-consuming and labor-intensive to directly solve the problem of drain pump flow from the hardware, and there are many inconveniences.

In view of the shortcomings of the prior art, a clear and convenient method for calculating the water and steam inlet flow of the low-pressure heater with drain pump is provided. This method is based on the heat balance method and uses the cycle iteration function of excel to calculate the water inlet and steam inlet flow of the low-pressure heater with drain pump conveniently and simultaneously.

2. Low pressure heater system with drain pump

In order to improve the regenerative effect of the heater, at present, drain pump is usually set for the heater. Figure 1 shows a low-pressure heater system with drain pump. The drain water from #5 and #6 low-pressure heaters flows automatically to #7 low-pressure heater stage by stage. The drain pump of #7 low-pressure heater pumps the drain water from #5, #6 and #7 three-stage low-pressure heaters to the outlet of #7 low-pressure heater. Therefore, after the drain pump is added, the condensate flow in and out of the heater is no longer the same. The outlet flow of #7 low-pressure heater is equal to the main condensate flow at the inlet of the deaerator. The water inlet flow and steam inlet flow of the #7 low-pressure heater are unknown.

* Corresponding author: zhulingkai@woyoxin.com

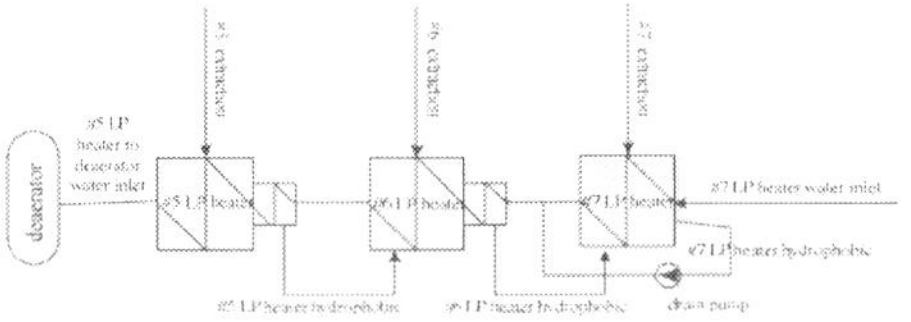

Figure 1. Overall strategy system of security protection.

3. Calculation Method for Inlet Steam and Inlet Water Flow of Drain Pump of Low Pressure Heater

There are n-stage low-pressure heaters in series. The outlet of the first stage low-pressure heater is connected to the deaerator. The n-stage low-pressure heater is equipped with a drain pump. The drain water of the n-stage low-pressure heater is mixed with the feed water at the outlet of the n-stage low-pressure heater by using the drain pump. The drain water of other low-pressure heaters flows automatically step by step. Its characteristics are: the heat balance equation is established for the 1st, 2nd ... and n-1 low-pressure heaters, and the extraction flow of the 1st, 2nd ... and n-1 low-pressure heaters is calculated. Wherein, according to the extraction pressure, extraction temperature, inlet water temperature, outlet water temperature and drain water temperature of the first stage low-pressure heater and the main condensate flow from the first stage low-pressure heater to the deaerator, the heat balance equation of the first stage low-pressure heater is established; According to the steam extraction pressure, steam extraction temperature, inlet water temperature, outlet water temperature and drain water temperature of the 2nd ... and n-1 stage low-pressure heaters and the main condensate flow from the 1st stage low-pressure heater to the deaerator, the heat balance equation of the 2nd ... and n-1 stage low-pressure heaters is established.

According to the steam extraction flow of the 1st, 2nd ... and n-1 stage low-pressure heaters and the main condensate flow from the 1st stage low-pressure heater to the deaerator, the heat balance equation and flow balance equation are established for the nth stage low-pressure heater at the same time. Based on the iterative method, the steam extraction flow and water inlet flow of the nth stage low-pressure heater are calculated, that is, the steam inlet flow and water inlet flow of the drain pump of the low-pressure heater are respectively.

The specific method for calculating the extraction steam flow of the 1st, 2nd ... and n-1 low-pressure heaters is as follows:

Obtain the extraction pressure, extraction temperature, inlet water temperature, outlet water temperature and drain water temperature of the first stage low-pressure heater, and calculate the extraction enthalpy, inlet water enthalpy, outlet water enthalpy and drain water enthalpy of the first stage low-pressure heater using the thermodynamic property model of water and steam. Obtain the main condensate flow from the first stage low-pressure heater to the deaerator. According to the above parameters, establish a heat balance equation for the first stage low-pressure heater, and calculate the extraction steam flow of the first stage low-pressure heater.

Obtain the extraction pressure, extraction temperature, inlet water temperature and drain water temperature of the 2nd ... and n-1 low-pressure heaters, and calculate the extraction enthalpy, inlet water enthalpy and drain water enthalpy of the 2nd ... and n-1 low-pressure heaters using the thermodynamic property model of water and steam. According to the main condensate flow from the first stage low-pressure heater to the deaerator, the extraction enthalpy, inlet enthalpy and drain enthalpy of the 2nd ... and n-1 low-pressure heaters, the heat balance equation is established for the 2nd ... and n-1 low-pressure heaters, and the extraction flow of the 2nd ... and n-1 low-pressure heaters is calculated.

Obtain the extraction steam pressure, extraction steam temperature and inlet water temperature of the nth stage low-pressure heater, and calculate the extraction steam enthalpy and inlet water enthalpy of the nth stage low-pressure heater using the thermodynamic property model of water and steam. Assumed water inlet flow of the nth stage low-pressure heater is set, and the heat balance equation of the nth stage low-pressure heater is established by combining the extraction flow of the 1st, 2nd ... and n-1 stage low-pressure heaters, the drain enthalpy of the n-1 stage low-pressure heater, the inlet enthalpy of the n-1 stage low-pressure heater, the main condensate flow from the first stage low-pressure heater to the deaerator, the extraction enthalpy of the n-stage low-pressure heater, and the inlet enthalpy of the n-stage low-pressure heater. The extraction steam flow of the nth stage low-pressure heater is calculated on the premise of the assumed inlet water flow of the nth stage low-pressure heater, which is the inlet steam flow of the drain pump of the low-pressure heater

The flow balance equation is established for the nth stage low-pressure heater according to the steam extraction flow of the 1st, 2nd ... and nth stage low-pressure heaters and the main condensate flow from the first stage low-pressure heater to the deaerator, and the arithmetic water inlet flow of the nth stage low-pressure heater is obtained through reverse calculation, so that the assumed water inlet flow of the nth stage low-pressure heater is equal to the arithmetic water inlet flow of the nth stage low-pressure heater, which is calculated based on the iterative method, The final iteration value is taken as the water inlet flow of the nth stage low-pressure heater, that is, the water inlet flow of the drain pump of the low-pressure heater.

Excel is used for automatic calculation based on iterative method. IFC-97 industrial water and steam thermodynamic property model in Excel is used for calculation.

The main condensate flow from the first stage low-pressure heater to the deaerator is measured by measuring the flow differential pressure from the first stage low-pressure heater to the main condensate of the deaerator. The flow differential pressure from the first stage low-pressure heater to the main condensate of the deaerator is measured by ASME flow nozzle, and then the main condensate flow from the first stage low-pressure heater to the deaerator is calculated by using the ASME nozzle flow calculation model in Excel.

The extraction steam pressure of the 1st, 2nd ... and n-stage low-pressure heaters is measured by pressure transmitter.

The steam extraction temperature, water inlet temperature, water outlet temperature and drain water temperature of the first stage low pressure heater, the steam extraction temperature, water inlet temperature and drain water temperature of the 2nd ... and n-1 stage low pressure heaters, and the steam extraction temperature and water inlet temperature of the n-stage low pressure heater are measured by temperature elements.

The heat balance equation is established for the first stage low-pressure heater, and the extraction steam flow of the first stage low-pressure heater is calculated.

$$G_1 h_1 + G_{ns} h_{1i} = G_1 h_{1h} + G_{ns} h_{1o} \quad (1)$$

Wheres:G_1-steam extraction flow of the first stage low-pressure heater, G_{ns}-main condensate flow from the 1st stage low-pressure heater to deaerator, h_1-enthalpy value of stage 1 steam extraction, h_{1i}-enthalpy value of inlet water of the first stage low-pressure heater, h_{1h}-drain enthalpy of the 1st stage low-pressure heater, h_{1o}-enthalpy value of outlet water of the first stage low-pressure heater.

According to the heat balance equation, the steam extraction flow of the first stage low-pressure heater is obtained as follows:

$$G_1 = \frac{G_{ns}(h_{1o}-h_{1i})}{h_1-h_{1h}} \quad (2)$$

Establish the heat balance equation for the 2nd ... and n-1 low-pressure heaters, and calculate the extraction steam flow of the 2nd ... and n-1 low-pressure heaters.

$$G_{n-1}h_{n-1} + G_{n-2}h_{n-2h} + G_{ns}h_{n-1i} = (G_{n-2} + G_{n-1})\, h_{n-1h} + G_{ns}h_{n-2i} \quad (3)$$

Wheres:G_{n-1}-steam inlet flow of stage n-1 low-pressure heater, h_{n-1}-enthalpy value of steam inlet of stage n-1 low-pressure heater, h_{n-1i}-enthalpy value of water inlet of n-1 low-pressure heater, h_{n-1h}-drain enthalpy of stage n-1 low-pressure heater.

According to the heat balance equation, the steam inlet flow of n-1 low-pressure heater is obtained as follows:

$$G_{n-1} = \frac{G_{ns}(h_{n-2i}-h_{n-1i})+G_{n-2}\,(h_{n-1h}-h_{n-2h})}{h_{n-1}-h_{n-1h}} \quad (4)$$

The heat balance equation of the nth stage low-pressure heater is as follows:

$$(G_{n-2} + G_{n-1})h_{n-1h} + G_{ni}h_{ni} + G_n h_n = G_{ns}h_{n-1i} \quad (5)$$

Wheres:G_n-steam inlet flow of the nth stage low-pressure heater, G_{ni}-assumed inlet water flow of the n-stage low-pressure heater, h_n-enthalpy value of steam inlet of the nth stage low-pressure heater, h_{ni}-enthalpy value of inlet water of the nth stage low-pressure heater.

According to the heat balance equation, the steam inlet flow of the nth stage low-pressure heater is obtained as follows:

$$G_n = \frac{G_{ns}h_{n-1i}-G_{ni}h_{ni}-\,(G_{n-2}+G_{n-1})\,h_{n-1h}}{h_n} \quad (6)$$

The flow balance equation of the nth stage low-pressure heater row is as follows:

$$G_{ni}' = G_{ns} - G_1 - G_2 -、、- G_{n-1} - G_n \quad (7)$$

Wheres:$\mathbf{G_{ni}}$'-arithmetic water inlet flow of the nth stage low-pressure heater.

4. Conclusion

The calculation method of inlet steam and inlet water flow of drain pump of low-pressure heater proposed in this paper is based on the heat and flow balance of the heater. The formula for calculating the inlet steam and inlet water flow is input into the excel table. The automatic iteration function of excel is used for automatic calculation, and the inlet steam and inlet water flow can be calculated at the same time. It is simple, intuitive and accurate, especially suitable for the thermal test practitioners of power plants to conduct thermal calculation

References

1. Xiaolong Guo. Optimization of steam turbine regenerative system in coal power plant[J]. Mechanical management development, 2022,(2):169-170.
2. Lujiang Li, Jian Wang. Improvement and economic analysis of drainage system of low-pressure heater [J]. Thermal power generation, 2008, 37 (12): 60-64.
3. Qindao Li. Economic Analysis of Drainage System of Low Pressure Heater of Domestic 300 MW Unit [J]. China Power, 1993, (3): 41-45.
4. Yingchun Zhao, Chaofan Zhang, Qi Duan. Optimization of regulation and control strategy for low pressure heater drain pump system [J]. Electric power survey and design, 2021, (11): 41-44
5. Guo Zheng. Application of Frequency Conversion Speed Regulation Technology in Drainage System of Low Pressure Heater [J]. Auxiliary equipment of power station, 2002, (02): 3-6.

A Novel Transmission Network Planning Framework under the New Power System

Chaochen Yan[1], Ye Tao[2], Yu Zhao[1], Jiaxing Wang[1], Yuqing Wang[1,*] and Jiang Zhu[3]

[1]North China Electric Power University (Baoding), Department of Economics and Management, 071003 Baoding, China
[2]Jiangsu Keneng Electric Power Engineering Consulting Co., Ltd., 210036 Nanjing, China
[3]Nyocor Co., Ltd, 100032 Beijing, China

Abstract. With the continuous development and construction of the new power system, there would be a large number of renewable energy connected to the electric network in the future. And the uncertainty of renewable energy would have a greater impact on the power grid, such as excessively large investment and low utilization rate. In view of this, this paper fully considers the adjustable comprehensive resources of the power supply and the load side, and proposes a novel transmission network planning framework under the new power system. Further, from the aspects of economy and reliability, an evaluation system of novel transmission network planning framework is established. At last, taking a municipal 220 KV power grid as an example, the novel transmission network planning framework is applied, and the effectiveness of the proposed planning framework is verified.

1. Introduction

With the rapid development of the new power system, a high proportion of renewable energy on the grid would become a fundamental feature of the future power system[1]. However, the randomness and uncertainty of renewable energy such as wind and photovoltaic power have a profound impact on the planning of power system[2]. The traditional grid planning method is based on the annual maximum load and does not consider the grid connection of renewable energy and other resources, possibly resulting in large investment and low utilization rate. Therefore, there is an urgent need for a better planning framework to meet the new needs of network planning under the development of the new power system.
In recent years, there has been a growing interest in the novel transmission network planning. The literature [3,4] analysed the uncertainties in the new power system, and used robust theory for optimization modelling. The literature [5,6] demonstrated the impact of high percentage of renewable energy connected to the network, and provided an outlook on transmission network planning. The literature [7,8] considered the uncertainty of renewable energy and the characteristic of bidirectional power flow of battery energy storage system, and a transmission grid planning approach included energy storage was proposed. The literature [9,10] constructed a transmission network planning model considered demand response, which analysed the influence of demand response resources on the network. In summary, the existing studies are limited to consider the impact of a single adjustable resource on transmission network planning, while few studies have combined renewable energy, energy storage and demand response resources as adjustable comprehensive resources into the novel transmission grid planning.
Therefore, this paper fully considers the need of transmission network planning mode change under the development of adjustable comprehensive resources. Meanwhile, a novel transmission grid planning framework is established, which focuses on two key methods in novel transmission grid planning, including confidence capacity evaluation of renewable energy and load forecasting. Further, an evaluation system of novel transmission network planning framework is constructed. Finally, the effectiveness and feasibility of the proposed planning framework are evidenced through example analysis.

2. A novel transmission network planning framework considering adjustable comprehensive resources

2.1 Demand analysis of transmission network planning improvement under new power system

Under the construction of new power system, the planning of the transmission grid is significantly impacted by the rapid growth of renewable energy, energy storage, demand response resources and other comprehensive resources, in particular: (1) The renewable energy resources connected to network leads to enhanced random fluctuations in tide and increased planning targets, so that network planning probably increase reliability investment while considering environmental protection requirements;

* Corresponding author: wangyuqingncepu@foxmail.com

(2) The application of energy storage transforms some renewable energy with random output into relatively stable adjustable power supply, which can reduce the investment in reliability of power grid, but also makes the power boundary of power grid planning change from static to dynamic interval; (3) The introduction of demand response can improve the optimal allocation of resources, and decrease the investment of the network because it can effectively reduce the peak load and compensate for the fluctuation of renewable energy output.

Traditional power planning is based on the current and planned conventional power supply scale and annual maximum load, carrying out the power balance and the substation capacity demand calculation of the planning year. However, it is simple to cause the problem of high investment and poor utilization rate of the grid. Because the supply side only considers the traditional power supply without taking into account the renewable energy output during planning, and the role of demand-side resources is not considered during load forecasting. Therefore, the impact of adjustable resources on power network planning is analyzed to improve: (1) In terms of planning ideas, the impact of adjustable comprehensive resources on the power system is fully considered; (2) In terms of planning process, a new planning mechanism is explored to form a planning process that fully considers market factors; (3) In terms of planning methods, considering economy and reliability, the transmission network planning theory and method for adjustable comprehensive resources are proposed.

2.2 A novel transmission network planning guidelines considering adjustable comprehensive resources

To build a novel transmission network planning model considering adjustable comprehensive resources, the basic goal should be to construct the new power system and adapt to economic and social development. Reasonable arrangements for the construction of power grids at all levels and promote the optimization of grid structure, in order to achieve reliable power supply and efficient economic operation of the power grid.. Specific ideas are as follows.

2.2.1 Consider the characteristics and impacts of adjustable comprehensive resources

Given that the combination of "renewable energy + energy storage" can provide relatively stable power supply, it should be included in the consideration of regional electric power and energy balance. Likewise, the renewable energy and energy storage should be included in the transmission grid planning as conventional generating sets using confidence capacity evaluation. In addition, load forecasting is carried out for demand response resources in adjustable comprehensive resources, and the results are used as a reference for transmission network planning.

2.2.2 Select the reasonable time sections to carry out safety checks

Considering the fluctuation and uncertainty of renewable energy power, the node with the worst power stability and the lowest power is selected as the time sections to carry out the network security verification according to its power curve. For example, summer peak load and renewable energy full output, summer peak load and renewable energy zero output, winter trough load and renewable energy full output, winter trough load and renewable energy zero output and other extreme conditions.

2.2.3 Demand response cost should be considered in the economic evaluation of transmission network planning scheme

Considering that the introduction of demand response resources into grid planning affects participant costs and system costs, the cost of demand response should be taken into account in the economic evaluation of transmission grid planning framework. In order to more reliably measure the effectiveness and economy of the new planning model.

3. Key methods for the novel transmission network planning considering adjustable comprehensive resources

3.1 Confidence Capacity Evaluation of Renewable Energy

Confidence capacity of renewable energy refers to the capacity of conventional generating sets that can be replaced by renewable energy generators under the premise of equal reliability. It can make the intermittent wind and photovoltaic power and conventional generating sets installed capacity at the same level to compare. In this paper, the equivalent capacity of conventional unit is used to assess the confidence capacity of the wind-PV-ES hybrid power system, that is, the unit with random outage rate is used to replace the renewable energy. When the reliability of replaced system is equal to the system with renewable energy, the capacity of replaced unit with random outage rate is the equivalent capacity of conventional unit.

Considering the complementary benefits between wind, photovoltaic and energy storage, under the new grid planning framework, the wind power, photovoltaic power and energy storage connected to each 220KV electric substation are regarded as a wind-PV-ES hybrid power system. The confidence capacity is defined as the size of the capacity of conventional unit that the wind-PV hybrid power system can be equivalent to under the same reliability. When the confidence capacity is defined under the same reliability, the capacity of wind-PV hybrid power system can be equivalent to the conventional unit. The calculation method is as follows.

Assuming that the output power of the wind-PV-ES hybrid power system is P_h.

$$P_h = P_W + P_{PV} + X_{dc} \quad (1)$$

Where P_W, P_{PV} and X_{dc} are the output power of wind turbine, photovoltaic generator set and energy storage plant respectively. If equation (2) is satisfied, the confidence capacity of the wind-PV-ES hybrid power system C_{un} is considered to be equal to the installed capacity of the equivalent unit, C_{na}.

$$f\{C_r, C_W, C_{PV}\} = f\{C_r, C_{na}\} \quad (2)$$

$f\{C_r, C_W, C_{PV}\}$ indicates the reliability of the system with a conventional unit capacity of C_r, a wind power capacity of C_W and a photovoltaic power capacity of C_{PV}. And $f\{C_r, C_{na}\}$ indicates the reliability of the system when the capacity of the conventional unit C_r is equivalent to the installed capacity of the unit C_{na}. At this point, the capacity reliability of the wind-PV hybrid power system can be defined as

$$\lambda = \frac{C_{na}}{C_{PV} + C_W} \quad (3)$$

In terms of system reliability metrics, this paper uses the expected energy not served (EENS) as a reference indicator. The expected energy not served (EENS) of the system varies depending on the load P_L and unit status as follows.

$$f_{EENS} = \begin{cases} P_L - \int_{P_{h\min}}^{P_{h\max}} P_h \cdot f_{P_h}(P_h)dP_h, P_L > P_{h\max} \\ \int_{P_{h\min}}^{P_L} (P_L - P_h) \cdot f_{P_h}(P_h)dP_h, P_{h\min} \le P_L < P_{h\max} \end{cases} \quad (4)$$

Where $P_{h\max}$ and $P_{h\min}$ represent the maximum and minimum values of power generation of wind-PV-ES hybrid power system.

3.2 Load forecasting correction considering demand-side resources

Incorporating demand response resources into power grid planning is conducive to improving the optimal allocation ability of resources and reducing system costs. Based on the traditional load forecasting, this paper proposes a correction model of load forecasting considering the load reduction effect of demand-side resources to realize regional maximum load forecasting considering demand-side resources.

The regional maximum load forecast considering demand side resources is

$$P_{DR}^0 = (1 - \xi) \cdot \gamma \cdot P_{nor}^0 \quad (5)$$

$$\xi = P_d / P_{nor}^0 \quad (6)$$

Where: P_{DR}^0 represents the maximum load forecast considering demand-side resources, MW; ξ represents the demand response rate, %; γ represents the load simultaneous rate, %; P_{nor}^0 represents the maximum load forecast without considering demand-side resources, MW; P_d represents the demand response load forecast, MW.

In summary, through the change of the maximum load before and after considering the demand-side resources, the substation capacity that can be avoided or delayed to construct can be calculated, as shown in the following formula.

$$\Delta C = R\left(P_{nor}^0 - P_{DR}^0\right) \quad (7)$$

Where: ΔC indicates the substation capacity that can be avoided or delayed after considering demand-side resources, MVA; R indicates the capacity-load ratio.

4. Evaluation of novel transmission network planning based on adjustable comprehensive resources

4.1 Reliability evaluation of novel transmission network planning based on adjustable comprehensive resources

The reliability evaluation of transmission grid is mainly used to evaluate the reliability of power supply, including two aspects: security and reliability evaluation of power supply. For power supply security evaluation, the novel transmission network planning framework takes into account the uncertainty of renewable energy output, mainly selects the extreme conditions for security verification. Through the power flow calculation, short-circuit current calculation and "N-1" security check of the planned grid, the security and stability analysis are carried out. For the reliability evaluation of grid, the target grid is evaluated by setting reliability evaluation indexes of power supply. For instance, insufficient load of system, insufficient energy of system, average interruption frequency of system, average interruption duration of system, average reliability of power supply of system, and average unavailability of power supply of system.

4.2 Economic evaluation of novel transmission network planning based on adjustable comprehensive resources

The economic evaluation is mainly based on two aspects: investment cost of construction and whole life cycle cost. Among them, the investment cost of construction aspect is mainly selected the increase power supply of unit investment and increase load of the unit investment as the main indicators to measure the economy of the scheme; The whole life cycle cost aspect mainly considers the total life cycle cost including the construction investment, operation and maintenance cost, and the call cost of adjustable comprehensive resources.

The economic evaluation of the construction investment dimension mainly selects two indicators, including incremental power supply of unit investment and incremental load supply of unit investment, which are calculated as follows.

$$IS_{UI} = (S - S_0)/C_C \tag{8}$$

$$IL_{UI} = (L - L_0)/C_C \tag{9}$$

Where IS_{UI} and IL_{UI} are the incremental power supply of planned unit investment and the incremental load of planned unit investment, respectively; S and L are the power supply of planned end-of-period and the maximum load of planned end-of-period; S_0 and L_0 are the power supply of end-of-period of the previous plan and the maximum load of end-of-period of the previous plan; C_C is the construction investment.

The whole life cycle cost needs to consider the construction cost, the grid operation and maintenance cost, and the call cost of adjustable comprehensive resources, etc., calculated as follows.

$$C = C_C + C_O + C_L + C_D - C_R \tag{10}$$

Where: C_C indicates the project construction investment, including the investment of line and equipment, C_O indicates the operation and maintenance cost, C_L indicates the network loss cost, C_D indicates the demand response cost, and C_R indicates the depreciated residual value of fixed assets at the end of the operation. Each cost is calculated as follows.

4.2.1 Project construction investment

Project construction investment C_C includes investment in line construction C_{line}, and investment in equipment C_{equ}, which is related to public network construction works, as indicated below.

$$C_C = C_{line} + C_{equ} \tag{11}$$

4.2.2 Operation and maintenance cost

The operation and maintenance cost consider the line and main transformer.

$$C_O = \sum_n^N (1+i)^n R_O \cdot A \tag{12}$$

Where: A is the total fixed assets for the year, including the line and main transformer; N is the length of the operating period, and i is the annual interest rate.

4.2.3 Network loss cost

Network loss cost C_L includes line loss cost C_{L-line} and the loss cost of main transformer C_{L-equ}.

① Line loss cost

$$C_{L-line} = \sum_n^N (1+i)^n N_L \cdot \tau \cdot \Delta P_{max} \cdot P_{flat} \tag{13}$$

$$\Delta P_{max} = 3I_L^2 \cdot R = \left(\frac{L_{max}}{N_L \cdot U \cdot \cos\varphi}\right)^2 \cdot L_L \cdot R_L \tag{14}$$

Where: ΔP_{max} is the line loss at maximum load; N_L is the number of lines in the year; I_L is the current value on a single line; L_{max} is the maximum load in the year.

② Main transformer loss cost

$$C_{L-equ} = \sum_n^N (1+i)^n N_T \cdot \left[\Delta P_0 \cdot T + \Delta P\left(\frac{L_{max}}{N_T \cdot S \cdot \cos\varphi}\right)^2 \cdot \tau\right] \cdot P_{flat} \tag{15}$$

Where: N_T is the number of main transformers for the year; T is the transformer operating time, assuming T =8760 h.

4.2.4 Adjustable comprehensive resources utilization cost

Demand response costs C_D include load transfer cost C_{D-S} and load shedding cost C_{D-C}.

$$C_D = C_{D-S,fix} + C_{D-S,price} \tag{16}$$

①Load transfer cost

Load transfer cost C_{D-S} includes fixed investment costs $C_{D-S,fix}$ and reduction of electricity revenue caused by the transfer of load $C_{D-S,price}$.

$$C_{D\text{-}S} = C_{D\text{-}S,\ fix} + C_{D\text{-}S,price} \tag{17}$$

$$C_{D\text{-}S,price} = \sum_n^N (1+i)^n L'_{max} \cdot R_{D\text{-}S} \cdot T_{D\text{-}S} \cdot (P_{peak} - P_{valley}) \tag{18}$$

Where: L'_{max} is the maximum load forecast for the year.

②Load reduction cost

The load reduction cost $C_{D\text{-}C}$ is the load-shedding compensation cost.

$$C_{D\text{-}C} = \sum_n^N (1+i)^n L'_{max} \cdot R_{D\text{-}C} \cdot P_{D\text{-}C} \tag{19}$$

4.2.5 Depreciable residual value of fixed assets

$$C_{LR} = R_{LR} \cdot C_{line} \tag{20}$$

$$C_{\mathrm{TR}} = R_{\mathrm{TR}} \cdot C_{equ} \quad (21)$$

The depreciable residual value of fixed assets C_{R} is the residual value of fixed assets at the time of reaching the depreciable life, which is equal to the depreciation residual ratio multiplied by the investment in fixed assets (subscript L represents lines and subscript T represents main transformers, respectively).

5. Example analysis

5.1 Basic data

Taking a municipal 220 kV power network as an example, based on the load and power supply level of the original grid plan, the impact of demand response on the load of this grid and the impact of renewable energy power plants and energy storage plants on the power supply output is considered. It can be divided into five scenarios according to the proportion of energy storage allocation and the proportion of demand response participation, details of which are shown in Table 1.

Table 1. Scenario Settings.

Planning Scenarios	Energy storage configuration ratio	Demand-side resources ratio
Scenario 1 - Baseline Scenario	10%	10%
Scenario 2 - Low energy storage scenario	5%	10%
Scenario 3 - High Energy Storage Scenario	15%	10%
Scenario 4 - High Demand Response Scenario	10%	12.5%
Scenario 5 - Low demand response scenario	10%	7.5%

5.2 Analysis of planning results

Figure 1 and Figure 2 represent the number of capacity projects and the substation capacity demand under the original power grid planning and five scenarios, respectively. From the figures, it can be seen that after considering the adjustable comprehensive resources, the number and capacity of grid planning under the five scenarios have been improved. The highest construction number of planned project is 7 for scenario 5 and the lowest is 2 for scenario 4, which is less than half compared with 14 in the original planning report. And the planned project capacity of the grid for Scenario 1, Scenario 2, Scenario 3, Scenario 4, and Scenario 5 is reduced by 79%, 78%, 83%, 93%, and 75%, respectively, compared to the original planning capacity.

5.3 Evaluation of planning results

5.3.1 Reliability evaluation

The grid planning solutions under five different scenarios were subjected to tidal currents and short-circuit calculations, and "N-1" safety and stability analyses. The result shows that there are no line or equipment overloads occurred under the "N-1" calibration of the five scenarios; the short-circuit current levels of each voltage level were within the rated range. All five scenarios meet the reliability requirements.

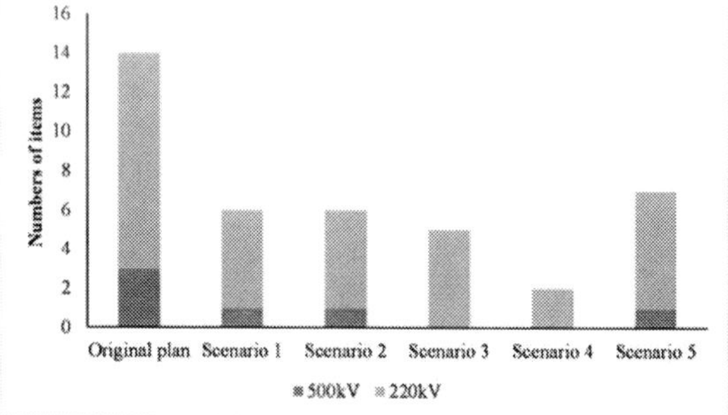

Figure 1 Number of network projects under different scenarios.

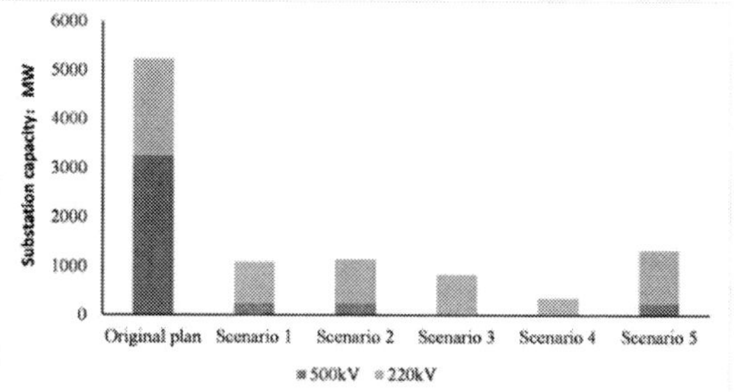

Figure 2 Capacity of power grid planning projects under different scenarios.

5.3.2 Economic evaluation

The economics of the grid planning scenarios under five different scenarios were evaluated, setting the operation period as 30 years, the conversion rate of fixed asset as 95%, the grid operation and maintenance rate as 2%, the compensation price of load shedding as 10 yuan/kw, and the discount rate as 8%, where the utilization cost of annual adjustable resources was mainly considered as the demand response cost. The economic benefits of each scenario were calculated and comparatively analyzed using the total investment during the planning period as well as the whole life cycle cost indicators. The results are shown in Table 2.

Comparing the total cost of the 5 scenarios and the original planning report in the planning period, it can be found that the total cost of the five scenarios considering adjustable comprehensive resources is lower than that in the original plan. Therefore, the integration of adjustable comprehensive resources into transmission network planning can reduce the grid investment cost in the planning period to a certain extent. And there is a more obvious improvement in investment efficiency, in which the most obvious improvement is scenario 4, followed by scenarios 3 and 1.

The sensitivity analysis of different scenarios shows that in this transmission network planning, for every 1 MW increase in energy storage capacity, the average reduction in construction investment is 390,800 RMB and the average reduction in total whole life cycle cost is 465,500 RMB. For every 1 MW increase in demand response capacity, the average reduction in construction investment is 872,100 RMB and the average reduction in total whole life cycle cost is 911,200 RMB. Overall, the effect of demand response resources utilization on grid planning investment is better than that of energy storage configuration, and almost no preliminary project investment is required.

In summary, the planning method of considering adjustable comprehensive resources is conducive to improving the economics and reliability of network operation. In particular, scenario 4 with the highest proportion of demand response participation has the best economy, which can bring higher investment efficiency and greater economic benefits to power grid companies.

Table 2. Economic comparison of various scenarios planning schemes.

Planning Scenario	New load during planning period (MW·h)	Construction investment (10^4 yuan)	Unit investment to increase the supply load (MW·h / 10^4 yuan)	Annual Operations and Maintenance Cost (10^4 yuan)	Annual adjustable comprehensive resources utilization cost (10^4 yuan)	Total Life-Cycle Cost (10^4 yuan)
Original plan	1591.2	118352.8	0.0134	2248.70	0	140980.73
Scenario 1	1435.2	32032.8	0.0448	608.62	431.81	42502.28
Scenario 2	1435.2	34832.8	0.0412	661.82	431.81	45837.61
Scenario 3	1435.2	28775.2	0.0499	546.73	431.81	38621.86
Scenario 4	1396	15975.2	0.0874	303.53	630.11	25370.14
Scenario 5	1474.4	36432.8	0.0405	692.22	332.65	46745.75

6. Conclusion

This paper proposed a novel transmission grid framework considering adjustable comprehensive resources under the new power system, and used a municipal grid as an example for planning and evaluating the novel transmission network. The results showed that all types of adjustable comprehensive resources have a certain degree of substitution for conventional power in grid planning, and in particular, demand response resources have the most significant available potential. Therefore, this paper made the following recommendations. (1) Consider the role of adjustable resources in grid planning and use them as an alternative resource on the generation side for transmission grid planning; (2) Explore user-oriented demand response models and improve the subsidy mechanism to guide users to actively participate in demand response; (3) Promote the application of big data in transmission grid planning, strengthen data governance, and improve the load classification statistics function of the mining system. Additionally, use data-driven statistics to depict the uncertainty factors, operation modes and scenarios in the planning process.

Acknowledgements

This paper is supported by Natural Science Foundation of Hebei Province: G2022502004.

References

1. C. KAN, L. YAO. Automation of Electric Power Systems, 41(9), 2-11 (2017)
2. W. SHI, J. QU, K. LUO, Strategic Study of CAE, 24(06), 52-63 (2022)
3. Q. HUANG, X. WANG, S. QI. Electric Power Construction, 39(06), 63-70 (2018)
4. S. Dehghan, N. Amjady. IEEE Transmission Transactions on Sustainable Energy, 7(2), 765-774 (2016)
5. J. YANG, F. ZHANG, W. CUI. Chinese Journal of Power Sources, 44(05), 757-761 (2020)
6. L. LIU, H. CHENG, Y. WU. Automation of Electric Power Systems, 45(13), 176-183 (2021)
7. H. SHI, Z. SHU, T. CHEN. A multi-scenario transmission network planning method with wind power and battery energy storage.
8. A. Moreira, D. Pozo, A. Street. IEEE Transactions on Power Systems, 32(4), 3246-3257 (2017)
9. H. SHI, North China Electric Power University, (2015)
10. W. NIU, J. TAN, H. LI, Distribution & Utilization, 35(04):32-36+59 (2018)

Day-ahead load optimal distribution of thermal power coupled electric energy storage frequency regulation system considering the effect of overtemperature

Xichao Zhou*, Nan Wang, Zhen Li, Pengxiang Zhao, Lin Cong

State Grid Integrated Energy Service Group Co., Ltd, Beijing, 100000, China

Abstract: Coupling energy storage devices on the generation side can significantly improve the AGC frequency regulation performance of thermal power units and bring frequency regulation benefits. In the traditional joint frequency regulation mode, energy storage is generally used to compensate the deviation between thermal power output and dispatching command, without considering the deep coordination between thermal power units and energy storage system. Thermal power units are still in fast variable load operation state, and the potential of coupled energy storage in improving the safety and other comprehensive performance of the units has not been explored deeply. Therefore, in this paper, the energy storage system is used to actively cooperate with the unit for secondary frequency regulation, and the load command is optimally distributed by minimizing the total cost function, including the variable load cost of the unit. The maintenance cost caused by overtemperature is considered in the cost model, and is analyzed through a simulation example. The results show that the total cost, considering the influence of over temperature on the optimal load distribution, is lower than that when the unit responds at a constant high load changing rate. The active load bearing of energy storage alleviates the cost increase caused by rapid load changing of the unit, making the unit benefit from energy storage and the total cost tends to be optimal.

Key words: joint frequency regulation, cost evaluation model, overtemperature cost conversion, variable load rate, load optimal distribution.

1. Introduction

Traditional coal-fueled thermal power generation has caused many problems for the environment, and many countries strongly support the proposal of carbon peaking and carbon neutrality for energy restructuring[1]. Renewable energy generation, such as photovoltaic and wind power generation, is strongly supported and developed, and the proportion of its connection to the power system is on the rise. However, renewable energy is highly stochastic. For example, wind power may have a large power output during the low power consumption period of the day, showing a reverse regulation characteristic. This stochasticity not only poses a challenge to the safe and stable operation of the power system, but also has a greater impact on thermal power units [2]. The power system requires that the active power at the generation side and the customer side can be kept in real-time balance, so that the frequency of the whole power system can be kept stable and the quality of electricity can be guaranteed. However, the random fluctuation of renewable energy output will affect the grid frequency, so it is necessary to adjust the output of power sources such as thermal power units, to maintain the power balance [3]. When the new energy output is high, the more controllable thermal power generation will generate as little power as possible, allowing the new energy to bear the load demand. And vice versa, when the new energy output is small, the generation sources such as thermal power units have to enhance the output to stabilize the load demand [4].

It can be seen that the large proportion of new energy power generation has brought greater frequency regulation pressure to traditional thermal power units, and frequency regulation units need to increase and decrease load frequently to cope with changing user load and new energy output. The power system requires real-time balance of active power, however, the response power units is difficult to keep up with the change of dispatching, limited by the large unit inertia. Electric energy storage devices can be a good aid to thermal power units for frequency regulation because the storage devices can be charged and discharged precisely in a very short time (millisecond level) [5]. This also gained further research and significantly improved the frequency regulation performance of the unit [6].

The traditional coordination mode of thermal power units and energy storage does not fully consider the synergy between them. After receiving the command to adjust the load, the units still rapidly increase and decrease the load. Energy storage is just to make up for the deviation between the unit output and the dispatching command. On

* Corresponding author: zhouxichao@sgecs.sgcc.com.cn

the one hand, the rapid load change of the unit will bring losses to the unit, such as the increase of total coal consumption, large temperature fluctuation of the unit, large life loss, etc. On the other hand, the rapid charging and discharging capacity of energy storage under this mode is not fully utilized, which has the problem of idle capacity [7]. Therefore, it is necessary to improve the traditional coordination mode, so that the energy storage device can actively bear part of the load demand when allowed, and allocate the dispatching instructions according to the minimum cost target of the frequency regulation process, so that the unit can better perform the secondary frequency regulation with the help of energy storage [8].

In the cost model of the combined unit and energy storage frequency regulation system, the cost of coal consumption of the unit needs to be considered, the configuration of energy storage and other costs are also indispensable, and the frequency regulation revenue should be calculated according to local policies. On this basis, adding the cost increase caused by overheating can make the overall cost model more complete. Related research [9] shows that the overall temperature fluctuation of the unit becomes more drastic when the load changes rapidly, and the losses of each part of the unit will increase because of the excessive temperature and the drastic temperature fluctuations, which on the one hand will increase the life loss of the unit itself, and on the other hand, the maintenance cycle of the unit will become shorter when the losses increase, and the normal power generation revenue will be damaged if the unit is shut down for maintenance. Therefore, it is necessary to consider the effect of overtemperature when optimizing the allocation of load orders, and the load allocation will be more meaningful if the results of the optimal load allocation can make the total cost model considering overtemperature tend to be minimal.

2. Total cost model of the joint frequency regulation process

In the process of joint frequency regulation, the costs of the units and energy storage system need to be characterized and used as the objective function for subsequent optimization. For thermal units, the cost of coal consumption, the cost of life loss caused by overtemperature and the cost of shorter maintenance cycles are mainly considered. The cost of energy storage system, such as configuration, is considered and its life loss is described. Frequency regulation gain is the main source of revenue. The more detailed the cost model portrayal is, the better understanding and calculation of the length of its payback period can be obtained [10].

2.1 Coal consumption cost model

2.1.1 *Steady-state coal consumption model*

A thermal power unit will have a corresponding supply coal consumption value for a specific load output. Generally, five sets of supply coal consumption tests are performed on the units and the supply coal consumption versus load curves are obtained by regression analysis. In order to obtain the supply coal consumption curve more conveniently and accurately, [10] conducted data regression analysis for various types of units and found that the specific load and specific supply coal consumption of most thermal power units can be approximated in the following form:

$$\mu = ae^{-\beta/t} + b \qquad (1)$$

where μ is the specific supply coal consumption, defined as $\mu = b_g / b_{g0}$, b_{g0} is the coal consumption of the unit under THA condition, and b_g is the supply coal consumption at a certain load P, and the unit is g/(kWh). β is the specific load, defined as $\beta = P / P_0$ and P_0 is the unit lode under THA condition, with *a=0.798, t=0.245, b=0.988.*

The model has high accuracy and is generally used as a generic specific coal consumption curve. Since turbines of different capacities can be considered as similar models with similar relative trends of internal efficiency with main steam flow, the specific coal consumption characteristic curve of the target unit can be considered to have the same shape as the general specific coal consumption characteristic curve. Since the shape is the same, it only needs to be shifted up and down to get as close as possible to the two or more test specific coal consumption data points of the unit. Since at least two points are required to determine the exact location, the "two-point method" is used to calculate the target unit. The steady-state coal consumption model of the unit used in this simulation can be calculated by the "two-point method" as follows:

$$b_g = b_{g0}\left[ae^{-(P/P_0+\beta_1)/t} + b - \mu_0\right] \qquad (2)$$

where $b_{g0} = 312.02g/(kWh), P_0 = 300.014MW, \beta_1 = -0.0348, \mu_0 = 0.0035$

2.1.2 *Transient coal consumption model*

When the initial and final steady-state conditions of the unit are the same but the variable load rate (transient process) is different, the total coal consumption will also be different during the whole variable load process. It was found in [12] that the transient coal consumption increment values were different when the unit raised or lowered the load at different variable load rates, and showed a certain variation pattern. The relationship between the variable load rate and the incremental transient coal consumption can be obtained by segmental fitting of the thermal test data.

$$\delta B = \frac{p_1 x + p_2}{x + q_1} \qquad (3)$$

where δB is the transient coal consumption increment in *kg/s*, *x* is the unit variable load rate, and the parameters of its segment fitting are shown in the following table.

Table 1 Table of model parameters

Calculated working condition	*p1*	*p2*	*q1*
50%-75% load up	1.313	0.9506	-1.181
75%-100% load up	1.182	0.3521	-1.192
75%-50% load down	-1.689	0.8297	0.2797
100%-75% load down	-1.102	-0.4988	1.226

The general idea of its cost conversion is as follows:

$$C_{coal} = \sum_{t=1}^{24} CB \times f^{t}_{coal}(P^{t}_{power}) \quad (4)$$

where CB is the price of coal combustion, in million RMB/kg; f^{t}_{coal} is the coal consumption of thermal power unit at moment t, in kg/h; t denotes the time period serial number, in h. P^{t}_{power} is the electric power output of thermal power unit, in MW.

In this paper, the price of coal taken in the simulation is 469 RMB/ton and its calorific value is 4700 kcal/kg, which is converted into the price per kg of standard coal. For a specific variable load process, the steady-state coal consumption and transient coal consumption values of each process are added up and multiplied by the converted price to accumulate the coal consumption cost.

2.2 Unit life loss cost

Related studies have shown that in the process of joint frequency regulation, if the fluctuation of load is small, the life loss caused by it is a very low, which is approximately negligible. However, when the unit is loaded and unloaded at a large variable load rate, the life loss will increase significantly. For example, when the unit is variable load at 5MW/min, the turbine rotor life loss is 4-7 times higher than that at 3MW/min. Therefore, the unit should reduce the variable load rate with the help of energy storage to reduce the life loss [13].

The life of a unit is inevitably lost when the unit is over-tempered and over-pressurized under rapidly varying load conditions. However, it is very difficult to characterize the life loss of a unit to a very precise degree, it is possible to approximate the life loss percentage of different variable load rates by an approximate order of magnitude. The life of a normal unit can be considered as 100%, and different life loss percentages are generated at different variable load rates. A model fit to the relevant experimental estimation data yields the following estimation model:

$$L = p_1x^3 + p_2x^2 + p_3x + p_4 \quad (5)$$

where L is the approximate percentage of life loss for one round-trip regulation, x is the unit variable load rate, and p1=4.329e-05, p2=-0.0002897, p3=0.000636, p4=-0.0002896.

When performing cost translation, it can be converted to opportunity cost C_{rep}:

$$C_{rep} = C_{jz} * L \quad (6)$$

where C_{jz} is the unit's cost in million RMB. This equation characterizes that the greater the variable load rate, the greater the potential lifetime loss opportunity cost. In the day-ahead optimal allocation, this cost is summed up over the lifetime loss opportunity cost generated by all variable load round-trip processes in a day.

2.3 Maintenance costs due to overtemperature

A related study [14] showed that in the process of joint frequency regulation, the temperature variation of different heating surfaces of the boiler in the unit increases with the increase of variable load rate. It selected five locations for dynamic process analysis: water-cooled wall outlet, screen superheater outlet, horizontal low-temperature reheater inlet, vertical low-temperature reheater outlet, and coal-saving wall outlet. It was found that the greater the variable load rate, the more drastic the temperature fluctuation of its heating surface. According to the ASME stress calculation standard, the alternating thermal stress can be corresponded to the maximum temperature deviation of the heating surface. In addition, the heat transfer capacity of the mass side is much stronger than that of the flue gas side, so the maximum temperature deviation of the mass can be used to reflect the maximum temperature deviation of the heating surface. In this way, the maximum temperature deviation of working medium is corresponding to the alternating thermal stress.

The larger the variable load rate is, the more drastic the fluctuation of the temperature of the mass on the heating surface is, and the greater the alternating thermal stress is, the greater the losses suffered by the unit, and the maintenance cycle will be shorter accordingly. First of all, according to the data of thermal test, it is found that the maximum temperature deviation of the working mass at each location of the unit has different deviations at different variable load rates. Therefore, the maximum temperature deviation of the mass at each position is averaged to measure the maximum temperature deviation of the mass at a specific variable load rate, i.e., the magnitude of alternating thermal stress. The following relational model is obtained by fitting the average value:

$$T_{dif} = p_1x^3 + p_2x^2 + p_3x + p_4 \quad (7)$$

where T_{dif} denotes the maximum temperature deviation of the average unit mass, which reflects the magnitude of the alternating thermal stress. x is the variable load rate. The lift-load process is fitted separately to improve the model accuracy. When x is positive, it indicates the lift-load, in which p1=0.001013, p2=-0.04737, p3=1.37, and p4 =6.788; when x is negative, it means decreasing load, and then the absolute value of input x is found and then

https://doi.org/10.1051/e3sconf/202337503005

substituted into equation, where p1=-0.003387, p2=0.1266, p3=-0.4742, p4=12.02.
It is difficult to portray very precisely the value of the variation of the maintenance cycle time caused by different variable load rates. In this regard, in the simulation, the cost per day was estimated based on engineering experience and shrunk to an order of magnitude consistent with engineering experience by multiplying it by an estimation factor:

$$C_{gs} = C_{wx} * T_{dif} * k \quad (8)$$

where C_{gs} is the estimated cost of maintenance corresponding to the variable load process at different variable load rates, and C_{wx} is the cost corresponding to one unit repair, in million RMB, taken as 10 million RMB in the simulation. T_{dif} is the alternating thermal stress term related to the variable load rate, this multiplied by the maintenance cost of a single time has initially reflected the trend of the maintenance cost becoming larger as the variable load rate becomes larger, but the cost value obtained from this trend is often not consistent with the actual engineering experience, therefore, the estimation factor k is multiplied at the end to shrink this cost to a quantitative level consistent with the engineering experience.
In this way, when the AGC load command and energy storage state change, the variable load rate of the unit will also change, so that its reasonable maintenance single-day commuted cost will also change between the quantity levels in line with the engineering experience. For the unit selected in the simulation, when it changes between the minimum and maximum variable load rate allowed, the single-day cost will fluctuate between 1000 and 8000 RMB. Of course, the cost will fluctuate depending on the specific maintenance situation due to the degree of wear and tear and different maintenance components of different units, but the trend of changing with the variable load rate is in line with the actual situation. In addition, when conducting the optimal allocation of load for thermal storage, if the result of the allocation can make its reasonable order of magnitude tends to the minimum cost, then the result of the optimal load allocation is also meaningful, it gets the lowest possible unit variable load rate, which makes the total cost tends to the minimum and the investment payback period tends to the shortest.

2.4 Energy Storage Costs and FM Benefits

When energy storage is first put into operation, its configuration costs should be considered. Then, its operation and maintenance (O&M) costs need to be considered. The cost of its configuration can be characterized by the following equation:

$$P_{storage} = C_b \frac{E_b}{\eta} + C_{pac} P_b + C_a E_b \quad (9)$$

where C_b is the unit energy price of the battery body, in million RMB/MWh; E_b is the rated capacity of the energy storage, in MWh. η is the total charging and discharging efficiency of the energy storage; C_{pac} is the unit power price of the converter of the energy storage system, in million RMB/MW. P_b is the charging and discharging power of the energy storage, in MW; C_a is the unit energy price of auxiliary facilities, in million RMB/MWh.
The O&M cost of energy storage system can be characterized by the following equation:

$$P_{oc} = c_{fix} P_b \quad (10)$$

where c_{fix} is the fixed operating cost per unit of power, in million RMB/MW.
When performing the day-ahead load optimization allocation, the total annual cost of energy storage is divided by 365 equally to a specific day as the cost of energy storage for a single day.
The frequency regulation gain can be characterized by the following equation:

$$gain_{tp} = k_p D' Q_{price} \quad (11)$$

where k_p is the comprehensive performance index of frequency regulation, which is taken as 2.4 when energy storage is involved in frequency regulation compensation and 1.1 when energy storage is not involved (based on historical data statistics). D' is the approximate frequency regulation mileage, and the difference between the two load points before and after the dispatch phase is used as the equivalent mileage calculation, in MW. Q_{price} is the frequency regulation compensation price, which is taken as 10 RMB/MW in the simulation. The total revenue of this load distribution can be obtained by summing up the revenue generated by all the variable load processes in a single day [15].

3. Analysis of the day-ahead optimization case

The capacity of the thermal power unit analyzed in this case is 330MW, and the storage capacity is 4.5MWh, and the rated charging and discharging power is 9MW.
According to the unit load command and unit output power history data for load optimization distribution simulation, the energy storage output and unit output should meet the load balance constraint. The load command fluctuation of a typical day is shown as follows.

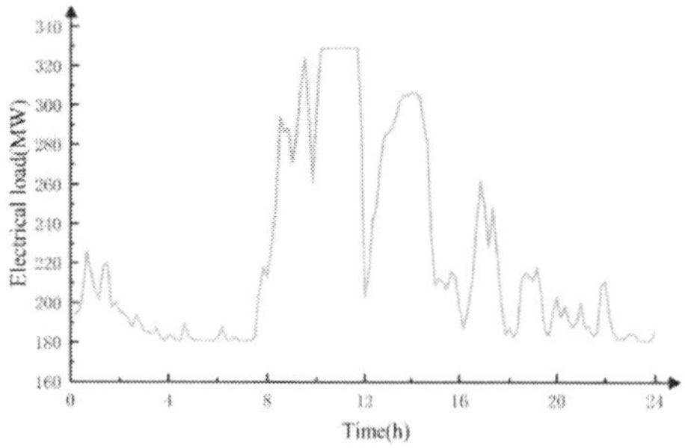

Figure 1 Typical daily electricity load demand (AGC command) graph

3.1 The objective function

The objective function of the day-ahead optimization is the total cost function of the joint frequency regulation system, and the load distribution of the unit and storage should make the total cost function tend to minimize and reduce its payback period. It can be expressed by the following equation:

$$OBJ = C_{coal} + C_{bat} + C_{gs} - gain_{tp} \qquad (12)$$

where C_{coal} is the total cost of coal consumption obtained by accumulating all variable-load processes in a single day, in million RMB. C_{bat} is the total cost of the energy storage unit levelized to a single day. C_{gs} is the cost of life loss plus the estimated cost of maintenance due to overtemperature spread to a single day. $gain_{tp}$ is the total frequency regulation gain obtained by accumulating all variable load processes in a single day.

3.2 Constraints

The most fundamental constraint in the joint frequency regulation system is the power balance, as shown in the following equation.

$$P^t_{power} + P^t_{bat} = P^t_{load} \qquad (13)$$

where P^t_{power} is the power output of the unit at a certain moment *t*, in MW. P^t_{bat} is the power output of energy storage at the same time. P^t_{load} is the load command value at the same time, i.e. the load required by the power grid.

Further, there is a unit output constraint as follows:

$$P^{min}_{power} \le P^t_{power} \le P^{max}_{power} \qquad (14)$$

where P^{min}_{power} is the minimum power output value; and P^{max}_{power} is the maximum power output value that can be achieved by the unit.

Upper and lower limit constraints of energy storage system:

$$P^{min}_{bat} \le P^t_{bat} \le P^{max}_{bat} \qquad (15)$$

where P^{min}_{bat} and P^{max}_{bat} are the maximum charging and discharging power of the energy storage system respectively.

For the sake of operation safety, there are the following residual energy constraints for energy storage devices:

$$E^{min}_{stor} \le E^t_{stor} \le E^{max}_{stor} \qquad (16)$$

where E^{min}_{stor} is the minimum required residual energy of energy storage, which ensures that the energy storage power cannot be too low, and its corresponding SOC value is generally 0.2. The residual energy of the highest requirement E^{max}_{stor} is to ensure that the energy storage cannot be overcharged, and its corresponding SOC value is generally 0.8. The residual energy is limited in order to make the energy storage work in a reasonable energy range and prolong its service life.

3.3 Optimization algorithm and results

In this paper, the artificial bee colony algorithm (IGABC) [16, 17] is used to solve the optimal load distribution problem. For the day-ahead optimization, the optimization variables selected are the power output values of the units and energy storage. The scheduling cycle is 5 minutes. Thus, there are 576 variables to be optimized. At the very beginning, 100 sets of random solutions (corresponding to the number of employed bees or onlooker bees) are generated in its solution space. A more optimal solution is continuously searched according to its solution process, in which the scout bees will avoid the solution process from falling into a local optimum. Its detailed solution flow chart is shown below.

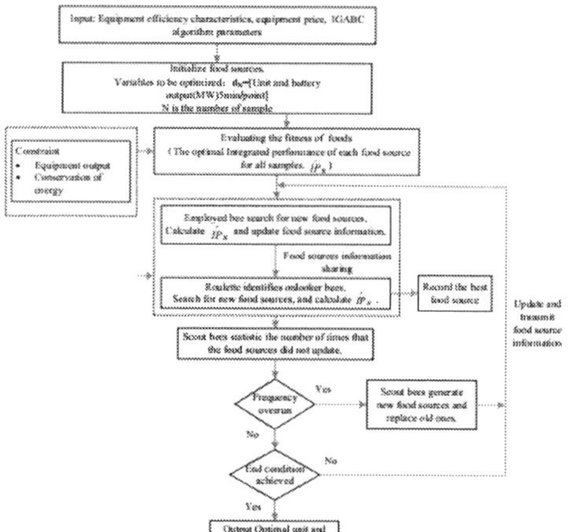

Fig. 2 Flow chart of artificial bee colony algorithm

After loading the load demand for a typical day in Figure 1 into the program, running the program according to the steps of the IGABC algorithm yields the following graph of convergence results.

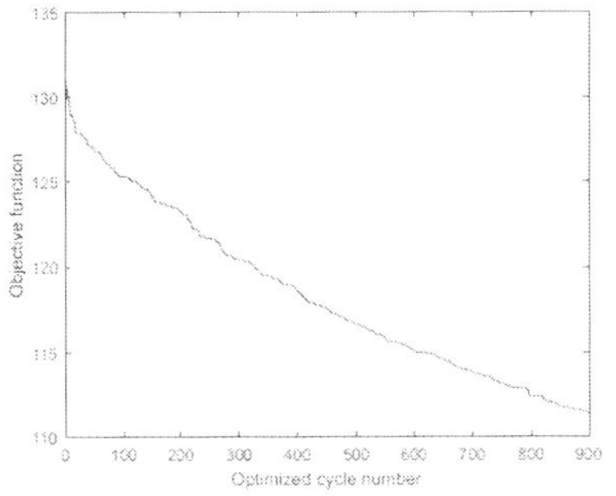

Fig. 3 Convergence diagram of the optimization search process

As can be seen from the diagram of the optimization search process, the optimal solution that makes the total cost tend to be smaller is refreshed as the solution is updated iteratively. After the maximum cycle of 900, the minimum cost of this example day is 1,115,160,000 RMB. Further, the output of thermal units and energy storage for a typical day of load demand in Figure 1 can be obtained as follows.

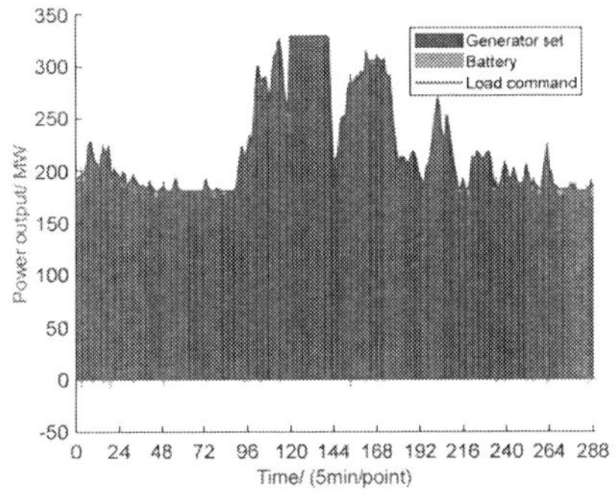

Figure 4 Thermal power unit and energy storage output map

Here the electrical load command is divided into 5 minute points, for a total of 288 load demand points in a day. Further, the variation of the energy storage SOC value (reflecting the residual energy) can be obtained as shown below.

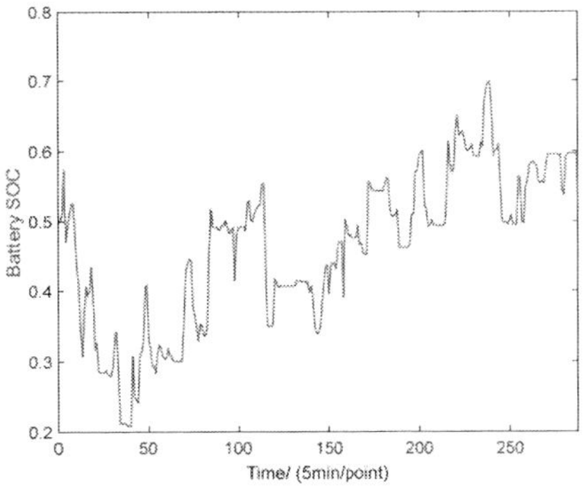

Figure 5 Graph of SOC variation of energy storage device

During the peak load hours, the energy storage will be discharged more to match the load of the unit, and its SOC value starts to show a decreasing trend. When the electricity consumption comes to the valley hours, the load of the unit starts to drop, and more energy is charged to the energy storage, and the energy storage SOC value starts to show an increasing trend.

Further analysis of each cost, comparing the optimized cost with the cost of the unit when the maximum variable load rate fast response command is applied, can be shown by the following table.

Table 2 Cost comparison table

Contrast Methods	Total Cost / (million RMB)	Coal consumpt-ion cost	Frequency regulation gain	Energy storage cost	Lifetime and maintenance costs
Unit optimized rate response	111.516	110.0854	12.95245	13.37955	1.0035
Unit constant and fast response	112.829	109.9505	11.0465	12.09516	1.82984
Relative change (%)	-1.1637	0.1227	17.254	10.619	-45.159

It can be seen that the total cost is reduced by operating under the optimally assigned load command. Especially for the life and maintenance costs, the relative change is large. This means that the unit is operated at the lowest possible variable load rate with the help of energy storage, so that the temperature fluctuation of the unit's own heating surface is not particularly sharp. The alternating thermal stress on the heating surface is lower than that at fast variable load, so that the calculated life time and maintenance cost estimates are lower.

4. Conclusion

By incorporating the estimated maintenance cost and life depreciation cost caused by overtemperature into the traditional cost model, the total cost of combined thermal unit and storage system is more detailed and complete, which is closer to the real cost of the frequency regulation process. This provides a certain reference value for the cost calculation and cost recovery period estimation of the combined thermal storage frequency regulation project.

In the optimal distribution of AGC load commands, the results show that with the help of energy storage, the response speed of the units to the commands is improved, and the overall frequency regulation performance index is improved. Therefore, compared with the fast response of the units, the energy storage only compensates for the deviation, and the frequency regulation benefit is increased, which is one of the main objectives. Let thermal power units benefit from the rapid load changing capability of energy storage and better meet the requirements of frequency regulation. It can also be seen that the coal consumption cost occupies a large percentage of the total cost, because coal consumption is needed every moment to maintain the unit's output value with the variable load rate value for lifting and lowering the load. This cost is subject to a range of fluctuations in the SOC status of energy storage, depending on the change in coal price and the size of the load command demand.

In addition, when the unit is frequency regulated at an optimized rate, energy storage can help the unit take on some of the load, and the unit can lift the load at the lowest allowable variable load rate in this case. In this way, the life loss of the unit will be greatly reduced, which can effectively extend the service life of the unit and reduce the commutation cost. After the load change rate is reduced, compared with the quick response command, the temperature fluctuation and alternating thermal stress of the internal boiler and turbine become smaller. Therefore, the maintenance cycle of the unit is extended. The relative change in its cost term (-45.159%) illustrates the necessity for optimal distribution of load. The deep coordination of thermal power and energy storage can not only improve the AGC frequency regulation gain, but also reduce the total cost, shorter the return period of investment.

Acknowledgements

This work was financially supported by State Grid Comprehensive Energy Service Group's science and technology project "Research on fire storage joint frequency regulation load distribution strategy and system development".(SGTYHT/20-JS-223)

References

1. Ma, X.J., et al., Carbon emissions from energy consumption in China: Its measurement and driving factors. SCIENCE OF THE TOTAL ENVIRONMENT, 2019. 648: p. 1411-1420.
2. Chen, Z., et al., Phase compensation based active disturbance rejection control for high order superheated steam temperature system. CONTROL ENGINEERING PRACTICE, 2022. 126.
3. Wu, X., et al., Fuzzy modeling and predictive control of superheater steam temperature for power plant. ISA TRANSACTIONS, 2015. 56: p. 241-251.
4. Chunmei, X., Research on using electric energy storage to improve frequency regulation performance of thermal power units. Thermal Power Generation, 2021. 50(06): p. 98-105.
5. RAO Yufei, et al., Operational benefit evaluation for frequency regulation application of large-scale battery energy storage. Energy Storage Science and Technology, 2020. 9(06): p. 1828-1836.
6. Xie, X.R., et al., Improving AGC Performance of Coal-Fueled Thermal Generators Using Multi-MW Scale BESS: A Practical Application. IEEE TRANSACTIONS ON SMART GRID, 2018. 9(3): p. 1769-1777.
7. Wang, F., Application Research of Large—Scale Energy Storage Technology Participation in AGC Frequency Regulation of Power System. 2020, North China Electric Power University. p. 61.
8. Bingying, S., Research on Operation Strategy and Capacity Configuration Method of Energy Storage Combined with Thermal Power Unit in AGC. 2018, North China Electric Power University. p. 70.
9. SHEN Qiankun, et al., Multivariate cooperative control for variable load process of supercritical units. Thermal Power Generation, 2020. 49(04): p. 119-124.
10. YANG Shuili, et al., Analysis and Enlightenment of AGC Modulation for Combined Fire and Storage system Based on Power and Capacity Compensation. Energy Storage Science and Technology, 2022: p. 1-13.
11. LIU Fuguo and Z. Xuhui, Determination of load-impacted curve on generation efnciency of coal-fired unit by two-point method. Thermal Power Generation, 2020. 49(09): p. 87-92.
12. WANG Chaoyang, et al., Study on Energy Consumption Characteristics During Transient Processes of a 660MW Coal-fired Power Plant With Different Load-Variation Rates. Proceedings of the CSEE, 2017. 37(19): p. 5665-5673+5843.
13. Yin, L., Influence of Thermal Power Unit Participating in Automatic Generation Control Operation on Unit Life. Thermal Power Generation, 2004(06): p. 6-10+87.
14. Yongliang, Z., Simulation Study on Dynamic Characteristics of 660 MW Supercritical Coal Fired Unit in Variable Load Process. ELECTRIC POWER, 2019. 52(05): p. 13-20.
15. CHEN Lijuan, JIANG Yuxuan and W. Chun, Strategy, and capacity of energy storage for improving AGC performance of power plant. Electric Power Automation Equipment, 2017. 37(08): p. 52-59.
16. Karaboga, D. and B. Basturk, A powerful and efficient algorithm for numerical function optimization: artificial bee colony (ABC) algorithm. JOURNAL OF GLOBAL OPTIMIZATION, 2007. 39(3): p. 459-471.
17. Hao, Y., et al., Multi-Objective Intelligent Optimization of Superheated Steam Temperature Control Based on Cascaded Disturbance Observer. Sustainability, 2020. 12(19).

Overview of landfill leachate treatment technology

Ruqiong Qin

Guangxi Polytechnic of Construction, Nanning 530003, China

Abstract: Household waste is directly related to people's quality of life and urban environment. As people pursue higher and higher spiritual quality, people pay more and more attention to the treatment of domestic garbage leachate. At present, although China has made some achievements in the treatment of domestic waste leachate, there are still many problems, which need to be solved. In this paper, the treatment of domestic waste leachate will be elaborated.

Key words: Landfill leachate; Treatment technology; Chemical method; Biological treatment method.

1. Introduction

With the rapid development of our economy, the amount of urban garbage also increases with it. Most cities in China use sanitary landfill or incineration to dispose of waste, resulting to a large amount of landfill leachate. Landfill leachate contains a variety of pollutants, such as heavy metal ions and organic matter, which not only exist in the water for a long time and a wide range, but also do great harm. If not properly treated, it will cause serious pollution to the environment. The effective collection and treatment of landfill leachate has become an urgent problem to be solved in urban environment. The treatment technology of landfill leachate has become a hot spot and difficult point for researchers.

2. Production and characteristics of landfill leachate

Landfill leachate, also known as leachate or leachate, is an inevitable secondary pollutant in landfill[1], mainly from precipitation, water contained in garbage and organic wastewater generated by anaerobic decomposition of microorganisms [2].Landfill leachate is a highly concentrated organic wastewater. If it is discharged directly without treatment or fails to meet the discharge standards, it will cause serious pollution to the surrounding groundwater, surface water and soil.

The pollutant content of landfill leachate is affected by many factors, such as waste composition, landfill years, climatic conditions, and landfill design [3]. The characteristics of landfill leachate water quality can be summarized as follows:

There are many kinds of pollutants, complex components and high concentration.Liu Jun et al. used GC-MS to analyze the organic components in the landfill leachate, and founded that there were 63 organic compounds, most of which were difficult to biodegrade, such as phenols, heterocyclic, heterocyclic aromatic and polycyclic aromatic compounds, accounting for more than 70% of the organic components in the leachate [4].High concentration of organic matter, high concentration of COD and BOD_5, up to tens of thousands of mg/L.

The changes of water quality and quantity are complex.The composition and yield of landfill leachate also change with different hydrologic and climatic conditions, geological conditions, geographical location, structural mode and landfill time.The biodegradability decreased with the increase of landfill age.

Imbalance of nutrition ratio.The content of ammonia nitrogen in the leachate is high, the C/N value is often maladjusted, and the p is deficient. The proportion of micronutrients cannot meet the requirements of water treatment.

3. Disposal method of landfill leachate

Landfill leachate, as a by-product produced in landfill, has always been the focus and difficulty in domestic sewage treatment field because of its high concentration of organic matter and ammonia nitrogen.At present, the treatment of landfill leachate in China can be roughly divided into physical and chemical methods, biological methods and membrane methods. The three types of leachate treatment technologies are generally introduced as follows:

3.1 Physicochemical method

Physicochemical method has a good treatment effect on refractory organic matter, atmospheric nitrogen and chroma in leachate. The commonly used physicochemical method to treat landfill leachate includes adsorption method, chemical precipitation method, REDOX method, blowing, membrane treatment and so on.

3.1.1 Chemical precipitation method

Chemical precipitation method is to add chemicals to the landfill leachate, so that it and some soluble substances in the leachate replacement reaction, generate insoluble salt precipitation, so as to reduce the dissolved pollutants in the water.The added chemical is called a precipitator.According to the properties of the insoluble salt produced, the chemical precipitation mainly includes hydroxide precipitation and sulfide precipitation.The commonly used precipitators are aluminum sulfate, ferric chloride and polyaluminum chloride.It is mainly used to remove chroma, heavy metal ions and turbidity of landfill leachate.

3.1.2 Chemical oxidation method

In the process of waste water treatment, the use of toxic and harmful substances dissolved in the waste water can be oxidized in the REDOX reaction, to transform it into non-toxic and harmless new substances, this method is called chemical oxidation method.The chemical oxidation method is mainly to remove the chroma and sulfide of the leachate and decompose the refractory organic matter in the leachate, so as to improve the biodegradability of the wastewater.Commonly used oxidants include ammonia, ozone, hydrogen peroxide, potassium permanganate and sodium hyponitine.

3.1.3 Adsorption

Adsorption method is a kind of processing technology that uses porous solid material with adsorption ability to remove trace dissolved impurities in water.The COD and ammonia nitrogen in leachate can be removed by adsorption. The commonly used adsorbents are granular activated carbon and powder activated carbon.The use of activated carbon on the leachate of pollutants treatment efficiency is high, the treatment effect is good, but the treatment equipment is simple, compact structure, but the treatment cost is high, easy to block, adsorption saturated adsorption material is difficult to regenerate, treatment, at present, most of the adsorption method as the end of the leachate advanced treatment process.

3.1.4 Blow off

Since ammonia nitrogen in the water mostly exists in the state of ammonia ion (NH^{4+}) and free helium (NH_3), and the two maintain a balance, ammonia escapes from the leachate by applying the physical effect of aeration and blowing after the pH value is increased.Blow-off is mainly used to remove high concentration of ammonia nitrogen in landfill leachate, so as to ensure the normal operation of subsequent biological treatment.The blown NH_3 should be recovered to prevent secondary pollution to the air.

3.2 Biological treatment

Compared with conventional physical and chemical methods, biological treatment has a higher removal effect on dissolved and colloidal organic compounds.The operating cost is relatively low;The good settling performance of the residual sludge is conducive to further dehydration.Therefore, at present, the treatment of landfill leachate mainly adopts biological method.Biological treatment can be divided into aerobic biological treatment method, anaerobic biological treatment method and aerobic - anaerobic combined treatment method.

3.2.1 Aerobic treatment

Aerobic treatment can effectively reduce COD and ammonia nitrogen, but also remove some pollutants such as iron and manganese.Aerobic treatment processes include: sequencing batch reactor (SBR), cyclic circulating activated sludge process (CASS), oxidation ditch, oxidation pond, etc.

Ehrig et al. [5]used activated sludge method to treat landfill leachate. The experiment showed that when BOD_5 load was 0.1kgBOD_5/kg MLSS/d, the effluent BOD_5 was less than 25mg/L.In the experiment, the high ammonia nitrogen load caused the accumulation of nitrite. When the ammonia nitrogen load was less than 0.03kg·NH^{4+}-N/kg MLSS/d, the nitrification process could be completely carried out. With the adaptation of microorganisms to the high ammonia nitrogen load, the ammonia nitrogen load could be increased to 0.1 kg BOD_5/kg·MLSS/d.

Mard et al.[6]found that for landfill leachate with COD of 4000~13000mg/L,BOD of 1600~11000mg/L and atmospheric nitrogen of 87~590mg/L, the removal rate of COD can be stabilized above 90% by aerobic activated sludge process.

Robinson[7] used continuous flow activated sludge to treat landfill leachate, and found that when the sludge age was less than 5 days, the treatment effect fluctuated greatly, and there were problems such as sludge swelling.When the operating temperature drops from 10^{0}C to 5^{0}C, the treatment effect is greatly affected.When the sludge age is more than 10 days, the BOD and COD of the treated effluent can be lower than 20mg/L and 150mg/L, respectively.However, it was found that when the ratio of N/BOD exceeded 36:100, the excess ammonia nitrogen could not be utilized by microorganisms and remained in the treated effluent.

Ying et al. [8]used the SBR reactor to treat the landfill leachate from a landfill in the United States, and found that the removal rate of total carbon in the leachate by the SBR reactor was above 95%.Harry et al. used the two-stage SBR method to treat landfill leachate. The BOD removal rate of the first stage SBR reactor could reach 98%, and the COD of the influent decreased from 5800mg/L to 112mg/L.After secondary SBR treatment, the effluent quality can meet the discharge standard.

Yalmaz et al. [9]conducted SBR biological denitrification of effluent from a landfill site in Istanbul after anaerobic biochemistry, and found that the ammonia nitrogen concentration could decrease from 1000mg/L in the inlet to 5mg/L in the effluent within a 24h cycle.

Wang Xiansheng et al. [10]adopted the two-stage contact oxidation process to treat landfill leachate, packed the soft packing inside, the aerobic volume was 6.71L, the influent concentration of the contact oxidation tank was 3100~6100mg/L, the secondary effluent concentration was maintained at 1100mg/L~3000mg/L, and the COD removal rate was ensured at about 10%.

In general, aerobic treatment has poor performance when toxic metal substances exist, and when the external temperature decreases rapidly, it is difficult to maintain a high removal rate.Energy consumption and other shortcomings.

3.2.2 *Anaerobic treatment*

Anaerobic treatment has a good effect on the treatment of high concentration organic wastewater, which has the advantages of high treatment load, low sludge production rate, low energy consumption and less land area.Anaerobic biological treatment technology mainly includes: UASB reactor, anaerobic biological filter, anaerobic pond, anaerobic contact method and new anaerobic baffle reactor, etc.The Water research Center in the United Kingdom uses UASB to treat leachate with COD >10000mg/L. When the load is 3.6~19.7 kg/(m^3·d), the average mud age is 1.0~4.3 d, and the temperature is 3^0C, the removal rates of COD and BOD_5 are 82% and 85%, respectively.

Xu Zhu [11]used upflow anaerobic biofilter to conduct continuous dynamic experiment on leachate from Chengdu landfill site.The results show that the upflow anaerobic biofilter is effective in treating landfill leachate.At medium temperature, the COD removal rate of landfill leachate with COD of 3000~8000mg/L can reach 95%, even at room temperature, the COD removal rate can reach 90%.The COD volume load of the reactor can reach more than 5kg/(m^3·d).

J.G. enry et al.[12], University of Toronto, Canada, used an anaerobic filter at room temperature to treat leachate with landfill age of 1.5 years (COD of 14000mg/L,B/C of 0.7) and 8 years (COD of 4000mg/L,B/C of 0.5), respectively.When the volume load is 1.26~1.45kg/(m^3 d) and the HRT is 24~96h, the COD removal rate can reach more than 90%.

Shen Yaoliang et al. [13]used an ABR reactor to treat the mixed liquid of leachate and municipal sewage from Qizishan domestic waste landfill in Suzhou.The results show that ABR can effectively improve the biodegradability of mixed wastewater.When influent BOD5/COD is 0.2~ 0.3, effluent BOD5/COD can be increased to 0.4~0.6.When the volume load is 4.71kgCOD/(m^3·d), the rod-shaped granular sludge with good settlement and particle size of 1~ 5mm can be formed.The sludge concentration in each compartment was 20~38g /L.The pretreatment of mixed wastewater by ABR greatly improves the operational stability of the subsequent aerobic treatment facility.

The anaerobic treatment has the advantages of low energy consumption and simple operation, so the investment and operation cost are low. At the same time, the anaerobic treatment produces less residual sludge and requires less nutrients such as nitrogen and phosphorus. Many polymer organic matter which is difficult to be treated under aerobic conditions can be biodegraded during anaerobic treatment.But anaerobic biological method can't remove the nitrogen and phosphorus in the waste water, the boot process, operation and management more complex, poor health condition, especially the single use of anaerobic treatment of high concentration organic wastewater, treatment is not completely, the water of the organic matter concentration are still high, should not be directly discharged into rivers or lakes, so in the biological treatment technology of landfill leachate,Anaerobic processes alone are rarely used.The combined anaerobic and aerobic mode is often used to treat high concentration organic wastewater.The anaerobic section is used to remove most of the organic pollutants in the water, and the polymer organic matter that is difficult to degrade is decomposed at the same time. The aerobic section is used for further treatment of organic matter and treatment of hydrogen and phosphorus.

Physicochemical method is mainly used to remove chroma, SS, ammonia nitrogen, heavy metal ions, refractory COD and other pollutants in leachate.Compared with the biological method, the physicochemical method has the ability to withstand the impact load, and has a better effect on the removal of precious metal ions and organics which are difficult to be degraded by the biological method.In particular, it has a good removal effect on the "aged" leachate with high ammonia nitrogen content and low BOD_5/COD ratio, which is difficult for biochemical treatment.However, the cost of physicochemical treatment is high, and the treatment process needs to be further optimized. At the same time, the physicochemical treatment process such as adsorption and precipitation only transfers the pollutants, and the pollutants will eventually return to the landfill in the form of physicochemical sludge, resulting in the circulation of pollutants.Therefore, physicochemical method is often used in the pretreatment and advanced treatment of landfill leachate, and the main process of leachate is usually biological method.

3.3 Membrane treatment technology

Membrane technology is a water treatment method that uses a membrane to separate the solvent from the solute and particles. According to the force of solute or solvent passing through the membrane, membrane separation can be divided into reverse osmosis, ultrafiltration and microporous filtration.In recent years, in order to reduce the degree of water pollution as much as possible, membrane method has also been applied in the field of leachate treatment, among which there are many applications and studies abroad.Thmos of Germany used reverse osmosis and microfiltration for the purification of leachate. The characteristics of reverse osmosis membranes under different conditions were studied and it was indicated that more than 95% osmosis removal rate could be achieved by using high-pressure reverse osmosis up to 120bar and microfiltration combined with a controlled crystallization process.A reverse osmosis unit was successfully used to continue the treatment of biochemical water at the Damsdorf landfill in Germany, and membrane separation technology has been used in several leachate treatment plants in the Netherlands and Switzerland.Foreign practice has proved that membrane technology is efficient and reliable to treat landfill leachate.Due to its extremely high cost, membrane technology can not be widely used in the treatment of landfill leachate at present stage.

3.4 Combined process flow

At present, due to the increasing environmental pollution, from the perspective of strengthening environmental protection, the state has issued the Pollution Control Standard of Household Waste Landfill; Quasi-Emission Standard (GB16889-2008), in which the total effluent nitrogen becomes an important index (40mg/L in non-sensitive areas, 20mg/L in sensitive areas).In order to meet the total nitrogen requirements in the new landfill leachate discharge standard, the original MBR process was further optimized and A secondary nitrification and denitrification process was added. As shown in Figure 1, the MBR process was optimized to A/O/O + A/O+ external ultrafiltration membrane (UF) to ensure the total nitrogen discharge of effluent reached the standard.
To sum up, the leachate treatment process is mainly "biological method + membrane treatment", and the leachate treatment technology can meet the discharge requirements of the 2008 "Domestic Waste Landfill Pollution Control Standard Discharge Standard".Among them, the biochemical treatment process can effectively degrade and eliminate the pollutants, and the membrane separation process can effectively separate and remove the non-biodegradable residual pollutants.

4. Conclusions and recommendations

Landfill leachate is a kind of high concentration organic wastewater with complex composition, and its treatment technology has advantages and disadvantages. It is difficult to make leachate discharge standard by using any treatment technology alone.Therefore, the treatment process must be developed from the single to the diversified, through the combination of processes to give full play to the advantages of each process, in order to achieve satisfactory results."Biological method + membrane treatment" process technology can meet the discharge requirements of 2008 "Domestic Waste Landfill Pollution Control standard Discharge Standard", but there are still some problems in the process of landfill leachate treatment.

Acknowledgements

This research is supported by Notice of 2021 Guangxi University Young and middle-aged Teachers Basic Research Ability Improvement Project "Current status and countermeasures of landfill leachate treatment under the background of garbage classification" (No.: 2021KY1165)

References

1. Li Chengjun, Liu Yongqiang, Liu Xiaojuan. Leather Production and Environmental Protection Technology. 2021(21);
2. Application of membrane separation technology in the treatment of landfill leachate [J]. Resource Conservation and Environmental Protection. 2021(12);
3. Study on the treatment status of landfill leachate in China [J]. Liu Wei. 2021(06);
4. Analysis of the status quo and existing problems of landfill leachate treatment [J]. Zhang Junjiao. Resource Conservation and Environmental Protection. 2022(03);
5. Research and application of anaerobic ammonia oxidation (ANAMMOX) in the treatment of landfill leachate [J]. Li Zhirong. Environmental Science and Technology. 2022(03);
6. Study on the treatment process of landfill leachate [J]. Dong Qingzhi. Henan Science and Technology. 2022(12);
7. Wang Qing, Jin Jing, Wang Dianer, Feng Xu, Hu Kailiang. Green Science and Technology. 2020(24);
8. Yu Zhibo, Zhang Kaijun, Jiang Hao. Research Progress of leachate treatment technology of domestic garbage [J]. Guangdong Chemical Industry. 2021(01);
9. Research progress on the treatment of landfill leachate in China [J]. Xu Qian, Ling Ting. Science and Technology Horizon. 2021(27);
10. Analysis on the treatment of landfill leachate in sanitary landfill [J].[Peng Lei. Comprehensive Utilization of Resources in China. 2021(09);
11. New process for the treatment of leachate from domestic garbage [J]. Chemical Engineering and Equipment. 2021(06)
12. Xu Changwen, Wang Shengdong. Environmental and Sustainable Development. 2020(05);
13. Chang Hongchen, Zeng Wenchao, Gong Xiwei, Li Panpan, Li Gencan. Environment and Development. 2020(09).

Highly active Cu-MFI catalyst for conversion of furfuryl alcohol to pentanediols

Dengfeng Dai [1], Dandan Liu [1, 2, *] and Yunqi Liu [1]

[1] State Key Laboratory of Heavy Oil Processing, China University of Petroleum (East China), Qingdao 266580 (China)
[2] College of New Energy, China University of Petroleum (East China), Qingdao 266580 (China).

Abstract. It is of great significance to use biomass-based furfuryl alcohol to produce oxygenated chemicals to replace petroleum-based chemicals. In this paper, a series of bifunctional Cu-MFI catalysts were developed, and the properties of biomass-based furfuryl alcohol to pentanediol, including 1,5-pentanediol and 1,2 pentanediol, were investigated. These catalysts were synthesized by ammonia evaporation method by loading copper nanoparticles on MFI molecular sieve systems with different silicon aluminum ratios. Among them, Cu-MFI (60)-AE catalyst containing abundantly Brønsted acid protons shows excellent performance, achieving 99.7% furfuryl alcohol conversion and 91% pentanediol selectivity under mild reaction conditions. The high catalytic activity can be attributed to the highly dispersed Cu species and Brønsted acid protons. Brønsted acid protons play a decisive role in the highly selective formation of 1,5-PDO. This paper can develop an economical and feasible path for converting furfuryl alcohol into high value-added fine chemicals.

1. Introduction

The preparation of fine chemicals from biomass as a substitute for petrochemical resources has attracted increasing attention[1]. Furfuryl alcohol (FFA), as a semi hydrogenation product of furfural, is an important biomass resource and has been widely used in industry[2,3]. FFA is an important renewable chemical, which can be transformed into a variety of high value-added chemicals through chemical paths[4,5]. The ring opening products of FFA are important diols, such as 1,5-pentanediol (1,5-PDO) and 1,2-pentanediol (1,2-PDO). Because C5 petroleum based raw materials are difficult to obtain[6,7], the route to obtain pentanediol from FFA is a sustainable reaction, which has significant economic prospects.

Copper based catalysts have been used in various oxygenated chemical transformations[8,9]. Chen's team explored the catalytic performance of Cu-$LaCoO_3$, Cu-LDO and Cu/Al_2O_3 bifunctional catalysts[10-12]. Cu/Al_2O_3 catalyst achieved 85.8% FFA conversion and 70.3% PDO selectivity under 8.0 MPa H_2. Jones research group realized 62% PDO yield by using reduced Cu-Co-Al mixed metal oxides[10-12]. These reactions are generally carried out under intermittent conditions of high hydrogen pressure, and the selectivity of 1,5-PDO is far from satisfactory.

Herein, we developed a bifunctional Cu-MFI catalyst by loading uniform copper nanoparticles on MFI zeolite by ammonia evaporation, which achieved 99.7% FFA conversion and 91% PDO selectivity under continuous conditions. The excellent activity achieved can be attributed to the cooperation between highly dispersed Cu species and Brønsted acid protons.

2. Experimental

2.1 Synthesis of Cu-MFI (60)-AE, Cu-MFI (130)-AE and Cu-MFI (Si)-AE

Three catalysts adopted HZSM-5 (SiO_2/Al_2O_3= 60, 130) and silicalite-1 as support and synthesized by ammonia evaporation methods with 3.0 wt.% Cu content, respectively. denoted as Cu-MFI (60)-AE, Cu-MFI (130)-AE and Cu-MFI(Si)-AE, respectively. The three samples were reduced under a flow of H_2 at 300°C for 2 h, and named as Cu-MFI (60)-AER, Cu-MFI (130)-AER, and Cu-MFI(Si)-AER, respectively.

2.2 Catalyst characterization

The catalyst characterization was carried out on X'pert Pro MPD X-ray diffractometer, ASAP 2020 microporous physical adsorption analyzer, JEOL JEM-2100F microscope, escalabmk X-ray photoelectron spectrometer, vertex 70 Fourier transform infrared spectrometer, AutoChem 2950 HP instrument, and IRIS Intrepid II XSP instrument.

2.3 Catalytic testing and product analysis

The products were analyzed by Agilent 7820a GC with HP-5 column; the conversion of FFA and pentanediol selectivity was calculated by the following formula:

$$\text{Conversion} = \frac{[FFA]f - [FFA]p}{[FFA]f} \times 100\% \qquad (1)$$

* Corresponding author: liudandan@upc.edu.cn

$$\text{Selectivity} = \frac{[PDO]p}{[FFA]f - [FFA]p} \times 100\% \qquad (2)$$

[FFA]f is the FFA content in the feedstock (%) and [FFA]p is the FFA content in the products (%); [PDO]p is the 1,2-PDO and 1,5-PDO content in the products (%).

3. Results and discussion

3.1 Characterization of copper-based catalysts

Table 1. Basic physical properties of catalysts

Catalyst	^{a}Cu/wt%	$^{a}SiO_2/Al_2O_3$	$^{b}S/m^2\cdot g^{-1}$	$^{c}V/cm^3\cdot g^{-1}$	$^{d}D_{Cu}/nm$
Cu-MFI (60)-AE	2.6	57	323	0.20	2.5
Cu-MFI (130)-AE	2.7	129	306	0.18	--
Cu-MFI(Si)-AE	2.9	--	320	0.18	--

(a) Obtained by ICP-OES. (b) Total surface area. (c) Total pore volume. (d) Average particle size of Cu species from HRTEM

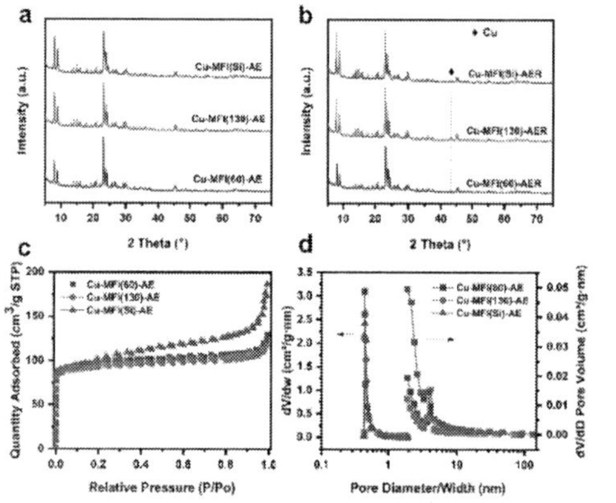

Figure 1. X-ray diffraction patterns of the (a) calcined catalysts, (b) reduced catalysts. (c) Nitrogen adsorption-desorption for calcined catalysts, (d) corresponding pore size distribution.

The XRD of calcined catalysts were shown in Figure 1a. All samples show a typical structure of MFI. No diffraction peak of CuO was found at 35.5o and 38.7o, indicating that CuO was evenly distributed on the surface of zeolite. After reduction, a weak diffraction peak attributed to Cu (111) appears at 43.3o, indicating that the species of Cu were evenly distributed. Nitrogen adsorption-desorption curves (Figure 1c) show that adsorbed rapidly at low pressure, and no hysteresis loop was found at 0.4-1.0P/Po, indicating that all three catalysts were typical microporous structures[14]. It was also confirmed by the corresponding pore size distribution (Figure 1d). The corresponding porous property data were also summarized in Table 1. In addition, ICP-OES showed that copper loading was close to theoretical value.

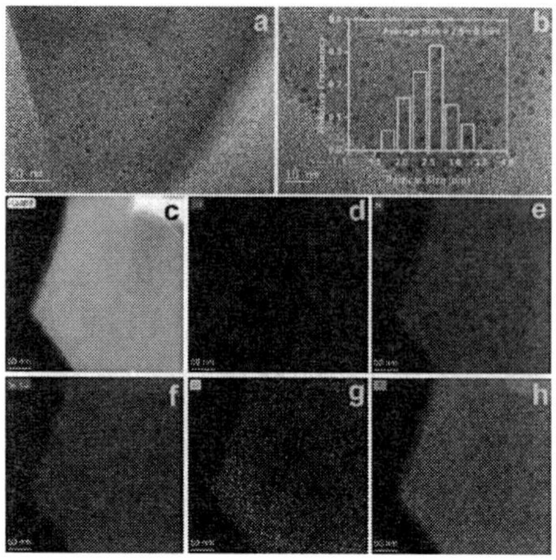

Figure 2. (a, b) HRTEM images for Cu-MFI (60)-AER, with corresponding size distributions of Cu species. (c-h) HAADF-STEM images, and corresponding EDS-mapping of Cu, Si, Si + Cu, and Al, O element.

HRTEM and HADDF-STEM were used to investigate the metal distribution on the catalyst surface, as shown in Figure 2. HRTEM images (Figure 2a, 2b) show that the distribution of copper particles is uniform, and no obvious particle agglomeration is found, which is consistent with the results of XRD. The average particle size of copper particles is 2.5nm. HADDF-STEM images and the corresponding EDS-mapping images show that copper and other elements are interlaced, indicating that Cu is evenly distributed on the surface of zeolites. This can improve the utilization of metal and promote the reaction.

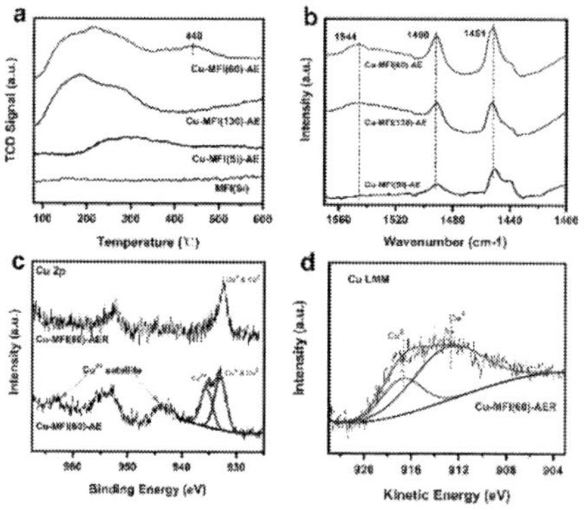

Figure 3. (a) NH3-TPD profiles of different catalysts and MFI(Si) support. (b) Pyridine infrared (Py-IR) spectrum profiles of catalysts. (c) Cu 2p XPS spectrum of Cu-MFI (60)-AE and Cu-MFI (60)-AER. (d) Cu LMM XAES spectrum of Cu-MFI (60)-AER.

To reveal the effect of surface acidity on ring opening of FFA, NH_3-TPD and Py-IR were used to investigate the surface acidity of the catalyst. The results are shown in Figure 3a and 3b. Cu-MFI (60)-AE shows a continuous desorption peak of NH_3 at 100-500 °C. The NH_3 desorption peak at 440 °C is attributed to the strong interaction between NH_3 and Brønsted acid protons. Cu-MFI (130)-AE shows a continuous desorption peak at 100-380 °C, and its strong acid content is lower than that of Cu-MFI (60)-AE, which is determined by the number of framework Si $(OH)^+$ Al unit[15]. However, Cu-MFI(Si)-AE shows a small desorption peak, and the corresponding MFI(Si) has no NH_3 desorption peak, indicating that the NH_3 desorption peak of Cu-MFI(Si)-AE is mainly caused by the interaction between NH_3 and copper species.

The three samples showed obvious characteristic bands at 1451 cm^{-1} and 1490 cm^{-1}, which were attributed to Lewis acid and Brønsted + Lewis acid, respectively. The difference was that Cu-MFI (60)-AE and Cu-MFI (130)-AE present a characteristic bands of 1544 cm^{-1}, corresponding to Brønsted acid[16], indicating that there were Brønsted acid protons in the catalysts. This was determined by the density of framework Si $(OH)^+$ Al unit of the support. and there is no Brønsted acid protons in Cu-MFI(Si)-AE.

The surface metal valence plays an important role in the ring opening of FFA. It was reported that unsaturated Cu^+ acts as anchoring sites for $-CH_2OH$, while Cu mainly plays the role of hydrogen activation. XPS and Cu LMM XAES were used to analyze the surface chemical state, and the results are shown in Figures 3c and 3d. After calcination, an obvious satellite peak appeared at binding energies of 943 eV for Cu-MFI (60)-AE catalyst, indicating the presence of Cu^{2+} species in the sample[17]. In addition, two overlapping peaks are found at 935.5 eV and 933.0 eV, corresponding to Cu^{2+} and Cu^0 and/or Cu^+ species[8], respectively. These indicate that part Cu^{2+} was reduced to Cu^+ and/or Cu during the roasting process, which may be exerted by the decomposition gas of NH_4^+ in the zeolite during the roasting process. The decomposition gas caused the reduction of copper, this is similar to that reported. After the reduction, the satellite peak disappeared, indicating that Cu^{2+} species were reduced to Cu^+ and/or Cu. The species composition was further analyzed using Cu LMM XAES, and an asymmetric peak was detected at 915.9 eV and 913.0 eV, which belong to Cu^0 species and Cu^+ species, respectively. This illustrated the coexistence of Cu^+ and Cu^0 species on the surface.

3.2 Catalysts performance for the FFA ring-opening to pentanediols

Scheme 1. Reaction pathway of FFA for different catalysts

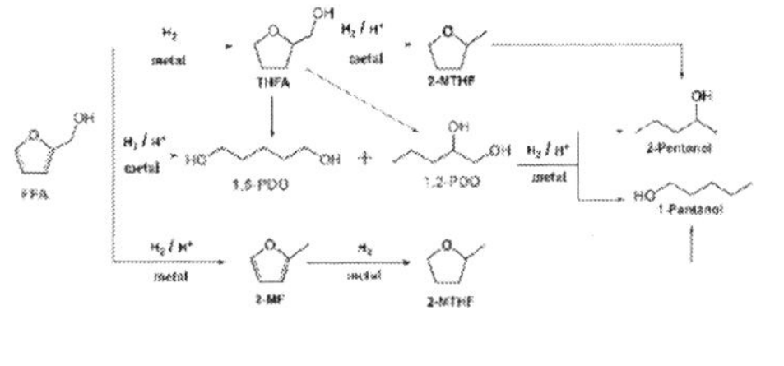

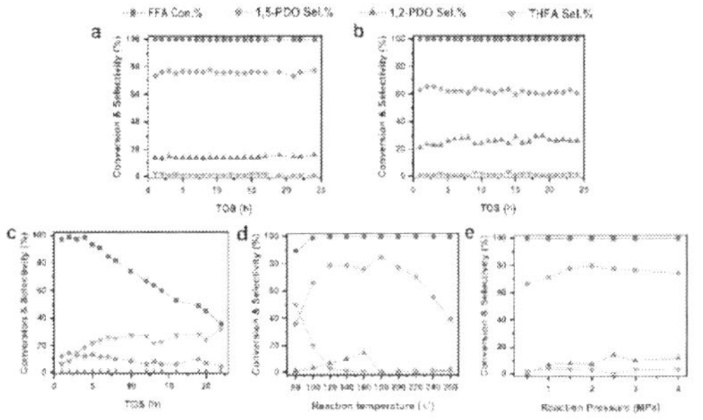

Figure 4. Catalyst performance test of (a) Cu-MFI (60)-AE, (b) Cu-MFI (130)-AE, (c) Cu-MFI(Si)-AE, 160 °C, 2.5 MPa H2. (d) catalyst performance test at different reaction temperature. (e) catalyst performance test at different H2 pressure.

The highly selective ring opening of FFA is an ideal path to obtain PDO (Scheme 2). The activity of Cu-MFI (60)-AE, Cu-MFI (130)-AE, and Cu-MFI(Si)-AE was tested at 160 °C and 2.5 MPa H_2, respectively. The results are shown in Figure 4. Cu-MFI (60)-AE catalyst exhibits excellent selectivity for PDO, which achieved a nearly 100% FFA conversion with a 91% selectivity for PDO, including 76% selectivity of 1,5-PDO and 15% selectivity of 1,2-PDO. Cu-MFI (130)-AE catalyst also achieved 87.7% PDO selectivity at nearly 100% conversion. The selectivity of 1,5-PDO reduces from 76% to 62%, while the selectivity of 1,2-PDO increases from 15% to 26% with the increase of SiO_2/Al_2O_3 ratios. This is closely related to the density of Brønsted acid protons, demonstrating that $Si(OH)^+Al$ could promote the formation of 1,5-PDO. Cu-MFI(Si)-AE catalyst showed low activity and selectivity. The conversion of FFA gradually decreased from 96% to 36%, while the highest selectivity of 1,5-PDO was only 14%, and no 1,2-PDO was observed. The difference in activity and selectivity mainly comes from the difference in Brønsted acid protons in the catalyst. This means that Brønsted acid protons are the key factor to promote the ring opening of FFA

In addition, the effects of reaction temperatures and pressures on the ring opening of FFA were also investigated. With the increase of reaction temperature, the selectivity of 1,5-PDO increased first and then decreased. The selectivity of 1,5-PDO reached the maximum value of 84.9% at 180 °C, surpassing the previously reported non-noble metal catalysts. The selectivity of 1,2-PDO was reduced to <2% after the temperature exceeded 180 °C, this may be caused by further dehydration of 1,2-PDO under high temperatures. Hydrogen pressure has little effect on the selectivity of FFA ring opening. nearly 100% conversion and 66% selectivity of PDO can still be achieved even at a low pressure of 0.5 MPa, demonstrating that acidity is the main factor determining the ring opening rather than hydrogen pressure.

4. Conclusions

In conclusion, a bifunctional Cu-MFI catalyst with high activity and selectivity was prepared for the ring opening

 https://doi.org/10.1051/e3sconf/202337503007

of FFA to pentanediol. The catalyst with abundantly Brønsted acid protons and highly dispersed copper species exhibits excellent performance. Among them, Brønsted acid protons are stem from molecular sieve support, and highly dispersed copper species derived from the ammonia evaporation preparation process. The yield of pentanediol is more than 91% in 24 h TOS. The excellent activity is superior to the previously reported non-noble metals, and the activity selectivity is comparable to that of noble metal catalysts. This study provides a new idea for the low-cost synthesis of pentanediol from biomass-based materials.

Acknowledgements

This work has been financially supported by the National Natural Science Foundation of China (22005342).

References

1. T. Zhang, W. Li, H. Xiao, Y. Jin, S. Wu, Bioresour. Technol. 2022, 354, 127126.
2. Z. Chen, X. Bai, A. Lusi, W. A. Jacoby, C. Wan, Bioresour. Technol. 2019, 289, 121708.
3. S. Liu, N. Govindarajan, K. Chan, Acs Catal. 2022, 12902-12910.
4. X. Li, P. Jia, T. Wang, Acs Catal. 2016, 6, 7621-7640.
5. R. Mariscal, P. M. Ojeda, I. Sádaba, M. López Granados, Energy Environ. Sci. 2016, 9, 1144-1189.
6. G. Hayes, M. Laurel, D. MacKinnon, T. Zhao, H. A. Houck, C. R. Becer, Chem. Rev. 2022.
7. D. Sun, S. Sato, W. Ueda, A. Primo, H. Garcia, A. Corma, Green chemistry. 2016, 18, 2579-2597.
8. Z. Zhang, Z. L. Wang, K. An, J.. Song, Y. Bando, Y. Yamauchi, Y. Liu, Small. 2021, 17, 2008052.
9. D. Dai, C. Feng, M. Wang, Q. Du, D. Liu, Y. Pan, Y. Liu, Catal. Sci. Technol. 2022, 12, 5879-5890.
10. H. Liu, Z. Huang, F. Zhao, F. Cui, X. Li, C. Xia, J. Chen, Catal. Sci. Technol. 2016, 6, 668-671.
11. Fangfang Gao, H, J. Chen, Z. Huang, C. Xia, Chinese Journal of Catalysis. 2018, 39, 1711-1723.
12. H. Liu, Z. Huang, H. Kang, C. Xia, J. Chen, Chinese Journal of Catalysis. 2016, 37, 700-710.
13. T. P. Sulmonetti, B. Hu, S. Lee, C. W. Jones, Acs Sustain. Chem. Eng. 2017, 5, 8959-8969.
14. M. Thommes, K. Kaneko, A. V. Neimark, J. P. Olivier, F. Rodriguez-Reinoso, J. Rouquerol, K. S. W. Sing, Pure Appl. Chem. 2015, 87, 1051-1069.
15. P. Arudra, T. I. Bhuiyan, A. M. Aitani, S. S. Al-Khattaf, H. Hattori, Acs Catal. 2014, 4, 4205-4214.
16. F. Yaripour, Z. Shariatinia, S. A. Irandoukht, Microporous Mesoporous Mat. 2015, 203, 41-53.
17. R. Ye, L. Lin, C. Chen, J. Yang, F. Li, X. Zhang, D. Li, Y. Qin, Z. Zhou, Y. Yao, Acs Catal. 2018, 8, 3382-3394.

A Real-time Corrosion Monitoring Method Based on Quartz Crystal Microbalance

Chaoxuan Qin [1,*], Lu Dai [1], Yuqin Zhu [1]

[1] Southwest Technology and Engineering Research Institute,401329 Chongqing, China

Abstract. This paper put forward a kind of fast, high precision, high sensitivity of corrosion monitoring method and developed a portable corrosion monitoring device with 6 road crystal vibration sensor and a temperature and humidity based on QCM theory. The device can work independently for a long time in harsh outdoor environment without relying on external power supply. The quantitative relationship between quartz crystal vibration frequency and corrosion quality was determined by calibration tests in the laboratory. The device is placed in a certain environment to monitor the corrosion. The results show that the device can effectively monitor the weak corrosion phenomenon, with real-time and high sensitivity.

1 Introduction

Metal is one of the most important materials for industrial equipment. In the process of service, metals are prone to corrosion under the action of environment and corrosive medium, the equipment is easy to form thinning, mechanical performance reduction, cracking, punch, etc., leading to equipment failure and safe hidden trouble, cause environmental pollution, and may even bring huge economic losses or safety accident. So, it is very important to monitor, prevent and warn the corrosion. At present, the main corrosion monitoring methods include metal hanging sheet method and resistance probe method, etc. These methods can only monitor uniform corrosion, and will destroy the integrity of the equipment itself during installation. Quartz Crystal Microbalance (QCM) is a new type of mass sensing and analysis technology with nanogram sensitivity[1]. It can reflect the mass change of adsorbable substance on the crystal surface by monitoring the resonant frequency change of quartz crystal. It has the advantages of high sensitivity, and real-time monitoring. QCM as a kind of high precision, high accuracy of the testing instrument is widely used in chemical, biological medicine, food, aerospace and other fields.

Corrosion monitoring is a hot research topic. The physical approach is to assess the corrosion of rebar by detecting changes in physical properties, including resistance, electromagnetism, heat conduction and sound propagation. These methods include time domain reflectometry (TDR)[2], X-ray diffraction and atomic absorption probe techniques, infrared (IR) thermography[3], and fiber corrosion detection[4]. Based on the modal analysis theory of elastic materials, Chen et al[5]. established a modal analysis model for corroded pipes and analyzed the relationship between the modes of corroded pipes and quality loss factors. The results show that the model can be used to evaluate the corrosion rate of corroded pipelines by monitoring and analyzing the modes. Dai [6]conducted in situ simulations using the HYSYS program to study the effects of operating conditions, inhibitors, pipeline parameters and flow modes on CO2 corrosion in order to predict CO2 corrosion in natural gas collection pipeline systems. The study showed that the corrosion rate was influenced by working pressure and carbonate concentration, and that there was an optimal value for the promotion of corrosion by temperature. By comparing the simulation data with the corrosion rate data in the field, the feasibility of the numerical simulation method is proved.

Based on QCM technology, a portable corrosion quality monitoring device is developed in this paper, which provides an efficient method for online corrosion monitoring and real-time early warning.

2 QCM theory

QCM [7]works by applying an alternating electric field through two electrodes to a quartz wafer, which is subjected to periodic mechanical deformation due to the inverse piezoelectric effect. This mechanical vibration propagates from the wafer in the form of thickness shear sound waves, known as thickness shear (TSM) oscillations. Generally speaking, the amplitude of this mechanical vibration is very small, but when the frequency of the alternating electric field is a certain value, the amplitude will suddenly increase, resulting in resonance, that is, piezoelectric oscillation.[8]

Under the alternating electric field, QCM generates a shear wave of a certain wavelength due to the piezoelectric effect. The shear wave propagates in the crystal along the direction parallel to the alternating electric field and is completely reflected at the crystal

* Corresponding author: qinchaoxuan1991@163.com

interface, that is, it generates a standing wave.[9] The conditions are as follows:

$$h_q = n\lambda_0 / 2 \quad (1)$$

Where h_q is the thickness of the quartz crystal oscillator and λ_0 is the shear wave length.

Assuming that the properties of QCM[10] surface material are similar to QCM -- and uniformly attached to QCM surface, the relation between frequency change and thickness change can be obtained by setting its thickness as Δh:

$$\Delta f = -\frac{\Delta m}{A\rho_q h_q} f_0 = -\frac{\Delta m}{A\rho_q h_q} \frac{f_0^2}{\frac{1}{2h_q}\sqrt{\frac{u_q}{\rho_q}}} \quad (2)$$

There are the following restrictions: (1) the frequency offset should be within 2% of the fundamental frequency; (2) The material on QCM surface is evenly distributed; (3) The crystal works in a resonant state.

According to Sauerbery's equation, when a QCM sensor plated with metal electrodes is working, the metal electrode will be corroded by corrosive gases in the air, resulting in an increase in the surface mass of the QCM sensor. The increase in mass Δm can be calculated from the output frequency change of the QCM sensor Δf. Then according to the corrosion time and the area of the metal electrode, the metal corrosion rate can be calculated.[11]

The crystal oscillator structure used in this paper is shown in Fig. 1. There are metal coatings on both sides of the quartz sheet, and the metal coatings are connected to the pins through the electrodes. The crystal vibrates steadily when a specific voltage is applied to both ends of the electrode. When the metal coating is exposed to the environment, corrosion will occur under the influence of the environment, and then the quality of the metal film will change, and the frequency of crystal vibration will change accordingly. By measuring the frequency of vibration, the corrosion of the metal can be monitored.

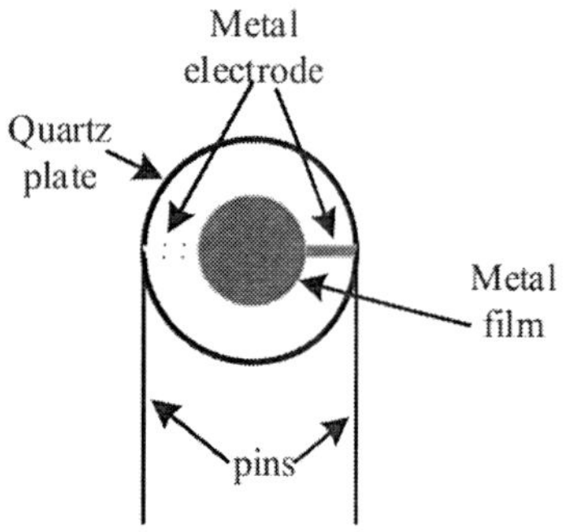

Figure 1. structure of quartz crystal

3 Portable corrosion monitoring device

Based on QCM principle, the corrosion test device was designed, its structure is shown in Fig. 2, the device is the core of the circuit by epoxy potting process, through the large capacity polymer lithium battery for power supply, makes the device work normally in harsh outdoor environment, through the low power design allows the device to get longer working hours. Each crystal oscillator is connected to a resonant circuit, through which the crystal oscillator generates stable vibration, and the vibration signal is transmitted to the main control circuit. The calibration and conversion algorithm of QCM is realized in the main control circuit, which measures the frequency of vibration and converts the frequency into corrosion weight gain or weight loss. The clock source provides the running clock for the master circuit. The temperature and humidity sensor collects the ambient temperature and humidity, and compensates the temperature and humidity through the main control circuit to improve the test accuracy and make up for the drift caused by the temperature and humidity. The measurement results of the main control circuit are stored in the flash memory chip. Through the upper computer can read and display the measured data.

Fig 3. is the physical picture of the device. There is a temperature and humidity sensor and 6 quartz crystals at the top, which are removable and replaceable. When not working, protective covers can be installed for protection. The working indicator light displays the working state of the device. When the corresponding sensor is working, the indicator light will blink, and the power indicator light displays the remaining power of the device.

Device installation program flow chart as shown in Fig 4., after work, first of all determine whether need to transmit the data to upper computer, after the initialized clock, timer, flash, etc, when reach the setting of sampling time, began to each sensor data acquisition and calculation during the frequency writing flash, after quality of converting frequency to corrosion.

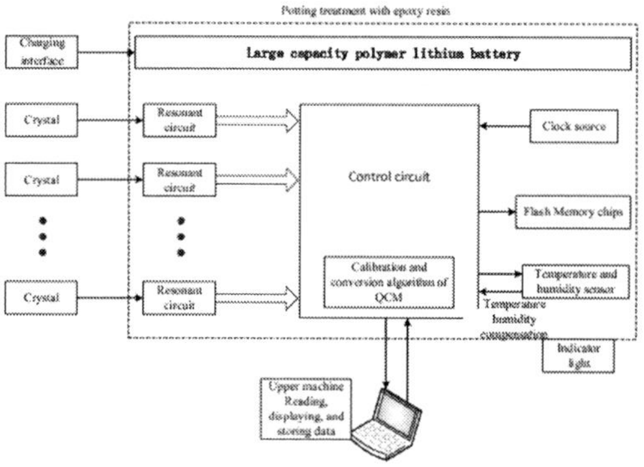

Figure 2. Structural composition diagram of the corrosion test device

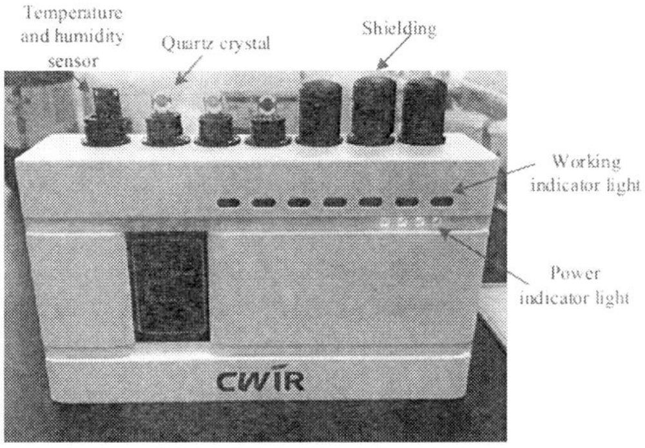

Figure 3. Real figure of the corrosion test device

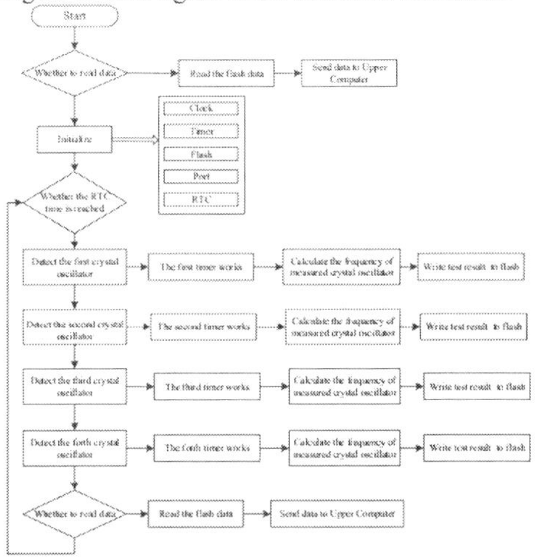

Figure 4 Program flow chart of the corrosion test device

4 Experiment

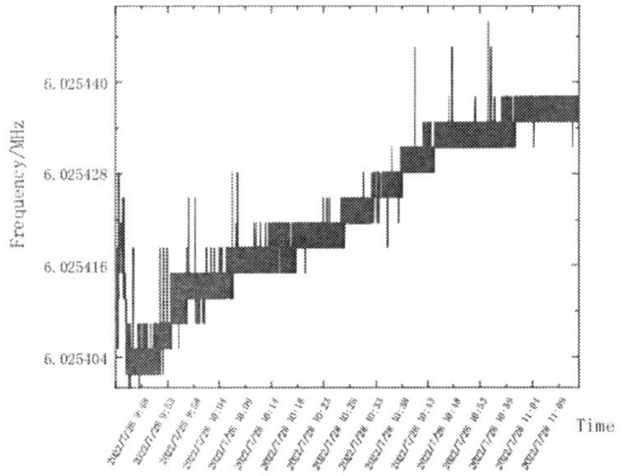

Figure 5 Corrosion test results

A large number of calibration tests were carried out in the laboratory. The relationship between frequency change and corrosion quality change was obtained by using AgNO3, KCl and Na2SO4 solutions for solution calibration experiments in the laboratory. Finally, the relationship was obtained as follows:

$$\Delta m = \Delta f / 644.2 \quad (3)$$

To test the corrosion monitoring effectiveness of the device, place the device under the environment of a corrosion monitoring, the result is shown in Fig. 5, the curve shows an hour or so of the frequency change, by type 1 can be calculated at this time of corrosion quality 0.056 ug, traditional methods are difficult to corrosion status of monitoring to the product in a short time, and the method can real-time monitor the corrosion condition. It has extremely high sensitivity.

5 Conclusion

To solve the real-time monitoring for corrosion problems, in this paper, based on the QCM design principle, put forward a kind of fast, high precision, high sensitivity of corrosion monitoring method and developed a portable corrosion monitoring device with 6 road crystal vibration sensor and a temperature and humidity. The device can work independently for a long time in harsh outdoor environment without relying on external power supply. The quantitative relationship between quartz crystal vibration frequency and corrosion quality was determined by calibration tests in the laboratory. The device is placed in a certain environment to monitor the corrosion. The results show that the device can effectively monitor the weak corrosion phenomenon, with real-time and high sensitivity.

References

1. Sun H Y, Wang Z F, Huang Y L. Progress and Trend of Marine Corrosion Monitoring[J]. Sea-lake Salt and Chemical Industry, 2005.
2. Bang S S, Yoon S H, Lee Y H, et al. Local corrosion monitoring of prestressed concrete strand via time-frequency domain reflectometry, 2018[C]. IEEE.
3. Simonov D, Vavilov V, Chulkov A. Infrared thermographic detector of hidden corrosion[J]. Sensor Review, 2020,40(3):283-289.
4. Fan L, Bao Y. Review of fiber optic sensors for corrosion monitoring in reinforced concrete[J]. Cement and Concrete Composites, 2021,120:104029.
5. Chen Y, Ming T, Shen B, et al. Evaluation Method of Pipe Corrosion Based on the Natural Frequency II: Corrosion of internal surface of pipe, 2021[C]. IOP Publishing.
6. Dai M. In situ mathematically simulation for CO2 internal corrosion in wet natural gas gathering pipelines system by HYSYS[J]. Engineering Failure Analysis, 2021,122:105265.
7. Mccubbin G A, Praporski S, Piantavigna S, et al. Adelaide Research and Scholarship: QCM-D fingerprinting of membrane-active peptides[J]. Springer-Verlag, 2011.
8. Bayramoglu, Gulay, Yilmaz, et al. Lysozyme specific aptamer immobilized MCM-41 silicate for single-step purification and quartz crystal microbalance (QCM)-based determination of lysozyme from chicken egg white[J]. Microporous

and mesoporous materials: The offical journal of the International Zeolite Association, 2015.

9. Sartore L, Barbaglio M, Borgese L, et al. Sensors and Actuators B : Chemical Polymer-grafted QCM chemical sensor and application to heavy metal ions real time detection[J]. 2011.
10. Wang L J, Wu C S, Hu Z Y, et al. Sensing Escherichia coli O157:H7 via frequency shift through a self-assembled monolayer based QCM immunosensor[J]. Journal of Zhejiang University-Science B, 2008.
11. Wang L. D.R.S Cumming, All-digital interface ASIC for a QCM-based electronic nose[J]. Sensors and Actuators B-Chemical, 2004,103:31-36.

The Detection Methods of Hydroxyl Radical: A Review

Chenkun Huang, Qiuyue Chen, Shengqiang Li and Jian Wang*

School of Urban Construction, Wuhan University of Science and Technology, Wuhan, China

Abstract. Hydroxyl radical (·OH) can rapidly react with biological macromolecules, organic and inorganic substances on account of its strong oxidizing property, so it is widely used in sewage treatment, chemiluminescence analysis, biological macromolecule modification and other fields. The strengthening effect of Hydrodynamic cavitation on chemical reactions depends on the number of free radicals initiated by cavitation. These free radicals are highly reactive and react rapidly with pollutants and other species in wastewater. Therefore, the systematic summary of detection methods of hydroxyl radical and the description of its advantages and disadvantages can provide technical support for scientific researchers, and guides the research in the future.

1. Introduction

In recent years, sewage treatment gradually become a hotspot. Hydrodynamic cavitation is an important means to develop intensive and miniaturized wastewater treatment technology. It is a greener and more efficient method to treat industrial wastewater. Occuring with the collapse of bubbles, hydrodynamic cavitation can break the strong chemical along with the high temperature and high pressure[1-3]. Among the free radicals formed during the collapse process, hydroxyl radical (·OH) has very strong oxidation property[4]. Hydroxyl radical is a kind of strong oxidizing radical, which can initiate and strengthen the chemical reaction process. The quantity produced in the cavitation process represents the intensity of cavitation[5-7]. Therefore, it is very important to detect the amount of hydroxyl radicals produced by Hydrodynamic cavitation for explorering the occurrence and strengthening effect of cavitation.

The hydroxyl radical has an unpaired electron, which is highly reactive and has a strong oxidation capacity. In addition, ·OH has a high reaction rate constant and a short life[8-9]. ·OH is a powerful oxidant with high efficiency and no secondary pollution[10]. However, for the detection of ·OH, there is no accurate and effective direct determination method at present. The only way to detect the amount of ·OH is to capture ·OH by agent and then detect the adducts indirectly. Therefore, such detection methods have high requirements on the capture efficiency of the agent, the stability and reactivity of the adduct, but the capture efficiency of the agent for ·OH is low in actual reactions. At present, the main methods include spectro-photometry, fluorescence spectrophotometry (FD), electron spin capture (ESR), high performance liquid chromatography (HPLC), chemiluminescence (CL), electrochemical detection (ECD) and so on.

2. Free radical detection technique

2.1 Spectrophotometric method

The ESR method and HPLC method are expensive and difficult for general laboratories to bear. Therefore, photometric method is commonly used by domestic researchers. Spectrophotometry is efficient, fast and practical, and the instrument is cheap. Through the development of technology in recent years, spectrophotometry has become a mature method for the detection of ·OH. Spectrophotometry uses the strong oxidation of ·OH to change the structure, properties and color of the probe capture agent, so as to change the absorption of the specific spectrum of the capture agent and indirectly determine ·OH.

Li Jiwu et al. [11] used bromophenol blue (BPB) as a liquid radical probe catcher to quantitatively detect the production of ·OH in the discharge process of high voltage corona liquid film. Yan Jun et al. [12] adopted visible spectrophotometry and used salicylic acid as the capture agent of hydroxyl radical ·OH produced by Fenton reagent reaction, making $FeSO_4$, H_2O_2, and salicylic acid-ethanol into a reaction system. They investigated the influencing factors of hydroxyl radical ·OH capture by salicylic acid, and used Fenton's reaction system as antioxidant filter. Yang Chunwei et al. [13] took methylene blue indicator as an example to capture hydroxyl free radicals generated by Fenton reagent reaction. With its high sensitivity and stability, it is suitable to be applied to study the mechanism of advanced oxidation method of environmental aqueous Fenton reagent. Wang Jingang et al.[14] used methylene blue spectrophotometry to detect hydroxyl radical by the change of absorbance of methylene blue before and after oxidation.

* Corresponding author: okwj@wust.edu.cn

2.2 Fluorescence photometric method

Fluorescence spectrophotometry has higher sensitivity than spectrophotometry and is not easily affected by other ions at specific wavelength. Although the accuracy and sensitivity of this method is less than that of ESR method, fluorescence photometry is simple to operate and easy to implement, so it is the most widely used method at present. The disadvantage of fluorescence photometry is that the adduct product of hydroxyl or the catcher itself needs characteristic fluorescence.

Liu Ting et al.[15] prepared Fe_2O_3-TiO_2-Al_20_3 catalyst and established a three-phase fluidized bed photoassisted heterogeneous Fenton reaction system. In the experiment, coumarin was used as ·OH capture agent, and the ·OH generated in the photoassisted heterogeneous Fenton reaction system was detected by fluorescence spectroscopy. Yang et al.[16] use quantitative DMSO to capture hydroxyl radicals and generate methyl radicals. The methyl radicals react with NitroXide-Linkednaphthalence, resulting in enhanced fluorescence intensity. The fluorescence intensity is proportional to the concentration of hydroxyl radicals and 0.1-2.0 μmol/ L of H_2O_2 can be detected, the lowest detection limit of H_2O_2 is 0.04μmol/L. Ding Haiyang et al.[17] studied the generation of ·OH in the degradation process of Ti/SnO_2 electrode by fluorescence photometry, and discussed the feasibility of the application of this method in the electrocatalytic oxidation process. At the same time, Ti/RuO_2 electrode was used as comparing electrodes, so as to study the different properties and mechanisms of the two electrodes in electrocatalytic oxidation activity.

2.3 Optical fiber spectral system measurement method

It is different from the existing spectrophotometry and fluorescence spectro-photometry. In order to detect radical concentration rapidly and accurately at low cost, Liu Huimin et al.[18] proposed to couple the online detection technology of ·OH radical and ·OH radical concentration optical fiber spectrum by UV-excited hydrogen peroxide production, and set up the online detection system of ·OH radical concentration optical fiber spectrum. The response characteristics of optical fiber spectra to methylene blue, and the effects of light conditions and initial concentration of hydrogen peroxide on the production and detection performance of ·OH radical were investigated. The results show that the concentration of ·OH radical can be accurately measured online by using the optical fiber spectrum system based on MB as ·OH radical catcher.

2.4 Electron spin capture method

Electron spin resonance (ESR), also known as electron paramagnetic resonance, was first discovered in 1944. It uses the characteristics that the material with unpaired electrons absorb the energy of electromagnetic wave under the action of magnetic field to make the electron transition between energy levels to detect and analyze paramagnetic substances. The method of spin capture is to add unsaturated diamagnetic compounds (spin capture reagent) into the reaction system, and combine them with all kinds of high activity and short life free radicals produced in the reaction system to form relatively stable spin adduct, which is suitable for ESR detection. The principle is to combine appropriate spin capture reagent and active short life free radicals to generate relatively stable spin adducts, the amounts of spin adducts can be detected by electron spin spectroscopy, and the amounts of spin adducts can be used to calculate the number of original free radicals. ESR is the most direct and effective technology for studying free radicals, but these free radicals must be relatively stable and reach to a certain concentration before they can be detected and studied by ESR technology. However, most of the free radicals produced in biological systems are unstable, which cannot be detected by conventional ESR spectroscopy. In order to overcome this limitation of ESR technology, spin capture technology has been developed, which is the most widely used and successful method for studying active free radicals in biological and medical systems [19].

Typical spin capture agents are nitrite or nitroxide compounds. When sufficient amounts of spin capture agent are added to the free radical producing system, the spin capture agent will react rapidly with any free radical present, producing stable and detectable nitroxide free radical adducts. The ESR spectrum of the free radical adduct formed has hyperfine cleavage given by the free radical trapping gene, which can identify the free radical adducts formed by the free radical and trapping agent which is very sensitive to the free radical structural changes[20]. There are three main types of spin capture agents, namely nitroso compounds, circular nitrones and linear nitrones.

Ding Qinxue et al. [21] used electron spin capture technology to detect hydroxyl radical (·OH) produced by the antitumor antibiotic boamycin in vitro. 5,5'-dimethylpivrlin-n-oxide was used as hydroxyl radical catcher, and the free radical signal was detected by electron spin resonance (ESR) spectrometer at room temperature. It was found that if nitrogen was added to the solution, no free radicals were produced when boamycin or ferrous sulfate were added to phosphate buffer alone or both. At the same time, if boamycin and ferrous sulfate were added to the buffer and oxygen, the signal appeared, and the strength of the signal was positively related to the concentration of boamycin. Cong Jianbo et al. [22] used liquid nitrogen (770 K) to preserve spin adducts to prolong the life of spin adducts, and the results showed that the preservation of short-lived free radical spin adducts in liquid nitrogen could prolong their life. For the short-lived free radicals that cannot be measured without ESR measurement conditions, this method can be used to select appropriate trapping agent, which can be stored in liquid nitrogen immediately after capture, and then removed and thawed for measurement when necessary.

Using free radical trap agent - the ESR method detecting hydroxyl radicals is simple and effective, but expensive and low cost. Since the spin adducts are unstable, their life is very short from only a few minutes or tens of minutes, must be measured immediately after capturing free radicals, so its quantitative analysis is not very accurate, which limits many experimental studies and industrial applications.

2.5 High performance liquid chromatography

High performance liquid chromatography (HPLC) is a chromatographic analysis technique used to separate mixtures in order to confirm and quantify the proportions of the individual components. It relies on a pump to pressurize the sample through a pressure column filled with an adsorbent, causing the components of the sample to separate. Free radicals should be captured by a trapping agent, which can react with free radicals to form a stable conjugate and is easy to be separated and detected. General ·OH capture agents are benzoic acid/salicylic acid (SA), phthalic acid, dimethyl sulfoxide (DMSO), etc. ·OH reacts with dimethyl sulfoxide to form mesylate, and its content can be determined by HPLC to infer that of ·OH.

Xue Juanqin et al.[23] used high performance liquid chromatoid-ultraviolet detection (HPLC-UV) to determine ·OH generated in the electrochemical oxidation system by using salicylic acid (SA) and 4-hydroxybenzoic acid (4-HBA) as capture agents. The results show that the electrochemical oxidation reactions of SA and 4-HBA with ·OH are in accordance with the first-order reaction kinetics, and the reaction rate constants K are $2.1833\times10^{-4}s^{-1}$ and $1.3500\times10^{-4}s^{-1}$, respectively. After three repeated experiments, it is found that the initial consumption of 4-HBA is 3 times of SA and the capture time is about 2.3 times of SA when the same ·OH capture amount is achieved. Through comparative study, it is found that SA has higher sensitivity and stronger capturing ability under UV detection conditions. Han Yao et al.[24] established a simple and sensitive method for the determination of ·OH produced by Fe(III) -catalyzed photoreaction. ·OH was captured by dimethyl sulfoxide (DMSO) to form formaldehyde, and then reacted with derivative reagent 2, 4-dinitrophenylhydrazine (DNPH) to form corresponding hydrazone (HCHO-DNPH), afterwards it was analyzed by high performance liquid chromatography. The effects of the concentration,pH, temperature and time of the derivatization on the derivatization reaction were studied. The optimal derivatization conditions were determined at 270μmol/L, pH=4, 50°C and 30 min. Yang Wei et al.[25] established a simple, efficient and rapid method for the determination of hydroxyl radical by using high performance liquid chromatography combined with Coulomb array multi-electrode detector system (HPLC-ECD).

The advantages of HPLC are convenient measurement, high efficiency and high sensitivity, the detection limit up is to 10^{-9}~10^{-11}g. However, it requires expensive equipment and the reaction process is relatively complex, besides, there are many intermediates and secondary products, so it is still unqualified in the accurate quantitative detection of free radicals.

3. Conclusions

Hydroxyl radical has always been a research hotspot. From the methods mentioned above, it can be seen that in recent years, the detection technology of hydroxyl radical is gradually mature, and there are a variety of detection methods at home and abroad. However, a few methods (such as ESR, laser fluorescence spectroscopy and long optical path absorption spectroscopy) can be directly measured, most of the determination methods are indirect. The indirect methods are affected by many factors, such as measuring system and measuring conditions. However, due to the characteristics of high activity, low concentration and short lifetime of ·OH, there are many interfering factors even for direct measurement. Therefore, the detection of ·OH should be combined with its own research characteristics, and try to use and combine different analysis methods to improve the analysis effects.

Acknowledgments

This work was financially supported by"Innovation and entrepreneurship training plan program for college students of Wuhan University of Science and Technology in 2021 (21ZA020) "and"Natural Science Foundation of Hubei Province (2021CFB495) ".

References

1. Gogate P R . Cavitation: an auxiliary technique in wastewater treatment schemes[J]. Advances in Environmental Research, 2002, 6(3):335-358.
2. Badve M , Gogate P , Pandit A , et al. Hydrodynamic cavitation as a novel approach for wastewater treatment in wood finishing industry[J]. Separation & Purification Technology, 2013, 106(Complete):15-21.
3. Mandar P. Badve and Parag R. Gogate and Aniruddha B. Pandit and Levente Csoka. Hydrodynamic cavitation as a novel approach for delignification of wheat straw for paper manufacturing[J]. Ultrasonics Sonochemistry, 2014.
4. Prasad S V , Thomas S A , Vinayak R V . Modelling of vortex based hydrodynamic cavitation reactors[J]. Chemical Engineering Journal, 2018, 377:S138589471831492X-.
5. Zhang Xiaodong, Yang Huizhong, Li Zhiyi. Relationship between Hydrodynamic cavitation strength and free radical production [J]. Journal of chemical industry, 2007,58 (1) : 27-32.
6. Yan Jun, Gou Xiaojun, Zou Quanfu, et al. Spectrophotometric determination of hydroxyl radical produced by Fenton reaction [J]. Journal of Chengdu University: Natural Science Edition, 2009, 28(2):4.
7. Xu Xiangrong, Wang Wenhua. Fluorometric method for determination of hydroxyl radicals produced by Fenton reaction [J]. Chinese Journal of Analytical Chemistry, 1998, 26(12):4.
8. Esplugas S , J Giménez, Contreras S , et al. Comparison of different advanced oxidation processes for phenol degradation[J]. Water Research, 2002, 36(4):1034-1042.
9. Jia Zhi-Shen, WU Jian-min, TANG Meng-cheng. Colorimetric determination of hydroxyl radicals

produced by Fenton reaction [J]. Progress in Biochemistry and Biophysics, 1996, 23(2):4.

10. Shao-An, Cheng, and, et al. Optimizing electron spin resonance detection of hydroxyl radical in water[J]. Chemosphere, 2003.
11. Li Jiwu, Yu Jie, Fan Lei. Quantitative detection of hydroxyl radical produced by high voltage corona liquid film discharge using bromophenol blue [J]. High Voltage Technology, 2012, 38(7): 1576-1581.
12. Yan Jun, Gou Xiaojun, Zou Quanfu, et al. Spectrophotometric determination of hydroxyl radical produced by Fenton reaction [J]. Journal of Chengdu University (Natural Science Edition), 2009, 28(2): 91-93, 103.
13. Yang Chunwei, Wang Dong. Detection of hydroxyl radicals produced by Fenton reaction in environmental simulated water by visible spectrophotometry [J]. Environmental Technology, 2005, (1): 29-31.
14. Wang Jingang, Li Guifang, Wang Xikui, et al. Detection of hydroxyl radical and nitrite produced by cavitation effect [C]// Proceedings of the Third National Conference on Environmental Chemistry, 2005:29-30.
15. Liu Ting, You Hong, Chen Qiwei, et al. Determination of hydroxyl radical in photoassisted heterogeneous Fenton system by fluorescence spectrometry and its influencing factors [J]. Environmental Science, 2009, 30(9): 2560-2564.
16. X. F. Yang,X. Q. Guo.·Investigation of the anthracene-nitroxide hybrid molecule as a probe for hydroxyl radicals[J].·The Analyst,·2001,·126 (10):·1800-4.
17. Ding Haiyang, Feng Yujie, Lv Jiangwei, et al. Detection and electrocatalytic mechanism of hydroxyl radical in Titanium based tin dioxide Electrode electrolysis [J]. Chinese Journal of Analytical Chemistry, 2007, (10): 1395-1399.
18. Liu Huimin, Wu Yongwu, Xiang Wenhao, et al. Study on on-line detection method of hydroxyl radical concentration based on optical fiber spectrum [J]. Laser Journal, 2020, 41(5): 62-66.
19. Zhao B L. Application of electron spin resonance (ESR) technology in biology and medicine [J]. Chinese Journal of Spectroscopy, 2010, 27(1): 51-67.
20. Yang Chunwei, Wang Dong, Guo Jianbo, et al. Comparison of hydroxyl radical detection methods in advanced oxidation of organic compounds in water [J]. Environmental Pollution Treatment Technology and Equipment, 2006, (1): 136-141.
21. Ding Qinxue, Lin Futian, Lu Daohui, et al. Study on spin capture of hydroxyl radicals produced by antitumor antibiotic boamycin in vitro [J]. Chinese Journal of Antibiotics, 1997, (1): 49-53.
22. Cong Jianbo, Sun Cunpu, Mo Jian. Low temperature preservation of short-lived free radicals by spin trapping [J]. Progress in Biochemistry and Biophysics, 1993, (4): 326-327.
23. Xue Juanqin, Jiang Meng, Yu Lihua, et al. Effect of trapping agent on detection of hydroxyl radical in electrochemical oxidation system [J]. Chinese Journal of Analytical Science, 2015, 31(5): 606-610.
24. Han Yao, Li Jialong, Yang Yalei, et al. Determination of hydroxyl radical by dimethyl sulfoxide capture and high performance liquid chromatography [J]. Chinese Journal of Analytical Sciences, 2021, 37(2): 177-182.
25. Yang W, Deng YL. HPLC-ECD method for the detection of hydroxyl radicals [J]. Analytical Laboratory,2007,(09):120-122.

Research on time-varying economic dispatch of smart grid based on Lagrangian pairing

Ying Zuo *

School of Management, Tianjin University of Technology, Tianjin, China

Abstract. To design a class of time-varying economic dispatching algorithms for smart grids based on the Lagrangian dual idea for the time-varying load problems that arise in practical applications in smart grids, and to demonstrate the convergence of the algorithms.

1. Introduction

Currently, smart grid is a hot issue in power system research[1,2]. Smart grid, called "Grid 2.0", is an intelligent grid that combines the traditional grid with advanced information and communication facilities to achieve two-way information transfer between the grid and end users. It isolates faulty components of the grid and automatically restores the system to its normal state to ensure system stability; it facilitates user participation in managing the operation of the power system; it provides a fast solution for restoring power when the grid is attacked and fails; and it allows all types of power generation systems to be easily connected to the system[3]. All countries attach great importance to the construction and development of smart grids. For example, the U.S. is building a smart grid in three directions: 1) upgrading the grid infrastructure to improve the reliability of power supply; 2) applying advanced information, communication and computer technologies to the power system; and 3) enhancing the information transfer between customers and power companies through advanced meter improvements. The main focus of the European smart grid is to improve the utilization of new energy sources to improve environmental, climate and energy issues. In addition, the main focus of China's smart grid is to try to meet the energy supply needed for the country's economic development. The development of smart grid is accompanied by a series of new technologies, new equipment updates and breakthroughs, and will encounter many new problems[4]. Among them, the issue of economic dispatch is widely The main focus is on the power distribution of the power generation, which aims to save the cost of power generation and ensure the balance of the grid..

The existing economic scheduling studies of smart grids mainly focus on the case of constant load, which is too strict an assumption[5]. A new class of economic dispatching algorithms is designed considering the time-varying load case in practical applications, and the convergence of the algorithms is rigorously demonstrated.

2. Symbols and problem descriptions

2.1 Symbols

Let $x_i \in R^{ni}$, $i = 1,2,\dots,m$, $col(x_1,\dots,x_m) = [x_1^T,\dots,x_m^T]^T$, where x_i^T is the transpose of x_i. I_n is an $n \times n$ unit matrix. $\nabla_x f(x,t)$ and $\nabla_{xx} f(x,t)$ denote the first order as well as the second order partial derivatives of the function $f(x,t)$ with respect to the vector x. $\nabla_{xt} f(x,t)$ denotes the function $\nabla_x f(x,t)$ with respect to t for the first-order partial derivative.

2.2 Problem Description

Suppose there is a grid with N buses and each bus line contains a generator and a local load, and denote by $P_i(t) \in \mathbb{R}$ and $D_i(t) \in \mathbb{R}$ denote the active power generated by the generators on the bus and the active power required by the load on the bus, respectively, and assume that each bus has a local load. the active power generated by the generator on the bus and the active power required by the load on the bus, respectively, and assuming that each each bus is assigned a local cost function.

$$f_i(P_i) = a_i P_i^2 + b_i P_i + c_i \quad (1)$$

where $a_i > 0$, $b_i > 0$, $c_i > 0$. Next consider the following economic scheduling problem:

* Corresponding author: zuoying10@163.com

$$P_i^* = argmin \sum_{i=1}^{n} f_i(P_i) \quad (2)$$

$$s.t. \sum_{i=1}^{N} P_i = \sum_{i=1}^{N} D_i(t) \quad (3)$$

$$0 \le P_i \le P_i^{max} \quad (4)$$

where f_i is known only for the bus. The goal of this paper is to design an algorithm for the bus is to design an algorithm such that the system

$$P_i^* = u_i \quad (5)$$

realizes $P_i \to P_i^*(t)$, $t \to \infty$, where $P_i^*(t)$ is the above optimal solution of the optimization problem.

Assume 1. The first-order and second-order derivatives of the load D_t exist and are bounded.

Note 1. From $a_i > 0$ it can be inferred that the overall cost function $\sum f_i(P_i)$ is convex , so the optimal solution of problem (2) exists. Assumption 1 can be satisfied in many cases because the actual load demand always varies slowly.

3. Algorithm derivation

Theorem 1. Let $a_i^* = a_i + \frac{c(t)}{2(P_i - P_i^{max})^2}$, $c(t) = e^{-2t}$, $b_i^* = b_i + 2a_i P_i^{max}$, $\theta^* = \sum_{i=1}^{N} \frac{1}{2a_i^*}$. Consider the following The following economic scheduling algorithm is

$$\dot{P}_i = -P_i + P_i^{max} - \frac{b_i^*}{2a_i^*} + \frac{1}{2a_i^*\theta^*} \sum_{i=1}^{N} \left(D_i + \dot{D}_i - P_i^{max} + \frac{b_i^*}{2a_i^*} \right) \quad (6)$$

$$\dot{\lambda} = -\lambda + \frac{1}{\theta^*} \sum_{i=1}^{N} \left(D_i + \dot{D}_i - P_i^{max} + \frac{b_i^*}{2a_i^*} \right) \quad (7)$$

The algorithm can realize the time-varying economic dispatch problem of smart grid.

Proof: The economic scheduling problem is essentially a constrained optimization problem. problem, and according to the design ideas of typical optimization algorithms, the optimization algorithm can be designed by constructing Lagrangian function, based on the idea of duality to design the optimization algorithm. Specifically, the following Lagrangian function is defined that,

$$L_2 = \sum_{i=1}^{n} f_i(P_i) + \lambda\left(\sum_{i=1}^{N} D_i(t) - \sum_{i=1}^{N} P_i\right) - c(t) \sum_{i=1}^{N} log(P_i^{max} - P_i) \quad (8)$$

Consider the original optimization variables and the dyadic variables in combination that is, let $z = col(P_1, \ldots, P_N, \lambda)$. Thus the above economic scheduling algorithm can be written in a more concise form, by means of the Lagrangian function The expression can be obtained whose first-order gradient has the following form:

$$\nabla_z L_2 = \begin{pmatrix} 2a_1 P_1 + b_1 - \lambda - \frac{c(t)}{(P_1 - P_1^{max})} \\ \vdots \\ 2a_N P_N + b_N - \lambda - \frac{c(t)}{(P_N - P_N^{max})} \\ \sum_{i=1}^{N} D_i - \sum_{i=1}^{N} P_i \end{pmatrix} \quad (9)$$

$$\nabla_{zt} L_2 = \begin{pmatrix} (P_1 - P_1^{max}) \\ \vdots \\ -\frac{c(t)}{(P_N - P_N^{max})} \\ \sum_{i=1}^{N} D_i \end{pmatrix} \quad (10)$$

In addition, the Lagrangian function expression gives its second order gradient degree has the following form.

$$\nabla_{zz} L_2 = \begin{bmatrix} & \Delta_{11} & 0 & \cdots & 0 & -1 \\ & 0 & \Delta_{22} & \cdots & 0 & -1 \\ -1 & \vdots & \vdots & \ddots & \vdots & \vdots \\ & 0 & 0 & \cdots & \Delta_{NN} & -1 \\ & -1 & -1 & \cdots & -1 & 0 \end{bmatrix} \quad (11)$$

where $\Delta_{ii} = 2a_i + \frac{c(t)}{(P_t - P_t^{max})^2}$, $i = 1, \ldots, N$. To further investigate the convergence properties of the algorithm, by simplifying the calculation of the Lagrangian Langeland function first-order gradient and its partial derivative with respect to time has a special symmetric form. Specifically, it is not difficult to compute

$$\nabla_{zz} L_2 = \begin{bmatrix} & 2a_1^* & 0 & \cdots & 0 & -1 \\ & 0 & 2a_2^* & \cdots & 0 & -1 \\ -1 & \vdots & \vdots & \ddots & \vdots & \vdots \\ & 0 & 0 & \cdots & 2a_N^* & -1 \\ & -1 & -1 & \cdots & -1 & 0 \end{bmatrix} \quad (12)$$

$$\nabla_z L_2 + \nabla_{zt} L_2 = \begin{pmatrix} 2a_1^*(P_1 - P_1^{max}) + b_1^* - \lambda \\ \vdots \\ 2a_N^*(P_N - P_N^{max}) + b_N^* - \lambda \\ \sum_{i=1}^{N} D_i - \sum_{i=1}^{N} P_i + \sum_{i=1}^{N} \dot{D}_i \end{pmatrix} \quad (13)$$

Further, the inverse matrix of the Hessian matrix can be obtained as the following expressions are obtained:

$$\nabla_{zz}^{-1} L_2 = \begin{bmatrix} \Lambda_{11} & -\frac{1}{4a_1^* a_2^* \theta^*} & \cdots & -\frac{1}{4a_1^* a_N^* \theta^*} & -\frac{1}{2a_1^* \theta^*} \\ -\frac{1}{4a_2^* a_1^* \theta^*} & \Lambda_{22} & \cdots & -\frac{1}{4a_2^* a_N^* \theta^*} & -\frac{1}{2a_2^* \theta^*} \\ \vdots & \vdots & \ddots & \vdots & \vdots \\ -\frac{1}{4a_N^* a_1^* \theta^*} & -\frac{1}{4a_N^* a_2^* \theta^*} & \cdots & \Lambda_{NN} & -\frac{1}{2a_N^* \theta^*} \\ -\frac{1}{2a_1^* \theta^*} & -\frac{1}{2a_2^* \theta^*} & \cdots & -\frac{1}{2a_N^* \theta^*} & -\frac{1}{\theta^*} \end{bmatrix} \quad (14)$$

where $\Lambda_{ii} = \frac{1}{2a_i^*} - \frac{1}{4(a_i^*)^2 \theta^*}$, $i = 1, \dots, N$. The symmetry of the form as above is carefully studied, and the trajectory of the original optimal variables and the pairwise variables over time can be investigated. Based on the symmetric form of the first-order gradient of the Lagrangian function and its partial derivatives with respect to time, the following relationship exists between the original optimal variables and the trajectory of the dyadic variables with respect to time:

$$\dot{z} = -\nabla_{zz}^{-1} L_2 (S \nabla_z L_2 + \nabla_{zt} L_2) \quad (15)$$

where $S = I$. It is further obtained that

$$\dot{\nabla}_z L_2 = -\nabla_z L_2 \quad (16)$$

The equation further states that there exists $\|z - z^*\| \le ce^{-t} \|\nabla_z L_1(z(0), 0)\|_2$, that is, the variable z will converge to the optimal solution. In addition, it is known from the above equation that the convergence of z has an exponential rate of convergence, so it is robust to external perturbations, especially to perturbation signals of small amplitude.

4. Simulation Example

Considering the specific economic scheduling problem, the corresponding parameters are shown in Table 1 shown in Table 1, the load-time variable track line can be characterized as

$$D_i = 70 + 10isin(0.5t) \quad (17)$$

Table 1. IEEE-14 bus parameters.

Bus	a_i	b_i	c_i
1	**0.04**	**2.0**	**561**
2	**0.03**	**3.0**	**310**
3	**0.035**	**4.0**	**4.0**
6	**0.03**	**4.0**	**4.0**
8	**0.04**	**2.5**	**78**

Based on the algorithm in Theorem 1, the numerical simulation leads to the specific of the bus load trajectory, as shown in Figures 1 and 2. It can be seen that the algorithm can achieve the economic scheduling goal and the convergence rate has exponential characteristics.

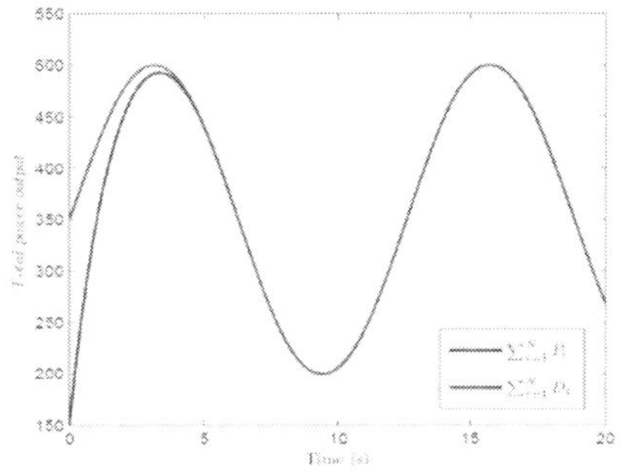

Figure 1. Time-varying load and power generation curves.

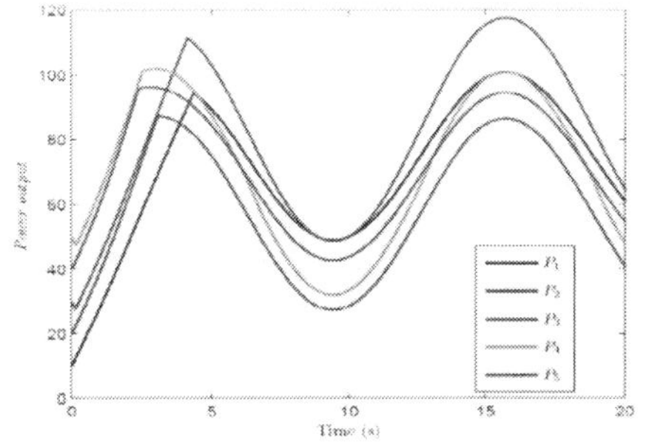

Figure 2. Different bus generation power rail lines.

5. Conclusion

The time-varying economic dispatching problem of smart grid is studied, and the corresponding algorithm is designed based on the Lagrangian pair idea, and the effectiveness of the algorithm is demonstrated. The designed scheduling algorithm is simple in form, low in complexity, and easy to calculate, which is well adapted to the actual grid system operation requirements; in addition, the algorithm has exponential convergence speed and is robust to external disturbance signals, so it is applicable to smart grids of different scales. Future research can turn to the economic dispatching problem of communication-constrained smart grid, the economic dispatching problem of information transmission with time delay and the economic dispatching problem of finite time convergence, etc.

References

1. Chowdhury B H, Rahman S. A review of recent advances in economic dispatch[J]. IEEE transactions on power systems, 1990, 5(4): 1248-1259.
2. Park J B, Jeong Y W, Shin J R, et al. An improved particle swarm optimization for nonconvex economic dispatch problems[J]. IEEE Transactions on power systems, 2009, 25(1): 156-166.
3. Zhenya Liu, Smart Grid Technology [M]. Beijing: China Electric Power Press, 2010.

4. Yumin Pan, Yonghong Deng, Quanzhu Zhang. Wavelet-chaotic time series prediction of short-term power load [J]. Journal of System Simulation, 2013,25(05):868-875.
5. Shengwei Mei, Jianquan Zhu. Some mathematical and control problems in smart grid and their prospects [J]. acta automatica sinica, 2013,39(02):119-131.

Application prospect of high efficiency and energy saving bromine extraction technology in the field of bromine-containing waste salt

Wenyan Zhang, Shujing Chai, Xiaocui Hao, Wei Liu, Yuqi Wang and Qi Zhang*

Institute of Seawater Desalination and Multipurpose Utilization, MNR (Tianjin), Tianjin 300192, China

Abstract. Bromine has a strong resource dependence. With the sharp decline of the total amount and grade of traditional brine resources, the difficulty of mining is increasing day by day, and the shortage of bromine supply and demand has become a normal situation. Therefore, it is urgent to find new sources of bromine resources and develop corresponding extraction technologies. Bromine-containing waste salt is a new source of bromine. The mature technology achievements of bromine extraction in underground brine, concentrated seawater and other water are transferred to the field of bromine extraction from bromine-containing waste salt, which realizes the closed-loop recycling of resources and extends the industrial chain, and has a good application prospect.

key word: Bromine, solid waste, bromine-containing waste salt, bromine extraction.

1. Introduction

Bromine is widely used in flame retardants, pharmaceutical intermediates, pesticide chemicals, oilfield chemicals and water treatment [1-3]. A large amount of bromine- containing wastewater will be produced in the production process.

Zero discharge of wastewater is an important way to realize the utilization of sewage resources. At present, the basic idea of "zero discharge" wastewater treatment process is to separate salt and water, to obtain recycled water and crystallized salt. For most industries, due to the complex salt composition of wastewater, it is very difficult to achieve "zero solid waste" discharge by simply evaporating. Due to the immature process of salt separation, most enterprises get bromine-containing waste salt after treating high-salt sewage containing bromide, which is temporarily stored as solid waste or dangerous waste. This not only causes resource waste, but also aggravates potential environmental protection and safety risks. We should also attach great importance to the prevention and control of solid waste pollution. In the work plan of "waste-free city" construction during the 14th Five-Year Plan Period, the comprehensive utilization of solid waste has been mentioned for many times, and the overall goal has been clearly stated to promote the construction of "waste-free city" in about 100 cities at prefecture level and above. By 2025, the production intensity of solid waste will be reduced rapidly, the comprehensive utilization level will be significantly improved, and the solid waste treatment system and capacity will be significantly improved. It is pointed out that the green and low-carbon development of industry should be accelerated, the pressure of industrial solid waste treatment should be reduced, and the comprehensive utilization of industrial solid waste should be strengthened.

As a potential source of solid bromine, bromine-containing waste salt has not been used to extract bromine on a large scale. The high value utilization of bromine containing waste salt can greatly develop circular economy, realize the win-win situation of closed loop utilization of resources and environmental governance, and effectively promote the economical and intensive recycling of resources. It is of great significance to safeguard national resource security and support the construction of ecological civilization at a high level.

2. Current situation of bromine resources

2.1 Current situation of traditional bromine resources

The total amount of bromine reserves is small, the grade is low, and the resources are exhausted day by day. Chinese bromine resources mainly come from the underground brine of Laizhou Bay, Shandong Province, but the average concentration of bromine is only 200 g/m^3, and the rate of decrease is 10g/m^3 per year. In addition to underground brines, bromine extraction from medium brines in the salt-making process from seawater accounts

* Corresponding author: zqhaoyun@163.com

for about 20% of the annual output. However, due to the policy restriction of salt production, it is difficult to increase the output of bromine extraction from medium brine effectively. After 2013, the production of bromine gradually decreased. As shown in Figure 1, in 2018, due to storm surge, safety inspection and other factors, many bromine plants in China stopped production and the annual output of bromine was about 85,000 tons. In addition to 2020, our annual output of bromine rebounded to 125,000 tons, other years the annual output of bromine about 70,000-90,000 tons.

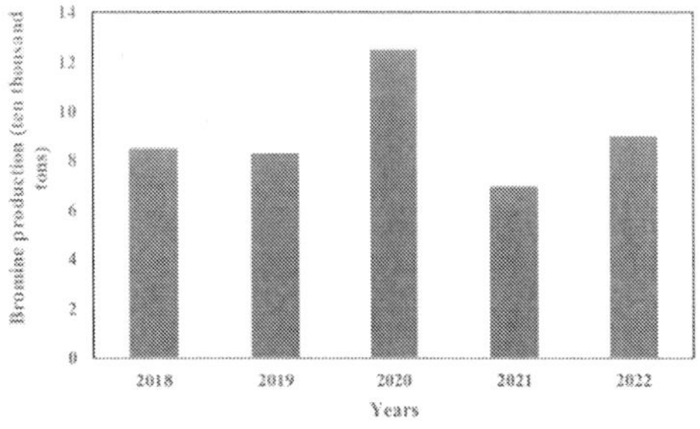

Figure 1. Variation of bromine production in China.

As shown in Figure 2, the average annual price of bromine has shown an oscillating upward trend in the past five years. The average annual price in 2021 is 45,502 yuan per ton, an increase of 53% compared with the previous year. The average bromine price in 2022 remained unchanged from 2021.

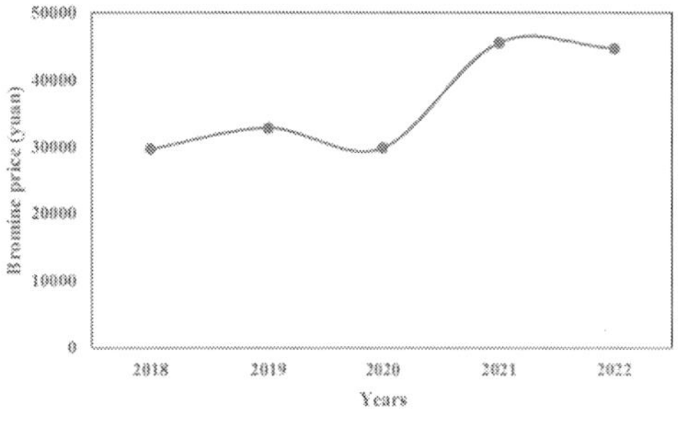

Figure 2. China's average annual bromine price in recent 5 years.

In terms of import, the accumulated import volume of bromine from January to November in 2022 is about 54,000 tons. Imported bromine is still an important supplement for domestic bromine. According to the import volume of bromine in previous years, the annual import volume of bromine is expected to exceed 60,000 tons in 2022.

Some places, such as Tianjin, have begun to extract bromine from desalinated concentrated seawater with a bromine content of only 80 g/m^3. Some domestic power plants and chemical plants have also planned to use warm seawater (50 ~ 60 g/m^3) discharged for bromine extraction. Relevant feasibility studies have been conducted, but due to the restriction of environmental protection policies, the construction has not started yet. Now, our country has become a net importer of bromine. The tension of bromine supply and demand caused by the shortage of resources has become normal. Many Chinese enterprises have gone to Africa, South America, Southeast Asia and other countries with insufficient supporting facilities to purchase brine resources to extract bromine. Therefore, developing new bromine resources is the necessary way to ensure the safe supply of bromine and realize the sustainable development of bromine and bromine industries.

2.2 Current situation of bromine resources in waste salt resources containing bromine

Bromine and bromides are widely used in flame retardants, medical intermediates, pesticide chemicals, oilfield chemicals and water treatment[1-3]. For example, bromoaniline as an important chemical intermediate, is widely used in medicine, dyes, pigments and other fine chemical products, and the synthesis of bromoaniline has a step of bromination reaction mainly using bromine as raw material. Therefore, in the synthesis process, it is inevitable to produce synthetic wastewater, which contains a large number of bromine ions not involved in the reaction. Direct discharge will cause serious harm to the environment and waste resources.

For most industries, due to the complex salinity of wastewater, it is very difficult to achieve "zero solid waste" discharge by simply evaporating. In the context of the state's efforts to control the discharge of wastewater, due to the immature process of salt separation, most enterprises get bromine-containing waste salt after treating high-salinity wastewater, which is temporarily stored as solid waste or dangerous waste. This not only causes a waste of resources, but also aggravates the potential risks of environmental protection and safety. The large-scale bromine-containing waste salt extraction project is still blank. According to the survey, the total amount of bromine-containing waste salt produced by 15 enterprises in a city can reach 15,000 tons/year, with an average bromine content of 13% (sodium bromide). If the resource treatment is carried out in a concentrated way, the annual output of bromine is expected to be about 1,500 tons. The annual average operating income is 66.9 million yuan (calculated at the average bromine price of 45,000 yuan/ton in 2022).

3. Current status of bromine extraction from bromine-containing waste salt

3.1 Current status of bromine-containing waste salt treatment technology

Bromine-containing waste salt is a potential source of solid bromine, and no large-scale bromine extraction projects have been carried out using bromine-containing waste salt. The literature reports mainly on how to treat bromine-containing wastewater to meet the emission standards[4]. Some bromine demand enterprises treat their own bromine-containing wastes mainly as bromine-containing wastewater and on a small scale. Most of the solid waste treatment enterprises collect bromine-containing waste salt and do not achieve the recovery of high-valent elements (bromine), and only use the form of salt separation for harmless treatment. The reports on bromine extraction process of bromine-containing wastewater are mainly focused on patents [5]. There is no research on the impurity removal process of bromine-containing waste salt reported in the literature.

3.2 Problems need to be solved in extracting bromine from bromine-containing waste salt

Bromine-containing waste salt are characterized by a wide range of sources, variety and complex composition. There are two forms of bromine in bromine-containing waste salt, namely organic and inorganic salts. The source of bromine-containing waste salt may be organic product manufacturers such as pharmaceutical intermediates, pesticide chemicals, dyestuffs, etc., which may increase the chlorine consumption in the oxidation stage due to high COD and affect the quality of bromine products due to volatilization of organic matter in the distillation process. Therefore, the pretreatment process of bromine-containing waste salt needs to remove COD firstly, and the common method to remove COD is calcination, and the bromine distribution law after calcination is one of the key issues in the study of bromine extraction from bromine-containing waste salt. Theoretically, the bromine in organic matter will enter the absorption solution with the tail gas, while the bromine in inorganic salts (e.g., sodium bromide, etc.) will be retained in the solid phase waste salt. However, whether there are other influencing factors to change the bromine distribution in the actual production process, practical verification of the bromine distribution pattern after calcination is needed.

After calcination, whether the bromine is distributed in the tail gas absorption solution or in the waste salt, the next process step is to convert it into a bromine-containing waste salt dissolution with a certain salt concentration. The bromine-containing waste salt is mainly inorganic salts such as sodium bromide, and the proportion of bromine in the waste salt is high after calcination, so most of the cases need to get the bromine containing waste salt dissolution by salting. If the calcium and magnesium scale forming ions in the bromine containing waste salt are too high, it will affect the equipment of the subsequent water distillation process, and the scaling problem will occur easily. Therefore, the pretreatment process should also include the removal of calcium and magnesium plasma.

Although there are many kinds of bromine-containing waste salts, inorganic salts are still mainly sodium bromide. Considering the high value of bromine, the bromine-containing waste salt as solid waste will not have too high bromine concentration, and the bromine concentration in the raw material solution of bromine-containing waste salt obtained through the above pretreatment process may range from several hundred to several ten thousand g/m^3. Because of the wide range of concentration of raw material liquid, the bromine extraction process route needs to be determined according to the actual situation.

3.3 Prospects of high efficiency and energy saving bromine extraction technology in the field of bromine-containing waste salt bromine extraction

Air blow-out method and water vapor distillation method are the most widely used bromine extraction processes in domestic and foreign industries [6]. The air blow-out method can extract bromine from underground brine (about 100-300 g/m^3) or seawater (50-65 g/m^3). This method has been adopted by almost all the enterprises using underground brine or (concentrated) seawater to extract bromine in Shandong province. According to the different absorbents, this method is divided into two types: air-blown acid absorption method and air-blown alkali absorption method. Although the former method has relatively low steam and electric power consumption, it needs to prepare sulfur dioxide gas with sulfur incinerator; although the latter method has relatively high electric power and acid-base consumption, it is easy to control the absorption by lye only, and the chlorine consumption can be reduced by about 50%. Water vapor distillation method is suitable for the raw material liquid with bromine content above 2~5 kg/m^3, which is basically used in foreign countries with high bromine grade such as the United States and Israel. In addition, after the concentration of the finished bromine solution is enriched to 20~80 kg/m^3 in the air blowing method, water distillation method must be applied to finally produce liquid bromine.

Analyzing from the technical point of view, our team has been continuously researching in the field of high-efficiency separation of bromine with multiple magnitudes, wide sources and low concentrations for more than ten years, and has comprehensively mastered the bromine extraction technologies such as air blow-out and water distillation, and has achieved many research results in bromine extraction from seawater, concentrated seawater and underground brine, and realized the transformation of the results. In response to the difficulties of low resource utilization and high energy consumption in China's bromine extraction industry, the project team has built indoor R&D platforms of 30m^3/d and 500m^3/d and on-site pilot production units with the continuous support of several national and provincial vertical research projects. The project team developed the industrialization technology and key core equipment of high efficiency and low energy consumption of bromine extraction from concentrated seawater by air blowing method, which broke through the bottleneck of low yield and high energy consumption of this process. During the research and development process, many national invention patents have been applied, covering the whole process of bromine extraction, including new process, key tower core components, tail gas purification and emission, etc. A relatively complete patent protection system has been formed.

Based on the above results, in 2019, the project team presided over and completed the first domestic 1,000-ton concentrated seawater high-efficiency and low-energy bromine extraction industrialization demonstration plant project in Tianjin. After the transformation, the bromine content of the bromine extraction mother liquor is 30%

lower than that before the transformation, and the production capacity is increased by 7.5%; the resistance loss of the components in the tower of the energy-efficient bromine extraction plant is about 60% of that in the tower of the conventional bromine extraction plant, and the energy consumption per ton of bromine is reduced by 10% on average. The project has passed the process checking in winter and summer organized by the manufacturer and the acceptance of relevant experts, and has been operating stably for three years with high reliability and good economic benefits. While realizing the industrialized operation of bromine extraction from concentrated seawater, the project has expanded its achievements to the field of bromine extraction from underground brine and sun-dried seawater step by step.

On the other hand, the bromine-containing waste salt needs to be transformed into bromine extraction raw material liquid before bromine extraction, which involves treatment processes such as dissolving salt and removal of calcium and magnesium scale forming ions. Our team has long been committed to the research on resource-based treatment of high-salt wastewater, and has made significant technical breakthroughs and fruitful research results in the research fields of crystallization technology, equipment and high-salt wastewater treatment, and formed advantageous technologies with independent intellectual property rights, which have been successfully applied in more than 10 demonstration projects in the fields of high-salt wastewater treatment and resource-based utilization in electric power, metallurgy, chemical industry and mining industry. Among them are 4 projects of high calcium and magnesium wastewater disposal (concentration, calcium and magnesium removal).

Coupling our team's mature technology of bromine extraction in underground brine, concentrated seawater and other water and high calcium and magnesium wastewater disposal technology in the new field of bromine-containing waste salt, we have better technical support for the application of high efficiency and energy saving bromine extraction technology in the field of bromine-containing waste salt bromine extraction.

From the perspective of application prospect, the bromine market in China has long been in short supply and the price continues to run high, so the economic benefits of the bromine extraction project are extremely good. The bromine extraction technology is transformed in the new field of bromine-containing waste salt. After the successful indoor research and development, it can be directly promoted and applied with the mature equipment and operation process.

In summary, the transformation of high efficiency and energy saving bromine extraction technology in the field of bromine-containing waste salt bromine extraction is technically feasible and has good application prospects.

4. Conclusion

Bromine-containing waste salt is a potential source of solid bromine, and no large-scale bromine extraction projects have been carried out using bromine-containing waste salt. Bromine-containing waste salt are characterized by a wide range of sources, variety and complex composition. The key technologies for extracting bromine from bromine-containing waste salt include pretreatment process and determination of bromine extraction process route. By coupling the mature technology of bromine extraction in water and the disposal technology of high-salinity wastewater in the new field of bromine extraction from bromine-containing waste salt, the transformation and industrialization of high efficiency and energy saving bromine extraction technology will be realized. Construction of bromine extraction project of bromine-containing waste salt can realize high-value utilization of high-value element (bromine) recovery and low-value element conversion, and achieve a win-win situation of resource utilization and environmental governance.

Acknowledgments

This work was financially supported by Key R&D Program in Shandong Province (2019JZZY010807), National Key Research and Development Program of China (2022YFC290400503) and Special basic research fund for central public welfare institute of China (R-JBYWF-2021-T01).

References

1. Y. S. Zhang , C. J. Gao, R. H. Cai, Application of brominated flame retardants and its development, Chem. Ind. Eng. 2009.
2. B. Rajender, S. Palaniappan. Efficient bromination of polyaniline base to poly(2-bromoaniline)-bromide salt and its application as a recyclable catalyst for the synthesis of 2-methyl-4-anilino-1,2,3,4-tetrahydroquinolines, Catal. Commun. 57(2014)94-98.
3. W. Sun, Y. T. Li, Sodium bromide acid dye intermediate synthesis improvement products. J. Salt Chem Ind. 2010.
4. H. Wang, W. Li, W. Li, A study on bromine wastewater treatment by two-uasb technology, Technol. Water Treat. 2013.
5. X. L. Zhou, Z.Y. Ji, J. Wang, Highly efficient recovery of bromine from shale gas wastewater by selective electrochemical oxidation, J. Environ. Chem. Eng. (2022) n. pag.
6. Y. N. Zhang, S. J. Chai, D. Song, Development status and future development trend of bromine, IOP Conf. Ser. Earth Environ. Sci. 300(2019) 032018.

 https://doi.org/10.1051/e3sconf/202337503012

Research on Energy-fed Suspension Control System of New Energy Vehicles Based on Energy Flow Analysis

Jiageng Ding

Case Western Reserve University, Cleveland, OH 44106, USA

Abstract. Suspension is the general name of the device that connects the conductive force between the car body and the wheels. Among them, operational stability and ride comfort are two contradictory requirements in actual driving. In the traditional vehicle suspension system, in order to maintain the ride comfort of the vehicle, the vibration energy caused by uneven road surface is usually dissipated by passive shock absorber in the form of heat. If this part of energy can be recovered in an effective way, the energy demand of the whole vehicle system can be reduced. Usually, the stiffness of semi-active suspension is constant, and the damping coefficient will change automatically with the change of driving conditions. Its advantages are simple structure, no need to consume automobile power, and the damping coefficient will change with the change of automobile driving conditions, which can effectively improve automobile riding comfort and handling stability. This paper expounds the problems encountered by traditional suspension, and analyzes the semi-automatic control strategy of regenerative suspension of new energy vehicles from the perspective of energy flow.

Key words: New energy vehicles; Energy-fed suspension control; Energy flow

1. Introduction

Suspension is the general name of the device that connects the conductive force between the car body and the wheels. As one of the most important car body components, suspension should have the following functions: supporting the whole car body or the quality of the car body; Transfer all the forces and moments between the car body and the wheels, so that the tires are as close to the road as possible, so as to minimize the change of the balance posture of the car body caused by external factors and provide good operational stability; The axle and the car body are elastically connected together, and the impact from the uneven road surface can be effectively attenuated, suppressed and isolated, so as to provide good riding comfort [1]. When the vehicle is running, the vibration caused by the uneven road surface excitation is mainly attenuated by the suspension and then transmitted to the body [2]. Although the suspension of modern automobile has various structural forms, it is generally composed of three parts: elastic element, shock absorber and guiding mechanism. In the traditional vehicle suspension system, in order to maintain the ride comfort of the vehicle, the vibration energy caused by the uneven road surface is usually dissipated by the passive shock absorber in the form of heat [3]. If this part of energy can be recovered in an effective way, the energy demand of the whole vehicle system can be reduced. The vibration energy recovery of suspension is an effective means of automobile energy saving, and it is also one of the research hotspots of suspension technology at present. The recovered energy can not only be used to drive the vehicle, reduce fuel consumption and increase driving mileage, but also be directly used to control the suspension and reduce its energy consumption [4].

As an important part of automobile, the performance of suspension system will have a vital impact on the ride comfort and handling stability of automobile [5]. Therefore, the requirements for automobile suspension are getting higher and higher. The traditional shock absorber mainly produces damping force through the friction of internal hydraulic oil to do work to attenuate vibration [6]. When the vehicle frame and the axle move relative to each other, the hydraulic oil in the shock absorber repeatedly passes through the damping hole with the movement of the piston, and at this time, the friction between the hole wall and the oil and the internal friction of the liquid molecules form a damping force, so that the vibration energy is converted into heat energy, which is absorbed by the oil and the shock absorber shell, and then dispersed into the atmosphere [7]. Usually, the stiffness of semi-active suspension is constant, and the damping coefficient will change automatically with the change of automobile driving conditions [8]. Its advantages are simple structure, no need to consume automobile power, and the damping coefficient will change with the change of automobile driving conditions, which can effectively improve automobile riding comfort and handling stability.

* Corresponding author: davyding65@gmail.com

Energy-feeding active suspension can realize active control of suspension, recover the vibration energy dissipated by shock absorber in the form of heat energy and store it in vehicle-mounted battery, which solves the contradiction between the economy and overall performance of suspension system to some extent, making it possible for it to be widely used in the future [9]. This paper expounds the problems encountered by traditional suspension, and analyzes the semi-active control strategy of regenerative suspension of new energy vehicles from the perspective of energy flow.

2. Problems encountered by traditional suspension

At present, most vehicles mainly use passive suspension due to technical and price reasons, and research and experiments have long shown that active suspension has superior performance, which can not only control the posture and height of the car body to adapt to various road conditions, but also improve the driving stability and ride comfort of vehicles by changing the damping characteristics of the suspension. Faced with the increasing shortage of petroleum energy and soaring prices, energy waste is widespread in the automobile industry. It has become an inevitable trend to develop environmental protection and energy-saving technologies in the automobile industry. Then the effective way to reduce the energy consumption of automobiles is to recycle these wasted energy, so as to achieve the purpose of saving energy [10]. If the motor can only run in a single quadrant, then the energy can only flow in one direction, that is, forward electric or reverse electric, and it can't be changed from power generation operation to electric operation, and vice versa. If you want to make the energy flow in two directions, you can only use a four-quadrant motor, that is, you can run in the starting state or the motor state. In the traditional passive vehicle suspension system, vibration energy is damped and converted into heat energy in the form of friction, and this heat energy is finally dissipated into the atmosphere and wasted. The amount of energy dissipated by the damping element is not only related to its friction coefficient, but also closely related to the vehicle speed, road bump and body weight. Then in order to recycle this part of energy, the necessary suspension system uses an energy recovery device to replace the damping element on the traditional passive suspension, and connects it in parallel with the elastic element to finally form the vehicle suspension.

3. Control system structure and energy flow analysis

The motor actuator is in the form of ball screw combined with motor. This kind of motor actuator has the advantages of low cost, high energy feeding efficiency, good reliability, high adaptability and simple structure, which meets the requirements of suspension system. In terms of energy storage, it adopts an energy storage device that combines a large-capacity battery with an energy storage electrolytic capacitor. Different from the ordinary active suspension, the electromagnetic energy-feedback active suspension uses the energy-feedback damper as the active actuator, and controls the actuator through the executive circuit, so as to realize the function of active control and energy recovery of the energy-feedback suspension [11]. The electric energy recovered by the motor can be quickly stored in the electrolytic capacitor and then released into the storage battery, where the electrolytic capacitor plays a buffering role and avoids charging the storage battery with a large instantaneous current. When the vehicle driving environment is poor, it is hoped that the controller will restrain the vibration caused by uneven road surface to the greatest extent, so as to achieve good vehicle driving smoothness. Therefore, at this time, the actuator control current covers all its realizable areas, and this mode is called full active energy consumption mode. The energy flow process of regenerative suspension system is shown in Figure 1.

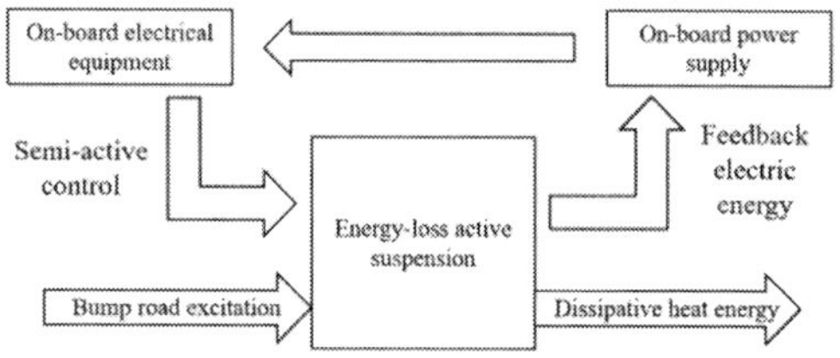

Figure 1 Energy flow process of regenerative suspension system

When the motor runs electrically, the electric energy needed is released from the electrolytic capacitor first, because the instantaneous discharge capacity of the battery is not enough to drive the motor to start with a large torque, which can not only prolong the service life of the battery, but also improve the driving capacity of the motor. By analyzing the simplified control circuit, it is known that different regions of the realizable range of control current correspond to different motor running states and energy flow states respectively. The motor used for the motor actuator of the regenerative suspension needs to meet the normal operation requirements of the whole suspension system, and the motor needs to move in the limited space between the car body and the tire, so the volume should be small. The rotating direction of the motor needs to be changed frequently and repeatedly, which requires it to have a strong ability to change direction. Because the weight of the suspension is changing, the motor must have a strong overload ability. The vibration of the car body causes the rotation of the motor to start and brake frequently, so the motor should have a small moment of inertia. When the on-board battery is too low and the requirement for road vibration suppression is not very high, the control current can be limited to be stable, so that the motor can work as a generator and charge the battery. This mode is called semi-active energy feeding mode.

4. Semi-active control of regenerative suspension

When the vehicle is running, the vibration of the vehicle is mainly caused by the irregular shape of the road surface. The greater the roughness of the road surface, the more intense the vibration of the vehicle and the worse the posture of the vehicle body. When analyzing the dynamic performance of active suspension, we must first determine what kind of road excitation to use. According to the different types of pavement, it can be divided into discrete pavement input and random pavement input. For the suspension system, the disturbance input can be expressed as pulse and step signals from the road surface, which expresses a fixed pothole [12] on the random road surface. According to common sense, this kind of disturbance signal we analyze has the characteristics of transient and sudden, so we can directly regard this kind of disturbance signal as pulse or step signal. When the car is running, the tire bumps and deforms due to the uneven road surface, which leads to the vibration of the whole car. At this time, the shock absorber produces vertical displacement, and the ball rolls on the threaded track of the lead screw, which converts the vertical displacement caused by road surface excitation into rotational displacement. At this time, the vehicle-mounted ECU calculates the ideal operating force through the controller, and the motor circuit supplies power to the motor according to the weak signal sent by the processor, so that the motor is in an active rotation state, so that the movement direction of the lead screw is opposite to the vibration direction, and the ball is in a follow-up state during this process.

Active suspension adopts active control mechanism, and the control law calculated by on-board ECU is transmitted to the actuator, so that it can provide driving force to resist road vibration. Energy-regenerative suspension is one of the active suspensions, which has the characteristics of energy recovery besides the related performance of active suspension. The main components of the energy-feeding active suspension structure are spring, suspension dynamic deflection sensor, drive energy storage circuit, main controller, vehicle battery and motor actuator, as shown in Figure 2.

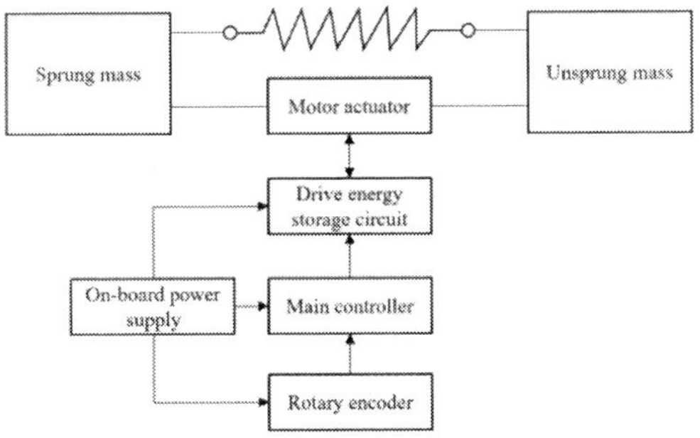

Figure 2 Energy-feeding active suspension structure

When the vehicle runs on uneven road, the energy-feeding suspension converts the linear motion between the sprung mass and the unsprung mass into the rotary motion of the energy-feeding motor, and the energy conversion device composed of the ball screw mechanism and the rotary motor can recover the vibration energy and deliver it to the battery or capacitor for use in other parts of the automobile. The relative motion of the spring-loaded mass, at this time, the energy feeding mechanism will convert the kinetic energy of this relative motion into electric energy recovery; However, excessive vibration of the car body will make the rider feel uncomfortable, and the ride comfort and handling stability will also be affected. At this time, the controller calculates the required damping force according to the car body state and the pre-designed control algorithm, which is obtained by the current value of the motor control circuit. When the vehicle is running, the road roughness, as the vibration input of the vehicle, will cause the relative motion between the sprung mass and the unsprung mass. At this time, the energy feeding mechanism will convert the kinetic energy of this relative motion into electric energy recovery; However, excessive vibration of the car body will make the occupants feel uncomfortable, and the ride comfort and handling stability will also be affected. At this time, the controller calculates the required damping force according to the state of the car body and the pre-designed control algorithm, and the damping force is obtained by the current value of the motor control circuit.

5. Conclusions

With the growth of automobile industry, people have higher and higher requirements for the ride comfort and handling stability of automobiles. Suspension system is an important part of automobile, and its performance will have a vital impact on the ride comfort and handling stability of automobile. As an important part of automobile suspension, shock absorber has great influence on automobile performance. Due to the energy and environmental crisis, electric vehicles will become the future development trend of the future automobile industry. However, electric vehicles have the problem of short driving range, so how to improve the driving range of electric vehicles is also very important. The actual output control force of the actuator should always be consistent with the ideal active control force calculated by the controller. The direct current controller controls the energy feedback loop to adjust the motor winding current, so that the electromagnetic damping force of the linear motor can track the ideal electromagnetic damping force required by the system, and the motor damping force can be adjusted in real time. The vibration suppression ability and energy recovery ability of electromagnetic suspension system are mutually restricted. In full active mode, the controller can get better ride comfort on the premise of consuming battery energy. In semi-active mode, energy can be recovered to the battery, but the improvement level of ride comfort is reduced.

References

1. Ricciardi V, Ivanov V, Dhaens M, et al. Ride Blending Control for Electric Vehicles[J]. World Electric Vehicle Journal, 2019, 10(2):36.

https://doi.org/10.1051/e3sconf/202337503012

2. Jiang B, Fitzgerald B. growth of a Rotary Screw Compressor for Electric Truck Air Brake and Suspension System[J]. SAE International Journal of Commercial Vehicles, 2021(3):14.
3. Hu X, Sun J, Chen Y, et al. Considering Well-to-Wheels Analysis in Control Design: Regenerative Suspension Helps to Reduce Greenhouse Gas Emissions from Battery Electric Vehicles[J]. Energies, 2019, 12(13):2594.
4. Urrea C, Garrido F, Kern J. Design and Implementation of Intelligent Agent Training Systems for Virtual Vehicles[J]. Sensors, 2021, 21(2):1-25.
5. Lu Y, Khajepour A, Liu Y, et al. Adaptive cabin suspension systems of commercial vehicles: a review of the state-of-art and future trends[J]. International Journal of Heavy Vehicle Systems (IJHVS), 2022(1):29.
6. Lv X, Ji Y, Zhao H, et al. Research Review of a Vehicle Energy-Regenerative Suspension System[J]. Energies, 2020, 13(2):441.
7. Kazemian A H, Fooladi M, Darijani H. Non-linear control of vehicle's rollover using sliding mode controller for new 8 degrees of freedom suspension model[J]. International Journal of Heavy Vehicle Systems, 2019, 26(5):707.
8. Dubrovskiy A, Aliukov S, Dubrovskiy S, et al. Adaptive suspension of vehicles with new principle of action: Theoretical bases and experimental investigations[J]. Engineering Letters, 2018, 26(4):526-534.
9. Liu Z, Hao H, Cheng X, et al. Critical issues of energy efficient and new energy vehicles development in China[J]. Energy Policy, 2018, 115(4):92-97.
10. Tan R, Lin B. Public perception of new energy vehicles: Evidence from willingness to pay for new energy bus fares in China[J]. Energy Policy, 2019, 130(7):347-354.
11. Hovgard M, Jonsson O, Murgovski N, et al. Cooperative energy management of electrified vehicles on hilly roads[J]. Control Engineering Practice, 2018, 73(4):66-78.
12. Liu Z, Liu X, Hao H, et al. Research on the Critical Issues for Power Battery Reusing of New Energy Vehicles in China[J]. Energies, 2020, 13(8):1932.

Evaluation model and promotion based on the ecological characteristics of Saihanba

Bingqi Wang

College of Science, Beijing Forestry University, Beijing, China.

Abstract. Saihanba Mechanical Forest Farm is an important sand source control project area in China.In order to promote the excellent case of Saihanba to other regions of the country and even the Asia-Pacific region, we have established a Saihanba ecological environmental impact model and ecological engineering construction assessment model, to provide suggestions for the establishment of ecological protection areas in China and even countries in the Asia-Pacific region. This paper determined the index of Saihanba ecological influence evaluation system using frequency analysis method and theoretical research method, determined the TOPSIS weight based on entropy weight method, to establish a complete ecological and environmental influence evaluation system. Subsequently, the paper introduced the data of relevant indicators around 1962 into the evaluation system, which got good scores, indicating that Saihanba has a great positive impact on the ecological environment. On the basis of problem 1, this paper had redefined the index and weight, established the index system based on the sandstorm background in Beijing, and brought the relevant data of Beijing in recent years into the model, and the establishment of Saihanba had played a relatively significant optimization effect on all the indicators of the sandstorm in Beijing. In view of problem 3, this article was first divided into 9 different ecological system service functional zones according to the National ecological function zoning (revised version) proposed by the Ministry of Environmental Protection and the Chinese Academy of Sciences. Meanwhile, based on the TOPSIS model, proposed a modified model based on DPSIR. The model fully omitted the impact of ecological, economic, policy and other factors on the construction of ecological protection areas .Meanwhile, multi-objective planning method based on genetic algorithm was used to solve the above problems to determine the best number and scale of ecological reserves in the region.

Key words: Entropy power method TOPSIS principal component analysis genetic algorithm multi-objective planning BP neural network DPSIR.

1. Restatement of the problem

China has always maintained that clear waters and green mountains are the ecological development force of gold and silver mountains.In China's state-owned forest farm, Saihanba Forest Farm, as an environmental model, has an extremely high ecological civilization value.According to the relevant situation of Saihanba, a mathematical model analysis was established to study the following problems:

1. According to the natural environmental conditions of Saihanba and relevant data and information, establish the Saihanba ecological environmental impact assessment model, analyze the impact of Saihanba on the local ecological environment, and quantify and compare the results before and after the recovery of Saihanba.
2. According to the model established in Question 1, study the wind and sand fixation in Beijing.According to the existing data and established indicators, the impact of Saihanba on Beijing against sandstorm capacity is evaluated, and the role of Saihanba in Beijing against sandstorms is evaluated.
3. Promote the ecological model of Saihanba, and multiple social and economic factors should be considered.By establishing which geographical locations are in China, ecological areas, ecological reserves are needed to be established, and the quantity and scale that need to be established.

2. Model establishment and solution

2.1 Problem 1: The Ecological Environment Impact Assessment model of Saihanba

2.1.1 *Study Area Overview*

Saihanba Forest Farm is located in the northernmost part of Weichang Manchu and Mongolian Autonomous County, Hebei Province. It is located in the semi-arid and

* Corresponding author: 798460040@qq.com

semi-humid zone. It belongs to the transition zone of the Inner Mongolia Plateau and the Greater Hinggan Mountains, and the terrain is relatively complex.Saihanba machinery forest farm in the national Beijing-Tianjin sand source management engineering area, after more than 50 years of continuous cultivation and protection, part of the forest has been included in the national and provincial scope of forest ecological benefits compensation (Saihanba machinery forest farm forest resources quality analysis and evaluation), it has high ecological benefits, in wind resistance, environmental protection, maintaining ecological balance and stability have played an important role.

2.1.2 Construction of an ecological function evaluation system in Saihanba

3.1.2.1 Design and analysis of Saihanba ecological function evaluation system

In order to evaluate the impact of Saihanba on the local environment before and after its construction is a relatively complex process, which involves many kinds of influencing factors. Therefore, we especially combine the function of the ecosystem and mainly consider the following aspects in the process of designing the Saihanba ecological function evaluation system:

Tab 1. Selected indicators

Target layer	Code layer
Forest natural ecological space	vegetation coverage
	Total proportion of wetland area
	Forest land area (forest accumulation) accounted for the total ratio
water conservation	Forest area
	annual precipitation
	The annual steam dispersion of forest
	Annual water conservation
Conservation of soil	Woodland soil organic matter content
	Forest maintain organic matter content every year
	Forest fixed soil quantity in different years
carbon fixation and oxygen release	Net productivity
	Annual carbon fixation of soil per unit area
	Annual oxygen release per unit area
wind prevention and sand fixation	wind speed
	Solid sand quantity
bio diversity protection	Shannon-Wiener Diversity index

1. Forest natural ecological space.Forest natural ecological space is the most basic condition for forestry ecosystem consideration, mainly including plant coverage rate, wetland area, nature reserve area and woodland area (accumulation area).

2. water conservation.Water conservation is particularly important for the treatment of Saihanba wasteland, which refers to the ecosystem interception, penetration and accumulation of precipitation through its unique structure and water interaction, and realizes the regulation of water flow and water cycle through evaporation.

3. Soil conservation.The benefits of conservation soil in the recovery of Saihanba are obvious.

4. carbon fixation and oxygen release.Absorbing carbon dioxide and releasing oxygen is the most basic biological function of plants and an important ecological function of forest plants.

5. wind prevention and sand fixation.Wind protection and sand fixation is a kind of ecological construction activity carried out in arid and semi-arid in areas to maintain soil and water and prevent dust-arid storms.

6. Biodiversity maintenance.Biodiversity is the sum of the ecological complexes formed by organisms and the environment and the various ecological processes associated with this, including three levels: ecosystem, species, and genes.

3.1.2.2 Selection method of indicators

The establishment of the function index of the forest ecosystem system is the key to affect the ecological environment evaluation system. The relevant literature and data foundation of Saihanba in the forest evaluation system are relatively weak, so the determination of the evaluation index is particularly important.In this paper, theoretical analysis and frequency statistics are used to determine the ecological function evaluation indicators.

2.1.3 Saihanba ecological function evaluation indicators are determined

For the selection of evaluation indicators, we fully consider the impact of the system on the system, taking into account the data that we can find, and the indicators that we choose are as follows:

2.1.4 Determine the weight of Saihanba ecological function evaluation index

3.1.4.1 Computational steps of the TOPSIS model based on the entropy weights method

The TOPSIS method is a common comprehensive evaluation method that can make full use of the raw data information, and its results can accurately reflect the gap between the evaluation schemes.The entropy method is an objective empowerment method, which determines the weight of the index by calculating the index according to the influence of the relative degree of change on the system overall. The index with a large degree of relative change has a large weight.Organic combination of the two contributes to the more scientific determination of the weight.

2.1.5 Assessment level and classification of Saihanba ecological function level

According to the range of ecosystem functional level numerical scores derived from the TOPSIS model, the functional level evaluation level can be divided into 6 levels (A is the best, F is the worst), and the classification is divided in the table:

Tab 2. Evaluation level and classification of ecological function level

grade	A	B	C	D	E	F
score	>80	60-80	45-60	30-45	20-30	<20
classify	Supreme function	Good function	Intermediate function	Primary function	Low-level function	Lack of function

2.1.6 Environmental impact and its evaluation before and after Saihanba recovery

3.1.6.1 Environmental evaluation before and after Saihanba recovery

Tab 3. Level of ecological function after and before the recovery of Saihanba

metric	2021year		1962year	
	Comprehensive weight	fraction	Comprehensive weight	fraction
Forest natural ecological space	10.76%	9.46	16.51%	1.37
water conservation	11.19%	9.84	16.57%	1.38
Conservation of soil	13.74%	16.26	15.7%	1.31
carbon fixation and oxygen release	2.11%	7.4360	16.8%	1.39
wind prevention and sand fixation	36.35%	31.99	12.4%	1.03
bio diversity protection	9.79%	8.61	15.6%	1.30
total points		88		8

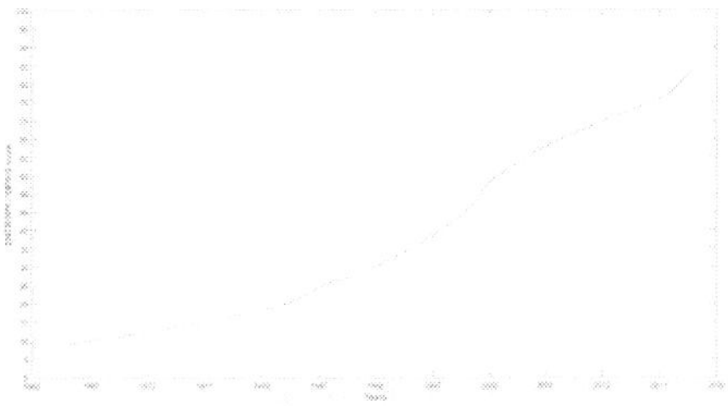

Figure 1 Distribution map of ecological function scores from 1962 to 2021 based on topsis calculation

2.2 Question 2: Instance study of windproof and sand fixation in Beijing by Saihanba Forest Farm

2.2.1 Construction of Saihanba to Beijing

3.2.1.1 The index of Saihanba to Beijing

Problem 2 is an example analysis of problem 1, and in combining the indicators in the problem 1 windproof and sand fixation function, we expand the index of problem 1 through the theoretical analysis method and the frequency statistical method. The specific analysis is as follows:

First of all, according to the sandstorm cause, we choose temperature, wind speed, annual sandstorm frequency and vegetation coverage, in order to improve the representative of sandstorm, meteorological and vegetation elements, we according to the seasonal distribution of sandstorm (Beijing and Tianjin sandstorm source management engineering area sandstorm spatial and temporal change and its vegetation recovery relationship), the occurrence of Beijing sandstorm has obvious seasonal, namely spring sandstorm frequency.In spring, the surface is bare and the vegetation is sparse, which provides sand source material for the sandstorm. At the same time, the air is dry and windy in spring, which provides conditions for the occurrence of dust storms, so we have narrowed the scope of meteorological indicators to spring.

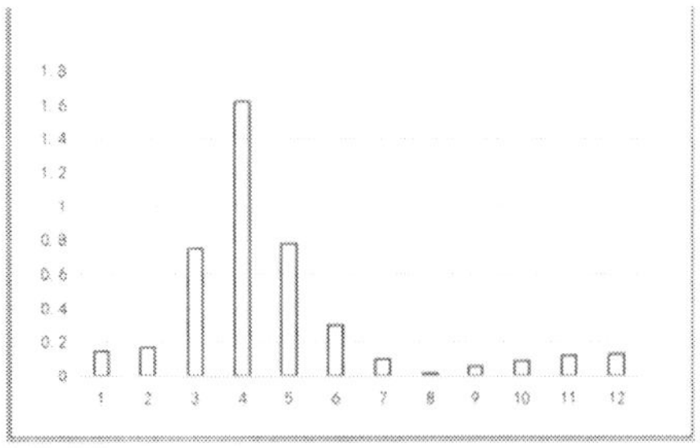

Figure 2 Distribution of Beijing sandstorm season in 1962-2021

3.2.1.2 Dust storm resistance to Beijing based on correlation analysis and nonlinear fitting

Correlation analysis is the analysis of two or more variable elements with correlation to measure the closeness of the two variable factors.The influence icon was quantified by establishing the cause of sandstorm in

https://doi.org/10.1051/e3sconf/202337503013

Beijing and the vegetation coverage data of Saihanba forest farm.
Nonlinear fitting mathematically fits the vegetation coverage rate of Saihanba with the Beijing meteorological indicators, determines the correlation by fitting the slope of the fitted curve, and then quantitatively analyzes the impact effect

2.2.2 Longitudinal impact assessment of Saihanba on sandstorm resistance in Beijing

3.2.2.1 Data
The meteorological data came from the Climate Data Center of the National Meteorological Administration, including the monthly average temperature, precipitation, average wind speed and frequency of dust storm of multiple weather stations.Vegetation coverage data were obtained from the Saihanba Mechanical Forest Farm.Data sources are accurate and reliable.

3.2.2.2 Relevant judgment of annual dust storm frequency and vegetation coverage rate in Saihanba
According to the established fitting equation, the annual dust frequency is negatively associated with the Saihanba vegetation coverage rate, that is, the increased vegetation coverage rate in Saihanba contributes to the decrease of the annual dust frequency.

3.2.2.3 Relevant judgment of spring average temperature and vegetation coverage rate in Saihanba
According to the established fitting equation, the annual temperature is positively associated with the Saihanba vegetation coverage rate, that is, the increased vegetation coverage rate in Saihanba contributes to the annual temperature increase.

3.2.2.4 Relevant judgment of spring monthly wind speed and Saihanba vegetation coverage rate
According to the established fitting equation, the monthly wind speed in spring is negatively associated with Saihanba vegetation coverage, that is, the increased vegetation coverage in Saihanba contributes to the decrease of wind speed.
According to the comprehensive evaluation of the above data, Saihanba is of great help to the wind and sand resistance capacity in Beijing.

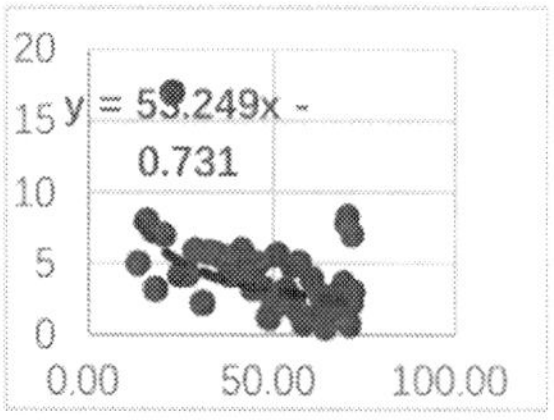

Figure 3. Plot of data fitting between sandstorm frequency in 2-2 years and vegetation coverage in Saihanba

y = 53.249x-0.731

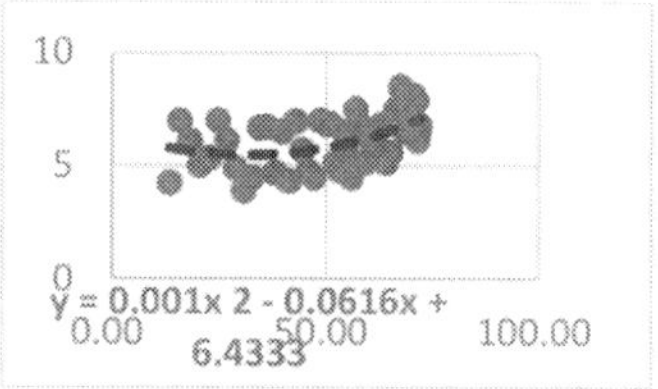

Figure 4 Plot of data fitting between mean air temperature in spring and vegetation coverage in Saihanba

y = 0.001x2 - 0.0616x + 6.4333

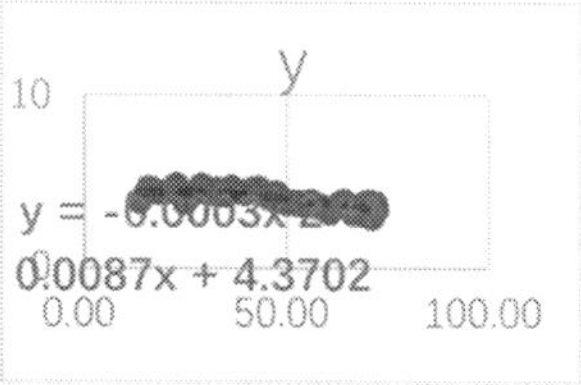

Figure 5 Plot of data fit for spring monthly wind speed and Saihanba vegetation coverage

y = -0.0003x2 + 0.0087x + 4.3702

2.2.3 Horizontal impact assessment of Saihanba on the sandstorm resistance capacity in Beijing

Subsequently, we plan to evaluate the evaluation of the dam on the sandstorm resistance ability of Beijing in the form of a pie chart. The predicted data results are as follows:

Figure 6 Preast of sand resistance of three lines of defense

Although Saihanba is of great help to Beijing against wind and sand, it is still relatively weaker than other lines of defense.

2.3 Problem 3: Saihanba mode expansion model and carbon neutralization evaluation analysis

2.3.1 National ecological function division

The content studied in this section, to explore which areas of the country need to establish nature reserves, in order to simplify the model and let the model has wider applicability, we according to the national ecological function division (modified), the national divided into nine kinds of ecological service system, and select each functional area representative 9 regions, analysis and evaluation.

Table 4 The National Ecological Function Division System

Ecological function category (Category 3 categories)	Ecological function type (Category 9)	Example of function zones
Ecological adjustment	Water conservation	Daba Mountain Water conservation functional area
	bio diversity protection	Biodiversity protection functional area of the Lesser Hinggan Mountains
	soil conservation	Soil protection functional area of loess hills and ravines in northern Shaanxi
	wind prevention and sand fixation	Horqin Sandy Land windproof and sand fixation functional area
	Flood regulation and storage	Wanjiang Wetland flood regulation and storage functional area
Product delivery	Supply of agricultural products	Sanjiang Plain agricultural products to provide functional areas
	Forest products are provided	Lesser Xingan Mountains mountain forest products provide functional areas
Living security	Metropolis group	The Yangtze River Delta metropolitan group functional area
	Key town groups	Wuhan urban group functional area

2.3.2 *TOPSIS ecological environment evaluation system with carbon neutralization index*

Table 5 A TOPSIS model introducing carbon neutralization metrics

Target layer	Code layer
Forest natural ecological space	vegetation coverage
	wetland area
water conservation	precipitation
	surface runoff
	Lake area
Conservation of soil	Soil contains nitrogen
	Soil contains phosphorus
carbon fixation and oxygen release	Vegetation solid carbon
	Soil solid carbon
wind prevention and sand fixation	wind speed
	Flow sand index
bio diversity protection	Total biomass
Carbon neutralization	carbon emission
	Carbon absorption

2.3.3 *Gray forecast for correlated data forecast for the next 20 years*

3.3.3.1 Implementation steps for gray prediction

1. Data inspection

Modeling using GM (1,1) requires testing of the data, first calculating the ratio of columns

$$\lambda(k) = \frac{x^0(k-1)}{x^0(k)}, 其中k = 2, 3, ..., n$$

$$X = (e^{\frac{-2}{n+1}}, e^{\frac{2}{n+1}})$$

If all level ratios fall within the allowable coverage interval, the columns x (0) can build a GM (1,1) model for gray prediction.Otherwise, you need to transform the data appropriately, such as translation.

2. Build a gray model

$$u = \begin{bmatrix} a \\ b \end{bmatrix}, Y = \begin{bmatrix} x^0(2) \\ x^0(3) \\ \dots \\ x^0(n) \end{bmatrix}, B = \begin{bmatrix} -z^1(2) & 1 \\ -z^1(3) & 1 \\ \dots\dots & \\ -z^1(n) & 1 \end{bmatrix}$$

Define the gray derivative, define the gray differential equation, define the bleaching background, gray action quantity, development coefficient; listed according to the matrix, GM (1,1) can be expressed as, Y=Bu, the next is to seek the value of a aa and b bb, can use linear regression or formal equation to find the value according to the least squares principle.

3, calculate

$$\frac{dx^1(t)}{dt} + ax^1(t) = b$$

$$x^1(t) = (x^0(1) - \frac{b}{a})e^{-a(t-1)} + \frac{b}{a}$$

The corresponding whitening model gives the solution of x 1 (t) to make t + 1 = t have

$$x^1(t+1) = (x^0(1) - \frac{b}{a})e^{-a} + \frac{b}{a},$$

That is the value that we predicted.

4, checkout

Table 6 Data predictions for the next two 0 years

Ecological function category (3)	Ecological function type (9)	Example of function zones	2020	2030	2040	2050
Ecological adjustment	Water conservation	Daba Mountain Water conservation functional area	11.52909446	10.54505972	9.645014611	8.821790421
	biodiversity protection	Biodiversity protection functional area of the Lesser Hinggan Mountains	7.906167409	5.92852619	4.44557027	3.333559538
	soil conservation	Soil protection functional area of loess hills and ravines in northern Shaanxi	52.55053724	50.63856949	48.79616565	47.02079474
	wind prevention and sand fixation	Horqin Sandy Land windproof and sand fixation functional area	19.12840272	18.6780674	18.23833422	17.80895357
	Flood regulation and storage	Wanjiang Wetland flood regulation and storage functional area	17.2614061	13.91840389	11.22283814	9.049320383
Product delivery	Supply of agricultural products	Sanjiang Plain agricultural products to provide functional areas	17.1578097	12.97170143	9.806906638	7.414248498
	Forest products are provided	Lesser Xingan Mountains mountain forest products provide functional areas	7.906167409	5.92852619	4.44557027	3.333559538
Living security	Metropolis group	The Yangtze River Delta metropolitan group functional area	187.1247155	162.4923874	141.1025576	122.5283971
	Key town groups	Wuhan urban group functional area	84.2788419	80.18846679	76.29661326	72.59364629

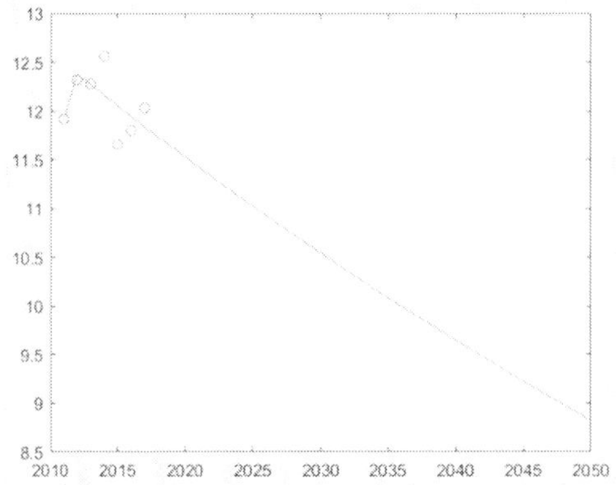

Figure 7 Grey prediction renderings

The data before the construction of the ecological reserve were brought into the model to get the corresponding scores.

From the above data, we obtain that the region should establish nature reserves, and that the establishment of ecological reserves has had a good impact on carbon neutrality.

3.3.3.2 A multi-objective planning model based on the genetic algorithm solves the number and model of ecological reserves

Taking into account the economic, political and ecological elements, we use the multi-objective planning method based on genetic algorithms to achieve the solution.We propose to establish [3], an evaluation system based on the DPSIR model.DPSIR, that is, driving force (D), pressure (P), state (S), impact index (I), response (R), which covers social, economic, environmental, human beings and many other factors, and more fully reflects the interaction relationship between human activities and the natural ecological environment.

The construction of the ecological protection area has its particularity relative to the construction of other projects. We obtain authoritative and specific calculation formulas by consulting the relevant information, [4], and supplement the calculation formulas of several other dimensions:

net present value (NPV)

$$NPV = \sum_{i=1}^{m} A_i \left[\sum_{t=0}^{n} \left(P_t Q_t + \sum E_j + \sum S_k - IP - IC \right)_t (1+i_0)^{-t} \right]$$

Internal rate of return (IRR)

$$NPV(IRR) = \sum_{i=1}^{m} A_i \left[\sum_{t=0}^{n} \left(P_t Q_t + \sum E_j + \sum S_k - IP - IC \right)_t (1+i_0)^{-t} \right] = 0$$

Dynamic investment payback period

$$\sum_{i=1}^{m} A_i \left[\sum_{t=0}^{T_p} \frac{P_t Q_t + \sum E_j + \sum S_k}{(1+i_0)^t} \right] = \sum_{i=1}^{m} A_i \left[\sum_{t=0}^{T_p} \frac{IP + IC}{(1+i_0)^t} \right]$$

Profit cost ratio

$$R = \frac{\sum_{i=1}^{m} A_i \left[\sum_{t=0}^{T_p} \frac{P_t Q_t + \sum E_j + \sum S_k}{(1+i_0)^t} \right]}{\sum_{i=1}^{m} A_i \left[\sum_{t=0}^{T_p} \frac{IP + IC}{(1+i_0)^t} \right]}$$

This model comprehensively considers the three ecological, economic and social benefits of ecological protection area construction, reflects the unique attributes of ecological engineering construction, and is of good help to the solution of this problem.Based on this, we solve the multi-objective planning problem described above using a genetic algorithm.

References

1. Code for Functional Assessment of Forest Ecosystem Services, State Forestry Administration of China, 2008.
2. The National Ecological Function Division (Revised Edition), the Ministry of Environmental Protection and the Chinese Academy of Sciences, 2015.
3. Research on ecological and environmental carrying capacity evaluation and obstacle factors of Hebei urban agglomeration based on DPSIR-TOPSIS model *, Kong Jiangwei, Zeng Jian and Gao Mengxi.
4. Forestry Bureau of 18 Station of Daxing'anling Forestry Bureau, Li Bo, Northeast Forestry University, Zhao Zhengyong.
5. Saihanba Forest Farm Research (1962-2017) —— and and ecological environment perspective
6. Trade-offs and synergies of ecosystem services in forest barrier belt of Northeast China.QI Lin1,2, ZHANG Yue2,3, XU Dong4, ZHU Qi2,3, ZHOU Wang-ming2, ZHOU Li2, WANG Qing-wei2, YU Da-pao2* ,2021
7. Ecological Environment Carrying Capacity Evaluation and Obstacle Factors ofHebei Province Urban Group Based on DPSI R -TOPSIS ModelKONG Jiang-wei, ZENG Jian, GAO Meng-xi ,2019

The Relation Between the CO_2 Concentration Levels and the Temperature

Yuyuan Lin, Zidong Zeng, Lang Chen, Zifeng Deng

Guangdong Experimental High School AP Curriculum, Guangzhou 510000, Guangdong, China

Abstract: Carbon dioxide can change the heat balance of the atmosphere. To study the relationship between CO_2 and temperature change, we use the given CO_2 concentrations to implement linear regression model, gray time series forecasting model, back-propagation model, and auto-regressive and moving average model to establish the growth function of CO_2 concentration. Errors are evaluated to choose the most suitable model. then we use the most suitable model in step one to further predict future land-ocean temperature. Finally, we use grey relational analysis to analyze the relationship between CO_2 concentrations and land-ocean temperatures since 1959.

Keywords: Global warming; prediction and fitting.

1. Introduction

Carbon dioxide can change the heat balance of the atmosphere. It can absorb infrared radiation from the Earth, causing an increase in near-surface atmospheric temperature. The warming of the near-surface atmosphere causes enhanced evaporation from the ground, resulting in an increase in atmospheric water vapor. The increase of water vapor in the atmosphere will further enhance the absorption of infrared radiation from the Earth by the near-surface atmosphere. As a result of this interaction, the increase of carbon dioxide in the atmosphere will change the original atmospheric heat balance and cause global warming.

To study the relationship between CO2 and temperature change, we use the given CO2 concentrations to implement linear regression model, gray time series forecasting model, back-propagation model, and auto-regressive and moving average model to establish the growth function of CO2 concentration. Errors are evaluated to choose the most suitable model. then we use the most suitable model in step one to further predict future land-ocean temperature. Finally, we use grey relational analysis to analyze the relationship between CO2 concentrations and land-ocean temperatures since 1959.

2. Assumptions of the Models and Notation

Assumption 1: We assume that there will be no future extreme natural conditions that would cause dramatic changes in carbon dioxide.

Justification 1: Extreme situations which dramatically increase the concentration of CO2, including volcanic eruption, or mountain fire of which the former has tiny probability of occurrence and the later has similar trend of increasing occurrence from 1959-2021 to the more distant future. Therefore, it is reasonable to ignore these extreme factors when establishing a model in this paper.

Assumption 2: We assume that recent climate patterns will continue in the future.

Justification 2: We anticipate that it will take a while for these initiatives to progress from research to mature operation and considerably benefit the ecological environment, even though changes in government regulations and technical advancement may lower future greenhouse gas emissions. There won't be any significant shifts in climatic patterns at least between now and 2100.

Assumption 3: We assume that the concentration of carbon dioxide is the only factor affecting global warming.

Justification 3: Apart from carbon dioxide; methane, nitrous oxide, hydrochlorofluorocarbons (HCFCs), hydrofluorocarbons (HFCs) and ozone are also common greenhouse gases that contribute to the overall warming effect. However, according to observations by the NOAA Global Monitoring Lab, in 2021 carbon dioxide alone was responsible for about two-thirds of the total heating influence of all human-produced greenhouse gases. [3] Therefore, for the sake of simplicity, we only take the effect of CO2 into consideration in terms of temperature change.

Assumption 4: We assume that the rising temperature will not decrease the solubility of CO2 in ocean to the extent which largely affects carbon dioxide level in the atmosphere.

Justification 4: The solubility of CO2 in water has been the subject of much investigation since its discovery in 1803. A model based on Henry's rule could be able to predict how much CO2 will be absorbed by water as a function of temperature and pressure. The Henry's constant (MPa/mol percentage), according to a critical examination of 91 publications in prestigious journals from around the world, grows linearly as temperature rises from 0 to 150 °C. [4] However, since the standard deviation of given temperature data calculated by excel is 0.335466113, a rather small value, it is reasonable to ignore the effect of temperature change on CO2 solubility.

Table 1 Notation

t	discrete time variable for years
$C(t)$	history data of CO2 concentration level
a	for cubic exponential smoothing model
p	prediction for auto regressive integrated
e	white noise

3. Model Theory, Implementation and Result

3.1 CO2 Concentration Changed over Time Elapsed

3.1.1 Data Processing

We use two methods to detect the outliers which may affect the model and prediction. Firstly, we draw a best fitting curve based on scattered data using excel. We choose to use polynomial function for curving fitting, and the R-squared values=0.9181 indicates the data are to the fitted regression line. Since no value deviates from the trend line in particular, we assume that there is no outlier.

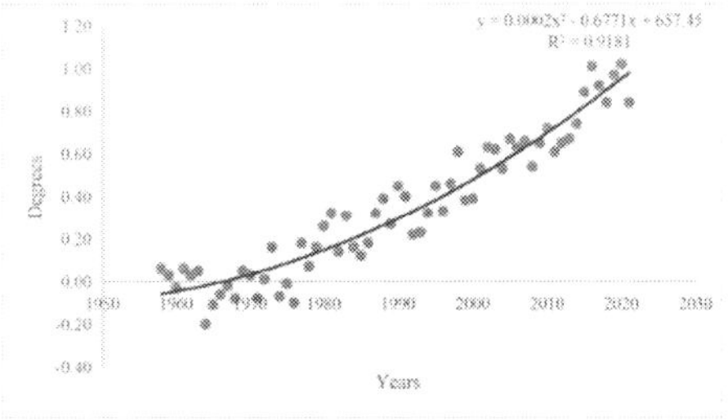

Figure 1 Fitted Regression Line of co2 Concentrations from 1959 to 2021

3.1.2 Polynomial Fitting

Let the polynomial be $y(w,x) = w_0 + w_1 x + w_2 x^2 + \cdots w_M x^M = \sum_{j=0}^{M} x_j x^j$. and we transform the polynomial into a mean-square error function:

$$E(w) = 1/2 \sum_{n=0}^{N} \{y(x_n, w) - t_n\}^2$$

We need to fit a curve such that the value of E takes a minimal value, where E is the mean-square error function. When the number of parameters is too large, because the noise in the dataset is given excessive importance, the fitting results will tend to include the noise in the model, and over-fitting may occur, i.e., the model and parameters fit well for a given dataset, but cannot give suitable estimates of the dependent variable for the new independent variable. Therefore, we can use the regularization method to add a penalty term to the error function to obtain:

$$E(w) = 1/2 \sum_{n=0}^{N} \{y(x_n, w) - t_n\}^2 + \lambda/2||x||^2$$

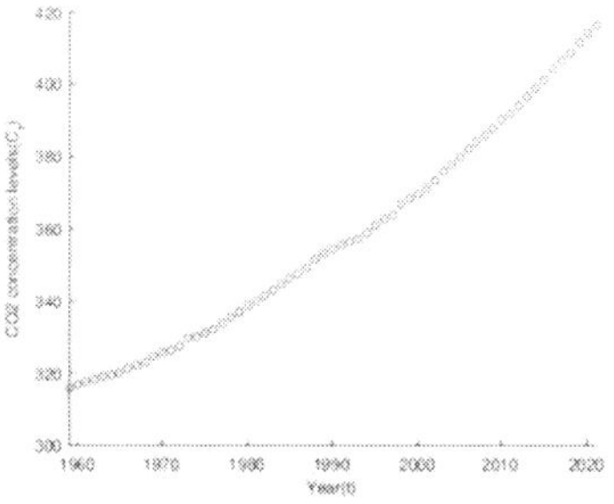

Figure 2 Measured and Predicted Carbon Dioxide Concentrations from 1959 to 2100, Generated by Polynomial Fitting

3.1.3 Grey GM (1,1) Model

First of all, in order to confirm the feasibility of the data using the GM (1,1) model, we should calculate $\lambda(k) = \frac{x_0(k-1)}{x_0(k)}$, k=2, 3..., n_0. if $\lambda(k)$is in the interval ($e^{\frac{-2}{n+1}}$, $e^{\frac{2}{n+2}}$), the GM (1,1) model can be used.

Secondly, we let the original sequences for the 63 years 1959-2021 be respectively $x_1^{(0)}$, $x_2^{(0)}$, $x_3^{(0)}$ ... $x_{63}^{(0)}$. Then we make an accumulation to generate a new sequence $x_1^{(1)}$, $x_2^{(1)}$, $x_3^{(1)}$ ... $x_{63}^{(1)}$, where there are

$$x_k^{(1)} = \sum_{i=1}^{k} x_i^{(0)}, k = 1,2,3 \dots 63$$

Afterwards, the mean series is generated:

$$z_k^{(1)} = \alpha x_k^{(1)} + (1-\alpha) x_{k-1}^{(1)}$$

where $0 \leq \alpha \leq 1$, are the weights. Usually $\alpha = 0.5$. Since the new sequence generated looks like a straight line, we can approximate this new sequence with an expression for a straight line.

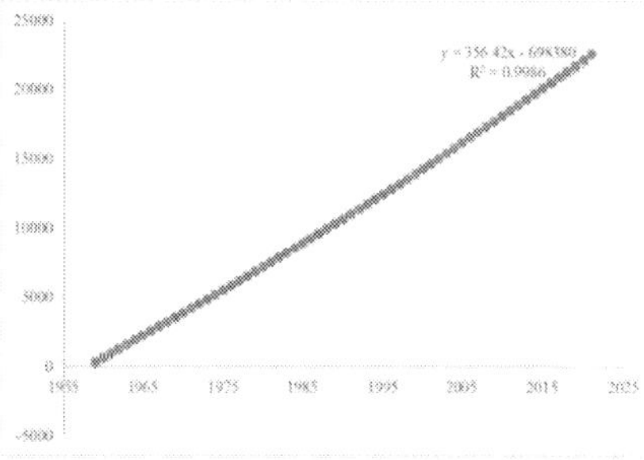

 https://doi.org/10.1051/e3sconf/202337503014

Figure 3 Function of $x_k^{(1)} = \sum_{i=1}^{k} x_i^{(0)}$

Therefore, we construct a first-order ordinary differential equation to solve for the functional expression of the fitted curve. From this figure, the gray differential equation is established as

$$x_k^{(0)} + az_k^{(1)} = b$$

The corresponding GM (1, 1) whitening differential equation is

$$\frac{dx^{(1)}}{d_t} + ax_t^{(1)} = b$$

Shifting the term of the grey differential equation gives.

$$-az_k^{(1)} + b = x_k^{(0)}$$

a,b are parameters to be determined, the above equation can be written in the form of a matrix in the form of $X\beta = Y$

Using least squares method, we can determine the parameter matrix β:

$$\hat{\beta} = (X^T X)^{-1} X^T Y$$

The resulting estimates of the parameters a,b are brought into the whitening equation to obtain the sequence $x_k^{(1)}$:

$$\widehat{x_k^{(1)}} = (x_1^{(0)} - b/a)\, e^{-a(k-1)}(1 - e^{a}), k = 1,2,3 \ldots 63$$

when k=1,2,3...63, we can get the fitted value; when k > 63, we can get the predicted value. We realize this model by using MATLAB. Carbon dioxide concentrations predicted from year 2022 to year 2100 are as follows:

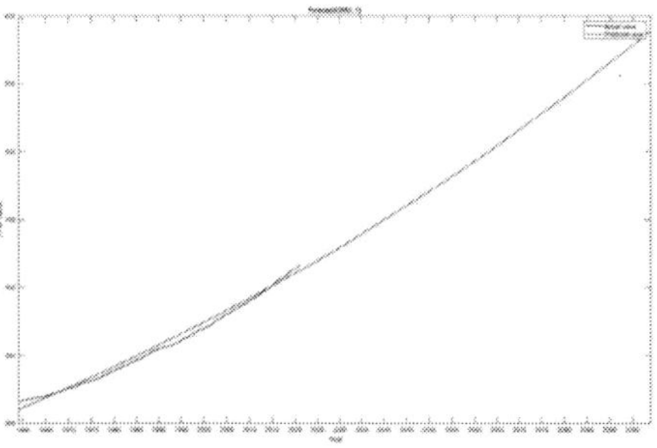

Figure 4 Measured and Predicted Carbon Dioxide Concentrations from 1959 to 2100, Generated Using Grey GM (1,1) model

3.1.4 Autoregressive Integrated Moving Average Model

From the fitted curves of the sample time series of carbon dioxide concentrations and years, it is clear that carbon dioxide concentration cannot continue "inertially" in its current state for some time in the future. Therefore, the series needs to be differenced.

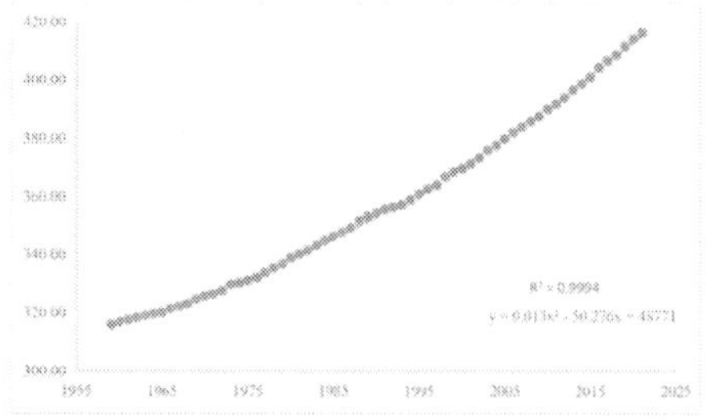

Figure 5 Fitted Curves of the Time Series of Carbon Dioxide Concentrations and Years

The model was initially determined as ARIMA(p,1,q) due to the first-order differencing of the original data. Based on the characteristics of the differenced PPM data, the order of the model can be determined based on the autocorrelation and then autocorrelation plots (ACF & PACF plots). The autocorrelation and partial autocorrelation plots are plotted with the help of the functions acf() and pacf() in MATLAB as follows:

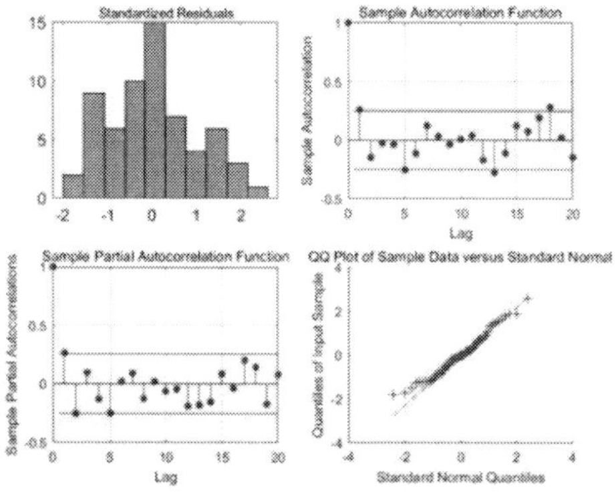

Figure 6 The autocorrelation and partial autocorrelation plots

According to the autocorrelation and partial autocorrelation plots, judging the stage of the model generally requires subjective guesses. From the information in the graphs, we can conclude that p=2 and q=1 in the ARIMA model.

According to the graph of residual test results, standardized residuals is to see if the residuals are close to normal distribution. According to the QQ plot below, the blue points are near the red line, so the residuals are close to the normal distribution. Therefore, the model ARIMA (2,1,1).Based on the ARIMA (2, 1, 1) model, the carbon dioxide concentration values can be predicted for 2022-2100. The prediction can be done by MATLAB function, results are shown in the following figure.

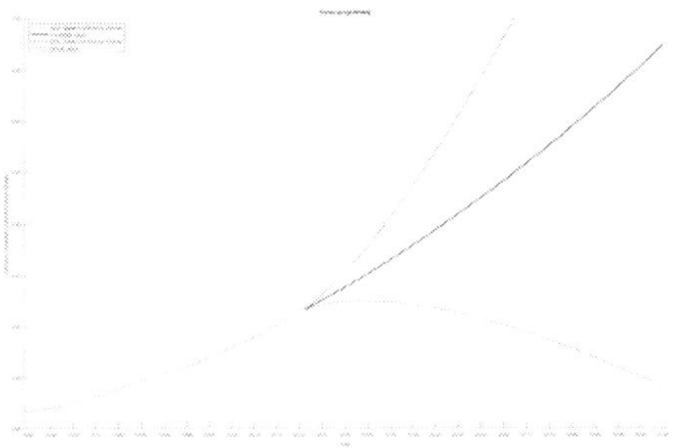

Figure 7 Measured and Predicted Carbon Dioxide Concentrations from 1959 to 2100, Generated by Autoregressive Integrated Moving Average Model

3.2 Comparison

We use three models in total to describe the changes in carbon dioxide concentrations over time and predict future carbon dioxide concentrations from 2022 to 2100. R2 is a measure of the goodness of fit of a model. We compare the R2 value for each model, it shows that the R2 values of polynomial fitting, grey GM (1,1) model, and autoregressive integrated moving average model are 0.8912,0.9867, and 0.9994, respectively, which proves ARIMA the most accurate model.

3.3 Model Future Land-Ocean Temperatures Changes

To determine future land-ocean temperatures changes, we use autoregressive integrated moving average model again due to its relative accuracy. Since we have discussed the principle of the model in the previous part, we will emphasize on the processes for predicting future land-ocean temperatures changes, which we have not covered before.

First of all, since the temperature for 1951-1957 were not found, we first used fitting to graph the data for 1958-1980 and expressed them as a function, The function gives y = 11.688ln(x) -88,639, where x represents the year and y represents the temperature value. the year is very unstable and irregular in the data co-fitted to the temperature. Therefore, a 1st order difference treatment is required for this series. In this case, the time series is weakly stationary. On the basis of the first-order difference of the given data, it was first determined that the model should be ARIMA (p, 1, q), and then the p and q values were determined based on the autocorrelation and then autocorrelation plots (ACF & PACF), respectively. Since the p, q values are subjective, we still set p=2 and q=1 here.

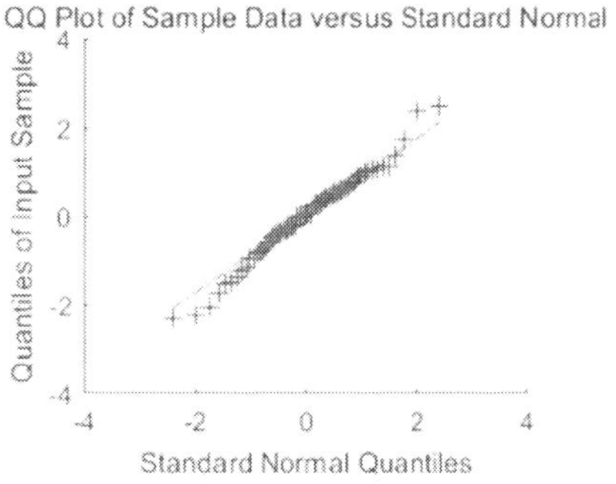

Figure 8 QQ Chart (Sample data-Standard Normal Distribution)

In this figure, almost all the blue points are very close to the red line, which means that the residuals are close to the normal distribution. Therefore it is feasible to use ARIMA (2,1,1) to predict future temperature.

4. Relationship Between CO2 Concentrations and Land-Ocean Temperatures Since 1959

The first method of relationship analysis in this paper is Grey relational analysis. Using excel, we can obtain matrix of the CO2 concentration and the temperature. after nondimensionalization. As calculated by MATLAB, the value of gray correlation is 0.9298, which shows high correlation between carbon dioxide concentration and temperature.

We also use linear regression model to further confirm the correlation between carbon dioxide concentration and temperature. By using excel, we can find that carbon dioxide concentration and temperature shows a linear correlation where the goodness of fit R2=0.9216, approximately the same as the value of gray correlation, which proves that either of the method is feasible.

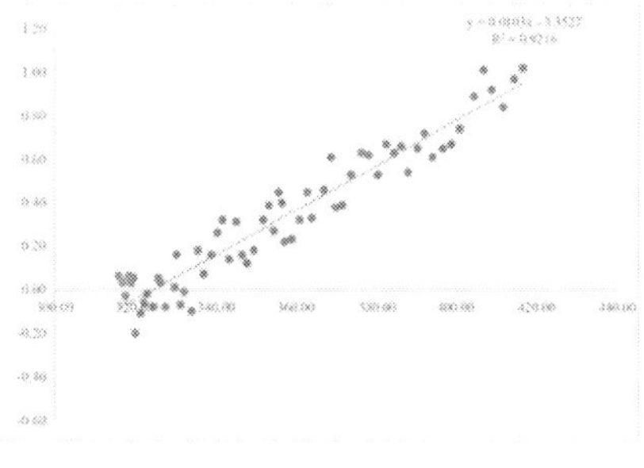

Figure 9 Correlation between Carbon Dioxide Concentrations and Temperatures

5. Conclusion and Possible Recommendations for the Future

We obtain the prediction data of CO2 concentrations and land-ocean temperatures by ARIMA model and compute the Spearman rank correlation coefficient of our two time series as 0.9989. Thus we can conclude that our model is reliable into the future by the probability which is close to 1. Notably, we have not used the history data, which are given in stable conditions and environment. Therefore, if the climate conditions and environment change acutely, our model's ability to predict future CO2 concentration levels and land-ocean temperatures may be worse, i.e., when extreme weather appears our model requires more improvement.

References

1. ZHANG Zhenghai, ZHANG Yayu, LEI Huixia, SHAO Cai, WU Chen, QIAN Jiaqi. Effect of light intensity-temperature-humidity and carbon dioxide on photosynthetic rate of ginseng[J]. Specialty Research,2020,42(02):41-46.
2. Xiao Bo. A coupled calculation method of temperature-pressure field for supercritical carbon dioxide dry fracturing[J]. China Science and Technology Paper,2020,15(01):119-124.
3. Li Yaohua. Effects of elevated temperature and carbon dioxide concentration on the biological characteristics of Aphis minor wasps[D]. Northwest Agriculture and Forestry University, 2018.
4. Zhang YJ. Experimental study on the effects of temperature and carbon dioxide on human comfort in a small space[D]. Chongqing University,2018.
5. Liang W B. Analysis of high exhaust temperature of CP-11 type carbon dioxide high-pressure compressor[J]. Light Industry Science and Technology,2017,33(04):50-52+60.
6. Feng Mengmeng,Tan Yingxin, Zhang Xiqing, Yang Yajun. Experimental study on the effect of initial temperature on the effect of carbon dioxide explosion suppression[J]. Fire Science and Technology,2015,34(11):1423-1425.
7. Wu Xiaohong, Xiao Min,Yang Wenyuan, Ren Bin, Ma Jingang. Experimental design of using temperature sensor to investigate the relationship between carbon dioxide and greenhouse effect[J]. Reference for teaching secondary school chemistry, 2015, (11):57-58.
8. Coco Zhang, Qi Y, Fu X, Miao B, Bai Xuejiao. Carbon dioxide temperature measurement based on TDLAS primary harmonics[J]. Optoelectronics, 2014, 34(04):300-303+307+312.
9. Translated with www.DeepL.com/Translator (free version) Climate Change, N. G. (2022, September 1). Carbon Dioxide Concentration | NASA Global Climate Change. Climate Change: Vital Signs of the Planet. Retrieved November 15, 2022, from https://climate.nasa.gov/vital-signs/carbon-dioxide
10. United Nations. (n.d.). The ocean – the world' s greatest ally against climate change. https://www.un.org/en/climatechange/science/climate-issues/ocea

A light pollution risk model based on improved assessment and prediction methods

Ziyan He, Ruikun Guo and Haiwei Yang

School of Jinan University, Guangzhou 519070, China.

Abstract: As light pollution becomes increasingly severe, its impact is becoming more widespread for instance affecting human health, social stability, and the ecological environment to varying degrees. In order to measure the risk level of light pollution and develop related measures, we propose the Light Pollution Risk Index (LPRI) and create various assessment, intervention and prediction models around it. It include three models: The LPRI Scoring System, The Light Pollution Risk Classification Model and the HSE Intervention Strategy & Potential Impact Prediction Model. Firstly, we use the improved EMW-AMP to determine the weights and elicicit the concepts for next models. We select 6 representative areas in different kinds of locations, combine the integrated weights with the Topsis method to score and rank. Followed by invoking K-Means cluster analysis, we reselect 108 areas and the consequence classifies the light pollution risk level into three levels: level A significant risk, indicator range from 0 to 4.23, level B average risk, indicator range from 4.23 to 7.64, and level C low risk, indicator range from 7.64 to 10. Then, followed by prediction of LPRI with a Grey linear regression combination prediction model. The predicted results can accurately and clearly reflect that the application of HEIS in Wuhan and HSIS in Los Angeles, which is the most effective. At the same time, we find that intervening in one or two of H, S, and E must have a non-positive effect on the risk indicator of the other side. Finally, we briefly discuss how the intervention strategies in Wuhan and Los Angeles affect the individual indicators and thus the level of risk.

Keywords: Light pollution; LPRI; improved EMW-AHP; Topsis; Grey linear regression combination prediction model

1. Introduction

Light pollution refers to the phenomenon of excessive light sources generated in human activities. It interferes with natural night environments and even damages biological clocks. According to the research, more than 90% of light pollution harms human life, destroys ecological environment, and causes social impact[1].

Worse still, a study published recently in the academic journal Science Advances reveals that more than 80% of the global population currently resides under the debilitating impact of light pollution, with one-third can no longer able to see the Milky Way. If the problem of light pollution continues to worsen, the entire population of the Earth may be deprived of the awe-inspiring sight of the starry sky one day.

Therefore, we urgently need a pinpoint model to appraise and forecast light pollution conditions in the present and the future. At the same time, reasonable intervention strategies are proposed to alleviate the crisis of light pollution and improve people's awareness of the impact of its impact.

2. Light Pollution Risk Indicator

2.1 Indicator selection

By referring to the papers and data, as well as experts' comments, we finally establish a model to measure the risk level of light pollution, and select 10 indicators from three aspects: human, ecological and social[2]. They are the impacts on physiology, psychology, GDP, energy, traffic, astronomical observation, animals, plants, climate, night sky.

2.2 Improve EWM-AHP

Because of the shortcomings of using entropy or hierarchical analysis alone, based on these two methods, we propose an improved entropy-based Analytic Hierarchy Process (AHP) method. It combines the subjective weights θ_i obtained through the AHP method with the objective weights β_i obtained through the entropy weighting method. In this way, the comprehensive weights ω_S can be obtained. Through several operations such as weighting and normalization, we can get the weights of each indicator as shown in Table 1.

Table 1: Indicators and their final synthetical weights

Upper-level		Sub-level		Normalized composite weights	Improved synthetical weight
Indicator	**Weight**	**Indicator**	**Synthetical weight**		
Human	0.169	Physiology	0.067	0.551	0.093
		Psychology	0.054	0.449	0.076
Ecology	0.443	Animals	0.162	0.323	0.143
		Plants	0.143	0.286	0.127
		Climate	0.082	0.163	0.072
		Night sky	0.113	0.227	0.101
Society	0.388	Traffic	0.102	0.269	0.104
		Energy	0.088	0.232	0.090
		GDP	0.078	0.205	0.080
		Astronomical observations	0.111	0.293	0.114

3. LPRI Scoring System Based on Topsis

In this section, we intend to combine the combined weights obtained above and evaluate three aspects of human risk index, ecological risk index, and social risk index based on the Topsis method to facilitate subsequent cluster analysis, and finally combine the weights of the criterion layer to obtain the light pollution risk index LPRI.
First of all, we take 24 regions and collected their indicators, positive and orthogonalize each indicator to eliminate the influence of dimension. Then, calculate the distance between the evaluation objects and the best value and the worst value. Ultimately, we can get their results: a score for each area in three aspects.
Combining the individual weights of HSE obtained by hierarchical analysis, the light pollution risky index calculation formula can be constructed as follows:

$$LPRI=\omega_{A1}HRI+\omega_{A2}SRI+\omega_{A3}ERI \tag{1}$$

where HRI is the human risk index, SRI is the social risk index, and ERI is the ecological risk index.

4. Light Pollution Risk Classification

In this chapter , we improved the traditional clustering method and created a 3D clustering model for forecasting the degree of light pollution. We mainly use k-means clustering analysis method to classify the light pollution risk level. Data sets of three dimensions include ***HRI***, ***SRI*** and ***ERI***, and the levels are classified into significant risk ***A***, general risk ***B*** and low risk ***C***.
We select **108 regions** for clustering, and the clustering results are shown below, with the three clusters having the centers of mass (3.6,2.7,3), (6,5.5,6.5) , (8.7,8.7,9.5) .

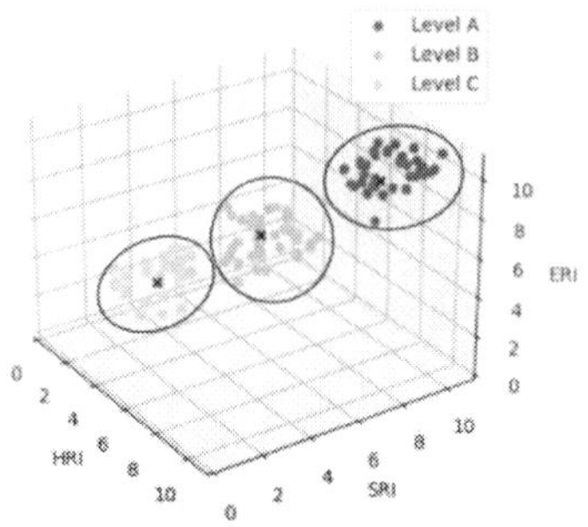

Figure1: The clustering results

We then invoke the light pollution risk index formula to calculate LPRI and quantitatively classify the risk level, and the classification results are as follows:

Table 3: Classification results

Level A	Level B	Level C
0~4.23	4.23~7.64	7.64~10

With the classification range, we can roughly divide the risk levels of the four areas. It is not difficult to see that the risk of protected land areas belongs to grade C, the risk of suburban and rural communities except Lusaka belongs to grade B, and the risk of urban areas belongs to grade A. This further proves the accuracy and feasibility of the evaluation model.

5. HSE Intervention Strategy & Potential Impact Prediction

5.1 HSE Intervention Strategy

According to the requirements of the title, three specific and stable intervention measures are proposed in this section for ***HRI***, ***SRI*** and ***ERI*** in ***LPRI*** to reduce the harm and influence caused by light pollution, namely ***HSIS***, ***HEIS*** and ***SEIS***.
The three intervention strategies are as follows:

1. Human health
(1) Reduce the irrational use of light.

2. Ecological sustainable development
(1) Strictly limit the use of glass curtain walls.
(2) Setting up protection areas for organisms sensitive to light environment.

3. Social stability
(1) Make relevant regulations on electricity consumption.
(2) Strengthen publicity about light pollution.
(3) Allocate light sources wisely.

 https://doi.org/10.1051/e3sconf/202337503015

5.2 The Principle of Gray Linear Regression Combination Forecasting Model

The gray linear regression combined forecasting model is an implicit gray combined model, which can improve the lack of exponential growth trend in linear regression and make up for the lack of linear growth in the gray forecasting model, which can improve the stability and accuracy of model forecasting. The combined forecasting model can be expressed by the formula.[6]

$$\hat{x}^{(1)}(k) = c_1 e^{vk} + c_2 k + c_3 \qquad (2)$$

where v, c_1, c_2, c_3 are coefficients to be determined.

The derivation leads to a proposed estimate of the parameter v

$$\hat{v} = \frac{\sum_{m=1}^{n-3} \sum_{t=1}^{n-2-m} v_m(t)}{\frac{(n-2)(n-3)}{2}} \qquad (3)$$

The estimation of parameters c_1, c_2, and c_3 can be estimated using the least squares method, and the derivation process is as follows. We let

$$\boldsymbol{c} = \begin{bmatrix} c_1 \\ c_2 \\ c_3 \end{bmatrix} = \left(\boldsymbol{A}^T \boldsymbol{A}\right)^{-1} \boldsymbol{A}^T \boldsymbol{X}^{(1)}$$

where:

$$\boldsymbol{A} = \begin{pmatrix} e^{\hat{v}} & 1 & 1 \\ e^{2\hat{v}} & 2 & 1 \\ \vdots & \vdots & \vdots \\ e^{n\hat{v}} & n & 1 \end{pmatrix} \quad \boldsymbol{X} = \begin{pmatrix} X^{(1)}(1) \\ X^{(1)}(2) \\ \vdots \\ X^{(1)}(n) \end{pmatrix}$$

Thus, the predicted values of the generated sequence are:

$$\hat{x}^{(1)}(k) = c_1 e^{vk} + c_2 k + c_3 \qquad (4)$$

From one single cumulative reduction of the equation above, we can get the predicted value of the original series $\widehat{X}^{(0)}(k)$:

$$\widehat{X}^{(0)}(k+1) = \widehat{X}^{(1)}(k+1) - \widehat{X}^{(1)}(k) \qquad (5)$$

Figure 2. Reduce the irrational use of light

5.3 Model testing

The combined model can be evaluated for good or bad prediction results by using a posteriori difference test, calculated as follows:

First calculate the mean of the original series as well as the standard deviation:

$$\bar{x}^{(0)} = \frac{1}{n} \sum_{k=1}^{n} x^{(0)}(k) \qquad (6)$$

$$S_1 = \sqrt{\frac{\sum [x^{(0)}(k) - \bar{x}^{(0)}]^2}{n-1}} \qquad (7)$$

And calculate the mean value of the residual series:

$$\Delta^{(0)} = \frac{1}{n} \sum_{k=1}^{n} \left[x^{(0)}(k) - \hat{x}^{(0)}(k) \right] \qquad (8)$$

Then calculate the standard deviation of the absolute error series:

$$S_2 = \sqrt{\frac{\sum [\Delta^{(0)}(k) - \Delta^{(0)}]^2}{n-1}} \qquad (9)$$

After that, calculate the value of the check ratio:

$$C = \frac{S_2}{S_1} \qquad (10)$$

The small error probability is:

$$P = p\{|\Delta^{(0)}(k) - \Delta^{(0)}| < 0.6745 S_1\} \qquad (11)$$

where the magnitude of C and P can classify the prediction accuracy of the model into the following classes, as shown in the following figure 2.

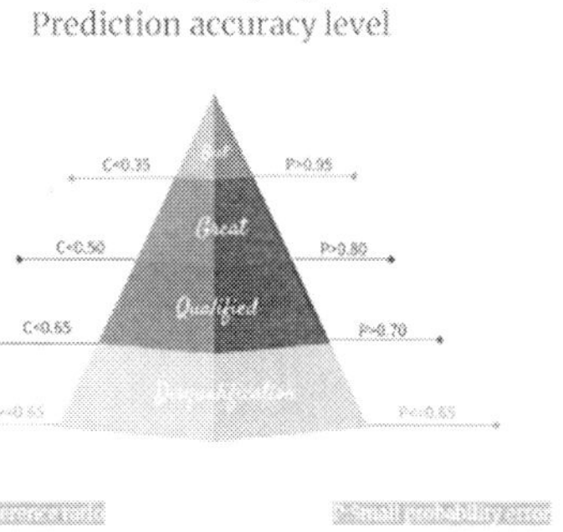

Figure 2: The prediction accuracy level

After validation and calculation, the posterior differences C and the probability of small errors P of the prediction results for the two regions are shown in the following table 4:

Table 4: The prediction results for the two regions

Area	Post-test difference ratio C	Small error probability P	Prediction accuracy level
Los Angeles	0.389	0.814	Great
Wuhan	0.576	0.725	Qualified

6. Conclusion

For the **Topsis** model based on the **improved EMW-AHP**: It integrates the weights from subjective and objective perspectives, which can effectively avoid large differences in weights owing to specific subjective factors or objective factors. Compared with the simple **EMW-AHP**, the improved version can integrate the subjective and objective weights more reasonably What's more, it can solve the problem that Indicator weights do not conform to reality.

For the **k**-means **three-dimensional cluster analysis model**: It has the advantages of combining information from multiple variables to classify samples, and the analysis results are simple and intuitive, which is mainly utilized in this paper for light pollution classification.

For the **Grey** linear **regression combination forecasting model**: The combination is suitable for both series with exponential growth trend and series with linear trend. It exploits the useful information of both single models to improve the accuracy of the model predictions by overcoming the shortcomings of each.

To sum up, light pollution is a global problem and has different degrees of harm to human, ecology and society. Aiming to solve this problem, we proposed the concept of Light Pollution Risk Index ***(LPRI)***, which contains three major aspects: Human Health Risk Index ***HRI***, Social Risk Index ***SRI***, and Ecological Risk Index ***ERI***.

Firstly, we build an ***improved EMW-AHP*** model. Through carefully screening the parameters of the model and determining the weights of the 10 selected indicators, the model was made as close to reality as possible. Secondly, we make full use of the Topsis model to score and rank the 24 cities under 4 different locations, and accurately derived the scores and rankings. Next, we select 108 cities randomly as the training set to classify them into three classes A, B and C with the **K-means 3D clustering model**. Finally, we test the 24 cities mentioned in the previous paper as the validation set. As a result, it shows that the K-means 3D clustering model is established scientifically and reasonably.

With the goal of reducing a city's future Light Pollution Risk Index (LPRI) and mitigating the effects of light pollution on various aspects, We propose a Grey linear regression combination forecasting model and three intervention strategies: ***HSIS***, ***HEIS***, and ***SEIS***. According to the question, we select two cities, Los Angeles and Wuhan to conduct the experiment. The results indicate that for Wuhan, the HEIS intervention strategy is optimal, however, for Los Angeles, the HSIS intervention strategy is optimal.

LPRI's research shows that the harm caused by light pollution should not be underestimated. We have to pay great attention to this problem and adopt practical solutions to combat it. In general, we should do our best to establish an eco-friendly and sustainable world for all mankind.

References

1. Xin Yu The speed of light pollution submerging the starry sky is beyond imagination [N] China Science Daily, 2023-01-31 (002).DOI:10.28514/n.cnki.nkxsb.2023.000194
2. Li Zhenfu. Research on urban light pollution [J]. Industrial Safety and Environmental Protection, 2002 (10): 23-25
3. Guo Jinwei, Pu Xuqiang, Gao Xiang, Zhang Yong'an. An improved method for calculating the weight of multi-objective decision-making indicators [J]. Journal of Xi'an University of Electronic Science and Technology, 2014,41 (06): 118-125
4. Huang Ruixue, Guo Yijiao, Zhang Dingling, et al. Current situation and suggestions of light pollution at night in residential areas in the main city of Chongqing [J]. Light Source and Lighting, 2022, No.173 (11): 5-7
   Wang Aiying, Shi Gang. Light pollution in urban nightscape lighting [J]. Urban Planning, 2003 (04): 95-96+99
5. Wang Lidong, Zheng Zhaorui, Fu Jiawang, Wu Binglin, Duan Yaoyong, Tang Baohua. Drug seizure prediction based on improved grey linear regression combination prediction model [J]. Journal of the People's Police University of China, 2022,38 (12): 22-25
6. Hu Yonghong. Improvement of TOPSIS method for comprehensive evaluation [J]. Practice and understanding of mathematics, 2002 (04): 572-575

Energy efficiency optimization of ships based on particle swarm optimization

Ailin Zhang[1,*], Zhengxu Liang[2]

[1]Economic Management College, Sichuan Vocational College of Chemical Industry, Sichuan, China
[2]COSCO SHIPPING Logistics (Xiamen) Co., Ltd., Siming District, Xiamen, Fujian, China

Abstract. In response to the severe energy-saving and emission-reduction situation, it is of great practical significance to reduce the fuel consumption of ships' main engines and carbon dioxide emissions, and to realize the energy-saving and emission reduction of ships. In this study, We use the particle swarm optimization algorithm to optimize and analyze the energy consumption of ships based on the energy consumption data of ships during navigation, and test the energy efficiency of ships through the ship energy efficiency operation index. The results of the present study show that in order to reduce the energy consumption of ships and ultimately achieve the purpose of energy saving and emission reduction, it is theoretically feasible to use particle swarm optimization algorithm to optimize the speed of ships.

Keywords: Energy saving and emission reduction; Ship energy efficiency improvement; particle swarm optimization.

1 Introduction

In the initial strategy of the International Maritime Organization (IMO), the goal of gradually reducing greenhouse gas emissions from international shipping was reached by the end of this century. However, the use of fossil fuels cannot meet the goals set out in the initial strategy of the International Maritime Organization. In order to achieve the goal of zero-carbon ships and promote innovation in sustainable technologies and alternative fuels, low-carbon shipping needs more comprehensive and in-depth research.

IMO statistics show that maritime transport accounts for 90% of the total global trade, and while bringing high production value, maritime transport is also the most unfavorable mode of transportation for the marine environment. On July 15, 2011, IMO added a new ship energy efficiency regulation within the framework of the MARPOL Convention, and put forward limits and management requirements for the energy efficiency design index (EEDI) of newly built ships and the energy efficiency operation index (EEOI) of ships in service[1]. EEOI measures the energy efficiency of existing ships in operation by the ratio of total CO_2 emissions to cargo turnover during actual voyages. In order to achieve this goal, scientific researchers improve the energy efficiency of ships through technical measures such as hull line optimization, lightweight design, energy-saving attachment design, high-efficiency diesel engine technology, and high-performance hull paint. Ship management personnel improve ship energy efficiency through operational management measures such as improved voyage planning, weather navigation, speed optimization, best trim, best ballast, hull maintenance, and propulsion system maintenance[2].

In view of the management and control of ship energy efficiency is one of the necessary measures to improve the marine economy and reduce CO_2 emissions. At the same time, in the process of ship operation, the problems of CO_2 emissions, sulfur emissions, nitrogen oxide emissions, chlorofluorocarbon emissions and other aspects are becoming more and more serious, Environmental problems caused by ship transportation have also attracted more and more attention around the world. It is particularly important to improve ship energy efficiency, improve fuel efficiency, and reduce pollutant emissions[3].

2 EEOI

2.1 EEOI formula analysis

According to EEOI Guidelines for Voluntary Application of Ships, ship energy efficiency operating index can be calculated as[4]:

$$\mathrm{EEOI} = \frac{\sum_{j} FC_j \times C_{F_j}}{m_{c\arg o} \times D} \quad (1)$$

The average ship energy efficiency operating index of multiple segments can be calculated as:

* Corresponding author: 250061294@qq.com

$$\text{AverageEEOI} = \frac{\sum_{i}\sum_{j} FC_{ij} \times C_{Fj}}{\sum_{i}(m_{cargo} \times D_i)} \quad (2)$$

where C_{Fj} is the CO_2 emission factor, which is the quality of CO_2 emitted by consuming 1g of fuel. Where FC_{ij} is the total fuel consumption of the ship, and m_{cargo} is the cargo capacity, and D is the voyage distance of the ship, and i is the number of flight segments, and j is the fuel type.

According to the above calculation formula of ship energy efficiency, it can be known that when the cargo capacity and the navigation distance are constant, the less fuel the ship consumes, the lower the EEOI value, which indicates the higher the ship's energy efficiency level.

2.2 CO_2 emission factor

The actual measurement of CO_2 emissions from ships has low operability and poor accuracy. However, the fuel consumption of ships is easy to obtain. Generally, the corresponding relationship between fuel consumption and CO_2 emissions is found to calculate CO_2 emissions, which is an international practice.

Two CO_2 emission detection schemes, Tire1 and Tire2, are proposed in IPCC GHG Inventory and Good Practice Guide for global greenhouse gas emission detection.

2.3 Statistical interval of oil consumption

Obtained by the EEOI definition formula, a single voyage is used as a single interval for statistical calculation of fuel consumption. According to the Guidelines for the Voluntary Application of EEOI in Ships, oil consumption refers to the sum of all fuel oil consumed by a ship in the course of port and navigation, including the fuel consumed by the ship's main engine, auxiliary engine and boiler. The ship's berthing process includes loading and unloading at the port. Therefore, although no cargo turnover is generated, the fuel consumption at the port must also be included in the statistics. The unloading time of this voyage is the start time of the next voyage[5]. Therefore, when calculating the EEOI, the start time is usually the time when the loading and unloading task of the cargo on the previous voyage is completed.

3 Particle swarm optimization algorithm

The basic concept of particle swarm optimization(PSO) algorithm originated from the research of bird swarm foraging behavior. Search for the individual closest to the food in the flock[6]. The PSO algorithm is inspired by this kind of biological population behavior characteristics and used to solve optimization problems.

Use a kind of particle to simulate the above-mentioned individual bird. Each particle can be regarded as a search individual in the N-dimensional search space. The current position of the particle is a candidate solution of the corresponding optimization problem. The flight process of the particle is the searching process of the individual. The particle velocity can be dynamically adjusted according to the historical optimal position of the particle and the historical optimal position of the population. PA particle has only two properties: speed, which represents how fast it is moving, and position, which represents the direction it is moving. The optimal solution searched individually for each particle is called the individual extremum, and the optimal individual extremum in the particle swarm is the current global optimal solution. Keep iterating, updating speed and location. Finally, the optimal solution satisfying the termination condition is obtained[7].

The algorithm flow is as follows:

1. Initialization

First of all, the maximum number of iterations, the number of independent variables of the objective function, and the maximum velocity of the particles are set. The position information is used as the entire search space, and the speed and position are randomly initialized in the speed range and search space. Set the particle swarm size as M, and each particle randomly initializes a flying speed.

$$N_i = [X_i, V_i] \quad (3)$$

$$X_i = [x_1, x_2, x_3 \ldots\ldots x_m] \quad (4)$$

$$V_i = [v_1, v_2, v_3 \ldots\ldots v_m] \quad (5)$$

2. Individual extremum and global optimal solution

The fitness function is defined, and the individual extreme value is the optimal solution found for each particle. From these optimal solutions, a global value is found, which is called the global optimal solution. Update with historical global optimum comparison.

3. Update speed and position formulas

$$V_{id} = wV_{id} + C_1 random(0,1)(P_{id} - x_{id}) + C_2 random(0,1)(P_{gd} - x_{id}) \quad (6)$$

$$x_{id} = x_{id} + v_{id} \quad (7)$$

where w is called the inertia factor, and C_1 and C_2 are called the acceleration constants, which usually $C_1 = C_2 = 2$. Where $random(0,1)$ represents a random number on the interval, and P_{id} represents the d-th dimension of the individual extreme value of the i-th variable, and P_{gd} represents the first dimension of the global optimal solution.

4. Termination conditions

Reach the set number of iterations or the difference between algebras to meet the minimum limit.

4 Establishment and optimization of ship energy consumption model

4.1 Ship fuel consumption statistics

As an important means of cargo transport in China, the energy consumption of ships can not be ignored. It is of great significance to control, manage and evaluate the energy consumption of ships.

(1) Calculation method of Marine ships

The fuel consumption of the ship's voyage can be calculated according to Equation (8).

$$Q = Q_z + Q_f + Q_g \quad (8)$$

where Q is the fuel consumption of ship voyage, Q_z is the fuel consumption of ship's main engine, Q_f is the fuel consumption of generator set, Q_g is the fuel consumption of the boiler.

(2) Calculation method of inland ships

The fuel consumption of the ship in operation can be calculated according to Equation (9).

$$Q=Q_1t+Q_2A+Q_f \tag{9}$$

where Q is the fuel consumption on voyage, Q_1 is the basic fuel consumption per hour, t is the standard voyage time, Q_2 is the unit turnover additional fuel consumption, A is the actual conversion turnover of voyage, Q_f is the fuel consumption for auxiliary production.

According to the research direction of this study, Marine ship calculation method will be used to calculate the fuel consumption of ship voyage. In view of the difficulty of data acquisition at present, only the fuel consumption of ship's main engine will be considered in this study.

4.2 Building ship energy consumption model

4.2.1 Route segment

By referring to the literature[8], an international route from Dalian to Kuala Lumpur, Malaysia is taken as the research object, and the relevant research on speed optimization is carried out in this study. The entire voyage is divided into 12 segments, namely L_1, L_2...L_{12}. The mileage information of each segment is listed in detail according to table 1. In addition, the entire voyage of the research subject ship is 3019.345 nm, and the voyage time of the ship is 296 h.

Table 1. Track section information table.

Segment number	mileage (nm)	Segment number	mileage (nm)
L_1	120.372	L_7	394.028
L_2	163.422	L_8	197.014
L_3	179.548	L_9	493.998
L_4	215.162	L_{10}	364.551
L_5	205.967	L_{11}	116.343
L_6	228.904	L_{12}	240.027

4.2.2 Objective function establishment

The ultimate goal of ship energy efficiency optimization is to achieve the least total fuel consumption when the ship completes a given voyage. Therefore, the fuel consumption of the entire voyage can be used as the objective function of the particle swarm optimization algorithm.

$$\text{Min W} = \min \sum_{i=1}^{12} w_i = \sum_{t=1}^{12} f_i(v_{si}) \tag{10}$$

where W represents fuel consumption for the entire voyage, and f_i represents the relationship function between diesel engine fuel consumption and ship speed in each flight segment.

According to the modeling process of diesel engine fuel consumption model in Literature[9], the fuel consumption of ships in each voyage can be calculated according to Equation(11).

$$W_i = g_{ei} \cdot p_{ei} \cdot t_i \tag{11}$$

where W_i represents the fuel consumption within each segment, g_{ei} and p_{ei} respectively are the diesel engine fuel consumption rate and output power of each segment. Where t_i is the voyage time of each segment, which can be obtained by dividing the voyage L_i of each voyage segment by the ship speed V_{si} according to Equation(12).

$$t_i = \frac{L_i}{V_{si}} \tag{12}$$

The trajectory segment of each segment has been given in Table 1. According to the parameters of the relationship between the speed of the ship and the speed of the ship in the reference literature[10], combined with the energy consumption of the ship in this voyage, the corresponding relationship between the speed of the diesel engine and the speed of the ship can be calculated as follows (taking segment 1 as an example) :

$$n_{e1} = 1.78V_{s1} + 40 \tag{13}$$

Therefore, The fuel consumption of the ship in each voyage is calculated.

$$W_1=(0.0558\times(1.78V_{s1}+40)^2-7.4382\times(1.78\times V_{s1}+40)+448.691\times0.0488\times(1.78V_{s1}+40)^3\times120.372)/V_s(14)$$

The relationship between the fuel consumption of the ship in the L_1 segment and the speed in the segment as Equation(14). For other flight segments, the calculation method is similar to this, so this study will not go into details. By calculating the fuel consumption of each flight segment, the objective function for the particle swarm optimization algorithm can be calculated.

4.3 Optimization of ship energy consumption model

4.3.1 Solving process

According to the above objective function, the particle swarm optimization algorithm is used to optimize it.

(1) Initialize the particle population and give the initial velocity and position to each particle. And initialize the learning factor C_1 and C_2, the maximum number of iterations, particle swarm size and other parameters.

(2) Calculate the fitness value of the particle, which is determined by the objective function. After the algorithm is initialized and updated, the constraint conditions of the two objective functions can be satisfied by limiting the control variable to the specified value range.

(3) Compare the fitness of the particles. The optimal position of the individual particles is updated according to the dominance relationship, and the global optimal position of the particles is randomly selected from it.

(4) Update the particle speed and position according to the particle update formula until the global optimal solution is found or the maximum number of iterations is satisfied. The specific flow chart of the solution is shown in Figure 1.

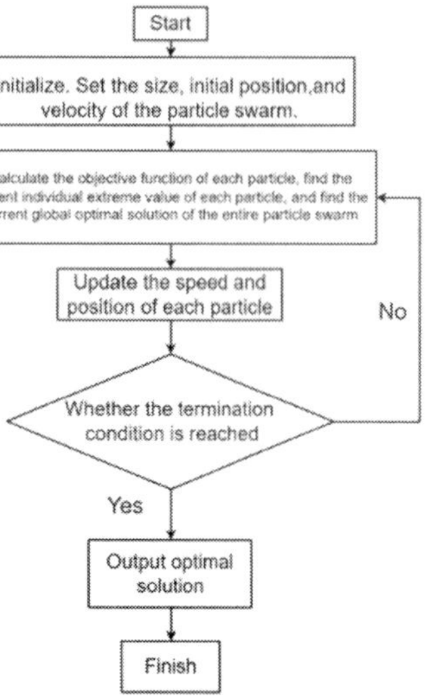

Fig. 1. PSO flow chart.

4.3.2 Running results

During the operation of the algorithm, the initial population number is set to 50, the space dimension is 1, the maximum number of iterations is 100, the inertia weight is 0.8, and the self-learning factor and the group learning factor are 2. The result of the operation is shown in Figure 2.

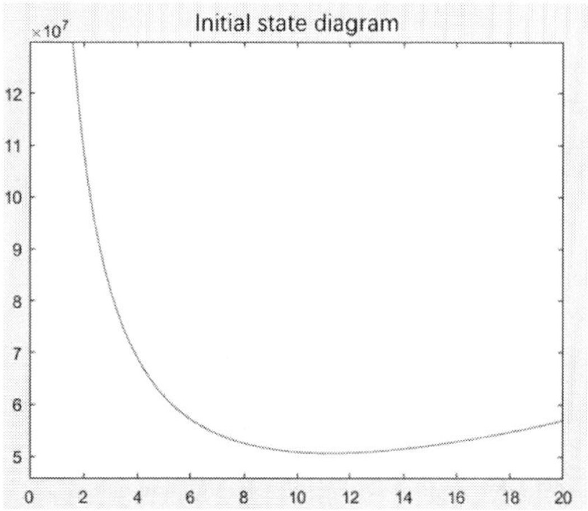

Fig. 2. Initial state diagram.

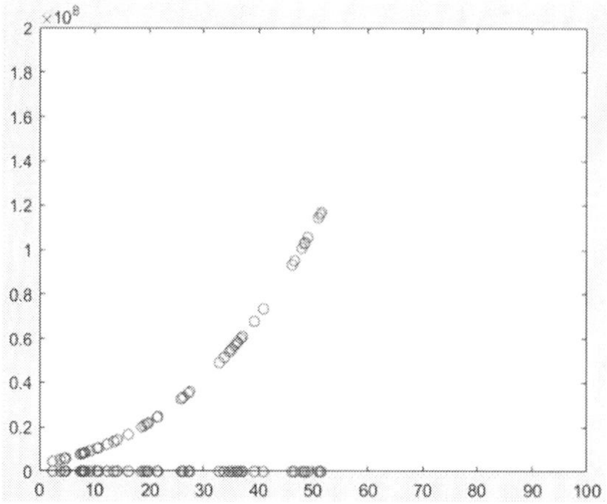

Fig. 3. Final position diagram.

The algorithm ran 10 times, the calculation results are consistent, and the algorithm has good stability. The convergence curve of a certain running program obtained randomly is shown in Figure 2, which shows that the optimized value of the objective function has converged. It can be seen from Figure 3 that the optimal solution is 1.23×108. According to the results of optimized energy consumption, calculate the ship energy efficiency operation index, evaluate and analyze the current ship energy efficiency, and finally realize the optimization of ship energy efficiency through speed control, engine speed or power control.

5 Conclusion

In today's era of vigorously advocating energy saving and emission reduction, green and low-carbon life, the study of ship energy consumption has important practical significance. In this study, we studied the realization and solution of the particle swarm optimization algorithm on the ship energy consumption optimization problem, which based on the relevant energy consumption data of ship navigation in the existing literature. Through the analysis of ship energy consumption optimization results, it can be concluded that in order to reduce ship energy consumption, it is feasible to use particle swarm optimization algorithm to optimize ship speed theoretically. And according to the results of algorithm optimization, the fuel consumption of the entire voyage is gradually reduced under the condition that the sailing time of the ship is not timeout.

References

1. F. Liu, *Analysis and Research on the Influence of EEDI on Ship's Overall Design* (DUT, Dalian, 2011)
2. J. Ni, *Research on Ship Energy Efficiency Operation Index* (SJTU, Shanghai, 2010)
3. G. Li, *Research on the selection of ship's main power plant based on energy efficiency design index* (DMU, Dalian, 2014)
4. T. Anh Tran, J. Oc. Eng. Sci, **4**, 3 (2019)

https://doi.org/10.1051/e3sconf/202337503016

5. N. Acomi, O. C. Acomi, Procedia Soc. Behav. Sci, **138** (2014)
6. L.G. Tonutti, B.O. Dalla Costa, H.P. Decolatti, G. Mendow, C.A. Querini, Chem. Eng. J, **424** (2021)
7. J. Hu, W. Xiong, IE. CP. APPL, **11**, 6-11 (2021)
8. L. Sun, *Research on Ship Energy Consumption Based on Speed Optimization* (HEU, Harbin, 2019)
9. Z. Liang, *Research on Ship Energy Efficiency Optimization Based on EEOI* (JMU, Xiamen, 2020)
10. Q. Chen, X. Yan, Tfc. Info. S, **32**, 87-91(2014)

https://doi.org/10.1051/e3sconf/202337503017

Diagnosis and Analysis of Thermal Power Plant Operation in Heating Season

Yanpeng Zhang *, Lingkai Zhu, Si Li, Panfeng Shang

State Grid Shandong Eectric Power Research Institute, Jinan, China

Abstract: In order to further improve the operation level of the thermal power units of the provincial grid thermal power plant, optimize the data quality of the online monitoring system of the provincial thermal power units, through the diagnosis and analysis of the operation data of the thermal power online monitoring system, and from an all-round perspective, analyze and evaluate the peak shaving and frequency modulation capabilities of the heat supply units of the thermal power plant from the aspects of the load carrying capacity of the heat supply units, the optimization of the startup mode of the whole plant, and the heating capacity, to provide support for the accurate scheduling of the heat supply units. This paper analyzes the heat supply, peak regulation and frequency regulation capacity of the whole thermal power plant through the operation data, draws the peak regulation interval and heat supply capacity interval, and provides support for the optimal operation of the whole plant.

key word: Thermal power plant, Heating capacity, Peak shaving capacity, diagnostic analysisg

1. Introduction

In order to further improve the operation level of the thermal power units of the provincial grid thermal power plant, optimize the data quality of the online monitoring system of the provincial thermal power units, through the diagnosis and analysis of the operation data of the thermal power online monitoring system, and from an all-round perspective, analyze and evaluate the peak shaving and frequency modulation capabilities of the heat supply units of the thermal power plant from the aspects of the load carrying capacity of the heat supply units, the optimization of the startup mode of the whole plant, and the heating capacity, to provide support for the accurate scheduling of the heat supply units.[1-2] This paper analyzes the heat supply, peak regulation and frequency regulation capacity of the whole thermal power plant through the operation data, draws the peak regulation interval and heat supply capacity interval, and provides support for the optimal operation of the whole plant. [3]

2. Summary of Thermal Power Plant

There are 4 units in the thermal power plant, including # 4, # 5, # 6 and # 7 units, all of which are condensing units. The data quality of the thermoelectric system is good, and the parameters of the measuring points meet the requirements.

The No. 4 and No. 5 units in the second phase of the thermal power plant are 330MW subcritical, primary intermediate reheat, single shaft, double cylinder and double exhaust, reaction condensing steam turbines, with the model of C330-16.7/0.8/538/538. The No. 6 and No. 7 units in the third phase of the thermal power plant are the first million kilowatt units in the world to apply the secondary intermediate reheat technology. The steam turbine adopts the ultra supercritical secondary intermediate reheat condensing steam turbine, which is arranged in series with ultra-high pressure cylinder, high pressure cylinder, intermediate pressure cylinder and two low pressure cylinders. The parameter is 31MPa/600°C/620°C/620°C. The design index is about 1% higher than the world's best secondary reheat generator unit, and the generating efficiency of the unit is more than 47.95%, about 2.2% higher than the average efficiency of domestic conventional ultra supercritical primary reheat unit. Under BRL working condition, the boiler's thermal efficiency is guaranteed to be no less than 94.65%, and the steam turbine's heat consumption is guaranteed to be no less than 7051kJ/kWh, making it the most efficient thermal power generation unit in the world.

In 2020, the power generation will reach 9913 million kwh (including 4.314 billion kwh for Unit 6 and 5.599

* Corresponding author: zhangyanpeng19883@163.com

https://doi.org/10.1051/e3sconf/202337503017

billion kwh for Unit 7), the accumulated utilization hours will be 4812 hours, the auxiliary power consumption rate will be 2.74%, and the coal consumption for power supply will be 265.58g/kwh, reaching the level of national thermal power benchmark units.

In 2020, Unit 6 and Unit 7 have successively participated in in-depth peak shaving for 306 times. The peak shaving time is up to 1051 hours, 50-40% peak shaving time is 1022 hours, and 40-30% peak shaving time is 29 hours. The deep regulation output of the unit reaches 36% of the rated capacity.

2.1 Heating condition of thermal power plant

According to the design data, the maximum steam extraction flow of # 4 and # 5 units is 240t/h, and according to the operation data, the maximum steam extraction flow is 300t/h# 6. The rated steam extraction flow of # 7 unit is 350t/h, and the maximum steam extraction flow is 400t/h; In 2019, 1 million square meters of heating area will be added, and 60t/h of steam extraction will be added. The newly increased area must be brought by Phase II, so a 1 million unit must be opened. The heating return water in the whole phase has a connecting valve. Due to different pressures, for safety reasons, the heating return water will not be mixed and will return to each initial station. At present, 4 million square meters are undertaken in Phase I and 4.5 million square meters in Phase II. The steam extraction capacity of small units needs to reach 240 t/h, which has reached the rated steam extraction capacity of the unit. Theoretically, the heat supply can be operated in the middle period of one big and one small, but the load of 330MW unit is required to reach about 200MW.

The thermal power plant has a design heating area of 4.5 million square meters for high-temperature hot water, and 100 tons of industrial steam per hour. The actual heat supply capacity of high-temperature water is 6 million square meters. Due to the large leakage loss of the heat supply pipeline from the industrial steam supply to the hot urban area, the steam supply was stopped in 2014. In 2016, the industrial steam was supplied to industrial users within 3 kilometers around, with 35 tons/hour. High temperature water heating management mode is wholesale, and industrial steam management mode is self operated. In 2015-2016 heating season, the heating capacity of high temperature water is 1.1655 million GJ, and the heating area is 4.1 million square meters.

2.2 Industrial steam extraction system of thermal power plant

(1) Thermal system of high pressure industrial steam supply pipeline

The high pressure industrial steam supply pipe is connected to the primary reheat low temperature steam pipe and supplied to the user after temperature and pressure reduction. The desuperheating water for high-pressure industrial steam supply comes from the primary intermediate tap of the feed pump and is connected to the desuperheating and pressure reducer after pressure reduction. The electric shut-off valve, pneumatic control valve, manual shut-off valve, filter screen, check valve and flow measuring device shall be set on the desuperheating water pipeline in order of flow direction.

(2) Thermal system of low pressure industrial steam supply pipeline

There are two steam sources for the low-pressure industrial steam supply pipeline, which are respectively connected from the primary reheat low-temperature steam pipeline and the secondary reheat low-temperature steam pipeline. The desuperheating water for low pressure industrial steam is from the condensate pipe, connected from the condensate booster pump, and then connected to the desuperheating and pressure reducer after pressure reduction. The electric shut-off valve, pneumatic control valve, manual shut-off valve, filter screen, check valve and flow measuring device shall be set on the desuperheating water pipeline in order of flow direction.

2.3 Description of heat supply calculation

The thermal power plant uses industrial extraction flow and heating extraction flow to calculate heat supply. The calculation formula of all units is the same, so it will not be repeated here. The calculation formula of # 4 unit is as follows:

#Heating and heating capacity of unit 4=# 4 heating and steam extraction flow × (# 4 heating extraction enthalpy - # 4 heating extraction drainage enthalpy);

#Industrial heat supply of unit 4=# 4 industrial steam extraction flow 2 (cold standby) ×# 4 Industrial steam extraction enthalpy 2 (cold standby)+# 4 industrial steam extraction flow 1 (cold standby) ×# 4 Industrial extraction enthalpy 1 (cold reheat).

3. Thermal Power Plant Capacity Summary

3.1 Summary of heating capacity of thermal power plant

The statistics shall be carried out according to three stages of heating: initial stage, middle stage and final stage. Statistics of heat extraction capacity, heat supply capacity, thermoelectricity ratio, power generation capacity, power supply capacity, converted heating area and other indicators of a single unit and the whole plant, the converted heating area is calculated as 40W/m^2, and the above thermoelectricity parameters are calculated as per unit operating hours.

Statistics of residential heating capacity and heating area in the last three heating seasons of the thermal power plant: 2977722.51 GJ of residential heating capacity and 7062400 m^2 of residential heating area in the heating season from 2019 to 2020; In the heating season from 2020 to 2021, the residential heating capacity will be 3559045.76GJ, and the residential heating area will be 8.4411 million m^2; In the heating season 2021-2022, the annual residential heating capacity is 299982042GJ, and the residential heating area is 7114800 m^2. The thermoelectric data statistics are normal, and the heating area calculation conforms to the actual situation. According to the rated external heating capacity of the power plant, the alternative heating area of the unit is

 https://doi.org/10.1051/e3sconf/202337503017

analyzed, and the results are shown in Table 1. According to the report of the power plant that the rated heating steam extraction flow of one large and one small is 240t/h and 400t/h respectively, the alternative heating area margin of the thermal power plant is 17.7742 million m^2.

Tab 1 Analysis on heating area of thermal power plant

Thermal power plant	Rated heating steam extraction flow (t/h)	Rated heating area (10000 m^2)	Actual heating steam extraction flow(t/h)	Actual heating amount(GJ)	Actual heating area(10000 m^2)	Alternative heating area allowance (10000 m^2)
#4 Units	240	466.67	85.86	623981.68	147.99	318.68
#5 Units	240	466.67	65.90	445836.55	105.74	360.93
#6 Units	400	777.78	180.03	1393382.56	330.47	447.31
#7 Units	400	777.78	93.46	536619.63	127.27	650.51
Whole plant	1280	2488.90	365.90	2999820.42	711.48	1777.42

3.2 Analysis and summary of power load adjustment capacity of thermal power plant

During heating period, the adjustable output range of # 4 unit under heating and heating is 178.48MW~301.79MW, and the adjustable output range under comprehensive heating is 180.24MW~296.26MW.# 5. The adjustable output range of the unit under heating and heating is 173.03MW~296.77MW, and the adjustable output range under comprehensive heating is 175.48MW~289.13MW. # 6. The adjustable output range of the unit under heating and heating is 583.39MW~995.88MW, and the adjustable output range under comprehensive heating is 594.32MW~951.76MW.# 7. The adjustable output range of the unit under heating and heating is 614.28MW~1021.15MW, and the adjustable output range under comprehensive heating is 619.60MW~999.69MW.

Analysis of deep peak shaving and derating of thermal power plant units during heating period. See Table 2 for specific data.

Tab 2 Deep peak shaving and derating of thermal power plant units

Thermal power plant	**Unit rated capacity (MW)**	**Minimum output of unit (MW)**	**Minimum adjustable output of integrated heating (MW)**	**Maximum adjustable output of integrated heating (MW)**	**Whether it has the ability of deep peak regulation**	**Deep regulation load (MW)**	**Reduced output load (MW)**
#4 Units	330	165	180.24	296.26	no	0.00	33.74
#5 Units	330	165	175.48	289.13	no	0.00	40.87
#6 Units	1030	500	594.32	951.76	no	0.00	78.24
#7 Units	1030	500	619.60	999.69	no	0.00	30.31

4. Conclusion

(1) Through calculation, the residential heating area of the thermal power plant in the three heating seasons since 2019 is 7062400 m^2, 8441100 m2 and 7114800 m^2 respectively. At present, the designed heating area of the thermal power plant is 24.889 million m^2, and the margin of alternative heating area is 17.7742 million m^2.
(2) At the beginning, middle and end of the 2021-2022 heating season, the actual average heating flow for people's livelihood is 350.84t/h, 383.50t/h and 327.24t/h respectively, and the actual average industrial heating flow is 167.76t/h, 164.16t/h and 150.84t/h respectively.
(3) During heating period, the minimum adjustable output of Unit # 4 under comprehensive heating is 180.24MW, which does not have deep peak shaving capacity. The maximum adjustable output under comprehensive heating is 296.26MW, which reduces the rated output by 33.74 MW.# 5 The minimum adjustable output under comprehensive heating is 175.48MW, which does not have deep peak shaving capacity. The maximum adjustable output under comprehensive heating is 289.13MW, which reduces the rated output by 40.87 MW.#6 The minimum adjustable output under comprehensive heating is 594.32MW, which does not have deep peak shaving capacity. The maximum adjustable output under comprehensive heating is 951.76MW, which reduces the rated output by 78.24MW.# 7 The minimum adjustable output under comprehensive heat supply is 619.60MW, which does not have deep peak shaving capacity. The maximum adjustable output under comprehensive heat supply is 999.69MW, which reduces the rated output by 30.31MW.

Acknowledgements

This work was supported by the scientific and technological project of State Grid Shandong Electric Power Research Institute "Research on Thermoelectric Cooperative Regulation Technology of Nuclear Power Units" (ZY-2022-10).

 https://doi.org/10.1051/e3sconf/202337503017

References

1. Shi Qingli, Wen SheJiao.Application of thermal test in thermal power plants [J]. Qinghai Electric Power, 2004, 23 (4): 4-10.
2. Weng Sicheng.Thermal performance test of SAIC 300MW steam turbine (Wangting Power Plant Unit 12) [J]. Shanghai Turbine, 1983 (01): 5-20.
3. Ji Ansen.Results and Analysis of Thermal Performance Test of Unit #12 in Wuhu Power Plant before Overhaul [J]. Anhui Electric Power Technical Information, 2000, 000 (002): 7-8.

Study on the relationship between tea concentration and Tyndall effect

Yifan Yang[1], Xiaofeng Su[2], Yongke Xu[2], Weigang Jin[2], Bangxing Shen[2], Xinfeng Zhang[2], Ruibo Lv[2], Yi He[2], Chunfeng Guo[1*], Yuefeng Li[1*], Jun Zou[1*]

[1]School of Science, Shanghai Institute of Technology, Shanghai, 201418, China
[2]Yantai Huachuang Intelligent Equipment Co., Ltd., Shandong, 265502, China

*Corresponding author: Chunfeng Guo (Email:cfguo@sit.edu.cn); YuefengLi(Email:yf_li@sit.edu.cn); Jun Zou (Email: zoujun@sit.edu.cn);

Abstract: The quality of tea is relative with its concentration. In this paper, we present a method to identify the concentration of tea based on Tyndall effect. The tea is illuminated by 3-watt LED light sources with the main wavelength in ultraviolet and visible bands. The intensities of transmitted light are detected using the distributed photometer and the images of light distribution are captured by CCD camera at vertical direction to observe the Tyndall effect. The intensities and images for different light sources are compared. The experimental results show that there exist Tyndall phenomenon in tea. In addition, the Tyndall effect becomes weaker as the tea concentration increases when the tea is illuminated by the LED light sources with the main wavelength in the ultraviolet band. The method presented in this paper should be helpful for rapid identification of the tea quality

1. Introduction

After thousands of years of development, tea has become one of the three major non-alcoholic beverages and tea is widely loved by people all over the world because of its excellent flavor and health benefits. The flavor and health care effect of tea is the result of the joint action of taste and aroma. Its components mainly include amino acids, sugars, fats, vitamins, chlorophyll, carotene, tea pigments, tea polyphenols, tea polysaccharides, caffeine, etc. These taste and aroma substance determines the quality of tea[1, 2].

At present, the research on the quality of the tea leaves is mainly divided into two types. The first is to study the tea leaves directly, and the second is to study the tea. Less research has been done on tea leaves. Bacterial species[3, 4], pest and disease disasters[5] are detected to study the tea leaves . More research has been done on tea soup as follows. The function of genetic traits[6] is studied by the use of molecular marker technology. and the structure of active ingredients[7]is determined by the use of liquid chromatography. Aroma substances[8] is analyzed by the use of mass spectrometry, and raw material varieties is identified by the use of near-infrared[9]. .In addition, absorption spectrum of tea is analyzed by using ultraviolet spectrophotometer[10]. To sum up, the current research on both tea leaves and tea requires professional equipment, and the operation is difficult. Therefore, there is an urgent need for a way to quickly identify tea. It was found that tea contains a variety of compounds, and the particle size of these compounds is generally in the range of colloids. When light passes through these colloidal particles, the Tyndall effect occurred. The concentration of tea is closely related to the content of ingredients in tea. Therefore, the relationship between the tea concentration and the Tyndall effect is explored in the paper for the first time.

In this article, in order to explore the relationship between tea and the Tyndall effect, two types of samples and light sources provided by SMC (Shenzhen) Co., Ltd. have been prepared for testing. In the section 1, the propagation mechanism of light in tea are introduced. In the section 2, the material preparation and experimental procedure of this study are introduced. In the section 3, the experimental data is analyzed and discussed.

2. Principle

When light encounters an object, Reflection, Scattering, Absorption and Transmission will occur. Reflection of light occur when light changes the direction of propagation at the interface between two substances and returns to the original substance. Scattering of light refers to the effect that part of the light travels away from the original direction when light passes through an inhomogeneous medium[11]. Transmission of light is that incident light is refracted through an object. The absorption of light refers to the effect that atoms absorb the energy of photons and transition from a low-energy state to a high-energy state when exposed to light. In the solution, the particle diameter of the dispersoid is less than

https://doi.org/10.1051/e3sconf/202337503018

1 nm. When the visible light enters the solution, the solution has a very weak scattering effect on the light, which mainly produces the transmission of light. While in the turbid liquid, the particle diameter of the dispersoid is generally larger than the wavelength of visible light. Therefore, when visible light enters the turbid liquid, the phenomenon of reflection mainly occurs. In colloids, the diameter of colloidal particles is less than the wavelength of visible light. When visible light enters the colloid, light scattering will occur to produce the Tyndall effect[12].

The Tyndall effect is formed by the scattering of visible light (wavelength 400 ~ 700 nm) [13]. In the process of light propagation, reflection or scattering of light will occur when light irradiates the particle. If the particle is many times larger than the wavelength of the incident light, light reflection occurs. If the particle is smaller than the wavelength of the incident light, light scattering occurs.

The diameter of colloidal particles is between 1nm and 100nm. It is smaller than the wavelength of visible light (400nm ~ 700nm), it will produce obvious scattering effect when visible light passes through the colloid. For a true solution, although the molecules or ions are smaller, the intensity of scattered light decreases significantly with the decrease in the volume of the scattering particles. Thus there is a very weak scattering effect when the true solution is illuminated by light. In addition, the intensity of scattered light also increases with increasing particle concentration in most cases. It is worth noting that the frequency of scattered light does not change in the Tyndall effect. And the Tyndall effect in the pure water and colloidal solution is shown in Figure 1.

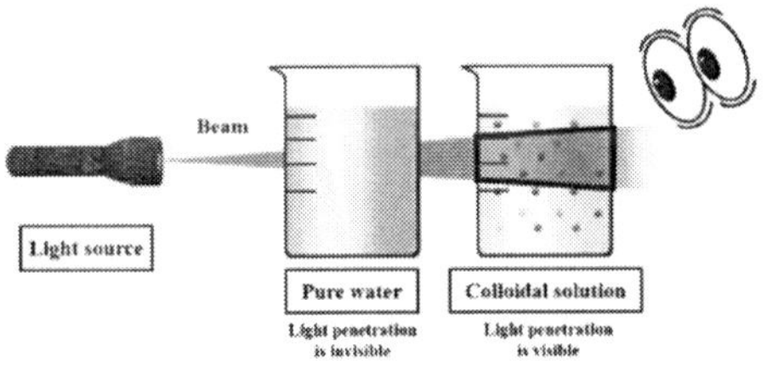

Figure 1. Schematic diagram of the Tyndall effect.

Luminous intensity is a physical quantity used to express the intensity of luminescence within a unit solid angle and a given direction of a light source. In this experiment, the intensity of light reflection in visible band is very small and can be ignored. Therefore, the intensity of light scattering in visible band is indirectly defined by testing the Luminous intensity of transmitted light, which is the intensity of Tyndall effect. The distributed photometer was used to test the Luminous intensity. The specific test principle is shown in Figure 2.

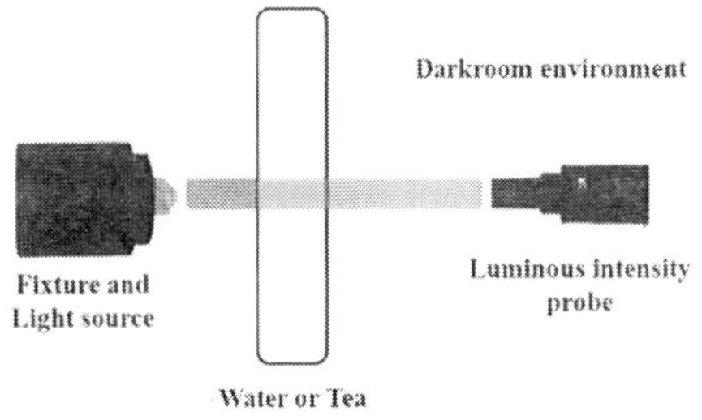

Figure 2. Luminous intensity test principle.

3. Experiment

3.1 Material preparation

The light sources used in this experiment are all 3535 vertical structure light sources. The power of these light sources is 3w and the divergence angle of these light sources is controlled within 120°. The specific wavelengths are shown in Table 1, the picture of 3535 vertical structure light source is shown as Figure 3.

Table 1. Light source specifications.

Band	Wavelength	Color
Ultraviolet	345nm 365nm 395nm	Not defined
Visible	405nm	
	440nm	
	470nm	
	505nm	
	530nm	
	560nm	
	605nm	
	630nm	
	660nm	

Xinyang Maojian Tea and Raw Pu'er Tea are choosed. Xinyang Maojian Tea is the representative tea in green tea, and Raw Pu'er Tea's fermentation degree is close to that of green tea. The fermentation degrees of Xinyang Maojian Tea and Raw Pu'er Tea are both ≤ 5%. The color of the Xinyang Maojian Tea is bright green, while the color of the Raw Pu'er Tea is light red. In this experiment, the tea water ratio of both two teas is 1g tea: 100ml water. Both teas were soaked in water at 80°C and placed at room temperature for 4 to 6 hours to fully dissolve the soluble molecules in the tea leaves, and then the tea leaves were taken out to obtain a filtrate. The filtered pictures of Xinyang Maojian Tea and Raw Pu'er Tea are shown in Figure 4.

 https://doi.org/10.1051/e3sconf/202337503018

Figure 3. 3535 vertical structure light source.

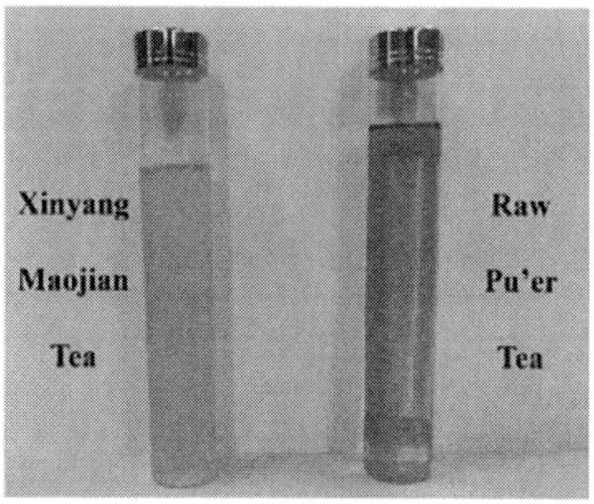

Figure 4. The filtrate of Xinyang Maojian Tea and Raw Pu'er Tea.

3.2 Methods

1、 Use the 12 groups of light sources in Table 1 to irradiate Xinyang Maojian Tea and Raw Pu'er Tea in a darkroom environment, and visually observe whether there is Tyndall effect and the intensity of Tyndall effect.
2、 Test the absorbance between 200nm and 800nm to explore whether the Tyndall effect is biased due to absorption. At the same time, use a distributed photometer to test the Luminous intensity of water and tea respectively in the band which almost no absorption, and obtain the Luminous intensity value of the Tyndall effect.
3、 Use the light sources for the best Tyndall effect which observed with naked eyes in darkroom environment to observe the Tyndall effect in a daylight environment, and select the best light source for the Tyndall effect in the daylight environment.
4、 Change the concentration of the tea, use the light source in the daylight with the best Tyndall effect for irradiation, observe the Tyndall effect result with the naked eye, and test the Luminous intensity data.

4. Results and discussions

Under the darkroom environment, the light sources in Table 1 are used to irradiate the tea in Xinyang Maojian Tea and Raw Pu'er Tea, and their Tyndall effect presented are shown in Figure 5 and Figure 6. According to Figure 5 and Figure 6, it is found that the Tyndall effect is existed in Xinyang Maojian Tea and Raw Pu'er Tea between 345nm to 660nm. Xinyang Maojian Tea and Raw Pu'er Tea have very obvious Tyndall effect at 345nm, 365nm, 395nm which are ultraviolet bands. At the same time, Xinyang Maojian Tea also has obvious Tyndall effect at 405nm, 470nm and 560nm, while Raw Pu'er Tea is at 470nm, 530nm and 560nm.

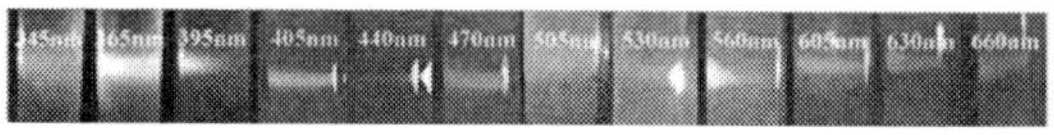

Figure 5. Tyndall effect of Xinyang Maojian Tea at different light sources.

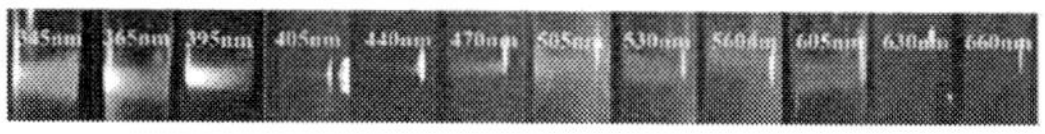

Figure 6. Tyndall effect of Raw Pu'er Tea at different light sources.

According to the definition of Tyndall effect, there should be no Tyndall effect in the ultraviolet band. However, it can be clearly found that not only the Tyndall effect exists in the ultraviolet band, but also the intensity of the Tyndall effect is very obvious through the naked eye observation. There are two reasons for this problem. One is that the ultraviolet light sources will be added with light in other wavelength bands during production to ensure the safety of the use of ultraviolet light sources. The second is that due to the emission of the tea in the ultraviolet band, the photons of the higher band excited show the Tyndall effect. Therefore, for 345nm, 365nm, and 395nm, the test of the light source spectrum and the test of the emission spectrum are carried out in order to distinguish between the scattering of visible light and the emission of tea. This part will be discussed after the analysis of the Tyndall effect caused by visible light.
The absorption of Xinyang Maojian Tea and Raw Pu'er Tea was tested, and the results are shown in Figure 7. As shown in the figure, the absorbance of Xinyang Maojian Tea and Raw Pu'er Tea is basically the same between 400nm and 700nm, the variances of Xinyang Maojian Tea and Raw Pu'er Tea are 0.004 and 0.01, respectively. Therefore, the absorbance in this between 400nm and 700nm can be ruled out as an average error when testing Luminous intensity. Since there is no reflection in tea and tea is only excited in the ultraviolet band, after excluding the average error of absorption, the result of the Luminous intensity test represents the intensity of the Tyndall effect to a certain extent.

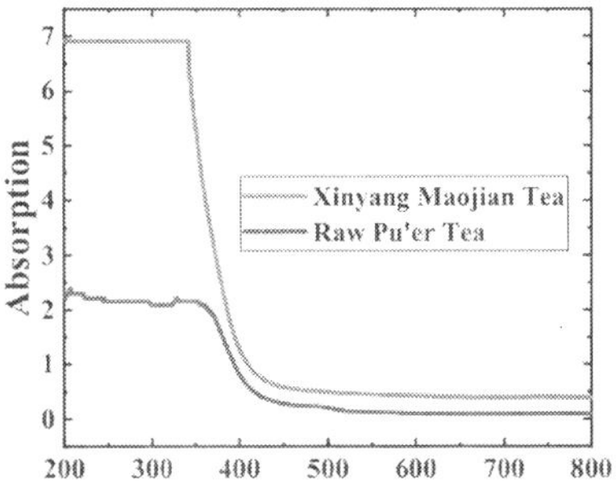

Figure 7. UV-vis diffuse reflectance spectra of two kinds of teas.

The results of the Luminous intensity test in visible light using the LED 626 are shown in Figure 8. Figure 8 shows the difference between the Luminous intensity passing through the water and the Luminous intensity passing through the teas.

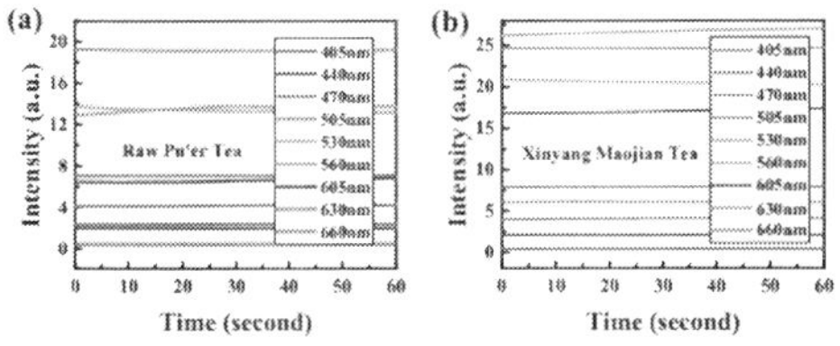

Figure 8. Luminous intensity of different light sources after pass through the two kinds of teas with Tyndall effect.

As shown in Figure 8, the Tyndall effect exhibited by 560nm light source in the Xinyang Maojian Tea is the strongest, followed by 505nm, and then 530nm, which is basically consistent with the effect observed by the naked eye. At the same time, the best Tyndall effect exhibited by Raw Pu'er Tea is at 505nm, followed by 530nm and 560nm. The bands with the strongest Tyndall effect values of different teas have a slight deviation, but they are generally in the green light band.

For the ultraviolet band, spectral tests and emission tests are added. The spectral test is from 420nm to 700nm, and the band in this range is tested to determine the visible light band mixed in the ultraviolet light source. The emission test tests the emission spectrum of Xinyang Maojian Tea and Raw Pu'er Tea at 345nm, 365nm, and 395nm respectively, to determine the emission band produced by ultraviolet to tea. The result of the spectrum is shown in Figure 9, the color coordinates and color temperature are shown in Table 2 and the result of emission spectrum is shown in Figure 10.

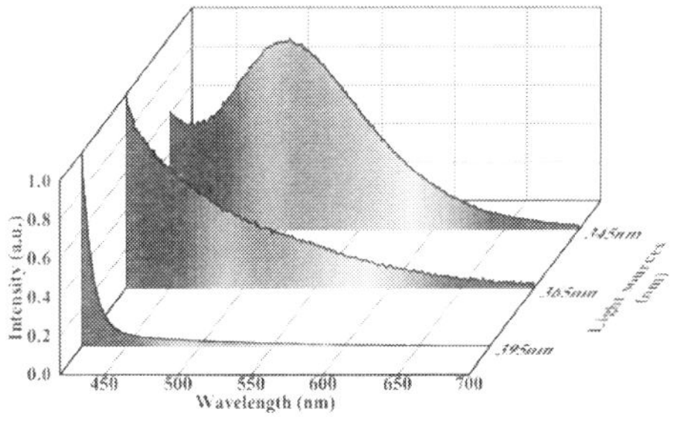

Figure 9. The Electroluminescence Spectra of stray lights in 345nm, 365nm and 395nm light sources.

Table 2. Color coordinates and color temperature.

Wavelength (nm)	Color coordinates (X, Y)	Color temperature (K)
345	0.2273, 0.3497	11717
365	0.2089, 0.2116	100000
395	0.1770, 0.0889	100000

As shown in Figure 9, it can be observed that there are visible light bands in 345nm, 365nm, and 395nm. According to the spectrum and color coordinates, the visible light band mixed with 345nm is biased towards green-white light, 365nm is biased towards blue-white light, and 395nm is biased towards violet light. In addition to the scattering in the visible light band, the Tyndall effect produced by the ultraviolet band also has an emission effect.

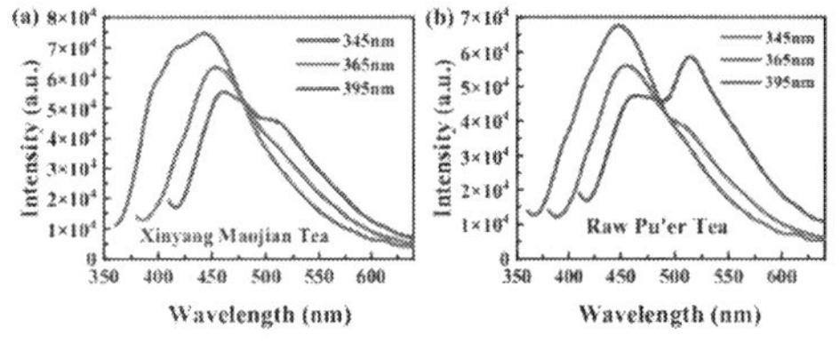

Figure 10. The PL emission spectra of two kinds of tea.

The stray light at 345nm is mainly green, so it can be clearly observed in Figure 5 and Figure 6 that the color of the Tyndall effect is mainly green, and according to the emission spectrum of Figure 10, the emission peak at 345nm is at 450nm. Therefore, the Tyndall effect observed in Figure 5 and Figure 6 includes the scattering of stray light and the emission effect at 345 nm. So the color of the Tyndall effect at 345nm in Xinyang Maojian Tea and Raw Pu'er Tea are green with a little blue. The color of stray light at 365nm is mainly white. Due to the emission peak in 365nm is at 455nm, so the color of Tyndall effect at 365nm in Xinyang Maojian Tea and Raw Pu'er Tea are white with a litter blue. The stray light in 395nm is purple with a litter white. According to the emission spectrums of 395nm, the emission peaks are at

525nm in Raw Pu'er Tea and at 460nm in Xinyang Maojian Tea. Therefore, the color of Tyndall effect at 395nm in Xinyang Maojian Tea is white with a litter blue while the color of Tyndall effect in Raw Pu'er Tea is green and purple mixed in white.

So, for the Tyndall effect in the ultraviolet band, it should be a combination of the scattering of the stray light in the visible band and the emission effect of ultraviolet band to tea. In addition, it can be observed from Figure 5 and Figure 6 that the Tyndall effect of a single band of visible light on tea is obviously not as strong as that of the ultraviolet band. There are two reasons for this. The first is that there is no obvious contrast between the background color generated by the visible light band and the color presented by the Tyndall effect, so the Tyndall effect is not obvious to the naked eye. Secondly, the wavelength of the light frequency in the visible light band is too narrow and the energy is low, compared with the emission and stray light in the ultraviolet band, its color rendering is slightly inferior. Based on the above two reasons, the Tyndall effect produced by the ultraviolet band in tea is the most obvious by naked eye.

Since the Tyndall effect is a physical phenomenon, the Tyndall effect generated in the ultraviolet band should be observed in the daylight environment. The Tyndall effect in daylight environment in ultraviolet band is shown in Figure 11. In Figure 11. As shown in the figure, the Tyndall effect of 395nm is the most obvious to the naked eye. Therefore, 395nm light has the strongest correlation with substances in tea.

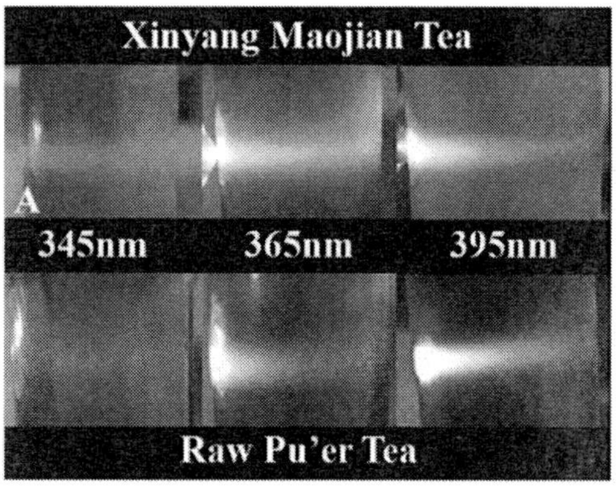

Figure 11. Tyndall effect of different teas at 345nm, 365nm and 395nm light sources in daylight environment.

In order to test the influence of tea concentration on the intensity of Tyndall effect, the 395nm light source with the best Tyndall effect was used to test the Luminous intensity of Xinyang Maojian Tea with different concentrations. The concentration of Xinyang Maojian Tea is controlled by changing the ratio between Xinyang Maojian and water. The specific ratio is shown in Table 3.

Table 4. Specific tea to water ratio.

Name	Xinyang Maojian	Water
0.5	0.5g	100ml
0.75	0.75g	100ml
1	1g	100ml
1.25	1.25g	100ml
1.5	1.5g	100ml

In a daylight environment, 395nm light source was used to irradiate Xinyang Maojian Tea with different ratios of tea to water, and the Tyndall effect was observed with the naked eye, as shown in Figure 12. Use the same method as above to test the Luminous intensity of Xinyang Maojian Tea with different concentrations. This time, the test Luminous intensity only discusses the relationship between concentration and Luminous intensity, so the scattering and emission effects in tea will affect the results, which is different from the visible light band. The test results are shown in Figure 12.

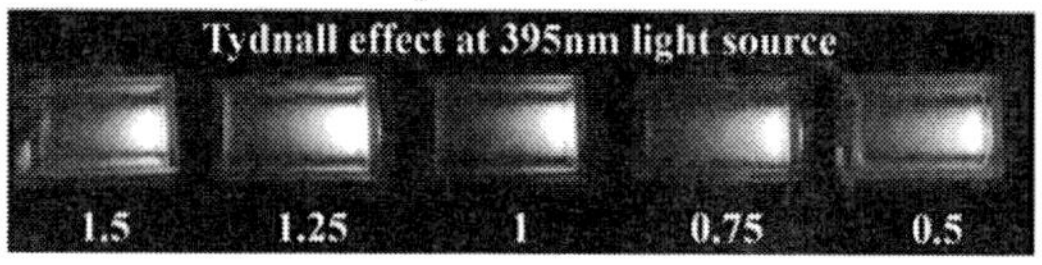

Figure 12. Tyndall effect of different concentrations of Xinyang Maojian Tea at 395nm light source.

As shown in Fig 12, it can be clearly observed that the Tyndall effect becomes weaker as the concentration of tea decreases, vice versa. Especially the intensity of the Tyndall effect at the two ratios of 0.5 and 1.5.

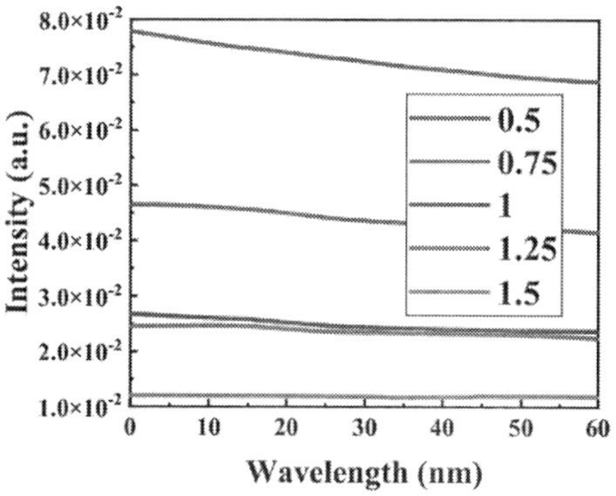

Figure 13. Luminous intensity of 395nm light source after passing through the different concentrations of Xinyang Maojian Tea.

As shown in Figure 13. It can be seen that the higher the concentration of tea, the smaller the Luminous intensity transmitted. This result proves that as the concentration increases, the transmitted light intensity gradually becomes weaker. The weaker the transmitted light, the stronger the Tyndall effect to a certain extent. This shows that there is a correlation between the Tyndall effect and tea concentration. The more flavor substances in the tea, the greater the concentration of the tea. And the flavor substances in the tea are the main factors that lead to the taste of tea. Therefore, the Tyndall effect can also be used to identify the taste and flavor of tea.

In addition, the Tyndall effect light path of different concentrations of tea is analyzed by using grayscale with MATLAB. One of the biggest benefits of using grayscale for Tyndall effect light path is that it can eliminate chromatic aberrations, exposure and some other inevitable matters caused by hardware during shooting. As shown in Figure 14, the light path of Tyndall effect after grayscale processing is very obvious, and the interference factors such as chromatic aberration, exposure are removed.

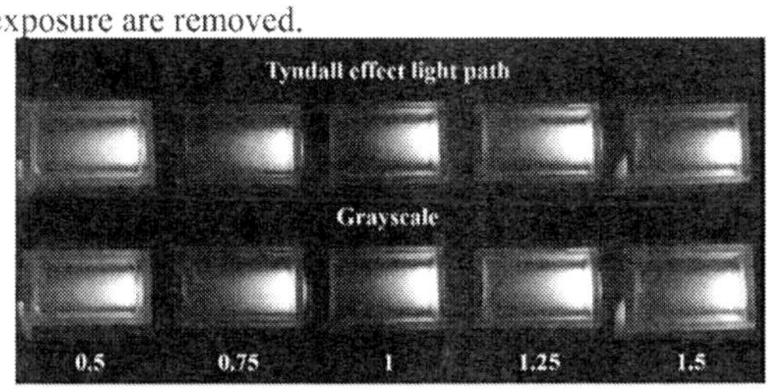

Figure 14. Grayscale image of tea with different concentrations.

As shown in Figure 14 and Figure 15, it can be clearly observed that the higher the concentration of tea, the weaker the scattered light (transmitted light) in the horizontal direction of the light source. The gray values of the red areas in Figure 15 are 107.3467 (0.5) and 88.3922 (1.5), respectively. This conclusion also confirms that the higher the concentration of tea, the weaker the transmitted light.

Figure 15. The gray area corresponding to the gray value.

5. Conclusion

In conclusion, an optical method based on the Tyndall effect is proposed to explore the relationship between tea concentration and scattered light. By testing the visible-ultraviolet spectrum and light intensity data, the conclusion was obtained that the Tyndall effect in the ultraviolet band is easier to observe than the visible band by using LED light sources with power 3w, because the ultraviolet LED has the scattering of visible light and the excitation of ultraviolet light. In addition, according to the results of image processing, the higher the tea concentration, the weaker the Tyndall effect. The changes in the concentration in tea leaves are mainly due to changes in the content of flavor substances and some compounds that determine the flavor of tea leaves. Therefore, the Tyndall effect can also be used to indirectly explore scattered light and tea flavor. This paper proposes a new research idea for using optical methods to study tea. In the future, the relationship between light and matter in tea will be studied.

Acknowledgements

1). National Key R&D Program of China (Grant No. 2021YFB3501700).
2). Shanghai Science and Technology Committee (STCSM) Science and Technology Innovation Program (Grant No. 22N21900400).
3). National Natural Science Foundation of China (Grant No. 12104311).
Conflict of Interest
The authors declare that they have no conflict of interest.

References

1. DAVID ROLL, JOANNA MALICKA, IGNACY GRYCZYNSKI, et al. Metallic colloid wavelength-ratiometric scattering sensors [J]. Analytical chemistry, 2003, 75(14): 3440-3445
2. ZHANG S., XUAN H., ZHANG L., et al. TBC2health: a database of experimentally validated health-beneficial effects of tea bioactive compounds [J]. Brief Bioinform, 2017, 18(5): 830-836
3. ZU-TONG QI, ZENG-MEI SUN. Identification of predominant species in brick tea [J]. Acta Mycologica Sinica, 1990, 03): 176-179
4. GAO W. J., YANG M., XIAO Y., et al. Rinsing Tea before Brewing Decreases Pesticide Residues in Tea Infusion [J]. Journal of Agricultural and Food Chemistry, 2019, 67(19): 5384-5393
5. WANG Z. H., ZHAO Z. X., HONG N., et al. Characterization of Causal Agents of a Novel Disease Inducing Brown-Black Spots on Tender Tea Leaves in China [J]. Plant Disease, 2017, 101(10): 1802-1811
6. ZHOU LIN. Development and application of EST-SNP molecular markers in tea and coffee [D]; Nanjing Agricultural University, 2017.
7. CEN XIONG, YEJING LIN, WEIGANG LI, et al. UHPLC-MS/MS Determination of 10 weight Loss Compounds in Slimming Health Foods [J]. PTCA(PART B: CHEM ANAL), 2017, 53(07): 780-786
8. SHENG-JIE ZHU. Studies on the Chromatography-Mass Spectrometry Analysis and Isolation of Glycoside Aroma Precursors in Tea [D]; Anhui Agricultural University, 2013.
9. JIAN ZHOU, HAO CHENG, JIAN-MING ZENG, et al. Study on Identification and Traceability of Tea Material Cultivar by Combined Analysis of Multi-

Partial Least Squares Models Based on Near Infrared Spectroscopy [J]. Spectroscopy and Spectral Analysis, 2010, 30(10): 2650-2653

10. MEI-HUA HONG, SAN-HE PENG, HONG-ZHI TANG, et al. Ultraviolet spectrophotometric method for quality identification of broken black tea [J]. Journal of Tea Communication, 1980, 02): 48-53
11. LEI H. B., HUANG X. Y., LUO J. J., et al. Rapid construction of superhydrophobic fabric coating assisted with 'Tyndall Effect' [J]. Surface Engineering, 2021, 37(10): 1277-1281
12. LIU H. Z., YANG X. Z., SU X. F., et al. Quality Identification of Sauce-Flavor Liquor Based on the Tyndall Phenomenon [J]. Applied Sciences-Basel, 2022, 12(1): 53
13. SUN Y., YUAN K. J., MO X. M., et al. Tyndall-Effect-inspired assay with gold nanoparticles for the colorimetric discrimination and quantification of mercury ions and glutathione [J]. Talanta, 2022, 238(part 1)

Basic Characteristic Analysis of Self-provided Power Plant in Iron and Steel Industry

Lingkai Zhu[1], Haijing Zhang [2], WeishuaiWang [2], QianWang [1], Wei Zheng [1], Ziwei Zhong [1]

[1] State Grid Shandong Electric Power Research Institute, Jinan, China
[2] State Grid Shandong Electric Power Company, Jinan, China

Abstract. Self-provided power plants are set up in order to maintain normal operation in some high energy-consuming industries in the era of power shortage in China. In a specific historical period, it effectively alleviated the problems of insufficient power supply and low reliability of the power grid, and promoted the rapid development of the national economy. The construction of self-provided power plants in the steel industry is to save electricity costs and improve the operational reliability of important units. This paper analyzes the basic production electrical characteristics and equipment load characteristics of the self-provided power plant in iron and steel industry, and summarizes the adjustment ability of the equipment from two aspects of power generation side and load side.

1. Introduction

China's self-provided power plants are produced in a specific historical period and under corresponding policies. Generally, large energy-consuming enterprises such as electrolytic aluminum, calcium carbide, steel, and cement invest in the establishment of self-provided power plants to provide self-balanced power supply systems within the enterprise, while providing relatively stable and low-cost power supply for end users[1].

Another purpose of the iron and steel industry to spend a lot of money to build its own power plant is to avoid the irreversible damage to the important units of the enterprise caused by the sudden power outage when the power off-grid accident occurs. More steel plant equipment, electricity load is more concentrated and the density is large, most of the electrical equipment is high voltage, high power, electrical equipment contains most of the high precision, continuous production load, so the continuity of power supply requirements higher[2]. Entering the 21st century, China's steel industry energy consumption accounted for one-tenth of the total energy consumption, electricity consumption is about ten percent of the total electricity.

For a long time, self-owned power plants have played an active role in comprehensive utilization of resources, alleviating the contradiction between power supply and demand, and improving the economic benefits of enterprises. However, some problems such as disorderly expansion, environmental protection supervision and safe operation have gradually been exposed in the development and operation. Therefore, it is urgent to carry out orderly rectification[2]. It is beneficial to improve the safety and reliability of the operation of the self-contained power plant and the power grid to carry out the interactive strategy of the self-contained power plant participating in the power grid[5]. Mastering the basic characteristics of the industry self-contained power plant is the premise of formulating a clear and reasonable and reliable interactive strategy of the self-contained power plant participating in the power grid.

2. Electrical characteristics of power plant basic production

Steel industry generally adopts twenty-four hours three shifts continuous working system, the annual load fluctuation is not big, no obvious peaks and troughs. In its process production, there are many continuous production equipment such as coke oven, sintering machine and blast furnace. In addition to the maintenance period, all kinds of equipment are usually running at full load, with high load rate and high requirements for power quality. Therefore, in the process of implementing demand response, fine management of adjustable load resources is needed. Typical daily load characteristic curve of iron and steel industry is shown in Figure 1.

* Corresponding author: zhulingkai@woyoxin.com

https://doi.org/10.1051/e3sconf/202337503019

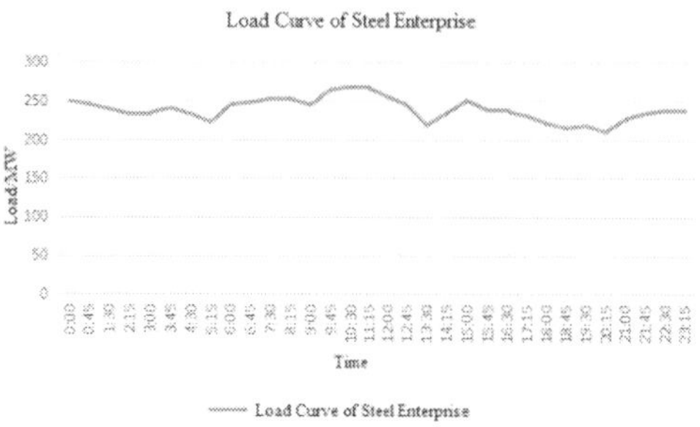

Figure 1. Typical Load Curve of Steel Enterprise.

3. Load Characteristics of Power Plant Equipment

The production load of iron and steel enterprises accounts for more than 70%of the total load, which is mainly composed of blast furnace, converter, continuous casting machine and other equipment. The auxiliary production load accounts for more than 8% of the total load, which is mainly composed of bar, wire fan, electric furnace fan, driving and other equipment. Safety load accounts for about 10% of the total load, which is mainly composed of exhaust gas, dust suction fan, circulating cooling water pump, fire and security equipment ; this kind of load loss may cause injury to personnel and loss of enterprise property.

Non-production load accounts for about 2%-5% of the total load, which is mainly composed of office equipment, central air conditioning and other equipment.

Steel enterprise equipment classification table is shown in Table 1 below.

Table 1. Iron and steel enterprise equipment classification table.

Load Category		Load Ratio	Major Installation
Production Load	Safety Load	10%-15%	dust recovery fan, circulating cooling water pump, Fire control and other equipment
	Main Production Load	55%-60%	electric furnace, refining furnace, oxygen machine, blast furnace, converter, etc
	Auxiliary Production Load	5%-10%	bars, wire fans, electric furnace factory fans, traffic, etc
Non-production load		2%-5%	office lighting, office appliances, central air conditioning, domestic electricity, etc

4. Analysis and Summary of Equipment Adjustment Ability

4.1 Power generation side potential analysis

Iron and steel industry power generation equipment is divided into pure coal-fired units, cogeneration units, waste heat and pressure utilization unit. Pure coal-fired power plant to follow the lack *of grid supply, surplus power grid* design principles, according to the basic power needs of enterprise production unit capacity selection. According to the agreement, the insufficient part of the peak power consumption is supplied by the power grid, and the excess part is transmitted to the power grid. When the load of electrical equipment on the production process becomes smaller, the power generated by the cogeneration unit may be reversed to the external power grid. However, the heat load in the production process of iron and steel enterprises will not change, so the change from the power grid to the power supply is only determined by the change of the load of the electrical equipment. The power generation of the waste heat and pressure unit is related to the characteristic production link in the production link, and the regulation performance is not high.

The power generation resources of the self-owned power plant can participate in the power generation right transaction, and the power generation capacity of the coal-fired unit is traded with the power generation right of the new energy enterprise, which greatly reduces the power generation cost and saves the coal consumption. At the same time, it reduces pollutant emissions and increases the consumption of new energy. The cogeneration unit has good regulation performance and low production cost. It can participate in peak shaving and valley filling, maintain the balance between supply and demand of the power grid, improve the frequency of the power grid, and promote flexible interaction with the power grid.

4.2 Load Side Potential Analysis

Iron and steel industry mainly ironmaking, steelmaking, cast steel, steel rolling four processes. The main equipment used in the process are reclaimer, ball mill, coke oven, sintering machine, blast furnace, converter and electric arc furnace, continuous casting machine, hot rolling mill, cold rolling mill and other auxiliary equipment.

The controllable load of iron and steel industry is mainly divided into productive load and non-productive load. The productive load includes electric furnace, refining furnace, oxygen making machine, steel rolling, bar and wire production line, among which electric furnace load accounts for the largest proportion, about 40%. The non-productive load includes office lighting, split and central air conditioning system and domestic electricity, accounting for a small proportion. The control mode is direct control (flexible), the preparation and recovery time can reach second level, and the response time is 0.5-2h. Adjustable productive load ratio is 19%, non-productive load is about 1%. Based on the analysis of typical user production data, the comprehensive regulation load accounts for about 20 % of the total production load in the steel industry when the production conditions permit. Controllable load can participate in the demand side response, participate in the power grid peak shaving interaction. Peak shaving is mainly by electric furnace, refining furnace, oxygen machine and non-productive load participation, reduce the peak load ; the load scheduling production is mainly carried out by the rolling steel, bar and wire production line, and the composite is transferred to the low load period to reduce the production cost. Load control is shown in Table 2 below.

Table 2. Load Control in Steel Industry.

Load Category	Major Installation	Load Ratio	Regulatory Mode	Regulatory Time: Setup Time	Regulatory Time: Response Time	Resumption Operation Time	Adjustable Load Ratio: Individual Effect	Adjustable Load Ratio: Overall Proportion	Total
Main production load	electric furnace	40 %	automatic	30 min	0.5h	30 min	10 %	19 %	20 %
	finery	3 %	automatic	30 min	0.5h	30 min	1.5 %		
	oxygen concentrator	2 %	automatic	10 min	0.5h - 1h	10 min	1 %		
	rolling production line	15 %	direct (rigid)	second grade	0.5h - 1h	second grade	5 %		
	bar production line	5 %	direct (rigid)	second grade	0.5h - 1h	second grade	2 %		
	rod production line	5 %	direct (rigid)	second grade	0.5h - 1h	second grade	2 %		
Non-production load		< 1 %	direct (rigid)	second grade	0.5h - 2h	second grade	<1 %	<1 %	
		< 1 %	direct (flexible)	second grade	0.5h - 2h	second grade	<1 %		
		< 1 %	direct (flexible)	second grade	0.5h - 2h	second grade	<1 %		

5. Generalize

To sum up, the self-owned power plant of iron and steel enterprises side coal-fired units and cogeneration regulation potential is excellent, can participate in new energy trading and peak shaving work. Electric furnaces, refining furnaces, steel rolling production and non-productive loads on the demand side have adjustment potential and can respond to electricity price incentives and compensation mechanisms to participate in grid interaction.

6. Acknowledgments

This work was financially supported by The Science and Technology Project of State Grid Corporation of China - Research on Flexible Interactive Technology of Autonomous Power Plants for Market-oriented Supply and Demand Adjustment (SGSDDK00YJJS1900143).

References

1. Yue Zhang, Zhongxian Dai, Jinli Lv, etc. Analysis on Related Policies and Development Problems of Self-provided Power Plant under Energy

Transformation[J]. Electrical Technology and Economics, 2020, (4):65-66.

2. Jianle Han. Research on the Characteristics of Power Supply and Distribution in Steel Plant and Its Integrated Automation System [J]. Electronic Science, 2012, (23):143.
3. Jiajia Yang. Main Problems and Policy Suggestions of Enterprise Owned Power Plant [J]. China Price Supervision and Inspection, 2012, (8):13-14.
4. Fenghu Wu. Summary of Some Problems in Development of Self-provided Power Plant[J]. Technological Diffusion, 2011, (11):97-98.
5. Lingkai Zhu, Weishuai Wang, Haijing Zhang, etc. Application Prospect of Self - provided Power Plant in Future Integrated Energy Field [J]. Shandong Electric Power Technology, 2022,49(291):12-16.

 https://doi.org/10.1051/e3sconf/202337503020

On the Importance of Self owned Power Plants Participating in the Adjustment of Supply and Demand of Power Network under the Reform of Power System

Lingkai Zhu [1], Haijing Zhang [2], Weishuai Wang [2], Qian Wang[1], Wei Zheng[1], Ziwei Zhong[1]

[1] State Grid Shandong Electric Power Research Institute, Jinan, China
[2] State Grid Shandong Electric Power Company, Jinan, China

Abstract. Centralized power generation by self owned power plants of high energy consumption enterprises will lead to insufficient peak shaving capacity of the power grid. Once the self owned power plants have operational problems, they will easily lead to large-scale blackouts, which will have a great impact on power supply reliability. The national electric power system reform opinion points out that the regulation focus of ensuring the balance of power supply and demand in the power grid should be shifted from the generation side to the load side, and the potential advantages of self owned power plants that combine power generation resources with large loads can be utilized. Self owned power plants participate in the equal management and fair competition of public power plants, can better play its role while improving the supervision of self owned power plants, which is conducive to maintaining the order of the power market and improving the stability of the implementation of power reform.

1. Introduction

As the name implies, self owned power plant enterprises generate electricity for their own use, and generally do not enter the national grid to provide electricity for residents. Most of these enterprises are energy consuming enterprises, which generally have surplus heat energy or provide steam for power generation[1]. Self owned power plants are produced in a special historical period, which meet the economic development needs of industrial policies in a specific era. The self owned power plant has both power generation resources and a large number of loads, but it has been in the state of self use for a long time. Its potential as an available resource for the balance of power supply and demand of the grid has not been tapped. At the same time, the centralized power generation of the self owned power plant will lead to insufficient peak shaving capacity of the grid. Once the operation problem occurs, it will have a disastrous impact on the power supply reliability. Therefore, the operation supervision of the self owned power plant must be continuously strengthened.

The promotion of new power transformation has laid a strong policy basis for scientific guidance and standardization of the development of self owned power plants. *Several Opinions of the CPC Central Committee and the State Council on Further Deepening the Reform of the Electric Power System* clearly puts forward that "control the middle, open up both ends", and will introduce more market competition at the power sales side[2]. The implementation of power sales side liberalization has become the focus of power generation enterprises. Enterprise owned power plants are an important part of China's power generation industry, and will also become a member of the power market reform to participate in market competition. *Guiding Opinions on Strengthening and Standardizing the Supervision and Management of Self provided Coal fired Power Plants* is a supporting document for the country to further deepen the reform of the power system and strengthen and standardize the supervision and management of self owned coal fired power plants. The introduction of this document is conducive to scientific guidance and standardization of the development of self owned power plants, promoting the equal management of self owned power plants and public power plants, achieving fair competition, effectively maintaining the order of the power market, and steadily promoting power reform[3].

In recent years, with the deepening of power system reform and the continuous improvement of marketization construction, some self owned power plant policies put forward by the state are being tightened step by step. The relevant guidance points out that no matter what the historical causes and institutional background of self owned power plants, they must stop their disorderly development, guide and standardize the development of self owned power plants, promote the joint management of all types of power plants, and popularize the auxiliary services of self owned power plants[4]. Deeply explore the regulation potential of self owned power plants, and carry out the strategy of self owned power plants participating in grid electricity and power interaction, which is conducive to strengthening the connection between the grid and self owned power plants, and

* Corresponding author: zhulingkai@woyoxin.com

improving the safety and reliability of self owned power plants and grid operation. The reasonable and reliable self owned power plant participating in the grid interaction strategy can strengthen the grid interaction and promote the safe and stable operation of the grid.

2. Development history and existing problems of self owned power plants

The development of self owned power plants in China can be divided into four stages: the first batch of self owned power plants were built by the government for specific industrial enterprises during the expansion of production scale; The second batch is at the weak stage of China's power infrastructure construction. In order to alleviate the problem of power shortage for large industrial users, the government encourages multi-channel financing for power generation and the development of local small thermal power plants; The third stage is to promote economic growth and create employment opportunities. The local government encourages the integration of self owned power plants and regional high energy consuming industries with investment promotion policies to make full use of local low-cost coal and other energy resources; The fourth stage is to standardize the operation and management of self owned power plants under the background of power reform and energy transformation[5,6].

Self owned power plants generally have low power generation efficiency of coal-fired units, idle environmental protection facilities and illegal operation. Taking Hebei Province as an example, according to *the Supervision Report on Air Pollution Prevention and Control Sites of Electric Power Enterprises in Hebei Province* issued by the National Energy Administration, in 2013, the average coal consumption of self owned generating units of 100000 kW and below in Hebei Province was up to 384.5 g/kWh, much higher than the average coal consumption of 321 g/kWh of coal-fired units nationwide. Some self owned power plants adopt intermittent operation of environmental protection facilities to reduce operating costs; Some self owned power plants for comprehensive utilization of resources adopt coal blending incineration measures to increase power generation and utilization hours; Some cogeneration units fail to meet the heat to power ratio and thermal efficiency standards set by the state[7]. These illegal means enable some enterprises with their own power plants to obtain lower power consumption costs than similar enterprises, gain unfair competitive advantages, hurt the enthusiasm of public power plants and the whole society to implement energy conservation and emission reduction, and will have a negative impact on the country's prevention of air pollution and building a harmonious society of ecological civilization.

Self owned power plants in some regions also have some outstanding problems, such as large proportion of power generation, bad operating environment and low level of safe operation. On the one hand, a large number of self owned power plants generate electricity in a centralized manner, resulting in insufficient peak shaving capacity of the power grid. Once operating problems occur, it is very easy to cause large-scale blackouts, which will have a significant impact on social stability. In some regions, for example, the installed capacity of self owned power plants in Xinjiang accounts for 26% of the total installed capacity in the region, and the average annual utilization hours of power generation is nearly 1493 hours higher than that of public units; In 2014, in Gansu Province, two thirds of power generation enterprises shut down in turn, while self owned power plants were basically fully operated; Heilongjiang and Jilin have insufficient peak shaving capacity due to their own power plants. Especially in the winter heating period, it is difficult to arrange the minimum startup mode, which reduces the safe operation level of power supply and consumption[8,9]. In some regions, enterprises rely on their own power plants to build power grids in violation of regulations, or even cross with public power grids in violation of regulations, which brings serious hidden dangers to the safe operation of power grids. On the other hand, enterprises with their own power plants are often important users above the national level II, and have a large number of important loads that can not be interrupted to supply power. However, the operation level of their own power plants is relatively weak, especially the self owned power plants operating in isolated networks. Without the backup of the power grid enterprise system, the safety problems of their own power plants are extremely prominent.

For the government funds and surcharges unpaid by the self owned power plants, the collection of system reserve fees needs to be further standardized. In order to promote the development of public utilities and maintain a fair market order, the State makes reasonable compensation for the power consumption of key people's livelihood such as agricultural production, residents' lives and remote areas, and clearly stipulates that self owned power plants need to bear funds and surcharges, and pay system reserve capacity fees. The actual implementation effect of self owned power plants is not good. Some self owned power plants with comprehensive utilization of resources and cogeneration refuse to pay on the pretext that the relevant national ministries and commissions have differences in the project and scope of government fund collection for such self owned power plants; Some provinces and cities have even made it clear in the local implementation rules that they will not levy or only levy a few of them, especially for self owned power plants operating in isolated networks, and will not pay any government funds; Some enterprises refuse to pay the reserve capacity fee of their own power plant system[10]. The behavior of these enterprises and self owned power plants has destroyed the market order of fair competition, evaded the national obligation to subsidize the people's livelihood, damaged the national interests, and played a negative role in guiding, which has led some public power plants to request to turn to self owned power plants to avoid fees.

 https://doi.org/10.1051/e3sconf/202337503020

3. Necessity Analysis of Self owned Power Plant Participating in Supply and Demand Regulation of Power Grid

The self owned power plant has now reached a critical period of decision. It will continue to face challenges in environmental protection supervision, supplementary payment of energy efficiency supervision funds, safe operation and other aspects. If it is closed for operation, the business development of the enterprise will be changed or even difficult to sustain. Self owned power plants are mostly multi energy cogeneration, with the effect of comprehensive utilization of resources. They play an active role in comprehensive utilization of resources, easing the contradiction between power supply and demand, and improving the economic benefits of enterprises. In the face of the continuous promotion of power system reform and the continuous improvement of market-oriented construction, self owned power plants need to be rectified in an orderly manner. Their participation in power grid supply and demand regulation is of great significance in improving the efficiency of energy comprehensive utilization, improving the supply and demand balance of regional power grids, and improving the energy consumption and environmental protection level of self owned power plants.

It is conducive to tapping the huge potential of self owned power plants in supply and demand regulation and improving the comprehensive utilization efficiency of energy resources. The research on the behavior characteristics and flexibility evaluation methods of self owned power plants can effectively tap and utilize the potential of self owned power plants in supply and demand regulation. The comprehensive utilization of resources can be realized by mastering the operation technical characteristics and comprehensive utilization of resources of the self owned power plant, formulating the grid safe operation management strategy, analyzing the behavior characteristics of the self owned power plant based on the collected data and using the data analysis method, and establishing the flexibility evaluation index system and evaluation method for the self owned power plant to participate in the grid interaction. It will lay a foundation for self owned power plants to orderly participate in project market and auxiliary service transactions. To some extent, it can also smooth the power demand curve of the power grid and improve the economic operation level of the power grid.

It is conducive to promoting the government to establish a mature and perfect market mechanism, assisting in solving the imbalance between supply and demand of regional power grids, and providing relevant policy recommendations to the government. The establishment of the market mode for self owned power plants to participate in grid auxiliary services and the proposal of the technical specification for self owned power plants to participate in flexible interaction can effectively guide self owned power plants to reasonably participate in power market transactions, and promote self owned power plants to participate in the real-time supply and demand balance of the grid by market-oriented means. By proposing the price mechanism that the power grid provides reserve capacity, reserve electricity and reliability for users of self owned power plants, the self owned power plants are encouraged to actively participate in the interactive market of the power grid. At the same time, in case of large renewable energy generation in the distance, greater renewable energy can be absorbed through the interaction between the self owned power plant and the power grid to help solve the imbalance between regional supply and demand.

It is conducive to improving the professional operation and management level of self owned power plants, improving the high energy consumption level of self owned power plants, and improving the environmental protection level of the whole society. The establishment of market mechanism and platform for flexible interaction between self owned power plants and power grid supply and demand can improve the environmental protection level of self owned power plants. By strengthening the overall planning and access standards of self owned power plants, strengthen operation management and energy efficiency and environmental protection supervision, use national industrial policies and energy efficiency and environmental protection standards to force self owned power plants to form a benign development mechanism, improve the professional operation management level and supervision ability of coal-fired self owned power plants, improve the energy consumption level of self owned power plants and reduce environmental emissions; By promoting power generation enterprises to participate in direct transactions with users, and on the basis of implementing the social responsibilities of self owned power plants such as government funds, policy based cross subsidies, and pollutant emission control, we will give full play to the power generation cost advantages of clean and efficient units, and improve the environmental protection level of the whole society.

It is conducive to improving the overall power service quality of the public and promoting social and economic development. The establishment of a flexible interactive platform between self owned power plants and power grid supply and demand can effectively improve the energy efficiency of self owned power plants. The establishment of an interactive potential assessment model can effectively monitor the environmental friendliness of self owned power plants and improve the environmental protection level of self owned power plants. At the same time, the research on the market mechanism of flexible interaction between self owned power plants and power grids can promote self owned power plants to participate in the real-time supply and demand balance of power grids through market-oriented means, It ensures the overall power service quality of the public under fault and stress conditions. The establishment of market mechanism can also provide relevant policy recommendations for relevant government departments to formulate policies on self owned power plants.

4. Summary

This paper describes the development history and existing problems of self owned power plants, and analyzes the importance of self owned power plants participating in the supply and demand regulation of the power grid in combination with the new power reform policy. The development direction of the supply and demand interaction between self owned power plants and the electricity market is clear, but through the analysis of existing research results and the investigation of several self owned power plant projects, it is found that due to the imperfect method system of flexible interaction design and operation optimization between self owned power plants and market-oriented supply and demand regulation, the actual operation effect of most demonstration projects has not reached the expected goals. The main reasons are as follows: the theoretical research on the behavior characteristics and flexibility evaluation methods of self owned power plants is still insufficient, and there is a lack of detailed operation technology characteristic model of multi type self owned power plants, leading to the lack of planning means; Inadequate evaluation and analysis of power generation and consumption behavior characteristics of self owned power plants, lack of analysis of online monitoring technology and external related data, resulting in unclear objectives for flexibility improvement; The research on the operation method of market mechanism coordination and optimization is insufficient; Inadequate policy guidance and low willingness of self owned power plant users to participate in market-oriented supply and demand interaction services. Therefore, it is the focus and direction for self owned power plants to participate in the research on power grid supply and demand regulation to carry out research on the operating characteristics, evaluation, market mechanism, etc., and to form an operable and scalable evaluation system and market mechanism based on the results of theoretical and technical research of demonstration projects.

Acknowledgments

This work was financially supported by The Science and Technology Project of State Grid Corporation of China - Research on Flexible Interactive Technology of Autonomous Power Plants for Market-oriented Supply and Demand Adjustment (SGSDDK00YJJS1900143).

References

1. Fenghu Wu. Summary of some problems in the development of self owned power plants[J]. Science and technology communication, 2011,(11):97-98.
2. Juan Yu. Deepen the reform of power system and establish a new power system with new energy as the theme[J]. Price in China, 2021,(11):82-84.
3. Shijun Liu, Qing Cai, Yuhui Bi, et al. Research, Judgment and Development Direction of Restrictive Policies for Enterprise Owned Thermal Power Units[J]. Baosteel Technology, 2016,(5):6-12.
4. Shuang Weng. Self owned Power Plant: How to Realize Nirvana and Rebirth[J]. China Electric Power Enterprise Management, 2018,(11):22-24.
5. Yue Zhang, Xianzhong Dai, Jinli Lv, et al. Analysis on relevant policies and development issues of self owned power plants under energy transformation [J]. Electrical technology and economy, 2020,(4):65-66.
6. Yifan Huang , HaijingZhang, Lei Wang, et al. Evaluation method for flexibility regulation of self owned power plants based on enterprise production characteristics[J]. Demand Side Management, 2021, 23 (1): 61-66.
7. Jianlin Shao, Minghui Zheng, Chenghao Guo, et al. Analysis on the application of energy storage technology for coal-fired cogeneration units under the dual carbon target[J]. Southern Energy Construction, 2022,9 (3): 102-110.
8. Lei Ma, Chao Zhang, Bo Yuan, et al. Analysis and Suggestions on the Development of Xinjiang Self owned Power Plants [J]. Electric Times, 2019 (10): 34-37.
9. Gaofeng Qin, Haitao Liu, Xu Li, et al. Research on the trading mode of self owned power plants participating in new energy consumption power generation rights[J]. Journal of Nanjing Institute of Engineering (Natural Science Edition), 2018,16 (3): 7-13.
10. Jiajia Yang. Main Problems and Policy Suggestions of Enterprise Owned Power Plants[J]. China Price Supervision and Inspection, 2012, (8): 13-14.

Test and Research on Load Capacity of a 210MW Unit in Heating State

Ziwei Zhong[1,*], Lingkai Zhu[1], Wei Zheng[1], Zhiqiang Gong[1], Panfeng Shang [1], Junshan Guo[1], Jun Liu [1], Yanpeng Zhang[1]

State Grid Shandong Electric Power Research Institute, Jinan, China

Abstract. As large-scale photovoltaic and wind power are integrated into the power grid, new energy power generation has strong randomicity, volatility and anti-peak shaving characteristics. Both the power supply side and the load side of the power grid will have random fluctuations, which brings huge challenges to the safe and stable operation of the power grid. The power grid will need more capacity and more flexible peak shaving resources as the basis for ensuring the smooth and orderly power supply. During the heating period, the thermoelectric units are faced with double pressures of power consumption peak and heating peak, and are forced to further increase and reduce output to meet the requirements of heating. Therefore, carrying out load capacity test on heating units is conducive to finding out the real output range of units and providing data support for power grid departments.

key word: Deep peak shaving; Heating period; Electric load adjustment section

1. Introduction

The turbine model of a power plant is K-210-130-1 super high pressure, one intermediate reheat, three cylinder double exhaust condensing type. The unit was put into operation in December 1989. In 2001, Alstom technology and equipment were used to retrofit the flow path of the low pressure part. After the retrofit, the original Bowman stage was cancelled, the last stage blade was lengthened from 765mm to 916mm, and the stages of the low pressure rotor remained unchanged at 2×Level 4, the nameplate of the unit is increased from 210MW to 225MW. The rated main steam pressure of this type of turbine is 12.75MPa, the rated main steam temperature is 540℃, and the rated reheat steam temperature is 540℃. The unit is equipped with 7-stage regenerative heater, including 3 high-pressure heaters, 1 deaerator and 3 low-pressure heaters. The design cooling water temperature is 20℃, and the rated back pressure is 5.0 KPa. Table 1 is the main technical specifications of steam turbine.

Table 1. Main design technical specifications of steam turbine

Project	Design parameters	Unit
Rated power	225	MW
Rated steam flow	646	t/h
Main steam pressure	12.75	MPa
Main steam temperature	540	℃
Reheat steam pressure	2.44	MPa
Reheat steam temperature	540	℃
Rated back pressure	5.0	kPa
Stages of regenerative steam extraction	3JG+1CY+3JD	
Cooling water temperature	20	℃
Feed water temperature	248.8	℃
Unit heat consumption	8700	kJ/kWh

* Corresponding author: dkyzhongziwei@163.com

https://doi.org/10.1051/e3sconf/202337503021

In November 2018, the heat supply transformation of the low pressure cylinder optical shaft back press of Unit # 3 was implemented, and the heat supply capacity was increased by 334MW, which can give full play to the heat supply capacity of Unit 3, minimize the impact of heat supply on the power generation capacity of Phase III units, improve the comprehensive energy utilization efficiency, energy conservation and consumption reduction, and have good social and economic benefits.

The smooth shaft transformation of the low pressure cylinder of the steam turbine adopts the double rotor scheme. The low pressure rotor of Unit 4 is processed into a smooth shaft, which is used for heating in winter. The original low pressure rotor of Unit 3 is used for pure condensation operation in summer. At the same time, the diaphragm in the low pressure cylinder is disassembled and assembled accordingly. The installation elevation of the front and rear bearings of the low pressure cylinder is adjusted. The half couplings at both ends of the rotor can be interchanged, The accuracy of the bolt holes of the coupling shall be completely consistent in terms of centering or two halves. The optical axis rotor of the double rotor scheme can meet the safe operation of the unit under back pressure during the heating period, and the pure condensing rotor can ensure the load carrying capacity and operating economy of the pure condensing operation in summer.

The steam exhaust from the intermediate pressure cylinder of the steam turbine is no longer discharged into the low pressure cylinder through the connecting pipe of the intermediate and low pressure cylinder, but is discharged to the newly built initial heating station of unit 3. The original connecting pipe of the intermediate and low pressure cylinder is replaced with a newly designed exhaust short pipe, and a connecting flange is added. The newly added heat supply pipeline is connected from the connecting flange, and then led to the two new heat supply network heaters through the regulating valve, safety valve, etc. After the smooth shaft transformation of the low pressure cylinder of the turbine, the amount of steam entering the low pressure cylinder is zero. In order to ensure the impulse starting of the unit, a new ventilation pipe system from the intermediate pressure cylinder to the condenser is added, which is connected to the DN500 pipe behind the original large bypass pressure reducer, and then connected to the condenser.

Double-rotor scheme is adopted for the reconstruction of the light axis of the low pressure cylinder of the steam turbine. The low pressure rotor of Unit 4 is processed into a light axis, which is used for heating in winter. The low pressure rotor of the original Unit 3 is used for condensing operation in summer. At the same time, the diaphragms in the low pressure cylinder are disassembled and disassembled accordingly. The installation elevation of the front and rear bearings of the low pressure cylinder is adjusted. The half couplings at both ends of the rotor can be interchanged, The accuracy of the bolt holes of the center or the two halves of the coupling shall be completely consistent. The optical-shaft rotor of the double-rotor scheme can meet the back-pressure safe operation of the unit during the heating period, and the pure condensing rotor can ensure the load capacity and operation economy of the pure condensing operation in summer.

The exhaust steam from the intermediate and low pressure cylinder of the steam turbine will no longer be discharged into the low pressure cylinder through the connecting pipe of the intermediate and low pressure cylinder, but will be discharged to the new initial heat supply station of Unit 3. The original connecting pipe of the intermediate and low pressure cylinder will be replaced with the newly designed exhaust short pipe, and the connecting flange will be added. The newly added heat supply pipeline is connected from the connecting flange, and then led to two new heating network heaters through the regulating valve, safety valve, etc. After the modification of the light axis of the low pressure cylinder of the steam turbine, the amount of steam entering the low pressure cylinder is zero. In order to ensure the need of impulse starting when the unit is started, a new ventilation pipe system from the intermediate pressure cylinder to the condenser is added, which is connected to the DN500 pipe behind the original large bypass pressure reducer, and then connected to the condenser.

2. On load capacity test process under heating condition

2.1 Test conditions

Before the test, due to the high temperature in this winter and the limitation of the current heating area, the high load capacity of the unit is limited. According to the actual situation of the unit and in combination with the typical working conditions of the unit's designed heating working condition diagram, the tester determined the following three working conditions. The operating modes of each working condition are rated parameters, and the power is determined by heat. After the subsequent conditions are met, the maximum electrical load working condition of the unit can be verified again:

Under working condition 1: Under 110MW working condition, adjust the heating butterfly valve to make the steam extraction pressure (intermediate exhaust pressure) close to the rated value 0.16MPa, and measure the main steam flow and external steam extraction flow of the unit.

Working condition 2: Under 90MW working condition, adjust the heating butterfly valve to make the steam extraction pressure (intermediate exhaust pressure) close to the rated value of 0.16MPa, and measure the main steam flow and external steam extraction flow of the unit.

Working condition 3: Under 77MW working condition, adjust the heating butterfly valve to make the steam extraction pressure (intermediate exhaust pressure) close to the rated value of 0.16MPa, and measure the main steam flow and external steam extraction flow of the unit.

DCS historical data are used for generator active power, auxiliary power rate, main steam flow, main steam pressure, main steam temperature, reheat pressure, reheat temperature, environmental protection parameters, condenser vacuum, heating extraction flow, heating extraction pressure, heating extraction temperature and other parameters. The collection period is 30s, and the test duration is 2 hours. Steam and water analysis and coal quality analysis shall adopt the statements of the power plant.

2.2 Test process

Working condition 1

Slowly bring the load to 110MW, adjust the main steam pressure and temperature as close as possible to the design value, adjust the opening of the heating butterfly valve to make the intermediate exhaust pressure close to the rated exhaust pressure of 0.16MPa, and keep the unit load, main steam pressure, main steam temperature and other parameters stable for 30 minutes. The effective time of the test under this working condition is 14:30-15:30 on January 9, 2019. During the test, the average active power of the generator of Unit # 3 is 113.35MW, the average main steam flow is 485.786t/h, and the external steam extraction flow (calculated by using the drainage flow of the heating network heater) is 345.554t/h. Under this working condition, the unit operation parameters are normal; No overheating of heating surface is found in the boiler; The auxiliary machine works normally; The environmental protection device works normally, the environmental protection indicators are qualified, the water supply temperature of the heating network is limited, and the main steam flow has reached the maximum value that can be reached at present, so it is determined that this working condition is the current maximum electrical load working condition of the unit.

Working condition 2

Slowly bring the load to 90MW, adjust the main steam pressure and temperature as close as possible to the design value, adjust the opening of the heating butterfly valve to make the intermediate exhaust pressure close to the rated exhaust pressure of 0.16MPa, and keep the unit load, main steam pressure, main steam temperature and other parameters stable for 30 minutes. The effective time of the test under this working condition is 11:30-15:30 on January 8, 2019. During the test, the average value of the active power of the generator of Unit # 3 is 90.79MW, the average value of the main steam flow is 399.702t/h, and the external steam extraction flow (calculated by using the drainage flow of the heating network heater) is 282.569t/h. Under this working condition, the unit operation parameters are normal; No overheating of heating surface is found in the boiler; The auxiliary machine works normally; The environmental protection device works normally and the environmental protection indicators are qualified.

Working condition 3

Slowly bring the load to 77MW, adjust the main steam pressure and temperature as close as possible to the design value, adjust the opening of the heating butterfly valve to make the intermediate exhaust pressure close to the rated exhaust pressure of 0.16MPa, and keep the unit load, main steam pressure, main steam temperature and other parameters stable for 30 minutes. The effective time of the test under this working condition is 10:30-12:30 on January 9, 2019. During the test, the average value of the active power of the generator of Unit # 3 is 78.36MW, the average value of the main steam flow is 357.342t/h, and the external steam extraction flow (calculated by using the drainage flow of the heat network heater) is 259.227t/h. Under this working condition, the unit operation parameters are normal; No overheating of heating surface is found in the boiler; The auxiliary machine works normally; The environmental protection device works normally, the environmental protection indicators are qualified, and the main steam flow has been 374t/h lower than the minimum designed stable combustion flow of the boiler. Therefore, it is determined that this working condition is the lowest electrical load working condition of the unit.The test points are arranged according to the requirements of ASME PTC 6-2004 test procedure.

2.3 Main limiting conditions for the test of extraction condensing unit:

1) Different thermal users have different requirements for steam extraction parameters, and the steam extraction parameters of the unit shall not exceed their limits.
2) The minimum electrical load of the unit is limited by the minimum steam intake of the low-pressure cylinder, the minimum stable combustion capacity of the boiler, the inlet flue gas temperature of the denitration device, the wall temperature of the heating surface, and environmental protection parameters.
3) For the reheater unit with steam extraction from the HP cylinder steam exhaust pipe, when the electrical load is low and the heat supply steam extraction volume is large, it is easy to occur that the reheater wall temperature exceeds the temperature, thus limiting the minimum electrical load of the unit.
4) For units that extract steam from the connecting pipe of the intermediate and low pressure cylinders, the exhaust pressure and temperature of the intermediate pressure cylinder shall not exceed the limits of the manufacturer.
5) The maximum electrical load of the unit is limited by the maximum evaporation capacity of the boiler, the wall temperature limit of the boiler heating surface, the quality of steam and water, the design output of the boiler denitration, desulfurization, dust removal and slag removal devices, and the design output of the main auxiliary equipment, including the pulverizing system, the six major fans, environmental protection parameters and the maximum steam inlet flow of the steam turbine.
6) The monitoring parameters affecting the unit safety shall not exceed the alarm value.

3. Conclusion

During the test of the unit #3, the minimum external heating steam extraction flow is 259.227t/h, and the corresponding unit electrical load is 78.36MW MW; The maximum external heating extraction flow is 345.554t/h, and the corresponding unit electrical load is 113.35MW.The load capacity test of the heating unit is conducive to understanding the actual output range of the unit and providing data support for the power grid department.

https://doi.org/10.1051/e3sconf/202337503021

Acknowledgements

This work is supported by the science and technology project of Electric Power Research Institute of State Grid Shandong electric power company(ZY-2022-15), which studies Research on peak capacity improvement of thermal power unit based on electric heating coordination optimization under guaranteed supply demand.

References

1. Shi Qingli, Wen SheJiao.Application of thermal test in thermal power plants [J].Qinghai Electric Power, 2004, 23 (4): 4-10.
2. Weng Sicheng.Thermal performance test of SAIC 300MW steam turbine (Wangting Power Plant Unit 12) [J]. Shanghai Turbine, 1983 (01): 5-20.
3. Ji Ansen.Results and Analysis of Thermal Performance Test of Unit #12 in Wuhu Power Plant before Overhaul [J].Anhui Electric Power Technical Information, 2000, 000 (002): 7-8.
4. Wang Xuedong, Wang Xuetong, Yang Jianzhu, etc.Energy saving transformation and thermal performance assessment test of flow passage of Unit 2 in Laiwu Power Plant [J].Shandong Electric Power Technology, 1998 (4): 4-12.
5. Wang Jizhou Study on thermal system mechanism model and off design operating characteristics of coal-fired power generation units [D].Huazhong University of Science and Technology, 2015.
6. Wang Wei.Analyze the application of thermal test in thermal power plants [J].China Science and Technology Wealth, 2012, 000 (014): 214-214.

A 1000MW ultra-supercritical secondary reheat unit Energy consumption analysis of thermodynamic system

Hu Zhang *, Jinxu Lao, Huitao Cheng, Kai Liang

State Grid Shandong Electric Power Research Institute, Jinan250003, China

Abstract: Taking the thermodynamic system of a 1000MW ultra-supercritical secondary reheating unit as the research object, the modeling energy consumption analysis was carried out, and the spatial distribution of energy consumption was determined, so as to provide a scientific basis for the operation optimization and energy-saving transformation of the unit. The analysis results show that: Boiler is one of the largest equipment factory irreversible loss, make up the loss by 85.1% of the total loss of the system, although the number of its external loss is not big but internal loss is very big, and the loss is high grade of energy, the number of the steam turbine cold source loss although large irreversible loss is not big but the grade is not high, its loss make up only 2.32% of the total loss system.

Key words: ultra-supercritical, thermodynamic system, secondary reheat unit, energy consumption analysis, Non reversible loss, Purpose (exergy) efficiency, Exergy loss coefficient.

1. Introduction

Under the guidance of national energy conservation and emission reduction policy, high-efficiency and low-pollution power generation technology has become the mainstream. Improving steam parameters and increasing reheating times are effective ways to improve unit efficiency, and ultra-supercritical secondary reheating technology is an important technical means to improve thermal economy of generating units[1].Compared with the supercritical unit, the thermal efficiency of the ultra-supercritical unit can be increased by 3%-4%, and the thermal efficiency of the whole plant is higher than 50% by adopting the secondary reheating technology[2-4].The waste heat recovery system is generally designed for the secondary reheating ulra-supercritical units, which further optimizes the thermal system, boiler efficiency, pipeline efficiency, turbine flow efficiency and plant power consumption rate[5].

The traditional method of heat balance analysis is based on the first law of thermodynamics. The inner efficiency of boiler and the absolute inner efficiency and cycle efficiency of steam turbine are used to describe the closeness of the actual steam cycle to the ideal cycle.However, the traditional thermal balance analysis method only focuses on the quantity balance of energy without considering the difference of energy in "quality", only calculates the external loss of the system but ignores the internal loss, and ignores the internal loss caused by the irreversibility of the process.Therefore, this method only shows the result of energy conversion, but cannot reveal the essential cause of energy loss, cannot accurately point out the location of energy saving potential, and sometimes even gives false appearance[6].For example, in the analysis of the condenser as a circulating cold source, the traditional heat balance analysis method thinks that most of the system loss occurs here, and it seems that the main energy saving space of the power plant is also here,But exergy analysis, based on the second law of thermodynamics, suggests that the loss here is not great, because the amount of heat emitted from the condenser is large but the energy of that heat is not high.In contrast, the conventional thermal balance analysis method calculates that the loss of the boiler is not large, and the efficiency of the boiler in the power plant can reach 90% or more, but the exergy analysis method considers the loss of the boiler to be large, because the amount of energy lost by the boiler is small but the loss of high grade energy[7-10].In this paper, a 1000MW ultra-supercritical secondary intermediate reheating unit is taken as the research object. Based on the first and second laws of thermodynamics, the loss of work capacity due to irreversibility in the cycle process is analyzed, and the location of energy loss is accurately located, so as to provide a reference for targeted energy saving and potential exploration.

* Corresponding author: 13075336627@163.com

2. Unit energy consumption analysis evaluation index

2.1 Purpose (exergy) efficiency of a device or system:

$$\eta_p = \frac{\Delta E_{out}}{\Delta E_{in}} \quad (1)$$

Where, ΔEout is the net income related to the purpose exergy and ΔEin is the net consumption to achieve the purpose exergy.The above equation is a general expression for exergy efficiency,The expression of specific exergy efficiency varies depending on the purpose for which the system or device is used.For example, ΔEout of the boiler plant is the feed water and reheated steam obtained in the boiler exergy, ΔEin is the fuel chemistry exergy.For the engine island, ΔEout is the output shaft work and ΔEin is the net input exergy.ΔEout of the heat exchanger device is the cold fluid obtained in the heat exchanger exergy, ΔEin is the hot fluid reduced in the heat exchanger exergy.

2.2 Exergy loss factor of a device or system:

The thermal system is divided into several subsystems, in which the ratio of the exergy loss of any subsystem to the total exergy consumption of the system is the exergy loss coefficient of the subsystem.For a power plant, the total exergy consumption of the system is the chemistry exergy of the fuel.

3. Unit energy consumption analysis example

3.1 Maching series Data

The boiler of a 1000MW ultra-supercritical secondary reheating unit in China adopts Single furnace tower arrangement, Tangential burning at four corners, swing and tail flue baffle temperature regulation, balanced ventilation, steel frame suspension structure, semi-open air arrangement, and AIR-cooled dry slag removal system.The steam turbine adopts a single-axis five-cylinder four-row steam mode, and a two-stage external steam cooler is used in the high-pressure heater side to make cross-stage utilization of the superheat energy.The main design parameters of boilers and steam turbines are as follows:

Table1 Boiler design parameters

Items	Unit	THA
Superheated steam flow rate	kg/s	710.349
Outlet pressure of superheated steam	MPa	31.62
Outlet temperature of superheated steam	°C	605
Primary reheat steam flow rate	kg/s	632.624
Primary reheat steam inlet pressure	MPa	10.4
Primary reheat steam inlet temperature	°C	424
Primary reheat steam outlet pressure	MPa	10.19
Primary reheat steam outlet temperature	°C	623
Secondary reheat steam flow rate	kg/s	544.852
Secondary reheat steam inlet pressure	MPa	3.3
Inlet temperature of secondary reheat steam	°C	445
Secondary reheat steam outlet pressure	MPa	3.06
Secondary reheat steam outlet temperature	°C	623
Feed water temperature	°C	324
Economizer inlet pressure	MPa	35.17
coal	t/h	376.3
Low calorific value of fuel	kJ/kg	19850
Total fuel moisture	%	16.1
ow thermal efficiency	%	95.03

Table 2 Steam turbine design parameters

Items	Unit	THA
power	MW	1000
main steam flow	kg/s	710.349
main steam pressure	MPa	30.138
main steam temperature	°C	600
High pressure cylinder inlet steam flow rate	kg/s	632.624
Exhaust steam pressure of ultra-high pressure cylinder	MPa	10.643
Exhaust steam temperature of ultra-high pressure cylinder	°C	425.5
Steam inlet pressure of high pressure cylinder	MPa	10.004
Inlet steam temperature of high pressure cylinder	°C	620
Steam inlet flow of medium pressure cylinder	kg/s	544.852
Exhaust steam pressure of high pressure cylinder	MPa	3.455
Exhaust steam temperature of high pressure cylinder	°C	447
Steam inlet pressure of medium pressure cylinder	MPa	3.041
Inlet steam temperature of medium pressure cylinder	°C	620
Back pressure	kPa	3.8
Feed water temperature	°C	325
Feed water pressure	MPa	36.095
Steam turbine heat consumption rate	kJ/kW.h	7038

Relevant (exergy) parameters are calculated according to the unit design parameters as shown in the following table:

https://doi.org/10.1051/e3sconf/202337503022

Table 3 THA unit exergy analysis related parameters table

Items	number	flow rate F kg/s	enthalpy h kJ/kg	Entropy s kJ/kg.°C	exergy e kJ/kg
superheated steam	1	710.349	3447.59	6.21762	1627.795
One reheat steam outlet	2	632.624	3680.83	6.95838	1643.881
Secondary reheat steam outlet	3	544.852	3734.75	7.55987	1521.474
A reheat steam inlet	4	632.624	3161.63	6.29218	1319.977
Secondary reheat steam inlet	5	544.852	3329.24	7.02136	1273.828
Furnace side feed water	6	710.349	1450.48	3.37178	464.943
fuel	7	104.528			20242.518
The main steam turbine side	8	710.349	3445.6	6.23412	1620.968
Inlet steam of high pressure cylinder	9	632.624	3674.91	6.9599	1637.515
Exhaust steam of ultra-high pressure cylinder	10	632.624	3161.51	6.28276	1322.619
Inlet steam of medium pressure cylinder	11	544.852	3728.05	7.55521	1516.14
Exhaust steam of high pressure cylinder	12	544.852	3331.64	7.00425	1281.244
turbine side feed water	13	710.349	1455.04	3.37725	467.899
NO.1 extraction	14	72.473	3161.51	6.28276	1322.619
NO.2 extraction	15	51.394	3513.25	6.98505	1468.483
NO.3 extraction	16	38.486	3332.1	7.00489	1281.517
NO.4 extraction	17	19.209	3550.27	7.58889	1328.487
NO.5 extraction	18	11.546	3383.54	7.60378	1157.392
NO.6 extraction	19	19.299	3274.57	7.61681	1044.602
NO.7 extraction	20	29.240	3109.7	7.63388	874.728
NO.8 extraction	21	16.04	2877.75	7.72708	615.457
NO.9 extraction	22	18.74	2728.4	7.73638	463.38
NO.10 extraction	23	20.948	2581.63	7.78599	302.067
To Feed pump turbine steam extraction	24	35.787	3383.54	7.60378	1157.392
Steam turbine exhaust	25	376.453	2396.31	7.97461	61.453
Inlet steam of #2 High-pressure heater outside steam cooler	26	51.394	3513.16	6.99811	1464.564
Inlet steam of #4 High-pressure heater outside steam cooler	27	19.209	3550.33	7.60245	1327.572
Inlet steam of #1 High-pressure heater	28	72.473	3161.42	6.29446	1319.099
Inlet steam of #2 High-pressure heater	29	51.394	2970.51	6.21709	1150.87
Inlet steam of #3 High-pressure heater	30	38.486	3331.89	7.01767	1277.56
Inlet steam of #4 High-pressure heater	31	19.209	3088.13	6.94063	1056.384
Inlet steam of Deaerator	32	11.546	3383.55	7.626	1150.888
Inlet steam of #6 low-pressure heater	33	19.299	3274.5	7.63939	1037.913
Inlet steam of #7 low-pressure heater	34	29.240	3109.71	7.65584	868.3
Inlet steam of #8 low-pressure heater	35	16.04	2877.83	7.74926	609.034
Inlet steam of #9 low-pressure heater	36	18.74	2728.48	7.7589	456.858
Inlet steam of #10 low-pressure heater	37	20.948	2582.1	7.8095	295.645
outlet water of #1 High-pressure heater	38	710.349	1402.99	3.2894	441.602
drain of #1 High-pressure heater	39	72.473	1239.34	3.06349	344.178
Inlet water of #1 High-pressure heater	40	710.349	1206.85	2.94412	346.681
Inlet water of #2 High-pressure heater	41	710.349	1052.56	2.65369	277.531
drain of #2 High-pressure heater	42	123.867	1073.11	2.76451	265.594
Inlet water of #3 High-pressure heater	43	710.349	891.307	2.32875	211.534
drain of #3 High-pressure heater	44	162.354	902.314	2.43073	192.646
Inlet water of #4 High-pressure heater	45	710.349	810.628	2.15652	181.344
drain of #4 High-pressure heater	46	181.563	817.573	2.25616	159.08
outlet water of deaerator	47	710.349	764.404	2.14221	139.315
inlet water of deaerator	48	517.24	687.088	1.96829	112.984
Inlet water of #6 low-pressure heater	49	517.24	587.907	1.73446	82.35
drain of #6 low-pressure heater	50	19.299	611.739	1.7926	89.138
Inlet water of #7 low-pressure heater	51	517.24	432.009	1.33895	42.396
drain of #7 low-pressure heater	52	48.539	455.198	1.40188	47.137
Inlet water of #8 low-pressure heater	53	452.661	342.426	1.09372	24.702
Inlet water of #9 low-pressure heater	54	452.661	244.043	0.80678	10.435
drain of #9 low-pressure heater	55	18.74	353.425	1.12736	25.839
Inlet water of #10 low-pressure heater	56	452.661	136.27	0.46806	1.958
drain of #10 low-pressure heater	57	20.948	254.826	0.84222	10.829
The drain water after mixing	58	39.688	301.398	0.97933	17.206
outlet drain water of drain water cooler	59	39.688	144.999	0.49973	1.403
Inlet water of drain water cooler	60	452.661	122.485	0.42268	1.476
Inlet steam of feed pump turbine	61	35.787	3383.55	7.626	1150.888
exhaust steam of feed pump turbine	62	35.787	2468.52	8.15938	79.498
hot well water	63	451.928	117.833	0.4105	0.395
inlet water of Feed water pump	64	710.349	764.404	2.14221	139.315
outlet water of Feed water pump	65	710.349	810.628	2.15652	181.344
power P	1000000 kW				

Note: The ambient temperature is 20°C

3.2 Efficiency of a device or system:

The boiler, steam turbine and pipeline efficiencies are shown in the following table:

Table 4 THA operating device or system efficiency

Items	Unit	Exergy efficiency	Heat balance efficiency
boiler	%	55.1	95.03
pipe	%	99.6	99.89
Steam turbine generator	%	92.256	98.57
Turbine island	%	87.24	51.15
Plant efficiency (Positive balance)	%	47.26	48.2

The plant efficiency calculated by the exergy analysis method is slightly lower than the plant efficiency calculated by the thermal balance method. The difference between the two is due to the characteristics of the fuel itself, namely, the low calorific value of the fuel and its

 https://doi.org/10.1051/e3sconf/202337503022

chemistry exergy.Although the total plant efficiency calculated by the two methods is approximately the same, the loss distribution calculated by the two methods is quite different.The boiler efficiency calculated by heat balance method is 95.03% and that calculated by exergy analysis method is 55.1%.The thermal efficiency of the turbine island calculated by the thermal balance method is 51.15%, while the thermal efficiency of the turbine island calculated by the exergy method is 87.24%.The reason is that the two efficiencies stand on different ground. Thermal balance calculation focuses on the quantity of each loss, while exergy analysis focuses on the quality of each loss, that is, the loss of the work capacity of the working medium.The working medium absorbs the heat transferred by the flue gas in the boiler. Due to the huge heat transfer temperature difference, the heat transfer process is inevitably accompanied by a huge irreversible loss.Such processes as combustion, heat transfer with finite temperature differences, logistic mixing of different parameters, and frictional resistance are not considered losses in traditional thermal equilibrium calculations because they obey the first law of thermodynamics, which theoretically conserved energy in quantity.But in exergy analysis these are all losses, because the capacity of energy to do work is reduced.A large amount of effective energy in the energy is converted into invalid energy in the boiler, which cannot be utilized in the later energy conversion process and can only be released to the environment through the condenser and cooling tower to form waste heat. Therefore, the heat balance analysis method believes that the main energy loss of the system occurs at the turbine side. The exergy analysis says that the main energy loss of the system has occurred on the boiler side but that the emission of this energy is on the turbine side.The energy consumption distribution of the whole plant thermal system is shown in the following table:

Tabie 5 Exergy loss coefficient of a plant or system in THA condition

Items	Exergy loss coefficient
boile	0.499
pipe	0.009261
Steam turbine generator	0.03967
steam cooler	0.001226
#1 High-pressure heater	0.001526
#2 High-pressure heater	0.0009793
#3 High-pressure heater	0.001847
#4 High-pressure heater	0.000586
deaerator	0.000779
#6 low-pressure heater	0.00117
#7 low-pressure heater	0.00196
#8 low-pressure heater	0.000619
#9 low-pressure heater	0.000765
#10 low-pressure heater	0.001006
mixed drain water	0.00001333
drain water cooler	0.0001933
Feed pump turbine system	0.004011
condenser	0.01222
NO.1 extraction pipe	0.00012056
NO.2 extraction pipe	0.00009518
NO.3 extraction pipe	0.0000719636
NO.4 extraction pipe	0.0000355429
NO.5 extraction pipe	0.0001454904
NO.6 extraction pipe	0.0000610128
NO.7 extraction pipe	0.0000888236
NO.8 extraction pipe	0.0000486837
NO.9 extraction pipe	0.0000577613
NO.10 extraction pipe	0.000063579
Overall plant exergy efficiency (counterbalance)	47.2385%

Through calculation and analysis, the irreversible loss of the system mainly occurs at the boiler side, and its loss accounts for 85.1% of the total loss of the system.The loss is mainly caused by the irreversibility of combustion and heat transfer.In the process of fuel combustion, chemical exergy is transformed into the physical fire of flue gas, which will be accompanied by a large amount of available energy loss. The furnace is the place with the highest working temperature of the boiler, and the heat transfer temperature difference can reach thousands of degrees. The huge heat transfer temperature difference leads to a great heat transfer loss.The incomplete combustion of fuel, heat transfer loss, heat dissipation loss, smoke exhaust loss and other factors combine to cause a great loss of available energy at the boiler side.

The irreversible loss of the turbo-generator body is the second largest loss source in the power plant, and its irreversible loss accounts for 7.52% of the total loss of the system.This loss includes nozzle, rotor blade, blade height, sector, impeller friction, partial steam intake, steam leakage, wet steam, residual speed loss and other losses.

The turbine cold source loss is the third largest loss source in power plant, and its irreversible loss accounts for 2.32% of the total system loss.Although the amount of steam turbine cold source loss is large, the steam exhaust temperature of the steam turbine is only 28.1°C, which is not big difference with the ambient temperature, so the grade of this part of heat is very low.

When the working medium flows through the main steam pipeline, water supply pipeline, primary reheating cold and hot section pipeline and secondary cold and hot section pipeline, the irreversible loss caused by the heat dissipation and pressure loss of the pipeline accounts for 1.755% of the total loss of the system, which is the fourth largest loss source in the power plant.

In the high pressure heater system, the design end difference of #1 high pressure heater and #3 high pressure heater is the same. The temperature rise of water supply in #1 high pressure heater and #3 high pressure heater is 39.8°C and 36.5°C, respectively. According to the design data, the heat transfer temperature difference between #1 high pressure heater and #3 high pressure heater is 26.83°C and 30°C, respectively.Although the heat exchange of #3 high pressure heater is less than that of #1 high pressure heater, the irreversible loss of #3 high pressure heater is greater than that of #1 high pressure heater due to the large heat exchange temperature difference. Since the use of an

 https://doi.org/10.1051/e3sconf/202337503022

external steam cooler reduces the heat transfer temperature difference between the #2 and #4 high pressure heaters, the irreversible loss of the #2 and #4 high pressure heaters is lower than that of the #1 and #3 high pressure heaters.

The irreversible loss of the low pressure heater system is slightly higher than that of the high pressure heater system. In the low-pressure heater system, the design end difference of the #6 low-pressure heater and the #7 low-pressure heater is the same. The temperature rise of the condensate in the #6 low-pressure heater and the #7 low-pressure heater is 23°C and 36.7°C, respectively. According to the design data, the heat transfer temperature difference between the #6 low-pressure heater and the #7 low-pressure heater is 31.26°C and 29.34°C, respectively. Although the heat transfer temperature difference of the #7 low-pressure heater is less than that of the #6 low-pressure heater, the irreversible loss of the #7 low-pressure heater is greater than that of the #6 low-pressure heater due to the large heat transfer.

The water drain of the #8 low-pressure heater flows into the outlet of the #8 low-pressure heater through the water drain pump. This method reduces the heat transfer temperature difference of the heater and is conducive to reducing the irreversible loss of heat transfer.

4. Conclusion

Through modeling and analyzing the thermal system of the unit, the following conclusions are drawn:

The boiler is the equipment with the largest irreversible loss in the power plant. Although the amount of lung heat emitted by the condenser is very large, the heat temperature carried by the exhaust steam of the steam turbine is not different from the ambient temperature. The taste of the exhaust steam of the steam turbine is very low, and the loss of work capacity is not large.

The difference between the whole-plant exergy efficiency calculated by exergy analysis and the whole-plant thermal efficiency of the first law depends on the characteristics of the fuel consumed in the plant. It depends on the difference between the low calorific value of the fuel and the chemical exergy.

Based on the second law of thermodynamics of exergy analysis takes into account not only the traditional heat balance analysis method have noticed in the system caused by the external loss but also considers the internal process within the irreversible loss, help us to understand the nature of the energy loss reasons, focus on how to reduce all kinds of irreversible factors in the system of debasement of energy, In order to make better use of energy.

Acknowledgements

This work was supported by the science and technology project of State grid shandong electric power research institute (ZY-2022-16), which studied the peak shaving capacity of different types of cogeneration units in Shandong power grid.

References

1. Yaxiu Gu, SHengpeng Wang, Thermal Economic Analysis of a Double R eheat Ultra Supercritical Pressure Unit [J]Journal of Xi'an University of Technology(2013): 357-361
2. Yaning Yin, Application Status and Development of USC Unit with Double Reheat Cycles[J]Power System Engineering 2013 (02) : 37-38
3. Haotian Gao, Haojie Fan, Jiancong Dong,et al, Development of Ultra Supercritcal Unit with Double Reheat Cycle[J]Boiler Technology 2014 (04) : 01-03
4. Guanpeng Li, Yida Liu, Qiang An,et al, System Comparison and Economic Analysis of 1000 MW Ultra-supercritical Double Reheat Units[J]electric power survey&design 2018 (05) : 52-55
5. Xiaolin Yao, CHang Fu, Yanzhou SHi,et al, Overall economical performance of 1 000 MW class ultra-supercritical unit with double-reheat cycle [J]THERMAL POWER GENERATION 2017 (08) : 16-22
6. Thermal Power Plants (4th Edition)[M]China Electric Power Press
7. SHeng Su, Ningning Si, ZHigang ZHao,et al, Exergy Analysis of 1000MW Double Reheat Ultra-supercritical Powerplant [J]J. Huazhong Univ.ofSci.&Tech.(Natural Science Edition)2017 (03) : 94-98
8. ZHiping Yang,Yongping Yang , Energy consumption and distribution of 1000MW coal-fired power generation unitJournal of North China Electric Power University2012 (01) : 76-80
9. Zuting CHeng , Ling Yang, Energy and ExergyAnalysisofal000MW Ultra Supercritical PowerUni[J] TURBINE TECHNOLOGY 2012 (05) 385-388
10. Xiang Liu, Caoming Fang, Haojie Fan,et al, Exergy Analysis on Ultra Supercritical Power Unit with Double Reheat [J] BOILER TECHNOLOGY 2016 (03) 01-05

Analysis of heating transformation of different types of thermal power units

Jinxu Lao *, Hu Zhang, Wei Zheng, Panfeng Shang, Yue Han

State grid Shandong electric power research institute, Jinan 250003, China

Abstract. This paper introduces the total heating capacity, total installed capacity, heating area change trend and unit transformation of direct regulating heating units in Shandong power grid in recent years. According to the operation data of condensing pumping units, back pressure heating units and cylinder cutting units in previous heating seasons, the heating and peak shaving performance of different types of cogeneration units of the same type, grade and grade in heating seasons are analyzed, It is convenient for the power grid dispatching department to more reasonably and orderly arrange the unit mediation plan of the province's thermal power plants according to the online situation of new energy and external power. At the same time, it is a reference for power generation enterprises to carry out flexible transformation according to the actual situation of units during the 14th Five Year Plan period, and take into account the low load peak shaving performance of units or the whole plant while increasing heating capacity.

Keywords: cogeneration, Peak shaving, reform.

1. Background

Cogeneration unit mainly includes regulating extraction condensing heating unit (hereinafter referred to as extraction condensing unit), back pressure heating unit (including high back pressure circulating water heating unit, optical axis transformation [1] unit, pure back pressure heating unit, etc.), low-pressure cylinder zero output heating[2-3] unit (mainly cylinder cutting unit), etc. As of June 2021, Shandong power grid has directly transferred 154 public thermal power units with a rated capacity of 57305MW, including 137 thermal power units with a rated capacity of 49840MW, accounting for 86.97%. Among the thermoelectric units, the number of condensate pumping units, back pressure units and cylinder cutting units is 88, 33 and 16 respectively, accounting for 79%, 11.79% and 9.21% of the capacity; According to the capacity level, the number of units below 300MW, 300MW, 600MW and 1000MW is 46, 62, 21 and 8 respectively, accounting for 14.23%, 41.37%, 27.99% and 16.41% of the capacity.

After the "carbon peaking and carbon neutralization" double carbon goal was put forward, the power system, as an important link of energy transformation, has accelerated the clean trend of power generation side, and the installed capacity of renewable energy has been increasing; At the same time, Shandong Province issued a notice on flexibility transformation to improve the regulation capacity of power system through flexibility transformation. Under the heating pressure brought by the consumption of renewable energy, urban development and the shutdown of small thermal power, coal-fired power units are facing the dual test of heating and peak shaving[4-6].

2. Heating trend and transformation of thermoelectric units

In the heating season from 2018 to 2019, the total capacity of the public fire electric motor assembly unit directly transferred by Shandong Province is 48215MW, the total heating capacity is 158580701GJ, which is equivalent to 20490 t/h of heating extraction flow and 337084442 ㎡ of heating area; In the heating season from 2019 to 2020, the total installed capacity is 49450MW, the total heating capacity is 167966636GJ, which is equivalent to the heating extraction flow of 21703 t/h and the heating area of 357035499 ㎡; In the heating season from 2020 to 2021, the total installed capacity is 49840MW, the total heating capacity is 190548310 GJ, which is equivalent to 24621 t/h of heating extraction flow and 405035859 ㎡ of heating area.

In 2018-2019, 136 public thermal power units will be directly transferred in the heating season, including 102 condensate extraction units, 31 back pressure units and 3 cylinder cutting units. There will be 139 sets in the heating season from 2019 to 2020 (5 sets of back pressure and cylinder cutting units are reconstructed from condensate extraction units). Compared with the previous heating season, 2 sets of back pressure units will be added: 1 set of 350MW and 1 set of 150MW, 3 sets of cylinder cutting

* Corresponding author: 1686844029@qq.com

units will be added: 1 set of 350MW, 1 set of 330MW and 1 set of 160MW, and 3 sets of condensate extraction units will be added: 1 set of 600MW, 1 set of 335MW and 1 set of 300MW. From 2020 to 2021, there will be 137 sets in the heating season (11 sets of back pressure and cylinder cutting units are transformed from condensate extraction units), one 145MW back pressure unit will be added compared with the previous heating season, and one 135MW back pressure unit will be shut down; 7 sets of 300MW and 3 sets of cylinder cutting transformation units below 300MW are added; Two 680MW condensate extraction units were added, and two 350MW and one 135MW condensate extraction unit were shut down. From the transformation situation, the transformation of back pressure unit is slowing down, and the cylinder cutting transformation has doubled.

3. Performance analysis of different types of thermoelectric units

3.1 Condensate extraction heating unit

Condensate extraction heating unit can supply one or more parameters of steam to the outside world, and its pressure is controlled by the pressure regulating system. This unit can meet the different requirements of external heat load and electric load in a large range, and has flexible operation, so it is widely used.

The heating performance of condensate extraction unit can be determined by three methods: test method[7] , working condition diagram analysis method and thermal calculation method. The test method can accurately reflect the actual heating capacity of the thermal power unit. The unit needs to carry out the load capacity test in the heating season, but there are many test conditions and it is difficult to summarize and sort out due to the limitations of on-site conditions. Thermodynamic calculation method is to calculate the variable working conditions of thermal power units under specific constraints. The calculation process of this method is complex. The working condition diagram method is a method to query the heating capacity of the unit according to the design thermoelectric characteristic curve of the unit. This method is more intuitive and easy to obtain the corresponding results.

The working condition diagram is the relationship curve between the active power of steam turbine generation, main steam flow and regulated steam extraction under different combinations. By digitizing and dimensional conversion of the working condition diagram, the trend diagram of peak regulation upper and lower limits under different steam extraction of the unit can be obtained, and then the calculation model of peak regulation upper and lower limits formula and the calculation model of steam extraction upper limit formula of condensing unit can be obtained.The upper and lower limits of the unit's output under the current steam extraction and the upper limit of the unit's steam extraction under the current output can be obtained by selecting an appropriate section according to the formula[8] . According to the average steam extraction capacity and output of the unit in different heating periods, the electric load adjustment interval under the average steam extraction capacity of different capacity condensing heating units in the initial, middle and final heating periods and the upper limit of steam extraction capacity under 50%, 75% and 100% rated capacity can be obtained by substituting into the calculation model.

According to the analysis of the data of heating season from 2020 to 2021, most condensing units can meet the heating demand by operating at about 60% of the rated load under the average steam extraction capacity of the initial, middle and final stages of heating. Under 50% rated load or minimum heating load, the upper limit of steam extraction capacity of units below 300MW level is about 60 ~ 175 t/h, that of 300MW level units is about 200 ~ 350 t/h, that of 600MW level units is about 500 ~ 600 t/h, and that of 1000MW level units is very small in heating season due to economic operation.

3.2 Back pressure heating unit

The back pressure heat supply unit increases the exhaust pressure of the steam turbine by removing the blades at the end of the low-pressure cylinder of the steam turbine, and uses all the cold source losses for heat supply. Generally, it includes high back pressure circulating water heating unit, optical axis transformation unit and pure back pressure heating unit. The medium and high back pressure circulating water heating units of Shandong direct dispatching public units are generally transformed from units below 300MW level or 300MW level. There are only two optical axis transformation units, the proportion of pure back pressure heating units is very small, and there are many self owned power plants.

The steam extraction heating unit can be transformed into a high back pressure heating unit, and different technical routes can be adopted according to the differences between wet cooling unit and air cooling unit. The operating back pressure of wet cooling unit is generally low, about 5 kPa. Excessive increase of back pressure will affect the operation safety of steam turbine or fail to meet the heating demand. There are two methods for high back pressure transformation of wet cooling unit, that is, removing the last stage blade of low-pressure rotor and one-time transformation of low-pressure rotor. To transform the direct air cooling unit into a high back pressure heating unit, in addition to adding a heating condenser, it is also necessary to partially transform the circulating water pump of the heating network, the heating pipeline and valves in the plant. During the heating period, according to the heating demand, adjust part or all of the exhaust steam to enter the heating condenser to realize high back pressure heating; In the non heating period, close the valves on the heating pipeline, and all the exhaust steam of the steam turbine enters the air cooling island. The indirect air cooling unit is similar to the pure condensing unit. It has a condenser. The exhaust steam condenses in the condenser, and the circulating water exchanges heat through the air cooling tower. The indirect air cooling unit double temperature zone condenser heating technology can be used for heating, which does not change the steam turbine body and intercooling tower.During the heat supply period, the back pressure of the steam turbine shall be properly increased, the circulating water of the heat network shall be used, and the exhaust heat of the steam turbine shall be recovered through the condenser for heat supply; In the non heating period, switch to the pure condensation condition of intercooling tower. Another scheme for high back pressure heat supply transformation of wet cooling unit is to maintain the unit back pressure of 5 kPa in summer, and replace the low-pressure pure condensing rotor with the low-pressure high back pressure rotor in winter, so as to increase the heating back pressure, such as about 50 kPa. After the heating period, replace the low-pressure pure condensing rotor again. Because two different low-pressure cylinder rotors are used in heating period and non heating period, it is called double rotor double back pressure exchange heating technology. This technology is applicable to the reconstruction of circulating water waste heat recovery of large units. In terms of safety, it solves many problems of high back pressure heating, but at the same time, this technology also has corresponding defects, that is, it digests the huge residual heat entering the exhaust steam of condenser, which requires a large and stable heating area.

The electric load of high back pressure heating unit is determined by the heat load it undertakes. The greater the heat load, the greater the steam exhaust required by the unit, and the higher the corresponding main steam and electrical load. Since Shandong high back pressure heating units operate stably throughout the heating season, the load trend is almost a straight line. Select different levels of high back pressure units, and directly transfer the unit load, circulating water flow, inlet and outlet temperature and other relevant data in a certain period to obtain the heating capacity and converted steam extraction capacity of these units, easy to compare.

The operation load of the high back pressure unit in the heating season is generally 50% ~ 75% of the rated load. Because the high back pressure heating unit makes use of all the cold source losses of the unit, the heating capacity is higher than that of the condensate extraction unit of the same grade. Generally, when the high back pressure heating unit below 300MW is converted into steam extraction capacity, it is about 200 ~ 300t/h, which is equivalent to the heating capacity of 300MW condensate extraction unit. When converted into steam extraction capacity, 300MW high back pressure heating unit is about 350 ~ 600t/h, which is equivalent to the heating capacity of 600MW condensate extraction unit. Based on the high back pressure unit's efficient heating capacity and good economic benefits, it should give priority to bear the maximum heat load.

Like the high back pressure unit, the optical axis unit operates stably throughout the heating season with almost no adjustment. Among the two optical axis units, one unit has a rated capacity of 110MW, an operating load of 62MW in the heating season, and an average steam extraction capacity of 280t/h; The other unit has a rated capacity of 225MW, an operating load of 113MW in the heating season, and an average steam extraction capacity of 345t/h. After the optical axis transformation, the daily operation load of the unit in the heating season is at a low level, but it has strong heating capacity.

Like the high back pressure unit, the optical axis unit operates stably throughout the heating season with almost no adjustment. Among the two optical axis units, one unit has a rated capacity of 110MW, an operating load of 62MW in the heating season, and an average steam extraction capacity of 280t/h; The other unit has a rated capacity of 225MW, an operating load of 113MW in the heating season, and an average steam extraction capacity of 345t/h. After the optical axis transformation, the daily operation load of the unit in the heating season is at a low level, but it has strong heating capacity.

3.3 Low pressure cylinder zero output heating unit

The low-pressure cylinder zero output heating unit is mainly transformed by cylinder cutting. After the strength check of the last two stages of blades, the original steam inlet pipe of the low-pressure cylinder is cut off, a bypass pipe is added in the medium and low-pressure connecting pipe, and a small amount of steam after check and calculation is used to take away the blast heat generated by the rotation of the low-pressure rotor after "cylinder cutting". The "zero output" operation of the low-pressure cylinder is realized by cutting off the low-pressure cylinder, which greatly reduces the electrical load of the unit and meets the deep peak shaving demand of the power grid, so as to realize "thermoelectric decoupling". The operation data of some cylinder cutting units in cylinder cutting and non cylinder cutting states are shown in Table 5.

With the decrease of volumetric flow, the efficiency of the stage will become worse. When the volume flow rate decreases to a certain value, the wheel circumference power of the stage is equal to zero; If the volume flow is further reduced, the stage will change from the original work stage to the blast power consumption stage, and the stage efficiency will change from positive to negative. At this time, the stagnation enthalpy of exhaust steam will be higher than the inlet stagnation enthalpy, blast friction will heat the steam flow, and the temperature of blades will also rise, causing cylinder deformation and excessive blade stress. When the low-pressure cylinder enters the blowing condition, the exhaust water spray device is usually set to reduce the exhaust temperature of the last stage. Under small volume flow, the last blade root works with negative reaction, and the water spray from LP cylinder will be sucked back into the moving blade through the last blade root, and with the vortex movement, the water droplet will be sucked back to erode the last blade root; The small flow rate greatly increases the outlet angle of the stationary blade, resulting in water droplets scouring to the back of the steam inlet side of the moving blade, resulting in erosion. The above two flow modes of water droplets do great damage to the blade and weaken the strength of the blade. At the same time, under the condition of small volume flow, the large negative angle of attack at the inlet of the last stage moving blade causes large-scale steam flow separation and induces self-excited vibration, resulting in the sudden increase of dynamic stress level and stall flutter. In relative volume , In the process of flow reduction, when the relative volume flow is low to a certain value, the blade vibration stress begins to increase rapidly, then reaches the maximum value, further reduces the volume flow, the vibration stress decreases gradually, and the vibration stress changes non monotonically with the relative volume flow. Therefore, there is a large dynamic stress amplification interval for the last stage blade at the relative volume flow of 0.1 ~ 0.2. During operation, try to avoid long-term operation in this working condition interval.

Table 1 Operation data of cylinder cutting unit in cylinder cutting and non cylinder cutting state

Unit level	Unit capacity (MW)	No cylinder cutting state		Cylinder cutting state			
		Maximum extraction capacity (t/h)	Output range(MW)	Close to the maximum steam extraction without cylinder cutting (t/h)	Corresponding load(MW)	Maximum extraction capacity (t/h)	Corresponding load(MW)
Below 300 MW	160	241	110 ~ 126	233	82	314	104
300 MW class	300	433	171 ~ 221	434	118	651	178
	330	452	231 ~ 267	452	132	667	218
	330	305	151 ~ 286	305	90	642	192
	350	510	283 ~ 294	510	203	640	283

In the 300MW class, the cylinder cutting transformation unit occupies the mainstream position. The maximum steam extraction capacity of the unit after transformation is about 650t/h, which is about 140 ~ 340t/h higher than that before transformation. Under the same load, the heat supply capacity of the unit after cylinder cutting is greater than that of the high back pressure unit. The operation time of the reformed heating unit is divided into two states: no cylinder cutting and cylinder cutting. Under the condition of constant external heating load, cylinder cutting can reduce the power generation by about 30% compared with no cylinder cutting.

Among the grades below 300MW, the heating type is the most comprehensive. Take 160MW unit as an example, the two units are high back pressure and cylinder cutting heat supply units. Before the transformation, they are the same type units produced by the same manufacturer, and the model is N150/CC135-13.24/535/535/0.981/0.23. The maximum steam extraction capacity after cylinder cutting transformation is similar to that of high back pressure unit and 110MW optical axis unit under daily operation, and can expand the steam extraction capacity by about 70 t/h compared with that of condensate extraction unit at the same level; At the same time, under the same heating level, the optical axis unit has the lowest operating load, but has no peak shaving capacity under the condition of constant heating capacity.

4. Conclusion

During the "14th five year plan" period, Shandong direct coal-fired power generation unit is required to carry out flexible transformation according to 20% of the capacity every year. The heating season of the existing condensate extraction unit is limited by 50% of the thermal power ratio, and the minimum output should reach 30% and 40% of the rated capacity. With the increase of heat load and the fluctuation of coal price in recent years, the contradiction between unit heating and peak shaving in heating season in Shandong Province is gradually deepened, and there are more flexible modifications of unit heating. The power generation enterprise shall flexibly transform according to the actual situation of the unit, and take into account the low load peak shaving performance of the unit or the whole plant while increasing the heating capacity.

Acknowledgements

This work was supported by the science and technology project of State grid shandong electric power research institute (ZY-2022-16), which carries out the research on the heating and peak shaving capacity of different types of cogeneration units in Shandong power grid.

References

[1] Ronghe Hao, Zhuang Li, Liping Wang , et al. Economic Analysis of Smooth Shaft Operation of Low Pressure Cylinder in 210 MW Unit[J]. Journal of Shenyang Institute of Engineering (Natural Science), 2020,16(02):25-28+44.

[2] Jianxun Wang. Analysis of Flexible Peak-Load R egulation Capability and Economy on the Zero Output Technology of Low-Pressure Cylinder for 650 MW Supercritical Unit[J].Journal of Engineering for Thermal Energy and Power, 2021,36(02):18-23.

[3] Tianfu Liang, Weiyang Xie, Fei Wang, et al.The key technique research of removing the low pressure cylinder of steam turbine[J].Turbine Technology, 2019, 61(6): 471–472.

[4] Haisheng Yang, Tuo Zhang , GuangTong Tang, et al. Influence of Zero-Output Technology of Low-Pressure Cylinder on Deep Peak Regulation Performance of Heating Unit and Compensation Standard for Peak Regulation[J]. Journal of Engineering for Thermal Energy and Power,2020,35(06):268-273.T

[5] Qinpeng Zhang, Xuedong Wang, Feng Li, et al. Analysis of Heating Capacity and Peak-regulating Capacity of 330 MW Steam Turbine Unit With Low-pressure Cylinder off Operation[J]. Shandong Electric Power, 2020, 47(12):72-76.

[6] Puxin Shi, Peiran Shi, Peiwen Wang, et al. Analysis and Operation Evaluation of Power Peak-shaving Ancillary Service Market in North China[J]. Automation of Electric Power Systems, 2021,45(20):175-184.

[7] Xuhui Zhang, Zhonghua Zhao, Fuxing Cui, et al. Experimental Research on Deep Peak Regulation Characteristics of 1030 MW Ultra Supercritical Unit[J]. Shandong Electric Power, 2021, 48(02):58-62.

[8] Lingkai Zhu, Wei Zheng, Junshan Guo , et al.Application of Heat Load Distribution in Improving Peak-shaving Ability of Plant[J]. Power System Engineering, 2019,35(06):43-46+49.

Prospect of Peak Regulation Capacity Improvement through Flexibility Transformation of Combined Heat and Power Units in Shandong Province

Zhiqiang Gong *, Yue Han, Junshan Guo, Panfeng Shang, Lingkai Zhu

State Grid Shandong Electric Power Research Institute, Jinan, 250003, China

Abstract: With the high proportion of renewable energy connected to the grid, peak shaving demand surge, which needs to enhance the flexibility of coal power with a larger proportion to support the stable operation of the grid. At present, the capacity of the combined heat and power unit accounts for about 87 % of direct dispatching coal-fired power plants by Shandong power grid. The flexible operation ability of the combined heat and power unit is poor, and its flexible transformation is urgent. This paper summarizes the current relatively mature flexibility transformation technology of combined heat and power unit, including low pressure cylinder zero output transformation (cylinder cutting transformation), bypass load regulation, thermal energy storage and electrical energy storage transformation technology, and puts forward to encourage the flexibility transformation of in-service combined heat and power unit, and actively promote the research and application of the intelligent heating network, the integration of thermal and electricity, and the clean heating construction and clean heating technology.

Key words: Carbon peaking and carbon neutrality, combined heat and power unit, flexibility transformation, peaking regulation.

1. Introduction

Since the goal of "carbon peak by 2030 and carbon neutral by 2060" was put forward, China has proposed to build a clean, low-carbon, safe and efficient energy system, control the total amount of fossil energy, focus on improving utilization efficiency, implement renewable energy substitution action, deepen power system reform, and build a new power system with new energy as the main body. These policies and measures will lead to substantial adjustment of power supply structure, and the thermal power generation industry will face huge challenges. Under this background, the proportion of new energy power generation in China's power structure will continue to increase. With the significant increase of installed capacity of intermittent and random renewable energy, as well as the impact of trans regional power transmission, coal power units with large capacity will transform to flexible power supply adjustment and actively participate in deep peak shaving, which puts forward higher requirements for peak shaving capacity of thermal power units. On the other hand, central heating has the advantages of high energy comprehensive utilization efficiency, energy conservation and environmental protection. The proportion of heating units in northern China is high, but the proportion of large condensing heating units is too large to fully participate in deep peak shaving. Therefore, it is urgent to carry out efficient research on flexible peak shaving of heat supply units to improve the peak shaving capacity of heat supply units.

2. China's energy structure and carbon emission status

2.1 China's energy structure

Data statistics show that in the post epidemic situation, China's primary energy demand will increase by 2.1% in 2020, lower than the average annual growth rate of 3.8% in the past decade, but it is also one of the few countries with increased energy demand[1]. China continues to promote the optimization of energy structure, and the proportion of coal in China's primary energy consumption structure will be reduced to 57% in 2020. The growth of renewable energy in China accounts for one third of the growth of global renewable energy consumption. Among them, wind energy (+14%) and solar energy (+15.8%) increased significantly. In 2020, China's installed wind power capacity increased by 72 million kilowatts, which exceeded the total increase in the past three years.

* Corresponding author: zhqgong@foxmail.com

2.2 Current situation of carbon emissions in China

In 2020, China's coal demand will increase by 0.3%, driven by the demand growth of power, building materials, steel and chemical industries. Coal imports were the highest since 2014. In 2020, China's carbon emissions will increase by 0.6%, making it one of the few regions in the world to increase. China's carbon emissions have continued to grow for the fourth consecutive year, ranking first in the world, reaching about 10 billion tons (Figure 2). China's carbon emissions have increased to about 31% of the total global carbon emissions[2].

China's carbon consumption and CO2 emissions mainly come from coal-fired power plants. According to the data of the National Bureau of Statistics, China's total power generation in 2020 will reach 7417.04 billion kWh, of which the total thermal power generation will reach 5279.87 billion kWh, with a cumulative growth of 1.2%, accounting for 71.2%. In recent five years, the proportion of thermal power generation in China's total power generation has declined year by year. However, as the base of thermal power is still large, the total amount of CO2 generated during coal combustion remains at a high level. In 2020, the total amount of CO2 generated from coal utilization will be close to 8 billion tons.

2.3 Necessity of flexibility transformation of thermal power generator set

The proposal of the 3060 policy has brought great opportunities and challenges for China to continue to promote the transformation of the energy structure and build a new power system. Thermal power generation units are the main source of carbon emissions in China. The proposal of 3060 policy and the pressure of new energy consumption make thermal power generation units face great pressure of transformation. The installed capacity and power generation of new energy will continue to grow. It is estimated that in 2035, the installed capacity of non-fossil energy power generation will account for 68.4%, and the installed capacity of new energy power generation will account for 54.7%. Non-fossil power generation accounted for 56.4%, and new energy power generation accounted for 32.9%. Electric energy accounts for 39.69% of terminal energy consumption[1].

Although the energy structure situation dominated by coal will not change in the near future, under the background of "dual carbon", with the substantial increase in the installed capacity and power generation proportion of clean energy power, the peak shaving demand of the power grid also increases significantly. As the main peak shaving capacity in the recent stage, the direct dispatching coal-fired power plants will undertake an important peak shaving task, and the coal power units need to actively transform from the electricity based power supply to the electricity based and power regulated power supply. Therefore, coal power generation units need to be flexibly transformed to improve the new energy consumption capacity to meet the safe and stable operation of the power grid.

2.4 Current situation of heat supply units in Shandong Province

By December 2021, there were 161 direct dispatching coal-fired power plants with a total capacity of 60.92 million kilowatts in Shandong Province. Among them, there are 147 thermal power units with a capacity of 52.9 million kilowatts, accounting for 86.8%. Among thermal power units, there are 86 condensing units, 34 back pressure units and 27 low pressure (LP) cylinder cutting units. The back pressure unit basically loses the flexibility adjustment ability. The unit operated with LP cylinder cutting realizes the weak decoupling of thermoelectricity. The designed minimum technical output can be reduced to less than 50% of the rated capacity. The flexible operation ability of the condensate extraction unit is poor. The demand for flexibility transformation of heating units in Shandong Province is urgent.

By the end of the 14th Five Year Plan period, all the direct dispatching coal-fired power plants in Shandong Province need to complete the flexibility transformation. After transformation, the minimum technical output of the unit under the pure condensation condition reaches 30% of the rated capacity, and the adjustable range of the unit is expanded to 30~100% of the rated capacity. The heating unit shall meet the minimum technical output of 40% of the rated capacity under the heating capacity. Therefore, the flexibility transformation of heat supply units is an important work to improve the capacity of regulated power supply in Shandong Province[3].

3. Flexibility transformation technology of heat supply unit

3.1 LP cylinder zero output transformation (cylinder cutting transformation)

For the condensate extraction unit using bridge tube for heat supply, the maximum steam extraction volume is limited by the minimum steam inlet volume of the LP cylinder. For safety reasons, the manufacturer reserves a large margin for the minimum steam inlet volume of the LP cylinder, which cannot give full play to the heat supply capacity of the unit. However, through verification and calculation, the steam inlet volume of the LP cylinder with 3% of the rated main steam flow is generally enough to take away the blast heat. The technical transformation of zero output (cylinder cutting) operation of LP cylinder is applicable to units with large demand for heat supply. In the heat supply state of the unit, the LP cylinder is in a high vacuum environment by removing most of the steam inlet from the LP cylinder of the turbine, increasing the heat supply capacity of steam extraction, and improving the peak shaving depth and heat supply capacity of the unit.

The cylinder cutting technology does not need to modify or replace the rotor, only the strength check of last stage and secondary last stage blades is required. The cost of transformation is low. During cylinder cutting operation, it can freely switch between back pressure working condition and condensate extraction working condition. After the LP cylinder is cut, intermediate pressure (IP)

 https://doi.org/10.1051/e3sconf/202337503024

cylinder exhaust are completely used for heating, with strong heating capacity. However, the blade may be damaged due to water erosion and air blowing loss during long-term operation of LP cylinder cutting[4].

3.2 Bypass load regulation technology

The bypass load regulation technology is generally divided into two technical schemes: steam turbine high pressure (HP)/LP bypass combined steam extraction heating and LP bypass steam extraction heating[5]. Combined bypass steam extraction refers to the use of HP bypass to reduce the temperature and pressure of some main steam and then send it to HP cylinder for steam exhaust. After being heated by the boiler reheater, the steam is extracted from LP bypass (IP cylinder inlet) for external heating. LP bypass steam extraction is to directly introduce some reheated steam for external heating by using LP bypass pipeline. The bypass load regulation technology has a small investment and can increase the peak regulation capacity by 10-20% of the rated capacity.

3.3 Thermal energy storage transformation technology

3.3.1 Hot water tank energy storage technology

The hot water tank energy storage technology of thermal power plants has been widely used at home and abroad, mainly for thermoelectric decoupling of heating power plants, to improve the flexibility of heating unit operation. The steam turbine is used to extract steam, heat the return water of the heat network to the heating temperature, and store it in a large hot water tank, so that the unit can store hot water energy during the period of high power load and low heating load. When the power load is low and the heating load is high, the energy storage system can provide heat. On the basis of meeting the heating requirements, the flexibility of unit operation can be improved, so that the cogeneration unit can participate in peak shaving[6].

According to the different heat storage capacity, the hot water tank energy storage technology can increase the peak shaving space of the original thermoelectric unit by 30~50% of the rated capacity, so that the thermoelectric unit can be reduced from 60~70% of the load under the previous heating state to 30~40% of the rated capacity, which is basically equivalent to the unit state under pure condensation.

3.3.2 Electrode boiler heating technology

The electrode boiler is a kind of electric boiler that is widely used in the industrial and civil heating markets at present. Unlike ordinary electric heating equipment, the electrode boiler uses high-voltage three-phase electrodes to directly discharge heat in the conductive brine in the boiler, so that the electric energy can be converted into heat energy with high conversion efficiency, and then the heat in the furnace can be transferred to the heating network through the heat exchanger. It has the advantages of large power, fast and smooth adjustment, etc. As an electric energy consumption equipment, the electrode boiler can directly reduce the output of the thermal power plant and increase the heating capacity, which is an effective peak shaving technology[7].

At present, electrode boilers are widely used in industrial heating and civil heating. This technology has the advantages of high power, fast and smooth adjustment, etc. The electrode boiler consumes the output power of the generator. According to different configurations, the capacity of the original thermal power unit can be increased by more than 50~70% of the rated capacity peak shaving capacity, or even zero output operation. However, the electric boiler does not have the heat storage performance. Under the restriction of heat load, the adjustment power of the electric boiler needs to be coordinated with the boiler adjustment load, so the deep peak shaving is limited. In addition, in the process, electric energy and heat energy are transformed mutually, and the economy of energy utilization is poor.

3.3.3 Regenerative electric boiler

The regenerative electric boiler refers to the use of electric boilers to convert electric energy into heat energy of high-temperature solids, and use the sensible heat of high-temperature solids to store heat energy[8]. When heat energy is needed, the heat energy of the heat storage body is converted into hot water, steam and other forms of heat use. The boiler uses the electric heating wire inserted in the heat exchange air duct at the bottom to heat the electric heating wire. During the heating cycle, the heat accumulator absorbs the heat released by the electric heating wire and temporarily stores it. The heat release cycle is released through the secondary heat exchange for users to use.

The investment cost of electric boiler solid heat storage technology is high, and the process involves electro thermal conversion. High grade electric energy is converted into heat energy, which is less economical in energy utilization and high in operation cost.

3.4 Battery energy storage transformation scheme

Battery energy storage can effectively improve the peak shaving capacity of the unit. When the thermal power unit needs to reduce the load, only part of the generated energy needs to be charged to the electric energy storage facility to reduce the grid connected load. When the thermal power unit needs to carry full load, the electric energy storage facility can discharge the grid, which can improve the overall output of the thermal power unit. However, the peak shaving capacity of the thermal power unit itself (high and low load range) is far less than that of the pure condensing unit, and the cost of electric energy storage is significantly higher than that of thermal energy storage. Therefore, to achieve the perfect peak shaving state (100~50% or even lower), large capacity and high-power electric energy storage equipment need to be configured, with huge investment costs[9].

4. Prospects for flexibility transformation of heat supply units to improve peak shaving capacity

4.1 Flexibility transformation of in-service heating units

Shandong Province will continue to promote the implementation of flexible manufacturing and flexible transformation of coal fired power generation units, encourage the in-service heat supply units to give priority to the use of mature technologies to implement flexible transformation, improve the flexibility regulation level of heat supply units and release the peak shaving capacity on the basis of ensuring or further improving the heat supply capacity and meeting the needs of new heat load[10]. This will help promote the transition from basic power supply to regulatory power supply of coal power units.

4.2 Research and application promotion of intelligent heating network and thermoelectricity integration

During the centralized heating period in Shandong Province, the capacity of peak load regulation and frequency regulation of heating units is greatly limited. In order to increase the peak shaving margin of heating units, the current heating mode with constant heat flow of heating units should be changed. The peak shaving capacity of heat supply units shall be improved by taking advantage of the different heating demands in different periods of the heat network and the thermal inertia of the heat network. It is estimated that the heat users will not feel obvious abnormality if the heating network in large cities stops heating for one hour. However, for thermal power units, huge peak shaving space can be released within one hour to meet the short-term peak shaving demand of the power grid.

4.3 Clean heating construction and clean heating technology R&D and application promotion

It is necessary to fully respect the will of the masses and accelerate the increase of the proportion of clean heating. At present, the market mechanism of clean heating is not mature, and technology research and development and promotion are not enough. It is suggested that the government make top-level design and issue relevant policies, guide each city to actively select applicable clean heating technologies, mobilize the enthusiasm of enterprises and heat users to participate in clean heating projects.

5. Conclusion

Under the background of "double carbon", the proportion of new energy power generation will be greatly increased, and coal power generation units are required to actively participate in deep peak shaving. At present, the capacity of heating units accounts for about 87% of the direct dispatching public coal-fired power units in Shandong Power Grid, with poor flexibility and urgent transformation needs.

The flexibility transformation technology of heat supply unit mainly includes zero output transformation of LP cylinder (cylinder cutting transformation), bypass load regulation technology, thermal energy storage transformation technology and electric energy storage transformation technology. It is necessary to encourage the flexible transformation of in-service heating units, and actively promote the development, application and promotion of intelligent heating network, thermoelectricity integration, and clean heating technology.

Acknowledgements

This work was supported by the Independent Research and Development Project of State Grid Shandong Electric Power Research Institute (ZY-2022-17).

References

1. BP. BP World Energy Statistics Yearbook[J]. 2022.
2. https://www.bp.com/content/dam/bp/business-sites/en/global/corporate/pdfs/energy-economics/statistical-review/bp-stats-review-2022-full-report.pdf
3. IEA CO2 Emissions from Fuel Combustion.
4. https://www.iea.org/subscribe-to-data-services/co2-emissions-statistics.
5. Jixiang Liu, Haichao Ma, Qingsong Wang, Shu Tian, Yue Xu, Yujie Zhang, Xueliang Yuan, Qiao Ma, Yuan Xu, Shuo Yang. Optimization of energy consumption structure based on carbon emission reduction target: A case study in Shandong Province, China[J]. Chinese Journal of Population, Resources and Environment, 2022, 20(2): 125-135.
6. Xu Han, Qiuliang Zhu, Jiandong Guan, Zhonghe Han. Research on wet steam condensation flow characteristics of steam turbine last stage under zero output condition[J]. International Journal of Thermal Sciences, 2022, 179: 107691.
7. Qingyuan Wei, Puyan Zheng, Siyu Zou, Tianyu Bai. Research on the combined low pressure steam bypass and heat storage peak shaving for industrial extraction steam heating units[J]. Energy Reports, 2022, 8(2): 179-187.
8. Mauricio Carmona, Anggie Rincón, Luis Gulfo. Energy and exergy model with parametric study of a hot water storage tank with PCM for domestic applications and experimental validation for multiple operational scenarios[J]. Energy Conversion and Management, 2020, 222: 113189.
9. Mattia Manni, Andrea Nicolini, Franco Cotana. Performance assessment of an electrode boiler for power-to-heat conversion in sustainable energy districts[J]. Energy and Buildings, 2022, 277: 112569.

10. Junhui Li, Yingnan Fu, Cuiping Li, Jiang Li, Zhitong Xing, Teng Ma. Improving wind power integration by regenerative electric boiler and battery energy storage device[J]. International Journal of Electrical Power & Energy Systems, 2021, 131: 107039.
11. Qingsong Wang, Pengfei Yang, Giuseppe Buja. Design and analysis on different functions of battery energy storage system for thermal power units frequency regulation[J]. Energy Reports, 2022, 8:11981-11991.
12. Jianjun Wang, Shuo Zhang, Jikun Huo, Yan Zhou, Li Li, Taoya Han. Dispatch optimization of thermal power unit flexibility transformation under the deep peak shaving demand based on invasive weed optimization[J]. Journal of Cleaner Production, 2021, 315: 128047.

Research on operation of low-pressure cylinder zero output heat supply retrofit unit

Jinxu Lao*, Hu Zhang, Yue Han, Panfeng Shang

State grid Shandong electric power research institute, Jinan 250003, China

Abstract: This paper introduces the technical route of the zero output heat supply transformation of low pressure cylinders in Shandong Province, the operation after the transformation, the impact on power grid dispatching, and the constraints on the normal operation of some units after the transformation, providing technical support for power generation enterprises to accurately implement the "three transformation linkage" transformation scheme and the accurate dispatching of the power dispatching center

Keywords: Zero output of low-pressure cylinder, Heating, Peak shaving

1. Background

As of August 2022, a total of 28 low pressure cylinder zero output retrofits [1-3] have been completed for direct thermal power generation units in Shandong Province, including 6 in 2019, 10 in 2020, 10 in 2021, and 2 in the first half of 2022; The rated capacity is 9265MW, accounting for 17.91% of the total capacity of direct thermal power units in Shandong Province. There are 4, 22 and 2 units below 300MW, 300MW and 600MW, respectively. The rated capacity is 600MW, 7335MW and 1330MW, respectively, and the capacity accounts for 6.48%, 79.17% and 14.36% respectively.

In 2021~2022 heating season, there are a total of 23 units with normal cylinder cutting, and no abnormal conditions affecting the normal operation of cylinder cutting are found. In order to meet the requirements of heat supply or peak load regulation [4-6], 19 units have been put into normal cylinder cutting operation, and 4 units have been put into operation for a short time; There are 5 units not put into cylinder cutting operation.

2. Low pressure cylinder zero output heat supply transformation

2.1 Low pressure cylinder zero output heating technology

The zero output heat supply technology of the low pressure cylinder is that under the high vacuum operation condition of the low pressure cylinder, the original steam inlet pipe of the low pressure cylinder is cut off with a fully sealed hydraulic butterfly valve, and a small amount of cooling steam is introduced through the newly added bypass pipe to take away the air blast heat generated by the low pressure rotor after the zero output transformation of the low pressure cylinder. Compared with conventional heating methods, low pressure cylinder zero output heating technology has the following advantages:

1) Compared with that before transformation, the zero output heating technology of low pressure cylinder uses the steam made by the original low pressure cylinder for heating, reducing the loss of cold source of the unit, and reducing the coal consumption rate of generating unit; Under the same boiler heat load, the heat supply capacity of the unit can be improved; Under the condition of constant heat supply, the generating power of the unit can be reduced to a certain extent to achieve deep peak shaving.

2) Compared with optical shaft heating [7], high back pressure heating [8] and other heating technologies, low pressure cylinder zero output heating technology can realize the flexible switching between the condensing operation mode and the high back pressure operation mode of the heating unit, making the unit have the characteristics of high back pressure unit with large heating capacity and flexible operation mode of the condensing heating unit, The problem of replacing the low pressure cylinder rotor twice during the heating period and the problem of storing and maintaining the spare rotor in the high back pressure heat supply transformation (double rotors) and smooth axis transformation scheme are avoided, and the maintenance cost during unit operation is greatly reduced.

2.2 Main technical route of transformation

1) Transformation of medium and low pressure connecting pipe and butterfly valve at the inlet of low pressure cylinder;

2) Add LP cylinder cooling steam system;

* Corresponding author: 1686844029@qq.com

3) Supporting transformation of monitoring and measuring points for turbine proper operation;
4) Check the operation safety of the low pressure secondary and final stage blades and spray the anti water corrosion metal wear-resistant coating;
5) Improvement of exhaust cylinder spray desuperheating system;
6) Supporting transformation of heating system, air extraction system, condensate system, etc;
7) Reconstruction of supporting automatic control system.

The transformation of the supporting automatic control system mainly includes sorting out the control logic related to heating and steam extraction in the original control system, and canceling all the locking control logic corresponding to the closing of the low-pressure butterfly valve; Cancel the protection logic of low load input of heat supply; Cancel relevant control logic that conflicts with zero output heat supply of low pressure cylinder. Sort out the protection settings related to the operation of low pressure cylinder in the original control system, and confirm that each control logic is consistent with the requirements for zero output heating and heating operation of low pressure cylinder. Add zero output heat supply input/removal control logic of low pressure cylinder. In the transformation scheme, monitoring and measuring points are added to connect to the DCS control system.

There are three different design methods for the additional LP cylinder cooling steam system. First, small flow cooling steam pipeline is equipped, and the adjustable range of cooling steam flow is small; Second, large flow cooling steam pipe is equipped as "auxiliary connecting pipe", and the cooling steam flow can be adjusted in a wide range. When the low pressure cylinder operates at zero output, adjust the steam regulating valve of the cooling bypass of the low pressure cylinder to make the unit "flexible cylinder cutting operation"; Third, on the basis of the first design method, the cooling steam is desuperheated and depressurized, and the steam and water are separated, reducing the demand for cooling steam, but the steam temperature at the outlet of the steam water separator must be controlled, increasing the preparation time for cylinder cutting.

3. Enable/disenable technical parameters

3.1 Several operation modes of cylinder cutting

1) Conventional cylinder cutting operation. The butterfly valve opening of the connecting pipe of the middle and low pressure cylinder is 0, and the cooling bypass steam just takes away the blast heat generated by the rotation of the low-pressure rotor.
2) Flexible cylinder cutting operation. As the "auxiliary connecting pipe", the cooling bypass steam pipe has a wide adjustable range of cooling steam flow. When the butterfly valve opening of the connecting pipe of the middle and low pressure cylinder is 0, adjust the cooling bypass steam flow to enable the flexible cylinder cutting operation of the unit.
3) Half cutting cylinder operation. The butterfly valve opening of the connecting pipe of the middle and low pressure cylinder shall be maintained at 5%~14%, and the low pressure cylinder shall retain a certain amount of steam. In this state, the operation shall still be carried out in the manner of cylinder cutting.
4) Switch to single cylinder/double cylinder operation. For 600MW units and above, there are generally two low-pressure cylinders. After cylinder cutting, there are two operation modes: single cylinder cutting and double cylinder cutting.

3.2 Preparation for cylinder cutting

1) The heat supply input is normal, whether the exhaust parameters and heating pressure of the intermediate pressure cylinder are within the safe range, and whether the opening of the steam supply regulating valve is within the required range;
2) Fully open the bypass cooling steam regulating valve of LP casing, and check that the bypass cooling steam flow reaches the required value;
3) Gradually reduce the opening of butterfly valve of connecting pipe to 15% or a certain limit, and check that the heating pressure and flow are normal;
4) Some units need to cut off the steam inlet at the steam side of #5 and #6 low pressure heaters;
Fully open the desuperheating water regulating valve of low pressure cylinder.

3.3 Control of main parameters of cylinder cutting

1) During the normal operation of the unit, the boiler load is basically more than 50%, and can be switched at any time except for the restriction of people's livelihood heating;
2) Parameters such as on/off time (excluding heating of cooling bypass) and cooling water volume change are shown in Table 1.

Table 1 Parameters such as startup/shutdown time and cooling water volume change of some units

Unit No	Cylinder cutting input time (min)	Cylinder cutting exit time (min)	Cooling steam flow (t/h)	Low pressure cylinder spray water flow (t/h)	Maximum temperature limit of last stage and next last stage (°C)	Actual operating last stage and next last stage temperature (°C)
1	5 ~ 7	5 ~ 7	20	7 ~ 10	100/175	33/70
2	1 ~ 2	1	8+Desuperheating water below 1t/h	2 ~ 4	140/140	40/50
3	5	2	20	10	60/80	60/70
4	4	8 ~ 9	26 ~ 27	6 ~ 7	110/150	23/62
5	30 ~ 40	10	10+Desuperheating water below 2t/h	0.5 ~ 2	50/110	50/110
6	15	5	10 (including desuperheating water)	11 ~ 13	140/140	40/100

As shown in the above table, the operation time of the unit under normal cylinder cutting operation is generally within 10 minutes, and some units need to be cooled and depressurized before cylinder cutting. The operation time is long, about 30-40 minutes; The exit time is generally within 10 minutes.

The operation time for cylinder cutting of some units is relatively long, and the reasons are analyzed as follows:

1) The exhaust parameters of intermediate pressure cylinder are high. The steam exhaust of the intermediate pressure cylinder is four extraction, which is higher than the steam exhaust parameters of the unit with five extraction in the middle. Before cylinder cutting, it is necessary to reduce the temperature and pressure, which leads to a long preparation time before cylinder cutting.

2) The outlet temperature of steam separator is high. Temperature and pressure reduction and steam separator are added to the cooling steam bypass of some units. Before cylinder cutting, the outlet temperature of steam separator needs to be lowered, which leads to a long preparation time before cylinder cutting (the operation mode of some units is to cut the cylinder first and then lower the outlet temperature of steam separator, which shortens the time for cylinder cutting).

3.4 Change of main parameters during cylinder cutting

For 300MW units, the electrical load after cylinder cutting is about 10MW~20MW lower than that before cylinder cutting, and the thermal load is about 120t/h higher when the main steam flow is unchanged. Under the condition of constant external heat load, cylinder cutting can reduce the power generation by about 30% compared with no cylinder cutting.

3.5 Main risks or problems in cylinder cutting adjustment

3.5.1 The butterfly valve of connecting pipe of middle and low pressure cylinder is jammed or not tight.

When the butterfly valve of the connecting pipe of the middle and low pressure cylinder fails, the original steam extraction heating operation mode shall be restored immediately. The maintenance personnel and technicians of the butterfly valve manufacturer shall analyze the cause of the failure, and the cylinder cutting operation shall be carried out after the relevant failure is completely eliminated.

3.5.2 After cylinder cutting, when the external heat load is low, there is a risk of intermediate exhaust overpressure.

When the high exhaust pressure alarm of the intermediate pressure cylinder occurs, the exhaust pressure of the intermediate pressure cylinder can be reduced by increasing the opening of the quick closing regulating valve for heating and steam extraction, and appropriately reducing the boiler load or the circulating water supply temperature of the heating network; If none of the above measures can restore the exhaust pressure of the intermediate pressure cylinder to the normal operating range, quickly open the butterfly valve of the intermediate and low-pressure connecting pipe until the exhaust pressure of the intermediate pressure cylinder returns to the normal operating range.

3.5.3 TSI monitoring parameters may be abnormal.

3.5.4 The water spray of the rear cylinder of the low pressure cylinder is blocked or the linearity difference of the adjustment valve causes the high temperature of the last stage and the second last stage blades.

It is recommended to control the last stage temperature of the low pressure cylinder and the exhaust temperature of the low pressure cylinder not to exceed 80 °C during operation, especially during the zero output operation of the low pressure cylinder. When the unit is shut down, check the water spray blockage of the rear cylinder through the manhole of the low pressure cylinder and deal with it. Check the water spray blockage of the rear low-pressure cylinder before starting as a routine inspection item for unit startup.

When the temperature of the last stage and the second last stage blades is too high, the desuperheating water flow of the low pressure cylinder can be increased by increasing the opening of the desuperheating water regulating valve of the low pressure cylinder and the outlet pressure of the condensate pump, and the opening of the bypass regulating valve of the connecting pipe of the middle and low pressure cylinders can be appropriately increased to increase the steam inlet flow of the low pressure cylinder until the temperature of the second last stage of the low pressure cylinder returns to the normal level.

3.5.5 The rear cylinder spray opening is too large, and the last stage blades of the low pressure cylinder are at risk of water erosion.

As the opening of the water injection regulating valve of the low-pressure cylinder decreases, the water injection flow decreases accordingly. When the exhaust temperature of the low-pressure cylinder is stabilized at the saturation temperature corresponding to the exhaust pressure+5 °C, stop reducing the water injection flow of the low-pressure cylinder. In the process of adjusting the water spray control valve of the low pressure cylinder, the opening change each time is not more than 5%. When the opening of the water spray control valve of the low pressure cylinder is less than 50%, the opening change each time is reduced to 2%. After the adjustment of the opening of the control valve each time, the next operation can be started after the final stage and exhaust temperature of the low pressure cylinder are stable.

3.5.6 The water level of deaerator and low-pressure heater may fluctuate when the cylinder cutting is on or off.

When the deaerator water level alarm is high, open the deaerator overflow valve, close the stage by stage drainage of No. 3 high-pressure heater, and close the deaerator water feeding main control valve; Close the deaerator steam inlet valve and open the deaerator emergency drain valve.

3.5.7 The thermal logic control problem causes a sudden change of a parameter or the cylinder cutting cannot be switched on or off normally.

If the butterfly valve opening is not put into manual operation control when the low pressure cylinder zero output operation mode is exited, the butterfly valve opening is automatically tracked to calculate the butterfly valve opening of the automatic control circuit for heating and steam extraction pressure, which may lead to sudden change of the hydraulic butterfly valve opening when the low pressure cylinder zero output operation mode is exited, causing certain disturbance to the unit operation. Considering that the unit is in operation, there are certain safety risks in online modification and downloading of control logic. It is recommended that the thermal engineering discipline forcibly exit the zero output exit button of the low pressure cylinder.

3.5.8 The cooling tower freezes during cylinder cutting operation.

Measures such as closing the water pouring in the cooling tower, switching the main bypass and covering the canvas are mainly taken.
During actual operation, the operating personnel shall closely monitor and timely adjust the main steam pressure, water level of steam drum, water level of deaerator and condenser, etc., and monitor and record the vibration, differential expansion, axial displacement, secondary and final stage blade temperature, low pressure cylinder exhaust temperature and other parameters of the turbine shaft system. If the parameters change abnormally, suspend the operation in time and analyze the reasons. If the parameter exceeds the limit, the emergency response measures shall be followed.

4. Conclusion

This paper mainly introduces the zero output heat supply transformation of low pressure cylinder in Shandong Province, the basic route and operation of the transformation, as well as three different low pressure cylinder cooling steam systems, and introduces in detail the main risks and solutions in the cylinder cutting adjustment process for reference by power generation enterprises.

Acknowledgements

This work was supported by the scientific and technological project of State Grid Shandong Electric Power Research Institute "Research on Thermoelectric Cooperative Regulation Technology of Nuclear Power Units" (ZY-2022-10).

References

1. Ronghe Hao, Zhuang Li, Liping Wang, et al. Economic Analysis of Smooth Shaft Operation of Low Pressure Cylinder in 210 MW Unit[J]. Journal of Shenyang Institute of Engineering (Natural Science), 2020,16(02):25-28+44.
2. Jianxun Wang. Analysis of Flexible Peak-Load R egulation Capability and Economy on the Zero Output Technology of Low-Pressure Cylinder for 650 MW Supercritical Unit[J].Journal of Engineering for Thermal Energy and Power,2021,36(02):18-23.
3. Tianfu Liang, Weiyang Xie, Fei Wang, et al.The key technique research of removing the low pressure cylinder of steam turbine[J].Turbine Technology, 2019, 61(6): 471–472.
4. Haisheng Yang, Tuo Zhang , GuangTong Tang, et al. Influence of Zero-Output Technology of Low-Pressure Cylinder on Deep Peak Regulation Performance of Heating Unit and Compensation Standard for Peak Regulation[J].Journal of Engineering for Thermal Energy and Power, 2020, 35(06):268-273.T
5. Qinpeng Zhang, Xuedong Wang, Feng Li, et al. Analysis of Heating Capacity and Peak-regulating Capacity of 330 MW Steam Turbine Unit With Low-pressure Cylinder off Operation[J]. Shandong Electric Power, 2020,47(12):72-76.
6. Puxin Shi, Peiran Shi, Peiwen Wang, et al.Analysis and Operation Evaluation of Power Peak-shaving Ancillary Service Market in North China[J]. Automation of Electric Power Systems, 2021,45(20):175-184.
7. Xuhui Zhang, Zhonghua Zhao, Fuxing Cui, et al. Experimental Research on Deep Peak Regulation Characteristics of 1030 MW Ultra Supercritical Unit[J]. Shandong Electric Power, 2021,48(02):58-62.
8. Lingkai Zhu, Wei Zheng, Junshan Guo , et al. Application of Heat Load Distribution in Improving Peak-shaving Ability of Plant[J].Power System Engineering, 2019,35(06):43-46+49.

Medium- and long-term power load forecasting model

Yuning Zhu [1], Xiaoliang Xu [1], Ruoqi Lai [1], Hengqing Guo, Fan Jia [2]

[1] Yongchuan Power Supply Branch, State Grid Chongqing Electric Power Company, Chongqing, 402160, China
[2] North China Electric Power University, Beijing, 102206, China

Abstract. Power load is an important part of power system, and power load forecasting has an important impact on power system analysis, design and control. With the development of smart micro grid, load forecasting has gradually become an important module in the energy management system, It is "source, network, load and storage" "An important link in energy flow matching. The staged combined demand forecasting model of power grid based on neural network and polynomial regression is adopted, and judgment conditions are added to the neural network. If the training sample data does not converge in the neural network training process, the neural network forecasting is terminated, and the data is automatically transferred to the polynomial regression model to obtain the forecasting results. This method can be initially used for annual and monthly load forecasting. It is an intelligent micro grid The planning of has laid a certain technical foundation.

Keywords. Load forecasting; Neural network; polynomial regression.

1. Introduction

Load forecasting is to explore the internal relationship between power load and various related factors according to the change law of power load, so as to develop a theoretical method that can scientifically reflect the relationship between past and future loads, and finally obtain the load value at a specific time in the future [1]. Load forecasting plays an extremely important role in conventional power systems and smart microgrids. High precision load forecasting is convenient for scientific dispatching and optimal configuration of power networks, which can provide users with safe and reliable power and improve social and economic benefits. According to the length of time, load forecasting can be divided into long-term, medium-term, short-term and ultra short-term forecasting[2]. Medium and long term load forecasting is usually based on the year, which can provide data reference for micro grid planning and construction; Short term load forecasting can usually be carried out in monthly, weekly, daily and hourly units, which should provide basis for the adjustment of power balance between regions; Ultra short term load forecasting is usually used for real-time dispatching, system security analysis, control strategies, and improving economic benefits in minutes. Medium and long-term load forecasting methods include unit consumption method[3], electric power elasticity coefficient method[4], trend extrapolation method [5], regression analysis method [6] and artificial neural network method [7]; Short term and ultra short term load forecasting methods include time series method [8] ,grey forecasting method [9], support vector machine method [10], wavelet analysis method [11] and comprehensive model. In the actual load forecasting work, data anomalies may occur in the detection, recording, conversion, transmission and other processes of historical data, or load data may change abnormally due to sudden changes in weather, line maintenance, emergencies, etc. [13]. These abnormal data usually interfere with the load forecasting results. If they are directly used for load forecasting, they will have a certain impact on the forecasting accuracy, or they can not get an ideal forecasting result at all [14] This paper uses the forecasting method combining regression analysis and trend extrapolation to identify abnormal data through horizontal comparison of load data, and then correct the abnormal data through interpolation, and then analyze the load data of historical time series, Find the appropriate function model curve to fit, and then get the load forecast data.

2. Combined demand forecasting model based on neural network and polynomial regression

2.1 Neural network model

The advantage of neural network is that it can effectively adapt to the laws of nonlinear relations through a large number of learning and training processes, and well fit and map the nonlinear relations. The rules for learning and training are easy to understand and operate. Based on the above characteristics, the neural network model is applicable to the prediction of power economic indicators that require high prediction efficiency and are affected by complex factors. The basic structure of neural network is composed of input layer, hidden layer and output layer. The input layer and the hidden layer have N and L neurons respectively, and usually meet L>2N; The number of neurons in the output layer is M. The topology of BP neural network is shown in Figure 1.

The sample data is input into the input layer and the output layer, and the computer is used to learn the mapping relationship between the input and output parameters. The use of the model does not depend on the specific expression of the mapping relationship, but is based on the principle of the minimum error signal, using the back transmission error signal to adjust the threshold and link weight of the neuron. The error signal refers to a certain norm of the prediction result and the expected value. Based on the back transmission of the error signal, the weight vector under the minimum error condition is calculated. Since the error is a discrete function, the minimum value cannot be calculated simply by derivative, but the minimum gradient should be calculated by iterative process.

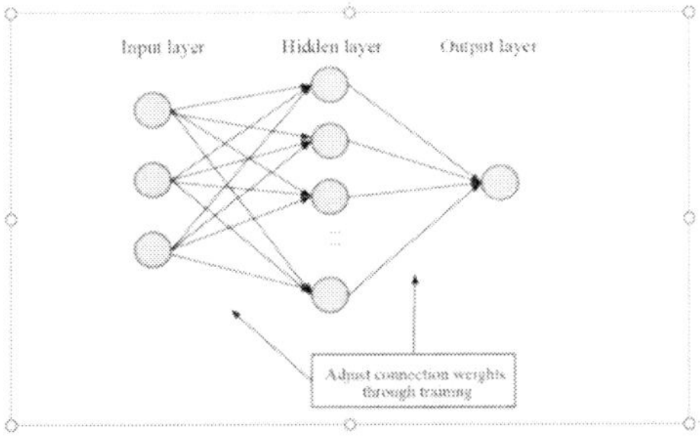

Fig 1 The topology of BP neural network

In practice, the number of iterations n and the error limit are set in the initial conditions. In the process of error back propagation, the weight correction can reduce the error, and the component change of the weight vector is consistent with the direction of gradient reduction. Theoretically, when the number of samples tends to infinity, the error should converge, but in practical application, when the number of iterations n reaches the maximum, the error still cannot meet the requirements of the error limit, and the network training fails, that is, it does not converge. If the goal is to solve the weight vector meeting the gradient limit requirements, the network training failure is that the gradient is still not less than the specified value after the completion of iterations, and the goal weight vector cannot be calculated.

2.2 Polynomial regression model

Regression analysis is used to analyze the functional relationship between dependent variables and independent variables. It is divided into polynomial regression and multiple polynomial regression. When there is only one independent variable, it is a single variable regression, and multiple independent variables are multiple regression. The regression equation of unary m-degree polynomial is:

$$\hat{y} = b_0 + b_1 x + b_2 x^2 + \cdots + b_m x^m$$

In the univariate regression analysis, if the relationship between the dependent variable and the independent variable is nonlinear, but no appropriate function curve can be found to fit, the univariate polynomial regression can be used. The greatest advantage of polynomial regression is that the measured points can be approximated by adding higher order terms of x until satisfactory results are obtained. Polynomial regression is a powerful tool to study the specific quantitative relationship between elements, which can establish a mathematical model reflecting the specific quantitative relationship between geographical elements. It can accurately measure the degree of correlation between various factors and the degree of regression fitting, and can accurately express the complex change rules of data series. It has been proved that continuous functions on a closed interval can always be approximated by a polynomial of a higher power under the specified accuracy. On the premise of large sample data, the use of polynomial regression can also achieve better prediction results. Regression analysis takes polynomial regression as a typical representative, and any continuous function can be approximated by this method by dividing the interval. Therefore, this method has been widely used in finance, voice and other fields.

2.3 A staged combined demand forecasting model based on neural network and polynomial regression

From the above analysis, we can know that the advantage of neural network is that it can effectively adapt to the laws of nonlinear relations through a large number of learning and training processes, and well fit and map nonlinear relations. In addition, its learning rules are simple and easy to implement. Therefore, it can effectively solve the problem of power demand forecasting. However, its shortcomings are also obvious. In the training process, it is difficult to ensure the global optimization, and it is easy to fall into the local. Therefore, in order to prevent the optimization from falling into the local, this paper adds a polynomial regression model. The specific solutions are as follows. After the convergence condition of BP neural network is set, for example, the relative error is not greater than 3% according to the power system regulations. In the training process, there may be non convergence, that is, when training and learning samples, it is difficult to ensure that the neural network is globally optimal, and it is easy to fall into the cycle of local optimal, thus affecting the prediction accuracy. At this time, in order to continue to obtain the prediction results of the given data, it is necessary to use other prediction methods. This paper adds judgment conditions to the neural network model, such as training 100000 times. When the sample data training does not converge, the sample data will be automatically filled into the polynomial regression model, so as to terminate the neural network prediction, jump out of the local minimum

 https://doi.org/10.1051/e3sconf/202337503026

loop, and build a polynomial regression model to obtain the prediction results. The combined model method is automatic intelligent training and learning, with high accuracy and generalization ability, and can be used in the prediction of various indicators of power economy.

The advantage of the neural network based on the neural network polynomial regression staged combination model for complex power grid demand forecasting is that it can effectively adapt to the laws of nonlinear relations through a lot of learning and training, and well fit and map the nonlinear relations. In addition, its learning rules are simple and easy to implement. Based on the above characteristics, the neural network model is applicable to the forecasting of power economic indicators that require high forecasting efficiency and are affected by complex factors, and has been widely used in power load forecasting. However, when the neural network is training and learning samples at the same time, it is difficult to ensure the global optimization, and it is easy to fall into the local optimal cycle, which may affect the prediction accuracy. Polynomial regression is a powerful tool to study the specific relationship between elements, which can establish mathematical models reflecting the specific relationship between geographical elements. It can accurately measure the degree of regression fit and correlation between various factors. It has been proved that continuous functions on a closed interval can always be approximated by a polynomial of a higher power under the specified accuracy. Therefore, this method has been widely used in finance, voice and other fields. Therefore, this paper establishes a neural network and polynomial staged combination forecasting model to forecast regional power grid load. In order to avoid the local minimum problem of the neural network, the solution given in this paper is to add judgment conditions in the neural network, such as the relative error of 0.3 and the training times of 100000. If the training sample data still does not converge during 100000 times of training, the data will be automatically transferred to the polynomial regression model, the neural network prediction will be terminated, the local optimization will be skipped, and the polynomial regression model will be established to obtain the prediction results. The staged combination method can be trained and learned automatically and intelligently, and it has high precision and generalization ability. In the established neural network model, there is only one hidden layer, and the number of neurons in the hidden layer adopts the trial and error method. First, set fewer neurons, and gradually increase the number of neurons in the training process until the training is successful and the error reaches the set range. If the sample is still not convergent after 100000 times of training, the data will be automatically filled into the polynomial regression model, the neural network prediction will be terminated, the endless loop will be jumped out, and the polynomial regression model will be established to obtain the prediction results.

3. Example analysis

In order to verify the validity of the model, this paper selects the load demand and influencing factor data of Chongqing, China, from 2012 to 2021 as the modeling data to forecast the medium and long-term load demand of power demand. Based on the field survey and data collection, we found that the high energy consuming industries in this region account for a large proportion of electricity consumption. By analyzing the correlation between independent variables and dependent variables, this paper finally selects four macroeconomic factors including regional gross product, regional population, industrial gross product and the whole society's fixed asset investment as influencing factors to forecast the annual maximum load demand of the power grid. The historical data of the maximum annual load and its influencing factors in the region from 2012 to 2021 are shown in Table 3-1. Because the dimensions of different factors are different, in order to facilitate comparison, the following formula is used for data standardization and normalization.

$$y_{i,j} = \frac{2*(x_{i,j} - x_{\min})}{x_{\max} - x_{\min}} - 1$$

Table 1 Historical Data of Maximum Load and Influencing Factors

particular year	Annual maximum load (10000 kW)	GDP (100 million yuan)	Regional population (10000 persons)	Regional population (10000 persons)	Investment in fixed assets of the whole society (100 million yuan)
2012	1320.00	11595.40	3343.44	5308.14	3193.31
2013	1300.00	13027.60	3358.42	5988.62	4236.20
2014	1430.00	14623.80	3375.20	6774.58	4566.35
2015	1324.00	16040.50	3371.84	7208.01	4904.05
2016	1676.00	18023.00	3392.11	7765.38	5325.94
2017	1842.00	20066.30	3389.82	8455.02	5706.71
2018	2048.00	21588.60	3403.64	8842.02	6076.38
2019	2138.00	23605.77	3416.29	9391.96	6307.39
2020	2188.00	25041.43	3514.23	9992.21	7048.92
2021	2350.00	27894.02	3546.23	11184.94	7353.26

Table 2 Normalized Historical Data

particular year	Annual maximum load	GDP	Regional population	Regional population	Investment in fixed assets of the whole society
2012	-0.962	-1.000	-1.000	-1.000	-1.000
2013	-1.000	-0.824	-0.852	-0.768	-0.499
2014	-0.752	-0.628	-0.687	-0.501	-0.340
2015	-0.954	-0.455	-0.720	-0.353	-0.178
2016	-0.284	-0.211	-0.520	-0.164	0.025
2017	0.032	0.039	-0.543	0.071	0.208
2018	0.425	0.226	-0.406	0.203	0.386
2019	0.596	0.474	-0.282	0.390	0.497
2020	0.691	0.650	0.684	0.594	0.854
2021	1.000	1.000	1.000	1.000	1.000

The normalized input and output data are brought into the neural network and polynomial regression staged combination prediction model proposed in this paper for training and learning. We set the convergence condition of the neural network as: relative error 0.3. When the number of training $N \geq 100000$ is still not convergent, stop training the neural network and establish a polynomial regression model. According to the prediction model combining neural network and polynomial regression, the results are shown in Figure 2. At the same time, this paper uses BP neural network and multiple linear regression model to forecast the load demand of the region, and makes a comparison. The prediction results of the comparison model are shown in Figure 2.

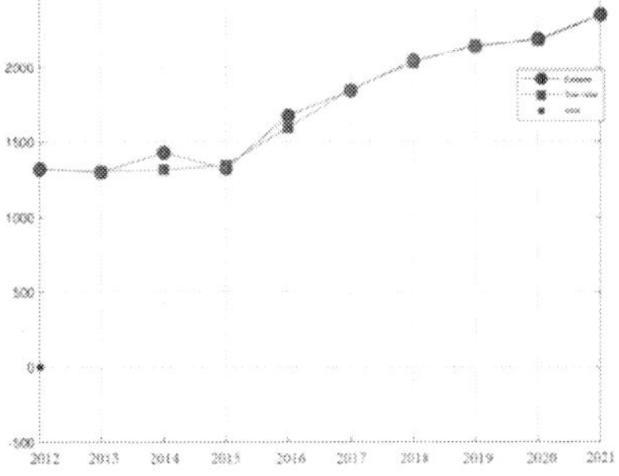

Figure 2 Comparison between predicted and actual values of test set

It can be seen from Figure 2 that the overall fitting effect of BP neural network and polynomial staged combination forecasting model is good, with the absolute error value MAE of 0.066 and the mean square error (MSE) of 0.430. The neural network multiple regression staged combination forecasting model has obvious advantages, can obtain higher forecasting accuracy, and its high accuracy and generalization ability can be applied to medium and long-term power demand forecasting.

4. Conclution

In this paper, a staged combined demand forecasting model of power grid based on neural network and polynomial regression is used to add judgment conditions to the neural network. If the training sample data does not converge in the neural network training process, the neural network forecasting is terminated, and the data is automatically transferred to the polynomial regression model to obtain the prediction results. The average prediction error is below 10, which fully demonstrates the practicability and effectiveness of this method, and can lay a certain technical foundation for the subsequent microgrid planning and configuration.

References

1. Zhou Hu, Jiang Yuechun, Chen Xu, et al. Short term load forecasting combining fuzzy clustering analysis and algebraic algorithm [J]. Journal of Power System and Automation, 2011,23 (3): 101-105.
2. He Miao, Ye Peng, He Ningning, et al. Overview of research on saturated load forecasting of power system [J]. Journal of Shenyang Institute of Engineering, 2017,13 (4): 340-345
3. Yang Shuangji, Dong Qingfeng. Application of output unit consumption method in load forecasting [J]. Journal of North China University of Water Resources and Hydropower, 2009, 30 (6): 73-76
4. Qian Wen. Load Prediction by Electric Power Elastic Coefficient Method [J]. Yunnan Electric Power Technology, 2001, 29 (1): 41-44
5. Ding Qia, Lu Jiangang, Qian Yumei, et al. A practical extrapolation method for ultra short term load forecasting curve [J]. Power System Automation, 2004,28 (16): 83-85
6. He Jing, Wei Gang, Xiong Lingling, et al. Fuzzy improvement of linear regression analysis method for load forecasting [J]. East China Power, 2003 (11): 21-23
7. Li Long, Wei Jing, Li Canbing, et al. Load Model Based on Artificial Neural Network [J]. Journal of Electrotechnics, 2015, 30(8):225-230.
8. Wang Qiumei. The principle and application of time series load forecasting [J]. East China Electric Power, 1993 (4): 37-39
9. Xiong Gang. Long term power demand forecasting based on grey system theory [J]. Journal of Shanghai Jiaotong University, 1993,27 (5): 15-18

10. Wang Ben, Leng Beixue, Zhang Xihai, et al. Overview of the application of support vector machine in short-term load forecasting [J]. Journal of Power System and Automation, 2011,23 (4): 115-121.
11. Tai Nengling, Hou Zhijian, Li Tao, et al. Short term load forecasting method of power system based on wavelet analysis [J]. Chinese Journal of Electrical Engineering, 2003,23 (1): 45-50
12. Cheng Xu, Kang Chongqing, Xia Qing, et al. Comprehensive model of short-term load forecasting [J]. Power System Automation, 2000 (9): 42-44
13. Chang Xianrong, Sun Jingwen. Load preprocessing based on improved fuzzy C-means clustering [J]. Journal of North China Electric Power University, 2014,41 (1): 27-32. [14] Kang Chongqing, Xia Qing, Liu Mei. Power system load forecasting [M]. Beijing: China Power Press, 2017
14. Shao, Z; Gao, F; Zhang, Q; Yang, SL. Multivariate statistical and similarity measure based semiparametric modeling of the probability distribution: A novel approach to the case study of mid-long term electricity consumption forecasting in China[J] . Applied Energy, 2015, 156: 502-518.

Study on the electrothermal coupling characteristics of different types of thermoelectric units

Ziwei Zhong[1,*], Lingkai Zhu[1], Wei Zheng[1], Zhiqiang Gong[1], Panfeng Shang [1], Junshan Guo[1], Jun Liu [1], Yanpeng Zhang[1]

State Grid Shandong Electric Power Research Institute, Jinan, China

Abstract. The electric heating characteristics of thermoelectric units refer to the coupling relationship between the generating power P and the steam extraction volume G of the unit for external heating. The heat supply forms of directly regulated public thermoelectric units mainly include steam extraction condensing heat supply and back pressure circulating water heat supply. Through the research on the electrothermal coupling characteristics of different types of thermoelectric units, we can master the electric load output range of thermoelectric units in the heating season, provide data support for the power grid dispatching department, effectively promote the consumption of new energy, and ensure the smooth and orderly supply of electric energy.

key word: Thermoelectric unit; Electrothermal coupling characteristic; Deep peaking.

1. Introduction

With the development of urbanization, the demand for clean and efficient central heating is becoming increasingly urgent[1]. Thermal power units can give consideration to both power generation and heat supply, with high energy utilization efficiency and significant economic benefits of energy conservation and emission reduction. Therefore, the proportion of thermal power units in coal power units is increasing year by year.During the "13th Five Year Plan" period, in order to comprehensively promote the implementation of the "three reform linkage" of coal power units, Shandong Province will complete the heat supply transformation of 30 sets of 19.9 million kilowatt units. In recent years, heat supply units have begun to develop towards large capacity and high parameters, and the capacity and heating area of thermoelectric units are increasing. In the future, with the reduction of coal consumption and the shutdown of small boilers and small thermoelectric units, the heating area and pressure of thermoelectric units will continue to increase[2].

Thermal power unit refers to a generator unit that not only produces electric energy, but also uses steam to heat users[3]. The thermoelectric unit is used to extract part of the steam that has been worked in the steam turbine and directly or indirectly supply it to the heat users, so as to reduce the cold source loss of the steam cycle (that is, the heat taken away by the circulating water in the condenser), so it has a good economy.

Thermal power units can be divided into back pressure units and condensate extraction units according to different heating modes. The heating modes and thermal characteristics of the two types of units are analyzed in detail below.

2. Thermodynamic characteristics analysis of different types of thermoelectric units

2.1 Back pressure unit

To increase the exhaust steam pressure of the steam turbine, the exhaust steam is used to provide heat directly to the heat users or the heat exchanger is used to provide heat indirectly to the heat users. The steam turbine that uses the exhaust steam to provide heat is called back pressure steam turbine. For back pressure steam turbine, the total heat of steam turbine exhaust is used for regenerative water supply and external heating. In actual operation, the electromechanical load of back pressure steam turbine is determined by the external thermal load, and its insufficient power is borne by the condensing steam turbine units operating in parallel in the power plant[4]. The back pressure turbine is also divided into a pure back pressure type B unit (with only one type of steam exhaust) and an extraction back pressure type CB unit (with two-stage regulating extraction of steam extraction with higher pressure and lower pressure). The exhaust steam of the back pressure turbine is used for heating, and the heat released from the exhaust steam is no longer an energy loss. The heat rate of the unit is independent of the cycle efficiency of the thermal system

* Corresponding author: dkyzhongziwei@163.com

 https://doi.org/10.1051/e3sconf/202337503027

and the relative internal efficiency of the unit, but only depends on the boiler efficiency, pipe efficiency, mechanical efficiency and generator efficiency. The heat supply mode of the back pressure machine is shown in Figure 1:

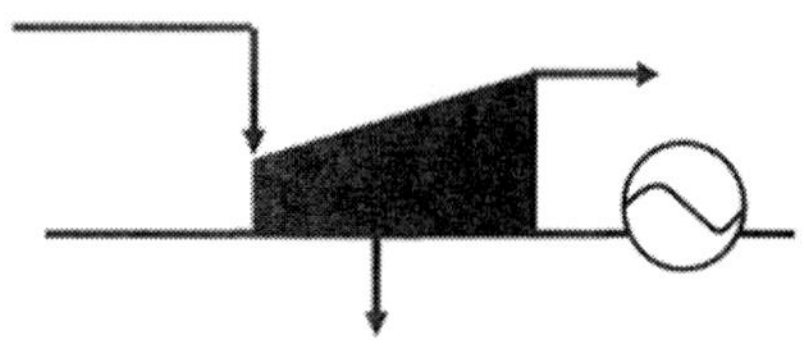

Fig. 1 Heat supply schematic diagram of back pressure machine

(1) Heating mode

In Shandong Province, pure back pressure units and extraction back units are generally small units below 50MW. Generally, this type of unit is used more in small local thermal power plants, and this type of unit has been basically eliminated in large direct dispatching power plants. Therefore, this project will not study the heating mode and thermal characteristics of this type of unit too much.

In recent years, with the increase of heat supply demand, some large capacity pure condensing or extraction condensing units in Shandong Province have also started to carry out heat supply transformation, including high back pressure transformation, optical shaft transformation and removal of low pressure cylinder operation. The heat supply mode of the transformed units is basically the same as that of the back pressure units.

The high back pressure heat supply unit is to improve the exhaust steam pressure of the low pressure cylinder rotor of the unit by retrofitting it, and directly heat the circulating water for external heat supply by using the exhaust steam. In this way, all the exhaust steam waste heat of the unit is taken away by the circulating water for heat supply, and the cold source loss is zero, which greatly improves the economy of the unit.

This type of unit is a retrofitted unit that has emerged in recent years to adapt to the heating and heating in the north. Most of them are retrofitted from pure condensing or pumped condensing units. It was used in Northeast China in the 1980s. In the 1990s, Shandong Huangtai Power Plant # 1 Unit (50MW) took the lead in reforming and applying high back pressure heat supply. In 2009, Yantai Power Plant carried out the transformation of high back pressure circulating water heating on the 150MW unit for the first time. The successful transformation of Yantai Power Plant lies in the elimination of the cold source loss of the unit during the heating period in winter, and the coal consumption for power generation is reduced to 150 g/kW H below, the overall operation is relatively normal. The main problem is that the efficiency of low pressure cylinder in non heating period is too low, about 16.8 percentage points lower than the design value, and the heat rate is far higher than the design value about 770kJ/kW h。 After further transformation, the problem of low efficiency under summer working conditions was basically solved by disassembling and assembling blades[5].

In order to meet the heat exchange requirements of the primary heating network and the secondary heating network as much as possible, the high back pressure circulating water heating adopts a tandem two-stage or three-stage heating system. The circulating water of the heating network first heats up through the condenser to absorb the waste heat from the exhaust steam of the low pressure cylinder, and then completes the second heating through the local extraction steam heater in the initial heating station[6]. In case of heat supply spikes, the temporary extraction steam heater is used for the third heating to generate high temperature hot water, The hot water is sent to the hot water pipe network to exchange heat with the circulating water of the secondary heat network through the secondary heat exchange station. The high-temperature hot water is cooled and then returned to the unit condenser to form a complete circulating water circuit.

(2) thermodynamic property

Without considering heat supply, high back pressure heat supply units are actually pure condensing units with high exhaust pressure. Therefore, from the perspective of power grid peak shaving, high back pressure heat supply units can participate in peak shaving. Their maximum electrical load is limited by the maximum evaporation capacity of the boiler, and the minimum electrical load is limited by the minimum stable combustion load of the boiler.

However, in the current dispatching mode of Shandong power grid, considering that the high back pressure heating units have no cold source loss and are more economical, the high back pressure heating units do not participate in the peak shaving of the power grid. The power plant reports the output curve of the units for the next day to middle note every day, and middle note dispatches according to the output curve reported by the power plant. However, as the number of units with high back pressure below 300MW in the province gradually increases, new energy grid connection and external power into Shandong increase, The peak shaving pressure of the power grid is increasing day by day, so the possibility of high back pressure heating units participating in peak shaving will not be ruled out in the future.

Under the current situation, for thermal power plants, the high back pressure heat supply units do not participate in peak shaving. Therefore, in actual operation, the high back pressure heat supply units are brought to full load to bear the main heat load. Other condensing units can bear less heat load, and the peak shaving capacity is enhanced, thus improving the overall peak shaving capacity of the whole plant.

2.2 Condensate extraction unit

(1) Heating mode

Condensate extraction unit refers to the steam extracted from an intermediate stage of the steam turbine to supply heat to external heat users, thereby reducing the amount of steam discharged into the condenser. According to the nature of heat users, it can be divided into single extraction (industrial steam), single extraction (heating steam) C-type units (steam pressure can be different), and CC type units that also provide two types of steam or two extraction parameters for industrial or heating steam. The condensing unit can meet the needs of both thermal and electrical loads changing at any time. Extraction steam is extracted from an intermediate stage of the turbine and sent to the heat user. After the heat user releases heat, the condensate is partially or completely recycled to the drain collector of the power plant, and then sent to the boiler by the water pump for recycling. If the heat load increases during operation, the steam turbine will increase the steam inlet according to the demand of heat load to meet the demand of external heat load increase. When the external heat load changes, the extraction capacity for heat supply and the total steam inlet capacity of the turbine can be adjusted at the same time to keep the power generation unchanged.

(2)Thermodynamic characteristics

For the condensing unit, the peak shaving capacity of the unit is directly related to the external steam extraction volume. When the steam extraction volume changes, the output range of the unit also changes. Generally speaking, the larger the steam extraction volume of the unit is, the smaller the adjustable range of the unit load is. The upper limit of the unit load is subject to the maximum steam inlet volume of the turbine, and the lower limit of the load is subject to the minimum steam inlet volume of the low pressure cylinder or the middle exhaust temperature limit. The relationship between the main steam flow, steam extraction volume and load under the steam extraction condition of the unit can be expressed by the heating working condition diagram. If two of the three parameters are known, the other one can be found according to the heating working condition diagram.

3. Limiting conditions for peak shaving capacity of thermal power units

The peak shaving capacity of thermal power units is the highest and lowest electrical load that the unit can reach under a certain external steam extraction capacity. The limiting conditions for different types of thermal power units to carry the highest and lowest electrical loads are also different.

3.1 Limiting conditions for peak shaving capacity of high back pressure heat supply units

1) The maximum electrical load of the unit is limited by the maximum evaporation capacity of the boiler, the wall temperature limit of the boiler heating surface, the quality of steam and water, the design output of the boiler denitration, desulfurization, dust removal and slag removal devices, the design output of the main auxiliary machines including the pulverizing system, six fans, environmental protection parameters, the maximum steam inlet flow of the turbine and the maximum exhaust temperature of the low-pressure cylinder.

2) The minimum electrical load of high back pressure heat supply unit with steam extraction is limited by the minimum steam inlet of low pressure cylinder.

3) For high back pressure heating units without steam extraction, the minimum electrical load is limited by the minimum stable combustion load of the boiler, flue gas temperature at the inlet of the denitration device, wall temperature of the heating surface, and environmental protection parameters.

4) The monitoring parameters affecting the safety of the unit shall not exceed the alarm value.

3.2 Limiting conditions for peak shaving capacity of condensate extraction unit

1) Different heat users have different requirements for steam extraction parameters, and the steam extraction parameters of the unit cannot exceed their limits.

2) The minimum electrical load of the unit is limited by the minimum steam inlet of the low-pressure cylinder, the minimum stable combustion capacity of the boiler, the flue gas temperature at the inlet of the denitration device, the wall temperature of the heating surface, environmental protection parameters, etc.

3) For the reheater unit that extracts steam from the exhaust pipe of the high-pressure cylinder, when the electrical load is low and the amount of heat and steam extraction is large, the reheater wall temperature is easy to overheat, which limits the minimum electrical load of the unit.

4) For the unit that extracts steam from the connecting pipe of the intermediate and low pressure cylinder, the exhaust pressure and temperature of the intermediate pressure cylinder shall not exceed the limit value of the manufacturer.

5) The maximum electrical load of the unit is limited by the maximum evaporation capacity of the boiler, the wall temperature limit of the boiler heating surface, the quality of steam and water, the design output of the boiler denitration, desulfurization, dust removal and slag removal devices, the design output of the main auxiliary machines including the pulverizing system, six fans, environmental protection parameters and the maximum steam inlet flow of the turbine.

6) The monitoring parameters affecting the safety of the unit shall not exceed the alarm value.

By studying the limiting conditions of peak shaving capacity of high back pressure heating units and condensing units in actual operation, it can provide reference for daily operation of thermal power plants.

4. Conclusion

The peak output capacity of the condensing unit is negatively related to the heat load. With the increase of heating and heating capacity of the unit, the output range of the unit becomes narrow and the peak output capacity is limited. The peak capacity of condensate extraction unit is limited by the main steam flow under TRL working condition (high back pressure working condition), and the main steam flow under TRL working condition (high back pressure working condition) reaches or approaches the

https://doi.org/10.1051/e3sconf/202337503027

main steam flow under TRL working condition (high back pressure working condition) when the unit has the maximum output.

The high back pressure heating unit operates in the mode of "determining power by heat" in the heating season. It does not undertake the peak shaving task in the heating season, and the load trend is almost a straight line. The greater the heating heat load of the unit, the higher the corresponding electrical load. The steam pressure after the high back pressure unit transformation increases. Therefore, the maximum output under the high back pressure operation state can only be brought to about 80% of the pure condensing state.

Acknowledgements

This work is supported by the science and technology project of State Grid Shandong Eectric Power Research Institute(ZY-2022-15), which studies Research on peak capacity improvement of thermal power unit based on electric heating coordination optimization under guaranteed supply demand.

References

1. Shi Qingli, Wen SheJiao.Application of thermal test in thermal power plants [J].Qinghai Electric Power, 2004, 23 (4): 4-10.
2. Lv Kai, Wang Hongyu, Zhou Jia, etc.Study on characteristics of electrothermal coal for cogeneration units [J].Thermal Power Generation, 2018, 47 (5): 7-12
3. Ji Ansen.Results and Analysis of Thermal Performance Test of Unit #12 in Wuhu Power Plant before Overhaul [J].Anhui Electric Power Technical Information, 2000, 000 (002): 7-10.
4. Zhang Guangcai, Zhou Ke, Lu Fen, etc. Discussion on deep peak shaving technology of coal-fired units [J].Thermal Power Generation, 2017, 46 (9): 7-15.
5. Wang Wei.Analyze the application of thermal test in thermal power plants [J].China Science and Technology Wealth, 2012, 000 (014): 214-214.
6. Li Hening.Optimal operation of electric heating combined system based on new energy consumption [D]. Shenyang University of Technology.

 https://doi.org/10.1051/e3sconf/202337503028

Introduction on Current Application of Voltage Transformer

Jun Yang[1, 2], Ziming Fan[1, 2], Yue Yang[1, 2], Xiayao Zhao[1, 2]

[1] Inner Mongolia Power (Group) Co., Ltd., Inner Mongolia Power Research Institute Branch, Hohhot 010020, China
[2] Inner Mongolia Enterprise Key Laboratory of High Voltage and Insulation Technology, Hohhot 010020, China

Abstract: As an important equipment, voltage transformer is widely used in power grid to obtain voltage signal and protect the system. In this paper, the voltage transformers in power system are classified according to the working principle. The working principle and characteristics of potential transformer (PT), capacitor voltage transformer (CVT), electronic voltage transformer (EVT) and optical voltage transducer (OVT) are briefly analyzed. Besides, this paper also introduces the principle, category and characteristics of voltage transformer, as well as the application and development of voltage transformer in power system.

Keywords: Potential transformer; Capacitor voltage transformer; Electronic voltage transformer; Optical voltage transducer

1. Introduction

Voltage transformer is an important equipment for power system to obtain voltage signals for metering and protect the power system. As a kind of voltage transformer, potential transformer (PT) is commonly used in low voltage, which is actually a transformer with very small capacity. Capacitor voltage transformer (CVT) is usually used in the high-voltage part of the current power system, which is mainly composed of capacitor voltage divider and electromagnetic part[1-4]. In which, the electromagnetic part is also a transformer, while the primary voltage is the lower voltage after the capacitance voltage divider. In recent years, with the rapid economic development of various countries, the demand for electric power is increasing day by day. The rated voltage level and rated current of the power system have been greatly improved, and many new and stricter requirements have been put forward for current and voltage transformers.

However, shortcomings are still existed: ① The insulation structure is complex, bulky and expensive, especially for ultra-high voltage systems and to meet the requirements of dynamic stability and thermal stability of large short-circuit capacity. ② Low linearity. It is easy to saturate in case of short circuit, small static and dynamic accuracy range. ③ Oil impregnated paper insulation is flammable, explosive and unsafe.

According to the principle of signal change, the existing voltage transformers can be divided into three types of electromagnetic voltage transformers (EVT), CVT, and optical voltage transformers (OVT) respectively. The EVT is a voltage transformer that converts the primary voltage into the secondary voltage proportionally through electromagnetic induction without additional electrical components that change the primary voltage. The essence of CVT is that the capacitor is connected to the electromagnetic unit after series voltage division, which has the advantages of better insulation performance and lower cost. And the higher the voltage level, the advantages are more. According to the statistics of the State Grid Corporation of China, the number of CVTs used in high-level power grids, especially in 330 kV and above voltage levels, has a large of advantages. The most important and core part of the optical voltage sensor is the voltage sensing unit, which converts the voltage information into optical information through certain effects from photoelectric part. In addition, the structure and performance of the voltage sensing unit also have an important impact on the structure and performance of the optical voltage transformer.

Potential transformer

PT is made of electromagnetic induction principle, which has the characteristics of small capacity, long-term stable operation, and nearly no-load state under normal operation. The winding impedance of voltage transformer is very small. When a short circuit fault occurs on the secondary side, the short circuit current will increase instantaneously, which will burn out the winding. According to different internal insulation materials, it can be divided into resin wound insulation transformer, oil immersed transformer and gas insulated voltage transformer.

The characteristic of the EVT is that the primary side and secondary side are relatively independent, and the key point is to ensure that the secondary side load cannot affect the grid voltage when the measuring instrument is connected to the secondary side.

https://doi.org/10.1051/e3sconf/202337503028

The classification of potential transformer is as followed. For the EVT closed with iron core, the box and iron core of the transformer are grounded. The primary coil and secondary coil are placed on the core column of the iron core. The windings are multi-layer, and the insulation between layers is made by oil impregnated paper. The structure can withstand impact and overvoltage with very high frequency. The heat dissipation performance can be improved by changing the structure, such as adding an oil gap. The electric field structure at the end of the winding must be considered during design and construction. Cascade EVT works with multiple transformers, which can not only ensure that the insulation performance of the transformer meets the requirements, but also achieve a very good heat dissipation effect. In high voltage power grid, the iron core of EVT with closed iron core is very easy to produce ferromagnetic resonance. Ferroresonance is formed by the resonant loop composed of winding inductance and circuit capacitance, which is very easy to form when switching or automatic is reclosing. Once ferromagnetic resonance is formed, over-voltage and over-current often make the instrument temperature too high, thus burning the instrument. For EVT with open iron core, the biggest feature of the transformer is that the iron core is open to withstand voltage value riseing to error, so that the shortcomings of the traditional electromagnetic voltage transformer are overcome.

2. Capacitor voltage transformer

Capacitor voltage transformer (CVT)[1-4] is widely used in 110 kV and above power system because of its good insulation performance and economy which is composed of capacitor unit and electromagnetic unit. High voltage capacitor and medium voltage capacitor are connected in series to form capacitor voltage dividing unit. Compensation reactor and intermediate transformer are composed of electromagnetic unit. The basic structure of CVT is shown in Figure.1. However, compared with PT[5], this type of voltage transformer is more complex in structure with low error stability, and is prone to out of tolerance in actual operation. The field operation experience[6-10] shows that the fault rate of CVT is about 5 times than that of PT and 10 times than that of electromagnetic current transformer (CT) in 110 kV and above voltage transformers.

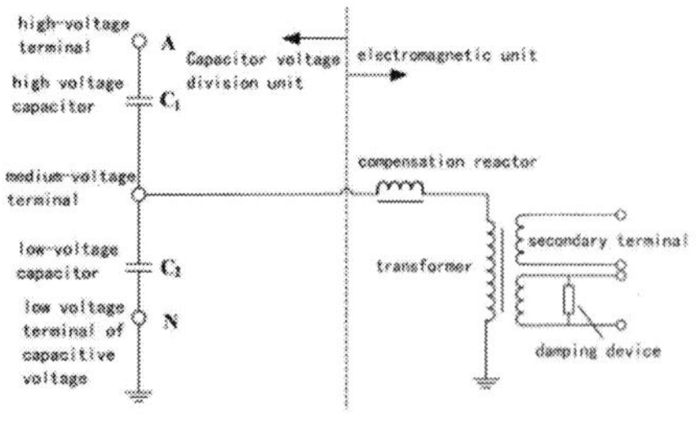

Figure 1 The basic structure of CVT

As a measuring equipment, the long-term stability of measurement error is one of the most important parameters to measure the operation performance of CVT. For the state evaluation of CVT measurement error, the general method is to use the standard to conduct error comparison detection with the verified CVT under power failure condition within a certain verification period[11-13]. However, due to the difficulty of power failure of high-voltage transmission lines, a large number of CVTs in the power grid are operating beyond the verification period, and there is a risk of measurement error exceeding the limit, which affects the fair trade settlement of electric energy. The existing measurement error state evaluation methods can not meet the operation requirements of intelligent substation for online monitoring of key equipment state. Therefore, it is necessary to carry out research on the evaluation and prediction methods of the measurement error status of CVT in operation under the condition of uninterrupted power supply, so as to grasp the measurement error status of CVT in real time, guide the operation and maintenance of CVT more pertinently, which is of great significance to ensure the safe, stable and economic operation of the power system. At the same time, the relevant technical routes and research methods for the evaluation and prediction of the measurement error status of CVT in operation can be extended to the research of other types of power transformers, which has important reference value for promoting the technical development of the industry.

3. Electronic Voltage Transformer

EVT is an important basic measuring equipment of smart grid, which can provide accurate and reliable secondary voltage information for smart grid measurement, metering control and relay protection, as shown in Figure.2. EVT is not an upgrade of traditional voltage transformer, but also a demand and reliable guarantee for the safe and stable operation of smart grid playing an important role in the construction of smart grid and smart substation. First of all, compared with traditional voltage transformers, EVT has the advantages of small size, light weight, no ferromagnetic resonance, good insulation performance, wide frequency response range, no requirement of short circuit at the secondary side, and digital output. Secondly, the digital signals output by EVT, electronic current transformer (ECT) and intelligent terminals form the layer network of intelligent substations, which enables intelligent substations to realize information digitization and information sharing, thus avoiding repeated sampling, simplifying secondary system wiring, and providing a solid foundation for safe operation of smart grids.Besides, acquisition of panoramic data of substation also provides conditions for new monitoring and protection methods of smart grid. Thirdly, the output of EVT reaches the process layer network through the merging unit, and the smart meter obtains the current and voltage data from the process layer for energy metering, which can eliminate the link of analog quantity transmission and secondary transformation of the meter and improve the accuracy of energy metering. Finally, EVT can meet the measurement requirements of PMU/WAMS, which is helpful to the realization of smart grid self-healing function, and will also play an important role in smart grid fast state estimation and grid transient stability monitoring. Therefore, the EVT has become a hot topic studied by scholars at home and abroad in recent years. So far, EVT developed by many companies has been put into grid test run or operation in the power system.

Figure 2 Basic structure block diagram of electronic voltage transformer

The existing EVT can be divided into passive type and active type according to the high-voltage side with or without power supply.. Passive EVT is usually an optical voltage transformer. Its primary sensing head uses the optical effect principle to convert the measured voltage into changes in optical quantities, and directly sends the optical measurement signal to the low voltage side through the optical fiber transmission system, then convers into corresponding electrical signals. There is no electronic circuit on the high voltage side, so there is no need for power supply on the high voltage side. The primary sensing head of the active EVT converts the measured voltage into an analog signal based on the electrical measurement principle, and further converts into a digital signal through an electronic circuit on the high voltage side. Then the electrical signal is converted into an optical signal through an electrical/optical conversion circuit, and transmitted to the low voltage side through optical fiber. Therefore, it is necessary to provide working power for the electronic circuit at the high voltage side. According to different installation methods, EVT can also be divided into two types of independent type and enclosed gas insulated combined electrical apparatus (GIS) matching type. The stand-alone EVT is used for outdoor independent installation, while the GIS equipped EVT is installed with GIS equipment, and the insulation is solved.

4. Optical Voltage Transducer

OVT uses optical materials to sense voltage and optical fiber to transmit signals with no iron core, inductance, ferromagnetic resonance and iron core saturation. It overcomes problems such as waveform distortion caused by magnetic saturation and measurement accuracy reduction caused by voltage drop of traditional voltage transformers[8-14]. Besides, the OVT also has the characteristics of compact structure, high security, wide measurement frequency band, large dynamic range, strong anti-interference ability, no secondary open circuit high voltage, easy to interface with computer, adapting to the requirements of power system automation and digitization, etc. The remote measurement and remote control of the system can be realized with the optical fiber transmission networking, which is easy to meet the requirements of miniaturization, intelligence and multi-function, and can not compare to traditional voltage transformers. OVT can also measure DC voltage, which is necessary in smelting and DC power system. As a result, OVT has attracted great attention in the industry since its inception. At present, Europe, America, Japan as well as other developed regions and countries have begun to use optical fiber sensors to develop new voltage transformers. The 161 kV voltage transformer of Tennessee Electricity Administration in 1989 was used to measure the voltage indirectly by current flowing through the capacitive voltage divider, as shown in Figure 3.

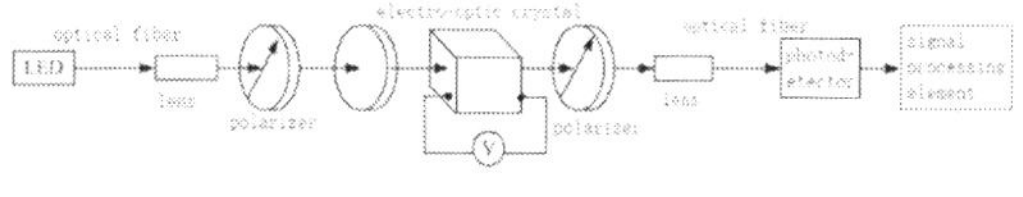

Figure 3 Principle diagram of optical voltage transformer

In terms of measurement principle, OVT can be roughly divided into three types based on Pockels effect (first order electro-optic effect)[15], Kerr effect (second order electro-optic effect) and inverse piezoelectric effect or electrostriction effect. In terms of type and structure, OVT can be divided into voltage division type and non voltage division type.

However, the OVT currently in the trial run stage is mostly based on the Pockels effect, which is required many optical components and difficult to calibrate and mass produce. In particular, the stability of the operation still needs to be improved. Therefore, countries around the world are constantly looking for and trying other more effective methods while improving the OVT based on the Pockels effect. In a word, OVT is still a long way from commercialization, and has great potential for performance improvement.

5. Conclusion

This paper classifies power system voltage transformers according to their working principles, and briefly analyzes the working principles and characteristics of PT, CVT, EVT and OVT. PT and CVT have a long history of application, and their implementation principles are simple and reliable, accounting for a large proportion in the power system. However, with the construction of ultra-high voltage, intelligent and digital power grids, ferromagnetic resonance, insulation difficulties, large floor space and other issues have become constraints to development. As new voltage transformers, EVT and OVT, can achieve real electrical isolation and eliminate the potential danger of ferromagnetic resonance by applying new measuring principles compared with traditional voltage transformers,. They have the advantages of small size, good insulation performance and strong anti-interference ability. The research of new voltage transformers is the need for the development of power system, but the research work and technology accumulation are obviously not enough at present, and the market share is still low. However, with the progress of science and technology, the improvement of production technology and the continuous efforts of researchers, EVT and OVT will be widely used in power system.

Acknowledgments

This research was financially supported by employee innovation studio innovation project of Inner Mongolia Power (Group) Co.,Ltd .

References

1. D. FERNANDESJ, W. NEVESA, J.C. VASCONCELOSB. Coupling capacitor voltage transformer: A model for electromagnetic transient studies [J]. Electric Power Systems Research, 2007, 77(2):125-134.
2. M. KEZUNOVIC, L. KOJOVIC, V. SKENDZIC. Digital models of coupling capacitor voltage transformers for protective relay transient studies [J]. IEEE Transactions on Power Delivery, 1992, 7(4):1927-1935.
3. F. GHASSEMI, P. GALE, B. CLEGG, et al. Coutts. Method to measure CVT transfer function [J]. IEEE Transactions on Power Delivery, 2002, 17:915-920.
4. M. GRAOVAC, R. IRAVANI, X. WANG, et al. Fast ferroresonance suppres-sion of coupling capacitor voltage transformers [J]. IEEE Transactions on Power Delivery,2003, 18(1):158-163.
5. Z.Y. ZHANG, F.C. LU, G.S. LIANG. High-frequency circuit model of a potential transformer for the very fast transient simulation in GIS [J]. IEEE Transactions on Power Delivery, 2008, 23(4):1995-1999.
6. D. A. DOUGLASS. Potential transformer accuracy at 60Hz voltages above and below rating and at frequencies above 60 Hz [J]. IEEE Transactions on Power Apparatus and Systems, 1981, 100(3):1370-1375.
7. K. SADIK, G. GEORGE, L. KARADY. Experimental comparison of conventional and optical VTs, and circuit model for optical VT. IEEE Transactions on Power Delivery, 2011, 26(3): 1571-1578.
8. W.X. SIMA, S.S. HUANG, Q. YANG, et al. Generation and distribution characteristics of space charge in liquid dielectric based on Kerr electro-optic effect[J]. High Voltage Engineering, 2014, 40(1): 16-23.
9. J.Y. WANG, Z.Z. GUO, G.Q. ZHANG, et al. Experimental investigation on optical current transducer's long-term operation stability[J]. Power System Technology, 2012, 36(6): 37-41.
10. Y.G. CHENG, H.j. LIU, Y.X. LI, et al. Basic principle and current development of optical voltage transformer[J]. Electric Power Automation Equipment, 2004, 24(5): 87-91.
11. C.S. LI, X. CUI. Summary of optical electric field sensor research[J]. High Voltage Engineering, 2000, 26(2): 40-43.
12. Z.H. XIAO. Study and comment of the optical transformers in power system[J]. Power System Protection and Control, 2014, 42(12): 148-154.
13. C.X. Zhang, X.J. Feng, S. Liang, et al. Quasi-reciprocal reflective optical voltage sensor based on Pockels effect with digital closed-loop detection technique. Optics Communications, 2010, 284(20): 3878-3883.
14. P. NIEWCZAS, L. DZITIALA, G. FUSIEK, et al. Design and evaluation of a pre-prototype hybrid fiber-optic voltage sensor for a remotely interrogated condition monitoring system. IEEE Transactions on Instrumentation and Measurement, 2005, 54(4): 1560-1564.
15. J.C. SANTOS, M.C. TAPLAMACIOGLU, K. HIDAKA. Pockels high voltage measurement system. IEEE Transactions on Power Delivery, 2000, 15(1): 8-13.
16. J. A. BRANDAO. An overview of electrooptic high voltage measuring systems. In: Aficon '92 Proceedings., 3rd Aficon Conference. Ezulwini Valley, 1992, 466-469.

AUTHOR INDEX

Ailin Zhang....425
Anqi Liu....310 314
Bangxing Shen....434
Bao Wang....274 278
Beibei Li....295
Biaoxian Chen....329
Bingqi Wang....409
Bo Yuan....310 314
Cai Li....286
Chao Zheng....254
Chaochen Yan....368
Chaoxuan Qin....33 389
Chengzhi Duan....18
Chenkun Huang....393
Chuanzhi Cui....223
Chunfeng Guo....434
Dandan Liu....385
Dengfeng Dai....385
Deyuan Lin....261
Dong Liu....237
Dongqi Kang....329
Fan Jia....472
Fang Zhang....1
Fangchao Zhao....27
Fangfang Jia....113
Fen Zhang....5
Fuji Yu....249
Guangyu Yuan....178 185
Guiyuan Shi....47
Guobiao Wang....223
Guowei Zhang....282
Guoyi Xu....295
Guoying Qin....153
Haijing Zhang....14 441 445
Haijun Zhou....5
Haiwei Wu....61
Haiwei Yang....421
Hang Gao....93
Haochen Zou....149
Haonan Xu....18
Hengqing Guo....472
Hong Lu....243
Hong Zheng....101
Hongbin Guo....208
Hongyu Xiao....228
Hongzhi Han....109
Houhua Lv....358
Hu Zhang....453 458 468
Hua Wang....274 278
Huitao Cheng....453
Jia Liu....329
Jiageng Ding....405
Jiameng Li....27 33
Jian Wang....393
Jiang Xiao....228
Jiang Zhu....368
Jiangwei Luo....47
Jianhui Ni....18
Jianjun Zheng....23
Jianjun Zhu....348
Jianting Qian....204
Jiaqi Zhang....295
Jiasu Li....213
Jiaxing Wang....368
Jiayi Li....208
Jie Liu....47
Jijiang Ge....208 213 218
Jin Chen....218
Jing Chen....133
Jing Fan....18
Jing Zhang....178
Jingbin Xi....318
Jinshan Wang....274 278
Jinxu Lao....453 458 468
Jiuhong Li....358
Juan Zhang....233
Jun Liu....10 335 339 449 477
Jun Yang....481
Jun Zou....434
Jundong Liu....204
Junfeng Liu....318
Junshan Guo....10 265 335 449 463 477
Kai Du....65
Kai Liang....453
Kang Liu....47
Kangwei Xiong....254
Ke Liu....233
Kexin Li....213
Kun Yang....286
Lang Chen....416
Le Zhang....254
Lei Han....295
Lei Wang....129
Leijing Zhu....329
Li Gao....137
Li Zhang....1 286

Lin Cong 348 374
Lin Li 141
Lingkai Zhu 10 14 265 335 339 364 430 441
.... 445 449 463 477
Linjie Guan 286
Linxi Hou 261
Liuyu Xu 358
Liwen You 300
Long Tian 162
Longjie Li 213
Lu Dai 27 33 389
Lujun Wen 18
Meiyi Jiang 158
Mengyuan Xu 358
Miaomiao Wang 318
Mingzheng Lu 170
Na Lei 43 191
Na Lv 69
Nan Wang 374
Nan Zhan 185
Nan Zhang 178
Ning Shi 228
Paisheng Xu 204
Pan Liu 125
Pan Wu 54
Panfeng Shang 10 335 430 449 458 463 468 477
Panpan Zhang 191
Peiyi Jia 89
Pengfei Chen 218
Pengju Chu 213
Pengxiang Zhao 348 374
Qi Liu 228
Qi Zhang 204 401
Qian Ma 310 314
Qian Wang 364 441 445
Qian Yao 5
Qiang Wang 348
Qingsong Fan 286
Qiuyue Chen 393
Qunyi Wang 237
Ren Yang 358
Ruibo Lv 434
Ruikun Guo 421
Ruoqi Lai 472
Ruqiong Qin 381
Senlin Gan 77
Shengqiang Li 393
Shima Chao 318
Shiping Liu 117
Shoulei Jiao 282
Shuai Wang 329
Shuiqingshan Lu 223
Shujing Chai 200 204 401
Si Li 430
Silu Tao 290
Song Han 185
Song Xia 310
Tao Li 200
Tiantian Tan 27
Tianze Liu 85
Tongfeng Cao 237
Wei Liu 200 204 401
Wei Wang 178 185 318
Wei Zhang 344
Wei Zheng 10 14 265 335 364 441 445 449 458 477
Weigang Jin 434
Weihao Liu 329
Weihua Zhang 43
Weikun Cheng 1
Weiqiang Zhang 358
Weishuai Wang 14 441 445
Weiwei Gong 23
Weixiong Chen 218
Weiyan Li 195
Wenhua Zhang 329
Wenhui Wang 218
Wenqiang Tian 274 278
Wenyan Zhang 200 204 401
Wenzhe Zhang 261
Xi Yan 178 185
Xian Yang 348
Xianbo Han 233
Xiang Li 97
Xiangnan Ding 145
Xiao Zhang 358
Xiaobao Liu 47
Xiaocui Hao 200 204 401
Xiaodan Yu 310 314
Xiaofei Zhang 47
Xiaofeng Su 434
Xiaojian Xia 261
Xiaojun Chen 310 314
Xiaoliang Xu 472
Xiaolong Zhang 204
Xiaoqi Chen 5
Xiaoxu Chen 282
Xiaoyan Wang 178 185
Xiaoyang Du 290
Xiaoyu Ren 47
Xiaoyu Yu 200
Xiayao Zhao 481
Xichao Zhou 374
Xidong Cai 185
Xincheng Yang 204

Xinfeng Zhang 434
Xing Wu 185
Xinrui Li 290
Xinyuan Wan 261
Xiping Huang 204
Xu Li 358
Xu Long 105
Xu Zhang 310 314
Xuefeng Lin 344
Yajun Jian 358
Yanbin Liang 185
Yanfeng Bao 166
Yang Zhang 237
Yanhua Han 358
Yanling Niu 5
Yanpeng Zhang 10 335 430 449 477
Yanru Zhang 348
Yantao Li 5
Yanxu Gou 274 278
Yaqiang Xu 323
Ye Liu 243
Ye Tao 368
Yi He 434
Yicheng Hong 261
Yifan Yang 434
Yinbao Miao 329
Ying Gao 81
Ying Zuo 397
Yisen Chen 269
Yiyang Chen 261
Yongbin Bi 237
Yongke Xu 434
Yongqiang Che 269
Yu Zhao 368
Yuanxian Liu 178
Yude Gao 121
Yue Han 458 463 468
Yue Yang 481
Yuefeng Li 434
Yueyang Guan 47
Yuming Ding 1
Yun Hu 243
Yunfang Liu 358
Yuning Zhu 472
Yunqi Liu 385
Yuqi Wang 204 401
Yuqin Zhu 27 33 389
Yuqing Wang 368
Yuwen Zhao 178
Yuxi Wang 314
Yuyuan Lin 416
Zejiang Wang 200
Zeyu Lin 178 185
Zhanlin Li 254
Zhaoling Yu 204
Zhe Chen 329
Zhehao Xu 23
Zhen Kang 318
Zhen Li 374
Zhengxu Liang 425
Zhenlu Zhang 318
Zhijun Wu 318
Zhiqiang Bao 274 278
Zhiqiang Gong 10 265 335 364 449 463 477
Zhonglin Li 73
Zhongmei Shen 38
Zian Wang 254
Zidong Zeng 416
Zifeng Deng 416
Ziming Fan 481
Ziwei Zhong 10 265 335 364 441 445 449 477
Ziyan He 421
Ziyue Wang 282